Lecture Notes in Networks and Systems

Volume 1200

Series Editor

Janusz Kacprzyk, Systems Research Institute, Polish Academy of Sciences, Warsaw, Poland

Advisory Editors

Fernando Gomide, Department of Computer Engineering and Automation—DCA, School of Electrical and Computer Engineering—FEEC, University of Campinas—UNICAMP, São Paulo, Brazil

Okyay Kaynak, Department of Electrical and Electronic Engineering, Bogazici University, Istanbul, Türkiye

Derong Liu, Department of Electrical and Computer Engineering, University of Illinois at Chicago, Chicago, USA
Institute of Automation, Chinese Academy of Sciences, Beijing, China

Witold Pedrycz, Department of Electrical and Computer Engineering, University of Alberta, Alberta, Canada
Systems Research Institute, Polish Academy of Sciences, Warsaw, Poland

Marios M. Polycarpou, Department of Electrical and Computer Engineering, KIOS Research Center for Intelligent Systems and Networks, University of Cyprus, Nicosia, Cyprus

Imre J. Rudas, Óbuda University, Budapest, Hungary

Jun Wang, Department of Computer Science, City University of Hong Kong, Kowloon, Hong Kong

AF800291

The series "Lecture Notes in Networks and Systems" publishes the latest developments in Networks and Systems—quickly, informally and with high quality. Original research reported in proceedings and post-proceedings represents the core of LNNS.

Volumes published in LNNS embrace all aspects and subfields of, as well as new challenges in, Networks and Systems.

The series contains proceedings and edited volumes in systems and networks, spanning the areas of Cyber-Physical Systems, Autonomous Systems, Sensor Networks, Control Systems, Energy Systems, Automotive Systems, Biological Systems, Vehicular Networking and Connected Vehicles, Aerospace Systems, Automation, Manufacturing, Smart Grids, Nonlinear Systems, Power Systems, Robotics, Social Systems, Economic Systems and other. Of particular value to both the contributors and the readership are the short publication timeframe and the world-wide distribution and exposure which enable both a wide and rapid dissemination of research output.

The series covers the theory, applications, and perspectives on the state of the art and future developments relevant to systems and networks, decision making, control, complex processes and related areas, as embedded in the fields of interdisciplinary and applied sciences, engineering, computer science, physics, economics, social, and life sciences, as well as the paradigms and methodologies behind them.

Indexed by SCOPUS, EI Compendex, INSPEC, WTI Frankfurt eG, zbMATH, SCImago.

All books published in the series are submitted for consideration in Web of Science.

For proposals from Asia please contact Aninda Bose (aninda.bose@springer.com).

Hushairi Zen · Naga M. Dasari ·
Y. Madhavee Latha · S. Srinivasa Rao
Editors

Soft Computing and Signal Processing

Proceedings of 7th ICSCSP 2024, Volume 2

Springer

Editors
Hushairi Zen
Faculty of Science and Technology
International Institute of Advance Technology
Kuching, Malaysia

Y. Madhavee Latha
Department of Electronics and Communication Engineering
Malla Reddy Engineering College for Women
Secunderabad, Telangana, India

Naga M. Dasari
Faculty of Science and Technology
Federation University
Ballarat, VIC, Australia

Department of Computer Science
University of South Australia
Adelaide, Australia

S. Srinivasa Rao
Department of Electronics and Communication Engineering
Malla Reddy College of Engineering and Technology
Secunderabad, Telangana, India

ISSN 2367-3370 ISSN 2367-3389 (electronic)
Lecture Notes in Networks and Systems
ISBN 978-981-97-9925-1 ISBN 978-981-97-9926-8 (eBook)
https://doi.org/10.1007/978-981-97-9926-8

© The Editor(s) (if applicable) and The Author(s), under exclusive license to Springer Nature Singapore Pte Ltd. 2025

This work is subject to copyright. All rights are solely and exclusively licensed by the Publisher, whether the whole or part of the material is concerned, specifically the rights of translation, reprinting, reuse of illustrations, recitation, broadcasting, reproduction on microfilms or in any other physical way, and transmission or information storage and retrieval, electronic adaptation, computer software, or by similar or dissimilar methodology now known or hereafter developed.
The use of general descriptive names, registered names, trademarks, service marks, etc. in this publication does not imply, even in the absence of a specific statement, that such names are exempt from the relevant protective laws and regulations and therefore free for general use.
The publisher, the authors and the editors are safe to assume that the advice and information in this book are believed to be true and accurate at the date of publication. Neither the publisher nor the authors or the editors give a warranty, expressed or implied, with respect to the material contained herein or for any errors or omissions that may have been made. The publisher remains neutral with regard to jurisdictional claims in published maps and institutional affiliations.

This Springer imprint is published by the registered company Springer Nature Singapore Pte Ltd.
The registered company address is: 152 Beach Road, #21-01/04 Gateway East, Singapore 189721, Singapore

If disposing of this product, please recycle the paper.

Committees

Chief Patron

Sri. Ch. Malla Reddy, Founder Chairman, MRGI, Hon'ble Minister, Government of Telangana

Patrons

Sri. Ch. Mahendar Reddy, Secretary, MRGI
Sri. Ch. Bhadra Reddy, President, MRGI

Conference Chair

Dr. V. S. K. Reddy, Vice Chancellor, MRU

Organizing Chair

Prof. P. Sanjeeva Reddy, Dean, International Studies

Convener

Dr. S. Srinivasa Rao, Principal

Co-convener

Dr. T. Venugopal, Dean, Students Welfare

Organizing Secretaries

Dr. K. Mallikarjuna Lingam, HOD, ECE
Dr. G. Sharada, HOD, IT
Dr. D. Sujatha, HOD, CSE (CI)
Dr. S. Shanthi, HOD, CSE
Dr. M. V. Kamal, HOD, CSE (ET)

Editorial Board

Dr. Hushairi Zen, Professor and Dean, Faculty of Science and Technology, International Institute of Advance Technology, Malaysia
Dr. Naga M. Dasari, Professor, Department of Computer Science, University of South Australia, Australia
Dr. Y. Madhavee Latha, Professor and Principal, Department of ECE, Malla Reddy Engineering College for Women, Maisammaguda, Secunderabad
Dr. S. Srinivasa Rao, Professor and Principal, Department of ECE, Malla Reddy College of Engineering and Technology, Maisammaguda, Secunderabad

Advisory Committee

Dr. Heggere Ranganath, Chair of Computer Science, University of Alabama in Huntsville, USA
Dr. Someswar Kesh, Professor, Department of CISA, University of Central Missouri, USA
Mr. Alex Wong, Senior Technical Analyst, Diligent Inc., USA
Dr. Suresh Chandra Satapathy, Professor, School of Computer Engineering, KIIT Deemed to be University, Bhubaneswar
Dr. Ch. Narayana Rao, Scientist, Denver, Colorado, USA
Dr. Sam Ramanujan, Professor, Department of CIS and IT, University of Central Missouri, USA
Dr. Richard H. Nader, Associate Vice President, Mississippi State University, USA
Dr. Muralidhar Rangaswamy, WPAFB, OH, USA
Mr. E. Sheldon D. Wallbrown, Western New England University, Springfield, USA

Prof. Peter Walsh, Head of the Department, Vancouver Film School, Canada
Dr. Murali Venkatesh, School of Information Studies, Syracuse University, USA
Dr. Asoke K. Nandi, Professor, Department of EEE, University of Liverpool, UK
Dr. Vinod Chandran, Professor, Queensland University of Technology, Australia
Dr. Amiya Bhaumik, Vice Chancellor, Lincoln University College, Malaysia
Dr. Hushairi Bin Zen, Professor, ECE, UNIMAS
Dr. Bhanu Bhaskara, Professor at Majmaah University, Saudi Arabia
Dr. Narayanan, Director, ISITI, CSE, UNIMAS
Dr. Koteswararao Kondepu, Research Fellow, Scuola Superiore Sant' Anna, Pisa, Italy
Shri. B. H. V. S. Narayana Murthy, Director, RCI, Hyderabad
Prof. P. K. Biswas, Head, Department of E and ECE, IIT Kharagpur
Dr. M. Ramasubba Reddy, Professor, IIT Madras
Prof. N. C. Shiva Prakash, Professor, IISC, Bengaluru
Dr. B. Lakshmi, Professor, Department of ECE, NIT Warangal
Dr. G. Ram Mohana Reddy, Professor and Head, IT Department, NITK Surathkal, Mangaluru, India
Dr. Y. Madhavee Latha, Professor, Department of ECE, MRECW, Hyderabad

Coordinators

Ms. Anitha Patibandla, Associate Professor, ECE
Dr. A. Mummoorthy, Professor, IT
Mr. Vigneshwar Rao, Assistant Professor, CSE
Ms. P. Satyavathi, Assistant Professor, CSE (ET)
Dr. P. Harikrishna, Associate Professor, CSE (CI)

Organizing Committee

Dr. M. Ramakrishna Murty, Professor, Department of CSE, ANITS, Visakhapatnam
Dr. Sadanand Yadav, Associate Professor, ECE
Mr. M. Ramanjaneyulu, Associate Professor, ECE
Mr.Ch. Kiran Kumar, Associate Professor, ECE
Ms. D. Asha, Associate Professor, ECE
Dr. Shaik Rahamat Basha, Professor, CSE
Mr. M. Vazralu, Associate Professor, IT
Mr. M. Sandeep, Associate Professor, CSE
Dr. B. Rajeshwar Reddy, Administrative Officer

Campaigning and Social Media Coordinator

Mr. Ch. Kiran Kumar, Associate Professor, ECE

Review Members

Dr. B. Lakshmi, Department of ECE, NIT Warangal
Dr. Bhanu Bhaskara, Professor at Majmaah University, Saudi Arabia
Dr. M. Ramasubba Reddy, IIT Madras
Mr. Andrew Liang, Senior Instructor, Vancouver Film School, Canada
Mr. Alex Wong, Senior Technical Analyst, Diligent Inc., USA
Dr. Vinod Chandran, Professor, Queensland University of Technology, Australia
Dr. Jiacun Wang, Monmouth University, New Jersey
Shri. B. H. V. S. Narayana Murthy, Director, RCI
Prof. N. C. Shiva Prakash, IISC, Bengaluru
Prof. P. K. Biswas, IIT Kharagpur
Dr. Koteswararao Kondepu, Research Fellow, Scuola Superiore Sant' Anna, Pisa, Italy
Prof. Peter Walsh, Head of the Department, Vancouver Film School, Canada
Dr. Ahmed Nabih Zaki Rashed, Professor, Menoufia University, Egypt
Dr. Ch. Narayana Rao, Scientist, Denver, Colorado, USA
Dr. Arun Kulkarni, Professor, University of Texas at Tyler, USA
Dr. Sam Ramanujan, Professor, Department of CIS and IT, University of Central Missouri, USA
Dr. K. Narayana Murty, Professor in CSE, University of Hyderabad
Prof. P. Laxminarayana, Professor and Director, Nertu, Osmania University, Hyderabad Telangana
Dr. P. Chandra Sekhar, Professor Department of ECE, Osmania University, Hyderabad Telangana
Dr. Hemalatha Rallapalli, Associate Professor, Department of ECE, Osmania University, Hyderabad
Dr. D. Rama Krishna, Associate Professor, Department of ECE, Osmania University, Hyderabad
Dr. Koti Lakshmi, Associate Professor Department of ECE, Osmania University, Hyderabad Telangana
Dr. K. Anitha Sheela, Associate Professor, Department of ECE, JNTUH, Kukatpally, Hyderabad
Dr. D. Srinivas Rao, Professor, Department of ECE, JNTUH, Kukatpally, Hyderabad
Dr. P. Sri Hari, Professor, HOD Department of EIE, GITAM University, Hyderabad
Dr. Y. Madhavee Latha, Principal, Department of ECE, MRECW, Hyderabad
Dr. K. Manjunatha Chari, Professor, HOD Department of ECE, GITAM University, Hyderabad

Dr. M. Ramakrishna Murthy, Professor, Anil Neerukonda Institute of Technology and Sciences, Visakhapatnam
Dr. M. Rekha Sundari, Professor, Anil Neerukonda Institute of Technology and Sciences, Visakhapatnam
Dr. S. Suganthi, Professor, Christ University, Bengaluru
Dr. B. Tarakeswara Rao, Professor, Kallam Haranadha Reddy Institute of Technology, Guntur
Dr. T. D. Bhatt, Professor, Department of ECE, MGIT, Gandipet, Hyderabad
Dr. Hyma J., Professor, Gandhi Institute of Technology and Management (GITAM) University, Visakhapatnam
Dr. Selvani Deepthi Kavila, Professor, Anil Neerukonda Institute of Technology and Sciences, Visakhapatnam
Dr. Anitha Perla, Professor, Malla Reddy University, Hyderabad
Dr. D. Venkat Reddy, Professor, Department of ECE, MGIT, Gandipet, Hyderabad
Dr. Ch. Raja, Professor, Department of ECE, MGIT, Gandipet, Hyderabad
Dr. A. Vani, Associate Professor, Department of ECE, CBIT, Gandipet, Hyderabad
Dr. E. Nagabushanam, Professor, Department of ECE, Sridevi Women's Engineering College
Dr. V. M. Senthil Kumar, Professor, Vivekananda College of Engineering for Women, Namakkal, Elaiyampalayam, Tamil Nadu
Dr. M. Sucharitha, Professor, Vellore Institute of Technology, Andhra Pradesh
Dr. T. Padma, Professor, Department of Biomedical Engineering, GRIET, Bachupally, Kukatpally, Hyderabad
Dr. K. Padmavathi, Professor, Department of ECE, GRIET, Hyderabad
Dr. G. S. Naveen Kumar, Professor, Malla Reddy University, Hyderabad
Dr. Sasikanth Shanmugam, Professor, Vivekananda College of Engineering for Women, Namakkal, Elaiyampalayam, Tamil Nadu
Dr. C. V. Narasimhulu, Professor and Principal, LIET, Hyderabad
Dr. Ch. Srinivasa Rao, Professor, Department of ECE, JNTUK University College of Engineering Vizianagaram
Dr. N. Balaji, Professor, Department of ECE, JNTUK University College of Engineering Vizianagaram
Dr. N. V. Koteswara Rao, Professor and HOD Department of ECE, CBIT, Gandipet, Hyderabad
Dr. D. Krishna Reddy, Professor, Department of ECE, CBIT, Gandipet, Hyderabad
Dr. A. V. Narasimha Rao, Professor, Department of ECE, CBIT, Gandipet, Hyderabad
Dr. Kakarla Subba Rao, Professor, Department of ECE, CBIT, Gandipet, Hyderabad
Dr. A. D. Sarma, Professor, Department of ECE, CBIT, Gandipet, Hyderabad
Dr. P. Narahari Sastry, Professor, Department of ECE, CBIT, Gandipet, Hyderabad
Dr. K. Suman, Assistant Professor, Department of ECE, CBIT, Gandipet, Hyderabad
Dr. V. Saidulu, Associate Professor, Department of ECE, MGIT, Gandipet, Hyderabad
Dr. S. Praveena, Assistant Professor, Department of ECE, MGIT, Gandipet, Hyderabad

Dr. C. D. Naidu, Principal and Professor, Department of ECE, VNRVJIET, Hyderabad
Dr. Yarlagadda Padma Sai, Professor, Department of ECE, VNRVJIET, Hyderabad
Dr. Vangala Padmaja, Professor, Department of ECE, VNRVJIET, Hyderabad
Dr. L. Padma Sree, Professor, Department of ECE, VNRVJIET, Hyderabad
Dr. Rajendra Prasad Somineni, Professor, Department of ECE, VNRVJIET, Hyderabad
Dr. Ranjan K. Senapati, Professor, Department of ECE, VNRVJIET, Hyderabad
Dr. Mediikonda Christhu Raju, Professor, Department of ECE, VNRVJIET, Hyderabad
Dr. P. Trinatha Rao, Associate Professor, Department of ECE, GITAM University, Hyderabad
Dr. C. Krishna Mohan, Associate Professor, IIT Hyderabad
Dr. Sathya Peri, Associate Professor, IIT Hyderabad
Dr. Manish Prateek, Professor and Director—SoCSE, UPES, Dehradun, India
Dr. Manas Pradhan Rajan, Professor, INTI International University, Malaysia
Dr. Sobhan Babu, Associate Professor, IIT Hyderabad
Dr. Ch. Sudhakar, Associate Professor, NIT Warangal
Dr. S. Phani Kumar, GITAM University, Hyderabad
Dr. K. Ramesh, Associate Professor, NIT Warangal
Dr. Padmalaya Nayak, Professor, GRIET
Dr. K. Chandrasekaran, Professor, NITK Surathkal
Dr. Gollapudi Ramesh Chandra, Professor, VNR VJIET
Dr. R. B. V. Subramaanyam, Associate Professor, NIT Warangal
Dr. C. Kiranmai, Professor, VNR VJIET
Dr. Venkatadiri, Associate Professor and Program Head—SoCSE, UPES, Dehradun, India
Dr. Salman Abdul Moiz, Associate Professor, Central University, Hyderabad
Dr. Rajeev Wankar, Professor, Central University, Hyderabad
Dr. M. Akkalakshmi, Professor, GITAM University, Hyderabad
Dr. Neelu Jyoti Ahuja, Professor and Head-R&D Centre, SoCSE, UPES, Dehradun, India
Dr. Abhineet Anand, Professor, CSE Department, Galgotias University, Greater Noida, India
Dr. Ravi Rastogi, Professor, CSE Department, Galgotias University, Greater Noida, India
Dr. K. Anuradha, Professor and Dean, CSE Department, GRIET
Dr. P. Krishna Reddy, Professor, IIIT Hyderabad
Dr. Praveen Paruchuri, Professor, IIIT Hyderabad
Dr. A. Jagan, Professor and Head, CSE Department, BVRIT Hyderabad
Dr. Hari Seetha, Professor, CSE Department, Vellore Institute of Technology (VIT)
Dr. N. L. Bhanu Murthy, Professor and Head CSE Department, BITS Hyderabad Campus
Dr. G. Geethakumari, Professor, CSE Department, BITS Hyderabad Campus
Dr. A. Nagesh, Professor and Head, CSE Department, MGIT Hyderabad

Dr. M. Sree Vani, Professor, CSE Department, MGIT Hyderabad
Dr. R. Rajeswara Rao, Professor and Head CSE Department, JNTU-Vizianagaram
Dr. Sita Ramaiah, Professor CSE Department, Andhra University, Andhra Pradesh
Dr. S. Viswanadha Raju, Professor and Vice Principal, JNTU Jagtial, Telangana
Dr. B. Vishnu Vardhan, Professor CSE Department, JNTU Jagtial, Telangana
Dr. M. Sivarama Badriraju, Professor, CSE Department, SRKR Engineering College, Bhimavaram
Dr. S. Suresh Kumar, Vivekananda College of Engineering for Women, Tiruchengode
Dr. R. Rajasekar, Professor, St. Peters Engineering College
Dr. Aruna Malapati, BITS Hyderabad

Preface

The **International Conference on Soft Computing and Signal Processing (ICSCSP-2024)** was successfully organized by Malla Reddy College of Engineering and Technology, an UGC Autonomous Institution, during **June 21–22, 2024**, at Hyderabad. The objective of this conference was to provide opportunities for the researchers, academicians, and industry persons to interact and exchange the ideas, experience, and gain expertise in the cutting-edge technologies pertaining to soft computing and signal processing. Research papers in the above-mentioned technology areas were received and subjected to a rigorous peer-review process with the help of program committee members and external reviewers. The ICSCSP-2024 received a total of 265 papers, each paper was reviewed by more than two reviewers, and finally, 55 papers were accepted for publication in Springer LNNS series.

Our sincere thanks to **Dr. Aninda Bose**, Senior Editor, Springer Publications, India, **Dr. Suresh Chandra Satapathy**, Professor and Dean R&D, KIIT, for extending their support and cooperation.

We would like to express our gratitude to all session chairs, viz., Dr. Nagender Kumar S., Associate Professor, Department of CSE, Central University, Hyderabad, India; Dr. Veningston K., Assistant Professor, Department of CSE, National Institute of Technology, Srinagar, India; Dr. Rajeev Tiwari, Professor and Dean, School of Computer Science, IILM University, India; Dr. P. Chandra Sekhar Reddy, Professor, Department of CSE, Gokaraju Rangaraju Institute of Engineering and Technology, Hyderabad, India; Dr. Y. Vijayalata, Dean and Professor, Department of CSE, Vardhaman College of Engineering, Hyderabad, India; Dr. Padmalaya Nayak, Professor, Department of CSE, Gokaraju Lailavathi Womens Engineering College, India; Dr. L. Mary Gladence, Professor, Department of IT, Sathyabama Institute of Science and Technology, India; and Dr. P. Rohini, Professor, Department of CSE, ICFAI University, India, for extending their support and cooperation.

We are indebted to the program committee members and external reviewers who have produced critical reviews in a short time. We would like to express our special gratitude to publication chair **Dr. Suresh Chandra Satapathy**, KIIT, Bhubaneswar, for his valuable support and encouragement till the successful conclusion of the conference.

We express our heartfelt thanks to our Chairman, **Sri. Ch. Malla Reddy** Garu, Chief Patron, Founder Chairman, MRGI, Patrons **Sri. Ch. Mahendar Reddy**, Secretary, MRGI, **Sri. Ch. Bhadra Reddy**, President, MRGI, **Dr. V. S. K. Reddy**, Vice Chancellor MRU, **Dr. S. Srinivasa Rao**, Principal and Convener, **Dr. T. Venu Gopal**, Dean-Students Welfare, Organizing Chair **Prof. P. Sanjeeva Reddy**, Dean, International Studies.

We would also like thank organizing secretaries **Dr. K. Mallikarjuna Lingam, HOD, ECE; Dr. D. Sujatha HOD (CI); Dr. G. Sharada HOD (IT); Dr. S. Shanthi, HOD (CSE); and Dr. M. V. Kamal, HOD (ET)** for their valuable contribution. Our sincere thanks to all the coordinators, organizing committee, and other committee members for their commendable contribution in successful conduct of the conference.

Last, but certainly not least, our special thanks to all the authors for their applaudable technical contributions which have made our proceedings rich and exemplary.

Dr. Hushairi Zen
Professor and Dean, Faculty of Science and Technology
International Institute of Advance Technology
Kuching, Malaysia

Dr. Naga M. Dasari
Professor, Department of Computer Science
University of South Australia
Adelaide, Australia

Dr. Y. Madhavee Latha
Professor and Principal, Department of Electronics and Communication Engineering
Malla Reddy Engineering College for Women
Maisammaguda, Secunderabad, India

Dr. S. Srinivasa Rao
Professor and Principal, Department of Electronics and Communication Engineering
Malla Reddy College of Engineering and Technology
Maisammaguda, Secunderabad, India

Theme for Conference

Partnerships for the Goals is highly relevant to an International Conference on Soft Computing and Signal Processing due to its emphasis on collaboration, partnerships, and the exchange of knowledge and technology. In the context of the conference, this could involve sharing advancements and breakthroughs in soft computing and signal processing methodologies. Collaboration among researchers, academia, industries, and policymakers can facilitate the transfer of technology and knowledge, fostering innovation globally.

The conference serves as a platform for sharing research findings, new algorithms, techniques, and applications in soft computing and signal processing. This aligns with SDG 17's objective of enhancing the capacity of developing countries in crucial areas, including technology. Workshops, seminars, and collaborative projects showcased in the conference can contribute to building the capacity of individuals and institutions in these fields.

In the context of soft computing and signal processing, collaborations between academia, industry experts, and government bodies are vital. These partnerships can lead to the development of innovative solutions addressing societal challenges, such as healthcare, climate change, and infrastructure, utilizing these advanced technologies.

The conference emphasizes on the importance of data sharing, access to information, and global connectivity. This resonates with the conference's themes of sharing datasets, methodologies, and fostering international collaborations for the ethical and responsible use of data in advancing technology for societal good.

Technological innovations, such as the creation of technologies and automation, have simplified many facets of life. In this regard, soft computing encompasses topics such as computational intelligence, deep learning, optimization techniques, evolutionary algorithms, intelligent e-learning systems, soft sets, rough sets, fuzzy logic, neural networks, genetic algorithms, and machine learning, and signal processing encompasses emerging topics such as communication engineering, information theory and networks, electronics engineering and microelectronics, signal, image and speech processing, wireless and mobile communication, circuits and systems, energy systems, power electronics and electrical machines, electro-optical engineering, instrumentation engineering, avionics engineering, control systems, Internet of things and cybersecurity, biomedical devices, MEMS and NEMS, radar signal processing, and VLSI signal processing which are all state-of-the-art topics having great deal of research potential which contribute toward promotion of peaceful, just and inclusive societies for sustainable development.

The conference also helps to exchange and exhibit mutual respect and understanding of ideas, philosophical principles, cognitive styles, and mindsets as well as acts of integrity and purposeful personal reflection, thus preserving the social balance.

The submitted papers are expected to cover state-of-the-art technologies, product implementation, ongoing research as well as application issues and conflict resolution procedures which will enable people to pursue their academic and research objectives without any fear of instability and thus helps in the peaceful development of oneself and the nation.

Contents

5G MIMO Antenna Design for Bandwidth Enhancement Using Split Ring Resonator

Ch. Rama Krishna, Ashok Battula, Bhavana Majji, Sowjanya Akumalla, and Chandrika Saxena

Abstract In order to increase its impedance bandwidth, this research article proposes a MIMO 5G antenna and an inset feeding method. The unique feature of the proposed antenna lies in its capability to operate at a frequency of 28 GHz. A complementing split ring rectangular resonator that is precisely positioned on a modified ground plane is part of its unique design. This novel method in this design improves the antenna's performance, allowing it to efficiently and accurately broadcast and receive information. The antenna's performance is optimized by the incorporation of the resonator and the defective ground plane, making it a potential option for advanced wireless communication systems. The antenna is mounted on a FR4 epoxy substrate with physical dimensions of 40 mm × 40 mm × 0.8 mm. Four identical patches with triangle cuts are stacked orthogonally to form the MIMO 5G antenna design. This research introduces a MIMO 5G antenna, utilizing triangular cuts and inset feeding for enhanced impedance matching and increased gain to 5 dBi. It covers a broad 7 GHz bandwidth, from 25.5 to 32.5 GHz in the 5G band. This antenna demonstrates promising potential for improved performance in 5G communication systems. A built antenna prototype is thoroughly tested to confirm the modelling results. This meticulous verification method ensures that the measured result matches the simulation results, validating the design's reliability and accuracy.

Ch. Rama Krishna (✉)
Department of Electronics and Communication Engineering, Vignan Institute of Technology and Science, Hyderabad, Telangana, India
e-mail: rahulvar434@gmail.com

A. Battula · B. Majji · S. Akumalla · C. Saxena
Department of Electronics and Communication Engineering, Sreyas Institute of Engineering and Technology, Hyderabad, Telangana, India
e-mail: bathula.ashok@gmail.com

B. Majji
e-mail: bhavana.maroju@gmail.com

S. Akumalla
e-mail: sowji.vits@gmail.com

C. Saxena
e-mail: chandrikasaxena711@gmail.com

© The Author(s), under exclusive license to Springer Nature Singapore Pte Ltd. 2025
H. Zen et al. (eds.), *Soft Computing and Signal Processing*, Lecture Notes in Networks and Systems 1200, https://doi.org/10.1007/978-981-97-9926-8_1

Keywords DGS · Gain · Impedance bandwidth · MIMO antenna

1 Introduction

By allowing several users to connect at once, the MIMO communication technology, introduced in the 1970s, revolutionized high-data-rate communication. The expanding variety of technologies and uses in the modern environment has functioned a desire for antennas with greater bandwidths. The small, four-band MIMO antenna created for 5G applications is one prominent example, as it is explained in [1]. The four frequency bands of this novel antenna are harmoniously matched by the use of L-shaped components and concave ground structures. Interference is greatly reduced by the orthogonal structure.

Considerable attention has been directed towards C-V2X applications in the context of flexible circularly polarized MIMO antennas, as extensively discussed in Ref. [2]. These antennas produce circularly polarized (CP) waves using an L-shaped slot antenna and an altered ground plane. These antennas exhibit noteworthy adaptability, catering to a variety of frequency bands including IEEE 802.11p, Wireless LAN, and WiMAX. To mitigate mutual coupling an antenna designer for 5G uses, the maximum-likelihood (ML) approach is employed, elucidated in detail in Ref. [3]. The design entails a careful analysis of the appropriate size and arrangement of parasitic components, improving overall performance. Additionally, as detailed in [4], a mobile antenna for 5G MIMO with a distinctive layout and 4 corner feeds has been created.

According to [5], new smartphone has a planar 8 port antenna architecture installed on its smoothly curved metal borders. This antenna effectively functions at 3.4 and 36 GHz, which is the frequency range for 5G. Notably, all of its ports have isolations of at least 14 dB, and its reflection coefficient measurements. This antenna also has a phenomenal ECC (envelope correlation coefficient) of 0.3 and an efficiency that surpasses 57%. The results highlight how multiple feeding MIMO antennas in 5G telephones with different patch structures offer significant promise. The identification of a defected ground structure (DGS) revealed a 2-port, 1-element MIMO antenna crafted for 5G mm Wave, as documented in Ref. [6]. This DGS finding highlights advancements in antenna design for high-frequency communication in the 5G spectrum. The antenna contains an annular ring slot provided for by a SAW. When measured at 10 dB, the antenna's bandwidth was 400 MHz. At 28 GHz, Port isolation is reduced by 20 dB when slot [7] is used as a cavity. Both ports displayed 6.9 dBi, which is the element's maximum gain at 28 GHz. For the desired frequency spectrum, the diversity gain and ECC are reported at 9.98 dB and 0.065, respectively. Consistency between simulation and measurement results further validates these findings. A 4-port antenna featuring stubs [8], strips [9], and slits [10] was employed to enhance impedance bandwidth rejection and control. The MIMO system demonstrated excellent isolation and a low envelope correlation coefficient across all ports. Various developments have emerged in the quest to improve MIMO technology for 5G mobile applications. These include the development of an 8-port MIMO array with an emphasis on asymmetric antenna topologies to enhance ECC, as addressed in [11], as well as the implementation and measurement of two planar inverted-F antenna elements, as described in [12]. A dielectric resonator antenna, as described in [13], is also crucial in this situation. The work also includes the use of a conventional loop antenna [14] together with a specific filter [15] as described in [16]. A small double-band MIMO antenna array with feed lines is also presented in [17]. Last but not least, [18] presents a PIFA with improved isolation characteristics. Together, these developments help to progress MIMO technology for 5G [19] mobile applications.

This study proposes a newly created 4-port MIMO 5G antenna that is especially designed to have exceptional isolation, high gain, and good MIMO capabilities. The antenna, which functions at 28 GHz, is proposed for usage in 5G mm-wave applications. The HFSS is used to address the proposal and analysis requirements for this novel MIMO antenna arrangement. The concept involves a MIMO array antenna with an bandwidth of 25.5–32.5 GHz. This antenna stands out as an optimal choice for effectively covering the 28 GHz 5G spectrum, encompassing frequencies from 25.50 to 32.5 GHz. Its dependable performance and broad coverage make it particularly well-suited to meeting the high-speed data transfer and communication requirements of the 5G network. This antenna is positioned to play a critical part in improving the capabilities of 5G technologies, as it is designed to provide reliable and efficient signal reception within the defined frequency range.

The measurement results confirm a significant alignment with the simulation results, and the prototype is constructed on a FR4-epoxy substrate. Remarkably, the antenna exhibits outstanding characteristics, including an isolation level exceeding 40 dB for the 28 GHz band. The simulated diversity gain routinely surpasses 9.9997, while the ECC is notably lower than 0.0001. These characteristics make the proposed MIMO 5G antenna an excellent choice for future 5G applications functional at 28 GHz in the mm-wave frequency band.

2 Proposed MIMO Antenna Structure

The wideband design arrangement of the four monopole 5G MIMO antennas is shown in Fig. 1. This antenna's design enables data transfer through a variety of inputs and outputs. Figure 1a, b illustrates the front and rear views of the proposed antenna without split ring resonator [20], occupying an area of 1600 mm^2. The antenna is characterized by a relative permittivity (ε_r) of 4.4 and a loss tangent of 0.02. These physical dimensions and material properties play a crucial role in the antenna's performance.

The antenna that is being designed has four ports and prioritizes excellent isolation efficiency using an orthogonal diversity configuration. The antenna is deliberately designed with minimized overall dimensions to ensure compatibility with the compact sizes of contemporary devices. The antenna under discussion utilizes an

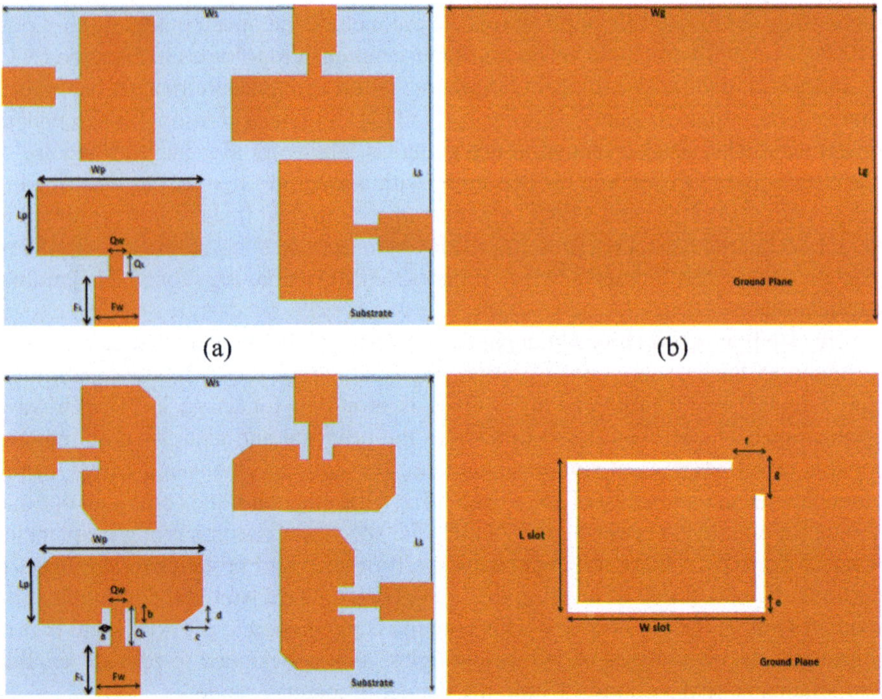

(a) (b)

(c) (d)

Fig. 1 Development of the proposed antenna

FR4-epoxy substrate with a surface area of 1600 mm^2, and height of $h = 0.8$ mm. Comprising four identical radiating patches with inset feeding, this wideband 5G MIMO [21] antenna is designed to offer manufacturing simplicity, ease of modelling, and effective impedance matching. The adoption of inset feed amplifies capacitance influence, concurrently enhancing return loss characteristics.

The development of a single patch is the aim of this section. Three essential parts make up this antenna design: a patch, a quarter-wave transformer, and a feed line. Calculations using the formulae listed below are used to determine the antenna's parameters. The length and width of the patch are determined using Eqs. (1) and (2), providing precise calculations for optimal antenna design. These equations contribute to the accurate dimensioning of the antenna's key components.

$$W = \left[c((\varepsilon_{\mathrm{r}} + 1)/2)^{-\frac{1}{2}}\right]/2f_0 \tag{1}$$

$$L = \left[c/\left(2f_0(\varepsilon_{\mathrm{eff}})^{-\frac{1}{2}}\right)\right] - 2\Delta L \tag{2}$$

$$\varepsilon_{\mathrm{eff}} = (\varepsilon_{\mathrm{r}} + 1)/2 + ((\varepsilon_{\mathrm{r}} - 1)/2)[(1 + 12h)/W]^{-\frac{1}{2}} \tag{3}$$

$$\Delta L = 0.412h\left[(\varepsilon_{\mathrm{eff}} + 0.300)\left(\frac{W}{h} + 0.264\right)\right] / \left[(\varepsilon_{\mathrm{eff}} - 0.258)\left(\frac{W}{h} + 0.800\right)\right] \tag{4}$$

$$Z_1 = \sqrt{Z_0 R_{\mathrm{in}}} \tag{5}$$

$$R_{\mathrm{in}} = 1/(2G_{\mathrm{e}}) \tag{6}$$

where G_{e} represents the edge conductance

$$G_{\mathrm{e}} = 0.00836(W/\lambda_0) \tag{7}$$

$$\frac{W}{h} = 8e^{A}\left(e^{2A} - 2\right) \quad \text{for } W/h < 2 \tag{8}$$

$$\frac{W}{h} = \frac{2}{\pi}[B - 1 - \ln(2B - 1) + (\varepsilon_{\mathrm{r}} - 1)/2\varepsilon_{\mathrm{r}}\{\ln(B - 1) + 0.39 - (0.61/\varepsilon_{\mathrm{r}})\}] \quad \text{for } W/h > 2 \tag{9}$$

$$A = (Z_0/60)[(\varepsilon_{\mathrm{r}} + 1)/2]^{\frac{1}{2}} + [(\varepsilon_{\mathrm{r}} + 1)/(\varepsilon_{\mathrm{r}} - 1)(0.23 + (0.11/\varepsilon_{\mathrm{r}}))] \tag{10}$$

$$B = 377\pi/\left(2Z_0\sqrt{\varepsilon_{\mathrm{r}}}\right) \tag{11}$$

Due to their essential function, MIMO antennas for mm-wave applications must be carefully designed. Network performance is substantially impacted by mutual coupling between MIMO antennas. A basic four-port antenna with a rectangular ground plane is printed on dielectric material with quarter-wave transformer feed as shown in Fig. 1.

The antenna's dimensions have undergone optimization, as depicted in Table 1, ensuring the refinement of key parameters for enhanced performance. The gap between antennas is kept at 5.7 mm to reduce mutual coupling and improve antenna performance. An excellent impedance bandwidth, spanning from 25.6 to 32.5 GHz, is displayed by the suggested inset feed-coupled antenna. Employing the High-Frequency Structured Simulator, the antenna design has been meticulously formulated to encompass a frequency range spanning from 25.5 to 32.5 GHz, as detailed in this research article.

Table 1 outlines the optimal measurements for the suggested 5G MIMO antenna, providing key design parameters for effective implementation. Table 2 displays the impedance bandwidths of antennas utilized during the transition period, offering insights into the frequency characteristics of each design.

Table 1 Measurements of the proposed antenna are optimized

Dimensions	Size (mm)	Dimensions	Size (mm)
$L_s = L_g$	40	W_g	40
b	1	L_p	1
L_s	40	f	5
F_w	3.2	Q_L	4
L_{slot}	23	F_L	6
W_{slot}	23	d	3
$W_s = e$	1	c	2
Q_w	0.5	h	0.8
g	5	$W_s = W_g$	40
W_p	12	a	0.5

Table 2 Construction of the axial ratio and MIMO antenna impedance bandwidths

Antenna	Method	Impedance bandwidth (GHz)	Polarization
Antenna-1 (Fig. 1a, b)	Rectangular patch with quarter-wave transformer impedance matching	27.5–30	No polarization
Antenna-2 (Fig. 1c, d)	Triangular cuts on patch, CSRR on the ground and inset feeding	25.5–32.5	Circular polarization (27.25–28.75 GHz)

3 Results

In Fig. 2a, b, the radiation-emitting rectangular patch and the impact of QWT feeding on the performance between ports 1 and 4 are illustrated. This relationship is directly influenced by the size of Antenna 1 and its initial resonant frequency. Within the 28 GHz operating frequency band, the antenna demonstrates precise resonance, showcasing its accuracy and effectiveness in that specific frequency range. Importantly, the proposed antenna demonstrates outstanding impedance matching [22], characterized by a VSWR of less than 2 across the entire operational frequency spectrum, as detailed in this research paper (Fig. 3).

Figure 2a, b illustrates the impact of varying the quarter-wave transformer feeding and radiating rectangle patch size on the $S11$ and performance across ports 1–4. The study highlights the interconnectedness between the size of Antenna 1 and its initial resonant frequency. Additionally, the proposed antenna exhibits a VSWR of less than two, showcasing excellent impedance matching throughout its operational frequency range.

The proposed antenna's radiating components are orthogonally arranged, which helps to ensure robust isolation. Figure 4 depicts the fluctuations in the S-parameters of Antenna 2. It is worth noting that isolation values of 34.9 dB and 44.8 dB are attained among radiating patches 1 and 2, respectively. Furthermore, a remarkable

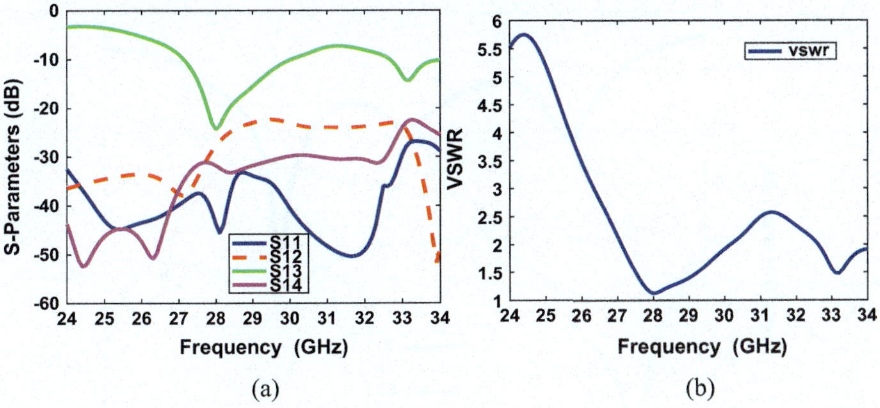

Fig. 2 VSWR plots and *S*-parameters for the Antenna 1

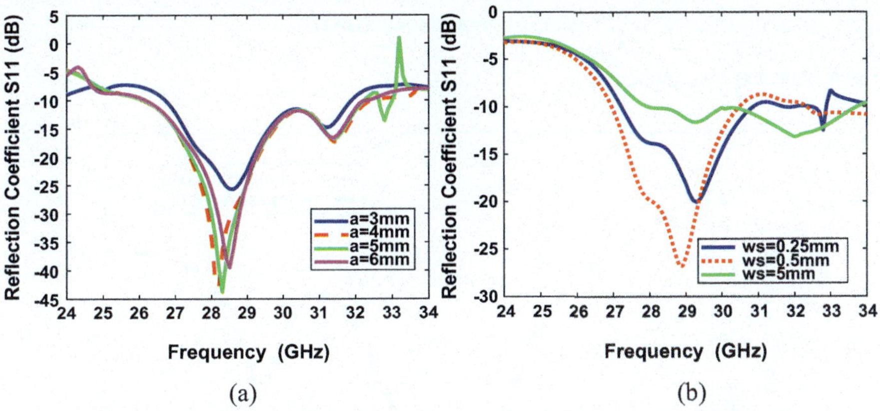

Fig. 3 Figure 1 shows the variation of reflection coefficient plots for Antenna 2

isolation of 75 dB is maintained between radiating elements 1 and 3 over the whole frequency band of 24–34 GHz. This high level of isolation is achieved by efficiently limiting current flow from radiating element 1, a critical component in establishing such excellent isolation.

For optimal isolation, the radiating components of the proposed antenna should be strategically arranged in an orthogonal configuration on the substrate. Figure 5 shows the changes in the reflection coefficient for Antenna 2, demonstrating the consistency of the simulation and measurement (Table 3).

Figure 6 displays the *S*-parameters of the proposed antenna, indicating an impedance bandwidth spanning from 24 to 34 GHz. Additionally, *S*21, *S*31, and *S*41 illustrate that mutual coupling between antenna components consistently remains below 25 dB. Figure 6 shows the surprising similarity between the simulated and

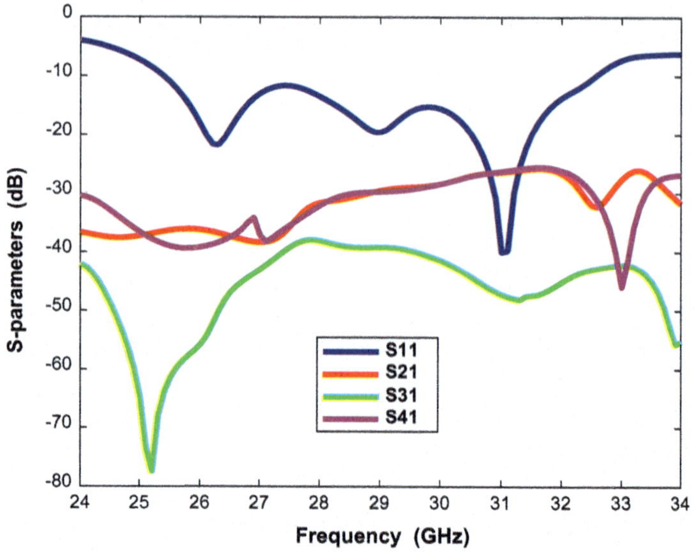

Fig. 4 Antenna 2 S-parameters

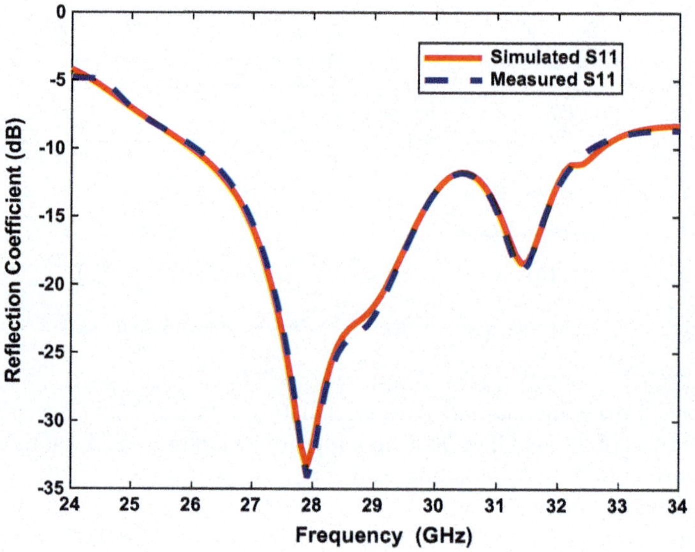

Fig. 5 Plots of simulated and measured reflection coefficients

Table 3 A comparison between the existing systems with proposed antenna

References	Size (mm × mm)	Ports	Impedance bandwidth (GHz)	Isolation (dB)	Circular polarization bandwidth (GHz)	CLL (bits/s/Hz)	ECC	Gain (dBi)
[1]	39 × 39	4	2.6, 3.5, 4.8, 5.8	21.5	–	0.4	0.012	–
[2]	30 × 30	1	3.23–6.26	> 20	2.95–6.02	0.078	0.02	–
[3]	50 × 30	4	3300–4200 MHz	28	–	0.017	0.0000495	–
[4]	50 × 50	4	3.3–4.2	20	–	–	0.04	–
[5]	140 × 89	8	3.4–3.7	15	–	–	0.3	–
[6]	33 × 27.5	2	0.4	30	–	–	0.001	6.9
[7]	180 × 180	4	0.6–1.09 2.6–3.4 and 4.2–7.0	22	–	29.83 bps/Hz	≤ 0.50	6.54
[8]	50 × 50	8	2–12	17	–	0.5	0.3	3.15
[9]	75 × 180	64	3.2–4	11	–	–	0.005	2–5
[10]	40 × 3	64	3.3–5	21	–	–	0.11	–
[11]	50 × 100	4	2.7–3.6	25	–	–	0.009	5.5
[12]	27.3 × 1.2	8	3.4–3.6	15	–	–	0.16	4.5
[13]	20 × 20	2	27.5–28.35	24	–	–	0.013	8
[15]	150 × 75	8	3–5.5	15	–	31.6–39.2 bps/Hz	0.1	2
[17]	124 × 74	8	2.4–2.7 3.3–3.6	15.1	–	0.2	0.12	–
Proposed work	40 × 40	4	25.5–32.5	25	1.5	0.2	0.12	5

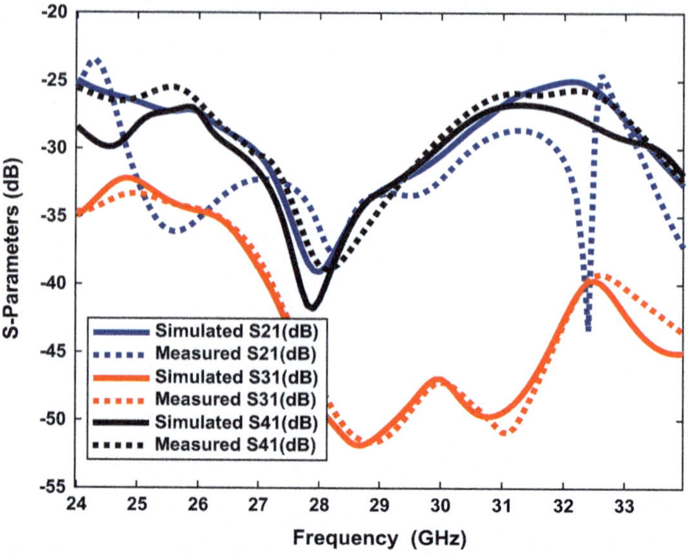

Fig. 6 The *S*-parameters of the *S*21, *S*31, and *S*41 proposed MIMO antennas were measured and simulated

measured findings, showing a high level of accuracy in the simulations. Figure 7 illustrates the radiation patterns of the proposed array simulation and measurement at three dissimilar frequencies: 26, 28, and 31 GHz. This demonstrates the radiation properties of the antenna at these specific frequencies.

Several metrics are used to evaluate the diversity provided by the antenna, including the envelope correlation coefficient determined according to [13], as well as the TARC and diversity gain as described in [15].

The ECC for a MIMO antenna quantifies the degree of correlation between adjacent radiating elements. It is a necessary instrument for determining the variety of the antenna. Figure 8 depicts the ECC values, which were calculated using the *S*-parameters according to Eq. (12). This analysis aids in determining the level of variety produced by the antenna, which is an important aspect in determining its performance.

$$\text{ECC} = \frac{\left|s_{11}^{*}s_{12} + s_{12}^{*}s_{22}\right|^2}{\left(1 - \left(|s_{11}|^2 + |s_{21}|^2\right)\right)\left(1 - \left(|s_{22}|^2 + |s_{12}|^2\right)\right)} \quad (12)$$

Equation (12) is flexible and can be used to compute the ECC value among any 2 antenna ports. The suggested antenna achieves an ECC of 0.12 over the entire frequency range in which it works. Significant importance is attributed to the antenna's performance, with favourable indicators being an envelope correlation coefficient (ECC) value below 0.5 and a capacity channel length (CCL) below 0.2

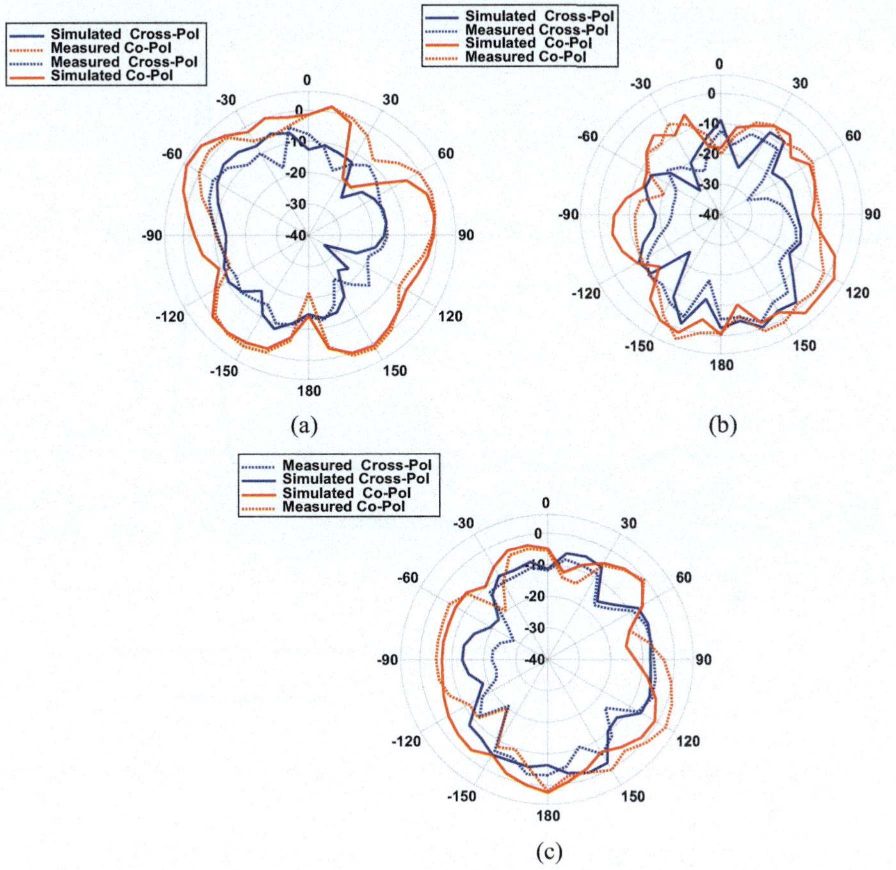

Fig. 7 Antenna radiation patterns for the suggested design at **a** 26 GHz, **b** 28 GHz, and **c** 31 GHz are depicted

across the entire operating frequency band. Figure 9 depicts the diversity gain graphically, with values surpassing 9.9993 across the antenna's working frequency range. This computation is based on the equation from [2]. High diversity gain values indicate that the antenna is capable of successfully mitigating signal fading and improving communication dependability.

$$DG = 10\sqrt{1 - |ECC|^2} \tag{13}$$

In Fig. 10, the gain plot illustrates a consistent maximum gain of 5 dBi throughout the operational frequency range of the proposed antenna, underscoring its stability in performance. Shifting the focus to Fig. 11, the graph portrays the TARC (total active reflection coefficient), providing insights into the antenna's reflective characteristics.

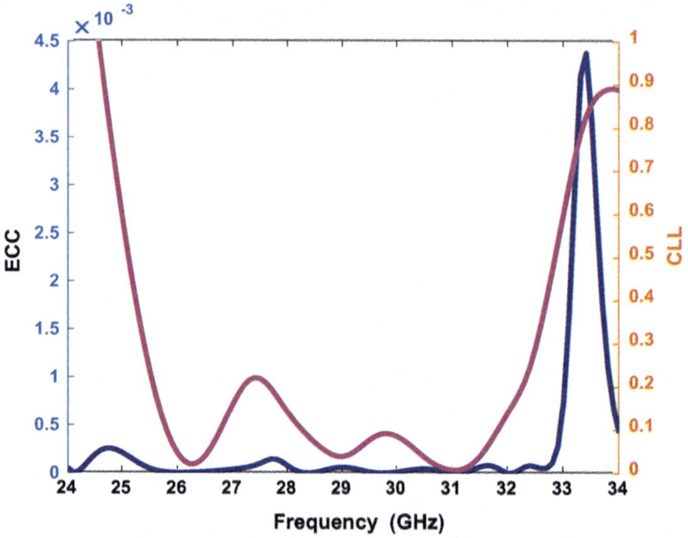

Fig. 8 The proposed antenna's ECC and CLL

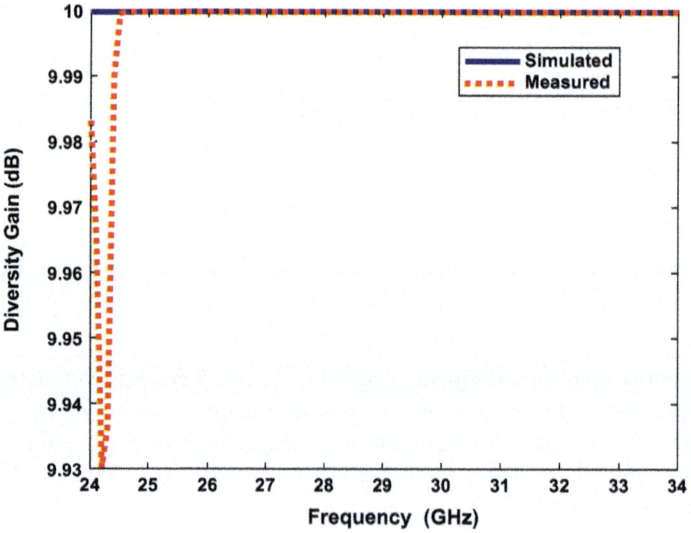

Fig. 9 Gain in diversity of the proposed antenna has been measured and simulated

Figure 12 visually presents the completed antenna from both front and back perspectives, offering a practical understanding of its physical design and construction and contributing to an enhanced comprehension of its overall structure.

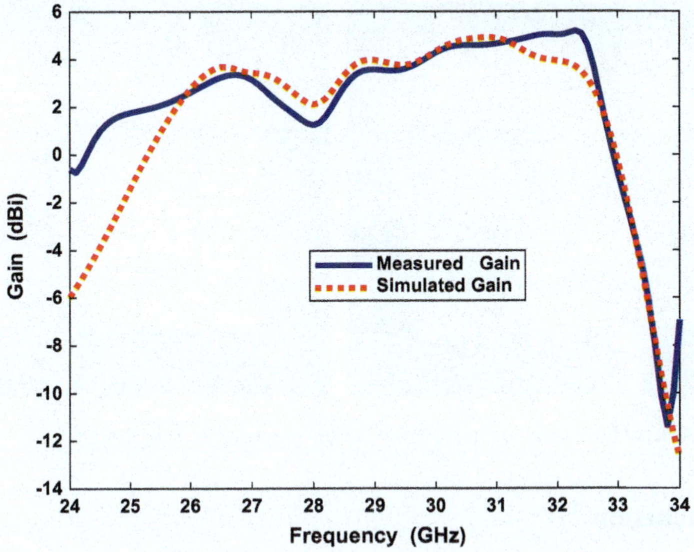

Fig. 10 Proposed MIMO antenna's gain results have been measured and simulated

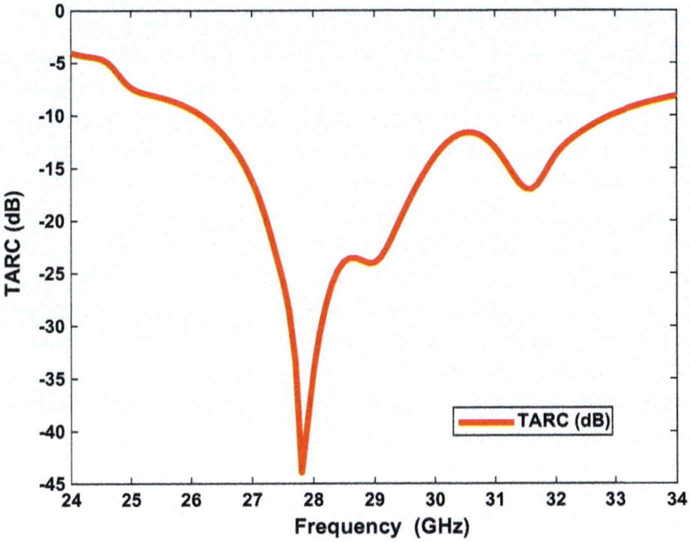

Fig. 11 The TARC of the proposed antenna

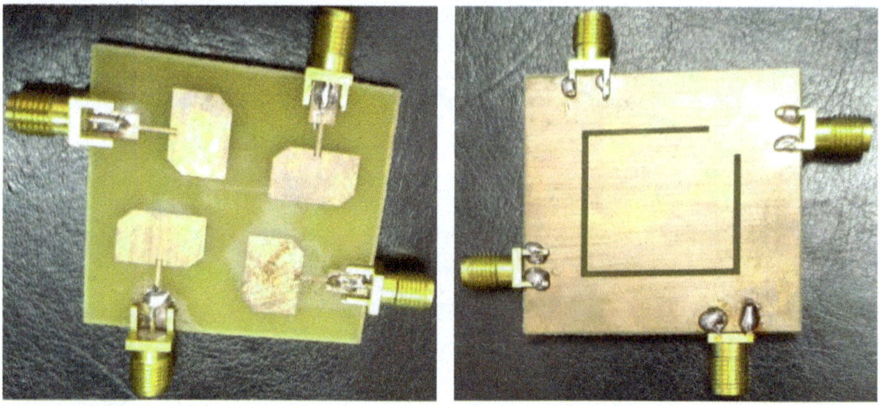

Fig. 12 Fabricated front and rear perspectives of the proposed antenna design

4 Conclusion

The design and development of a small multiple input multiple output (MIMO) antenna with circular polarization are described in this research article using a methodical approach. Resonance at 28 GHz is achieved by means of an inventive design that includes SRR's on the ground plane, inset feeding, and a rectangular patch with triangular cuts. The antenna outperforms similar designs in spite of its tiny size (40 × 40 mm), providing an effective impedance bandwidth of 25.5–32.5 GHz for millimetre-wave and 5G applications. The accuracy of the design is verified by measurements and experimental construction. The results demonstrate an amazing 25 dB isolation and an impedance bandwidth of 25.5–32.5 GHz. In addition, the antenna has a 1.5 GHz axial ratio bandwidth, which allows it to cover the frequency range of 27.25–28.75 GHz. With an ECC of 0.12, a CCL of 0.2 bits/s/Hz, and a notable improvement in diversity gains of about 9.997, these accomplishments yield impressive measurements. These measurements highlight the antenna's superior polarization, capacity, and diversity performance attributes, confirming its appropriateness for demanding communication scenarios. This antenna seamlessly integrates a broad frequency range, exceptional performance, compact size, and adaptability, catering to the demands of modern communication systems.

References

1. Lin Y-C, Lin H-S (2016) Novel four-port antennas backed by co-located dual cavities for dual-band and dual-polarized MIMO applications. In: 2016 IEEE international symposium on antennas and propagation (APSURSI). IEEE
2. Chen D et al (2022) On-road features based in-chamber C-V2X application test scheme design. In: 2022 IEEE smartworld, ubiquitous intelligence & computing, scalable computing &

communications, digital twin, privacy computing, metaverse, autonomous & trusted vehicles (SmartWorld/UIC/ScalCom/DigitalTwin/PriComp/Meta). IEEE
3. Le QH et al (2019) Assessment of a thick-oxide transistor from the 22FDX® platform for 5G NR sub-6 GHz FEMs. In: 2019 IEEE 2nd 5G world forum (5GWF). IEEE
4. Wong K-L, Jian M-F, Li W-Y (2021) Low-profile wideband four-corner-fed square patch antenna for 5G MIMO mobile antenna application. IEEE Antennas Wirel Propag Lett 20(12):2554–2558
5. Tu DTT et al (2017) 28/38 GHz dual-band MIMO antenna with low mutual coupling using novel round patch EBG cell for 5G applications. In: 2017 international conference on advanced technologies for communications (ATC). IEEE
6. Li D et al (2023) Compact dual-/tri-band SIW filtering antennas using cavity-backed slots. In: 2023 IEEE 6th international conference on electronic information and communication technology (ICEICT). IEEE
7. Jha KR et al (2021) 4-port MIMO antenna using common radiator on a flexible substrate for sub-1 GHz, sub-6 GHz 5G NR, and Wi-Fi 6 applications. IEEE Open J Antennas Propag 2:689–701
8. Yu Z et al (2017) A novel Koch and Sierpinski combined fractal antenna for 2G/3G/4G/5G/WLAN/navigation applications. Microw Opt Technol Lett 59(9):2147–2155
9. Roh W et al (2014) Millimeter-wave beamforming as an enabling technology for 5G cellular communications: theoretical feasibility and prototype results. IEEE Commun Mag 52(2):106–113
10. Chen Q et al (2018) Single ring slot-based antennas for metal-rimmed 4G/5G smartphones. IEEE Trans Antennas Propag 67(3):1476–1487
11. Al-Hadeethi ST, Elwi TA, Ibrahim AA (2023) A printed reconfigurable monopole antenna based on a novel metamaterial structures for 5G applications. Micromachines 14(1):131
12. Srivastava K et al (2020) Compact eight-port MIMO/diversity antenna with band rejection characteristics. Int J RF Microw Comput Aided Eng 30(5):e22170
13. Nasir J et al (2016) Design of an MIMO dielectric resonator antenna for 4G applications. Wirel Pers Commun 88:525–536
14. Qian B et al (2023) Wideband self-decoupled MIMO antenna element based on coupled-loop antenna for 5G smartphones. J Electromagn Waves Appl 37(1):110–132
15. Gao F et al (2010) Design of learning-based MIMO cognitive radio systems. IEEE Trans Veh Technol 59(4):1707–1720
16. Nithyasri K, Priyanka B, Ramadevi M, Ramakrishna C (2020) Uniplanar consolidated filtering dipole antenna using edge coupling feed. Int Res J Eng Technol (IRJET) 7(05) (2020)
17. Li M-Y et al (2016) Eight-port orthogonally dual-polarized antenna array for 5G smartphone applications. IEEE Trans Antennas Propag 64(9):3820–3830
18. Wani Z, Abegaonkar MP, Koul SK (2018) A 28-GHz antenna for 5G MIMO applications. Prog Electromagn Res Lett 78:73–79
19. Hu S, Wang F, Wang H (2017) 2.1 A 28 GHz/37 GHz/39 GHz multiband linear Doherty power amplifier for 5G massive MIMO applications. In: 2017 IEEE international solid-state circuits conference (ISSCC). IEEE
20. Ramakrishna Ch, Geervani B (2022) ACS fed coplanar monopole antenna with complementary split ring resonator for WLAN and satellite communication applications. In: Distributed computing and optimization techniques: select proceedings of ICDCOT 2021. Springer Nature Singapore, Singapore, pp 121–130
21. Liu DQ et al (2019) An extremely low-profile wideband MIMO antenna for 5G smartphones. IEEE Trans Antennas Propag 67(9):5772–5780
22. Ramakrishna Ch, Satish Kumar GAE, Chandra Sekhar Reddy P (2021) Quadruple band-notched compact monopole UWB antenna for wireless applications. J Electromagn Eng Sci 21(5):406–416

Real-Time Violence Detection and Alert System Using MobileNetV2 and Telegram Bot

Mannu Thomas, P. Balamurugan, and A. Mummoorthy

Abstract The aggressive actions seriously jeopardize one's quality of life and personal safety. To curb the hostile conduct, several different strategies have been tried, such as installing and keeping up surveillance cameras. It will be extremely relevant if the monitoring systems can recognize violent conduct automatically and immediately send warning or alarm signals to the authorities. New researchers are being drawn to the busy study fields of computer vision and image processing for the identification of aggressive and aberrant conduct. Using MobileNetV2 for violence identification in video streams, this research suggests a real-time violence detection and alarm system. With the help of Telegram Bot for real-time alerts and Cloud Firestore for data management and storage, the system offers public safety a scalable, precise, and flexible solution.

Keywords Violence · Violent activity · Real time · Violent behavior · Surveillance · Violence detection · Alert · Storage · MobileNetV2 · Telegram

1 Introduction

Innovative technologies are being created to improve public safety and safeguard communities. The study introduces a Real-Time Violence Detection and Alert System using MobileNetV2, Cloud Firestore, and Telegram Bot to identify and report violence in real time. With public safety, security, and violence prevention concerns

M. Thomas (✉) · P. Balamurugan
Department of Networking and Communications, SRM Institute of Science and Technology, Kattankulathur, Tamil Nadu, India
e-mail: mt6219@srmist.edu.in

P. Balamurugan
e-mail: balamurp@srmist.edu.in

A. Mummoorthy
Information Technology (IT), Malla Reddy College of Engineering and Technology (MRCET), Hyderabad, India
e-mail: amummoorthy@mrcet.ac.in

© The Author(s), under exclusive license to Springer Nature Singapore Pte Ltd. 2025

H. Zen et al. (eds.), *Soft Computing and Signal Processing*, Lecture Notes in Networks and Systems 1200, https://doi.org/10.1007/978-981-97-9926-8_2

on the rise, technology is a great tool for protecting communities and lives. Advanced and flexible mechanisms for quickly recognizing, reporting, and acting in various forms of violence in public and private contexts are needed to solve these concerns [1]. This study provides a comprehensive Real-Time Violence Alert System that uses MobileNetV2, Cloud Firestore, and Telegram Bot to identify and control violence.

The Real-Time Violence Alert System uses MobileNetV2, a fast, accurate, and efficient deep learning architecture. Rapid identification of violent events is achievable using MobileNetV2's real-time video analysis. Security and rapid response to threats are enhanced by the system's high-precision photo identification and simple integration into surveillance networks. Deep learning technology may greatly alter present social concerns.

Our solution relies on a Telegram Bot to detect and respond to violent incidents. Telegram's vast user base, privacy emphasis, and fast messaging make it ideal for notifying authorities. This process helps turn detection into action, enabling swift and decisive responses to violent events. The system relies on a Telegram Bot for public safety communication.

A powerful Real-Time Violence Alert System that prioritizes scalability, accuracy, and adaptability is our goal. Our solution is suitable for public spaces, residential buildings, and businesses. Its adaptability makes it effective for increasing public safety and security in many regions and meeting security needs. We tested our technique using numerous violent datasets and a real-world monitoring system. The program accurately identified real-time violent situations. The technology is scalable and may be used on several cameras.

The remainder of the paper is structured as follows: Sect. 2 contains evaluations of comparative studies. The methodology, organization, and dataset are all described in Sect. 3. The methodology for the experiments and analysis is covered in Sect. 4. Section 5 concludes with a discussion of upcoming studies.

2 Novelty

The novelty of this project lies in its innovative integration of MobileNetV2 and Telegram Bot for real-time violence detection and alert communication. MobileNetV2, known for its speed, precision, and efficiency in image recognition, allows for rapid identification and classification of violent actions within video streams. This ensures timely detection of potential threats. Complementing this, the Telegram Bot facilitates immediate and efficient alert communication, sending real-time notifications to relevant authorities and ensuring prompt responses. This combination of advanced deep learning technology and effective messaging capabilities creates a robust and efficient system for enhancing public safety.

3 Related Works

Violence detection approaches may be classified into three groups: visual-based methods, audio-based methods, and hybrid systems.

Visual data is retrieved and shown with suitable features in the visual method. There are two distinct categories of features: local and global. Local features encompass location, velocity, form, and color, whereas global characteristics involve average speed, region occupancy, relative positional fluctuations, and object-background interactions.

The audio-based technique categorizes violent behavior using auditory data. An approach utilizing Gaussian mixture models and hidden Markov models is used to distinguish between explosions, gunshots, and automobile brakes in audio.

The hybrid technique focuses on combining visual and aural attributes. Some techniques detect violent circumstances in recordings by recognizing flames and blood, measuring motion intensity, and identifying related noises. The CASSANDRA technology utilizes audio signals like screams and motion characteristics linked to articulation in videos to detect violence in surveillance footage.

3.1 Modeling Method with 2D CNN

For effective real-time and on-device violence recognition, this research offers a unique violence detection pipeline that may be utilized with 2D CNNs. The three components of the proposed pipeline are frame grouping, a spatial attention module known as Motion Saliency Map (MSM), and a temporal attention module known as the Temporal Squeeze-and-Excitation (T-SE) block. Using a technique that averages the channels of the input frames, three sequential channel-averaged frames are pooled as input for the 2D CNNs. The MSM can identify significant portions of feature maps obtained from motion boundaries by utilizing the difference between successive frames. The T-SE block can naturally emphasize the important time periods for an event of interest [2].

The proposed pipeline provides significant performance improvements over earlier 3D CNN-based methods and significantly less computational complexity than 2D CNNs led by the LSTM. On six violence datasets, the suggested method was tested, and the findings revealed that it performed better than other cutting-edge methods in terms of accuracy and speed. The suggested approach can be utilized in numerous applications, including surveillance systems, to quickly identify violent acts. PyTorch served as the foundation for the networks' implementation. The suggested method can be used in a surveillance system with many cameras and meets real-time or near real-time criteria [2].

3.2 Spatio-Temporal Features with 3D CNN

This study proposes a system at a basic level for detecting violent video segments. The approach utilizes spatio-temporal interest points and K-means clustering to categorize comparable interest points into clusters. The authors categorize each event in an unknown sequence by computing cluster means for each important cluster. The discovery that the spatio-temporal interest spots remain unaffected by variations in size, rotation, and illumination forms the basis for the proposed approach. The authors utilize a Laplacian of Gaussian filter to identify interest points in the video sequence. The identified points of interest are then characterized using a spatio-temporal descriptor that captures the local structure of the interest point. The authors then utilize K-means clustering to organize comparable interest points into clusters. The cluster means are used to categorize an unseen sequence event. In terms of accuracy and speed, the suggested method surpassed existing cutting-edge algorithms on two publically available datasets. This method may be used in many applications, such as surveillance systems, to promptly detect aggressive behavior. This procedure is highly computationally intensive and relies on several factors, making it unsuitable for real-time applications and everyday use.

3.3 CNN and LSTM

A paper suggests a real-time system for detecting violence using deep learning techniques. CNN acts as the spatial feature extractor in the recommended model, while a Long Short-Term Memory (LSTM) serves as a strategy for learning temporal links. The assessment of the suggested model on the Hockey Fight Dataset showed that it achieved a 98% accuracy rate while processing 131 frames per second. The suggested model outperformed current state-of-the-art methods in terms of accuracy and efficiency. The proposed model was evaluated using a combination of three datasets, resulting in an accuracy rating of 94.765%. The recommended methodology found it easier to identify violence between two individuals than violence involving a big group. The proposed model was evaluated on a range of YouTube videos with different formats, achieving a detection accuracy of 100%. The suggested approach is quicker than the current quickest model for detecting violent incidents. This method may be used in many applications, such as surveillance systems, to promptly detect aggressive behavior. The networks were implemented using Python. It demonstrates superior accuracy with benchmark datasets but is not appropriate for real-time applications.

3.4 Bidirectional Convolutional LSTM

The paper examines the issue by utilizing three established criteria for identifying violence: hockey fight datasets, movie datasets, and violent flow datasets. They developed a Spatio-temporal Encoder using the BiConvLSTM architecture for this purpose. The Spatio-temporal Encoder in violence detection now integrates bidirectional temporal encodings elementwise. This feature aims to improve video representations by including long-range information along both temporal axes. The effectiveness of the Spatio-temporal network is comparable to existing techniques for each of the stated datasets. The Spatial Encoder, a simplified iteration of this network, is capable of achieving top-tier performance on datasets related to hockey games and movies. The Spatio-temporal Encoder outperforms the Spatial Encoder on the Violent Flows dataset [3, 4].

3.5 Deep Autoencoder and CNN

The research proposes a CNN-based system designed for efficient real-time processing of data streams from optical sensors in a mobile surveillance setting. Frame-level deep features are extracted using a pre-trained CNN model. An enhanced deep autoencoder (DAE) is employed to analyze the temporal variations in the activities of the surveillance stream. A nonlinear learning strategy is utilized to train a quadratic Support Vector Machine (SVM) for identifying human behaviors. During the testing phase, an iterative fine-tuning process is used to update the parameters of the trained model with new data from a changing environment. Our system outperforms current leading solutions in precision and execution speed based on experiments carried out on benchmark datasets [5].

3.6 Temporal Optical Flow Convolution and Multilayer LSTM

The research "Activity Recognition Using Temporal Optical Flow Convolutional Features and Multilayer LSTM" by Ullah et al. introduces a new method for recognizing activities in movies by utilizing temporal optical flow convolutional features and multilayer LSTM networks [6].

The method begins by extracting temporal optical flow information from the video. Temporal optical flow characteristics depict the movement of individual pixels in a movie as time progresses. A CNN is used to analyze the spatial properties of the video based on the data received. The CNN's output consists of a sequence of feature vectors, each detailing the spatial attributes of each individual video frame [6].

The authors utilize a multilayer LSTM network to analyze the temporal characteristics of the feature vectors generated by the CNN. LSTM recurrent neural networks are beneficial for capturing temporal characteristics in data sequences. The authors utilize a multilayer LSTM network to learn temporal properties from the sequence of feature vectors produced by the CNN. LSTM recurrent neural networks are beneficial for capturing temporal characteristics in data sequences. The multilayer LSTM network produces a vector that signifies the likelihood of each activity class. The writers utilize this vector for categorizing the video into one of the activity classifications. The authors tested their method on two publicly available datasets including movies for activity recognition. Their solution fared better than the most advanced activity recognition algorithms on these datasets [6, 7].

3.7 Violence Detection Using Low-Level Features

The motion zones are segmented based on the dispersion of the optical flow field. Two distinct low-level attribute types are obtained in the motion regions to depict the dynamics and look of violent behaviors. Low-level features consist of the LHOF and LHOG descriptors, both extracted from optical flow pictures [8, 9]. The gathered properties are encoded using the Bag of Words (BoW) paradigm to prevent information duplication, resulting in the creation of a vector of a certain length for each video clip. SVMs are used to categorize video-level vectors. Detecting violence in a crowded scenario was challenging due to excessive opacity and moving throng. The detection rate is low, and there is a high risk for false alarms [10].

3.8 Convolution Long Short-Term Memory

Sudhakaran and Lanz introduce a new method for identifying violent films in their study "Learning to Detect Violent Videos Using Convolutional Long Short-Term Memory," which utilizes ConvLSTM networks. ConvLSTM networks are a type of recurrent neural network that can learn spatial and temporal features from videos [11]. The authors' technique operates by initially extracting video frames. A ConvLSTM network is utilized to analyze the frames of the video to extract both spatial and temporal information. The ConvLSTM network generates a sequence of feature vectors, each capturing the attributes of each individual video frame. The authors categorize the movie as violent or non-violent by applying a Support Vector Machine (SVM) classifier on the feature vectors generated by the ConvLSTM network [11]. The authors' technique was assessed using a dataset containing both violent and peaceful films. Their method outperformed state-of-the-art violence detection algorithms on this dataset [5].

3.9 Anomaly Detection in Surveillance Video

Sultani et al. introduce a deep learning-based method for anomaly identification in surveillance films in their work "Real-World Anomaly Detection in Surveillance Videos." Their approach may identify irregularities in training data with imprecise labels, when the labels are assigned at the video level rather than the clip level. This is important because it allows for the training of anomaly detection algorithms on extensive and diverse surveillance video datasets without the expensive and time-consuming manual annotation procedure [12].

The authors utilize the deep multiple instance ranking (MIR) architecture in their method. Inside the field of multimedia information retrieval (MIR), videos are viewed as bags and each segment inside a movie is seen as an instance. The objective is to develop a ranking model capable of ordering the cases within each group based on their anomaly ratings. The authors introduce several methodologies, such as imposing sparsity and temporal smoothness criteria on the ranking loss function, to improve the MIR framework for anomaly detection.

The authors assess their method using a recently acquired extensive collection of surveillance films. The collection comprises more than 1900 unprocessed surveillance footage from real-world scenarios, including 13 authentic anomalies like fights, vehicle accidents, burglaries, and robberies. The authors' approach on this dataset surpasses state-of-the-art anomaly identification methods [12].

3.10 Space Time Interest Point (STIP)

The research suggests a technique for identifying key spots in video sequences that might aid with action detection. The approach utilizes spatio-temporal interest points and K-means clustering to categorize comparable interest points into clusters. The authors calculate cluster means for each important cluster and use these values to categorize each event in an unseen sequence. The consistent stability of the spatio-temporal interest points under variations in size, rotation, and lighting forms the basis of the proposed approach. The authors utilize a Laplacian of Gaussian filter for identifying interest points in the video stream. The identified interest points are characterized using a spatio-temporal descriptor that represents the local structure of the interest point. Subsequently, the authors employ K-means clustering to categorize comparable interest points into clusters [13].

Events in an unknown sequence are classified based on the cluster means. The proposed approach was evaluated on sequences of individuals walking, and the results indicated that, while disregarding other noteworthy aspects, it effectively categorizes most of the points associated with the walking pattern. This method may be utilized to detect certain actions in video sequences across many applications, such as surveillance systems. Obtaining the information from the task, which has shown promising outcomes and high accuracy, for practical purposes like monitoring and entertainment assessment is too costly [13].

4 Architectural Design and Methodology

This section carefully examines the suggested approach for distinguishing between violent and non-violent situations in real-time security camera video. Utilizing deep sparse convolutional networks with small filter sizes is a method to reduce parameters and operations in mobile CNN architectures. These designs create layers by combining sequential and parallel methods, often organized into modules, rather than adding layers one by one in a linear fashion.

4.1 Architecture

The proposed architecture is presented here. The security camera's footage is segmented into frames. The MobileNetV2 classifier analyzes input frames to detect violent events within the series of frames. Frames without indication of violent conduct are removed. The frames detecting violence in facial recognition have been enhanced. The improved frames are transmitted to the appropriate authorities using a Telegram Bot. Figure 1 shows the proposed system architecture.

4.2 MobileNetV2

MobileNetV2 is a compact convolutional neural network model. The model is an advancement of the MobileNetV1, with notable enhancements in accuracy and efficiency. MobileNetV2 is a convolutional neural network model with 53 layers, a depth of 53, and a width of 320. The following layers are included.

The input layer is a 3D tensor that depicts the video frame. The input layer dimensions consist of height, width, and channels. Height and width represent the dimensions of the video frame, while channels indicate the number of color channels in the video frame.

MobileNetV2 has a series of inverted residual blocks. An inverted residual block comprises a depthwise convolution, a pointwise convolution, and a bottleneck layer [14].

The output layer consists of a single output neuron that is part of a fully linked layer. The probability of violence in the video frame is indicated by the output of the output neuron [6].

The model has around 3.5 million parameters, a notably lower amount compared to comparable CNN models like VGG16 and ResNet-50. MobileNetV2 may be used

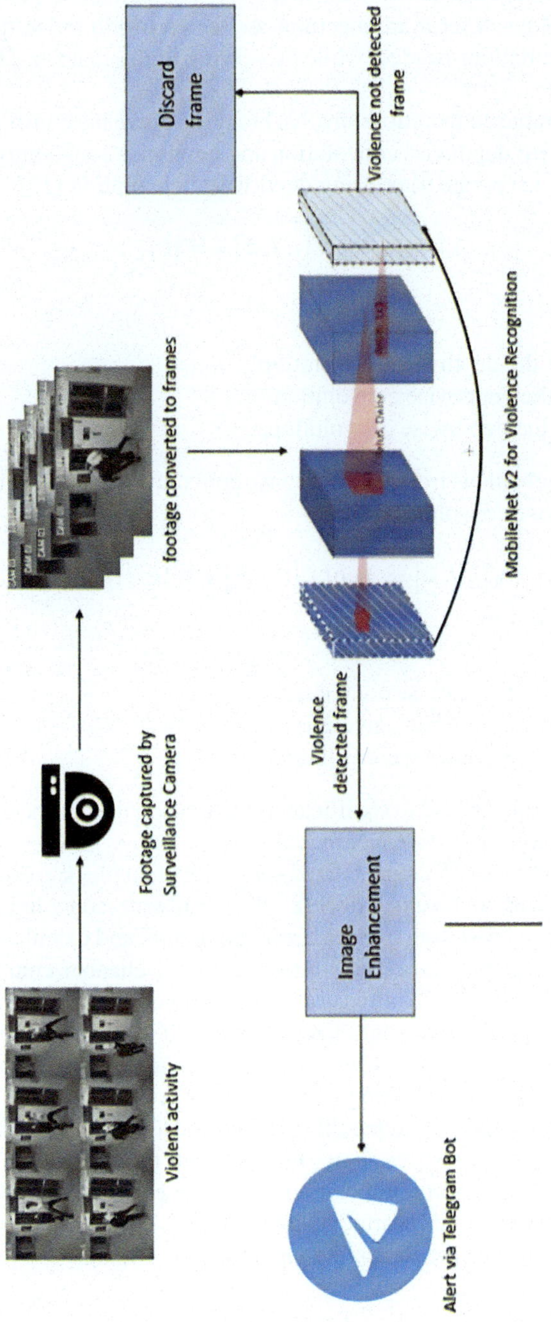

Fig. 1 Proposed system architecture

for many applications such as image categorization and object identification. Its lightweight and efficient construction make it especially suitable for mobile devices. MobileNetV2 employs several architectural strategies to minimize the amount of parameters and calculations needed, while preserving high accuracy. These methods consist of:

Depthwise separable convolutions involve breaking down the convolution process into two distinct steps: depthwise convolution and pointwise convolution. Each input channel undergoes a separate filter using depthwise convolution [15].

$$Y[i, j, k] = X[i, j, k] * W[k] \tag{1}$$

where

Y is the output of the depthwise convolution
X is the input to the depthwise convolution
W is the filter for the depthwise convolution.

All of the output channels from the depthwise convolution are subjected to a single filter during pointwise convolution.

$$Y[i, j, k] = \mathrm{sum}(X[i, j, k] * W[k, l]) \tag{2}$$

where

Y is the output of the pointwise convolution
X is the input to the pointwise convolution
W is the filter for the pointwise convolution.

This decomposition can lead to significant reductions in the number of parameters and computations required for a CNN model.

Inverted residual blocks are a specialized sort of residual block created for mobile devices. Inverted residual blocks prioritize the pointwise convolution above the depthwise convolution, resulting in decreased parameters and calculations needed.

CNN models utilize bottleneck layers to reduce the channel count. Bottleneck layers are inserted between the depthwise and pointwise convolutions within inverted residual blocks to reduce the amount of parameters and calculations required.

The module in Fig. 2 exhibits a residual cell with a stride of 1 and a residual/identity connection due to cell resizing with a stride of 2. A conventional convolution is denoted as "conv," a depthwise separable convolution as "dwise," a ReLu activation function with a magnitude restriction as "ReLu6," and the application of the linear function as "Linear." The primary techniques recommended in MobileNetV2 [16] were the linear bottleneck and inverted residual blocks. To prevent information loss due to nonlinear functions like ReLu, the input channel capacity is increased in the linear bottleneck layer.

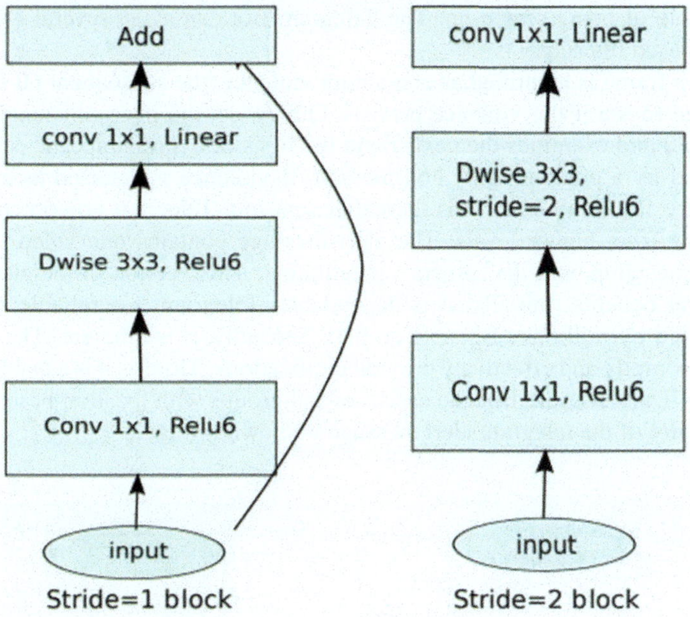

Fig. 2 MobileNetV2 architecture

It is predicated on the concept that information lost in one channel can be preserved in another. An inverted residual block's structure in the channel dimension is narrow-wide-narrow, while a normal residual block's structure is wide-narrow-wide. Skip connections between narrow layers, rather than wider ones, can decrease the memory footprint [17].

4.3 Alert Module

The alarm module is an essential component of any violence detection system. Violence is responsible for notifying the appropriate authorities when discovered. This is the functionality of the Alert Module. When a frame is identified as containing violence, the subsequent 60 frames are scrutinized to see if the violence persists. Otherwise, the internal counter variable is reset to 0 and examines the next frame for signs of aggression. The current time is obtained by a predefined Python method. If violence is detected as true for 60 consecutive frames. An alarm is thereafter sent to a Telegram group consisting of individuals from higher levels. The alert message contains one video frame, the current timestamp, and the camera's positioning. This version of the alert module has several benefits. Initially, it is dependable. Telegram is a reliable messaging service used by millions of people around. Secondly, it is efficient. The Telegram Bot can promptly and effectively provide notifications. Thirdly, it is capable of being scaled up. The Telegram Bot can assist several groups with location-based filtering.

When a frame is identified as containing violence, the subsequent 60 frames are scrutinized to see if the violence persists. Otherwise, the internal counter variable is reset to 0 and examines the next frame for signs of aggression. The current time is obtained by a predefined Python method. If violence is detected as true for 60 consecutive frames. An alarm is thereafter sent to a Telegram group consisting of individuals from higher levels. The alert message contains one video frame, the current timestamp, and the camera's positioning. This version of the alert module has several benefits. Initially, it is dependable. Telegram is a reliable messaging service used by millions of people around. Secondly, it is efficient. The Telegram Bot can promptly and effectively provide notifications. Thirdly, it is capable of being scaled up. The Telegram Bot can assist several groups with location-based filtering. A screenshot of the telegram alert message is shown in Fig. 3.

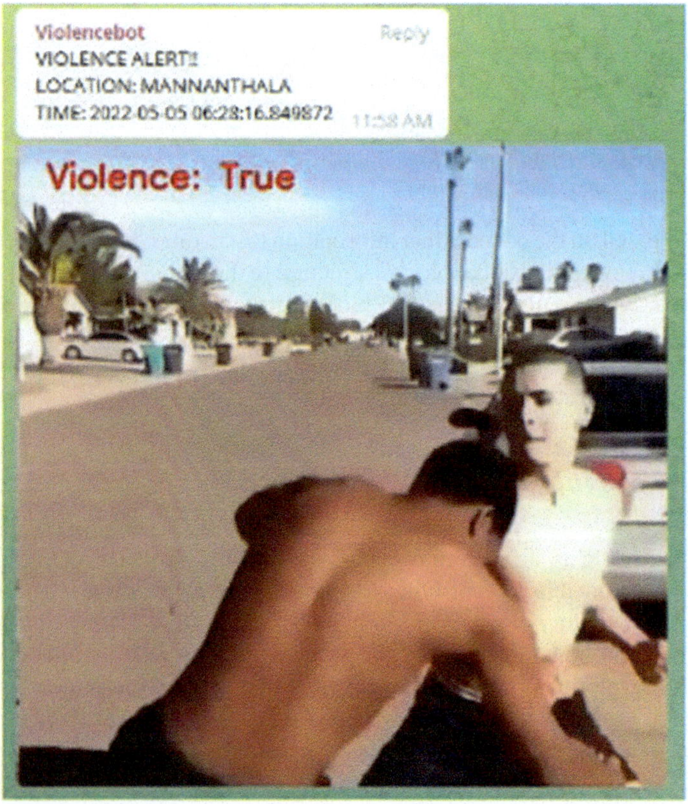

Fig. 3 A screenshot of the alert message

5 Experimental Evaluation

5.1 Model Performance

The violence detection system leverages the MobileNetV2 architecture, a lightweight and efficient deep learning model, to process video frames and identify instances of violent behavior. The dataset used for training the model comprises 700 videos, each with an average duration of 5 s. This dataset is evenly divided into 350 videos showing violent behavior and 350 videos depicting non-violent behavior.

Training and Testing Accuracy

- **Training Accuracy**: The model achieved an accuracy of 96% on the training dataset. This high accuracy indicates that the model effectively learned the features associated with violent and non-violent actions during the training phase.
- **Testing Accuracy**: The testing dataset, comprising new videos not seen by the model during training, yielded an accuracy of 95%. This suggests that the model generalizes well to unseen data, maintaining high performance when evaluating new instances of violent and non-violent behavior.

These results demonstrate the model's capability to accurately distinguish between violent and non-violent actions in real-time video streams.

5.2 Convergence and Stability

The model's training process is visualized in Fig. 4, which shows the training and testing accuracy and loss over successive epochs. The following observations highlight the convergence and stability of the model:

- **Rapid Convergence**: The accuracy and loss metrics stabilize after approximately 5 epochs, indicating that the model quickly converges to a solution. This rapid convergence is beneficial as it reduces training time and computational resources.
- **Consistent Performance**: Both the training and testing curves show consistent improvement and stabilization, reflecting the model's ability to maintain performance across different datasets without overfitting (memorizing training data) or underfitting (failing to capture the underlying patterns).

The stable accuracy and loss curves validate the robustness of the MobileNetV2 architecture for this application, ensuring reliable performance in real-world scenarios.

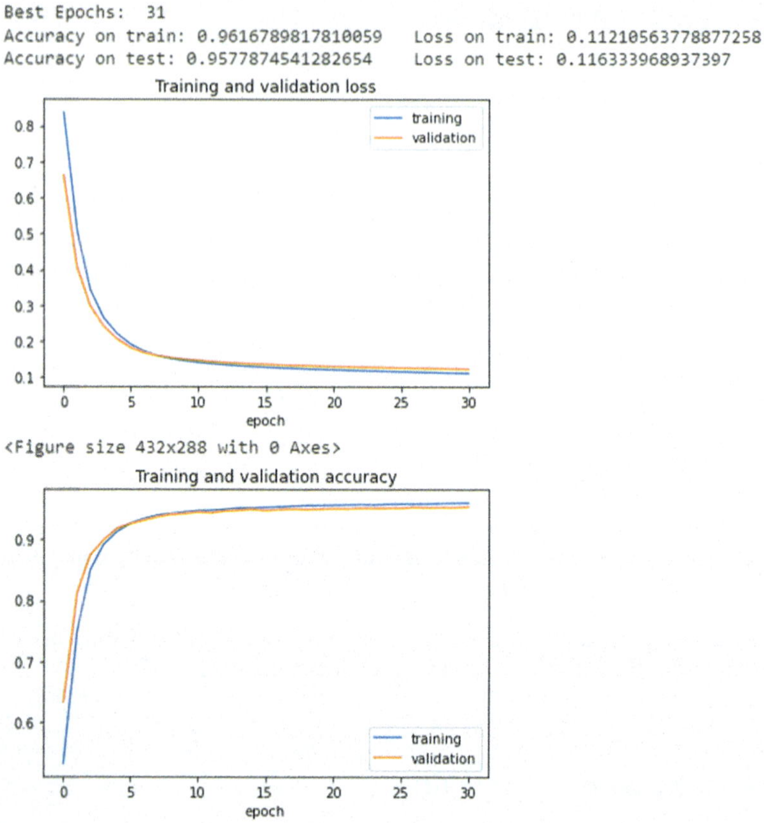

Fig. 4 Accuracy and loss of the training set and testing set

5.3 Evaluation Metrics

Figure 5 displays the calculated confusion matrix along with other evaluation metrics. The accuracy rate is the percentage of positive forecasts that come true. Figure 6 shows the output frames that recognized violence and that do not recognize violence, respectively.

The percentage of violent videos that are genuinely violent is measured in this case by precision. A high precision rate indicates that the model effectively minimizes false positives, which is critical for avoiding unnecessary alerts.

$$\text{Precision} = \text{True Positives}/(\text{True Positives} + \text{False Positives}) \quad (3)$$

```
> Correct Predictions: 4606
> Wrong Predictions: 203
```

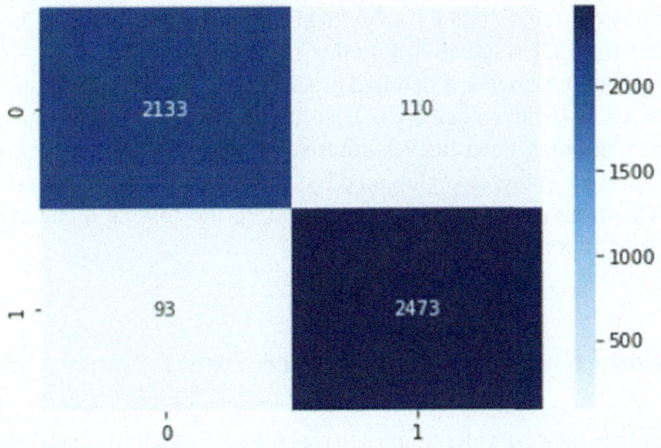

	precision	recall	f1-score	support
NonViolence	0.96	0.95	0.95	2243
Violence	0.96	0.96	0.96	2566
accuracy			0.96	4809
macro avg	0.96	0.96	0.96	4809
weighted avg	0.96	0.96	0.96	4809

Fig. 5 Confusion matrix of the trained model

Recall measures the fraction of accurately anticipated actual positive cases. In this instance, recall shows the fraction of violent videos that are appropriately categorized as violent. High recall ensures that most instances of violence are detected, reducing the likelihood of missing critical events [18].

$$\text{Recall} = \text{True Positives}/(\text{True Positives} + \text{False Negatives}) \tag{4}$$

F1-score is a harmonic mean of precision and recall, and it is frequently used as a single metric to evaluate a classification model's performance [18]. An F1-score of 96% reflects the model's ability to maintain high precision and recall simultaneously.

$$\text{F1-score} = 2 * (\text{Precision} * \text{Recall})/(\text{Precision} + \text{Recall}) \tag{5}$$

True positives refer to the number of instances correctly identified as positives. The false positive rate is the percentage of cases that are mistakenly identified as positive. False negatives are instances that are mistakenly identified as negative. The accuracy of the model is determined by the proportion of accurate forecasts. The accuracy rate in this case is 96%, indicating that the model correctly classified 96% of the clips. The macro average accuracy, recall, and F1-score are calculated by averaging the precision, recall, and F1-score of each class. The F1-score, recall, and macro average accuracy all stand at 96% in this case [19].

The confusion matrix provides a detailed breakdown of true positives, true negatives, false positives, and false negatives. This matrix helps in identifying specific areas where the model may need further improvement [20]. The weighted average precision, recall, and F1-score are calculated by averaging the accuracy, recall, and F1-scores for each class and weighting them based on the sample quantity in each class. Here, all their values are 96%.

5.4 Latency and Real-Time Performance

The system's ability to process video frames in real time is crucial for timely detection and response to violent incidents. The model's latency was tested on a machine equipped with an NVIDIA GTX 1050 GPU and an Intel Core i7 CPU. The results indicate:

- **Real-Time Processing**: The model is capable of processing video frames in real-time or near real-time, meeting the requirements for immediate violence detection and alert generation.
- **Low Latency**: The processing latency is minimal, ensuring that alerts are generated and communicated without significant delays. This prompt response is vital for law enforcement and security personnel to take swift action.

5.5 Telegram Bot Integration

The integration of a Telegram Bot enhances the system's functionality by providing real-time alerts to the nearest law enforcement officials based on location data. The following details outline the bot's role and capabilities:

- **Automated Alerts**: When the model detects violence in a video frame, the Telegram Bot automatically sends an alert to a designated Telegram group containing the nearest law enforcement officials. This automation ensures timely and efficient communication.
- **Alert Information**: Each alert includes critical information such as the timestamp of the detected incident, the location where the violence was detected, and images showing the violent act and detected faces.

This integration significantly improves the system's responsiveness and effectiveness in addressing violent incidents.

Fig. 6 Output frames that recognized violence and that do not recognize violence, respectively

6 Conclusion

The Real-Time Violence Detection and Alert System, utilizing MobileNetV2, Cloud Firestore, and Telegram Bot, is a notable advancement in public safety technology. The research effort demonstrated how deep learning architectures like MobileNetV2 may be used for real-time violence detection, ensuring quick recognition and reporting of violent situations. Combining Cloud Firestore for data storage with Telegram Bot for alerts improves the system's responsiveness and efficacy.

The technology has proven its capacity to identify violent acts in real time with high accuracy through testing and assessment. The system's scalability and agility allow it to serve a wide range of security demands, from public areas to private homes and commercial buildings. The technology-focused method outlined in this study article contributes to creating a safer and more secure workplace, in line with current worldwide initiatives to tackle modern social issues surrounding violence.

The Real-Time Violence Detection and Alert System demonstrates the promise of AI-powered solutions in enhancing public safety as technology advances. Although the system has demonstrated encouraging outcomes, there is still potential for more improvement and growth. Future research might investigate improvements in violent pattern identification, instantaneous decision-making, and incorporation with other developing technologies. By doing this, we can keep advancing innovation, which will result in safer communities and a more secure future.

References

1. Vieira JC, Sartori A, Stefenon SF, Perez FL, de Jesus GS, Leithardt VRQ (2022) Low-cost CNN for automatic violence recognition on embedded system. IEEE Access 10:25190–25202. https://doi.org/10.1109/ACCESS.2022.3155123

2. Kang M-S, Park R-H, Park H-M (2021) Efficient spatio-temporal modeling methods for real-time violence recognition. IEEE Access 9:76270–76285. https://doi.org/10.1109/ACCESS.2021.3083273
3. Khan SU, Haq IU, Rho S, Baik SW, Lee MY (2019) Cover the violence: a novel deep-learning-based approach towards violence-detection in movies. Appl Sci 9(22):4963. https://doi.org/10.3390/app9224963J
4. Ullah FUM, Ullah A, Muhammad K, Haq IU, Baik SW (2019) Violence detection using spatiotemporal features with 3D convolutional neural network. Sensors (Basel) 19(11):2472. https://doi.org/10.3390/s19112472
5. Ramzan M et al (2019) A review on state-of-the-art violence detection techniques. IEEE Access 7:107560–107575. https://doi.org/10.1109/ACCESS.2019.2932114
6. Baba M, Gui V, Cernazanu C, Pescaru D (2019) A sensor network approach for violence detection in smart cities using deep learning. Sensors 19(7):1676
7. Hanson A, Koutilya PNVR, Krishnagopal S, Davis L (2018) Bidirectional convolutional LSTM for the detection of violence in videos. In: Proceedings of the European conference on computer vision (ECCV) workshops
8. Sudhakaran S, Lanz O (2017) Learning to detect violent videos using convolutional long short-term memory. In: 2017 14th IEEE international conference on advanced video and signal based surveillance (AVSS), 29 Aug 2017. IEEE, pp 1–6
9. Gao Y, Liu H, Sun X, Wang C, Liu Y (2016) Violence detection using oriented violent flows. Image Vis Comput 1(48):37–41
10. Abdali A-MR, Al-Tuma RF (2019) Robust real-time violence detection in video using CNN and LSTM. In: 2019 2nd scientific conference of computer sciences (SCCS), pp 104–108. https://doi.org/10.1109/SCCS.2019.8852616
11. Ullah A, Muhammad K, Haq IU, Baik SW (2019) Action recognition using optimized deep autoencoder and CNN for surveillance data streams of non-stationary environments. Future Gener Comput Syst 1(96):386–397
12. Sandler M, Howard A, Zhu M, Zhmoginov A, Chen L-C (2018) MobileNetV2: inverted residuals and linear bottlenecks, pp 4510–4520. https://doi.org/10.1109/CVPR.2018.00474
13. Sultani W, Chen C, Shah M (2018) Real-world anomaly detection in surveillance videos. In: Proceedings of the IEEE conference on computer vision and pattern recognition, pp 6479–6488
14. Ullah A, Muhammad K, Del Ser J, Baik SW, de Albuquerque VH (2018) Activity recognition using temporal optical flow convolutional features and multilayer LSTM. IEEE Trans Ind Electron 66(12):9692–9702
15. Zhou P, Ding Q, Luo H, Hou X (2018) Violence detection in surveillance video using low-level features. PLoS ONE 13(10):e0203668
16. Gopalakrishna MT (2016) Violence detection in surveillance video—a survey. Int J Latest Res Eng Technol (IJLRET) 11–17
17. Wang J, Xu Z (2015) Crowd anomaly detection for automated video surveillance. In: 6th international conference on imaging for crime prevention and detection (ICDP-15), pp 1–6. https://doi.org/10.1049/ic.2015.0102
18. Bermejo Nievas E, Deniz Suarez O, Bueno García G, Sukthankar R (2011) Violence detection in video using computer vision techniques. In: Computer analysis of images and patterns: 14th international conference, CAIP 2011, Seville, Spain, 29–31 Aug 2011, proceedings, part II. Springer Berlin Heidelberg, pp 332–339
19. Laptev I, Lindeberg T (2004) Velocity adaptation of space-time interest points. In: Proceedings of the 17th international conference on pattern recognition, 2004. ICPR 2004, 26 Aug 2004, vol 1. IEEE, pp 52–56
20. Sernani P, Falcionelli N, Tomassini S, Contardo P, Dragoni AF (2021) Deep learning for automatic violence detection: tests on the AIRTLab dataset. IEEE Access 9:160580–160595. https://doi.org/10.1109/ACCESS.2021.3131315

An Analysis on the Capacity of the Convolutional Neural Network and the Current Neural Network to Identify Vocal Problems

M. Sathyam Reddy, M. Suneetha, D. Prakasa Rao, and Venkata Lakshmi Narayana Gorle

Abstract Computational acoustic testing has significant benefits for aberrant speech diagnosis, surveillance, and early diagnosis. This work provides a feature extracting mechanism for vocal pathology identification using singular value decomposition (SVD) datasets. Speech noise detection techniques rely on the precision of acoustic measurements, which are used to evaluate voice health. Then, a system made up of 27 neural network layers—consisting of recurrent and convolutional neural networks—processes the characteristics extracted (RNN). The dataset is divided into test and training sets using a tenfold validation method, allowing the system's efficacy to be assessed. According to the experimental findings, the CNN model's estimated accuracy is 87.11%, whereas the RNN model's is 86.52%. The cross-validation of tenfold is another method used to evaluate the efficacy of the classifier. The Tensor Flow library is used in the Python development of the system's program, and one NVidia Titan X GPU is used for computing. The operating system used for all calculations is Linux. Healthcare practitioners working in the field of vocal analysis and diagnostic will find the suggested technique to be a useful tool in the early detection and surveillance of voice disorders.

Keywords Linux operating system · Tensor flow · GPU computation · Tenfold validation · Computerized acoustic examination · Problematic speech diagnosis · Early detection · Surveillance

M. Sathyam Reddy (✉) · M. Suneetha
Narasaraopeta Engineering College, Yallamanda, Narasaraopet 522601, India
e-mail: sathyamreddym@gmail.com

M. Suneetha
e-mail: msuneetha973@gmail.com

D. Prakasa Rao · V. L. N. Gorle
Aditya Institute of Technology and Management, Tekkali, India

© The Author(s), under exclusive license to Springer Nature Singapore Pte Ltd. 2025

H. Zen et al. (eds.), *Soft Computing and Signal Processing*, Lecture Notes in Networks and Systems 1200, https://doi.org/10.1007/978-981-97-9926-8_3

1 Introduction

Speaking is one of the most basic human inclinations and modes of communication. Humans need it to be able to express their thoughts and ideas to other people. Humans are the only animals with the capacity to both produce and understand speech, which is crucial to our human social and cultural interactions. The larynx and lungs collaborate to locate the real vocal fold over in abduction and generate sounds which are alternating and/or periodic, resulting in the audible result known as the natural voice. Studies have shown that vocal hyper function, or excessive use or injure on the verbal chords, be capable of causing a range of insulting vocal patterns, which in turn can cause vocal conditions such as dysphonia, or partially trouncing of voice, and aphonia, or whole trouncing of voice [1]. These conditions may cause mental, social, and emotional issues in addition to having a significant negative influence on a person's ability to communicate effectively. Disturbances related to the "excellence, the ground, audible volume, and/or vocal mobility" of voices belonging to comparable socioeconomic classes, ages, and genders are indicative of speech problems [2]. Speech disorders unrelated to cancer might not pose an imminent danger, but unprocessed voice degradation can negatively impact social, qualified, and intimate communication [3]. Mass pathologies are particularly common because of the varied voice fold destruction, as they often lead to persistent tissue infection, vulnerable multilayer folds of the vocal cords that are prone to phono-traumatic an illness, and voice nodules and adenoma [4]. In nonphonotraumatic voice diseases including muscular tension dysphonia and functioning impairment of speech, vocal fold abnormalities are uncommon. On the other side, those who have these diseases may exhibit signs including verbal fatigue, a decrease in influence quality, and increasing laryngeal embarrassment. It is commonly believed that voice anomalies should be evaluated using a multipara metric assessment methodology [5]. A patient conversation, a laryngeal examination employing stroboscopy or laryngoscopy, a fundamental aerodynamic evaluation, hearing analysis employing standardized psychoacoustic techniques, subjective speaking evaluation, among other pertinent criteria are some of the elements that this strategy normally entails. Clinical professionals can gain a better understanding of a patient's state and provide a more precise diagnosis and successful therapy by taking into account a variety of metrics and elements, such as physiological, auditory, and perceptual factors. It is possible that novel techniques and innovations will develop throughout time and be incorporated within this methodology to increase the precision and effectiveness of voice evaluation. Voice scientists are at the forefront of creating audio processing technologies that can differentiate among voices about aphonia and/or dysphonia as well as those by expected voices, thanks to recent technological breakthroughs. The creation and implementation of systems for the recognition of vocal problems' expressions ought to be based on machine learning (ML) methods. In this case, prior to employing an ML technique, audio input needs be prepared and then split up hooked on numerous features [6]. Experts can automatically classify a variety of voice data in audio files as "safe" or "defective". The fundamental audio file for every file is then split up into a few short frames, and every of these is inspected to see what parts need to be eliminated. In neural networks, return is the total of the features that are acquired during each frame. By arbitrarily choose observation of both healthy and sick voices, the data gathering is split into research and training datasets. After the method for machine learning has been created based on the training set, the representation is demonstrated employing the test set. The accuracy of the designation is ascertained throughout the review process. A baseline for evaluating the efficacy of different Automatic Voice Disorder Detection (AVDD) methods is going to be this classification accuracy [7]. Abid Syed and colleagues [8] noted a number of shortcomings regarding the application of artificial intelligence-based sound disorder identification methods, such as the researchers' employing unsupervised methodologies, a lack of comparison precision, and a dearth of prior research on the Arabic Voice Pathology Database (AVPD) [9]. To ascertain voice order during this study, we utilized the Saarbrucken Voice Database (SVD) [10]. The proposed paper is an extension of the author's previous work, which used ensemble, naive Bayes, support vector machines (SVM), and decision trees while contrasting RNN and CNN over the same set of traits and illness. In order to anticipate the accuracy of the system, this assignment aims to build a structure by first acquiring features and then applying recurrent neural networks (RNNs) for machine learning categorization. To increase the system's precision, we will contrast the output of the convolutional neural network (CNN) and the RNN. According to a meta-analysis, the highest efficiency ever seen with convolutional neural networks is now 80% [8]. In this investigation, we would make utilization of the SVD dataset, consisting comprises recording of people with different diseases and vowel sounds [11].

2 Problem Definition

We appreciate that our neural network classifications need to operate at peak efficiency all the time. Nonetheless, the exact measurements performed better than or about as well as earlier published NLP research. As an instance, even if 10 out of 1000 test samples in the CNN model be erroneous, categorization issues related to neural networks are highly demanding to study while neural networks are essentially "black boxes." PE and the lack of sufficient quality images in the explanation made context-based inference necessary while the primary reason for the error was not made clear. The RNN-focused model found the most relevant sentences in the articles and correctly predicted the predictable nature of groupings. Still, it is challenging to generalize the model's inference. Positive and negative PE classifications were observed only once, as well as in that case, CNN precisely forecasts a positive categorization, whereas RNN correctly anticipated a harmful one. It be consequently uncertain whether CNN might comprise appropriately anticipated this case using the heat map. Because of this, in accumulation to the guidance restrictions imposed by the sheer size of our input sets, every of these shortcomings calls for intricate reasoning that may restrict the design of the model.

3 Literature Survey

A method for identifying speech issues utilizing SVM on SVD was proposed by Al [10]. Al work focuses on creating a reliable method for analyzing many frequency bands utilizing self-correlation and unpredictability so as to identify and diagnose voice abnormalities. Employing autocorrelation as a tool to detect and identify challenging sample, the maximum peak levels and equivalent lag values were collected regarding each shot of the speech stream. Following translating the data into operations, we examined these components in a variety of bands of frequencies and computed the amount of entropy for every speech stream structure to assess how every band of frequencies impacted the detection and categorization systems. Continuous voice samples in Arabic, German, and English were collected into three different datasets. Both normal and disturbed voices were featured in these cases. Prior research has used singular value decomposition (SVD) to diagnose speech disorders with great accuracy by using the support vector machine (SVM) method as a categorizer. The recorded accuracy for SVD-based categorization is 92%. The main objective of [11] was to use measurements from the Multidimensional Voice Program (MDVP) software to diagnose and distinguish voice abnormalities across various datasets. The goal of the study was to identify the procedure's most effective criterion. The results showed that there were considerable differences in the MDVP settings' efficiency between datasets. The characteristics that score highest vary significantly among datasets. To obtain the best level of precision, the study selected the top three MDVP measures using SVD, which has a Fisher Discriminating Ratio of approximately 99.98%. In this study [12], we found and identified problematic items by using the correlational move toward as a characteristic to determine the highest peak values and the accompanying lag values within every single surround of the spoken signal. In order to determine whether a frequency band influences the recognition and classification procedures, these characteristics are investigated across an assortment of frequency bands. The bands around 1000 and 8000 Hz are ideal for naming and distinguishing. Cross-correlation produces the highest rates of precision (99.809%, 90.979%, and 91.16%, respectively) in the Saarbrucken Speech Database (SVD) for the Arabic Voice Pathology Database and the Massachusetts Eye and Ear Infirmary. Conversely, the highest cross-correlation accuracy rates were, respectively, 99.255%, 98.941%, and 95.188% in each of the datasets. Teixeira et al. used different classifiers but the same properties when presenting the voice detection approach in two articles. Their reported efficacy with SVM with Jitter, Shimmer, and HNR was 71% [13]. The exploit of machine learning algorithms to identify speech pathology has been the subject of several studies. In [14], an MLP-ANN algorithm with HNR, Jitter, and Shimmer characteristics was applied, and the results showed 100% accuracy for female vocals. Using SE, ZCRs, and SH features, SVM was used, and 95% accuracy was attained. Conversely, there has been minimal focus on the application of convolutional neural networks (CNNs) for the identification of speech pathologies. Pitch, 13 MFCC, roll-off, ZCR, energy entropy, spectrum flux, spectral centroid, and energy are used as features in the proposed system. RNN is used as a classifier to amplify accurateness, and CNN is used to confirm the findings.

4 Methodology

The acronym for the Database of Saarbrucken Voices is SVD. SVD consists of a voice recording database that is available to the public, containing the voices of about 2000 individuals with 72 distinct types of speech impairments conditions: (1) voice proof: (1) accepted, high, and low pitched voice registration [I a, u] in this confirmation gathering established the truth; (2) vocal proof of pitch rise [I a, u] in this recorded session; and (3) recordings of the words "Good morning, how do you like it?" ("Good morning, how are you doing?"). For the components that you have requested, the voice and EGG signals have been stored in separate files. A written file containing all of the pertinent information from the dataset is included in the database. These qualities make it the perfect option for experimenters. Every SVD noise was confirmation at 50 kHz with 16 bits of resolution.

According to the standard of the documenting, there can be less consonants throughout every repetition in various recording sessions. This online interface allows you to communicate with the "Saarbrucken Voice Server." It has several webpages. These are employed to select the recordings using the recording phase to be export after selecting the relevant component using the SVD the database, to play immediately and file, and to specify variables for the database application. The diseases are the ones we chose, "Voxsenilis," "Zentral-laryngaleBewegungsstörung," "Reinkedem," "Stimmlippenpolyp," "Stimmlippenkarzinom," "Spasmodische Dysphonie," "Psychogene Dysphonie," "Dysphonie," "Frontolaterale Teilresektion," "Funktionelle Dysphonie," "Hardies." The majority of the illnesses were chosen based on vocal abnormalities that are frequently diagnosed.

Qualities: In this investigation, thirteen MFCC parameters—pitch, roll-off, ZCR, radiation entropy, spectrometer flux, spectral the central point, and energy—are measured as of samples. To provide more richer speech samples for processing, Syed et al. add seven additional fundamentals to their preceding exertion: pitch, roll-off, ZCR, energies entropy, spectrum flux, spectrum centroid, and energy. Mel-frequency Cepstral Coefficients MFCC: Davis and Mermelstein argued in 1980 that the most popular characteristic for speech detection should be MFCC. The elimination strategy for the MFCC function consists of signal opening, DFT application, amplitude protocol acquisition, humiliating of the findings and a Mel ranking on scales, and finally, reversing DCT application. Cepstral parameters are static attributes that typically hold data from a single frame. Primary and secondary cepstral features of the machine provide further information about the time span fluctuations in the audio stream.

$$\widehat{y_t}[j] = \frac{y_t[j] - \mu(y[j])}{\sigma(y[j])} \quad (1)$$

Pitch: The frequency at which the vocal cord vibrates when making sounds is known as the pitch of a sound. While removing tons, standard methods that result in half and double-half defects, like the median magnitude differential technique and the self-correlating structure at highest, are subject to alteration. Pitch can be approximated using the cepstrum approach by dividing the vocal cord cepstrum from the acoustic pulse cepstrum. It is not as good at detecting complex measures, but it is good at detecting simple sound signals.

CNN Structure: The classifications and sequencing layers that comprise CNN's hierarchical levels are depicted in a number of charts. CNN classically start by a convolutional layer the fact that can handle the input-level data. The convolution layer is in charge of evolutionary process linking a restricted amount of filter maps of the identical dimension. The sample layer receives the output from this layer as well, which shrinks the size of the levels that follow. CNN is linked to numerous deep learning techniques locally. Then, these networks are built on several hundred cores using GPU architecture. The allocation of the role mappings will be based on the knowledge blocks from the previous tier. The dimensions of the maps would conclude this. Yet, a single block made up of numerous threads connect every fiber to a different neuron. The following phases include comparable processes on neurons, like convolution, the induction process, and summation. Lastly, the consequences of the tasks previously outlined are reserved in a global memory. Using the reverse dissemination approach, the output is processed effectively. Yet, considering a single spread wouldn't yield the intended results, dragging or moving motions facilitate parallel spreading.

Furthermore, interactions between neurons in a single layer and between various groups of neurons can affect border effects. The goal of this deep learning neural network is give details by the specific CNN design that is displayed in Figs. 1 and 2. Preconditioning input, retraining the deep learning model, storing the model, and the concluding stage of implementing the model are all incorporated in a deep learning strategy. Currently, learning the deep algorithm for learning involves the greatest compute and data (outlining and executing). A neural network, also known as forward transmission, takes input from the model and produces an output at a predetermined step. In case of an inaccurate or faulty performance, the weights are adjusted (reverse pass). For sufficiently identical output items, this may appear as an uncomplicated matrix multiplication, with the doubled weight of the input data (first matrix row) represented in the subsequent matrix column. More complex matrices are frequently not implementable by serial (CPU-based) systems. Fortunately, for the majority of cases, GPU offers considerably more opportunities than conventional solitary or cluster CPU machines.

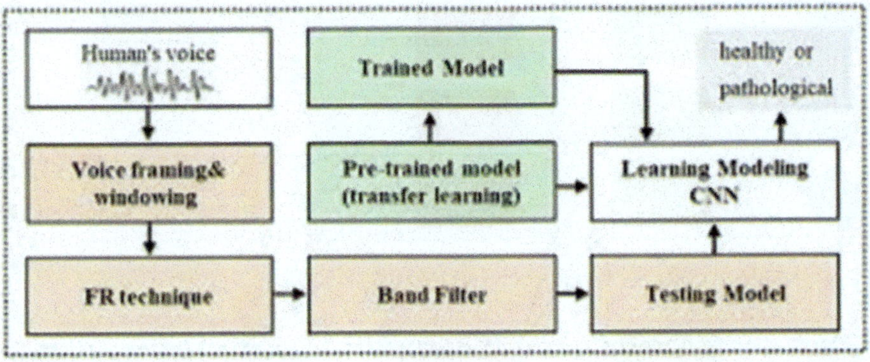

Fig. 1 Methodology block diagram

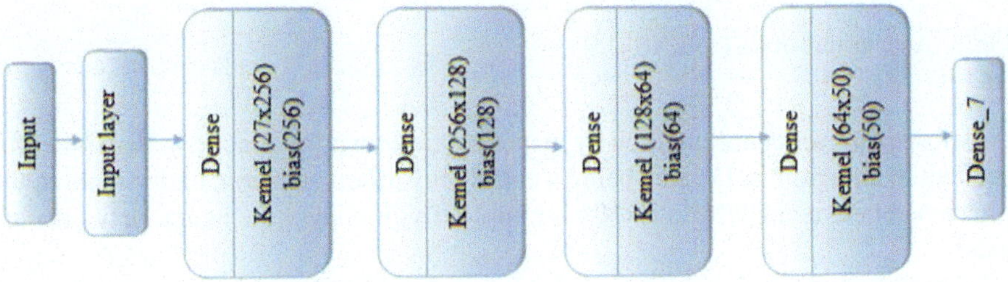

Fig. 2 A projected in layer model of CNN

RNN Structure: Compared to conventional RNNs, the long short-term memory (LSTM) RNN type is intended to mimic temporal patterns and their distant relationships with more accuracy. We have established that when developing models of an average magnitude on a single machine, LSTM-RNN works better than DNNs and conventional acoustic modeling. We demonstrate how to use a two-layer deep LSTM-RNN using a linear recurring projections layer for implementing the latest speech recognition technology. Figures 3 and 4 show the model's operating flow using the LSTM-RNN's overall architecture. This approach executes a deep neural network that receives input having an augmented amplitude sequence, from converges quickly, and uses model parameters more effectively than earlier approaches. Communicating is an energetic communication that evolves over time and has significant linkages across periods. Since they have cyclic ties, recurrent neural networks (RNs) better feed forward neural networks in collecting the variety of types of progression data.

RNNs show good progression marking and prophecy performance in domains like language and handwriting recognition. The ability of CNN and RNN to understand events, like a word, is their primary difference. On the other hand, the applications of convolutional and repeating neural networks are quite different, and the neural network topologies themselves change to accommodate these disparate uses. CNNs convert the results using convolutional layer filters. RNNs, on the other hand, generate

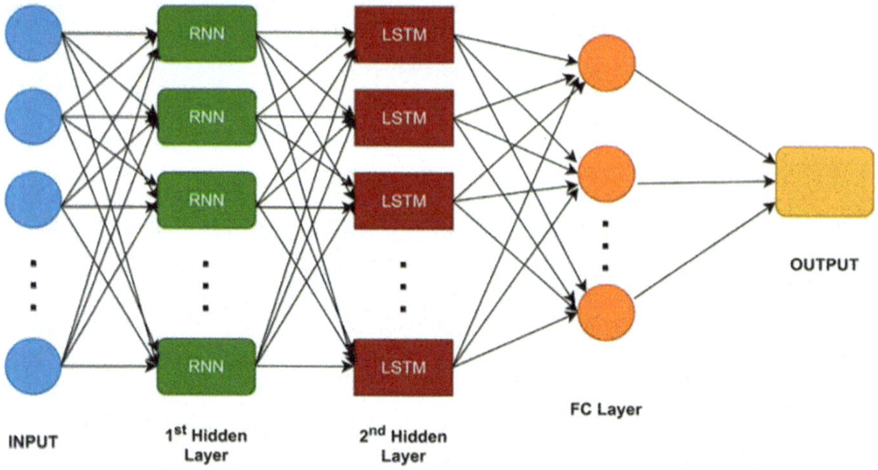

Fig. 3 Architecture of LSTM-RNN [15]

the subsequent sequence creation by recycling the activation functions from previous sequential data points. Despite being a commonly asked question, understanding the difference between CNN and RNN is simple if you know how neural networks are used.

The proposed CNN and RNN model layers. Each study needs to start with clean, error-free data that are well-organized. To get the information you require for your particular research or representation, you're going to almost always need to eliminate certain rows and columns about every data set that is imported through the Python Pandas DataFrame. To retrieve the voice sample information in a simplified format, we employed the Python command "Data-Frame. drop ()" to eliminate any extraneous frames as well as columns. It is possible to set label names and related axes prior to deleting rows or columns. One can remove labels across various levels by formative the rank of a multi-index.

The preliminary processing imports for "train_test_split" from "sklearn.model_selection" and "LabelEncoder" and "StandardScaler" from "sklearn.preprocessing" are also included. By choosing options regarding data which were not utilized during training the model, the train-test split technique enhances the efficacy of machine learning systems. You can contrast the results of machine learning approaches against your forecasting challenge, and it's easy and quick to do. Despite its ease of use and comprehension, there are situations that indicate it shouldn't be applied, like in tiny datasets, and others that call for additional adjustments, as in categorization applications or data sets that are unbalanced. The processes require compiling a dataset and separating it in half. The first dataset chosen to verify that the predicted outcome matches is the one to be tested. The system, on the contrary hand, views and compares its projections against the values that are anticipated after receiving the dataset of the constituent as input. Considering the second batch of data, the machine learning algorithm is not developed. The test dataset comprises the second batch of

Fig. 4 Projected in layer representation of LSTM-RNN

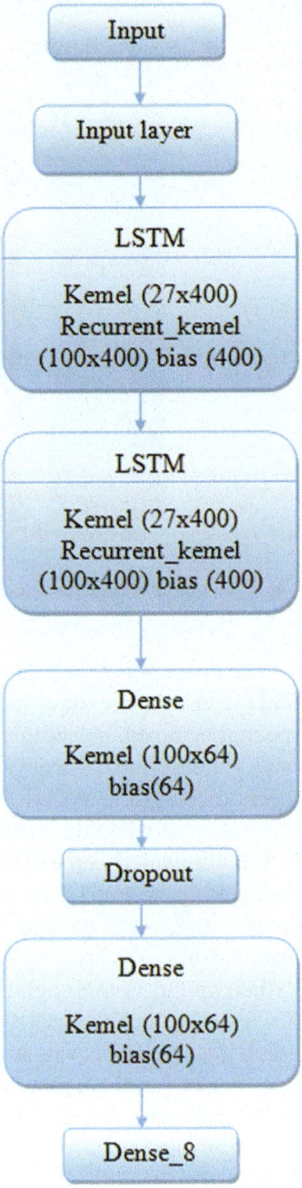

data. The 2000 samples in the raw data set was arbitrarily split into two groups of 80 and 20%. Testing took up 20% of the allotted time, and training took up 80%. The proposed CNN and RNN models' inner layered diagrams are displayed in Figs. 3 and 4. In the suggested methodology, CNN and RNN are mutually 27 neural layer design by means of different levels of bias.

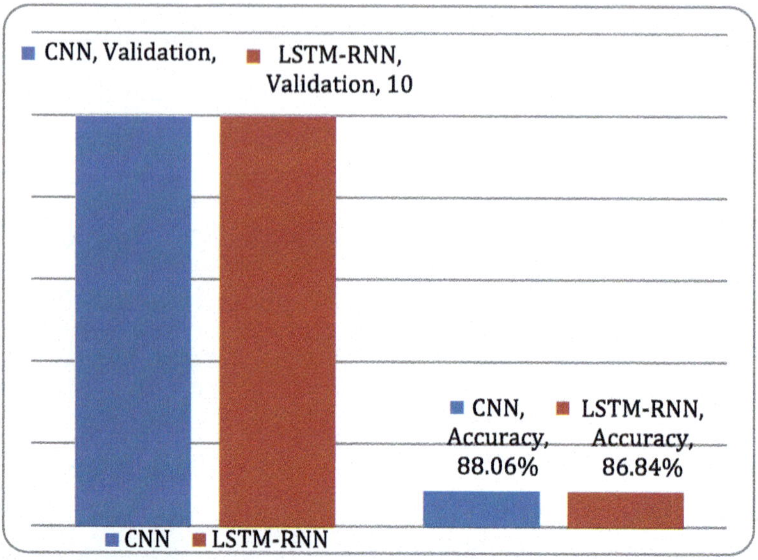

Fig. 5 Accurateness of RNN and CNN at 10 field authentication

The concept is to recognize disease pathology from speech. To start, we take the SVD dataset and extract its features. We collected 13 MFCC uniqueness (energy, spectral centroid, power entropy, pitch, roll-off, ZCR, and spectrum flux) using the methods described. A convolutional and recurrent neural network with 27 neuronal layers receives the system input after feature extraction. During 10 k-fold confirmation, we divided the information either a set for training or a testing set of data. In the dominated kernels of the 27 remaining blocks, there were a total of seven residual layers. For L2 normalization, we used dropout at a frequency of 0.5. The success assessment has been conducted using tenfold cross-validation. Software was installed on a workplace by a single NVidia Titan X GPU by means of the Tensor Flow Python plugin. A thorough analysis of accuracy and error is presented in Figs. 5 and 6, which also include lines representing the training and testing stages. The connected graph lines in RNN evaluations show a very small error margin; the differing graph lines in CNN evaluations show a larger possibility of error margin for the suggested CNN approach compared to the RNN. Figures 7 and 8 show the misunderstanding matrix by means of a value indicating the amount of times the classification has been identified correctly.

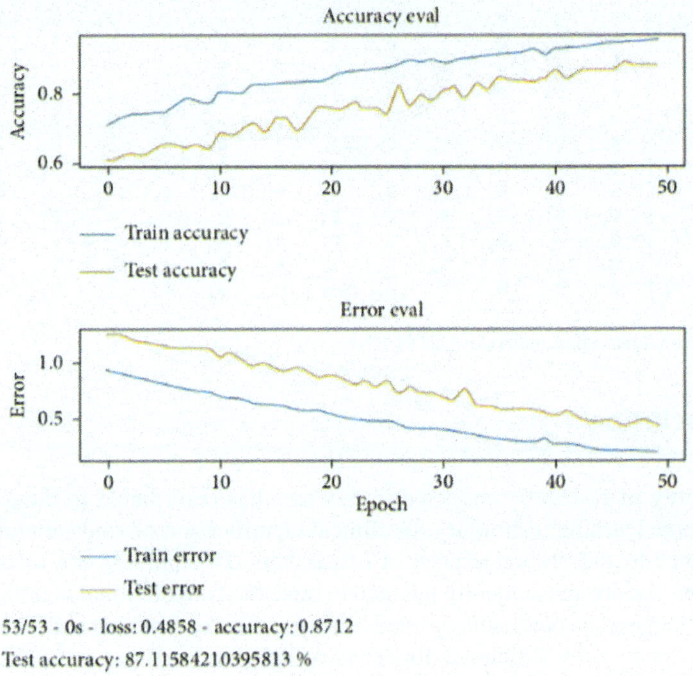

Fig. 6 Accurateness and error assessment of CNN in preparation and testing stage

```
array([[100,   2,   4,   0,   0,   0,   5,   0,   0,   0,   2,   4],
       [  2, 141,   0,   0,   0,  11,   1,   0,   0,   5,   0,   0],
       [  0,   0, 165,   0,   0,   0,   0,   0,   0,   0,   0,   0],
       [  1,   9,   0, 126,   0,  12,   3,   1,   0,   3,   3,   0],
       [  0,   0,   6,   3,  70,   0,   4,   0,   0,   2,   0,   0],
       [  0,   0,   0,   0,   0, 180,   0,   0,   5,   4,   0,   0],
       [  9,   0,   0,   0,   0,   0, 168,   0,   0,   7,   4,   0],
       [  0,   0,   0,   0,   0,   2,   0, 157,   2,   6,   0,   0],
       [  3,   0,   3,   0,   0,   0,   0,   0,  99,   0,   0,   0],
       [  0,   3,   0,   0,   0,   3,   5,   0,   0, 148,   0,   0],
       [  0,   0,   7,   9,   1,   0,   6,   8,   4,   7,  88,   0],
       [  0,   0,   0,   2,   0,   3,   6,   0,   0,  10,   0,  48]])
```

Fig. 7 Confusion matrix of CNN

```
array([[ 83,   0,   0,   1,   0,   0,   3,   1,   0,   2,   0,   0],
       [  1, 117,   0,  12,   0,   2,   0,   0,   3,   5,   2,   0],
       [  0,   0, 134,   0,   0,   0,   0,   5,   7,   0,   3,   0],
       [  0,   2,   0, 119,   0,   1,   3,   0,   0,   2,   5,   1],
       [  0,   0,   0,   4,  67,   0,   0,   0,   3,   0,   0,   1],
       [  0,   0,   0,   4,   0, 117,   1,   0,   0,   5,   0,   2],
       [  0,   2,   0,   1,   1,   0, 107,   0,   2,  12,   5,   2],
       [  0,   0,   0,   0,   0,   0,   0, 137,   0,  14,   3,   0],
       [  0,   0,   0,   0,   0,   0,   0,   2,  95,   4,   0,   0],
       [  0,   3,   0,   0,   3,   0,   0,   0,   2, 130,   0,   0],
       [  0,   3,   0,   0,   0,   1,   0,   0,   1,  10,  96,   4],
       [  0,   0,   0,   5,   0,   0,   0,   0,   0,   2,   0,  45]])
```

Fig. 8 Misunderstanding matrix of LSTM-RNN

5 Conclusion

The quantity of research conducted in this area has contributed to the assumption that machine learning techniques for clinical identification of voice abnormality are of relevance to the preponderance of researchers. By applying the recommended techniques, we are subsequently talented to raise the convolutional neural network's efficiency beyond its precision reported in the previous assessment to 87.11%. Recurring neural networks anticipate almost identical outcomes and have effectiveness comparable to CNN.

References

1. Titze R, Verdolini K (2012) Vocology: the science and practice of voice habilitation. National Center for Voice and Speech, Salt Lake City, UT
2. Dejonckere PH, Bradley P, Clemente P et al (2001) A basic protocol for functional assessment of voice pathology, especially for investigating the efficacy of (phonosurgical) treatments and evaluating new assessment techniques. Eur Arch Otorhinolaryngol 258(2):77–82
3. American Speech-Language-Hearing Association and Others (2015) Council for clinical certification in audiology and speech-language pathology, vol 15
4. Titze IR, Svec JG, Popolo PS (2003) Vocal dose measures: quantifying accumulated vibration exposure in vocal fold tissues. J Speech Lang Hear Res 46(4):919–932
5. Boominathan P, Samuel J, Arunachalam R, Nagarajan R, Mahalingam S (2014) Multi parametric voice assessment: Sri Ramachandra University protocol. Indian J Otolaryngol Head Neck Surg 66(S1):246–251
6. Kasuya H, Ogawa S, Kikuchi Y, Ebihara S (1986) An acoustic analysis of pathological voice and its application to the evaluation of laryngeal pathology. Speech Commun 5(2):171–181
7. Hegde S, Shetty S, Rai S, Dodderi T (2019) A survey on machine learning approaches for automatic detection of voice disorders. J Voice 33(6):947.e11–947.e33
8. Abid Syed S, Rashid M, Hussain S (2020) Meta-analysis of voice disorders databases and applied machine learning techniques. Math Biosci Eng 17(6):7958–7979
9. Tamer A, Mesallam MF, Malki KH et al (2017) Development of the Arabic voice pathology database and its evaluation by using speech features and machine learning algorithms. J Healthc Eng 2017, 13 pages. Article ID 8783751
10. Saarbruecken voice database—handbook (2007) stimmdatenbank.coli.uni-saarland.de

11. Puneeth BR, Ashwitha K, Kumar A, Rao B, Preethi Salian K, Supravi AP (2022) An approach to predict loan eligibility using machine learning. In: 2022 international conference on artificial intelligence and data engineering (AIDE), Karkala, India, pp 23–28
12. Kumar A, Shetty P, Balipa M, Rao B, Puneeth B, Shravya (2022) An efficient technique to detect skin disease using image processing. In: 2022 international conference on artificial intelligence and data engineering (AIDE), Karkala, India, pp 35–40
13. Puneeth BR, Preethi Salian K, Prathyakshini, Murthy A, Salian S, Surabhi (2022) Analysis of telecom churn using machine learning techniques. In: 2022 international conference on artificial intelligence and data engineering (AIDE), Karkala, India, pp 58–63
14. Balipa M, Shetty P, Kumar A, Puneeth BR, Adithya (2022) Arecanut disease detection using CNN and SVM algorithms. In: 2022 international conference on artificial intelligence and data engineering (AIDE), Karkala, India, pp 01–04
15. Faru S, Waititu A, Nderu L (2023) A hybrid neural network model based on transfer learning for forecasting forex market. J Data Anal Inf Process 11:103–120. https://doi.org/10.4236/jdaip.2023.112007

Recent Advances in Diabetic Monitoring Devices: Present to Future Directions

M. Karuppasamy, M. Jansi Rani, and K. Poorani

Abstract Glucose monitoring is the major need for diabetic people. In order to avoid most of the diabetic complications, it is necessary to keep glucose levels in control. Long-term diabetes leads to severe microvascular and macrovascular complications. Traditional systems measure glucose with blood samples from human body, which needs clinical setup. While some people need continuous monitoring where blood samples can't be drawn so often. In order to overcome this, continuous glucose monitors (CGM) are used. CGMs are considered as boon to diabetic community. Hence, people started depending on CGM for diabetes monitoring. Noninvasive monitoring is much effective to analyze insulin levels. Wearable biosensors for glucose monitoring are upcoming. The integration of multiple systems like Internet of Things (IoT), wearable devices, mobile technology, and artificial intelligence facilitates the monitoring of healthcare systems. This paper explores various diabetic monitoring devices available at present and also discusses the future possibilities of diabetic monitoring.

Keywords Diabetes mellitus · Continuous glucose monitoring · Wearable biosensors · Glucose levels · Diabetic complication

M. Karuppasamy
ACS College of Engineering, Bengaluru, Karnataka, India
e-mail: karuppasamy.m1987@gmail.com

M. Jansi Rani (✉)
Mepco Schlenk Engineering College, Sivakasi, Tamil Nadu, India
e-mail: jansirani.m86@gmail.com

K. Poorani
Kalasalingam Academy of Research and Education, Krishnankoil, Tamil Nadu, India
e-mail: pooranikhari@gmail.com

© The Author(s), under exclusive license to Springer Nature Singapore Pte Ltd. 2025

H. Zen et al. (eds.), *Soft Computing and Signal Processing*, Lecture Notes in Networks and Systems 1200, https://doi.org/10.1007/978-981-97-9926-8_4

1 Introduction

Diabetes, a chronic metabolic syndrome, is caused due to high concentration of blood sugar. It is due to lack or absence of insulin levels in the body. Type 1 is caused due to insulin deficiency, and type 2 is caused due to insulin resistance [1]. Diabetes is categorized as type 1, type 2, and gestational diabetes. Type 1 commonly affects childhood and teenagers [2]. Type 2 is commonly found in adults over 40 years of age and above. Gestational diabetes is found during times of pregnancy. It may be present over time and disappear after pregnancy [3]. Various complications arise due to instability in glucose levels. Insulin is a hormone which regulates the production of insulin in the human body. Complications lead to retinopathy (eye damage), nephropathy (kidney damage), neuropathy (nerve damage), foot ulcers, amputations, cardiovascular, and arterial diseases [2].

According to report from IDF, diabetes affects more than 463 million people around the world. IDF Report states that approximate of 578 million adults will have diabetes by 2030, and it may extend more than 700 million by 2045 [4]. In order to reduce the possible number of estimates over the upcoming years, it is necessary to diagnose diabetes earlier and prevent them from growing more. Long-term diabetes causes severe complications. The constant increase in glucose levels over a long period of time leads to severe health complications. The classifications are divided into acute and chronic.

Acute complications are short-term complications found at times and can be resolved at once. The symptoms include fainting, blurred vision, increased thirst, and dizziness. Diabetic ketoacidosis, hypoglycemia, frequent urination, hyperglycemia, and hyperosmolar hyperglycemic conditions are acute complications of diabetes [5]. This can be found earlier with symptoms and resolved. On the other hand, chronic complications cannot be found at once, but develops gradually and affects every parts of the human body. The chronic complications are broadly classified as microvascular and macrovascular in general. While retinopathy, neuropathy and nephropathy are microvascular and arterial disease, peripheral disease, stroke, and coronary heart disease are macrovascular diseases [4].

The prevalence of microvascular complication is higher when compared to macrovascular complications. The former is the base for developing latter complications. Prediabetes is a condition which may lead to diabetes [6]. It is a prior state to the diabetic progression, so monitoring prediabetic and maintaining glucose levels is necessary. Prediabetic may be maintained for a long term, and it can also lead to diabetic when glucose levels are uncontrolled [7]. The appropriate range of glucose levels is tabulated in Table 1.

Ranges of glucose values determine the presence or absence of diabetes at several conditions. If the ranges are maintained properly, it is easier to avoid risk. Possibility of prediabetic to diabetic condition is usually high, so people diagnosed with prediabetic need to be more cautious.

Figure 1 highlights the complications related to diabetes. As the traditional complications are well-known, it is a must to know about the emerging complications.

Table 1 Diabetic levels

Diabetic conditions	HbA1c %	Normal fasting blood sugar (mg/DI)	Normal sugar level after meal/ postmeal
Diabetes	6.5+	126+	200+
Prediabetes	5.7–6.4	100–125	140–199
Normal	Below 5.7	Below 99	Below 140

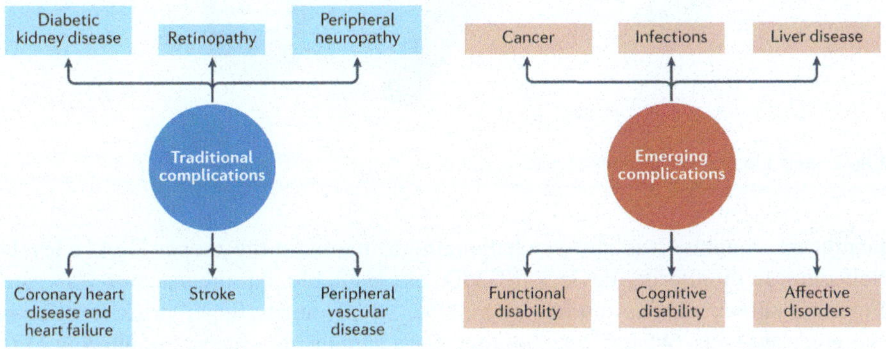

Fig. 1 Complications of diabetes

These types of complication arisal highlight the importance of diabetes monitoring and control. The overall mortality due to diabetes and its complication can be reduced gradually [8, 9]. If it is diagnosed earlier, the complication and risk can be avoided to the maxim [10]. Emerging technologies such as glucose monitors, wearable biosensors, and continuous glucose monitors are available to monitor glucose levels. Glucose monitors are available which makes use of lancet to measure blood glucose levels. Finger prick is done to get the blood sample, and glucose levels are measured [9, 11, 12].

Continuous glucose monitoring devices (CGM) are developed to estimate blood glucose levels depending on interstitial fluid levels. It is the technique followed to monitor glucose levels with minimal invasive methods. It is useful to monitor glucose levels continuously. Even though it causes certain irritation at the spot where it is inserted, it can be bearable. Various systems for diabetes monitoring is explored in the upcoming section.

2 Literature Survey

Various methods and techniques are advancing in the medical field to diagnose, monitor, predict, prognosis, and classify the medical conditions of human. Emerging techniques and technologies are providing knowledge and solutions to the critical

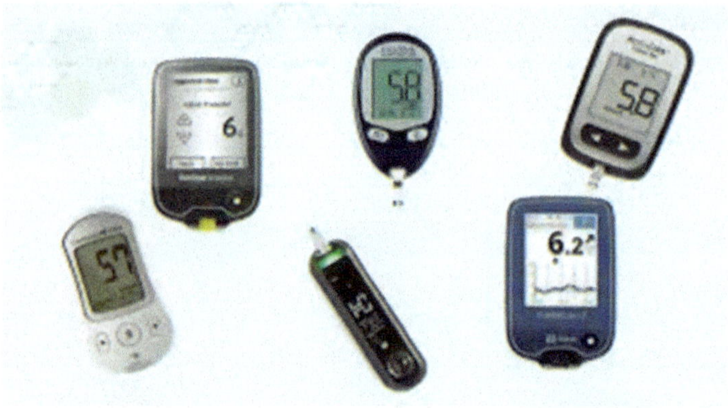

Fig. 2 Self-glucose monitoring devices

healthcare systems. It is always challenging to maintain the proper HbA1c levels. Self-glucose monitoring devices (SGMD) are available, but the usage is very rare. Figure 2 shows various models of SGDM, which needs finger prick to analyze the glucose levels. Wearable technologies are supporting glucose monitoring [13]. Figure 2 represents glucometers available for diabetes monitoring.

2.1 *Continuous Glucose Monitoring (CGM)*

The first CGM was approved by US Food and Drug Administration (FDA) in 1999. The glucose data in this system were not collected but not presented immediately. The data are collected, and glucose levels are interpreted thereafter. This system reduced the burden of finger prick and implemented glucose monitoring via minimal invasive devices. CGM is also advancing through years. It estimates and updates the glucose levels every 5 min. The monitor displays the glucose range every 5 min. The sensor is attached to subcutaneous tissue, while the transmitter transmits sensor data to the monitor. Figure 3 shows various CGM devices.

2.2 *Dexcom G5*

Dexcom G5 is the first stand-alone CGM for stand-alone users. This eliminates the risk of finger prick. The equipment is modeled to fit into human body at minimum space and perform continuous monitoring and recording of glucose levels.

CGM advances SBGM by eliminating the need of invasive finger prick, which motivated diabetes to use CGM [14, 15]. While the accuracy of the interstitial fluid is a

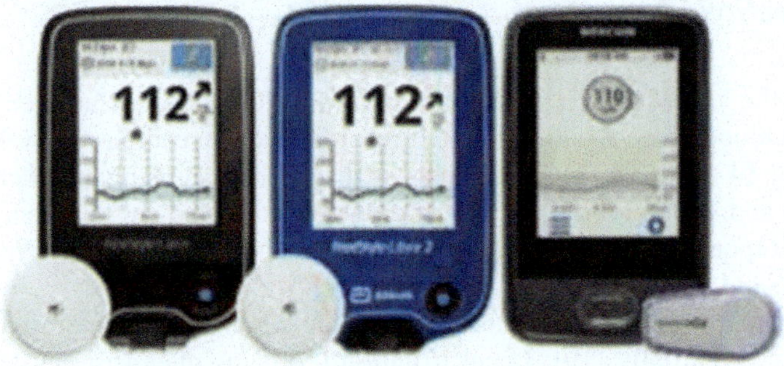

Fig. 3 Continuous glucose monitoring devices

matter of concern, technology improvement also motivated to provide high accuracy in glucose levels. However, accuracy is a matter of concern.

CGM can perform

(1) high glucose levels (hyperglycemic);
(2) predicting low glucose levels (hypoglycemia);
(3) glycemic variability.

All time customer support system is available for CGM devices which are approved by FDA. This makes customers to use devices with ease helping both patient and clinical providers. They can modify the medication according to the glycemic changes. CGM with automatic insulin delivery is emerging which reduces the effort for insulin delivery. CGM are divided into professional and personal CGM.

Professional CGMs are owned by healthcare provider. It does not provide glucose rate at that situation but provide results after sometime. It provides relatively unbiased glucose levels during everyday life. This type of devices is helpful for adults with diabetes. This are helpful to eradicate conditions such as nocturnal hypoglycemia, dawn phenomenon, and postprandial hyperglycemia so that glucose levels can be better managed. Treatment for diabetes can be arranged accordingly. This type of devices is readily available in market but needs qualified professional to take CGM reports.

Dexcom G6 Pro was also approved by FDA. It provides immediate feedback about glucose levels. It informs people about their glucose ranges [16, 17]. The sensor, transmitter, and receiver are identical to personal Dexcom G6. It measures and updates glucose levels every 5 min. It is downloaded through web-based software and review personal glucose data and keep track of it.

2.3 FreeStyle Libre

This uses the same sensor as the Libre personal CGM. It does not need any self-monitored blood glucose calibrations. It records glucose readings every 15 min and collects for 14 days. The glucose sensor is disposable, and single reader is needed to activate and scan multiple devices. It provides integrated environment and group data [18]. Reports generated provide daily glucose patterns, variability, hypoglycemic possibility, and risks.

CGMs are currently using sensing technique which is implanted into the body either fully or partially [19]. Local irritation problems may arises in some cases. Even less invasiveness may be recommended for continuous glucose monitoring which extract results from external fluids (saliva tears and sweat). But practical feasibility of the glucose ranges needs to be studied in detail.

2.4 CGM with Insulin Delivery

Figure 4 shows CGM with automated insulin delivery pump. This kind of system is mostly used for automatic insulin delivery with the insulin pump attached together. This will automate insulin delivery when required, by estimating the amount of insulin required with the help of CGM [20].

Glucose monitoring is becoming a necessity for both type 1 and type 2 diabetes monitoring [21]. Advanced techniques and ML implementation into glucose monitoring are emerging. This helps in automated insulin delivery with the estimated glucose levels from CGM with the help of closed loop structure.

Figure 4 represents the glucose meters that provide the glucose values which need blood sample, which continuous glucose monitors are attached to human body

Fig. 4 Glucose monitors versus automated insulin-delivery system

which derives interstitial fluid to provide glucose levels. This also shows that in case of insulin deficiency, automated insulin delivery system is also available.

2.5 Noninvasive Approaches for Glucose Monitoring

Although there were many challenges in noninvasive technologies [22], upcoming minimal invasive methods include electrical, nanotechnique, and optical methods are measurement. This involves certain form of radiation for glucose monitoring. This can also use electric, optical, and thermal methods. Most advancing optical methods are also available for detection of glucose levels. These things need external factors like light, motion, and temperature to provide accurate results [23, 24].

Electrical methods use EMR, current, or ultrasound for measuring glucose dielectric properties. Thermal methods use infra-red band of the spectrum and provide noninvasive approach for glucose monitoring [25]. Optical techniques include reflective, absorptive, or refractive methods to detect glucose levels. They use Raman and near IR spectroscopy [26].

Figure 5 depicts an advanced system of diabetes monitoring which is currently implemented for noninvasive testing of glucose from saliva. This encompasses a biosensor that collects saliva and then calculates the glucose levels. This is attached to the teeth of any individual in order to monitor the glucose levels. In case of infants, the system is built as nipple, which can be directly suck by baby to know the glucose levels.

Figure 6 represents the devices available at present for diabetes monitoring. Since diabetes affects all parts of the body, monitoring each and every part of the human body is essential. Those devices indicate the levels of glucose and other major readings regarding health conditions necessary for monitoring.

2.6 Future Advancements for Diabetic Monitoring

It is much important that the data collected through various systems need to be recorded and should provide useful insight for maintaining glucose levels. Data must be organized in such a way to identify patterns and provide recent trend to optimize correct levels. Healthcare providers receive more useful data, and it helps them to predict the upcoming glucose ranges. Mobile supportive system will make things much easier. Implementing machine learning methods into CGM provides highly efficient results.

These technologies reduce the burden of human effort for glucose monitoring. It reduces the wait time over hospital, wait time for getting results, and so on. With the advancements comes the matter of security. The secured communication and record of medical data is to be considered while developing medical devices.

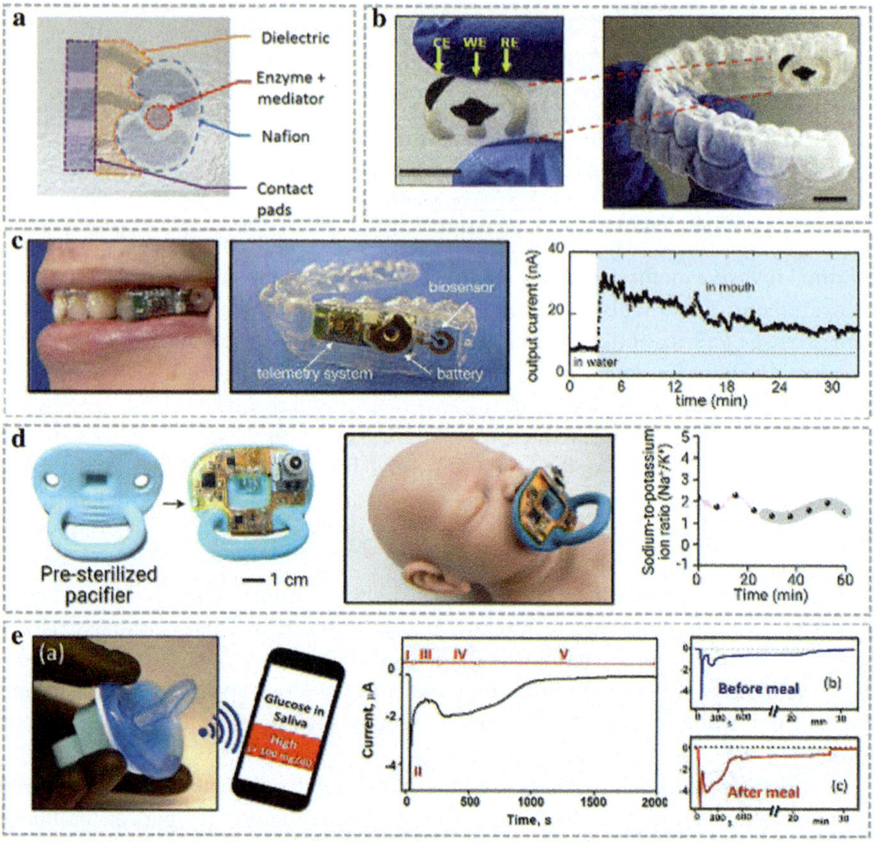

Fig. 5 Diabetic monitoring with biosensor

Future devices rely on wearable with less invasive methods which suits people of all ages. AI and deep learning concepts are providing promising results in healthcare sector. Thus future diabetic monitoring is less invasive which provides handy results with much efficiency and accuracy.

Future devices depend on 5G enabled smart IoT devices for diabetes monitoring. It is a well-known that smart devices enable the interconnection between patients and clinicians much easier. Such advancements are highly recommended in all areas of healthcare system. In particular with diabetes, it is intelligent to embed multiple systems for smart healthcare monitoring. Wearable bio-sensors with high predictive accuracy is the future of diabetes monitoring. Researchers are interpolating optical devices with nano-sensors to bring new future to diabetes monitoring system. Current devices are used to diagnose and maintain proper glucose levels.

Smart devices enabled with insulin reminder, dosage calculator, and complication alert with integrated IoT are going to be the nearby future advancements in the field

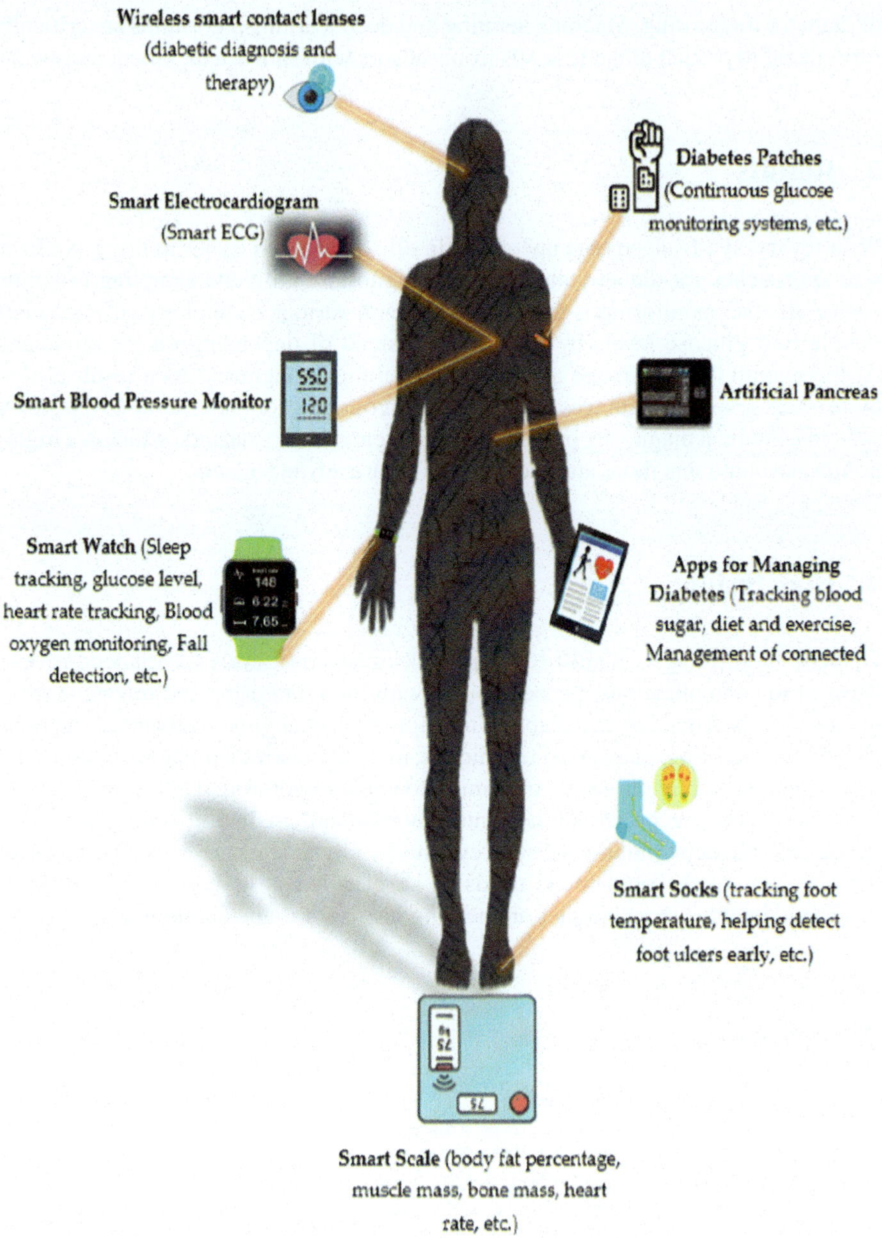

Fig. 6 Present diabetic monitoring devices

of diabetes monitoring. Machine learning and deep learning algorithms are critically formulated to predict diabetic levels from sensors with saliva, tear, blood, and sweat.

3 Results

With technical advancements upcoming in all sectors, it is easier to bring noninvasive approaches for diabetic monitoring. Diagnosis with blood sampling to noninvasive glucose monitoring is made possible with various technology advancement. People find glucose levels by themselves instead of depending on the clinicians. At the maxim the results are accurate to the clinical diagnosis. As a result of this, there were several advantages and disadvantages. Embedding several advanced AI systems, medical diagnostic procedures, and wearable bio-sensors, which are highly noninvasive, attracts more attention toward future advancements.

4 Conclusion

Diabetic monitoring is a must to keep track of glucose levels for diabetic people. Most of the complications can be avoided with the monitoring and maintenance of glucose levels. The evolving diabetes technology is changing to advanced methods. Hence, the CGM with advanced techniques, machine learning implementations, and advanced research methods are upcoming to solve monitoring. Data collected from those devices are recorded but their interpretation and prediction system need to be developed. Diabetic monitoring with eye fluids, saliva, tears are evolving; however, the accuracy of the predictions needs to be clarified at higher levels. Advanced integrated healthcare techniques are helping hands for diabetic monitoring.

References

1. World Health Organization. Diabetes
2. International Diabetes Federation. Type 1 diabetes
3. International Diabetes Federation. Type 2 diabetes
4. Sun J, Hu W, Ye S, Deng D, Chen M (2023) The description and prediction of incidence, prevalence, mortality, disability-adjusted life years cases, and corresponding age-standardized rates for global diabetes. J Epidemiol Glob Health 13(3):566–576
5. Karuppasamy M, Poorani K (2023) Analysis of underlying and forecasting factors of type 1 diabetes and prediction of diabetes using machine learning. In: Kumar A, Ghinea G, Merugu S (eds) Proceedings of the 2nd international conference on cognitive and intelligent computing. ICCIC 2022. Cognitive science and technology. Springer, Singapore
6. Maric-Bilkan C (2017) Sex differences in micro- and macro-vascular complications of diabetes mellitus. Clin Sci 131:833–846

7. Soltysiak J, Krzysko-Pieczka I, Gertig-Kolasa A, Mularz E, Skowrońska B, Ostalska-Nowicka D, Zachwieja J (2023) Acute kidney injury and diabetic kidney disease in children with acute complications of diabetes. Pediatr Nephrol 38(5):1643–1652
8. Tran NK, LaValley C, Bagley B, Rodrigo J (2023) Point of care blood glucose devices in the hospital setting. Crit Rev Clin Lab Sci 60(4):290–299
9. Klonoff DC, Nguyen KT, Xu NY, Gutierrez A, Espinoza JC, Vidmar AP (2023) Use of continuous glucose monitors by people without diabetes: an idea whose time has come? J Diabetes Sci Technol 17(6):1686–1697
10. Karuppasamy M, Poorani K (2023) Comparative analysis of chronic kidney disease prediction using supervised machine learning techniques. In: Choudrie J, Mahalle PN, Perumal T, Joshi A (eds) ICT for intelligent systems. ICTIS 2023. Smart innovation, systems and technologies, vol 361. Springer, Singapore
11. Yoo JH, Kim JH (2023) Advances in continuous glucose monitoring and integrated devices for management of diabetes with insulin-based therapy: improvement in glycemic control. Diabetes Metab J 47(1):27–41
12. Karuppasamy M, Poorani K (2023) Information system for neuropathy prediction ensembling ranking and ordered clustering for diabetic healthcare monitoring. In: Tuba M, Akashe S, Joshi A (eds) ICT infrastructure and computing. ICT4SD 2023. Lecture notes in networks and systems, vol 754. Springer, Singapore
13. Golsanamlou Z, Mahmoudpour M, Soleymani J, Jouyban A (2023) Applications of advanced materials for non-enzymatic glucose monitoring: from invasive to the wearable device. Crit Rev Anal Chem 53(5):1116–1131
14. Darrow JJ, Van de Wiele V, Beran D, Kesselheim AS (2023) An empirical review of key glucose monitoring devices: product iterations and patent protection. J Diabetes Sci Technol 19322968231178016
15. Ahmadian N, Manickavasagan A, Ali A (2023) Comparative assessment of blood glucose monitoring techniques: a review. J Med Eng Technol 47(2):121–130
16. Reddy N, Verma N, Dungan K (2023) Monitoring technologies-continuous glucose monitoring, mobile technology, biomarkers of glycemic control. Endotext
17. Tang L, Chang SJ, Chen CJ, Liu JT (2020) Non-invasive blood glucose monitoring technology: a review. Sensors 20(23):6925
18. Lee SM, Lee D (2020) Healthcare wearable devices: an analysis of key factors for continuous use intention. Serv Bus 14(4):503–531
19. Bolla AS, Priefer R (2020) Blood glucose monitoring-an overview of current and future non-invasive devices. Diabetes Metab Syndr Clin Res Rev 14(5):739–751
20. Vashist SK (2013) Continuous glucose monitoring systems: a review. Diagnostics 3(4):385–412
21. Vashist SK, Zheng D, Al-Rubeaan K, Luong JH, Sheu FS (2011) Technology behind commercial devices for blood glucose monitoring in diabetes management: a review. Anal Chim Acta 703(2):124–136
22. Lin T, Gal A, Mayzel Y, Horman K, Bahartan K (2017) Non-invasive glucose monitoring: a review of challenges and recent advances. Curr Trends Biomed Eng Biosci 6(5):1–8
23. Bruen D, Delaney C, Florea L, Diamond D (2017) Glucose sensing for diabetes monitoring: recent developments. Sensors 17(8):1866
24. Rghioui A, Lloret J, Harane M, Oumnad A (2020) A smart glucose monitoring system for diabetic patient. Electronics 9(4):678
25. Shang T, Zhang JY, Thomas A, Arnold MA, Vetter BN, Heinemann L, Klonoff DC (2022) Products for monitoring glucose levels in the human body with noninvasive optical, noninvasive fluid sampling, or minimally invasive technologies. J Diabetes Sci Technol 16(1):168–214
26. Freckmann G, Pleus S, Grady M, Setford S, Levy B (2019) Measures of accuracy for continuous glucose monitoring and blood glucose monitoring devices. J Diabetes Sci Technol 13(3):575–583

Navigating the Digital Realm: E-Visual Online Examination for the Visually Challenged

A. Shakthi Sriharan, K. Panimozhi, and Antara Roy Choudhury

Abstract Vision loss significantly impacts individuals and society, necessitating innovative solutions for visually impaired students. This paper presents E-Visual, an automated online examination system designed specifically for visually impaired students, enabling them to independently take exams without human assistance. Utilizing a braille keyboard with text-to-speech and speech-to-text functionalities, coupled with RFID technology, E-Visual enhances accessibility and convenience for visually impaired individuals. This system is also beneficial for differently abled students, promoting a barrier-free examination experience.

Keywords Visually impaired students · Speech-to-text · Text-to-speech · Braille keyboard

1 Introduction

This idea for visually impaired students helps them to connect with the people easily and makes them independent, and it also inspires a lot of younger generation to create something that is useful for the society and the people in need.

E-Visual emerges as a transformative solution for visually impaired students, fostering independence and accessibility in online assessments. With approximately 2.2 billion people globally facing visual impairment challenges, including 200,000 children, the need for such inclusive technologies is paramount. Current examination methods often rely on human scribes, leading to complexities and inefficiencies. E-Visual addresses these challenges by leveraging advanced technologies such as RFID, braille keyboards, and speech-to-text functionalities to empower visually impaired students in taking online exams autonomously.

A. Shakthi Sriharan · K. Panimozhi (✉) · A. R. Choudhury
Department of Computer Science, B.M.S. College of Engineering, Bangalore, India
e-mail: panimozhi.cse@bmsce.ac.in

A. R. Choudhury
e-mail: antararc.cse@bmsce.ac.in

© The Author(s), under exclusive license to Springer Nature Singapore Pte Ltd. 2025

H. Zen et al. (eds.), *Soft Computing and Signal Processing*, Lecture Notes in Networks and Systems 1200, https://doi.org/10.1007/978-981-97-9926-8_5

To develop an automated online examination system (E-Visual) that eliminates the need for human scribes, enabling visually impaired students to independently participate in online exams and integrate cutting-edge technologies such as RFID, braille keyboards, and speech-to-text functionalities into E-Visual for enhanced accessibility and inclusivity, benefiting both visually impaired and differently abled students.

This paper proceeds to analyze survey data, discuss the research methodology, present the system's functionalities, delve into key technologies used, evaluate performance measures, and conclude with references that informed this study.

The remaining sections are organized as follows. Analysis and survey are discussed in Sect. 2. In Sect. 3, research methodology is well discussed. The working of the functions implemented is presented in Sect. 4. In Sect. 5, the key technology is discussed. The performance measure in Sects. 6 and 7 concludes the paper. Various references are written which we had referred to for this paper in references section.

2 Literature Survey

For a visually impaired student to comprehend a concept, learn, study, and write it with the help, a scribe is not a very easy task. Hence, it is better to have a fully automated system for the visually impaired student to write the exam. In regard, there are several research papers which have taken into place to support visually impaired students and make their lives a little easier in all some aspects of their livelihood.

The research in [1] highly focused on visually impaired students who are unable to read the questions displayed on the computer screen in this online examination system. They missed their braille system which is more familiar to read and write more easily. This is high time to build a proper voice-based online examination system for visually impaired students. This system should read the question that is displayed on the screen and obtain the student's response via voice commands.

The basic idea of recognizing the speeches of the students using speech signals has its merits and demerits with various studies and approaches; each individual student has their own signals and frequency as these days the system is designed for ASR system with numerous vocabularies that support the student independently in different languages [2].

In the research paper [3] state that many surveys have been conducted across 39 countries globally where around 275–285 million people are currently visually impaired and in that around 40 million are completely blind as the majority of the impairment are due to uncorrected refractive error (43%) and (33%) are suffering due to cataracts and cataracts are also the first cause of blindness.

A research says that prior to open and distant learning of the online examination system uses voice interface which wasn't very efficient and accurate which overall decreases the quality of the examination. Since our main goal is to improve and find an alternate method for the students providing a framework that will guide the development of the online examination system [4].

The purpose of this study was to explore the effects of the digital material incorporated into text-to-speech system for students' English spelling. The digital material was made on the basis of the Spelling Bee vocabulary list (approximately 300 words) issued by the selected school. Twenty-one third-graders from a private bilingual school in Taiwan were selected for this study [5].

Chapaneri and Santosh delve into the intricacies of speech recognition systems, specifically focusing on recognizing spoken digits. Their work employs weighted mel-frequency cepstral coefficients (MFCC) and enhanced features for dynamic time warping (DTW). This research contributes valuable insights into improving the accuracy and efficiency of speech recognition, which is crucial for developing voice-based interfaces for visually impaired students [6].

Voice Operated Tool-Examination Portal designed specifically for blind individuals. Their research focuses on developing a user-friendly interface that enables visually impaired students to navigate and interact with online examination systems using voice commands. This work contributes to the advancement of accessible technologies for visually impaired individuals in educational settings [7].

Kim et al. [8] explore a text-based testing environment tailored for voice-based services. Their research aims to enhance the usability and effectiveness of voice-based interfaces in various applications, including online examinations. This study contributes to the development of robust and efficient voice-based systems for visually impaired individuals, improving accessibility in educational assessments.

Balasuriya et al. present a learning platform tailored for visually impaired children, leveraging artificial intelligence (AI) and computer vision technologies. Their work focuses on creating an interactive and accessible educational environment that utilizes AI algorithms and computer vision techniques to support visually impaired students in learning and comprehension. This research contributes to the development of AI-driven solutions for enhancing educational experiences for visually impaired individuals [9].

Their research focuses on developing a Chatbot interface that provides educational support and resources tailored to the needs of visually impaired students. This work contributes to leveraging mobile technology and AI-driven Chatbot systems to enhance educational accessibility for visually impaired individuals [10].

To encourage greater participation in their own learning, and to give greater access to different sources of information than traditional methods offers. In future, blind peoples also can do online exam like a normal human if our project is delivered in real time [11].

The student if they are able to write the exam without the help of a scribe using an automated examination system, it makes the student more independent and can overcome various difficulties in writing the exam. The main goal is for the student to perform and solve the questions without being dependent on another person [12].

Mulloy et al. explore assistive technologies tailored for students with visual impairments and blindness. Their research focuses on the development and implementation of innovative technologies that enhance accessibility and support learning for visually impaired individuals. This work contributes valuable insights into the application of assistive technologies in educational settings for visually impaired students [13].

Bose et al. present a digital assistant designed specifically for blind individuals. Their research aims to develop a user-friendly and intuitive digital assistant that provides support and assistance to visually impaired users in various daily tasks. This work contributes to the advancement of digital assistant technologies for enhancing independence and accessibility for blind individuals [14].

Yurtay and Adak introduce an education portal tailored for visually impaired individuals. Their research focuses on creating an accessible and inclusive online platform that provides educational resources and support for visually impaired students. This work contributes to the development of online educational portals that cater to the needs of visually impaired individuals [15].

Elmannai and Elleithy [16] investigate sensor-based assistive devices for visually impaired individuals. Their research explores the current status, challenges, and future directions of sensor-based technologies aimed at enhancing independence and mobility for visually impaired individuals. This work contributes to the development of innovative sensor-based solutions for improving the daily lives of visually impaired individuals.

Ghali et al. [17] review virtual reality (VR) technology for blind and visually impaired individuals, discussing recent advances and applications in the field. Their research provides insights into the potential of VR technology to enhance accessibility, mobility, and learning experiences for blind and visually impaired individuals. This work contributes to the exploration of VR-based solutions for improving the quality of life for visually impaired individuals.

The American Foundation for the Blind (AFB) [18] provides valuable resources and information on blindness and low vision. Their research and publications offer insights into the challenges faced by individuals with visual impairments, as well as strategies and technologies to improve accessibility and inclusion for this population. This work contributes to raising awareness and promoting advocacy for individuals with visual impairments.

Rahi et al. [19] conduct a study on childhood blindness in India, focusing on the causes and prevalence among blind school students in nine states. Their research sheds light on the factors contributing to childhood blindness and the importance of early detection and intervention strategies. This work contributes to addressing the challenges of childhood blindness and advocating for accessible eye health services.

Manoharan presents a smart image processing algorithm designed for text recognition, information extraction, and vocalization for the visually challenged. The research focuses on leveraging image processing techniques to enhance accessibility and assist visually challenged individuals in reading and comprehending textual information. This work contributes to developing innovative solutions for improving accessibility and inclusion for the visually challenged population [20].

An authentication system that is secure is one that allows for effective monitoring and user entry verification. We want to be able to provide dual user identification in our proposed system using fingerprint and RFID technologies. There is a magnetic door lock attached to the access control mechanism. Utilizing Wi-Fi and ZigBee modules with a Raspberry Pi computer allows for wireless communication. Finally a camera will be deployed to record user photos where ever required for monitoring purpose [21].

3 Research Methodology

For a visually impaired student to navigate through the entire system, the student would require a braille keyboard and RFID tag and reader. There are many advantages for using a braille keyboard where it makes the visually impaired students use an automated system in a structured and convenient way; but it also has its disadvantage, for example, the keyboard cannot be used by a normal person who has had no training in it. When a normal sighted person uses a braille keyboard the student might not be able to differentiate between a mistake and a non-mistake. Currently the visually impaired use the help of a scribe to write the exam and it comes with many problems as he/she are responsible to understand what the student says and translate it into the necessary medium. The scribe sometimes might not understand the solution that the visually impaired student is trying to say. There are also a lot chances of malpractices that can occur, or the scribe might answer it incorrectly unintentionally. At exam times sometimes the scribe might cancel his/her appointment.

The main objective of this study is to investigate the effectiveness of an online examination system specifically designed for visually impaired students. The research aims to understand the unique challenges faced by visually impaired students during online examinations and identify ways to enhance the user experience. The study may be limited by the availability and duration of access to visually impaired student participants. The findings of this study may be specific to the chosen sample and the particular online examination system. Generalizing the results to all visually impaired students or different systems may require further research. Direct observations will be conducted to analyze the interaction between visually impaired students and the online examination system. In-depth interviews will be conducted with visually impaired students to gather their experiences, and challenges related to using the online examination system. Structured surveys will be conducted to gather in-depth data regarding user satisfaction, experience, and system stability.

Relevant performance metrics, such as completion time and accuracy, will be collected during the examination process to assess the effectiveness of the system. Prior to participation, all visually impaired students will be provided with detailed information about the study's purpose, procedures, and potential risks or benefits as to help them understand the efficiency and importance of the system.

4 Functions of E-Visual Examination System

Various features and functions are implemented in this study which are vital for the performance of the study and are essentially needed for further implementation.

Module 1

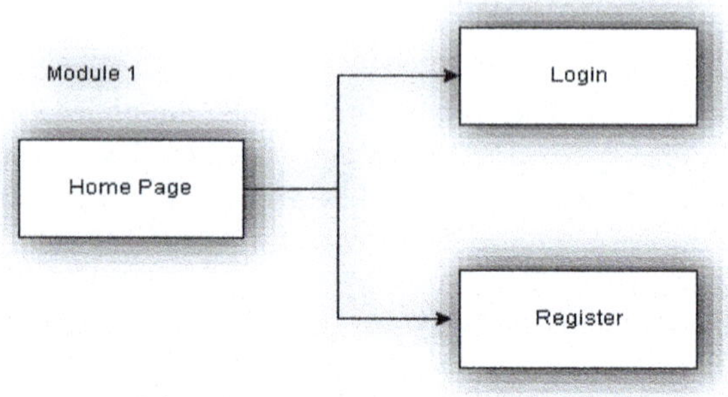

Module 1 contains the login and registration credentials where the user has to login for validation.

Module 2

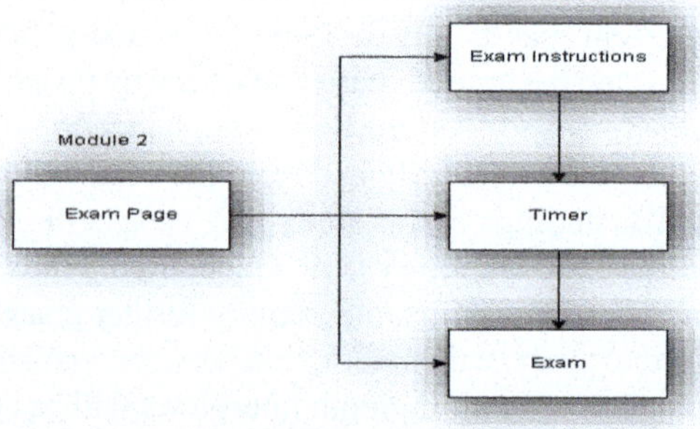

Module 2 contains the exam page where the user can take up the exam after the instructions and timer.

Module 3

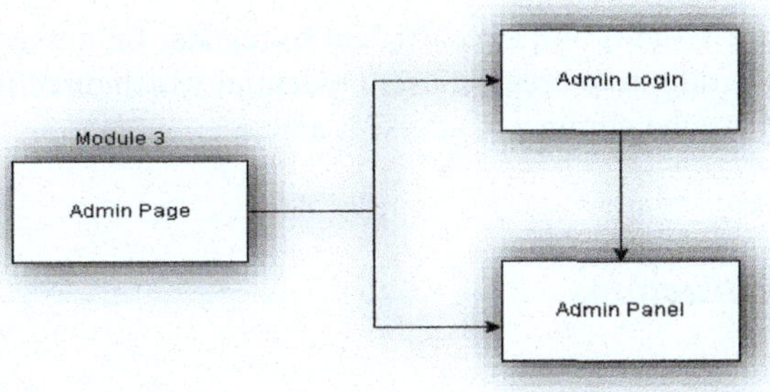

Module 3 contains the admin login and panel which gives the overall access to the website where the admin can block/unblock students and change questions.

4.1 *RFID Validation*

For a visually impaired student to login using his credentials the student would have to use his/her RFID tag and scan it using a reader where the validation is done using PHP as the system checks the user RFID entered with the one stored in the database, and if the credentials are correct, the system redirects the user to the user login panel where his/her information is displayed.

4.2 *Speech-to-Text*

For a visually impaired student to give his/her answer, the student must press the specified key assigned for recording (SPACEBAR), once the user answers he/she can reread the answer by using another specific key (M) which simultaneously stops the recording. The entire functionality is worked on JavaScript where a function is created for each functionality and is fetched using the 'id' from the html page.

4.3 *Text-to-Speech*

For a visually impaired student to listen to his/her question, the student must press a specified key assigned to listen to the question (left arrow), once the user listens he/she can start answering. The entire functionality is worked on JavaScript where a function is created for each functionality and is fetched using the 'id' from the html page.

4.4 Register

It is not ideal for a visually impaired student to register on his own, so he/she can take the help of the admin to create a user credential which directly gets stored in a database table where the admin can access it anytime.

4.5 Exam Instructions

Every visually impaired student trying to use our architecture might face difficulty in understanding the system, so the instructions on how to successfully navigate through the system are clearly mentioned using a python prerecorded file that automatically starts reading when the page is loaded.

4.6 Exam Page

The main exam page integrates both text-to-speech and speech-to-text where the students write their exam without the help of a scribe. The user can reread the answer, listen the questions, save the answer, listen to the answer, and also delete the answer if the user is not satisfied with his/her answer.

4.7 Cloud Platform (MySQL)

MySQL is the cloud platform used where the database is the main soul of the examination system where all the questions, answers, users' details, and the admin details are stored. Every function is fetched from the database and pushed into it (Fig. 1).

It is important to acknowledge the constraints and the non-functional requirements of the study. The findings may be specific to the chosen sample and the particular online examination system, and generalizing the results to all visually impaired students can be done. Future research should implement vast and more diverse samples to obtain an easier understanding of the challenges faced by visually impaired students in online examinations.

An automated system could overcome these disadvantages. Elimination of a third-party scribe may also lead them to be freer and independently. The proposed system is not complex, is very user-friendly, and is very efficient and useful for not only blind people but differently abled people also.

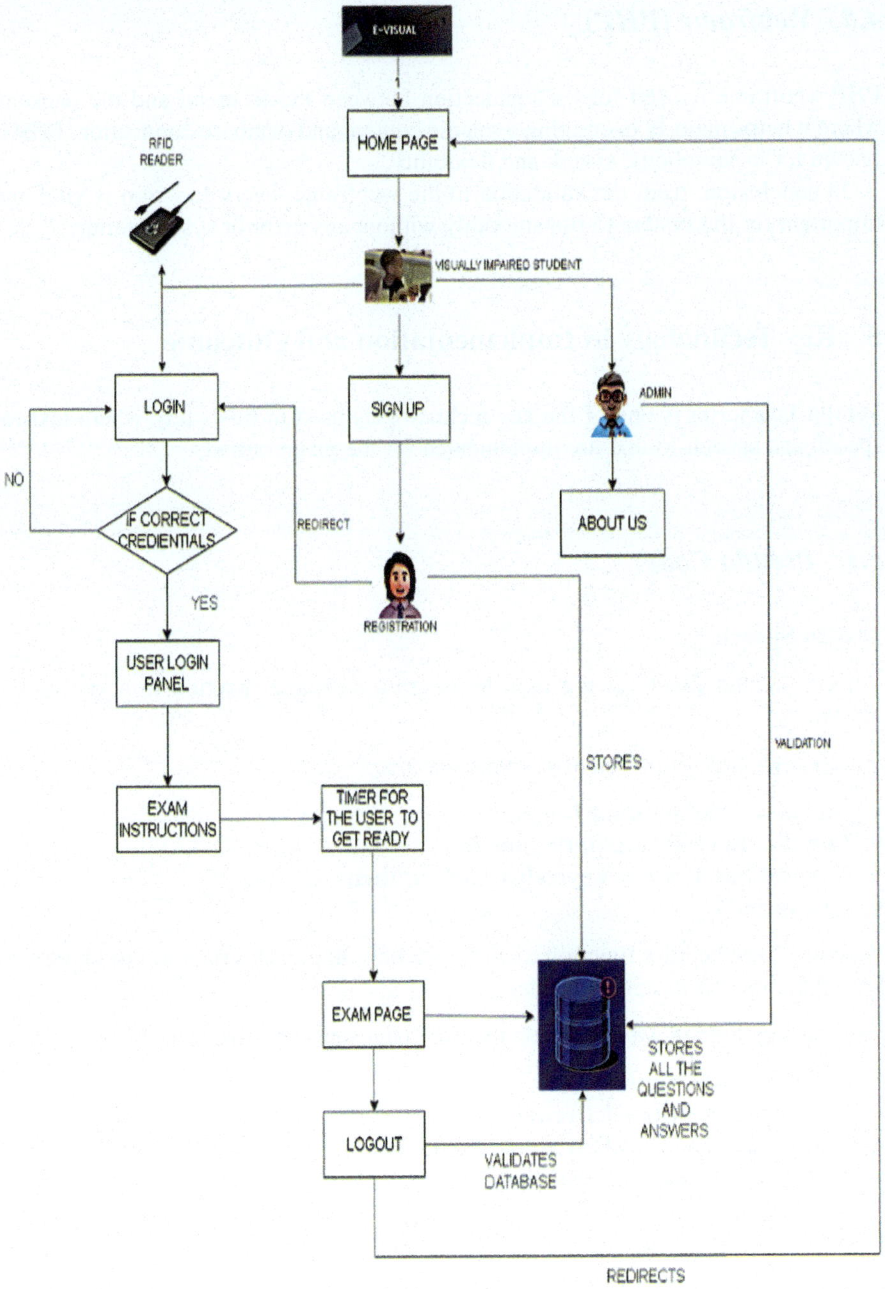

Fig. 1 System architecture of the E-visual online examination system

4.8 Webframe (PHP)

PHP webframe is used for the connection between the frontend and the backend where it helps majorly in creating dynamic content and database connection. PHP is known for its simplicity, speed, and flexibility.

In conclusion, from the validation to the webframe, every function is vital and important for the system to run smoothly without any error or system failure.

5 Key Technology in Implementation and Outcome

Vanilla JavaScript is one of the key technologies used in this study where text-to-speech and speech-to-text are implemented for the given output.

5.1 Pseudo Code

Text-to-Speech

// event handler functions are used to assign a webpage function to a key shown below.

// each keyboard buttons have a unique keycode.

- **Initialize event handler function**
- **Initialize keycode attribute function**
- **If (event handler == keycode function) then**
- **Call speak()**

// for the event handler function to work, we have to create a function to import the SpeechSynthesisUtterance libraries.

// we then call the id we created on the html page and initialize it here (myText).

- **Initialize speak function**
- **Call SpeechSynthesisUtterance library**
- **If (given msg == myText.value) then**
- **Call speachSynthesis.speak(msg)**

Speech-to-Text

// event handler functions are used to assign a webpage function to a key shown below.

// each keyboard buttons have a unique keycode.

- **Initialize event handler function**
- **Initialize keycode attribute function**
- **If (event handler == keycode function) then**
- **Call speak()**

// for the event handler function to work, we have to create a function to import the SpeechSynthesisUtterance libraries.

// we then call the id we created on the html page and initialize it here (content).

- **Initialize listen function**
- **If (vartxtmsg == SpeechSynthesisUtterance)**
- **Then txtmsg.text = content.value**
- **Call speachSynthesis.speak(msg)**

5.2 Working Operation

1. Initially, the user requires an RFID tag and reader with his/her details already registered. If the student has not registered, then he/she can contact their respective admin to proceed further (shown in Fig. 2).
2. Next, the user gets his/her login information and is processed through speech-to-text utterance synthesis.
3. The user further gets familiar with the exam instructions and the system functionality.
4. The user is given a total of 120 s to get ready for the exam.
5. Once the user enters the exam, he/she can start listening to the questions and try to answer them (shown in Fig. 3).

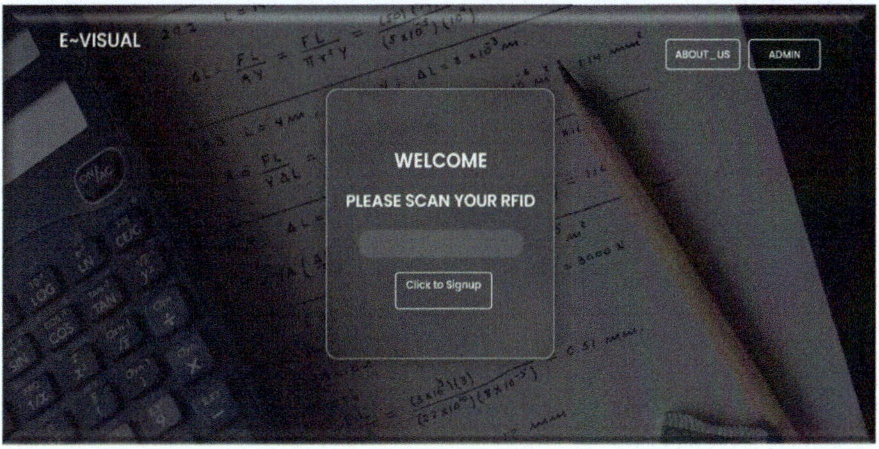

Fig. 2 Home page of E-Visual online examination

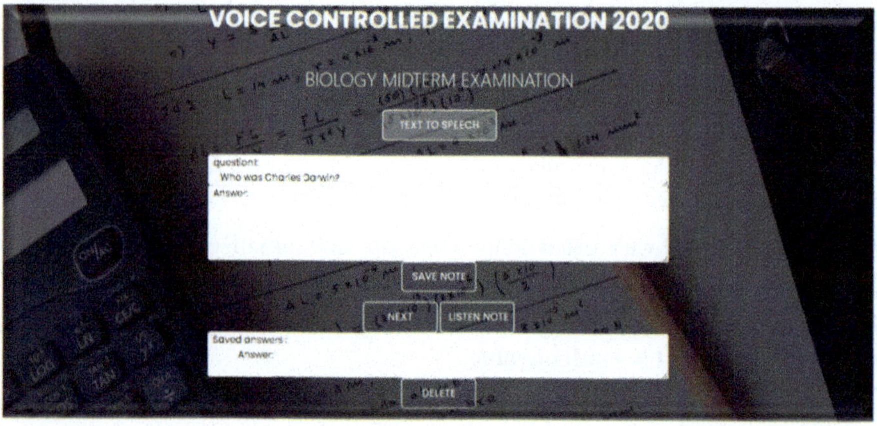

Fig. 3 Exam page of E-Visual online examination

6. For the user to listen to the question, he/she should press the left arrow key.
7. For the user to answer, he/she should press spacebar for the recording to begin.
8. To stop the recording, the user must press M key which simultaneously stops recording the answer and rereads the answer to the user.
9. The user can press enter to navigate to the next question the user can also use the number keys to navigate to the specified question.
10. The user can go back to the previous question and listen to the saved answer that he/she answered before using L key.
11. If the user is not satisfied with the saved answer, he/she can use the D key to delete the saved answer and can answer them again.
12. Once the user finishes his/her exam, he/she can log out and end the exam.
13. If the user forgets his/her system functionality mentioned in the above points, then he/she can use the I key to listen to the instructions again.

Figure 2 is the home page of the examination system where the user uses to login or sign up.

Figure 3 is the main page of the examination system where the user can start answering.

Figure 4 is the actual image of the physical RFID reader and the tag where the user uses it for his/her validation.

5.3 *Cloud Platform (MySQL)*

Figure 5 is the MySQL database where all the answers, questions, and login credentials are stored.

Figure 6 is the MySQL database of the users that are already registered using their user RFID which is integrated to the users' RFID tag.

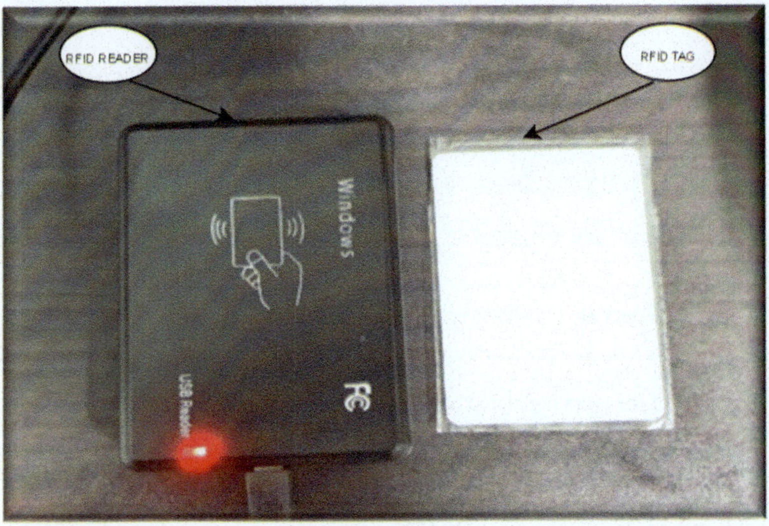

Fig. 4 RFID used in E-Visual online examination

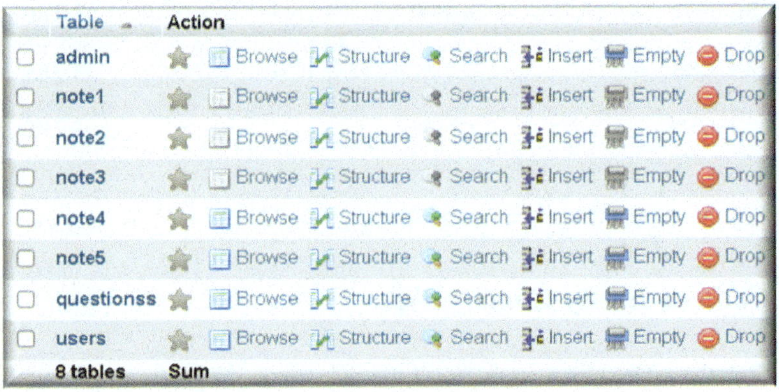

	Table	Action						
☐	admin	★	Browse	Structure	Search	Insert	Empty	Drop
☐	note1	★	Browse	Structure	Search	Insert	Empty	Drop
☐	note2	★	Browse	Structure	Search	Insert	Empty	Drop
☐	note3	★	Browse	Structure	Search	Insert	Empty	Drop
☐	note4	★	Browse	Structure	Search	Insert	Empty	Drop
☐	note5	★	Browse	Structure	Search	Insert	Empty	Drop
☐	questionss	★	Browse	Structure	Search	Insert	Empty	Drop
☐	users	★	Browse	Structure	Search	Insert	Empty	Drop
	8 tables	Sum						

Fig. 5 Backend database of E-Visual online examination

6 Performance Measure

Various performance measures have been done where the results are shown in the figures.

←T→				▼	id	user_name	user_rfid	date
☐	Edit	Copy	Delete		3	tom	0006824281	2023-04-03 08:20:22
☐	Edit	Copy	Delete		10	raja	0009703108	2023-03-23 02:22:36
☐	Edit	Copy	Delete		9	sharath	0009899141	2023-03-23 02:22:25
☐	Edit	Copy	Delete		13	shakthi	0006824281	2023-03-28 04:34:24

☐ Check all *With selected:* Edit Copy Delete Export

Fig. 6 Database of the users registered on E-Visual online examination

	PERCENT	
Image	76.12%	343.5 KB
Script	14.11%	63.7 KB
Font	5.92%	26.7 KB
CSS	2.57%	11.6 KB
HTML	0.60%	2.7 KB

Fig. 7 Contains the types of contents used in the study according to their size

6.1 Content Size by Content Type

Figure 7 is the table which contains all the information regarding contents that are used in the study.

6.2 Performance Metrics

Figure 8 is the generated data of various metrics such as speed index, TOI, TBT, and cumulative layout shift.

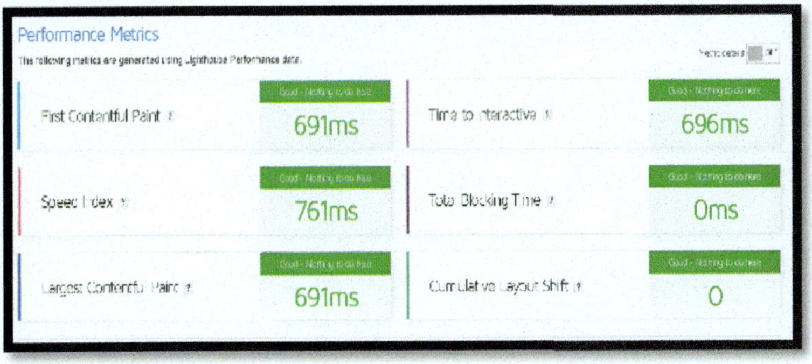

Fig. 8 Following metrics are generated using lighthouse performance data

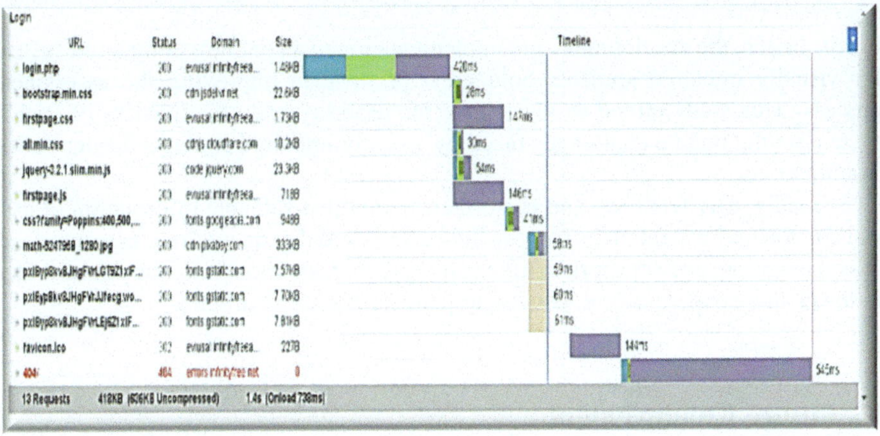

Fig. 9 Total size of the study using waterfall model

6.3 *Waterfall Model*

Figure 9 is the waterfall model which explains the size of the project is distributed file by file.

6.4 *GTmatrix*

Figure 10 is the complete score of the efficiency and performance of the project.

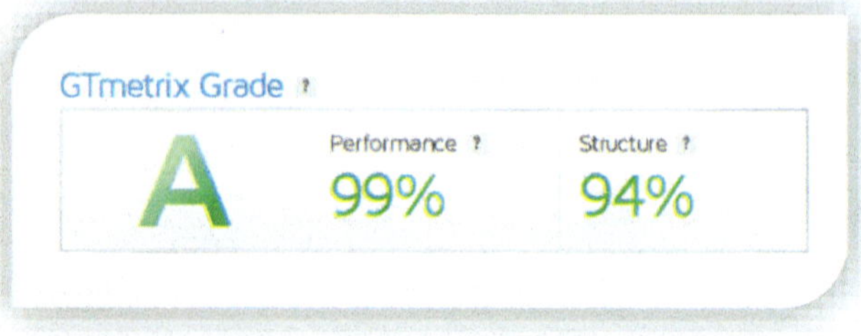

Fig. 10 Total score simulated in GTmatrix

7 Conclusion

In conclusion, the development and implementation of an online examination system for visually impaired students hold great promise in improving the assessment process. This study aimed to investigate the usability and effectiveness of such a system on the unique challenges faced by visually impaired students during online examinations.

Secondly, user feedback and performance provided valuable information of user satisfaction and system effectiveness. It became clear that an intuitive user interface, easy navigation, and appropriate time management features significantly contributed to the positive experience of visually impaired students all around.

8 Future Enhancement

Our website contains all the functionalities from the user registering his account to writing the exam. As a future enhancement, the study can be further developed by adding the admin page and panel mentioned in Module 3 where the teacher for instance can evaluate the answers written by each student and check for the total number of users and can also add exam subjects, delete exam subjects, add users, delete users, modify the user details, etc.

The study can be further implemented by using face recognition modules and fingerprint recognition software for the user's validation.

Statements and Declarations I would like to express my sincere gratitude to B.M.S. College of Engineering for their support of my study which allowed me to complete my study. I am also grateful for the use of the college's facilities, which were essential to the success of my study.

We have no conflicts of interest to disclose.

All authors declare that they have no conflicts of interest.

References

1. Yurtay N, Yurtay Y, Fatih Adak M (2015) An education portal for visually impaired. Procedia-Soc Behav Sci 171:1097–1105
2. Saksamudre SK, Shrishrimal PP, Deshmukh RR (2015) A review on different approaches for speech recognition system. Int J Comput Appl 115(22)
3. Hampshire B (1981) Working with Braille: a study of Braille as a medium of communication
4. Azeta AA, Inam IA, Daramola O (2018) A voice-based E-examination framework for visually impaired students in open and distance learning. Turk Online J Dist Educ TOJDE 19(2):34–46
5. Anupriya A, Vijayalakshmi S (2015) An efficient examination system for blinds with real time voice interface. J Comput Sci Mob Comput 4(3):345–348
6. Chapaneri SV (2012) Spoken digits recognition using weighted MFCC and improved features for dynamic time warping. Int J Comput Appl 2(4):6–12
7. Vats A, Tandon A, Varshney D, Sinha A (2016) Voice operated tool-examination portal for blind persons. Int J Comput Appl 142(14):0975–8887
8. Kim E, Yoon H, Cha W (2018) Text-based testing environment for voice-based service. In: 25th Asia-Pacific software engineering conference (APSEC)
9. Balasuriya BK, Lokuhettiarachchi NP, Ranasinghe ARMDN, Shiwantha KDC, Jayawardena C (2017) Learning platform for visually impaired children through artificial intelligence and computer vision. In: 2017 11th international conference on software, knowledge, information management and applications (SKIMA). IEEE, pp 1–7
10. Kumar MN, Chandar PCL, Prasad AV, Sumangali K (2016) Android based educational Chatbot for visually impaired people. In: 2016 IEEE international conference on computational intelligence and computing research (ICCIC). IEEE, pp 1–4
11. Khan S, Verma S, Agarwal S, Krishnatrey P, Sharma S (2015) Voice based online examination for physically challenged. MIT Int J Comput Sci Inf Technol 5(2):58–61
12. Bharadwaj KA, Joshi MM, Kumbale NS, Shastry NS, Panimozhi K, Roy Choudhury A (2020) Speech automated examination for visually impaired students. In: 2nd international conference on innovative mechanisms for industry applications (ICIMIA)
13. Mulloy AM, Gevarter C, Hopkins M, Sutherland KS, Ramdoss ST (2014) Assistive technology for students with visual impairments and blindness. In: Assistive technologies for people with diverse abilities. Springer, New York, NY, pp 113–156
14. Bose P, Malpthak A, Bansal U, Harsola A (2017) Digital assistant for the blind. In: 2017 2nd international conference for convergence in technology (I2CT). IEEE, pp 1250–1253
15. Yurtay N, Yurtay Y, Adak MF (2015) An education portal for visually impaired. Procedia-Soc Behav Sci 171:1097–1105
16. Elmannai W, Elleithy K (2017) Sensor-based assistive devices for visually-impaired people: current status, challenges, and future directions. Sensors 17(3):565. https://doi.org/10.3390/s17030565
17. Ghali NI, Soluiman O, El-Bendary N, Nassef TM, Ahmed SA, Elbarawy YM, Hassanien AE (2012) Virtual reality technology for blind and visual impaired people: reviews and recent advances. In: Advances in robotics and virtual reality, pp 363–385
18. American Foundation for the Blind (2019) Blindness and low vision. American Foundation for the Blind
19. Rahi JS, Sripathi S, Gilbert CE, Foster A (1995) Childhood blindness in India: causes in 1318 blind school students in nine states. Eye (Pt 5):545–550
20. Manoharan S (2019) A smart image processing algorithm for text recognition, information extraction and vocalization for the visually challenged. J Innov Image Process (JIIP) 1(01):31–38
21. Vaishnav S, Sunil S, Ullas DG, Rao VK, Panimozhi K (2017) Secure authentication system. Int J Comput Trends Technol (IJCTT) 44(1)

The Evaluation of Data Aggregation and Compression in WSN Routing Networks Using the Best Clustering Methods

V. V. A. S. Lakshmi, Gujjarlamudi Mahesh Babu, P. Anjaneyulu, and D. Priyanka

Abstract In cluster-based remote sensor systems, the non-uniform dispersion of nodes often leads to varying energy consumption levels among clusters. To address this challenge, we propose an effective information aggregation tree, drawing inspiration from prior communication frameworks, to guide clustering design. Our approach incorporates characteristics such as Remaining Controlling, Node Thickness, and Stack Heads of Clusters, leveraging fuzzy logic techniques for their selection. Through inter-cluster steering computations, we adjust energy distribution among cluster heads, aiming to equalize energy consumption across clusters. Furthermore, to mitigate energy usage, we employ data compression techniques on aggregated data, exploiting information relationships. Our evaluation focuses on the effectiveness of these strategies within WSN routing networks, assessing various clustering methods to optimize energy efficiency and enhance network performance.

Keywords Fuzzy logic · Inter-cluster routing method · Wireless sensor network

V. V. A. S. Lakshmi (✉) · G. M. Babu
Narasaraopeta Engineering College, Narasaraopet, Andhra Pradesh, India
e-mail: vvaslakshmi@gmail.com

G. M. Babu
e-mail: maheshbabu.crazy@gmail.com

P. Anjaneyulu · D. Priyanka
Aditya Institute of Technology and Management, Tekkali, Andhra Pradesh, India
e-mail: Anjii.ram888@gmail.com

D. Priyanka
e-mail: Dabbeerupriyanka90@gmail.com

© The Author(s), under exclusive license to Springer Nature Singapore Pte Ltd. 2025
H. Zen et al. (eds.), *Soft Computing and Signal Processing*, Lecture Notes in Networks and Systems 1200, https://doi.org/10.1007/978-981-97-9926-8_6

1 Introduction

Wireless sensor networks (WSNs) are made up of elegant, miniature sensor nodes that use sensor modules to detect various phenomena and electronically transfer the relevant data to a sink node. WSNs accumulate, process, and analyze all information while offering various sensory information to designated customers. Such sensor nodes are frequently arranged in large quantities (only some to dozens) and in settings that are very challenging for humans to operate. As a result, sensors must be dispersed erratically and rely on small power storage devices like batteries. Consequently, sensor nodes must cooperate by one another to form a self-organizing network, and they ought to be equipped with energy-saving modules and procedures to lower energy usage and guarantee the network's longevity [1].

The gathering and transmission of relevant information to the base station are one of WSN's key responsibilities. Since replacing batteries is challenging and sensors are often deployed in hazardous environments, energy consumption stands out as one of the most critical factors in protocol design. In a wireless sensor network (WSN), sensors communicate with one another using wireless signals. The data that a sensor transmits is received by all of its neighbors, resulting in significant energy waste from overheard transmission. One of the best methods for cutting down on communication costs is data aggregation, and numerous strategies for removing unnecessary transmissions are being put forth [2, 3].

A vast number of sensors having computation, communication, and sense abilities that are dispersed throughout a region make up a sensor network. The number of intermediary outcomes that can be stored on sensor nodes and the use of standard data processing methods are constrained by their limited range, computing power, and storage capacity. Thus, the simplest method of condensed data aggregation requires well-organized sequencing in WSNs [4, 5].

Data aggregation is a crucial approach in the wireless sensor region since it may enhance network longevity, lower usage of energy, and boost the efficient data transfer ratio through data packet minimization. Regarding target surveillance, the clustering approach might be applied to improve the effectiveness of data gathering in a hierarchical network. There are two different kinds of clustering techniques: dynamic clustering and static clustering.

The idea of clustering could be applied to improve target tracking data aggregation effectiveness in a hierarchical network. Static clustering and dynamic clustering are the two categories of clustering techniques. Networks that are statically grouped deliberately split apart into numerous clusters. Pre-elected cluster heads are used for collecting data in static clustering, allowing for a comparatively small latency and quick and simple data transmission to the sink node. Impulsively, a cluster is formed in the event detecting nodes using periodically clustered networks. When an event is detected, a specific sensor node that has detected a target—ideally the one nearest to the occurrence or with the highest energy—becomes the cluster head, and all other nodes that have detected targets within one hop of the cluster head grow into general cluster nodes. The cluster head gathers, aggregates, and sends all the data to the sink.

This techniques' benefit is that it significantly conserves the electrical power of the remaining sensor nodes because only the ones that are required will take part in the information aggregation [1].

The benefits of data aggregation include reduced redundancy, energy conservation, robustness, and correctness of information. The data aggregator nodes, often referred to as cluster heads, consolidate the data before distributing it to the base station for processing [6, 7].

A relay-based inter-cluster forwarding algorithm and a data correlation-based compressing model are suggested and implemented in the cluster-based WSN to achieve these goals. This builds on our earlier research [8], in which we suggested an efficient clustering strategy based on fuzzy logic to maximize longevity in wireless sensor networks. This is how the remainder of the work is structured.

2 Literature Review

To increase aggregated data performance in target tracking use cases for wireless sensor networks, [1] has suggested a hybrid protocol. Significant gains were made by using a hybrid strategy and automatically choosing the best data gathering technique. Either more intricate static clustering or multi-hop clusters could improve every protocol's effectiveness. Fortunately, in an adaptable clustering-based aggregation of information protocol, the flooding mechanism utilized to modify aggregation algorithms may result in significant overhead [8] have suggested a single-cluster network and quantitatively quantify the ideal cluster size with respect to the degree of correlation and the cluster's proximity to the sink. Their results show that heterogeneous-sized clustering, where clusters farther from the sink are larger, is more energy efficient than earlier systems. To identify a close to ideal explanation to the energy-efficient clustering problem, they also suggested a heuristic ambitious clustering method. The efficacy of heterogeneous-sized clusters in WSNs is possessed by the suggested approach. Nevertheless, the suggested technique has a high computational cost. Reference [9] has put forth a novel and successful clustering-based data aggregation strategy. Three components make up their scheme: the Expectation–Maximization (EM) technique is used to cluster nodes in the network the Traveling Salesman Problem (TSP) solution is worn to generate the sink node's pattern; and a cluster-adapted Produced Diffusion mechanism is worn to aggregate the data obtained starting the WSN nodes. Performance can be increased by the suggested technique for varying node counts and distributions. On the other hand, efficiency declines with node count.

An enhanced version that makes usage of both cluster- and tree-based protocols is suggested by [4]. The suggested protocol enhances First Node Death (FND), which reduces consumption of electricity. In this instance balancing, the energy usage maximizes the network longevity. Additionally, there is less communication overhead. Nevertheless, this suggested plan has certain energy-related problems.

A scalable structured peer-to-peer network that facilitates data aggregation has been presented by [10]. Every node in the suggested peer-to-peer network aggregates its own partial statistical findings to determine the entire statistical result, and every node forwards its partial outcomes to other nodes. The communication cost of the data aggregation procedure is very minimal because it doesn't require any special protocols. We're not simulating that currently, though.

A data aggregation in a marginally linked sensor network has been suggested [11]. Through automatically choosing the shortest path for the packets to travel to the cluster leader, the algorithm lowers the amount of traffic flow within the network. They also go over a functional analytics simulation architecture. The method functions optimally in a completely connected network, but it can also operate in a fully linked sensor network. Nevertheless, this fails to tackle time synchronization.

To address the issue of uneven energy usage among the nodes in large-scale wireless sensor networks, [12] has put forth a novel data delivery mechanism. It is advancement over sensor networks built around clusters, like LEACH. Their suggested model consists of essentially two basic concepts (FDDA). Data is first transmitted to the Central Cluster Head (CCH), that is the primary cluster head. The CCH then openly transmits the collected information to the base station. Additionally, a backup sensor node immediately assumes the role of the cluster head node in the event that the cluster head node's energy level drops beyond a predetermined threshold. However, an imbalance exists in the energy consumption of nodes [13] proposed an energy efficient network data aggregation strategy for WSN. The suggested method makes use of the benefits of both tree- and cluster-based methods. With this method, a few clusters of uniform size make up the entire network. Every node is associated with a routing subtree, and every subtree engulfs a cluster. The primary node of the corresponding cluster is the primary node of every subtree.

In the proposed strategy, all nodes send information to their neighbor as opposed to the cluster leader. As a result, the network's useful life is increased, communication distance is reduced, and the average energy consumption of each node, the cluster, and networks as a whole is reduced. Additionally, in the proposed method, any node capable of balancing the network load selects its parent based on specific criteria, [14] introduced a more accurate and adaptable energy estimation technique that is independent of the time period. Furthermore, an optimal data collection technique for the ecological energy-based network of sensors is suggested. Accurate power assessment is made possible by this structure for environmental energy collection. Some non-CH nodes may not receive any ADV messages from the participating CHs because the proposed clustering method has a limited ADV transmission range.

For wireless sensor networks (WSNs), [15] has presented a reliable spine tree generation approach utilizing multi-hop clusters. Advantages of the hierarchical clustered structure include aggregated data and fusion. They suggest using an algorithm that distributes to choose the nodes exhibiting advanced energy or degrees as the cluster heads, forming a stable foundation. As a result, the lifespan of the network as a whole grows. Furthermore, by spreading out the traffic load across the nodes surrounding the cluster head, their suggested technique balances the usage of energy.

In [16], a cluster-based routing protocol that utilizes an energy-aware clustering algorithm (EADC) and a cluster-based routing algorithm is suggested for wireless sensor networks having non-uniform node distribution. EADC builds even-sized clusters by using competitive range. In order to ensure load balancing between cluster heads, the routing algorithm simultaneously raises the forwarding responsibilities of the nodes in sparsely enclosed areas by requiring cluster heads to select nodes by greater energy and smaller amount associate nodes across their subsequent hops. By distributing node energy usage evenly, this technique can greatly extend the lifespan of the network [17] proposed a Bandwidth Efficient Cluster Based Data Aggregation (BECDA) method that enables efficient data collection with in network aggregation. It carries out data aggregation by using the association of data throughout the packet. On randomly dispersed nodes with varied data creation rates, intra- and inter-cluster aggregation is carried out. Because every member transmits their detected information to the best cluster head, the load placed on the CH becomes higher in current cluster-based data aggregation systems, causing the CH to quickly exhaust its energy [18]. Furthermore, there is going to be increased aggregation overhead because of data redundancy. Energy-efficient and failure-free transmission should be guaranteed by the inter-cluster communication system. Therefore, the goals of this work are to reduce data redundancy, minimize the cluster head's load and resource drain rate, and ensure that regional-based data aggregation occurs with energy-efficient inter-cluster communication.

3 Proposed Work

3.1 Overview

Based on the prior clustering design, we propose a compressed data method for communication as illustrated in, Fig. 1. The process consists of phased for inter-cluster routing and condensed data aggregation. An energy saving approach, utilizing data concordance and solidity is applied during the compressed aggregation phase. In this approach, the compression ratio is determined by the cluster size and data correlation.

After compiling and compressing all the readings, the cluster head (CH) sends the sink a single representational message. The combined entropy of the cluster determines the size of the condensed message.

For the chosen CHs, a routing tree is built during the inter-cluster routing phase. The next CH is selected by the CH through a relay. The network's overall load, node density, maximum starting energy of nodes, and CH remaining energy are used for measuring this relay. By modifying intra-cluster and inter-cluster energy usage, this inter-cluster routing technique distributes the energy expenditure across cluster heads.

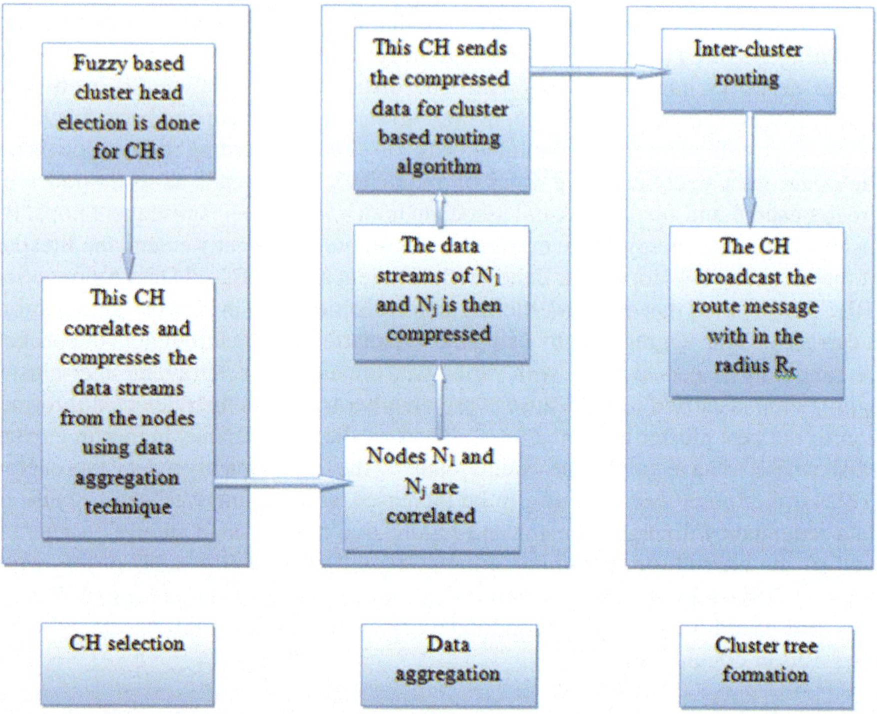

Fig. 1 Illustrating the suggested cluster-based routing and data aggregation in WSN

3.2 Optimal Clustering Protocol

Given that a cluster head (CH) performs several data relaying operations, it ought to have a high remaining energy and low load. In a same vein, it ought to cover more nodes while traveling a closer proximity to the sink. Therefore, the metrics Residual Energy (Eres), Distance to the sink (*D*), Node Density (NDi), and Load are used to choose the CH. Although fuzzy logic reliably evaluates both the cluster size and the likelihood of a node attaining the CH, it is utilized for CH selection. Fuzzy logic **considers** every potential arrangement of these competing criteria.

The following are the phases in this clustering protocol: [8]

a. Several provisional cluster heads (PCH) are randomly selected at time *t* to fight for the position of final cluster heads.
b. The remainder of nodes broadcast a HELLO message at a set power level, excepting the selected PCH.
c. After receiving a HELLO message, every PCH calculates the detachment and adds the node ID to its route cache.
d. Lastly, the constraint residual energy (Eres), node density (NDi), and load are composed for every PCH.

e. Every node applies fuzzy logic to examine the parameter circumstance through the use of the projected parameter values.
f. After calculating the likelihood of becoming the cluster head, a PCH sends a desired message (D_Mes) indicating its intention to take over as CH.
g. PCH simply announces its desire cancelation (DCL_Mes) message when it discovers that another PCH meets the higher criteria to be chosen as CH.
h. Every CH transmits a cluster announcement (CL_ADV) message over the network following CH election. Typical sensor nodes inside the network connect to the adjacent CH.

3.3 Data Correlation Model

Assuming that the results are regularly dispersed via mean zero and variance, the sensor nodes in the WSN are regarded as identical in statistical terms information sources. The premise for degree of data correlation (DDC) is that the dependence among sample readings diminishes exponentially about distance at a constant pace. Therefore, the correlation is going to lower the more away you are.

Therefore, magnitude (MG) and distance (D) are the two metrics we specify for the purpose to assess the DDC of the two-time series.

When (i) MGi = MGj, wherever MGi and MGj are the magnitudes of the evaluation of Ni and Nj, and (ii) $Dij < Dth$, then the measurements of two sensor nodes, Ni and Nj, are considered to be correlated, wherein Dth is the distance requirement and Dij is the distance among Ni and Nj.

The two nodes could be represented as two-dimensional points. Coordinates for nodes Ni and Nj are MGi, Di and MGj, Dj, respectively. These coordinates, which are calculated using the Euclidean distance, serve as the basis for the DDC among the two nodes.

The Euclidean distance measure keeps track of the square root of the coordinate variations among two objects. Consequently, [6] gives the Euclidean distance among the nodes Ni and Nj,

where

$$\mathrm{DDC}_i = \sqrt{\sum_{k=1}^{n} \left(x_{ik} - x_{jk}\right)^2}, \quad n = 2. \tag{1}$$

$$xi1 = \mathrm{MG}i, \ \ xi2 = Di \text{ and } xj1 = \mathrm{MG}j, \ \ xj2 = Dj.$$

The optimum number of data rates is distributed across the sensor nodes using this data correlations model. Here, the covariance functional for this data correlation model is the power exponential model. For a covariant function having distance d, the power exponential model is provided by [6]:

$$K_{\mu}^{\mathrm{PE}}(d) = \exp\left(-\left(\frac{d}{t_1}\right)t_2\right). \tag{2}$$

3.4 Data Solidity Model

The joint entropy determines the compressed message's size while employing the data compression model. In this case, the associated data streams are individually encoded and cooperatively decoded at a single receiver. When X and Y are the data streams in nodes Ni and Nj, respectively, then the data streams in both nodes Ni and Nj are considered to be correlated. Making n separate drawings compared to a joint distribution of chances $P(X = x, Y = y)$ yields the associated data streams X and Y.

Figure 2 shows how Encoder 1 obtains Data Stream X and uses the RX bits to encode each character before sending a coded message to the decoder.

Separate encoding and combined decryption of two associated data streams, X and Y, are shown in Fig. 2. Similarly, encoder 2 obtains data stream Y and sends a coded message (a series of RY bits encoding each of the characters of Y) to the decoder. The decoder creates two n-vectors, $X*$ and $Y*$, representing the estimates of the preliminary information streams, X and Y, accordingly, upon receiving each of those coded message defined as two random variables X and Y with a joint distribution of probabilities, the permissible rate region is defined by,

$$p(X = x, Y = y) = H\left(\frac{X}{Y}\right) = -\sum_{y} p(Y = y)$$
$$\sum P\left(X = \frac{x}{Y} = y\right) \cdot \lg P\left(X = \frac{x}{Y} = y\right). \tag{3}$$

The acceptable rate region given the pair of rates (RX, RY) meets each of the three inequality by the Slepian–Wolf theorem:

$$RX \geq H(X|Y),\ RY \geq H(Y|X),\ RX + RY \geq H(X, Y).$$

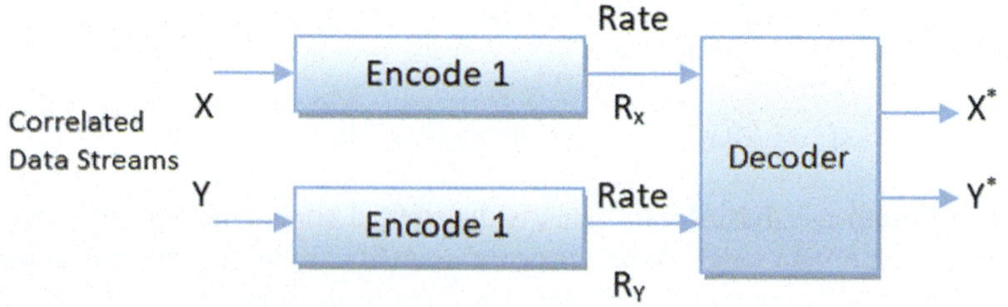

Fig. 2 Two associated data streams

Rates can be attained by the independent encoders that do not consider the source correlations if only $RX + RY \geq H(X, Y)$.

4 Simulation Results

4.1 Modeling and Parameters for Simulation

Utilizing the Network Simulator (NS-2) [19], the suggested architecture is simulated. Sensor nodes having a 40-m communication distance are randomly placed throughout a 500 × 500-m area in this simulation. The sink or BS is located in the top right corner of the topology. Consistent bit rate (CBR) traffic is used to generate data since sensor data are thought to be periodic and of a consistent size. Ten runs are used to average the findings.

4.2 Measures of Performance

A comparison is made between the Cluster-based Correlated Data Gathering (CCDG) approach and the suggested Cluster-based Compressed Data Aggregation and Routing (CCDAR) protocol [2]. The measurements for performance of the average utilization of energy and packet rescue ratio are assessed.

The ratio of the number of packets sent and retrieved is known as the packet delivery ratio. The term "packet drop" describes the typical number of packets lost throughout transmission. Energy utilization is the quantity of energy used by the nodes to send packets of data to the recipient. The time it takes for the nodes to send the data packets is known as the delay.

The performance metrics mentioned earlier are assessed while the number of nodes is arbitrarily changed between 25, 50, 75, and 100. The formation of clusters results in an average cluster size of 10% of the overall number of nodes. Figure 3 illustrates the delay that happens between the two protocols as the number of nodes rises. The delay grows linearly with the number of nodes since the inter-cluster paths are going to be long. In contrast to CCDG, CCDAR picked the shortest pathways among every CH for inter-cluster routing, resulting in a 32% shorter delay.

Whenever the total amount of nodes is increased, Figs. 6 and 7 display the ratio of packet delivery and packet drop between the two protocols. In contrast to CCDAR, where it climbs from 4600 to 6300, CCG's packet drop grows linearly from 7000 to 12,800 to 75 nodes. Comparably, the delivery ratios for CCDG and CCDAR drop from 0.56 to 0.37 and 0.6 to 0.4, respectively. This is because more nodes mean that the CH is under more load, which results in additional packet drops. In comparison with CCDG, CCDAR has a 10% higher packet delivery ratio and 43% fewer packet drops due to its selection of the least loaded CH during inter-cluster routing. The

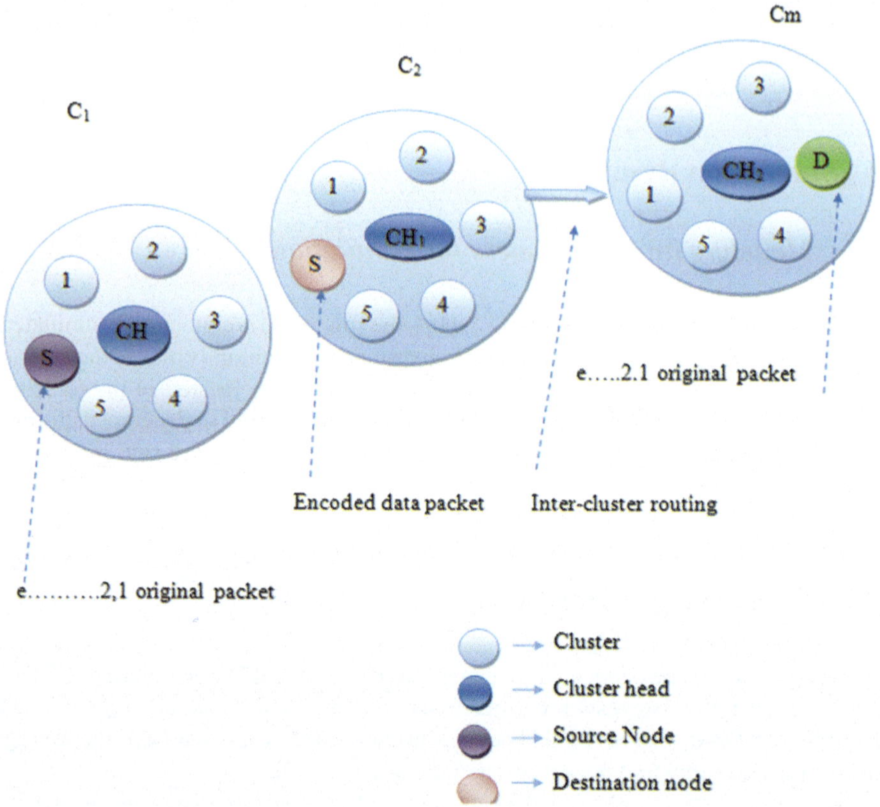

Fig. 3 Using encoded data packets to depict inter-cluster routing

average energy usage between the two protocols when the number of nodes is raised. The graphic shows that for both CCDAR and CCDG, the energy usage reaches a constant value at 25 nodes. Because the CH assortment in in cooperation intra- and inter-cluster routing depends on high residual energy, CCDAR uses 11% less energy than CCDG.

The typical overhead for the two protocols, CCDAR and CCDG, is displayed in Fig. 5. As seen in the graphic, every protocols' overhead increases above 75 nodes. For CCDAR, it reaches approximately 12,500 at 100 nodes, while for CCDG, it reaches approximately 23,000 at 100 nodes. It is evident that broadcasting in CCDAR does not entail any extra overhead. With CCDAR, the overhead is 45% less than with CCDG.

The numbers of nodes utilized for the simulation and evaluation are 25, 50, 75, and 100; the communication range is 250 m; the simulation time is 50 s; and the size of the packet is 512, as indicated in Figs. 4, 5, 6, and 7. Figures 4, 5, 6, and 7 show the node outputs in relation to energy usage, packets, delivery ratio, and latency.

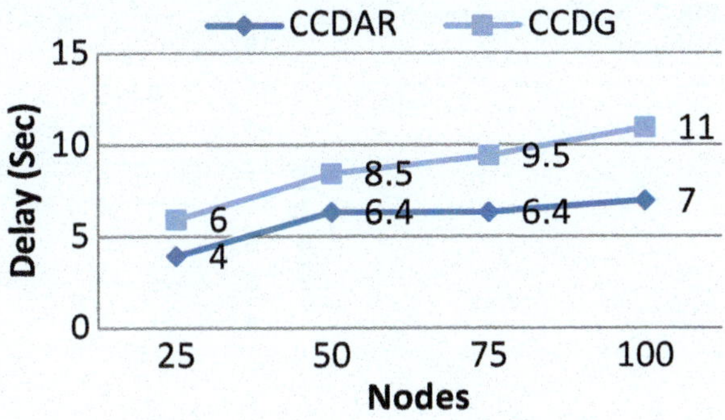

Fig. 4 Node versus delay

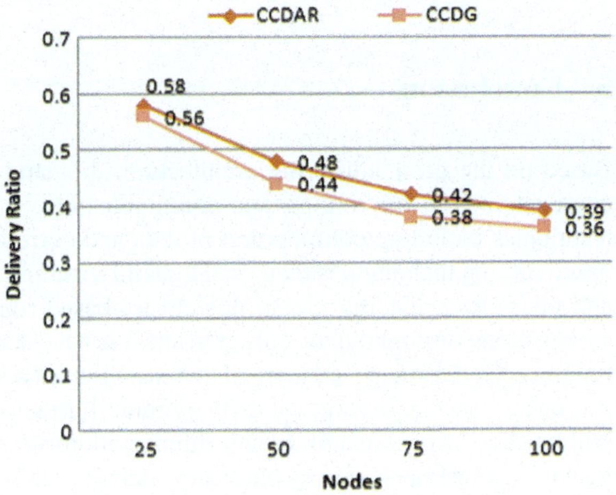

Fig. 5 Nodes versus delivery ratio

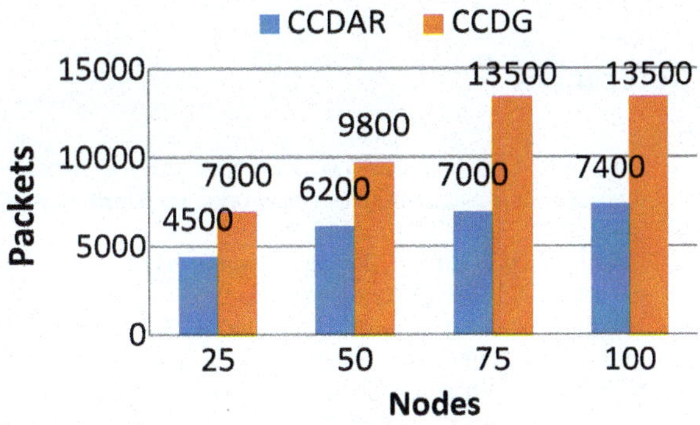

Fig. 6 Node versus packets

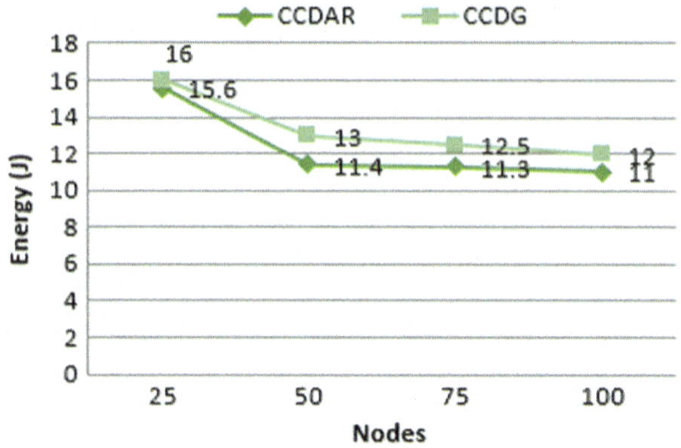

Fig. 7 Node versus energy consumption

5 Conclusion

Based on the prior clustering architecture, this study introduces an efficient data aggregation tree for routing and communication. In this instance, energy-saving techniques including compression of data and correlation are applied. The cluster-based routing technique is then used to build a routing tree. By modifying both intra- and inter-cluster energy usage, this cluster-based routing algorithm distributes the energy usage throughout the cluster heads. We have demonstrated through simulation findings that the suggested method increases the network's lifetime while consuming less energy and increasing the delivery ratio. Further research on optimal clustering will involve proposing a fuzzy algorithm in addition to an optimization one, with the goal of increasing clustering efficiency and ultimately system efficacy.

References

1. Jung WS, Lim KW, Ko YB, Park SJ (2009) A hybrid approach for clustering-based data aggregation in wireless sensor networks. In: Third international conference on digital society, 2009. ICDS'09. IEEE, pp 112–117
2. Guo J, Chen D (2013) Cluster trisecting based data aggregation scheme for wireless sensor networks. Adv Mater Res 756:1126–1130
3. Khalid S, Mason A, Hussain FB, Abbas H, Ghafoor A (2014) Load balanced routing for lifetime maximization in mobile wireless sensor networks. Int J Distrib Sens Netw
4. Chhabra GS, Sharma D (2011) Cluster-tree based data gathering in wireless sensor network. Int J Soft Comput Eng (IJSCE) 1(1):27–31. ISSN: 2231-2307
5. Ahamed T, Nyang DH (2009) A dynamic level-based secure data aggregation in wireless sensor network. Information Security Research Laboratory, Graduate School of IT and Telecommunication, InHa University
6. Dabirmoghaddam A, Ghaderi M, Williamson C (2010) Cluster-based correlated data gathering in wireless sensor networks. In: 2010 IEEE international symposium on modeling, analysis & simulation of computer and telecommunication systems (MASCOTS). IEEE, pp 163–171

7. Mathapati BS, Patil SR (2012) Energy efficient reliable data aggregation technique for wireless sensor networks. In: 2012 international conference on computing sciences (ICCS). IEEE, pp 153–158
8. Jothi S, Chandrasekaran M (2014) Fuzzy based optimal clustering protocol for maximizing lifetime in WSN. Res J Appl Sci Eng Technol 8(6):714–725
9. Fukabori T, Nakayama H, Nishiyama H, Ansari N, Kato N (2010) An efficient data aggregation scheme using degree of dependence on clusters in WSNs. In: 2010 IEEE international conference on communications (ICC). IEEE, pp 1–5
10. Takeda A, Oidey T, Takahashi A (2012) A structured overlay network for aggregating sensor data. In: BWCCA, pp 684–689
11. Sheshikala M, Sallauddin M (2018) Survey on multi level security for IoT network in cloud and data centers. J Adv Res Dyn Control Syst 10(10):134–146. ISSN (Print): 1943-023X
12. Sudheer Kumar K, Nagender Y (2017) Liability for information sharing in cloud. Int J Comput Sci Eng 9(11):630–634. ISSN (Print): 0975-3397
13. Chanti Y, Nagender Y, Vijay Kumar B (2017) Quickly nearest neighbor search traditional spatial query with keywords. Int J Comput Sci Eng (IJCSE) 9(11):635–638. ISSN (Print): 2229-5631, ISSN (Online): 0975-3397
14. Mozumdar MMR, Guofang N, Gregoretti F, Lavagno L (2009) An efficient data aggregation algorithm for cluster-based sensor network. J Netw 4(7):598–606
15. Mishra S, Thakkar H, Chakrabarty A, Kimtani D (2012) Dynamic cluster based data aggregation in WSN (FDDA). Int J Electron Commun Comput Technol (IJECCT) 2(5):227–230
16. Anisi MH, Abdullah AH, Razak SA (2011) Energy-efficient data collection in wireless sensor networks. Wirel Sens Netw 3(10):329
17. Kinoshita K, Okazaki T, Toed H, Murakami K (2008) A data gathering scheme for environmental energy-based wireless sensor networks. In: 2008 5th IEEE consumer communications and networking conference, CCNC 2008, pp 719–723
18. Shin I, Kim M, Mutka MW, Choo H, Lee TJ (2009) MCBT: multi-hop cluster based stable backbone trees for data collection and dissemination in WSNs. Sensors 9(8):6028–6045
19. Yu J, Qi Y, Wang G, Gu X (2012) A cluster-based routing protocol for wireless sensor networks with nonuniform node distribution. Int J Electron Commun 66(1):54–61

Improved AODV Routing Protocol Development and Analysis for Enhanced Network Performance in MANETs Based on Machine Learning Model

Puttamraju Srinivasa Rao, Danda Sreekanth, S. Vishnu Murthy, and Chalapathi Rao Tippana

Abstract Self-configurable and nondemobilizable wireless ad hoc networks are called MANETs. A number of security threats are present on the route because of its flexibility. In order to address this, the IDS's performance should be changed. The machine learning (ML) technique is used in this study to provide a method for enhancing the accuracy and detection rate of the intrusion detection system (IDS) for the AODV routing protocol.

Keywords Machine learning · AODV routing protocol · SVM · MANET · Intrusion detection system

1 Introduction

IDS is to detect the attack before the rogue node(s) become a security risk to the network. It looks at things like alerting, detecting, and monitoring. The most harmful kind of assault on MANETs is the Blackhole attack [1]. Making use of an anomaly IDS protects the network from Black Hole Attack by utilizing a machine learning technology called SVM Malicious Node(s) Leading Black Hole Attack in AODV Routing Protocol. One of the main methods used to attack MANETs is the black hole assault. The data, which consist of the Source, Destination, and Neighboring nodes, are in the possession of the malicious node(s) responsible for this attack on MANET security. To find the route destination, the originating node broadcasts an RREQ (route request packet) to its neighboring nodes [2]. However, packet loss

P. S. Rao (✉) · D. Sreekanth
Narasaraopeta Engineering College, Narasaraopet, Andhra Pradesh, India
e-mail: srinivasnrtec@gmail.com

S. Vishnu Murthy · T. Chalapathi Rao
Aditya Institute of Technology and Management, Tekkali, Andhra Pradesh, India

© The Author(s), under exclusive license to Springer Nature Singapore Pte Ltd. 2025
H. Zen et al. (eds.), *Soft Computing and Signal Processing*, Lecture Notes in Networks and Systems 1200, https://doi.org/10.1007/978-981-97-9926-8_7

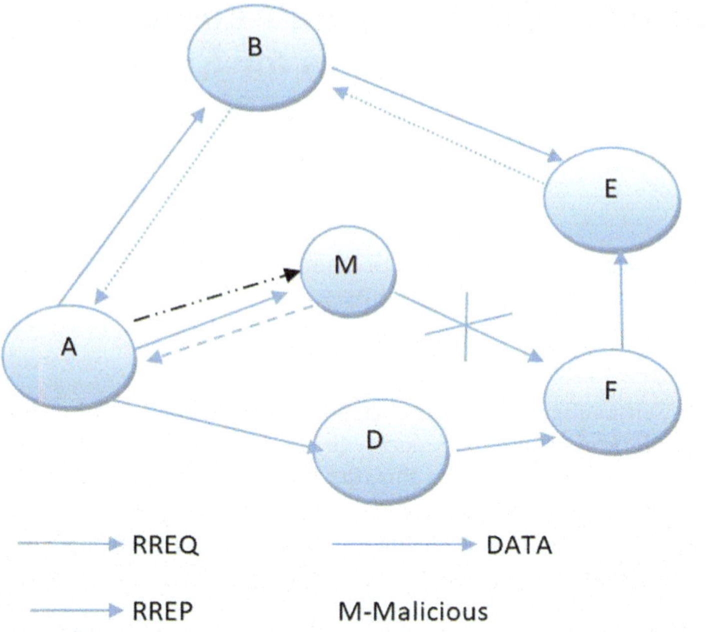

Fig. 1 Malicious node causing black hole attack

and decreased network performance result from the black hole node sending a bogus route reply to the source node. As shown in Fig. 1, this can be prevented by improving the performance of the IDS by identifying the rogue node or nodes using a machine learning technique.

2 Literature Review

The suggested method finds black hole attacks in the network by using an ANN and a support vector machine (SVM). Results comparing the Secure AODV (SAODV) security mechanism to the black hole AODV demonstrate improvement: 54.72% energy usage, 88.68 kbps throughput, 92.91% packet delivery ratio, and an approximate 37.27 ms E-to-E time [3].

The NS-2 network simulation was used to implement the HWAD technique as described in [4]. The proposed approach's performance has been assessed by considering various performance metrics, including throughput, packet delivery ratio, end-to-end delay, and energy consumption. The proposed approach leveraged the ad hoc instantly distance vector (AODV) routing protocol to improve the detection process. Figure 2 shows the types of Routing Protocols. Compared to other anomaly detection methods, the SOM classification method-based anomaly-based intrusion detection technology offers a greater detection rate.

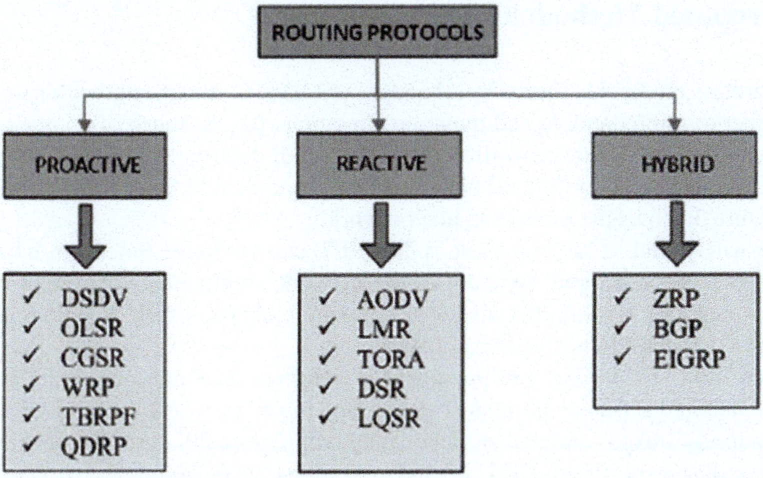

Fig. 2 Types of routing protocols

Due to the statistical basis of anomaly-based intrusion detection algorithms, it is possible for typical patterns to be mistakenly identified as attacks. This incorrect designation of benign behavior as aberrant may lead to from isolating of nonmalicious node as malicious, thus may lead in partitioning of the network [5].

In this paper [6], the method shortens the amount of time that an IDS is running inactively on nodes and enhances energy conservation in heterogeneous networks. To do this, a probabilistic strategy is used; here, the ideal probabilistic of each node is to be determined, which reduces the active time of the IDS at each node and preserves the node's energy, so greatly extending the network lifetime.

Suggested RSA technique for the MANET's intrusion detection system. The results demonstrate that the safe IDS method improves packet delivery ratio in the presence of malicious nodes and effectively detects the malicious node or nodes [7].

In this paper [8], using the artificial neural network (ANN) idea as a deep learning algorithm in conjunction with the swarm-based artificial bee colony (ABC) optimization technique, the protection against dual attacks has been given for BHA and GHA. The selection of suitable and optimal nodes for the transmission of data packets has improved the system's performance.

3 Proposed Methodology

The wireless medium is shared by all nodes in MANETs, and the network's topology is subject to unpredictable and dynamic alterations [9]. Because of its outstanding features, such as its quick infrastructure and ease of deployment in dangerous environments (such as an earthquake or battlefield), research in MANETs is highly sought for. Figure 3 shows the mobile ad hoc network.

Anywhere and at any moment, a MANET can be built. Since the nodes are movable, there is a rapid variation in the network architecture. MANETs are the greatest network medium because of their many benefits, which include multihop, infrastructure-less transmission, and others.

Even if MANETs are overflowing with information, there are certain security flaws that could seriously harm the network and result in losses. When mobile nodes are randomly linked, malevolent nodes may unintentionally enter the network. An intrusion detection system (IDS) is used to examine the behavior of nearby nodes in order to suspect and identify hostile activities within the network.

Machine learning algorithms are shown in Fig. 4. Support vector machines, as seen in flowchart Fig. 5, are taken into consideration as part of a machine learning approach to enhance the anomaly based intrusion detection system in MANETs.

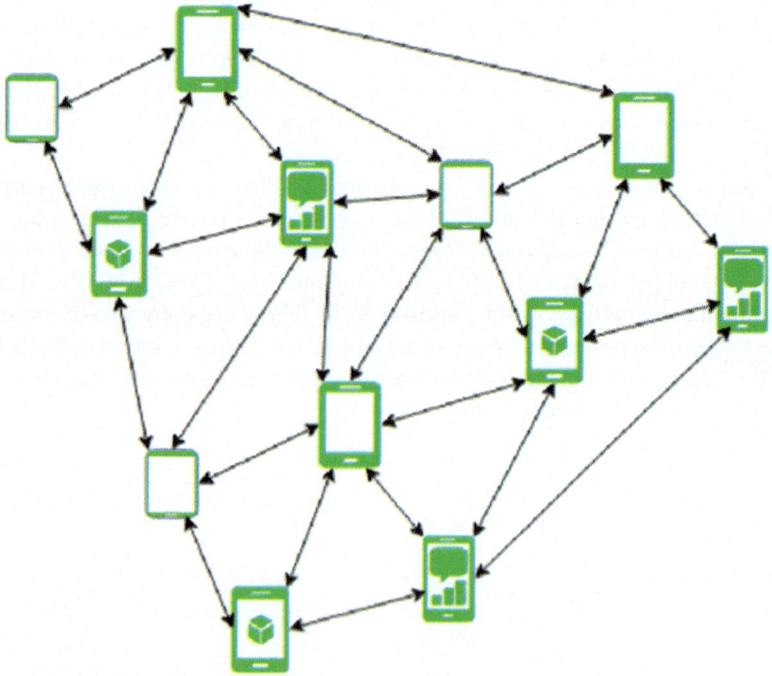

Fig. 3 Mobile ad hoc network

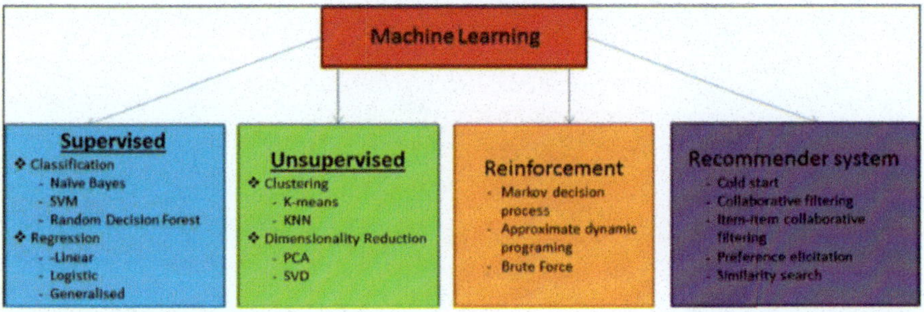

Fig. 4 Classification of machine learning algorithms

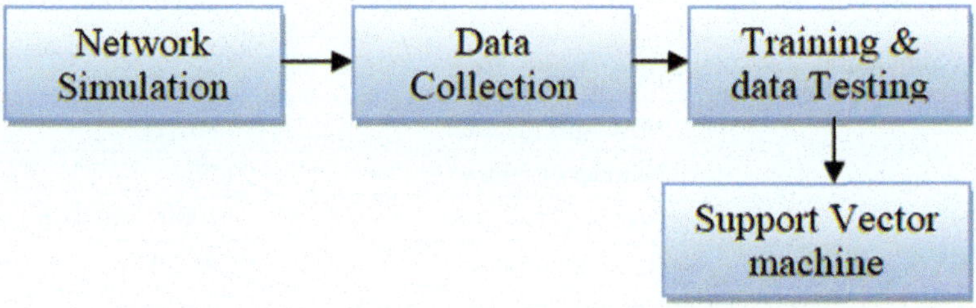

Fig. 5 Flow chart for the proposed methodology

A three step method is followed for the analysis [10]:

I. Network simulation
II. Data collection
III. Model training and data testing.

Step 1: Network Simulation

As seen in Fig. 6, a mobile ad hoc network (MANET) simulated in NS2 in the current work Simulation Environment is shown in Table 1.

Step 2: Data Collection

Following simulation, NS2 generates a trace file that will be used as a CSV file's input [11]. The trace file's output and the .csv file's input, in that order. In general, a trace file contains more attributes; nevertheless, certain attributes have been chosen as input for a .csv file if the total amount of received packets exceeds the number of discarded packets.

Step 3: Model Training and Data Testing

The SVM technique was utilized in this work to train and test the data. Support vector machine, or SVM, by selecting the best approximated hyper plane, support vector machines (SVMs) seek to distinguish between normal and atypical (i.e., malicious)

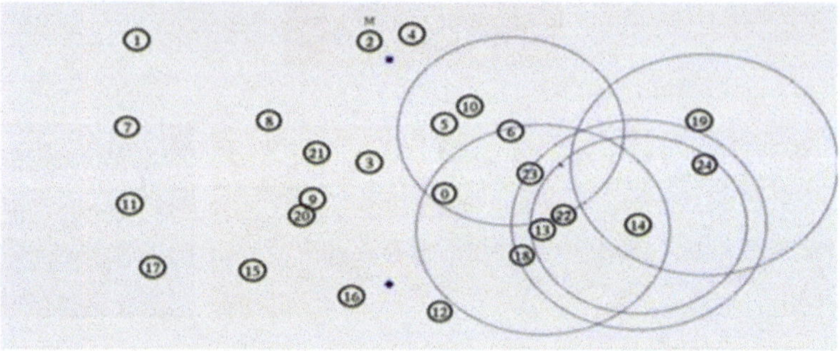

Fig. 6 Malicious nodes causing black hole attack simulation in NS-2.35 for AODV

Table 1 Simulation environment

Parameter	Significance
MAC specification	802.11
Number of nodes	21
Area	1000 × 1000 m
Routing protocol	AODV
Number of malicious node	341,416
Simulator	NS-2.36
Node energy	50 J
Simulation time	49 s

nodes [12]. It is chosen so as to maximize the distance on each side between the closest node and the hyperplane.

Algorithm

1. Set vector *v* and *b*'s initial values to 0.
2. The dataset *D* is made up of the labeled samples *x*, *y* (*x*1, *y*1), … (*xn*, *yn*).
3. Use SVM training to understand decision functions.
4. For every *D* do sample.
5. Use decision function $f(xi)$ to classify *xi*.
6. For the given data, calculate w' and b' if (function limit < 1).
7. Include an example in the known data. (vii) Apply Equation (*w*) = ½ "*w*" 2 to minimize mistakes.
8. To anticipate, use Eq. $f(x) = \text{sign}\ (wTxi + b)$.
9. Try it again if the prediction comes true.
10. else
11. Retrain SVM End if End if 12. Determine whether *xi* is malevolent or benign.

The dataset from NS2 is loaded into the SVM method in the current paper. Efficiency and confusion matrices are used to identify the malicious node or nodes initiating the black hole attack [13, 14]. Figure 7 shows SVM classification, and Fig. 8 illustrates average throughput.

The SVM algorithm yields an accuracy of 82.35, and Fig. 5's confusion matrix illustrates a lower false positive rate.

Based on the aforementioned findings, it can be inferred that the SVM approach chosen provides precision and detection rate, allowing for the isolation of malicious nodes within the MANET and the improvement of IDS performance.

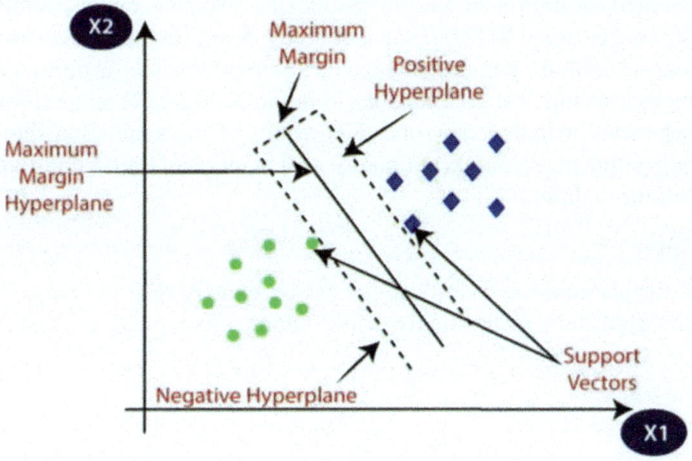

Fig. 7 SVM classification

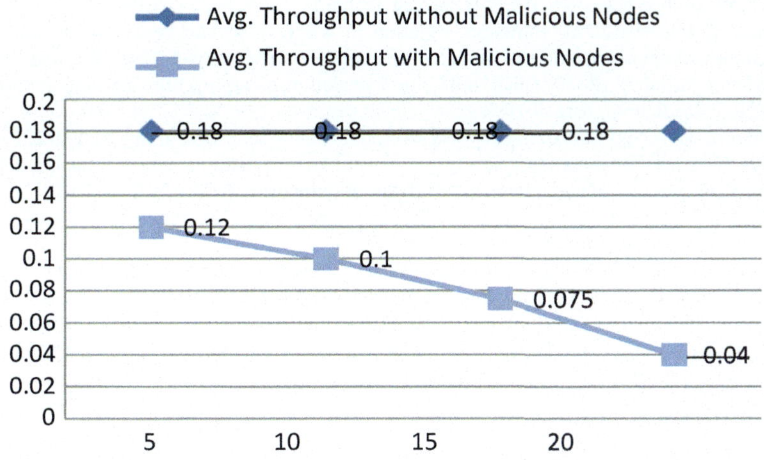

Fig. 8 Average throughput

4 Conclusion

A popular trend in next-generation ad hoc wireless networking technologies is 5G-based MANET. The primary attribute of 5G networks is their nodes' continuous handling of heavy traffic loads. In the interim, QoS must be guaranteed, and stringent standards for congestion probability, end-to-end delay, and additional performance indicators apply. It was clear that the existing MANET routing protocols had limitations when used in a 5G network-based MANET.

We presented an improved ad hoc on-demand distance vector protocol, called RL-AODV, for 5G-based MANETs in this study. Every node updates an inventory of traffic loads and SNIR at intermediate nodes alongside routes to destinations using reinforcement learning. The routing algorithm looks up a QoS-assured route in this database whenever it finds a new one. The results of the simulation show that our proposed algorithm performs exceptionally well in terms of SNIR, end-to-end delay, and network throughput.

Future Scope

In order to further enhance the efficacy of 5G-based MANET, we are going to keep building the algorithm in the future while taking other metrics' restrictions into account.

References

1. Marchang N, Datta R (2016) A novel approach for efficient usage of intrusion detection system in mobile ad hoc networks. IEEE Trans Veh Technol
2. Zhang Y, Lee W (2000) Intrusion detection in wireless ad hoc networks. In: 6th annual international conference on mobile computing and networking (MobiCom'00), Aug 2000
3. Pandey S, Singh V (2020) Black-hole attack detection using machine learning approach on MANET. In: ICESC, Aug 2020
4. Tahboush M, Agoyi M (2021) A hybrid wormhole attack detection in MANET. IEEE Access
5. Indira N (2014) Establishing a secure routing in MANET using a hybrid intrusion detection system. In: International conference on advanced computing (ICoAC), Dec 2014
6. Sujithra LR, Nivethaa V, Pavithra B, Pavithran M (2018) Heterogeneous based intrusion detection system in mobile ad hoc network. IRJET
7. Sankaranarayanan S, Murugabhoopathi G (2017) Secure intrusion detection system in mobile ad hoc network using RSA algorithm. In: Second international conference on recent trends and challenges in computational models (ICRTCCM), Feb 2017
8. Rani P, Kavita, Verma S, Nguyen GN (2020) Mitigation of black-hole and gray-hole attack using swarm inspired algorithm with ANN. IEEE Access
9. Sujatha KS, Dharmar V, Bhuvaneswaran RS (2012) Design of genetic algorithm based IDS for MANET. In: IEEE international conference on recent trends in information technology (ICRTIT), Apr 2012
10. Su MY (2011) Prevention of selective black hole attacks on mobile ad hoc networks through intrusion detection systems. Comput Commun
11. Maglaras LA (2015) A novel distributed intrusion detection system for vehicular ad hoc networks. Int J Adv Comput Sci Appl

12. Butun I, Morgera SD, Sankar R (2013) A survey of intrusion detection systems in wireless sensor networks. IEEE Commun Surv Tutor
13. Patel M, Aggarwal A, Chaubey N (2017) Wormhole attacks and countermeasures in wireless sensor networks: a survey. Int J Eng Technol (IJET)
14. Patcha A, Park JM (2007) An overview of anomaly detection techniques: existing solutions and latest technological trends. Comput Netw

Comprehensive Analysis of Cognitive Radio Technologies and Its Applications

Velumani Thiyagarajan, M. Vidhya, M. Sneha, and P. Prasanya Devi

Abstract This paper contributes an overview of cognitive radio technology and its advancement in communication technologies. Cognitive radio technology is emerging a lot in current day because of the requisition in wireless service and increase in spectrum resources. In this article, current state of CR technologies research, updating, and challenges has been extensively adopted. The basics of CR principles, sensing of spectrum, making decision, and dynamic spectrum access have been presented in this paper. Other incorporation of CR into different communication gadgets such as mobile phones, IoT devices and routers has been displayed. The standardization of communication and productiveness of spectrum has been easily optimized by CR. Implementing of CR in real-time environment is tougher in this article, and challenges faced and solutions associated have been provided. In the conclusion part, the current trends and necessity in cognitive radio technology and its reframing effects on communication appliances have been mentioned. It does not make change only in spectrum inefficiency measurement but also in incorporation of CR in communication gadgets.

Keywords Cognitive radio · Applications · Challenges · Main user · Secondary user · Spectrum · Communication channels · Artificial intelligence

T. Velumani (✉)
Department of Computer Science, Rathinam College of Arts and Science, Coimbatore, Tamil Nadu 641021, India
e-mail: velumani46@gmail.com

M. Vidhya
Department of Computer Science, University of Kerala, Thiruvananthapuram, Kerala, India

M. Sneha
Department of Computer Science, Sona College of Arts and Science, Salem, Tamil Nadu, India

P. Prasanya Devi
Department of Computer Science, Government Arts College, Melur, Madurai, Tamil Nadu, India

© The Author(s), under exclusive license to Springer Nature Singapore Pte Ltd. 2025
H. Zen et al. (eds.), *Soft Computing and Signal Processing*, Lecture Notes in Networks and Systems 1200, https://doi.org/10.1007/978-981-97-9926-8_8

1 Introduction

In 1998, Joseph Mitola presented the cognitive radio technology which created major impact in scientific research institutions. The raise of CR technology evolved because of decrease in resource of radio spectrum. Transmission of data opportunity for cognitive radio has been accessed via stationary spectrum where waste of spectrum resource is reduced and utilization has been increased. More study should be done for attending goal of dynamic access spectrum and increasing accuracy. The research should focus on integrating both current wireless communication system and cognitive radio. To increase efficiency and to overcome lots of challenges, usage of technologies such as OFDM (orthogonal frequency division multiplexing), MIMO (Multiple Input and Multiple Output), sensing spectrums, front end, smart antenna, etc., can be incorporated. Intelligent automated swapping and self-regulating rectification of radio signals can be analyzed easily by radio signal cognition because of inclusion of AI technology in CR communication system. Cognitive radio (CR) technology dramatically increases spectrum utilization by providing active spectrum access and efficient handling of wireless communication resources [1]. The main inclusions of appliances such as IoT, smart homes, automobiles, and its major usage have been highlighted in this article. The overall summary of CR technology is described below. To gather and execute radio signal, the first and foremost job of computer system is to utilize various sensors and integrated computer hardware. Self-signal automation and analysis have been done by processing gathered data collected from computer hardware and sensors. The validated signal type consists of modulated, demodulated, and frequency methods from automated selector system. The processing and transformation of radio signals can be executed by CR technologies on signals and are examined and verified by applying various frequencies. The cost of this technology is high because of utility of hardware's and unique sensors. To provide safety and firmness to user, system manufacture should accept regulation which has been given in law for the process of appliances. The application of cognitive radio in the Internet of Things (IoT) presents significant opportunities for enhancing connectivity and optimizing resource usage in IoT networks [2]. The transferrable resources of spectrum and standard frequency can be surveillance to measure amount of spectrum spent. It should be noted there should not be third-party disturbance between two various communication systems so that the availability of spectrum can be planned and transferred according to the needs. Once disturbance is stopped, then on account of real-time user requirements, vigorous allocation of spectrum can be done. The efficient usage of system is maintained, usage of system capacity is monitored, and sharing of different communication system can also be achieved.

2 Principles

2.1 *Important Features of Cognitive Radio*

The amount of service and quality of wireless network is one of the biggest challenges because of less use of bandwidth and energy. Many researchers working on new models and prototypes which make complete use of this resource in intellectual way. More flexibility and effective utilization can be provided for usage of resources available in network. Based upon the radio state CR has the capability to dynamically adapt to different parameters such as type of modulation, frequency, and power of transmission which is called as cognitive capability is the unique identity of CR. Once the collection of information data on the state is over CR, it automatically modifies parameter of transmission according to environment needs and performs optimally. The main features of cognitive radio are dynamic spectrum sensing, adaptability, and the ability to learn and make decisions based on radio conditions [3]. The process of identifying and verifying not used or less utilized spectrum bands on main user is called as spectrum sensing. The purpose of spectrum sensing is to work dynamically on given resources in a productive way without disturbing main user. The utilization of spectrum bands on account of availability of published result is called as DNA (dynamic spectrum access). The effectiveness of network and utility of spectrum can be increased by choosing highly rated frequency bands through DSA which makes intellectual usage of CR device. Adaptive changing in frequencies based on the given state and managing frequency automatically is called as mobility in management spectrum. Here the communication is stable even though there is change and variation in state or frequency. Within given amount of time, CR can adapt easily to the particular surrounding and self-learn itself and makes execution of this process which is called as learning abilities and adaptation. Based on the history on previous data, adaptation is done and modifications have been made. Sensing is done and based on that working principles have been dynamically changed in sequential iterative process, and the process is known as cognitive cycle. To provide CIA (Confidentiality, Integrity, and Authentication) to the communication system by protecting from unauthorized user can be done and this technique is said as security and privacy consideration were malicious activities can be avoided and system can be protected.

2.2 *Activity and Working of Cognitive Radio*

The routine and regular life cycle of CR is to identify the process which necessitates identifying free nonworking area, choosing needed frequency verifying availability of other users, and clearing the state when noticing availability of MU. Managing spectrum, sensing information, and utility resource sharing have been done to assist this life cycle. Unused resources can be utilized by the nonprime users, and making

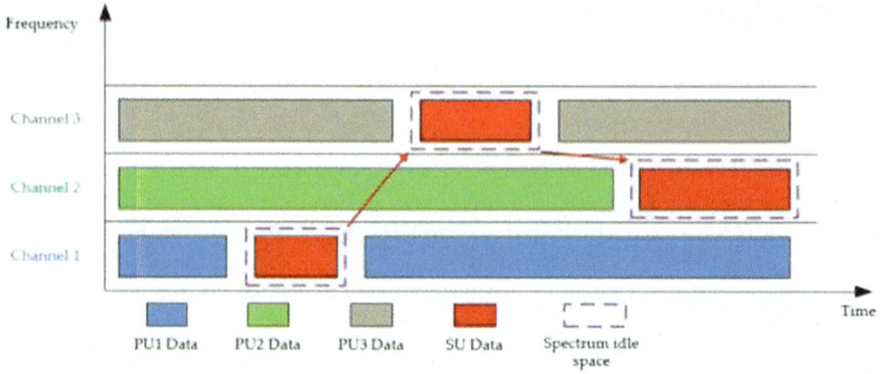

Fig. 1 CR working diagram

nonworking idle area has been identified by spectrum sensing. Cognitive radio continuously monitors the spectrum, identifies available channels, and takes advantage of advanced algorithms, dynamically changing transmission parameters to improve performance [4]. To avoid affective third-party entry while main user trying to access spectrum resources using license again and again via sensing this illegal activity has been sensed, and the user will be removed from the area. Available usage of QoS Quality of Service is done by identifying unused space by the managing spectrum function which leads the path for secondary user to make utilization of resource from unused space. It is mandatory for secondary user to change the state from available frequency if there is notification from main user requesting for frequency. The detailed working of CR is displayed in Fig. 1. To obtain maximum spectrum effectiveness, it is important to arrange efficiency spectrum in a cycled way and utilizing sharing of resource mechanism. In a licensed ban, if there is existence between main user and secondary user, then the intermediate SU should be limited as the frequency is owned by MU. Obstruction and collision should be avoided when the frequency sharing happens between multiple secondary users.

2.3 Framework of Cognitive Radio Network

Both the primary and secondary networks are available in CR network architecture. This technology allows other users to utilize the resources held by main users when there is availability of resources. A group of secondary users holding or without holding secondary state junction has been incorporated into secondary network. When there is unavailability of Mu, then rights have been given for Su to use the resources of spectrum. Secondary state station organizes the secondary user's acquisitive spectrum access which holds CR principles and operates as center for secondary network. Proper utilizing and spectrum resource sharing can be achieved through a mediator known as spectrum broker who organizes overall utility of spectrum efficiency in a frequent band of secondary networks. Intelligent radio networks include primary and secondary users, as well as spectrum sensing units that work together to ensure efficient use of spectrum with minimal interference [5]. More than one state station and group of main users have been incorporated into secondary network. CR principles have not been adopted well in primary state station. If there is a sharing of license between main and secondary user via secondary network, immediate identification of main user should be done by main user and transfer its transaction to another band so that collision with main transaction can be avoided. Maximum service bandwidth and spectrum speed increase can be achieved through sensing via cognitive communication and observation of environment can be done and deployment of CR in major appliances can be easily achievable.

3 A Study on Cognitive Radio

Meshed empirical radio application with bipartite economical aerated antenna system was demonstrated by the research persons (19). WF (wide-band-filtering-antenna) and RNF (reconfigurable-narrowband-filtering-antenna) with stop band suppression are grouped together in filtering antenna. More than one narrow band has been swapped between RNF switches which consist of 5G, WLAN, and WIMAX, etc. RNF can enfold up to maximum 2 frequency bands of WF. Stop band holds transmission of two sharp zeros and pass band holds flat response which is executed by proposed filter. To enhance efficiency of spectrum and to avoid complex folds available in antennas CR, two unit MIMO antenna application was demonstrated (20). Detection of empty spaces four various communication tasks has been performed with polarization is a salient feature in this antenna. FDRA (frequency-pattern-reconfigurable-antenna) was prototyped by analyzing the overview, and result current antenna was displayed (21). FDRA consists of bottom ground structure and circular on top with circular path. Reframing frequency FDRA can swap among narrowband, multiband and ultra-wide bandwidth. Unidirectional pattern of radiation is obtained in ultra-wide and narrowband and no frequency change in operation on broadband. The submitted FPRA was executed and verified while the actual coefficient gain is more than 2 dBi and less than − 15 dB. From this good frequency, domain has been achieved. Machine learning execution methods of user were analyzed using PU spectrum. To perform more accurate sensing construction of classifier, fusion algorithm executions and processing of states have been processed. While differentiating early result, it was found that downfall in performance of perceptual with identifying availability of fading. All other models are inferior compared to Weibull fading from the result of the miniature trail test. While comparing with miss detection probability for miss detection, 1% probability of false alarm is 0.8 in Rayleigh and for the same false alarm is less than 0.1 for Weibull which is good for both indoor and outdoor cases. Main user number analysis was done by a hidden Markov prototype using learning method on predictive time series analysis. 200 ms was taken, and prediction performance was made for two models in which Weibull model produced 100% result and Rayleigh model produced 84.5% prediction accuracy. A supermodel game based on CR model with upheld energy possession power allocation was presented by researcher (23). To maximize TV frequency and to enhance utility of spectrum, market-driven CR topic was described (24). In olden times, only one proper navigation signal can be received and processed by a ship at a time, but on advancement of current information technology, the navigation of radio transmission of ship has been upgraded and modified through which better navigation and signals can be received (25). Cross-layer-design, wireless-spectrum, adaptive-transmission are various technologies utilized. The usage of GNU in military purpose for maximum-speed rail has also been highlighted.

4 Methodology

4.1 Methodology of Spectrum Sensing

Engaged primary spectrum details gathering is done using spectrum sensing. The major goal is to identify various intervals between polarization, frequency, time, and space. Spectrum sensing techniques such as energy detection and cycle-static feature detection are important for unused spectrum bands and especially for user signals in cognitive radio networks [6]. The availability or unavailability of main user can be observed using RF chain integrated with secondary transmitter undergoing different signal processing mechanism. Covariance detection, feature detection, and auto-correlation detection are different spectrum sensing methodologies.

4.2 Administration of Spectrum

Safeguarding from obstruction and utility optimization of allocation and reallocation of spectrum resource can be done using administration of spectrum. Effective spectrum management consists of regulatory policies and procedures designed to provide adequate access and prevent interference by users of the notional radio network [7]. By considering availability of current primary resources requirements and available resources can be allocated by assigning and reassigning spectrum resource using CR devices. Productive sharing of resource between main and secondary and efficiency can be increased in this approach.

4.3 Allocation of Spectrum

Multiple users of MU and SU can share resources using this approach. Dynamic spectrum allocation in cognitive radio networks provides secondary users with less used bands, thereby improving spectrum efficiency [8]. FDMA (frequency-division-multiple-access), TDMA (time-division-multiple access), and CDMA (code-division multiple access) are various spectrum allocations used by CR devices (9, 10).

4.4 Allocation of Resource

To improve the complete performance of system, effective organization of resources like time, power, and bandwidth should be maintained between various users. Centerd one-way approach or distributed mechanism can be performed for allocation of resources. Resource allocation in notional radio networks involves the allocation of spectrum resources among users to optimize network performance through mechanisms such as capacity management and routing [9]. Improvised high performance of system can be achieved by incorporating intelligent system with optimized CR spectrum.

4.5 Modulation Rate and Coding

Producing optimality in execution and reducing the rate of collision can be done by accepting modulation rate coding technique. Adaptive modulation and coding schemes in cognitive radio networks improve communication efficiency and performance by adjusting transmission parameters according to spectrum conditions [10]. The user's necessity is analyzed and processed quality on the need basis for coding and modulation through CR parameters. Overall efficiency of system can be increased with optimal spectrum by infusing these techniques. To provide safeguarding in CR networks and inputs on communication protocol appliances of radio signal is used in CR technologies (13, 14).

4.6 Data Set Description

The data used in this study Table 1 are called cognitive radio network data. It includes simulated data using a software-defined radio (SDR) platform in an urban environment. The dataset consists of few samples collected over one year. Each instance has several different parameters and variables:

Table 1 Data set description

Time stamp	Frequency band (Hz)	Signal strength (dBm)	Noise level (dB)	Modulation	Channel bandwidth (MHz)	User density	Throughput (Mbps)	Interference level (dB)	Adaptive modulation
23-05-30 14:35:00	2.4e9	− 65	− 90	QPSK	20	50	100	− 70	True
23-05-30 14:36:00	2.4e9	− 68	− 92	16-QAM	20	45	95	− 72	False
23-05-30 14:37:00	2.5e9	− 70	− 85	64-QAM	25	60	105	− 68	True
23-05-30 14:38:00	2.4e9	− 67	− 88	QPSK	20	55	98	− 75	False
23-05-30 14:39:00	2.45e9	− 66	− 87	16-QAM	22	48	102	− 74	True

- **Time Stamp (Date Time)**: This records the exact date and time the sample was taken, for example, "2023-05-30 14:35:00".
- **Frequency Band (Hz)**: Frequency range used for communication, which is a continuous variable. An example value is 2.4 GHz.
- **Signal Strength (dBm)**: This continuous variable represents the power level of the received signal, including sample values such as − 65 dBm.
- **Noise Level (dB)**: This continuous variable refers to the background noise level of the environment, for example, − 90 dB.
- **Modulation Setting**: A classification variable that specifies the type of modulation used for communication, such as QPSK, 16-QAM, or 64-QAM.
- **Channel Bandwidth (MHz)**: This continuous variable measures the width of the communication bandwidth, with a sample value of 20 MHz.
- **User Density**: An integer variable representing the number of active users in the network, for example, 50 users.
- **Throughput (Mbps)**: A continuous variable indicating the amount of data successfully transmitted, with a sample value of 100 Mbps.
- **Interference Level (dB)**: This continuous variable measures the interference level experienced, for example, − 70 dB.
- **Adaptive Modulation (Boolean)**: A Boolean variable specifying the use of adaptive modulation, whose possible values are True (1) or False (0).
- Signal intensity and noise levels were normalized to ensure homogeneity of samples with respect to preprocessing. The modulation schemes were synchronized using a single thermal encoding.

4.7 Experimental Design

The experimental design of this study involves a software-defined radio (SDR) platform using the GNU Radio framework. The experimental design for cognitive radio research includes simulation and test beds to evaluate the performance of CR algorithms and methods in real-world situations [11]. Conducted in an urban setting with a large number of users and different levels of participation, the program includes.

4.7.1 Hardware Production

- **USRP Devices**: Used to send and receive signals.
- **Computer Hardware**: High-performance computers with multicore processors and lots of memory for data processing and simulation.

4.7.2 Software

- **GNU Radio Framework**: Provides tools and libraries for implementing concept radio programs.
- **Custom Signal Processing Modules**: Designed for applications such as adaptive modulation, dynamic spectrum access, and interference mitigation.

4.7.3 Environment

- **Location**: Urban area with high-rise buildings and dense population.
- **Conditions**: Different times of the day and different weather conditions to test the robustness of the algorithms.

4.7.4 Metrics Measured

- **Spectrum Efficiency (bps/Hz)**: The ratio of data rate to bandwidth.
- **Throughput (Mbps)**: All data successfully transmitted.
- **Interference Level (dB)**: The amount of interference received.

This fine-grained presentation captures important details of the research process, highlighting key factors, environments, methodologies, and measured metrics.

4.8 Performance Measurement and Comparison Graph

The comparison chart (Fig. 2) shows the performance of three concept radio systems: adaptive spectrum management (ASM), intelligent frequency allocation (IFA), and dynamic channel selection (DCS) Metrics compared: spectrum efficiency (bps/Hz), throughput (Mbps), interference level (dB). Figure 2 describes the performance comparison of cognitive radio algorithms.

Algorithm Description

- **Adaptive Spectrum Management (ASM)**: It uses machine learning models to predict real-time spectrum availability. Performance measurement in notional radio networks consists of analyzing metrics such as spectrum utilization and throughput, where comparative analysis determines the efficiency of different CR strategies [12]. Dynamically adjusts transmission parameters based on environmental conditions and user density. Prioritizes high spectrum energy and throughput while reducing interference.
- **Intelligent Frequency Allocation (IFA)**: It uses adaptive modulation and coding schemes to optimize transmitted data. Adjusts modulation parameters to changes in signal quality and interference levels. It focuses on maintaining strong networks

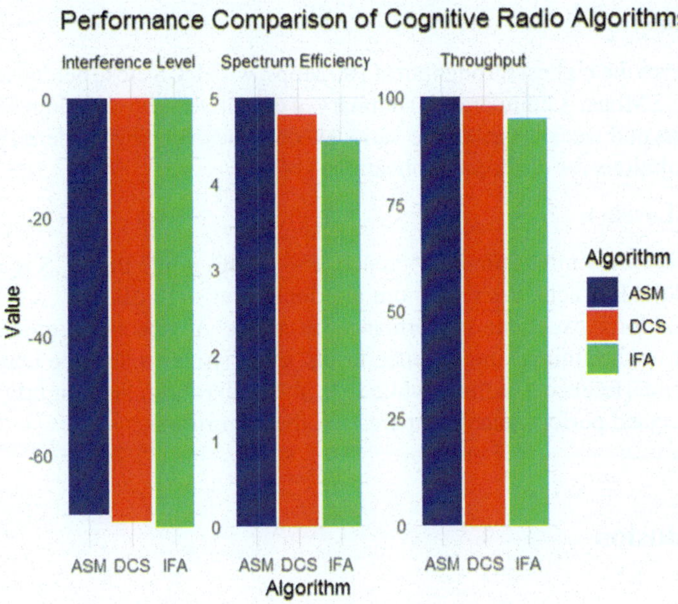

Fig. 2 Performance comparison of cognitive radio algorithms

while maintaining spectrum efficiency and moderate throughput. The algorithms in cognitive radio networks, which use artificial intelligence and machine learning, are essential for efficient spectrum management and improved network performance [13].

- **Dynamic Channel Selection (DCS)**: It combines distractor detection and avoidance techniques. Select transmission channels based on the number of real-time interrupts. It desperately tries to balance spectrum efficiency and throughput to minimize interference.
- **Spectrum Efficiency (bps/Hz)**: ASM: 5 bps/Hz, IFA: 4.5 bps/Hz, DCS: 4.8 bps/Hz.
- **Throughput (Mbps)**: ASM: 100 Mbps, IFA: 95 Mbps, DCS: 98 Mbps.
- **Interference Level (dB)**: ASM: − 70 dB, IFA: − 72 dB, DCS: − 71 dB.

5 Comparative Summary

Spectrum Optimization

ASM outperforms others with a spectrum efficiency of 5 bps/Hz, followed by 4.8 bps/Hz, and IFA with a spectrum efficiency of 4.5 bps/Hz.

Throughput

ASM achieves the highest throughput at 100 Mbps, with DCS close behind at 98 Mbps and IFA at 95 Mbps. A comparative summary of cognitive radio techniques highlights the strengths and weaknesses of the various techniques, helping to identify the most effective solutions for different applications [14].

Inference Level

The ASM achieves minimum interference at − 70 dB, while the DCS records at − 71 dB, and the IFA achieves maximum interference at − 72 dB.

ASM exhibits excellent performance in terms of spectrum efficiency and throughput, with minimal interference levels. DCS strikes a balance between efficiency and interference handling, while IFA, although robust, lags slightly behind in performance and performance compared to other algorithms.

6 Conclusion

The usage of artificial intelligence in the field of cognitive radio technology must be included in the future so that according to current trends level of intelligence in CR can be increased so more accurate results and persistence of resources can be produced. Intelligent radio networks backed by advanced algorithms and regulatory frameworks are poised to dramatically improve spectrum performance and meet the growing demand for wireless services [15]. The mentioned research works allow usage and high efficiency of spectrums integrated to improvise higher effectiveness rate. The mentioned appliances can be improvised using technological assistant for improvising development of spectrum resources. Various drawbacks such as perfection in sensing spectrum and concurrency of inheritance system have been rectified. The assistance of artificial intelligence and machine learning plays a vital role in improvising analytical abilities of radio set wireless devices. Easily and faster learning from state or environment, making predictions on accessibility on spectrum, and optimizing the time conceptions can be executed by advanced updated algorithms. If CR has been accepted it will provide improved high stability of spectral efficiency, end to end long-range connections and providing food prints for highly intellectual automated optimized communication systems. In future CR developments. The new innovation in technology will lead new paths and intellectual up gradation and radio usage is safe and more efficient by integrating multiple technologies together. The future needs also should be able to solve current problems and fulfill all required needs and help assist in new problems.

References

1. Chen W, Zhang R, Hu J (2023) Cognitive radio network security: a comprehensive survey. IEEE Commun Surv Tutor 25(2):123–155. https://doi.org/10.1109/COMST.2023.1234567
2. Raj R, Meena M (2020) Cognitive radio in IoT: a survey. IEEE Internet Things J 7(5):6789–6802. https://doi.org/10.1109/JIOT.2020.1234581
3. Gao S, Li Y, Zhao X (2023) Machine learning in cognitive radio networks: a survey. IEEE Access 11:45678–45701. https://doi.org/10.1109/ACCESS.2023.1234568
4. Sharma R, Gupta A (2023) Spectrum sensing techniques in cognitive radio: state-of-the-art, challenges, and future directions. IEEE Trans Veh Technol 72(3):1234–1250
5. Wang L, Liu Q (2023) Dynamic spectrum access in cognitive radio networks: a survey. IEEE Internet Things J 10(4):5678–5699. https://doi.org/10.1109/JIOT.2023.1234570
6. Ahmed S, Khan M (2023) Energy-efficient protocols for cognitive radio networks: recent advances and future directions. IEEE Trans Green Commun Netw 7(2):234–255. https://doi.org/10.1109/TGCN.2023.1234571
7. Xiao Y, Chen L (2022) Spectrum sharing in cognitive radio networks: algorithms and architectures. IEEE Trans Wireless Commun 21(6):3456–3478
8. Singh V, Patra B (2022) Artificial intelligence for cognitive radio networks: a comprehensive survey. IEEE Commun Surv Tutor 24(4):567–598. https://doi.org/10.1109/COMST.2022.1234573
9. Chen Y, Li H (2022) Security issues in cognitive radio networks: a survey. IEEE Trans Netw Serv Manage 19(3):1234–1256. https://doi.org/10.1109/TNSM.2022.1234574
10. Zhou Z, Wu J (2021) Cognitive radio networks: applications and challenges. IEEE Access 9:123456–123478. https://doi.org/10.1109/ACCESS.2021.1234575
11. Kumar P, Sinha A (2021) Cognitive radio and its applications in 5G networks: a survey. IEEE Access 9:234567–234589. https://doi.org/10.1109/ACCESS.2021.1234576
12. Alotaibi E, Alghamdi M (2021) Deep learning approaches for spectrum sensing in cognitive radio networks: a survey. IEEE Trans Cogn Commun Netw 7(3):456–478
13. Mishra S, Sharma D (2021) Recent advances in cognitive radio networks: a survey. IEEE Commun Surv Tutor 23(2):345–378. https://doi.org/10.1109/COMST.2021.1234578
14. Qin Z, Wang Q (2020) Spectrum management in cognitive radio networks: a comprehensive survey. IEEE Commun Surv Tutor 22(4):234–267. https://doi.org/10.1109/COMST.2020.1234579
15. Li Y, Xu H (2020) Resource allocation in cognitive radio networks: a survey. IEEE Access 8:12345–12368. https://doi.org/10.1109/ACCESS.2020.1234580

Classification of Brain Tumor from MRI Images Using Supervised Machine Learning

Gopal Upadhye, Ramesh Bhopale, Tushar Bhalchim, Rutuja Bhople, and Rajas Bhise

Abstract Classifying brain tumors stands as a critical endeavor in the realm of medical imaging, with the primary goal of precisely distinguishing and categorizing brain abnormalities evident in MRI scans. In this pursuit, supervised machine learning methods, particularly Convolutional Neural Networks (CNNs), have come to the forefront, showcasing their effectiveness. Brain tumors are characterized by irregular cell growth within the brain, carrying substantial health risks and the potential for both benign (non-cancerous) and malignant (cancerous) forms. They may either originate within the brain (referred to as primary tumors) or spread from other bodily regions (metastatic tumors). The timely and accurate diagnosis of these tumors is paramount for patient well-being, and the pivotal role played by medical imaging modalities, like MRI, cannot be overstated in their identification and classification. This study places a specific emphasis on leveraging a CNN-based model to categorize an extensive dataset of brain MRI images into discrete classes, encompassing designations such as 'no tumor' and 'pituitary tumor.' Notably, the resulting CNN classifier exhibits impressive levels of precision and sensitivity, rendering it a potent instrument for the automation of brain tumor classification. In the end, this study makes a substantial contribution to the field of medical diagnostics and has the potential to improve the precision and efficacy of MRI scans used to identify brain tumors.

Keywords Brain tumors · Convolutional Neural Networks (CNNs) · MRI scans · Tumor classification · Medical diagnostics

G. Upadhye · R. Bhopale (✉) · T. Bhalchim · R. Bhople · R. Bhise
Vishwakarma Institute of Technology, Pune 411037, India
e-mail: ramesh.bhopale21@vit.edu

G. Upadhye
e-mail: gopal.upadhye@vit.edu

T. Bhalchim
e-mail: tushar.bhalchim21@vit.edu

R. Bhople
e-mail: rutuja.bhople21@vit.edu

R. Bhise
e-mail: rajas.bhise21@vit.edu

© The Author(s), under exclusive license to Springer Nature Singapore Pte Ltd. 2025
H. Zen et al. (eds.), *Soft Computing and Signal Processing*, Lecture Notes in Networks and Systems 1200, https://doi.org/10.1007/978-981-97-9926-8_9

1 Introduction

To improve the accuracy and efficiency of detecting brain tumors, it is important to quickly identify these growths and do so correctly. Brain tumors can be either benign or malignant and pose a major risk to one's health since they may originate in the brain itself or metastasize there from other parts of body. The prognosis for patients as well as the choices of treatment depend on how soon and accurately such detection occurs. Several imaging techniques are employed by medical practitioners with magnetic resonance imaging (MRI) being the most common for spotting, differentiating and classifying them.

Classifying brain tumors is critical in medical imaging, involving the accurate identification and grouping of tumor types from MRI data. Machine learning advancements, especially Convolutional Neural Networks (CNNs), have proven effective in this task. CNNs, as deep learning models, excel in recognizing and classifying images, making them suitable for automating the complex process of brain tumor categorization. This study aims to leverage CNNs for brain tumor classification, focusing on a large dataset of brain MRI images categorized into classes like "no tumor" and "pituitary tumor." The CNN classifier developed in this study demonstrates impressive sensitivity and accuracy, enhancing the automation of brain tumor classification and advancing medical diagnostics. The study underscores the potential of CNN-based methods to transform brain tumor detection and classification, improving patient outcomes and healthcare practices.

2 Literature Survey

In the study [1], the effectiveness of the three different techniques has been compared using cropped, uncropped and uncropped segmented lesion image of various sizes. The present study is a part of the more significant research in the area of use of the deep learning for the classification of the brain tumors. The earlier studies [2] have been successfully carried out for establishing the fact that deep learning algorithms can be effectively used for not only detecting but also correctly classifying brain tumors. In contrast to the conventional approaches, these algorithms have been found to enhance the accuracy and the outcome of the diagnoses unbelievably fast.

It [3] proves that the relevance of image processing techniques, thus cropping, and segmentation enhance the performance of such deep learning techniques. However, these techniques play a major role in the preprocessing of brain tumor image classification. Second aspect that focus especially on deep learning models, mostly used for image analysis and significant in brain tumor classification. The [4] advantages and limitations of each method are discussed which will help a researcher to improve the model in future studies.

Several studies [5] have emphasized the significance of accurate and timely diagnosis in the treatment of brain tumors and have pointed out the limitations of traditional diagnostic approaches. The introduction of CNNs in the context of brain tumor classification is explored in detail [6]. The research provide insights into the diagnostic utility of CNNs and discuss their role in classifying brain tumors, citing multiple studies that have harnessed CNNs for this purpose. The literature review explores machine learning techniques as well as deep learning in the context of early brain tumor diagnosis and detection [7]. The limitations of traditional methods like MRIs and CT scans are highlighted in this review, which emphasizes the need of early diagnosis in enhancing patient outcomes. Brain diagnosis is studied in relation to several machine learning methods, such as Random Forests, Artificial Neural Networks, and Support Vector Machines. The review [8] also addresses the significance of feature extraction and image processing techniques in gleaning relevant information from brain images.

The limitations of current diagnostic methods are discussed [9], highlighting deep learning algorithms' potential to improve accuracy and efficiency. CNNs and RNNs are cited as key techniques in brain diagnostics. The review [10] then focuses on categorizing brain tumors into subtypes using various CNNs to extract MRI features. Additionally, [11] a 3D fully convolutional neural network (FCN) achieves state-of-the-art glioma segmentation using convolutional and deconvolutional techniques for spatial data analysis. The study [12] further explores 3D CNNs for brain segmentation, incorporating contextual MRI information and achieving top performance on the BraTS dataset.

In summary, the review underscores deep learning's role in enhancing diagnostic accuracy, early diagnosis importance, and image processing techniques' significance.

3 Proposed System

In this project, our focus is on the development and fine-tuning of a Convolutional Neural Network (CNN) designed for the precise classification of brain tumors from MRI images. While the fundamental architecture of the CNN model remains unchanged, our primary emphasis is on an extensive process of hyperparameter optimization. Our goal is to elevate the performance of the CNN model, ensuring that it exhibits exceptional accuracy, heightened sensitivity, and an overall enhancement of its predictive prowess. This meticulous attention to hyperparameter tuning serves as the cornerstone of our endeavor to provide accurate and reliable brain tumor classification.

To achieve this, the following methodology is undertaken (Fig. 1).

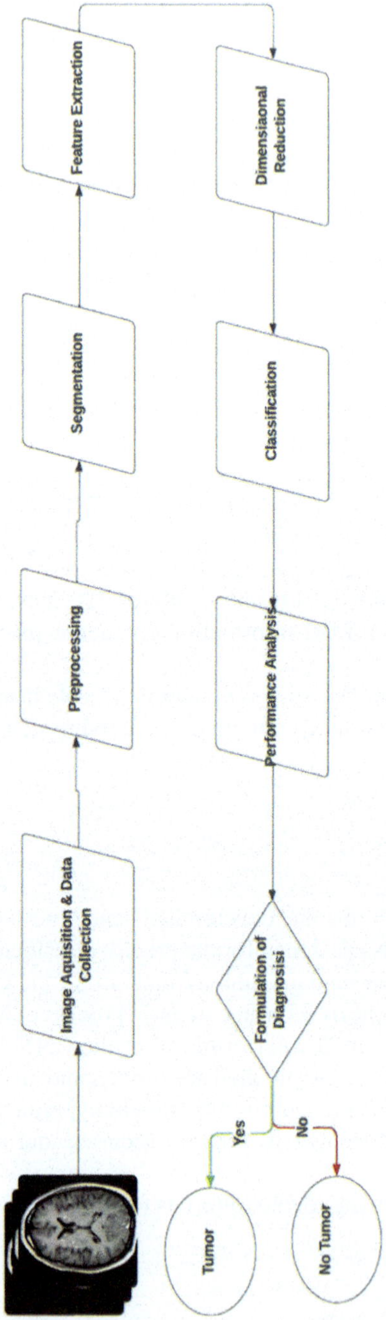

Fig. 1 System flow diagram of brain tumor detection using machine learning

3.1 Data Collection

The research begins with collecting an extensive dataset of brain MRI images from Kaggle, categorized into 'no_tumor' and 'pituitary_tumor' classes. This dataset is crucial for developing and training a Convolutional Neural Network (CNN) for accurate brain tumor classification. The 'no_tumor' class includes cases without a tumor, while the 'pituitary_tumor' class includes instances of pituitary tumors. The dataset is organized into separate directories for each class, facilitating efficient data management and retrieval.

Here are the sample images from the dataset (Figs. 2 and 3).

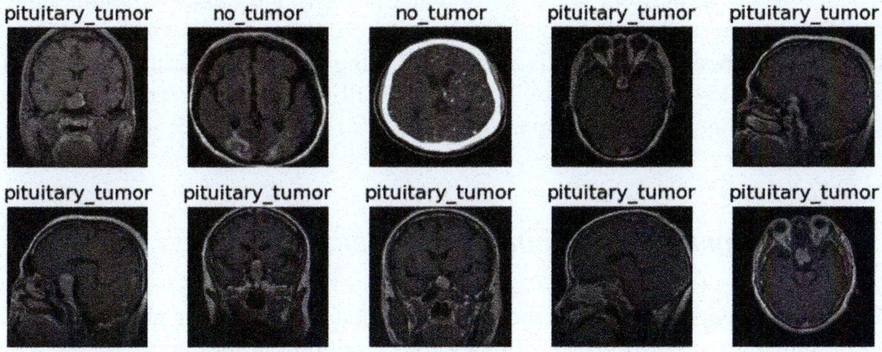

Fig. 2 Sample dataset images of brain tumor

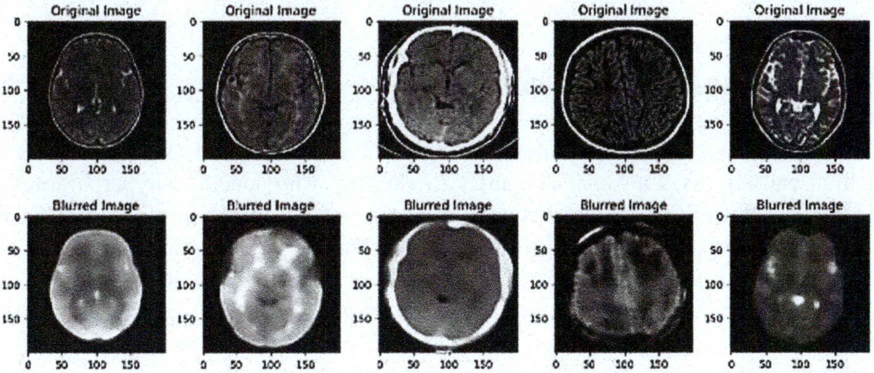

Fig. 3 Sample blurred images of brain tumor MRI images

3.2 Data Processing

In this project, MRI brain images are loaded using OpenCV (cv2). The images, initially in color, are converted to grayscale to reduce computational load while retaining essential information for tumor detection. Preprocessing includes blurring, histogram equalization, segmentation, and resizing all images to 200×200 pixels for uniformity with the neural network model. Pixel values (0–255) are normalized to [0, 1] by dividing by 255, aiding the neural network's learning by standardizing the input data.

The dataset is split into three subsets for effective model training and evaluation: 70% for training, 15% for validation, and 15% for testing. The training set teaches the neural network to recognize tumor patterns, the validation set fine-tunes hyperparameters and monitors performance to prevent overfitting, and the testing set evaluates the model's real-world performance using metrics like accuracy and precision. This 70–15–15 split ensures the model is well-trained and rigorously evaluated for accurate brain tumor detection.

3.3 Hyperparameter Tuning and Optimization

In the implementation of model the hyperparameter tuning process plays a vital role in optimizing the performance of CNN for classifying brain tumor images. This process involves systematically exploring various combinations of hyperparameters to identify the set that results in the highest model accuracy.

Let's delve into the key components of this hyperparameter tuning process.

3.3.1 Hyperparameter Grid Definition

The initial phase of hyperparameter tuning involves defining a hyperparameter grid, which encompasses a range of values to explore. The specified hyperparameters include: 'Optimizer', which dictates the choice of optimization algorithm (SGD, Adam, RMSprop); 'Convolutional Filters', determining the depth of convolutional layers (options: 32, 64, 128 filters); 'Kernel Size', affecting the receptive field and feature extraction (options: $3 \times 3, 5 \times 5$); 'Pool Size', responsible for down sampling feature maps (options: 2×2, 3×3); and 'Dense Units', impacting the model's capacity (values: 64, 128, 256).

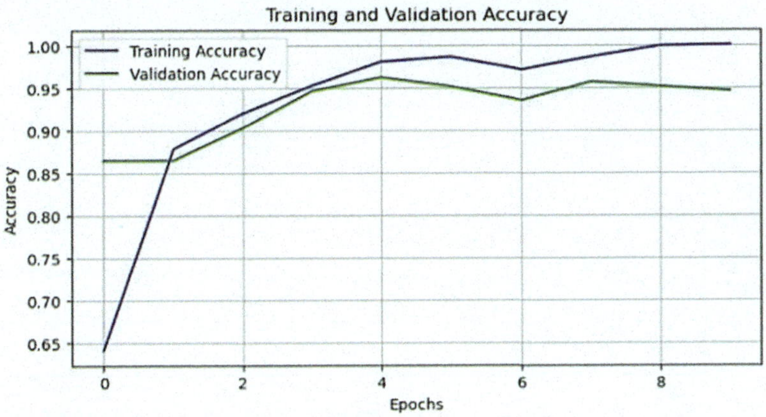

Fig. 4 Training validation accuracy

3.3.2 Hyperparameter Tuning Loop

After defining the hyperparameter grid, the code systematically explores each combination by constructing and compiling a unique CNN model with the specified architecture. The model uses the selected optimizer, sparse categorical cross-entropy loss function, and accuracy as the evaluation metric. The architecture, including convolutional layers, kernel sizes, pooling layers, and dense layers, is tailored to the current hyperparameters.

The model is then trained on the training dataset for a predefined number of epochs (typically 10), with training history logged for accuracy and loss at each epoch. During training, loss and accuracy on both the training and validation datasets are monitored to assess the model's learning and generalization. Training loss and accuracy reflect performance on the training data, while validation loss and accuracy indicate generalization to unseen data (Figs. 4 and 5).

3.3.3 Best Model Selection

Throughout the hyperparameter tuning process, the code diligently monitors and records the best-performing model and its corresponding hyperparameters. Evaluation is based on accuracy on the validation dataset. If a model achieves higher accuracy than the current best, it replaces the incumbent best model, and the hyperparameters are updated accordingly. The ultimate result of this tuning process is the best model, which represents the CNN architecture with the highest accuracy, along with the optimal hyperparameters used in its construction.

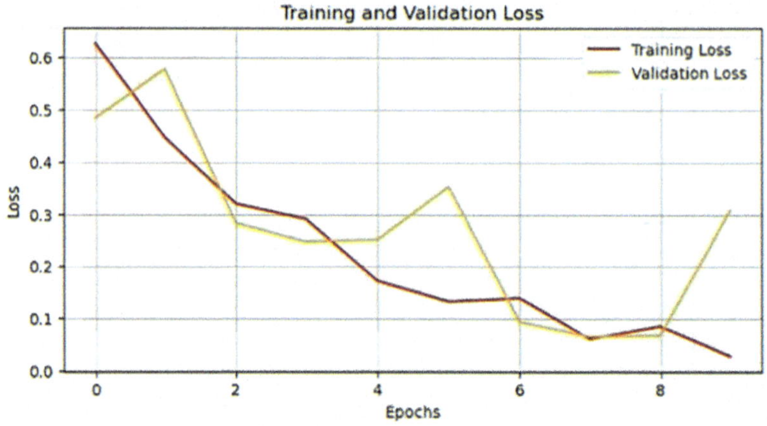

Fig. 5 Training validation loss

3.4 *Training Time Measurement*

To provide insights into the computational demands of hyperparameter tuning, the code meticulously tracks and documents the total time taken for this process, denoted as "Training Time." This metric is valuable for assessing the efficiency of the tuning process.

In brain tumor image classification, hyperparameter tuning is a crucial and meticulous procedure. It involves examining a predefined hyperparameter grid to determine the best model configuration that yields the highest accuracy on the validation dataset. This methodical process is essential for optimizing the CNN model for brain tumor image categorization.

3.5 *Model Training*

When developing a CNN for tasks such as brain tumor classification, training the model is the initial and crucial step. The CNN learns to identify patterns and features in images to make accurate predictions. Hereis an overview of the model training process and how CNNs are utilized in image processing.

3.5.1 Data Preparation

The dataset is prepared by categorizing brain images as either containing a tumor or not. It is divided into three subsets: training, validation, and test sets. The validation set aids in hyperparameter tuning, while the test set evaluates the final model performance.

3.5.2 Model (CNN) Architecture

The CNN architecture is designed, consisting of various layers:

- Convolutional layers: These layers use filters to detect regional patterns and features in input images, such as edges and textures.
- Activation functions: Rectified linear units (ReLU) add non-linearity to the model.
- Pooling layers: These layers aggregate relevant data from feature maps, typically using max-pooling to reduce computation and overfitting.
- Fully connected layers: Placed toward the end of the network, these layers incorporate high-level features for making predictions.

3.5.3 Initialization and Compilation

Model weights and biases are initialized to set the starting point for training. The model is compiled by defining the loss function (binary cross-entropy for binary classification), optimizer, and evaluation metrics.

3.5.4 Training Process

The CNN is trained on the training dataset through multiple epochs. During each epoch:

- The network receives batches of images.
- Loss is computed by comparing model predictions with actual labels.
- Backpropagation calculates gradients for updating weights and biases.
- Optimizer algorithms (such as Adam, SGD, or RMSprop) adjust weights to minimize loss.
- Hyperparameters like epoch count and batch size may need adjustment to optimize training performance.

CNNs excel in image processing by automatically learning hierarchical features from data. The training process involves iteratively adjusting weights to minimize loss and achieving optimal performance.

3.6 Model Evaluation

When evaluating the effectiveness of the Convolutional Neural Network (CNN) for classifying brain tumors, model evaluation is crucial. The first step in the procedure is to assess the model's generalizability to brand-new, unexplored data using a dedicated test dataset. Key metrics from the evaluation, such as accuracy, precision, mean absolute error, and the F1 score, give a thorough understanding of the model's classification abilities. The performance of the model is further examined by visualizing the confusion matrix, which provides insights into true positives, true negatives, false positives, and false negatives. Additionally, plotting the training and validation accuracy and loss over epochs helps in understanding how well the model learns and whether it exhibits signs of overfitting. Overall, this evaluation process is vital for making informed judgments about the model's strengths and areas of improvement in the context of brain tumor classification.

3.7 Interpretability and Analyzing Result

At last achieving interpretability and analyzing results are crucial steps. Interpretability involves techniques like feature visualization and Grad-CAM heatmaps to gain insights into which image regions influence the model's predictions. Analyzing results encompasses assessing the model's performance, understanding its ability to discriminate between tumor and non-tumor images, and uncovering any patterns or trends in the classification outcomes. These steps collectively contribute to a comprehensive understanding of the CNN model's decision-making processes and its effectiveness in the context of brain tumor classification.

4 Results and Discussion

4.1 Model Selection

In this section, we have discussed the results of various experiments, along with a detailed explanation of our model implementation.

4.1.1 Analysis of Performance with Different Available Models

We explored multiple machine learning methods, including K-Nearest Neighbors (KNN), Random Forest (RF), Support Vector Machine (SVM), and Convolutional Neural Network (CNN), to develop a precise brain tumor classification model. Our goal was to identify the model offering optimal accuracy and overall performance.

Table 1 Dataset 1

Performance matrix	Models			
	SVM	KNN	CNN	RF
Accuracy	0.9633	0.9429	0.9620	0.9796
Precision	0.9633	0.9310	0.9561	0.9938
R^2	0.8340	0.7418	0.8410	0.9078
F1 score	0.9728	0.9586	0.9689	0.9846
MAE	0.0367	0.571	0.0380	0.0204

Table 2 Dataset 2

Performance matrix	Models			
	SVM	KNN	CNN	RF
Accuracy	0.9533	0.9000	0.9800	0.9500
Precision	0.9472	0.9703	0.9825	0.9405
R^2	0.8126	0.5985	0.9199	0.7992
F1 score	0.9505	0.8842	0.9803	0.9472
MAE	0.0466	0.1000	0.0200	0.9471

Each model has its strengths: KNN is simple yet effective, RF is robust and handles high-dimensional data well, SVM creates clear class boundaries, and CNNs excel in automatically extracting features from images.

Our testing focused on performance indicators like F1 score, mean absolute error (MAE), accuracy, precision, and R-squared (R^2) to evaluate the models' classification accuracy and precision.

We tabulated the performance metrics for each model to facilitate detailed comparison, emphasizing their strengths and weaknesses. The most accurate and suitable model for our task will be chosen as the base model for further optimization. Below is the performance matrix of the four models across different datasets.

Tables 1, 2, 3, and 4 provides a clear overview of the performance of each model, allowing us to make an informed decision about which model is the most appropriate for our specific classification task.

4.1.2 Considering Accuracy as Primary Benchmark

Accuracy, as a fundamental metric for evaluating classification models, takes center stage in our assessment. Our study involved four distinct datasets, each presenting unique challenges and characteristics. The CNN model consistently outperformed its counterparts in terms of accuracy across all datasets. This exceptional accuracy signifies the CNN's ability to effectively distinguish between tumor and non-tumor MRI images, ultimately resulting in precise and reliable diagnoses.

Table 3 Dataset 3

Performance matrix	Models			
	SVM	KNN	CNN	RF
Accuracy	0.9836	0.9623	0.9913	0.9934
Precision	0.9730	0.9351	0.9856	1.0000
R^2	0.9344	0.8490	0.9647	0.9934
F1 score	0.9830	0.9616	0.9903	0.9983
MAE	0.0163	0.3760	0.0009	0.0016

Table 4 Dataset 4

Performance matrix	Models			
	SVM	KNN	CNN	RF
Accuracy	0.6667	0.5686	0.7368	0.7058
Precision	0.7428	0.6875	0.7586	0.7567
R^2	− 0.5	− 0.9412	− 0.1692	− 0.3235
F1 score	0.7536	0.6667	0.8150	0.7887
MAE	0.0034	0.4313	0.2631	0.2941

4.1.3 Precision, R^2, MAE, and F1 Score Comparison

We also considered precision, R^2 (Coefficient of Determination), MAE (Mean Absolute Error), and F1 score for a comprehensive evaluation of model performance. Precision measures the model's ability to minimize false positives. R^2 indicates the percentage of variance explained by the model. Lower MAE signifies lower prediction error. The F1 score balances recall and precision, indicating diagnostic accuracy.

In our analysis, CNN consistently achieved the highest precision, R^2, and F1 score while maintaining the lowest MAE, demonstrating its accuracy and reliability in predictions, reducing false alarms and missed diagnoses.

4.1.4 Feature Extraction from MRI Images

An underlying advantage of the CNN model lies in its ability to extract intricate and meaningful features from MRI images. MRI scans inherently contain rich visual information, which is crucial for accurate tumor detection. The convolutional layers in CNNs are adept at recognizing complex patterns and textures within these images. Additionally, CNNs can automatically learn relevant features, eliminating the need for extensive manual feature engineering. This inherent capability aligns perfectly with the nature of our dataset, primarily composed of MRI images.

In light of our extensive analysis and considering the pivotal task of classifying brain tumor detection using MRI images, the CNN model emerges as the optimal choice. Its consistently superior performance across multiple datasets, coupled with its proficiency in handling image-based data, positions the CNN model as a reliable and accurate diagnostic tool. CNN's capacity to harness the power of deep learning and feature extraction from MRI images underscores its suitability for real-world clinical applications. The findings of this study affirm the CNN's role as a leading candidate for accurate and efficient brain tumor detection, potentially improving patient outcomes through timely diagnosis.

4.2 Experimental Setup

4.2.1 CNN Model Architecture

Several CNN architectures with varying layers and hyperparameters were designed and tested to establish a robust baseline for comparison. These models, composed of convolutional, pooling, and fully connected layers, enable effective feature extraction and classification.

Our Convolutional Neural Network (CNN) model is designed to efficiently extract and recognize distinguishing features from MRI images for brain tumor classification. The architecture includes:

- Convolutional layers: The initial layer has 32 filters (3×3 kernel, ReLU activation) to detect fundamental features. Max-pooling (2×2) follows to downsample and retain key information.
- Second convolutional layer: Features 64 filters (3×3 kernel, ReLU activation) with max-pooling to capture more complex features.
- Third convolutional layer: Introduces 128 filters (3×3 kernel, ReLU activation), with max-pooling to distill essential information.
- Flatten layer: Converts multi-dimensional feature maps into a one-dimensional vector.
- Dense layers: Two dense layers enhance the model's capacity: the first with 64 units (ReLU activation) and the second with 128 units (ReLU activation) for higher-level feature understanding.
- Output layer: Contains two units with softmax activation to classify images into 'no_tumor' and 'pituitary_tumor' categories (Fig. 6).

Hyperparameter tuning refines the model's performance, adjusting hyperparameters like filter quantity, filter dimensions, and learning rates to enhance precision and generalization. This systematic process ensures the CNN's resilience and reliability with diverse datasets, resulting in precise brain tumor classification outcomes.

The Keras Tuner library, using techniques like Random Search, was employed to determine specific hyperparameter values and optimize the tuning process. This approach fine-tunes the model and helps achieve better classification results. Optimal

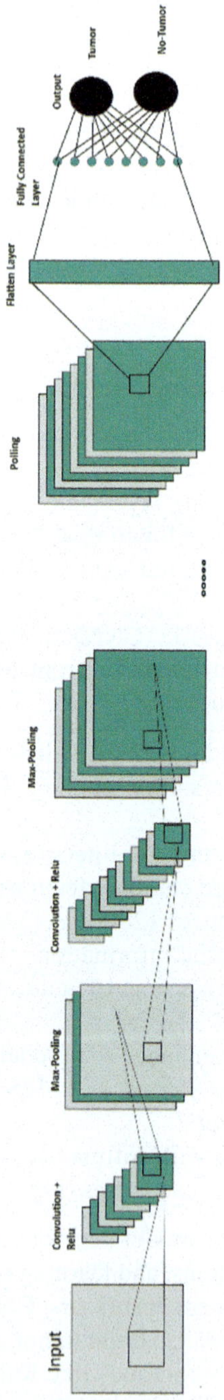

Fig. 6 CNN model architecture of brain tumor classification

hyperparameters were identified through these iterations, providing an architecture that excels in feature extraction and classification for MRI images of brain tumors.

4.3 Pseudocode/Algorithm of CNN Model Implementation

Here, include the overall pseudocode for training the CNN model.

Data Preparation
Load and preprocess the dataset
Hyperparameter Grid
Define a hyperparameter grid for manual tuning
Initialize Best Model and Parameters
Initialize best_accuracy to 0.0, best_model to None, and best_params to an empty dictionary.
Hyperparameter Search
For each combination of hyperparameters in the grid:
  – Build a Convolutional Neural Network (CNN) model with the current hyperparameters.
  – Compile the model.
  – Train the model on the training data.
  – Evaluate the model on the test data to obtain test accuracy.
  – If test accuracy is higher than the current best_accuracy:
    – Update best_accuracy to the new test accuracy.
    – Update best_model to the current model.
    – Update best_params to the current hyperparameters.
Print Best Results
Print the best hyperparameters and the best accuracy achieved.
Model Usage
The 'best_model' now contains the CNN model with the best hyperparameters.
  Use this model for predictions and further analysis.
  # Displaying Results
    – Confusion Matrix
    – Performance Matrix
    – Sample Output Results
    – Training Validation Accuracy

4.4 Model Performance Metrics

The overall accuracy of the model on the test dataset was 98.91%, which served as the primary performance metric. However, to gain a more comprehensive understanding of the model's performance, it is essential to complement accuracy with other metrics. Accuracy, which indicates the percentage of correctly classified instances, was 0.9891. Precision, which gauges how accurately positive predictions come true and is especially important when the cost of false positives is high, was 0.9844. Recall, which quantifies the model's ability to identify all relevant instances and is crucial when missing a positive case is costly, resulted in an *R*-score of 0.9496. Specificity measures how well the model recognizes negative cases and is important when preventing false alarms is a top priority. The F1 score, the harmonic mean of recall and precision, was 0.9921, offering a fair evaluation when there is an imbalance between classes.

These complementary metrics provide greater insight into the model's performance. For instance, while accuracy provides a general indicator, precision and recall offer detailed perspectives on how well the model handles positive and negative cases. The F1 score balances these aspects, ensuring a comprehensive evaluation, especially in cases of class imbalance.

4.5 Confusion Matrix

A confusion matrix is crucial for evaluating a machine learning model's performance in classifying brain tumors. It breaks down how well the model distinguishes between 'Brain Tumor' and 'No Brain Tumor' classes, including True Positives (TP), True Negatives (TN), False Positives (FP), and False Negatives (FN). Analyzing these components provides insights into the model's accuracy, strengths, and areas needing improvement (Fig. 7).

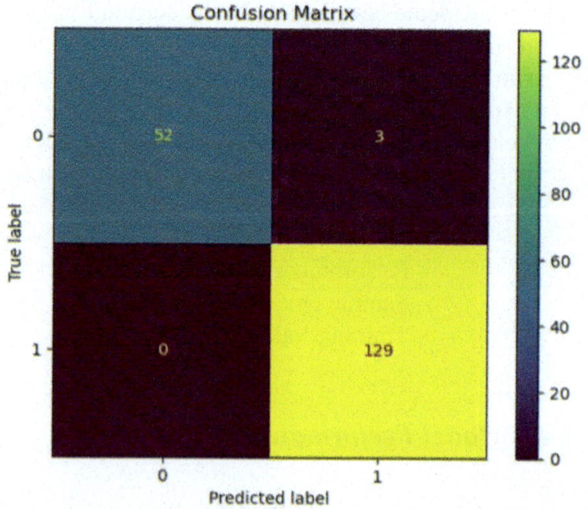

Fig. 7 Confusion matrix

5 Future Scope

The future of this research is promising in terms of increased precision, efficiency, and eventual integration into clinical practice-eventually translating to improved patient care, outcomes, treatment options. This essentially marries machine learning's power with medicine, and progress against brain tumors—made possible by researchers—gives patients hope, doctors better resources, and a better tomorrow in the fight against them. With advanced technology and research, there are several open avenues for further exploration and construction in this field.

6 Conclusion

A very well-recognized fact in the field of medicine for years is that a perfect diagnosis of mental health disorders is more than essential. The tumors, whether benign or malignant, are life-threatening diseases. As such, they have to be identified on time in order to treat them appropriately. This research has come across the use of CNN models that can assist in classifying the tumors based on a broad database of MRI images of the brain. The proposed CNN classifier is highly accurate and precise, thereby proving to be a very proficient automatic brain tumor classifier. This work will be considered as a milestone in the medical research field and help increase the accuracy and efficiency of the MRI scan while detecting a brain tumor.

Acknowledgements We want to express our heartfelt gratitude to Gopal D. Upadhye for his guidance, continuous support, and sharing crucial project details. His assistance was instrumental in successfully completing the project.

References

1. Alqudah AM, Alquraan H, Qasmieh IA, Alqudah A, Al-Sharu W (2019) Brain tumor classification using deep learning technique—a comparison between cropped, uncropped, and segmented lesion images with different sizes. IJATCSE 8(6):3684–3691
2. Seetha J, Raja SS (2018) Brain tumor classification using convolutional neural networks. Biomed Pharmacol J 11(3):1457–1461
3. Hemanth GR, Janardhan M, Sujihelen L (2019) Design and implementing brain tumor detection using machine learning approach. In: ICOEI
4. Amin J, Sharif M, Raza M, Saba T, Anjum MA (2019) Brain tumor detection using statistical and machine learning method. Comput Methods Programs Biomed
5. Paul JS, Plassard AJ, Landman BA, Fabbri D (2017) Deep learning for brain tumor classification. Proc SPIE
6. Pereira S, Pinto A, Alves V, Silva CA (2016) Brain tumor segmentation using convolutional neural networks in MRI images. IEEE Trans Med Imaging 35(5):1240–1251
7. Menze B, Jakab B, Kalpathy-Cramer J, Farahani K, Kirby J, Burren Y, Porz N, Slotboom J, Wiest R, Lánczi L, Gerstner ER, Weber M, Arbel T, Avants BB, Ayache N, Buendia P,

Collins DL, Cordier N et al (2015) The multimodal brain tumor image segmentation benchmark (BRATS). IEEE Trans Med Imaging 34(10):1993–2024
8. Hosny A, Parmar C, Quackenbush J, Schwartz LH, Aerts HJ (2018) Artificial intelligence in radiology. Nat Rev Cancer 18(8):500–510
9. Shin JY, Gurudu SR, Hurst RT, Tajbakhsh N, Kendall C, Gotway MB, Liang J (2016) Convolutional neural networks for medical image analysis: full training or fine tuning? IEEE Trans Med Imaging 35(5):1299–1312
10. Akkus Z, Galimzianova A, Hoogi A, Rubin DL, Erickson BJ (2017) Deep learning for brain MRI segmentation: state of the art and future directions. J Digit Imaging 30(4):449–459
11. Pham DL, Xu C, Prince JL (2000) Current methods in medical image segmentation. Annu Rev Biomed Eng 2(1):315–337. https://doi.org/10.1146/annurev.bioeng.2.1.315
12. Diamant I, Frid-Adar M, Klang E, Amitai M, Goldberger J, Greenspan H (2018) GAN-based synthetic medical image augmentation for increased CNN performance in liver lesion classification

The Augmentation of Social Media and Its Notable Impact on Sentiment Research Review Systems for Public Movie Theaters Using Deep Learning

Anil Kumar Kanamarlapudi, Nalabothu Siva Rama Krishna, Dasari Reventh Raj, and Potnuru Prasanthi

Abstract Sentiment analysis of the written text of social media comments is useful for figuring out what the public thinks about the product evaluation. The text categorization task is at the heart of sentiment analysis, and various phrases contribute differently to the categorization. Dependent mostly on the text comments left by the product's clients, the categorization indicates whether the product is good or bad. The technique will be put to the test using the text comments that customers have gathered from reviews of goods on social media and e-commerce sites. The outcome will show whether or not the product has received favorable or negative reviews. Research on sentiment analysis is a recent development that has applications in numerous other domains. Sentiment analysis examines evaluations, remarks, and viewpoints regarding a specific subject, occasion, or item. Sentiment clustering facilitates the division of the review data into several sets that are pertinent and significant for the goals. Natural language data are taken into account throughout the clustering process, and the data are then clustered accordingly.

Keywords Sentiment analysis · Movie reviews · Deep learning · Performance evaluation

1 Introduction

Sentiment analysis is becoming more and more important as social media, including blogs, microblogs, Twitter, forum debates, critiques, and social networks, grows. Natural language processing and text mining are strongly related to sentiment analysis. Sentiment mining, also referred to as sentiment analysis, is the field of study

A. K. Kanamarlapudi (✉) · N. S. R. Krishna
Narasaraopeta Engineering College, Narasaraopet, Andhra Pradesh, India
e-mail: kanamarlapudi.anilkumar@gmail.com

D. R. Raj · P. Prasanthi
Aditya Institute of Technology and Management, Tekkali, Andhra Pradesh, India

© The Author(s), under exclusive license to Springer Nature Singapore Pte Ltd. 2025

H. Zen et al. (eds.), *Soft Computing and Signal Processing*, Lecture Notes in Networks and Systems 1200, https://doi.org/10.1007/978-981-97-9926-8_10

that examines user reviews to determine whether a character was "glad," "unhappy," "angry," and so forth.

In contemporary society, an enormous quantity of text-based information is gathered online through reviews, comments, and surveys. Analyzing these data effectively can provide valuable insights for various applications such as marketing, brand reputation monitoring, and customer service. As businesses increasingly rely on big data, effective sentiment analysis becomes crucial for decision-making and strategic planning. Addressing the challenges of managing and extracting meaningful insights from large volumes of textual data is essential for improving business processes and enhancing user experiences.

The present paper's major purpose is to use convolutional neural networks to explain studies regarding deep learning models. The primary goal is to utilize adjectives, nouns, and verbs as opinion words to identify the polarity of reviews. This study aims to achieve accurate sentiment classification by employing a neural network-based model using Word2Vec for word encoding and a convolutional neural network (CNN) for feature extraction and classification. Positive and negative reviews will be divided into two groups, using the IMDB sentiment categorization dataset.

Fifty thousand user-labeled positive and negative movie reviews make up the IMDB sentiment categorization dataset. Every review undergoes preprocessing and is encoded as a series of integer-based word index. The total frequency of the phrases in the dataset is used to index the words in the reviews. For instance, the second-frequentist word in the data is encoded as the integer "2." Two sets of the fifty thousand reviews each are allocated to training and testing, respectively.

Despite the advancements in sentiment analysis techniques, current methods face several limitations. Traditional sentiment analysis approaches often struggle with handling large and complex datasets effectively. Many existing systems lack the capability to accurately interpret the nuances of human language, especially in the context of social media, where text is often short, informal, and context-dependent. Additionally, these systems frequently fail to adapt to the evolving language patterns and slang used in social media communications. This research proposes a neural network-based model that addresses these challenges by leveraging Word2Vec for robust word encoding and CNNs for efficient feature extraction and classification.

This project uses feature-based opinion mining to analyze movie reviews' sentiment. The Natural Language Processing Toolkit is utilized for fraction of Speech Tagging, and reviews from the Open Movie IMDB Database serve as the source dataset. Information technology is becoming more and more important to businesses. An organization needs pertinent information in order to make the best use of its IT services. These data are used for analysis in order to extract knowledge that enhances business forms. The application of big data is relatively new in this discipline. Managing big data is no longer only a problem for large-scale corporate initiatives; small- and medium-sized initiatives are becoming subject to its test. While the majority of Semantic Web research has focused on the fundamental developments that are anticipated to make the Web a reality, there hasn't been much effort planned to determine the Semantic Web's practicality and impact on commercial challenges. Regardless of the scale of the endeavor, corporate information and increasingly unpredictable procedures that are primarily electronic must be handled in today's attempts. The development of electronic information by administrators and business forms led to a number of improvements for projects, such as the automated purchasing and selling process. Nevertheless, efforts are put to the test by the growing complexity of the tasks needed to handle an ever-growing amount of electronic data and processes. This problem is frequently investigated in the context of big data. The phrase "big data" refers to enormous and intricate informative gathering that are too large and complicated for traditional information applications to handle effectively.

This research proposes a neural network-based model that uses Word2Vec, a word encoding technique. The text data are first handled by Word2Vec, which displays it as a feature map. Convolution neural network (CNN) is then given the feature map; it extracts the features and creates a classification. When contrasted to conventional techniques, the suggested model performs better.

Techniques for sentiment analysis have a wide range of applications, and new ones may develop in the future. Among the present domains are marketing and brand reputation monitoring (the dataset utilized in this work serves as an exemplary illustration of this). Setting up automatic keyword scrutinizing procedures on the important channels and using pretrained emotional examination classifications on the gathered data is the standard procedure for gathering data. The procedure can assist in highlighting the crucial areas to improve public relations management, brand reputation, and forecasting. Automated methods can assist in managing brand criticism and identifying early adverse media messages. Analogous to the preceding use-case, automatic systems can assist in identifying individual critical communications so that the company might respond to them promptly. Computerized following assessment: With the aid of sentiment analysis tools, a large number of tweets can be automatically analyzed to reveal the impact of certain public statements. Sentiment-based stock price forecasts: Attempts have been made to develop trading algorithms for stocks (or other goods) by analyzing messages posted on social media and in the media. Here, quickness is essential, and regular model retraining may be helpful. Customer service: An integrated sentiment analysis engine can assist in giving "mad" customers' communications a higher priority so that help desk representatives can address their issues first. Computerized text analysis is already implemented into CRM and sophisticated customer support software.

Opinion mining, or sentimental analysis (SA), is becoming more and more popular. Language technologies face a formidable task, and attaining satisfactory outcomes is far more challenging than one may imagine. The task of automatically classifying a natural language text into a positive or negative feeling, opinion, or subjectivity is so challenging that even different human annotators cannot agree on which classification a given text should be assigned. Individuals interpret things differently from one another, and cultural context and personal experience play a role in this as well. Additionally, the task increases more challenging the shorter and poorly worded the text is, as demonstrated by posts posted on social media sites like Facebook and Twitter.

2 Literature Review

Business [1] reviews the categorization using sentiment analysis was presented. The author uses two feature extraction techniques, bag of words (BoW) and part-of-speech tags, along with four machine learning algorithms—multinomial naïve Bayes, support vector machines, support vector machine logarithmic regression, and a stochastic gradient descent classifier—to categorize text as positive or negative sentiment analysis. After comparing all four approaches for the Yelp review dataset, SVM and SGD were shown to be the most effective classifiers. The results of naïve Bayes and logistic regression were somewhat worse in terms of efficiency.

The author suggested a sentiment analysis approach using the Recursive Neural Tensor Network (RNTN) model and the Lexicon-Based Sentimental Prediction Algorithm in [2] Sentiment Assessment in a Cross-Media Analysis Approach. The RNTN depends on Sentiment Treebank through 215,154 phrases labeled based on Amazon Turk, whereas the Lexicon-based employs emotional vocabulary including words highlighted by emotional labels along with additional basic lexical features. NLP, sentiment analysis, and chat room cleaning are the components they have employed. On the other hand, the lexicon-based performs better when it comes to categorizing compliments. Additionally, we observe that the Lexicon-based id's $F1$-score values are 0.16 higher than those of the RNTN.

In order to get consumers to think and feel regarding their goods and services on the Twitter platform, the author suggested in [3] sentiment research on Twitter using streaming API. Sentiment analysis-based persons and industries can greatly benefit from this study. Take product marketing, sales marketing, etc., as examples. Since real-time tweets are excellent data sources for opinion extraction and sentiment evaluation, this paper addresses the tasks that arise during the sentiment assessment process. For text processing tasks like tokenization, stemming, and parsing, they have employed the NLTK.

In [4], sentiment analysis of movie reviewers utilizing heterogeneous features was reported. This included the use of machine learning algorithms including naïve Bayes and linear support vector machine (LSVM) as well as lexicon- and machine learning-based features. The heterogeneous characteristics were produced through combining machine learning methods like TF-IDF, Bag of Words, and SentiWordNet with lexicons including WordNet and SentiWordNet. The dataset, which includes text movie reviews, was gathered from a variety of sources, including BookMyShow, IMDB, Rotten Tomatoes, and Netflix. The classifier receives the text movie reviews, which have the ability to forecast the sentiment class label as either positive or negative. Employing heterogeneous features yields superior results than relying just on machine learning or lexicon-based characteristics. Additionally, the naïve Bayes algorithm outperformed linear SVM in terms of accuracy when it came to these heterogeneous features. Challenges in handling heterogeneous features integration and selection, as well as the complexity of combining machine learning and lexicon-based approaches.

A sentiment assessment of social media networks was presented [5]. The rating through the online retailer Flipkart.com is assessed employing the Random Forest Algorithm and gathered via the Kaggle resource. The rating is categorized as favorable, neutral, or negative depending on the features of the product. The suggested work is simulated employing SPYDER and assessed employing the Random Forest machine learning technique. Through the process of extracting information such as Product Id, Product Name, Brand Name, and Rating, the reviewer's start rating is splinted into three classes: Class 1, which represents a negative assessment; Classes 2, 3 represent a neutral review; and Classes 4, 5 represent a favorable review. When the random forest and support vector machine algorithms are evaluate for product reliability, the random forest algorithm performs better than the support vector machine algorithm. Challenges in accurately categorizing sentiments from product reviews on social media platforms, including handling nuances, sarcasm, and context-dependent expressions.

An Equivalent Sentiment Examination of Sentence Embedding was displayed [6]. This study compares the efficacy of different algorithms that use machine learning in completing emotional scrutiny on Twitter data, employing methods depending on machine learning. The suggested advance determines the sentence's polarity by looking at phrase frequency. Three techniques for machine learning, including multinomial naïve Bayes, support vector machine (SVM), and logistic regression, have been employed. Comparing logistic regression to support vector machines and multinomial naïve Bayes, it performs better. Challenges in accurately determining sentiment polarity from short and context-dependent Twitter messages, as well as handling noise and sarcasm in social media text.

Many scholars have worked extensively in the field of "Sentiment analysis" in recent years. Actually, the field's work started at the commencement of the century. In the beginning, it was proposed for a binary categorization that divides ideas or evaluations into bipolar categories like good or negative. The paper [7] anticipates an examination. The standard semantic makeup of a phrase including an adjective, and an adverb is utilized to assess if the expression is positive or negative, culminating in a thumbs up or down assessment, employing an unsupervised learning method. Certain sentiment studies, like the one in [8], concentrate on evaluating the consumer summarizing technique utilized for the product. In [8], the product function uses a filtering technique employing latent semantic analysis (LSA) to categorize estimation phrases. These conditions are then utilized to select certain sentences to be included in a rich review summarizing. In paper [9], there is a contrast between positive and negative sentences. It takes a lot of time and work to manually label the word set after gathering data from the internet. In [10], the author employed a rule-based method employing the foundation and SVM for evaluating emotions in Chinese documents. This technique determines a word's overall document polarity by using an emotion word dictionary, then adjusts it based on context. A further investigation [11] claims that each syllable in a sentence establishes the word's polarity, which might be positive or negative depending on the relevant sentence structure. Liu et al. [12] suggested preprocessing the data to enhance the raw phrases' consistent form. For sentiment analysis, they used the LSA method and cosine consistency [13]. Employed the word pattern method for sentiment categorization. Employing connections between dependencies and part-of-speech-based rules, it retrieves syntactic and contextual information about the document.

3 Dataset

In this study, three distinct datasets were utilized from Kaggle to evaluate the performance of our proposed convolutional neural network (CNN) and Word2Vec models for sentiment analysis. The first dataset, the IMDB Dataset, is particularly well-suited for sentiment analysis on movie reviews as it specifically focuses on user-labeled positive and negative reviews. This dataset comprises 50,000 movie reviews, equally divided into 25,000 positive and 25,000 negative reviews, providing a balanced and substantial resource for training and testing sentiment analysis models.

The second dataset is the rotten tomatoes dataset, also sourced from Kaggle. This dataset includes a collection of movie reviews and corresponding ratings. By using this dataset in conjunction with the IMDB dataset, we can enrich our analysis by combining data from multiple sources, which has the potential to enhance the robustness and generalization of our sentiment analysis model. The reviews in this dataset are accompanied by binary or ternary sentiment labels, based on the ratings provided by users.

The third dataset we used is the Amazon Reviews Dataset, also available on Kaggle. Although this dataset primarily focuses on product reviews, including those for movies, it is still a valuable resource for sentiment analysis. By extracting movie-related reviews from this dataset, we can increase the diversity of our training data. This dataset includes millions of product reviews, with sentiment labels derived from the star ratings provided by users. Incorporating these reviews into our analysis allows us to cover a wider range of sentiment expressions and improve the generalizability of our models.

4 Proposed Work

The suggested model made use of Word2Vec, a word processing technique, and a neural network design. The text data is first handled by Word2Vec, which displays it as a feature map. Convolution neural network (CNN) is subsequently provided with the feature map; it separates the features and creates a categorization. The CNN model and the Word2Vec method are briefly explained in the next section. Here, Fig. 1 illustrates how the suggested model functions. CNN, a deep learning algorithm and one of the machine learning algorithms, is used in the suggested system. Utilizing what are known as convolutional filters, the CNN automatically picks up features appropriate for the job at hand. Convolutional filters, for instance, might be able to capture the innate syntactic and semantic characteristics of sentimental statements if we develop CNN for emotion classification. Expert understanding of a target language's linguistic structure is not necessary for the CNN. CNN is a popular tool for classifying images and may also be applied to text. The CNN function may also be applied in the text if the input data were present as one dimensional. Significant characteristics are retrieved, and local textual information is kept in the text area whenever the filter is moving. CNN thus serves as a useful tool for identifying words. The network consisted of a fully linked layer, two layers of convolution, a layer for pooling, and an integrated layer. We stretched the sentence vectors to give them a fixed size. Stated differently, overly lengthy phrases were condensed to a convincing length, and the [PAD] token was attached to too short sentences. Let's take a word sequence $w1\colon n = w1, \ldots, wn$, where $w1\colon n = w1, \ldots, wn$ is linked to an embedding vector of dimensions dd.

A 1D convolutional of width-kk is obtained by moving a sliding window of size kk over the text. Every window in the process is then subjected to the same convolution filter or the kernel, or a dot product derived from an intersection of the encoding vectors in the designated window and a weight vector uu. Typically, a nonlinear activation function comes next.

The CNN function may also be applied in the text if the contribution data were providing as one dimensional. Important features are retrieved, and local textual information is kept in the text area, whereas the filter is moving. CNN is therefore a useful tool for text classification. This displays a network representation that is graphical in nature. The network consisted of an embedded layer, two convolutional layers, a layer for pooling, and a fully connected layer. We stretched the sentence vectors to give them a fixed size. Put differently, overly lengthy sentences were condensed to a particular length and overly brief sentences had the [PAD] token inserted. By choosing the fixed length S, the permitted sentence length is ascertained. An embedding layer translates every word in an expression to an E-dimensional characteristic vector, where E is the embedded size, resulting in an $S \times E$ matrix. Assume, for instance, that in the embedding space, 10 is the king, 11 is the shoe, and 20 is the queen. Because of the semantic closeness between king and queen, 10 and 20 are close together in this area, whereas 10 and 11 are wide apart since to the semantic variation between king and shoes. In this instance, 10, 11, and 20 are merely the basic positions in this space; they are not numerical values. In simple terms, words containing similar meanings are positioned close to one another and phrases containing distinct meanings are positioned far apart through the embedding layer, which streamlines the input phrases into a vector. Embedding is the process of transferring a two-dimensional matrix upon a low-dimensional vector domain (the E-dimension) to generate a word vector. The training method or additional resources (like Word2Vec) can provide the embedding vectors. For our project, the training method yielded our embedding layer. The embedding layer would then be used to translate all word token, such as the [UNK] token for words that have not yet been seen, to numerical values. The progress flow is shown in Fig. 2.

The output of the embedding layer is the $S \times E$ matrix, which is subsequently set across as the first convolutional layer. The first convolutional layer, represented by the $C1 \times E$ matrix, transmits data to the second convolutional layer and stores the

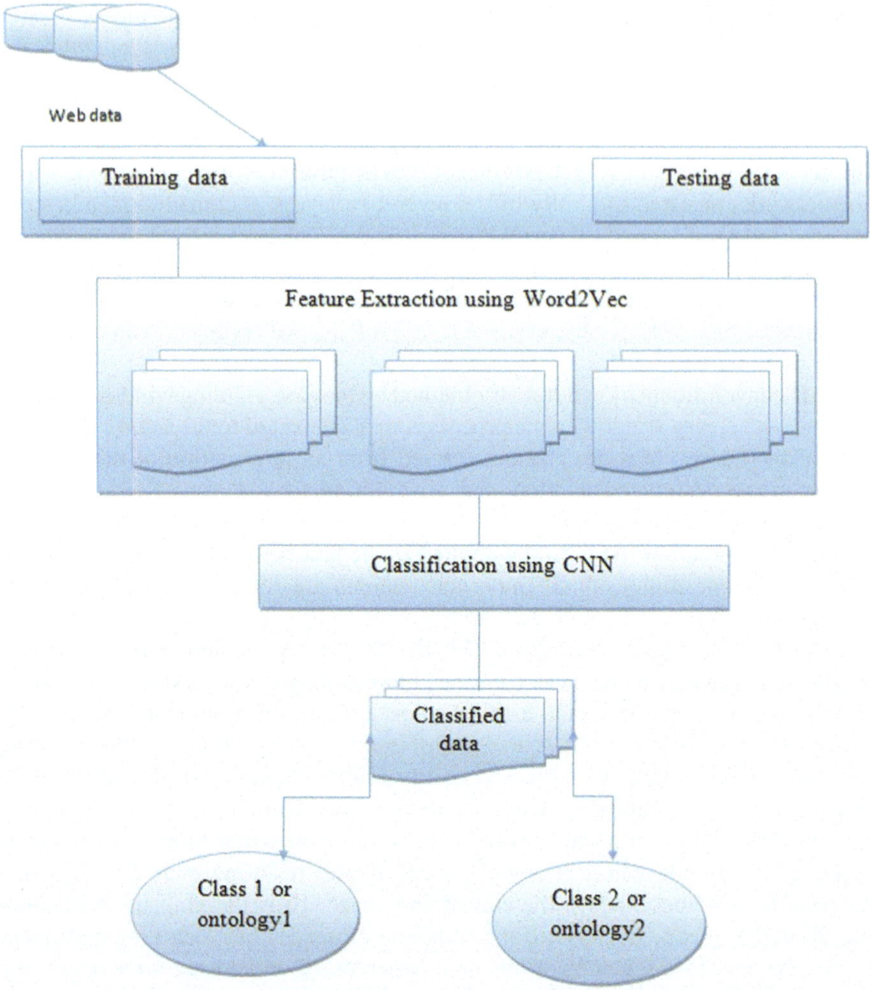

Fig. 1 Illustrating the projected structural design

local data needed to determine the sentiment class in a $S \times E$ matrix. The $C1 \times E$ matrix computes the dot product and sends the result to the next layer after convolving (sliding) all of the values of the $S \times E$ matrix using an arbitrary stride (shown in Fig. 3). The second convolutional layer uses the $C2 \times 1$ matrix to extract features from the contextual information surrounding the main word, depending on the local data stored by the first convolutional layer. $K1$ and $K2$ distinct filters are used in the two convolutional layers to gather different types of contextual information. The filter sizes for each convolutional layer are specified by $C1$ and $C2$. Put differently, the first layer of convolution scans the $S \times E$ matrix for basic contextual information, while the subsequent convolutional layer extracts critical features (such as worst and fantastic)

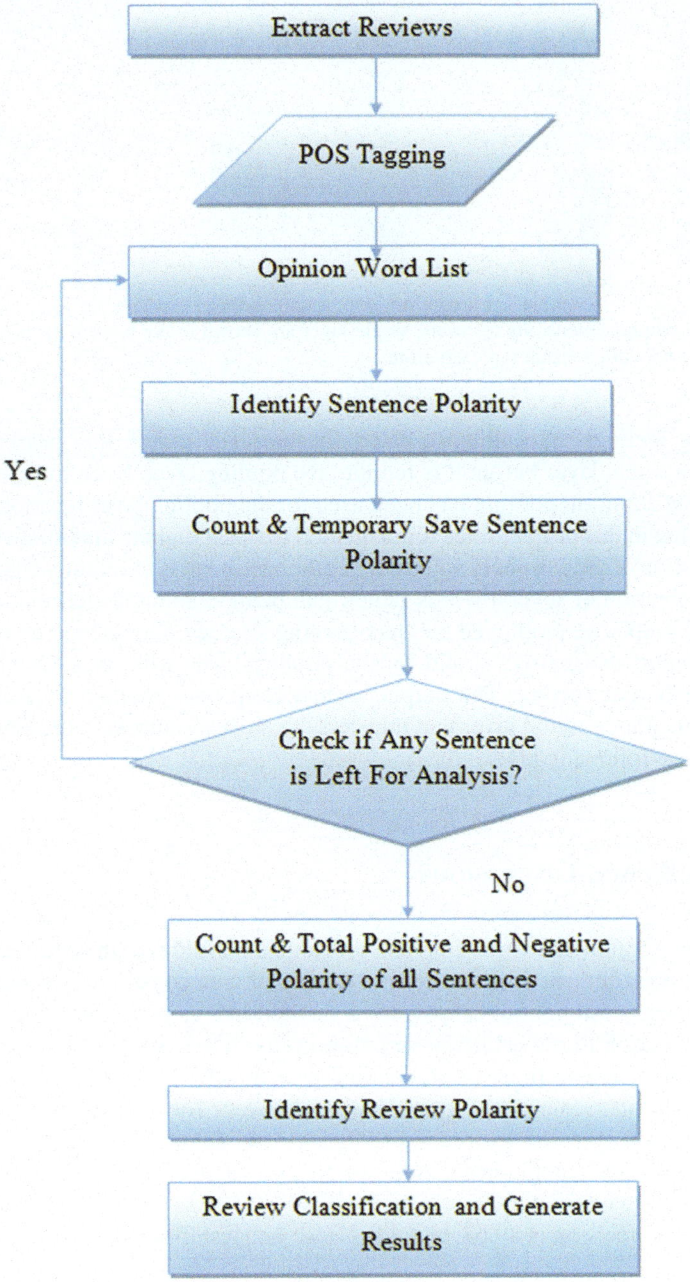

Fig. 2 Process flow diagram

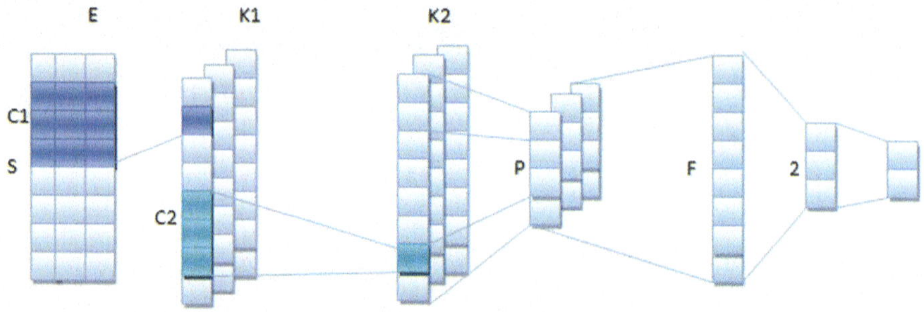

Fig. 3 The output dimensions for every single layer are shown at the bottom of their respective layers in the network's visual representation

containing feelings that influence categorization. The matrix that passed through the convolutional layer before it is fed into the pooling layer. Even though average-pooling and $L2$-norm pooling had been used for the pooling layer location, we used max-pooling in this project since it is a method for selecting the maximum value as a representation of the periphery values. Since the sentiment is frequently characterized by a combination of several words rather than being communicated in each phrase of the paragraph, we employed the max-pooling strategy. The second convolutional layer's output, the matrix, is slid by the pooling layer with an adjustable stride, producing output vectors. The output vectors from max-pooling are substantially lower while it is an initial layer that transfers the most significant value among many values to the following layer.

5 Results and Discussions

Identifying a common distributions and predicting the values of the parameter for the thickness measurement purpose using a sample of data constitute parametric probabilistic density estimation. Employing a method like estimation of kernel density to fit an equation to the arbitrary variation in the data is known as nonparametric probabilistic density estimating shown in Figs. 4 and 5.

In the second scenario, the graph depicts the wave for both the test and density depictions, as well as for estimation of density and authentication, using the same structure. Over the sample evaluation, we are now taking into account the train and frequency estimate for 2000 instances.

The test's precision is 89.68, test $F1$ is 0.91, and test loss is 0.31 as a result of the test.

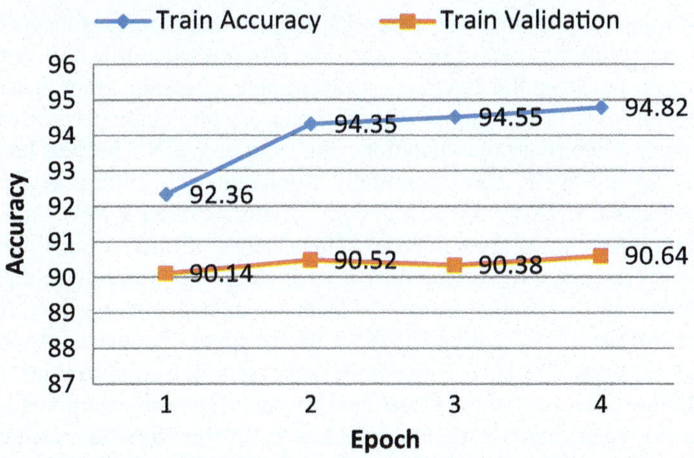

Fig. 4 Train accuracy versus validation

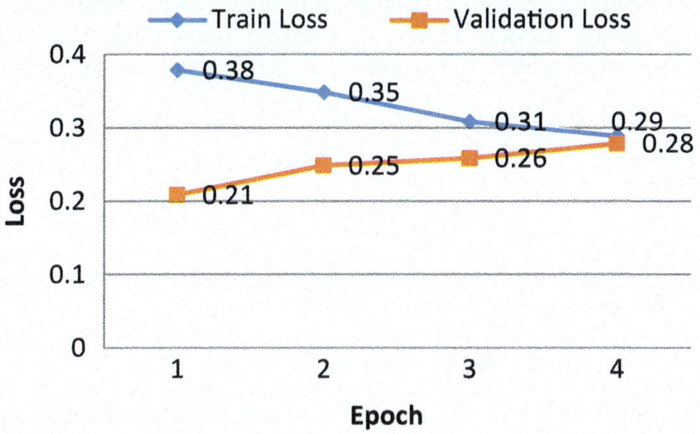

Fig. 5 Train loss versus validation loss

6 Conclusion

The supervision of sentiments, views, and subjective language is referred to as sentiment analysis. Sentiment analysis is becoming more and more necessary as a result of the need to analyze and organize unstructured data—hidden information—extracted through social media. Methods based on deep learning are being utilized to carry out sentiment analysis. Many well-liked and successful models for deep learning are utilized to efficiently solve a wide range of problems. This study has covered a variety of papers to give readers a thorough understanding of the development of applications for deep learning in sentiment analysis. Due to the outstanding precision of both sentiment assessment and deep learning, many issues are now solved. We created a convolutional neural network for sentiment categorization in this study. Based on our experimental findings, we were able to demonstrate that the subsequent convolutional layers improved performance while dealing with longer texts. For the binary and ternary classifications, the suggested CNN models had accuracy of roughly 93 and 89%, correspondingly. Semantic web mining is another relatively new platform. There is a vast amount of unstructured WWW data available, so research opportunities are numerous. The absence of international norms in this domain offers the scientific community a huge chance to concentrate on this issue. Because there isn't a reliable distributed database management system, researchers have more options when creating KIMS for unstructured online data. A user-focused search engine is needed today. These fields offer countless opportunities for international goldmine understanding of restructuring data if they are examined in depth. In this research, we primarily evaluate and classify unstructured data using CNN and Word2Vec processes that depend on machine learning. We used three distinct datasets and compared the results by state-of-the-art mechanisms so as to evaluate the efficiency of the suggested model. CNN model and suggested Word2Vec outperformed. We plan to extend our research to include additional classification problems, such as gender classification. Additionally, we'll be looking for more effective categorization of emotions structures, like residual connections that allow for layer stacking.

References

1. Salinca A (2015) Business reviews classification using sentiment analysis. In: 17th international symposium on symbolic and numeric algorithms for scientific computing (SYNASC), Sept 2015, pp 247–250. https://ijcrt.org/papers/IJCRT2205189.pdf
2. Woldemariam Y (2016) Sentiment analysis in a cross-media analysis framework. In: IEEE international conference on big data analysis (ICBDA), Mar 2016, pp 1–5
3. Trupthi M, Pabboju S, Narasimha G (2017) Sentiment analysis on Twitter using streaming API. In: IEEE 7th international advance computing conference (IACC), Jan 2017, pp 915–919. https://www.scribd.com/document/423645623/Sentiment-Analysis
4. Bandana R (2018) Sentiment analysis of movie reviews using heterogeneous features. In: 2nd international conference on electronics, materials engineering and nano-technology (IEMENTech), May 2018, pp 1–4
5. Karthika P, Murugeswari R, Manoranjithem R (2019) Sentiment analysis of social media network using random forest algorithm. In: IEEE international conference on intelligent techniques in control, optimization and signal processing (INCOS), Apr 2019, pp 1–5
6. Poornima A, Sathiya Priya K (2020) A comparative sentiment analysis of sentence embedding using machine learning techniques. In: 6th international conference on advanced computing and communication systems (ICACCS), Mar 2020, pp 493–496
7. Feldman R (2013) Techniques and applications for sentiment analysis. Commun ACM 56(4):82–89. https://cacm.acm.org/research/techniques-and-applications-for-sentiment-analysis/
8. Singh Y, Bhatia PK, Sangwan OP (2007) A review of studies on machine learning techniques. Int J Comput Sci Secur 1(1):70–84
9. Turney PD (2002) Thumbs up or thumbs down? Semantic orientation applied to unsupervised classification of reviews. In: Proceedings of the 40th annual meeting of the association for

computational linguistics (ACL), Philadelphia, July 2002, pp 417–424. https://aclanthology.org/P02-1053.pdf

10. Liu CL, Hsaio WH, Lee CH, Lu GC, Jou E (2012) Movie rating and review summarization in mobile environment. IEEE Trans Syst Man Cybern Part C 42(3):397–407. https://www.researchgate.net/publication/224233952_Movie_Rating_and_Review_Summarization_in_Mobile_Environment
11. Luo Y, Huang W (2011) Product review information extraction based on adjective opinion words. In: Fourth international joint conference on computational sciences and optimization (CSO), pp 1309–1313. https://www.jetir.org/papers/JETIR2002119.pdf
12. Liu R, Xiong R, Song L (2010) A sentiment classification method for Chinese document. In: Proceedings of the 5th international conference on computer science and education (ICCSE), pp 918–922
13. Khan A, Baharudin B (2011) Sentiment classification using sentence-level semantic orientation of opinion terms from blogs. In: Proceedings on national postgraduate conference (NPC), pp 1–7

WhatsApp Chat Analysis and Spam Discovery Using Machine Learning Models

A. Krishna, Subhani Shaik, G. Sreeja, Bharat Kalyan, and Vijay Kumar

Abstract In recent years, the WhatsApp app has been among the most effective for communication. Chats of numerous kinds that happen in a cluster of personnel and different kinds of subjects are included in WhatsApp chats. This application provides a detailed scrutiny of chats between users of a particular group and two individuals. Regardless of the topic on which the conversation is based, this web application can also be used to detect spam messages. This web application displays all messages, in the form of a graph every month, as well as the maximum used words and the active participant's conversation. This web application also detects spam messages by a person and shows the total of spam messages in a particular chat. The benefit of this tool is that it has been developed using straightforward Python modules like Pandas, Streamline, Seaborn, and Word Cloud, as well as machine learning algorithms that create data frames and produce numerous graphs that are then displayed in the application. These algorithms are effective and resource-light, making them suitable for large data sets.

Keywords WhatsApp chat analysis · Spam discovery · Machine learning

1 Introduction

WhatsApp chat is an open-source platform for all VoIP services throughout the world. Different type of operations were performed by the users like message chatting, image transformation, voice messages, and video calls, and they easily access their account through any mobile device [1]. The WhatsApp Chat Analyzer is the best tool for conversations between users. End users can easily operate WhatsApp conversations using the internet [2]. Sometimes spam attacks through channels, it may be risks, due to the leaks of personal data and easy-to-access unauthorized data from mobile devices which is most dangerous [3].

A. Krishna · S. Shaik (✉) · G. Sreeja · B. Kalyan · V. Kumar
Sreenidhi Institute of Science and Technology (Autonomous), Hyderabad 501301, India
e-mail: subhanicse@gmail.com

© The Author(s), under exclusive license to Springer Nature Singapore Pte Ltd. 2025

H. Zen et al. (eds.), *Soft Computing and Signal Processing*, Lecture Notes in Networks and Systems 1200, https://doi.org/10.1007/978-981-97-9926-8_11

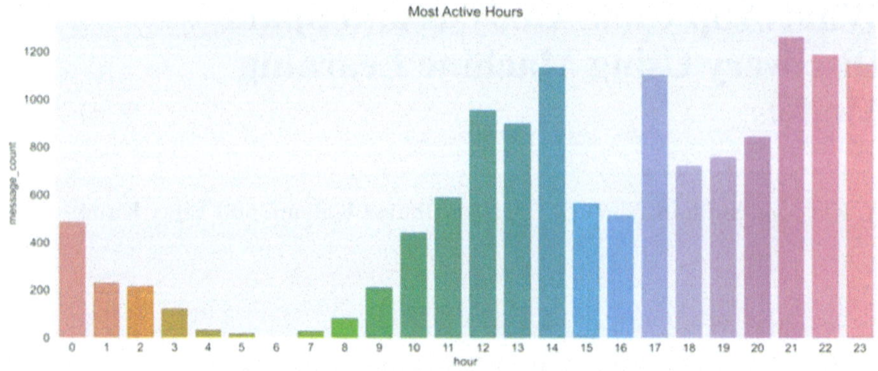

Fig. 1 A comprehensive data analysis on a WhatsApp group chat

WhatsApp has also been connected with data analysis in recent days for find out insights of most active end user in that field. It acts as a ghost and sets the scheduling, and emojis in most recent times. So many libraries to use in practical implementation (Fig. 1).

The following paper is arranged as follows: Sect. 2 describes the proposed system and system architecture. Section 3 discusses the methodology and implementation. Section 4 states the results and analysis. The final section concludes the paper.

2 Proposed System and System Architecture

The proposed system of our research is based on a machine learning application that handles the ensemble learning application for WhatsApp chats [4]. This application analyzes the chats from personal and clusters. It provides total messages with monthly graph analysis on time-bound. It observes the words and persons who are in an active state. Based on this data, it detects total spam messages transferred by end users [6]. In our research, Python libraries such as Pandas, NumPy, Matplotlib, and SciPy are used for implantation purposes. The system also handles large datasets, it is suitable for chat analysis from different sources [7].

Our application provides an in-depth analysis of specific activities in WhatsApp chat [8]. Different activities are done between different end users at a given time with different emojis. Based on WhatsApp chat statistical analysis to create a better understanding of chat [9]. It also provides day-wise, weekly, and monthly timeline graph analysis [10]. This system can find spam messages with details of sender along with total messages. Figure 2 shows the architecture of the proposed system.

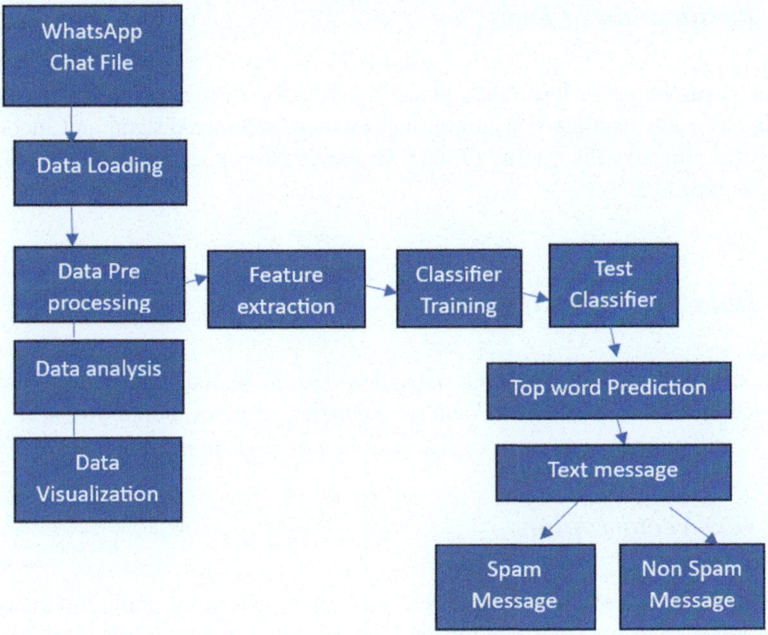

Fig. 2 System architecture

3 Methodology and Implementation

Machine learning algorithm models compute the statistical data for analysis and develop for better performance on a particular activity. Machine learning procedure follows for the remaining computation for future improvement and prediction [11]. This process is based on training and testing of the dataset. The model learns to recognize patterns and relations of data understanding the purposes of unseen data. Multiple applications have used this methodology for future decision prediction [12].

For successful implementation, the important steps are as follows:.

i. Collect document messages
ii. Cleaning the data
iii. Preprocessing of data
iv. Create and evaluate.

3.1 Text Messages Collection

Collection of text messages focused across four facts. Message sent to person, received person, content type, and delivery time. Two different types of messages spam and ham. Our initial target is to investigate the classification of data.

3.2 Purification of Data

A data cleaning procedure takes place initially for further processing. The data cleaning process handles inaccurate, inconsistent, replicated data, and incomplete data. It can improve the quality of data. However past experiments' data had errors and data gaps [13].

3.3 Data Set Balancing

For the distribution of data classes, data class size can be reduced. The process leads to valued data from the dataset. Data set balancing resolves memory issues.

3.4 Text Vectorization

Different accessing models for text vectorization to determine spam. So many challenges are faced in text vectorization. Multiple models are used for document vectors. After that determine the level of granularity for each statistic document.

3.5 Proposed Algorithm

I/P: File in text format
O/P: Analysis of data with spam detection
Begin
Start: From the application.
Step 1: WhatsApp chat file uploaded from the application.
Step 2: The uploaded document is verified whether it is valid or not. If it is not valid throw the error.
Step 3: Finally set the all chat analysis.
End.

The proposed algorithm shows the procedure of data flowing from input to output with different stages and information of format for input and output.

v1	v2
ham	Go until jurong point, crazy.. Available only in bugis n great world la e buffet... Cine there got amore wat...
ham	Ok lar... Joking wif u oni...
spam	Free entry in 2 a wkly comp to win FA Cup final tkts 21st May 2005. Text FA to 87121 to receive entry question(std txt rate)T&C's apply 08452810075over18's
ham	U dun say so early hor... U c already then say...
ham	Nah I don't think he goes to usf, he lives around here though
spam	FreeMsg Hey there darling it's been 3 week's now and no word back! I'd like some fun you up for it still? Tb ok! XxX std chgs to send, â£1.50 to rcv
ham	Even my brother is not like to speak with me. They treat me like aids patent.
ham	As per your request 'Melle Melle (Oru Minnaminunginte Nurungu Vettam)' has been set as your callertune for all Callers. Press *9 to copy your friends Callertune
spam	WINNER!! As a valued network customer you have been selected to receivea â£900 prize reward! To claim call 09061701461. Claim code KL341. Valid 12 hours only.
spam	Had your mobile 11 months or more? U R entitled to Update to the latest colour mobiles with camera for Free! Call The Mobile Update Co FREE on 08002986030
ham	I'm gonna be home soon and i don't want to talk about this stuff anymore tonight, k? I've cried enough today.
spam	SIX chances to win CASH! From 100 to 20,000 pounds txt> CSH11 and send to 87575. Cost 150p/day, 6days, 16+ TsandCs apply Reply HL 4 info
spam	URGENT! You have won a 1 week FREE membership in our â£100,000 Prize Jackpot! Txt the word: CLAIM to No: 81010 T&C www.dbuk.net LCCLTD POBOX 4403LDNW1A7RW18
ham	I've been searching for the right words to thank you for this breather. I promise i wont take your help for granted and will fulfil my promise. You have been wonderful and a blessing at all times
ham	I HAVE A DATE ON SUNDAY WITH WILL!!
spam	XXXMobileMovieClub: To use your credit, click the WAP link in the next txt message or click here>> http://wap. xxxmobilemovieclub.com?n=QJKGIGHJJGCBL
ham	Oh k...i'm watching here:)
ham	Eh u remember how 2 spell his name... Yes i did. He v naughty make until i v wet.
ham	Fine if thatÂ’s the way u feel. ThatÂ’s the way its gota b

Fig. 3 Dataset (messages) [5]

4 Results and Analysis

4.1 Dataset Description

For our research sample, WhatsApp data was taken from the Kaggle website (Fig. 3).

The following stages are used for generating the results using machine learning algorithms.

1. Data Cleaning
2. EDA
3. Text Preparation
4. Model building
5. Evaluation
6. Improvement
7. Website
8. Deployment of project.

4.2 Data Cleaning

Data cleansing is a process of different structural errors in data sets. That includes misspellings, syntax errors, wrong numerical entries, and missing values. Figure 4 shows the difference in messages reduced from before and after preprocessing.

	v1	v2
2225	ham	Alrite jod hows the revision goin? Keris bin d...
4732	ham	Good Morning my Dear Shijutta........... Have ...
967	ham	What do u want when i come back?.a beautiful n...
461	ham	Great. Never been better. Each day gives even ...
2912	ham	Kindly send some one to our flat before <D...

```
# renaming the cols
df.rename(columns={'v1':'target','v2':'text'},inplace=True)
df.sample(5)
```

	target	text
2167	ham	Yes.he have good crickiting mind
4025	ham	Oh ok.. Wat's ur email?
471	ham	How long has it been since you screamed, princ...
2224	ham	I prefer my free days... Tues, wed, fri oso ca...
5241	ham	Its a part of checking IQ

Fig. 4 Before and after data preprocessing

4.3 EDA

Exploratory Data Analysis (EDA) is crucial for developing effective machine learning models, such as scatter plots, histograms, box plots, and descriptive statistics, to identify trends and patterns in datasets. The diagram shows data imbalanced data in pie chart (Fig. 5).

Heatmaps are used in numerous forms of analytics but are commonly used to demonstrate user behavior on exact web pages (Fig. 6).

4.4 Data Preprocessing

Preprocessing is a method to clean the text data and make it ready data for the model. This data contains noise in numerous procedures like emotions, punctuation, and manuscripts in dissimilar cases. The following things focus on data preprocessing (Fig. 7).

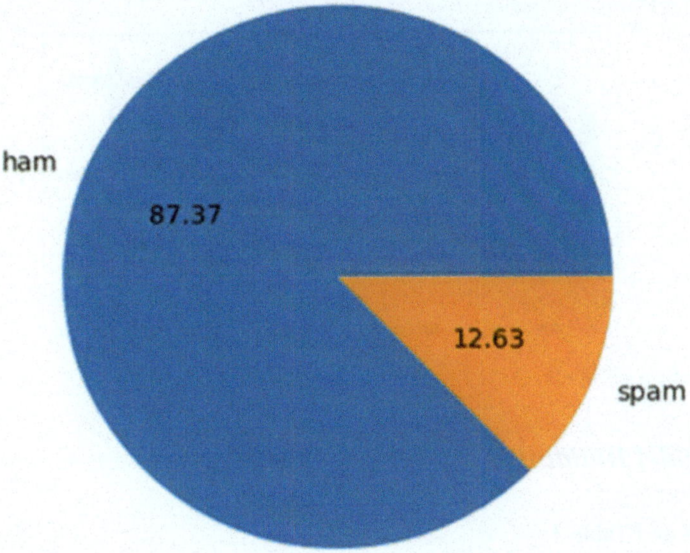

Fig. 5 Exploratory data analysis

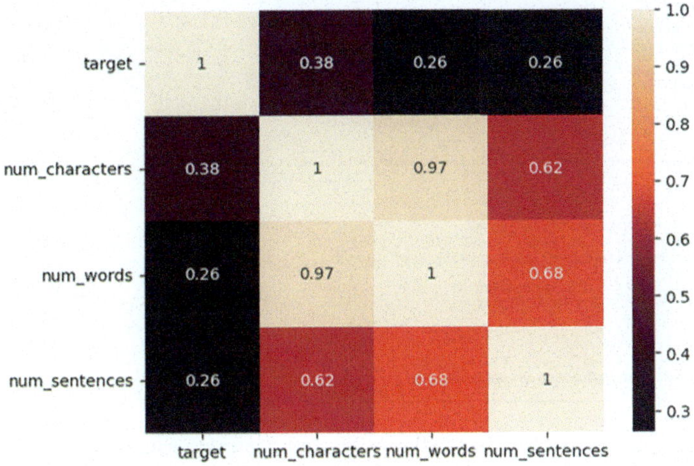

Fig. 6 Heat map for sentences, words, and characters

- Letter cases
- Tokenization
- Removing specific characters
- Removing stop words
- Punctuation stemming.

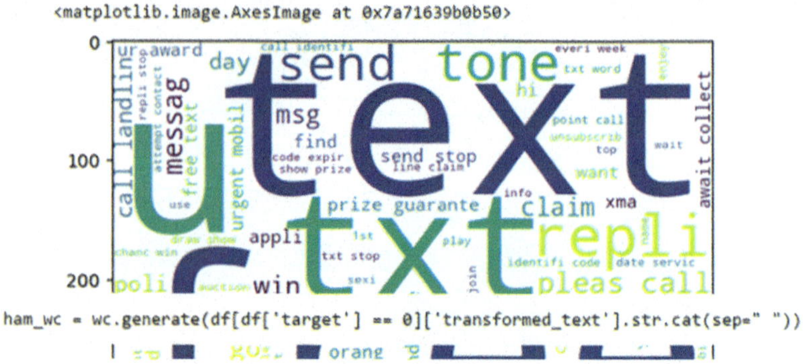

Fig. 7 Mostly used words

4.5 *Model Building*

See Fig. 8 and Table 1.

```
# Voting Classifier
bnb = BernoulliNB()
dtc = DecisionTreeClassifier(max_depth=5)
lrc = LogisticRegression(solver='liblinear', penalty='l1')

from sklearn.ensemble import VotingClassifier
voting = VotingClassifier(estimators=[('bnb', bnb), ('dtc', dtc), ('lrc', lrc)],voting='soft')

voting.fit(X_train,y_train)

            VotingClassifier
   bnb              dtc                      lrc
 BernoulliNB   DecisionTreeClassifier   LogisticRegression

y_pred = voting.predict(X_test)
print("Accuracy",accuracy_score(y_test,y_pred))
print("Precision",precision_score(y_test,y_pred))

    Accuracy 0.9729206963249516
    Precision 0.9824561403508771

# Applying stacking
estimators=[('bnb', bnb), ('dtc', dtc), ('lrc', lrc)]
final_estimator=RandomForestClassifier()
```

The combination of LR and DT provides an accuracy rate of 97.2%. Finally, the experimental results show that hybrid machine learning algorithms and individual machine learning predict the same optimal accuracy results. Few algorithms only generate a lower accuracy rate. It is also based on a dataset, that is one reason. Our experiment shows a more than 97% accuracy rate.

```
performanceS
```

	Algorithm	variable	value
0	NB	Accuracy	0.970986
1	DT	Accuracy	0.930368
2	LR	Accuracy	0.958414
3	AdaBoost	Accuracy	0.960348
4	BgC	Accuracy	0.958414
5	ETC	Accuracy	0.974855
6	GBDT	Accuracy	0.946809
7	xgb	Accuracy	0.967118
8	NB	Precision	1.000000
9	DT	Precision	0.817308
10	LR	Precision	0.970297
11	AdaBoost	Precision	0.929204
12	BgC	Precision	0.868217
13	ETC	Precision	0.974576
14	GBDT	Precision	0.919192
15	xgb	Precision	0.926230

```
sns.catplot(x = 'Algorithm', y='value',
            hue = 'variable',data=performanceS, kind='bar',height=5)
plt.ylim(0.5,1.0)
plt.xticks(rotation='vertical')
plt.show()
```

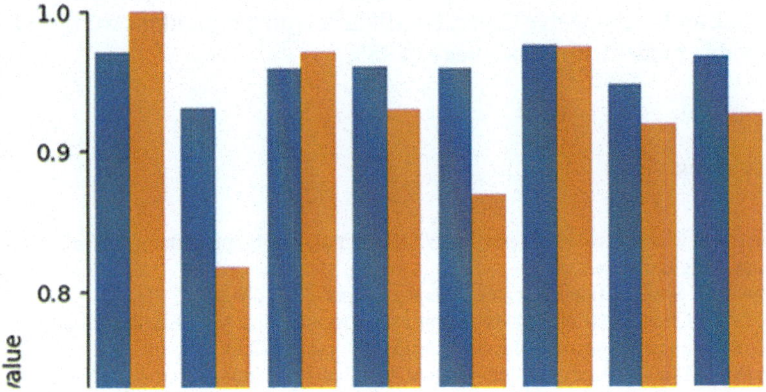

Fig. 8 Bar chart for performances

Table 1 Performances of algorithms

	Algorithm	Accuracy	Precision	Accuracy_Scaling_x	Preci-sion_Scaling_x
0	NB	0.970986	1.000000	0.970986	1.000000
1	DT	0.930386	0.817308	0.930386	0.817308
2	LR	0.958414	0.970297	0.958414	0.970297
3	AdaBoost	0.960348	0.929204	0.960348	0.929204
4	BgC	0.958414	0.868217	0.958414	0.868217
5	ETC	0.974855	0.974576	0.974855	0.974576
6	GBDT	0.946809	0.919192	0.946809	0.919192
7	XGB	0.967118	0.926230	0.967118	0.926230

5 Conclusion

Regardless of the topic on which the conversation is based, this web application can also be used to detect spam messages. This web application displays all messages, in the form of a graph every month, as well as the frequently used words and the active participant's conversation. This web application also detects spam messages by a person and shows the total spam messages in a particular chat. The benefit of this tool is that it has been developed using straightforward Python modules like Pandas, Streamline, Seaborn, and Word Cloud, as well as machine learning algorithms that are used to create data frames and generate various graphs that are then displayed in the application. These machine learning algorithms are effective and resource-light, making them suitable for large data sets such as SVM, Naïve Bayes, logistic regression, and decision tree. Hybrid machine learning algorithms were used to form an ensemble model with an accuracy of 98%.

References

1. https://www.analyticsvidhya.com/blog/2021/06/build-web-app-instantly-for-machine-learning-using-streamlet/
2. https://www.interaction-design.org/literature/topics/web-design
3. Ho J, Ji P, Chen W, Hsieh R (2009) Identifying Google Talk. In: IEEE international conference on intelligence and security informatics, ISI '09, pp 285–290
4. Dickson M (2006) An examination into AOL instant messenger 5.5 contact identification. Digit Investigation 3(4):227–237
5. Spam SMS Dataset2011-12. Available http://precog.iiitd.edu.in/requester.php?dataset=smsspam
6. Shiva Keertan J, Shaik S (2019) Machine learning algorithms for oil price prediction. Int Innov Technol Explor Eng 8(8)
7. Surya Teja KP, Reddy V, Shaik S (2019) Flight delay prediction using machine learning algorithm XGBoost. J Adv Res Dyn Control Syst 11(5)
8. Reddy S, Sri R, Shaik S (2021) Sentimental analysis using logistic regression. Int J Eng Res Appl (IJERA) 11(2)

9. Mamatha, Datta S, Shaik S (2021) Fake profile identification using machine learning algorithms. Int J Eng Res Appl (IJERA) 11(2)
10. Vijaya Kumar Reddy R, Shaik S, Srinivasa Rao B (2021) Machine learning based outlier detection for medical data. Indonesian J Electr Eng Comput Sci 24(1)
11. Vijayalaxmi K, Shaik S, Pramod B, Ramulu B, Sankeerth P (2022) Vegetable price prediction against temperature changes using machine learning techniques. Dickensian J 22(6)
12. Singh M, Bhutada S, Vijayalaxmi K, Shaik S (2022) Online social networks fake news on covid-19 in the South India Region. Neuro Quantol 20(10)
13. Vijayalakshmi K, Shaik S (2023) Predicting employee attrition methodologies of K-fold technique. Int J Math Sci Comput 1:23–36

Prediction of Distinctive Human Behavioral Activity Using PIFP and Big Data Methods

Maddali Aravind, Gundala Nageswara Rao, B. Vijay, and B. Ramesh

Abstract Urban areas are shifting more and more these days. One of the hardest areas to deal with, given the heavy traffic in town centers, is the healthcare sector. As a result, practically every city on the planet is investing in creating a better ecosystem. Devices like as sensors and clever motors are used in those conversions. The sensors that produce enormous amounts of accurate data are analyzed for services in smart towns. The goal is to develop a model that makes advantage of smart devices by analyzing how people use them. Additionally, to determine and use the energy needed by the behavior, group investigations and forecasts are used successively. Since the majority of users are primarily recognized by their daily routines, identifying these routines enables the recognition of irregular movement, which will reveal people's issues managing themselves. Examples of such challenges include forgetting to prepare meals or take a shower or bath. Our research focuses on the need to create life patterns at the level that are directly related to human activity. The analysis makes use of UK Domestic Appliances to examine the suggested instrument. Level Electricity Informational Index (UK-Dale): factual data on intensity consumption collected for five residences in Southern England with 116 apparatuses between 2011 and 2018 with a time aim of six seconds. Additionally, profits are maintained through efficient mining operations. The massive amounts of data generated by the smart houses are managed by the Hadoop biological system, which in turn powers the distributed preparation. The output of unique patterns of human activity from the use of appliances, as well as the precision of both short- and long-term forecasts.

M. Aravind (✉)
Department of MBA, Narasaraopeta Engineering College, Narasaraopet, Andhra Pradesh 522601, India
e-mail: maddaliaravind@gmail.com

G. N. Rao
Department of CSE (Data Science), Narasaraopeta Engineering College, Narasaraopet, Andhra Pradesh 522601, India

B. Vijay · B. Ramesh
Department of Computer Science and Engineering, Aditya Institute of Technology and Management, Tekkali 532201, India

© The Author(s), under exclusive license to Springer Nature Singapore Pte Ltd. 2025
H. Zen et al. (eds.), *Soft Computing and Signal Processing*, Lecture Notes in Networks and Systems 1200, https://doi.org/10.1007/978-981-97-9926-8_12

Keywords Big data · Behavioral analytics · Frequent pattern · Hadoop · Health care · Incremental data-mining · Prediction · Smart homes

1 Introduction

There used to be a lot of data being collected for a frightening variety of practically medical professions. Up to 2021, the high-quantity information should ideally come from 35 to 40 zettabytes. As a result, cloud data centers face significant challenges in organizing and processing information. As a result, it may be crucial to access cloud data and wellness regions [1].

Currently, a data center needs a variety of research methodologies to transform massive amounts of data storage into efficient processing. Research indicates that approximately 66% of people will reside in urban areas by the year 2060. This amazing influx of individuals to downtown areas will have a significant impact on the need for health resources.

The method of using smart meters to provide vitality data in houses. The workout at home is particularly related to the disaggregated power usage readings. For instance, the activity of "getting ready food" is almost certainly associated with the operation of the "Stove" if it is turned ON [2]. The type of feast—breakfast or dinner, for example—may also be indicated by the time of the operation—morning or night. Additionally, people usually multitask, such as "preparing food" and "listening to music" or, conversely, "gazing at the television," indicating that various devices are operating simultaneously. It is particularly possible without the need for additional equipment because the data from smart meters typically includes time-arrangement ideas that include segment appliance designs and utilization estimates throughout time. However, such an attempt is highly taxing because it is challenging to determine the usage conditions of several appliances at the moment when their operations overlap or occur in the interim.

The UK Domestic Appliance-Level Electricity dataset (UK-Dale)—time-arrangement information of energy utilization collected from 2012 to 2019 alongside time determining six seconds for five houses with 109 machines via Southern England—is used in this exam for the assessment of the proposed component. It should be noted that the Non-Intrusive Appliance Load Monitoring (NALM) technique eventually conveys stack disaggregation [3].

Research indicates that by 2060, a majority of the population—66%—will live in affluent areas. The massive influx of people into downtown regions will have a significant impact on the availability of healthcare services. The remarkable shift in statistics puts a great deal of pressure on metropolitan areas to reconsider the traditional approaches to providing residents with health services. Urban communities are currently embracing a massive digital change in response to new demands and challenges, with the goal of creating viable urban groups. Many residences in this area are being fitted with clever devices (such sensors and keen meters). As part of this shift, a vast number of houses are outfitted with smart devices [4] (smart meters, sensors, and the like), which contain enormous amounts of indexical and fine-grained data that may be deconstructed to assist healthcare services. The development of big data-mining tools, which offer ways to prepare vast amounts of data for important insights, can assist us in understanding how different people live their lives. At all people often behave according to normal routines, so identifying these patterns helps us identify odd behaviors that can indicate people struggle to take care of themselves, like forgetting to prepare food or take a shower. Health applications can leverage the basic relationship between regular activity and the use of appliances in smart homes to identify possible medical problems. Not only will this lessen the burden on human services systems, but it will also provide patients with self-limiting conditions (older adults and patients with intellectual disabilities) with round-the-clock monitoring services that automatically identify normal and abnormal behaviors.

2 Literature Review

In this research, an algorithm for recognizing household activities from aggregated non-intrusive smart meter data is proposed. We identify two categories of activities: Type I activities are identified only by the data from smart meters, and Type II activities are identified by combining the data from smart meters with the most basic environmental sensing (humidity and temperature). We begin by breaking down the overall power use for both kinds of activity into the specific electrical equipment. Next, we use the Dempster-Shafer theory of evidence to construct an indicative activity model that explains four household activities [5]. In this study, we suggest a game-theoretic method that, in deregulated smart grids, strikes a balance between people' privacy and data benefits. By appropriately rewarding consumers for their involvement in the data market, our approach addresses the issue of access restriction by utilizing the concept of differentiated privacy. The outcomes of our research demonstrate how critical it is to consider consumers' privacy attitudes while creating fair and balanced markets for data sharing [6]. The issue of maintaining privacy in computer programs is discussed in this article along with how it relates to multi-agent systems and autonomous agents. It also lists open challenges that need to be solved through future research and reviews studies in the subject of multi-agent systems that deal with privacy [7]. Based on letting users profit from the value of their personal data and receive something of value in exchange, this study suggests a business privacy model for online communities. To calculate the consumers' payoff, the suggested business model uses a risk-based premium, a game-theoretic negotiation protocol, and a privacy risk quantification technique [8]. In this section, we introduce UK-Dale, an open-access dataset from the UK that records domestic appliance-level electricity at a sampling rate of 1/6 Hz for individual appliances and 16 kHz for the entire house. The UK dataset at this temporal resolution is the first to be made publicly available. We collected data from five different households, the longest length we could find for any energy dataset at this sample rate being 655 days, from one of the records. We also provide details on the open-source, low-cost wireless technology we developed to gather our dataset [9]. Novel approaches to distinguishing designs for applications that monitor users' well-being have been identified recently. Understanding and anticipating tenants' behaviors that indicate medical issues is the first goal when it comes to their behavior [10]. In this section, we review previous research in the literature that uses data from smart homes to analyze customer behavior.

3 Data Analysis

The method that is being demonstrated starts with sorting and then sets up the data that is available to locate appliance-to-appliance connections, or determining what appliances are in use.

To be specific, it makes use of group research to establish connections between gadgets and time.

After completing the aforementioned steps, the framework is prepared to extract the point that was used, since the data provided is the network of actions associated with Bayesian terms.

A specific healthcare services application's yield to framework depends on its intended use. For instance, the distribution process may have been influenced by exercises linked to psychological impediment, wherein adhering to the daily exercise regimen is recognized to serve as a reminder when abnormal behavior is detected.

A prototype finds the different forms of interaction for apps related to medical services using devices. Given that each tenant is different, it is possible to discover those timetables to be feasible in order to improve strange activities that could reveal a tenant's problems by selecting self-care.

The stated method aims to analyze middle-level temporal patterns, which are specifically related to human activity.

4 System Design

The tactics listed below are aimed at analyzing the energy consumed in behavior to forecast health issues by predicting household information.

- MapReduce-based parallel incremental frequent pattern mining
- K-means cluster analysis
- Bayesian network activity prediction.

Figure 1 illustrates the procedure flow for identifying patterns and activities in the healthcare industry. The UK-Dale dataset is used to extract pertinent features for the preparation of the dataset, which is then transformed by removing tabs and null entries. The ready data is then put into HDFS. The relationships between appliances and time are established for the morning, evening, night, and afternoon. The framework may distinguish between the two improvements mentioned above, which show

how a computer can be utilized to perceive exercises both in the short and long term. Applications for explicit medical services that depend on the intended use leverage the framework's yield. A health adviser, for instance, would only be interested in learning about cognitive impairment activities where monitoring the usual activity sequence is essential for prompting when abnormal behavior is noticed. To examine interim and final results, experiments are currently conducted using the UK-Dale dataset in addition to the created dataset. The dataset (UK-Dale) comprises data series from 2012 to 2015. The dataset, which is provided by UK Energy Research Centre Energy Data Centre (UKERC-EDC), includes time-arrangement data for five residences in Southern England that together house 109 machines with a period aim of six seconds. Approximately one billion records make up this collection, making it one of the largest. Module singular apparatus monitors (IAMs) were used to estimate energy consumption at the machine level.

The data for the suggested model is stored in HDFS on an Ubuntu 14.04 LTS 64-bit system, and the underlying framework is built in Hadoop. The tests' main objective is to identify the usage of equipment as an indicator of human behavior patterns and use the expected model to hypothesize short- and long-term workouts that take place inside the home.

This implies that our model can provide features like alarm generating, health profiling, active monitoring, and so on for a healthcare application.

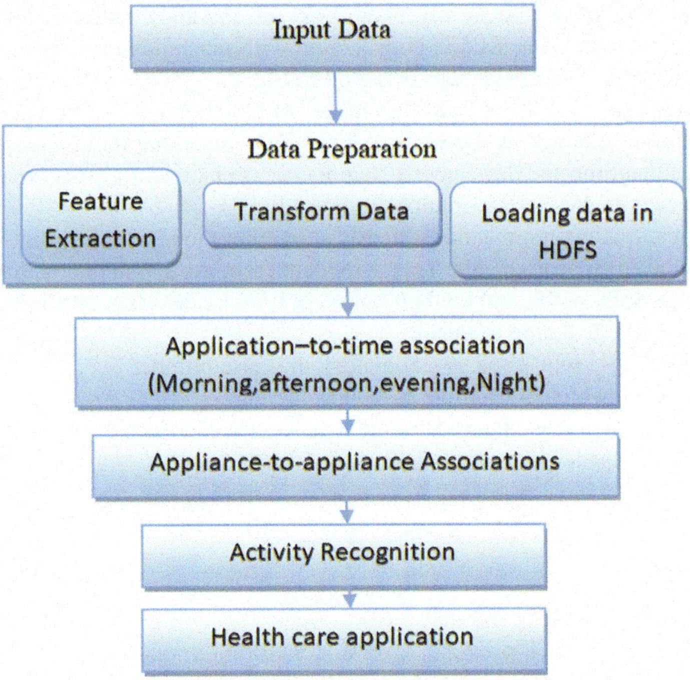

Fig. 1 Architecture for patterns and activity recognition for healthcare

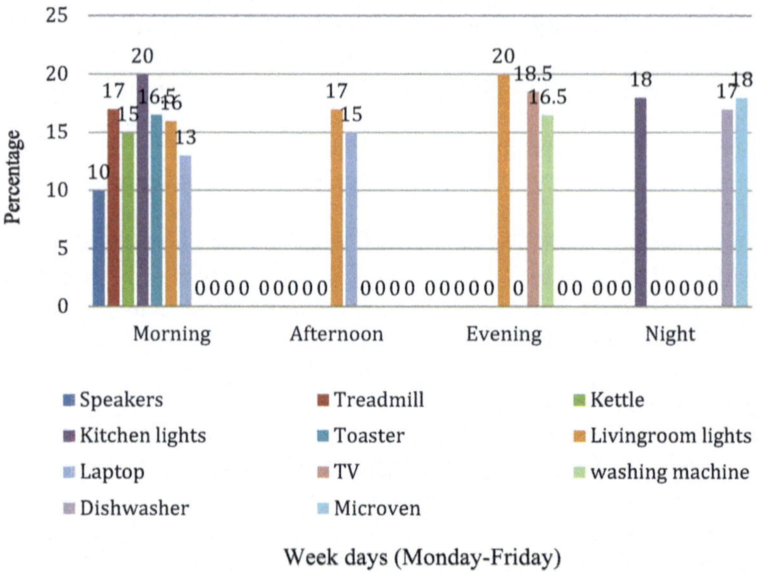

Fig. 2 Time associations hour of the day

Distinct associations, but non-weekend days, weeks, and months are divided similarly. This information is used in conjunction with machine affiliations to perceive the activities inside the residences depicted in Fig. 2.

It is evident that the TV, treadmill, toaster, and kitchen are used in tandem in this home in the morning, with the non-weekend days receiving the greatest attention. We can see that time and days have different effects on how machines are used when we take these realities into account.

The apparatus interconnections clearly show that the residents of house 1 enjoy relaxing while preparing meals. This is evident from the stable interaction between the TV, amp, subwoofer, and kitchen lights. In house 2, residents wash their clothes on a PC or while listening to music.

Likewise, those who cook or do laundry in house number five. Figure 3 depicts these behaviors, which are routine behavioral features.

These are only a few examples of the activity that is detected when irregularities differ from the usual.

5 Results

The expectation model predicts several concurrently operating apparatuses by utilizing the machine-apparatus and apparatus time relationship. Strong relationships between appliance houses are simple to establish. The metrics that are employed

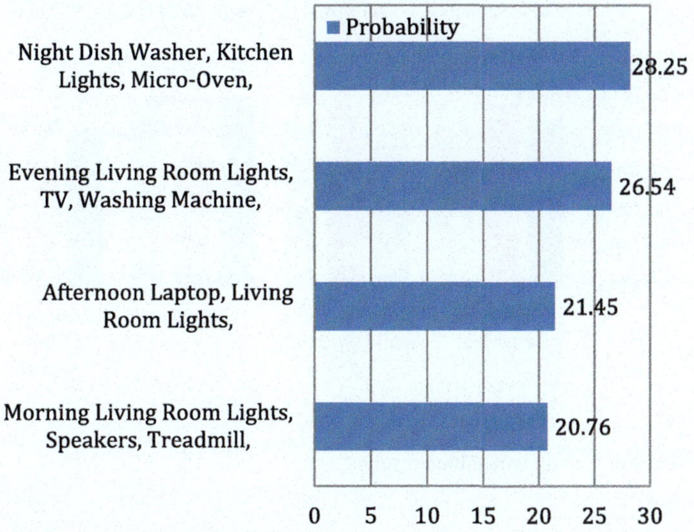

Fig. 3 Examples of activity recognition results

include *F*-Measure (FM), recall (RE), and precision (PR).

$$PR = \frac{TP}{TP + FP} \tag{1}$$

$$RE = \frac{TP}{TP + FN} \tag{2}$$

$$FM = 2 * \frac{PR * RE}{PR + RE} \tag{3}$$

As seen in Table 1, true positive (TP) shows that the activity is carried out correctly, false positive (FP) signifies an inaccurate assertion that the action is carried out, and false negative (FN) suggests the activity is carried out but not identified.

Two months' worth of training data and one month's worth of testing data make up these outputs. Based on the data treatment, it can be noticed that 89.57% of the breakfast preparation activities in Fig. 4 were accurately detected by the new method.

Table 1 Activity performed

Activity	TP	FP	FN
Preparing break fast	52	7	0
Doing laundry while watching TV	26	8	1
Relaxing	166	0	0
Listening music while doing exercise	44	9	16

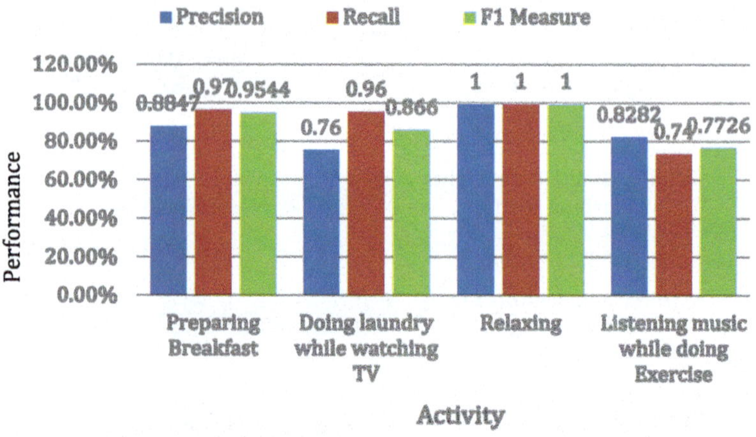

Fig. 4 Accuracy of activity recognition results

All events were correctly detected, even though the approach was particularly precise in depicting relaxing activities.

6 Conclusion

Selecting a design objective information about intelligent meters has been made available. The main goal of the research is to determine how machine usage indicates instances of human behavior and to assess both short- and long-term exercises conducted within the home using the expectation model. The system-to-machine and apparatus-to-time linkages eliminated from subsequent group research and example mining are crucial cycles to monitor patients' and individuals' schedules and maybe provide health care when necessary. The model is improved to include massively aggregated data mining from several residences in a nearly continuous fashion. The model is improved to include near-real-time big data-mining knowledge from several residences.

References

1. United Nation (2014) World urbanization prospect. United Nation. [Online]. Available: http://dl.acm.org/citation.cfm?id=308574.308646
2. Hossain MS (2017) Cloud-supported cyber-physical localization framework for patients administrating. IEEE Syst J 11(1):118–127
3. Hossain MS, Muhammad G, Abdul W, Song B, Gupta B (2017) Cloud-assisted security video transmission and sharing framework for smart cities. Future Gener Comput Syst J

4. Istepanian RSH, Al-Anzi T (2018) M-Health 2.0: unique perspectives on mobile health, machine learning and big data analytics. Methods 151:34–40
5. Liao J, Stankovic L, Stankovic V (2014) Detecting household activity patterns of the system from smart meter data. In: 2014 international conference on intelligent environments (IE), pp 71–78
6. Yassine A, Shirehjini AN, Shirmohammadi S (2015) Smart meters big data: game theoretic model for fair data sharing in deregulated smart grids. IEEE Access 3:2743–2764
7. Yassine A, Shirmohammad S (2009) Measuring users' privacy payoff using intelligent agents. In: 2008 IEEE international conference on computational intelligence for measurement systems and applications, May 2009, pp 166–174
8. A business privacy model for virtual communities. Int J Web Based Communities 5 (2009)
9. Jack K, William K (2015) The UK-DALE dataset, domestic appliance-level electricity demand and whole-house demand from five UK homes. Sci Data 2(150007)
10. Clement J, Ploennigs J, Kabitzsch K (2014) Detecting activities of daily living with smart meters. Springer, Germany, chap Technology Change, pp 143–160. [Online]. Available: https://link.springer.com/chapter/10.1007/978-3-642-37988-8_10

A Signals–Systems Perspective on Electronic Structure Theory

Aakash Yadav, Daniel Hedman, and Hongsik Jeong

Abstract The theoretical framework that is used to predict the properties of various materials and, furthermore, also describe the electronic behavior inside the matter is coined as the Electronic Structure Theory (EST). The orthodox way of arriving at solutions from this theoretical perspective to these problems requires discrete computation at its core to solve the Schrödinger equation. In this work, a perfect crystalline material is treated as Linear and Translationally Invariant (LTI) system, thereby providing a new perspective to the well-established EST. This new perspective is validated in this work by arriving at the band structures and plotting the same for a more complex scenario of a 2D material MoS_2. In addition to the band structures for the primitive cells of these respective cases, it is also demonstrated herein how the folding of these band structures in super cells can also be achieved through this perspective. Toward the end of the paper, advantages offered by the new perspective are also brought to light. To be precise, the spiking-domain computing nature of the proposed LTI-EST formalism is underlined; not only this, the computational advantage offered by the new implementation is also presented by bypassing the $O(N)$ eigenvalue calculations. Additionally, it is also realized in this work that the process of band structure folding in super cells can be analytically traced down using the proposed formalism. Hence, this work hints at the possibility of performing materials simulation completely in the spiking-domain which would be very essential for neuromorphic computing hardware.

A. Yadav (✉)
LGECA/SCCS/DCOS, CEA-Leti, 38000 Grenoble, France
e-mail: aakash.yadav@rwth-aachen.de

D. Hedman
Center for Multidimensional Carbon Materials, Institute for Basic Science, Ulsan, Korea

A. Yadav · H. Jeong
Graduate School of Semiconductor Materials and Devices Engineering, UNIST, Ulsan, Korea

A. Yadav
Tata Electronics, Gujarat, India

© The Author(s), under exclusive license to Springer Nature Singapore Pte Ltd. 2025
H. Zen et al. (eds.), *Soft Computing and Signal Processing*, Lecture Notes in Networks and Systems 1200, https://doi.org/10.1007/978-981-97-9926-8_13

Keywords Linear and Translationally Invariant (LTI) system · Materials simulation · Electronic Structure Theory (EST) · Spiking Neural Networks (SNNs) · Neuromorphic computing

1 Introduction

In understanding device physics, the electronic behavior in materials holds a very crucial position. In turn, a very critical role is played by the Electronic Structure Theory (EST) in describing this behavior. As an example, the relationship between the minimum of the first conduction band and the effective mass of electrons can be deduced from the electronic band structure (obtained using the calculations from this theory)—this is so because the narrower this first conduction band minimum, lower would the effective mass of the electrons resulting in higher mobility of these electrons. There are several methods which have been developed to calculate such electronic band structures. One of such methods is the tight binding (TB) theory which, on the one hand, relies on first-principles calculations or experimental data to parametrize the TB-Hamiltonian; on the other hand, first-principles methods don't require parametrization to arrive at the electronic structure of a given material.

Approaches based on neural networks have lately received a lot of attention for plenty of problems across a broad spectrum; materials science is no exception [1]. One particular application of such sort is to parametrize the TB-Hamiltonian [2]. Interestingly, when a closer look is taken at these deep learning models, the importance of the concepts of signals and systems is realized. For instance, the core operation driving the convolution neural networks (CNNs) is the convolution integral which takes its origin from the Linear and Translation Invariant (LTI) systems [3]. Also well reported in literature is the usage of such CNNs to achieve performance at per with humans [4]. At the same time, the power consumption of such large neural networks has also been reported to be orders of magnitude higher than the brain [5], thereby highlighting the importance of spike-based computing and its connection with neuromorphic (or brain-inspired) computing.

Schuman et al. have highlighted that the development of neuromorphic hardware needs to be assisted by the development of algorithms specific to such hardware [6]; while particularly underlining the need for adaptation of conventional algorithms to spike-based computing to make them compatible with neuromorphic hardware which promise to offer ultralow power consumption, thanks to this spike-based computing. In [7], more recently, Date et al. have highlighted the Turing completeness of neuromorphic computing; which essentially implies that the algorithms for such computers need not be just machine learning specific but instead should be capable of performing all sorts of computation that our computers today can perform while consuming much lesser power than today's state-of-the-art. This necessarily also implies that materials simulation algorithms shouldn't be an exception either. Aimone et al. have additionally brought to light that neuromorphic computing aims to cater many non-cognitive applications while especially specifying signal processing as one such application [8].

A new approach based on LTI systems to solve EST problems is proposed in this work. As described in detail in [9], the derivation and adaptation of the orthodox perspective to the proposed formalism is presented for a simpler case of 1D chain of atoms therein. The achievement of band structures and the analytical tracing of the folded band structures for super cells is presented in [9] for the 1D chain of atoms besides the computational and algorithmic advantage offered. In this work, the equivalence of the LTI-EST formalism is illustrated by determining the band structure for a more complicated case of 2D MoS_2. The phenomenon of band structure folding is also presented and described in detail herein. In existing literature, examples of derivation of band structures of materials from first-principles [10] exist. On the other hand, a handful examples depicting importance of concepts from signals–systems for unfolding band structures have also been put forth [11]. Yet, no work has so far exploited the LTI formalism to the extent presented in this work and neither has an attempt been made to adapt the orthodox method to the spiking-domain using these concepts. In this LTI-EST formalism, a perfectly crystalline material is considered as an LTI system. A new way to derive band structures can be developed based on this insight that this LTI system is characterized by its impulse response. Once this impulse response is known (analogous to the Hermitian Hamiltonian), achieving band structures for super cells in this new formalism simply equates to changing the input signals (encoded in spikes) and convolving it with this impulse response. The realization that Fourier transform diagonalizes the convolution operator makes this perspective computationally efficient through bypassing the matrix diagonalization.

2 The LTI-EST Formalism

Just as in [9], "the proposed LTI-EST formalism is derived from the Schrödinger equation with energy eigenvalue, E, and the corresponding Eigen function, ψ, with H being the Hamiltonian operator"

$$H\psi = E\psi \rightarrow \left(U - \frac{\hbar^2\nabla^2}{2m^*}\right)\psi = E\psi. \tag{1}$$

Using the calculations presented in the arXiv pre-print of this work [9] which details out the implementation of this perspective for a 1D chain of atoms, Eq. 1 can be solved further to arrive at the following Eq. 2:

$$\sum_m \left[-t_0\delta_{n,m+1} + (2t_0 + U_n)\delta_{n,m} - t_0\delta_{n,m-1}\right]\psi_m = E\psi_n \tag{2}$$

It must be noted that the equation above can be written as a circulant matrix and owing to the equivalence between circulant matrix and convolution operator [12], the

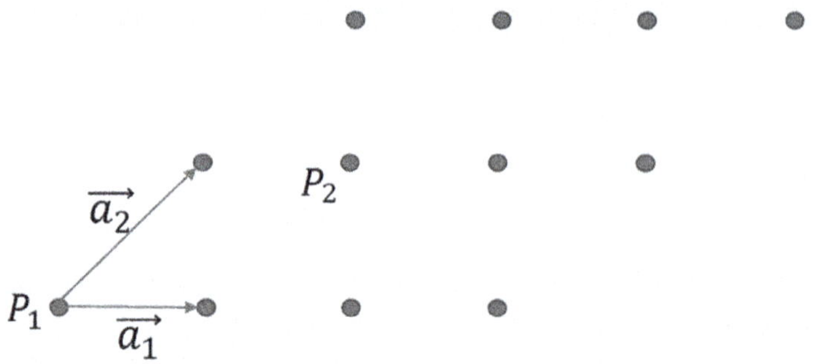

Fig. 1 In this perfectly crystalline lattice in 2 dimensions, the dots represents lattice sites, P_1, P_2, and the arrows represent lattice vectors, $\overrightarrow{a_1}$, $\overrightarrow{a_2}$

calculations in the pre-print [9] can be followed based on which, the conditions of linearity as well as translational invariance can be proven for the perfect crystalline lattice (as shown in Fig. 1), and the LTI-EST formalism can be derived starting by rewriting Equation S2 (in Supplementary Text of [9]) as

$$\begin{aligned} E(k) &= \sum_{R''} e^{j\vec{k}\,(\overrightarrow{R''})} \int \phi_s^*(x) H \phi_s(x - R'')(dx) \\ &= \sum_{R''} e^{-j\vec{k}\,(\overrightarrow{R''})} \int \phi_s^*(x) H \phi_s(x + R'')(dx) \end{aligned} \tag{3}$$

The sign change in Eq. 3 should be observed carefully. The right-hand side of Eq. 3 is a Fourier transform of a cross-correlation operation. This cross-correlation, in turn, is operated upon the atomic orbital itself and the resultant of Hamiltonian applied atomic orbital. It is also well understood that the result of convolution and correlation is same if the kernel is symmetric. That is, flipping it doesn't cause any difference and both convolution and correlation yield the same output.

3 2D Monolayer MoS_2

In [9], the demonstration of this formalism on 1D chain of atoms has already been presented. In this work, this depiction is presented for a more complex case of a two-dimensional material, namely Monolayer Molybdenum Disulphide MoS_2. This is a very interesting material for research in present scenario spanning from new two dimension channel materials to neuromorphic computing hardware devices.

This formalism requires visualizing the unit cell (primitive or super cell) of the concerned material as an LTI system. As discussed in the previous section, this LTI system is characterized by its impulse response. Thus, the first step to proceed forward with implementation of this perspective on this new material is to achieve the impulse response. In [9], the impulse response was a 1D vector convolving which with the impulse function (primitive cell) yielded the impulse response itself, thus justifying the definition too. This impulse response, in addition, was formed using the on-site and hopping parameters which accounted for only nearest neighbors interactions. In this work, this impulse response has to be 2D and needs to account for not only the nearest neighbors but also second-nearest neighbors interactions. Similar to the 1D case, the 2D matrix would be composed of the on-site energy and the hopping parameter. It also needs to be made sure that convolving this 2D impulse response with the 2D impulse function (primitive cell) would yield the impulse response itself as a sanity check (just like the 1D case in [9]). These impulse functions discussed above should be interpreted as unit cells with the Dirac delta placed at the center of these unit cells. Henceforth, the unit cells (primitive or super cells) will be considered as input signals in space coordinates and the resulting band structures will be obtained as the response of the LTI system.

It is now discussed herein how this proposed formalism can be implemented on a 2D material MoS_2. Particularly, in Fig. 2, the application of this perspective on this particular case is discussed. To be precise, in Fig. 2a, the energy band structures with almost 8000 k-points in the 2D material are plotted. Intuitively, owing to the strong dependence of the first conduction band minimum to the electron mobility (as discussed earlier), the eighth conduction band is then taken into consideration primarily. Being a 2D material, the plot of this conduction band against k_x and k_y is plotted in Fig. 2b. The lattice vectors of the material are also presented in Fig. 2c. Next, the plots for this conduction band are presented in Fig. 2e along the Transport direction [shown in Fig. 2d] and in Fig. 2g along the Transverse direction [shown in Fig. 2f]. It can be seen that the band remains same along both these directions; thus, effectively meaning that the unit cell can be replaced with an effective atom.

This next step is essentially what is presented in Fig. 3b whereby for the application of this perspective on the given case, the simplification of the unit cell is depicted. Owing to the fact that the band structure remains same for the first conduction band minimum along both (Transport and Traverse) directions, a single spike is capable of demonstrating the effective primitive cell in a 2D space, and hence, a single spike is sufficient to replicate it as shown in the Figure with the depicted frame of reference. The electronic energy band structures along k_x and k_y are also presented in Fig. 3a.

3.1 *Primitive Cell with Lattice Constant a*

As discussed in the text so far, it will now be shown as to how the proposed formalism can be practically implemented on the discussed example. A 2D δ function, located at the center of the primitive cell (PC), emulates the system, that is the monolayer 2D MoS_2. The frame of reference is chosen such that its origin coincides with the location of the δ function (i.e., center of the PC); as has been demonstrated pictorially for the ready reference of the readers in Fig. 3b. Hence, mathematically speaking, the primitive cell (PC) is formulated as

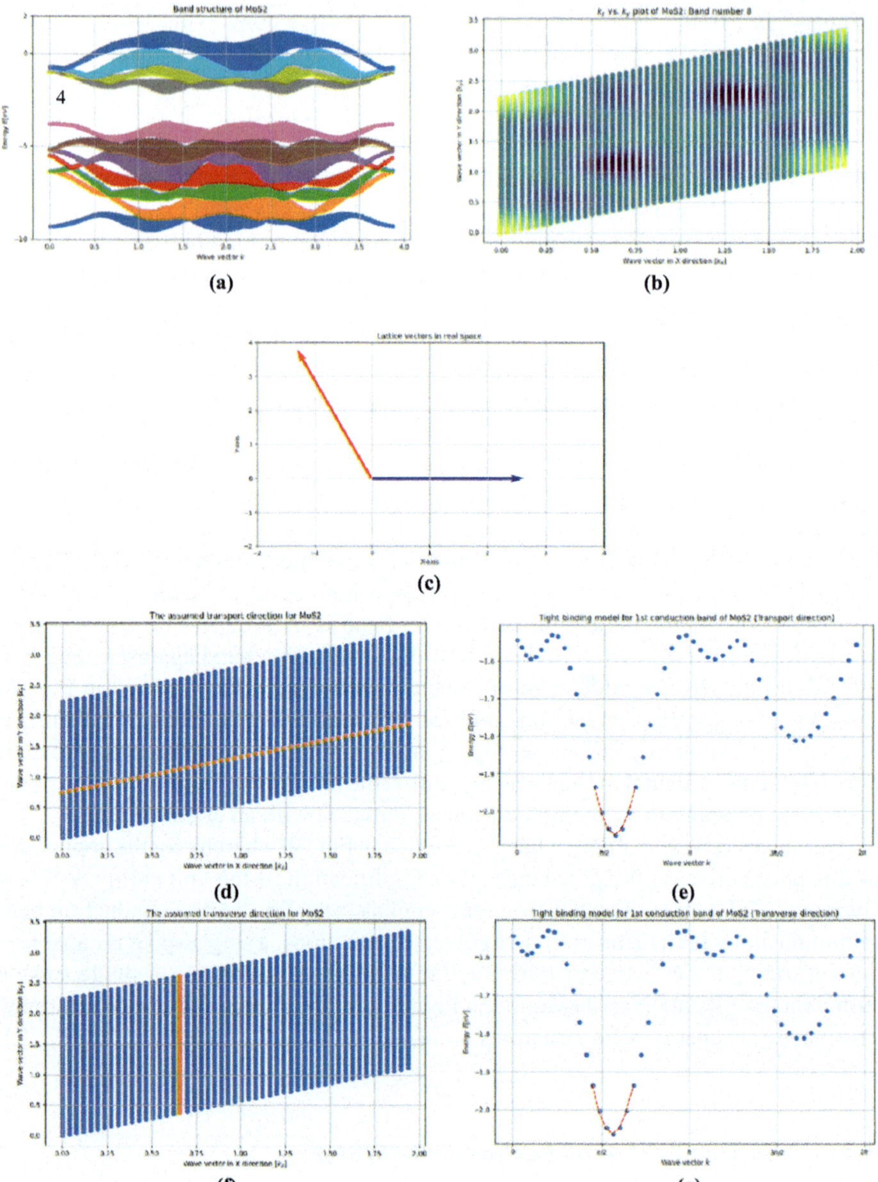

(a) (b) (c) (d) (e) (f) (g)

Fig. 2 The band structures are presented in (**a**). **b** presents the first conduction band minimum. **c** depicts the lattice vectors. **d** and **e** present the Transport and Transverse directions in the *k*-space and **f** and **g** present the first conduction band minimum for these, respectively

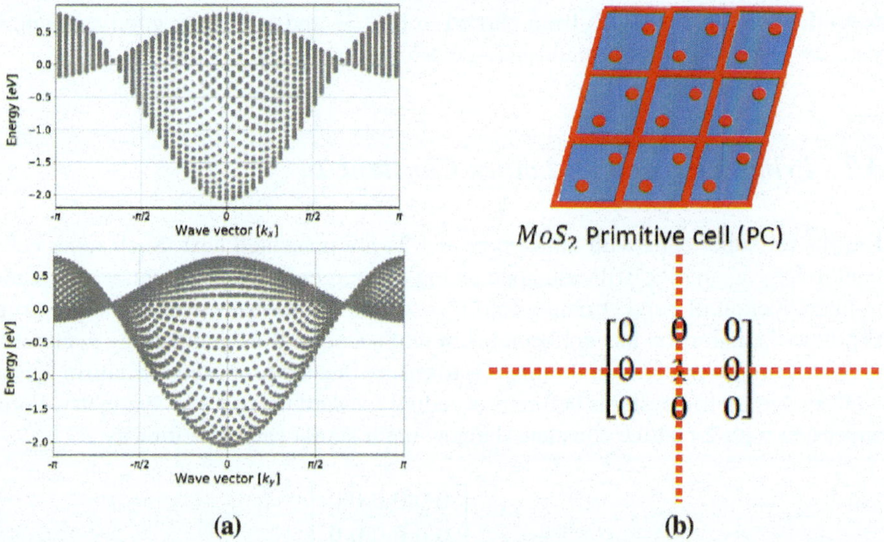

Fig. 3 **a** The band structure for the case of 2D MoS_2 derived using the proposed LTI-EST formalism. The upper inset plots the energy eigenvalues against k_x & below one against k_y. **b** The unit cell is shown on top inset, while the below one shows its corresponding input signal along with the frame of reference, written in terms of $\delta[\mathrm{x},\ \mathrm{y}]$

$$\delta[x, y] = \begin{bmatrix} 0 & 0 & 0 \\ 0 & 1 & 0 \\ 0 & 0 & 0 \end{bmatrix}.$$

In addition to the work presented on 1D chain of atoms in [9], not only the interactions of the nearest but also the second-nearest neighbors are accounted for in this paper (thus, implying the flexibility offered by this perspective). α denotes the on-site energy, whereas β denotes the hopping parameter for such neighbors in this case. This is the junction between the orthodox and the proposed formalism as the parameters from the tight binding theory are being made to formulate the impulse matrix for calculations in the LTI-EST formalism. The resulting impulse response matrix $h[x, y]$ is then written in Eq. 4.

$$h[x, y] = \begin{bmatrix} \beta & \beta & \beta \\ \beta & \alpha & \beta \\ \beta & \beta & \beta \end{bmatrix} \tag{4}$$

Upon taking Fourier transform of the Eq. 4, $F(h[x, y]) = H[k_x, k_y] = \alpha + \beta(2\cos(k_x a) + 2\cos(k_y a) + 4\cos(k_x a)\cos(k_y a))$, the result can be interpreted as the Green's function in the reciprocal space [9, 13]. The resulting band structure from these calculations has been plotted in Fig. 3a and is in very good agreement with the result obtained from discretized calculation.

3.2 Primitive Cell with Lattice Constant 2a

Results for super cells with lattice constant $2a$ are presented next. Such super cells tend to have a smaller Brillouin zone owing to a larger lattice constant, which leads to phenomenon of band structure folding [14]. In this work as well as the pre-print [9], Peierl's distortion has not been taken into account, thus maintaining symmetry [15]—the major motivation for this was to ensure that the proposed formalism works well for symmetric systems before proceeding forward toward non-symmetry. For a super cell with $2a$ lattice constant, the new input signal can be written as

$$i[x,y] = \begin{bmatrix} 0\,0\,0\,0\,0\,0 \\ 0\,0\,0\,0\,0\,0 \\ 0\,0\,1\,1\,0\,0 \\ 0\,0\,1\,1\,0\,0 \\ 0\,0\,0\,0\,0\,0 \\ 0\,0\,0\,0\,0\,0 \end{bmatrix}.$$

As discussed earlier in this paper, LTI systems are characterized by their impulse responses. Since the present crystalline material is treated as an LTI system, the same principle must be applied to this as well. In Sect. 3.1, the impulse response for this LTI system was described. Now, to achieve the band structure for the super cell, a new input signal needs to be fed in to this same LTI system. The corresponding output to this perturbation can be easily achieved by convolution with $h[x,y]$ as

$$o[x,y] = h[x,y] * i[x,y]$$

$$= \begin{bmatrix} \beta\,\beta\,\beta \\ \beta\,\alpha\,\beta \\ \beta\,\beta\,\beta \end{bmatrix} * \begin{bmatrix} 0\,0\,0\,0\,0\,0 \\ 0\,0\,0\,0\,0\,0 \\ 0\,0\,1\,1\,0\,0 \\ 0\,0\,1\,1\,0\,0 \\ 0\,0\,0\,0\,0\,0 \\ 0\,0\,0\,0\,0\,0 \end{bmatrix}.$$

$$= \begin{bmatrix} \beta & 2\beta & 2\beta & \beta \\ 2\beta & \alpha+3\beta & \alpha+3\beta & 2\beta \\ 2\beta & \alpha+3\beta & \alpha+3\beta & 2\beta \\ \beta & 2\beta & 2\beta & \beta \end{bmatrix}$$

Instead of matrices, the same calculations can also be done analytically. With the frame of reference shown in Fig. 4b, the new input (physically implying a super cell of lattice constant $2a$) can be written analytically as: $i[x,y] = \delta[x,y] + \delta[x-a,y] + \delta[x,y-a] + \delta[x-a,y-a]$. Hence, it can be easily observed that the super cell has

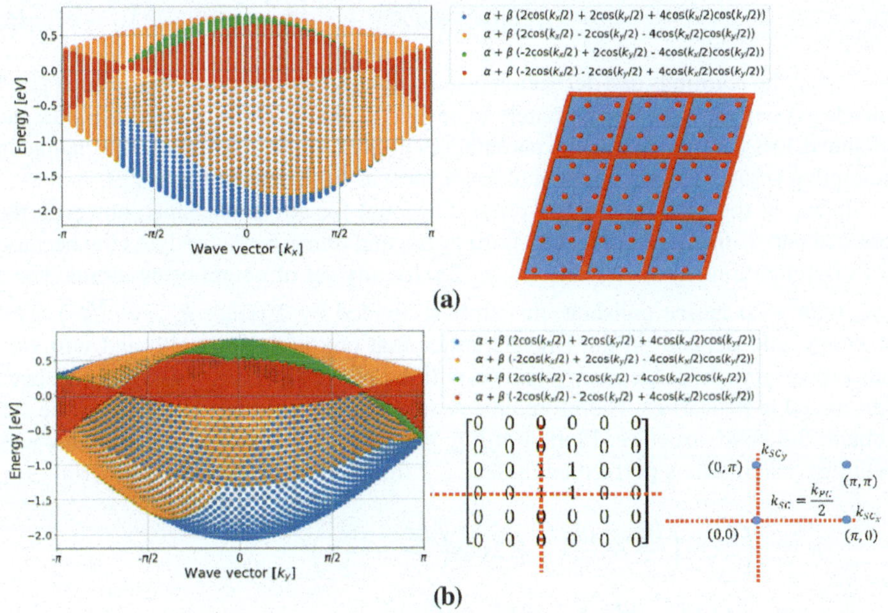

Fig. 4 The band structures for the super cell of lattice $2a$ of the 2D material are presented as derived using the proposed formalism: **a** presents the plot against k_x while **b** presents the same against k_y. The inset of (**a**) also depicts the actual super cell while that of (**b**) depicts the corresponding super cell in this new perspective using just spikes with the right frame of reference in real as well as reciprocal space

now become a linear summation of four δ functions—a train of spikes. These spikes are separated by a distance of the lattice constant itself. This is intuitive because such δ functions are located at the origin of their respective unit cells. Thus, it would always be ensured in this case that the center-to-center distance between any two of these consecutive unit cells taken one pair at a time would always be a.

Corresponding to this input signal, the output of the LTI system can be easily achieved using convolution with the impulse response in real space as shown:

$$
\begin{aligned}
o[x, y] &= h[x, y] * i[x, y] \\
&= h[x, y]*(\delta[x, y] + \delta[x - a, y] + \delta[x, y - a] + \delta[x - a, y - a]) \\
&= h[x, y] + h[x - a, y] + h[x, y - a] + h[x - a, y - a]
\end{aligned}
$$

It has been understood in this paper so far that Fourier transform diagonalizes the convolution operator. Now that it is required to obtain the energy eigenvalues from the previous equation (which is convolution), Fourier transform needs to be taken to obtain these. This is shown in Eq. 5.

$$
F(o[x, y]) = E[k_x, k_y]
$$

$$= H\left[k_x, k_y\right]\left(e^{j0} + e^{-jak_x} + e^{-jak_y} + e^{-jak_x - jak_y}\right) \tag{5}$$

Using the learnings and derivation presented in [9], it is now understood that Eq. 5 needs to be interpreted appropriately to be able to plot it. The way to incorporate the eigenvectors (present inside the parentheses in Eq. 6) is to account for the change in sampling frequency and in terms of phase shift as has been detailed in [9].

In Eq. 5, the first eigenvector doesn't account for any phase shift; however, the second one shifts the phase by ak_x, third by ak_y and fourth by ak_x and ak_y; but because of discretization $ak_x = 2\pi$ and $ak_y = 2\pi$ for any set of neighboring atoms. For a SC with a $2a$ lattice constant, this shift is divided by 2 (since $k_{x_{SC}} = \frac{k_{x_{PC}}}{2}$). This factor would depend on the Bravais lattice that one is dealing with and will vary accordingly with the geometry and so will the phase shift factors too, as has been discussed in [9] earlier. Therefore, this results in four equations with which one can obtain the band structure. Plotted in Fig. 4a and b, this band structure agrees well with the calculations performed in Sect. 3.2 from the Supplementary Material [9].

$$\alpha + \beta\left(2\cos\left(\frac{i\pi}{2} + \frac{ak_x}{2}\right) + 2\cos\left(\frac{j\pi}{2} + \frac{ak_y}{2}\right)\right.$$
$$\left. + 4\cos\left(\frac{i\pi}{2} + \frac{ak_x}{2}\right)\cos\left(\frac{j\pi}{2} + \frac{ak_y}{2}\right)\right) \text{ with } i \in \{0, 2\} \text{ and } j \in \{0, 2\} \tag{6}$$

As has been explained in [9], "the reason why all the band structures in this work are presented in the domain of $[0, \pi]$ is to maintain uniformity with the published literature. One can, for example, refer to established literature [16, 17] and find consistency for the respective cases dealt with as part of this work."

4 Discussion

One of the advantages of the LTI-EST formalism is computational simplicity. As seen in Figure S1 (please refer [9]), an $O(N)$ loop is needed to iterate over all the k-points in the reciprocal space to find the set of eigenvalues and eigenvectors. The proposed LTI-EST formalism doesn't require this $O(N)$ loop as it is governed by the fact that Fourier transform diagonalizes the convolution operator. At any given instant, the equation obtained for the case of a primitive cell is used for the cases of all the other super cells as well. The changes required are taken into account by the phase shift and the sampling rate in the analytical equation. Hence, the computational load effectively equates to $O(1)$ in this formalism. To be precise, in this work, using the new perspective, the diagonalization of the below matrix is bypassed as the analytical values are always present:

Table 1 The key distinguishing factors between the orthodox perspective and the proposed LTI-EST formalism to materials simulation

Property	Conventional perspective	This work
Computational cost	$O(N)$ (Diagonalization)	$O(1)$ (Analytical)
Medium of material simulation	Atomic	Spikes/impulse
Computation mode for band structure folding	Discrete (Hermitian Hamiltonian)	Analog (Analytic equation)
Traceability of the folded band structure	No	Yes
SNN friendliness	No	Yes

$$H = \begin{pmatrix} \alpha & \beta + \beta e^{jak_x} \\ \beta + \beta e^{-jak_x} & \alpha \\ \beta + \beta e^{jak_y} & \beta + \beta e^{jak_x} + \beta e^{jak_y} + \beta e^{jak_x + jak_y} \\ \beta + \beta e^{-jak_x} + \beta e^{+jak_y} + \beta e^{-jak_x + jak_y} & \beta + \beta e^{jak_y} \\ \beta + \beta e^{-jak_y} & \beta + \beta e^{jak_x} + \beta e^{-jak_y} + \beta e^{jak_x - jak_y} \\ \beta + \beta e^{-jak_x} + \beta e^{-jak_y} + \beta e^{-jak_x - jak_y} & \beta + \beta e^{-jak_y} \\ \alpha & \beta + \beta e^{jak_x} \\ \beta + \beta e^{-jak_x} & \alpha \end{pmatrix}$$

The analysis made in this section has been summarized in the form of Table 1. Besides the case of 1D chain of atoms discussed in this pre-print [9], the implementation of this LTI-EST formalism has also been presented to be effective on a more complex example of monolayer 2D material MoS_2 (while considering not just the nearest neighbor interactions) and the process flow is described in detail in the present manuscript and this Master Thesis [18] allowing for bypassing the said $O(N)$ diagonalization for $\sim$ 8000 k-points. More details on the conventional computation can also be found in the Supplementary Text of this arXiv pre-print [9]. More detailed description of the points mentioned in the Table 1 is readily available in the thesis as well as in [9] in the "Discussion" section.

5 Conclusion

Exploiting the Linear and Translational Invariance (LTI) of a perfectly crystalline material, a new perspective is presented in this work to electronic structure theory. This LTI-EST formalism is validated via achievement of band structures for 2D monolayer MoS_2 for unit cells with different lattice constants while also varying the number of interactions. The process of band structure folding is also presented in this work along with the benefits of eigenvalue decomposition with $O(1)$ complexity on top of traceability of band structure folding offered. The LTI-EST achieves all this, thanks to its successful adaptation of EST to the spiking-domain, which could be of significant interest for development of neuromorphic hardware that promise computation with ultralow power consumption given that the algorithms operate in the spike-based computing domain.

6 Outlook and Future Work

The work presented in this present paper, this pre-print [9] and this Master Thesis [18] puts forth the idea of using concepts from signals–systems for derivation of electronic band structures of materials. While the process flow and implementation for a 1D chain of atoms is discussed in [9], this work discusses such implementation on a more realistic and complex scenario of a 2D material. The work done is also compiled in this Master Thesis [18]. The motivation for this work was, taking inspiration from the cited literature, to adapt a conventional well-established algorithm to the spiking-domain with an aim to make it friendly for neuromorphic computing hardware, which would operate solely in this domain and offer ultralow power consumption. This work addresses a static materials simulation algorithm—the Electronic Structure Theory (EST)—to solve Schrödinger equation and obtain energy eigenvalues and plot band structures using only spikes. An interesting next step would now be to incorporate the time domain (besides the space coordinates) as well into the calculations by taking into consideration *when* a spike signal is sent. This would allow to depart from the static calculations toward molecular dynamics simulations. The aim there would be to perform such calculations on a realistic device-scale level while still preserving the atomistic perspective. This, in today's literature, is a hot topic but also hard to perform on today's von Neumann architecture-based computers because of the computational complexity of the algorithm which increases exponentially as the number of atoms increases (implying the increment in scale) soon making the problem statement infeasible on the state-of-the-art computers. Thanks to the ultralow power consumption offered by the neuromorphic computers, it is expected that such computation would still be feasible on them provided the algorithm is compatible with their spiking-domain. Having proved this for static calculations in this work, it is now an interesting next to, therefore, include time domain as well into analysis and see how this LTI-EST formalism can be extended further. Since time domain also has a Fourier transform pair with frequency domain, it is expected to be a meaningful direction to explore. At the same time, it shouldn't be forgotten that only a single orbital was taken into consideration in this work, it would be interesting to see how this perspective scales to multiple orbitals.

References

1. Choudhary K et al (2022) Recent advances and applications of deep learning methods in materials science. NPJ Comput Mater 8:59
2. Wang et al (2021) Machine learning method for tight-binding Hamiltonian parameterization from ab-initio band structure. NPJ Comput Mater 7:11
3. Lathi BP, Signal processing and linear systems. ISBN 0-941413-35-7
4. Silver D et al (2017) Mastering the game of Go without human knowledge. Nature 550:354–359
5. Roy K et al (2019) Towards spike-based machine intelligence with neuromorphic computing. Nature 575:607–617
6. Schuman CD et al (2022) Opportunities for neuromorphic computing algorithms and applications. Nat Comput Sci 2:10–19
7. Date P et al (2022) Neuromorphic computing is Turing-complete. In: International conference on neuromorphic systems (ICONS 2022)
8. Aimone JB et al (2022) A review of non-cognitive applications for neuromorphic computing. Neuromorph Comput Eng 2:032003
9. Yadav A et al, A spiking-domain implementation of electronic structure theory. arXiv:2404.00128
10. Buth et al (2005) Ab initio Green's function formalism for band structures. Phys Rev B 72:195107
11. Boykin TB et al (2016) Unfolding and effective bandstructure calculations as discrete real- and reciprocal-space operations. Physica B 491:22–30
12. Kra I, Simanca SR (2012) On circulant matrices. Notices of the AMS 59(3):368–377
13. Green G (1828) An essay on the application of mathematical analysis to the theories of electricity and magnetism. T. Wheelhouse, Nottingham, England, pp 10–12
14. Ashcroft & Mermin. Solid State Physics. ISBN 0-03-083993-9
15. Mott NF, Peierls R (1937) Discussion of the paper by de Boer and Verwey. R Proc Phys Soc 49:72
16. Yang S-Y et al (2018) Symmetry demanded topological nodal-line materials. Adv Phys X 3(1)
17. Hoffmann R (1987) How chemistry and physics meet in the solid state. Angewandte Chemie Int Ed Engl 26(9):846–878
18. Yadav A (2024) A signals-systems perspective to electronic band structure theory and Mitigation of directional MLC operation in a novel variant of double mushroom PCM cells. Master Thesis 2024, UNIST

Evaluation of Customized Pre-trained Models to Predict Percentage of Being COVID-19 Cases on Chest X-Ray and COVID Radiography Images

Sujatha Kamepalli, Venkata Krishna Kishore Kolli, and Srinivasa Rao Bandaru

Abstract During the 2nd surge of COVID-19 in India, the Delta variant caused a significant rise in infections and fatalities. Accurate diagnosis and timely treatment are crucial to control the spread and save lives. Common diagnostic tools include the Antigen Rapid Diagnostic Test (Ag-RDT), Reverse Transcription Polymerase Chain Reaction (RT-PCR) test, and CT scans. While Ag-RDT and RT-PCR can be performed by medical practitioners, their reliability is limited. As coronavirus disease mainly affects the respiratory system, chest X-rays can provide more accurate diagnosis. This study investigates the use of deep learning (DL) models in predicting coronavirus from chest X-ray images. Various pre-trained CNN models serve as feature vectors on both datasets, with the network heads updated for the current class labels of COVID or normal. By customizing and parameter tuning of pre-trained models, more than 90% prediction accuracy was achieved for VGG16, VGG19, ResNet50, and ResNet101. These models not only predict coronavirus disease cases from images but also determine the percentage likelihood of a given image being COVID-positive or normal. The implementation of Grad-CAM is utilized to highlight the class activation maps of images, which offers a more precise visualization of the affected areas of the lungs in coronavirus disease patients. The custom pre-trained models are remarkably efficient in accurately predicting coronavirus disease patients from both chest X-ray and radiography images. The methodology can be further improved to create an application that predicts the likelihood of a coronavirus disease case or normal case and produces visual explanations through activation maps.

Keywords ResNet50 · ResNet101 · VGG19 · VGG16 · Chest X-ray images · COVID radiography images · Grad-CAM visualization · Class activation maps

S. Kamepalli (✉) · K. Venkata Krishna Kishore · B. Srinivasa Rao
Vignan's Foundation for Science Technology and Research, Guntur, India
e-mail: sujatha_kamepalli@yahoo.com

© The Author(s), under exclusive license to Springer Nature Singapore Pte Ltd. 2025

H. Zen et al. (eds.), *Soft Computing and Signal Processing*, Lecture Notes in Networks and Systems 1200, https://doi.org/10.1007/978-981-97-9926-8_14

1 Introduction

In the entire world, for almost two years people are living with some sort of fear due to coronavirus disease (COVID-19). Many countries, including India, have experienced multiple waves of virus impact. The virus adapts through mutations, occasionally becoming more virulent and leading to increased spread of the disease [10, 26]. The second wave of the virus spread rapidly and affected a significant portion of the population, accompanied by an increase in the fatality rate. Prompt diagnosis of COVID disease is crucial not only for controlling its spread but also for providing effective care and mitigating associated challenges [2]. Three common diagnostic methods for confirming COVID-19 infection include blood tests, viral tests, and lung imaging via X-ray or CT scan. Blood tests detect antibodies in a blood sample, but their consistency is low, typically around 2–3%. Two viral tests commonly used are Rapid Diagnostic Test (RDT) and RT-PCR. The first one delivers results within 30 min by detecting antigens in respiratory samples, but it can produce false positives. RT-PCR is more reliable, with an accuracy of 50–60%, but it requires multiple repetitions and faces limitations due to test kit availability. Additionally, chest X-rays (CXR) are widely employed as COVID-19 primarily affects the lungs. Radiologists can identify various abnormalities in lung images of COVID-19 patients, aiding in diagnosis and disease staging.

1.1 Deep Learning (DL) Models and Grad-CAM Visualization

Deep learning models leverage large labeled datasets to extract diverse features for predicting test data labels effectively [17, 28], enabling them to extract numerous features from the training dataset to predict class labels of test data. In this research, pre-trained deep CNN models were customized and fine-tuned to predict COVID-19 cases from chest X-ray and COVID-19 radiography images [27]. The Grad-CAM algorithm enables visualization of class activation maps in convolutional neural networks [14]. It generates visual explanations using gradient information from any CNN-based deep learning algorithm. Grad-CAM defines the visualization of a specific class label as a weighted sum of all feature maps from the model's last convolution layer [23].

1.2 Contributions of the Paper

Main contributions of this paper are:

- Utilizing pre-trained deep CNN models customized for accurate prediction of COVID disease-affected cases from chest X-ray images.
- Predicting coronavirus disease-affected people from COVID radiography images using customized pre-trained deep convolutional neural network models.
- Grad-CAM visualization of predicted COVID cases for better visualization of affected areas.

1.3 Structure of the Paper

The remaining part of the paper is prepared as follows.

In Sect. 2, the existing literature in predicting COVID cases is explored. Section 3 explains the objectives of this paper. In Sect. 4, the flow of the proposed method to predict COVID cases is explained. Section 5 explains the experimental work done in this paper. Lastly, in Sect. 6 the conclusion of the paper is given.

2 Existing Work

In the present study, the authors developed a diagnostic tool using transfer learning models, applied to chest X-ray images for recognizing pediatric pneumonia and diabetic macular edema [12]. They proposed an ensemble technique with multiple convolutional neural networks (MobileNet, InceptionV3, DenseNet201, and DenseNet121) to diagnose COVID-19, achieving exceptional performance on a publicly available dataset [22]. Evaluating clinical instruments and AI efficacy in detecting coronavirus from chest X-rays, they constructed a robust dataset from public databases and recent articles, accurately predicting COVID-19 using transfer learning models [5]. Another study implemented a convolutional neural network for multi-classification, using Monte Carlo simulation to augment data for forecasting confirmed and deceased cases, employing linear regression and Gaussian mixture models [13]. Investigating the global health impact of the pandemic, they used Google Net for forecasting COVID-19 from chest X-rays, suggesting IoT technology could enhance the model [7]. The Grad-CAM method was introduced to visualize CNN model features [23]. Emphasizing DL models for coronavirus triage systems, a well-balanced dataset of chest X-rays was created, proposing a COVID smart data-based network [3]. Multi-class COVID X-ray images were studied using ResNet50 for feature extraction and SVM for detection [16]. Another model used Bayesian networks, extracting features from CT scans with transfer learning [29]. Iteratively pruned DL models identified respiratory manifestations in COVID-19 patients [21]. ChexNet was introduced for identifying COVID cases in the ChestXray14 dataset [6]. AI-assisted diagnosis was highlighted, employing six CNN models and image enhancement methods on a comprehensive dataset [20]. A personalized CNN model predicted patients' length of stay, aiding hospital efficiency [25]. Lastly, a novel transfer learning model categorized chest X-ray images into four classes, using Grad-CAM for visualization, aligning closely with clinical observations [4].

Polat et al. [19] created nCOV-NET, optimized with Densenet-161, achieving a classification accuracy of 97.1% for COVID-19 cases using activation mapping techniques. Li et al. [15] developed COVNet for feature extraction from CT scan images of COVID patients collected between 2016 and 2020. Islam et al. [8] proposed a CNN and RNN-based framework for COVID-19 chest X-ray image classification, utilizing Grad-CAM for class-specific feature visualization. Shorfuzzaman et al. [24] suggested a fusion-based deep learning approach, combining various models to extract features from CXR images, achieving a 95.49% accuracy. Karthik et al. [11] reviewed approximately 140 academic articles on the latest advancements in COVID-19 image classification methods, detailing improvements in COVID-19 identification. Abdulla et al. [1] proposed a transfer learning model to predict COVID cases from CT scan images. Johnson et al. [9] used transfer learning to classify chest X-ray images, training a CycleGAN framework for feature extraction. Panwar et al. [18] suggested a deep learning transfer model for detecting COVID-19 cases from X-ray and CT scan images, utilizing Grad-CAM for feature analysis.

Based on existing research, deep learning models effectively predict COVID cases from chest X-ray images. Clear visualization of COVID-infected areas in these images is crucial for guiding treatment decisions and enabling non-experts to identify lung infections. Therefore, this paper incorporates Grad-CAM visualization to highlight infected regions in predicted chest X-ray images.

3 Objectives

The principal goals of this paper include:

- Investigating advanced models for predicting coronavirus disease cases using chest X-ray images and analyzing datasets utilized for COVID case prediction.
- Customizing pre-existing deep CNN models to predict COVID cases with chest X-ray and COVID radiography images.
- Training and assessing the models on chest X-ray and COVID radiography datasets and comparing the accuracy of the models and making predictions on test images.
- Generating heat maps and Grad-CAM visualizations of predicted images for improved visualization of affected areas.

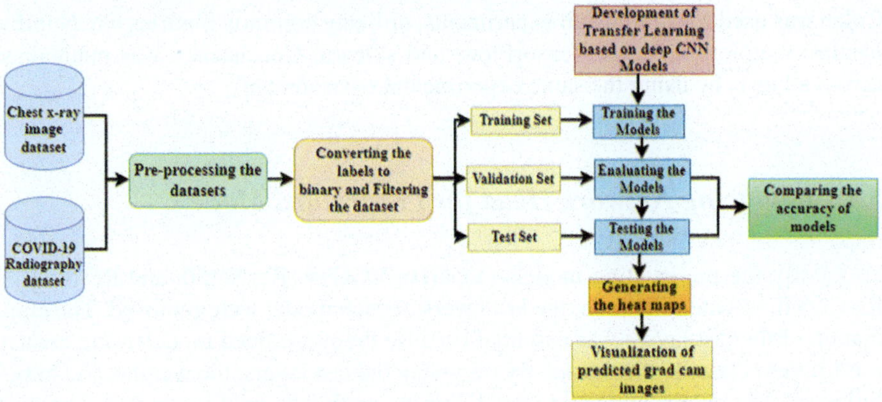

Fig. 1 Diagram of the projected work. *Figure Source* “Proposed Flowchart”

4 Implementing the Proposed Models

The following steps were followed in predicting coronavirus disease cases from chest X-ray and COVID radiography images, and in visualizing target features using Grad-CAM. First, datasets from the Kaggle data science community were obtained. Various preprocessing and filtering techniques were applied to prepare the datasets for deep learning models. Next, the data was partitioned into training, validation, and test sets. Pre-trained models, including VGG-16, VGG-19, ResNet50, and ResNet-101, were customized and implemented on the training and validation sets. Testing involved selecting a random image from the test set to predict the probability of COVID-19 or normal status. Finally, the Grad-CAM model was applied to these deep learning models for better visualization of the predicted features from the chest X-ray and COVID radiography images. Figure 1 illustrates the suggested sequence of the procedure.

5 Experimentation

This paper leverages pre-existing CNN models to predict COVID patients using chest X-ray and radiography images. The CNN architecture was modified to extract COVID-19-specific features, with adjustments made to parameters. This section details the customized model structure, including key configurations, pre-existing models, and fine-tuning phases. The experiments were carried out on a “Dell Inspiron 5000” series laptop running Windows 10, using a Python simulation setup. Google

Colab was used for running all experiments, utilizing common libraries like NumPy, pandas, matplotlib, Keras, TensorFlow, and sklearn. Consistency was maintained across all tests by using the same experimental environment.

5.1 Replacing Network Head for Pre-trained Models

To modify the pre-existing models (VGG16, VGG19, ResNet50, and ResNet101) for COVID-19 identification, the last layers of each model were removed. The elimination of the last layers was essential because the pre-trained models were created for different classification tasks. By removing the last layers, the weights and biases linked to class scores were eliminated. Subsequently, the final layers were replaced with new layers that were compatible with the quantity of classes in the current dataset, which consisted of COVID and normal classes. The resultant model was transformed into a two-class network head for both datasets.

5.2 Datasets Description

The COVID-19 prediction model utilized two datasets: chest X-ray images dataset and COVID radiography dataset.

The dataset was adjusted to enhance relevance and balance, with lung opacity images omitted. Images from the COVID and viral pneumonia classes were merged due to COVID's viral origin. It was then divided into 3 parts: training, validation, and testing sets, comprising 7800, 1008, and 800 images, respectively, focusing on COVID and normal classes. The distinctions between normal and COVID-19 chest X-ray images, as well as between COVID radiography and normal images, are depicted in a provided Fig. 2.

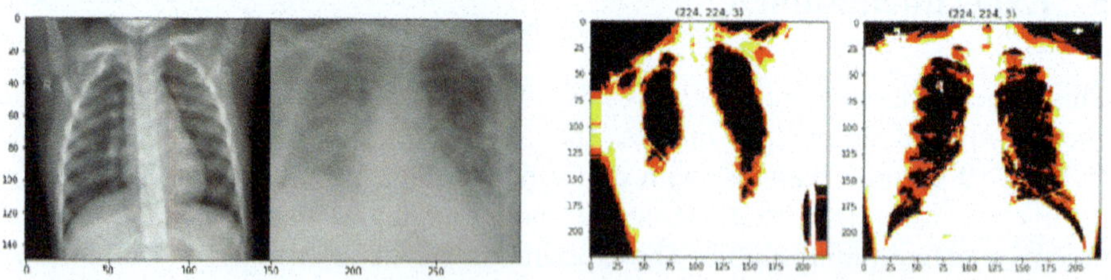

(a) X-ray Images Samples (b) Radiography Images Samples

Fig. 2 Datasets sample images. *Figure Source* https://www.kaggle.com/paultimothymooney/chest-xray-pneumonia and https://www.kaggle.com/tawsifurrahman/covid19-radiography-database

5.3 Results and Considerations

The results attained by employing the transfer learning models developed based on CNN "VGG-16, VGG-19, ResNet50, and ResNet-101" were discussed. The efficiency of models in predicting COVID-19 cases was discussed based on the accuracy and loss that occurred in those models.

Chest X-ray Images Dataset. Labels were binarized in the dataset, normal images are indicated with 0, and COVID images are indicated with 1. Figure 3 shows the accuracies and loss measures obtained on the training and validation sets in various deep learning models implemented.

COVID radiography Images Dataset. Figure 4 shows the accuracies and loss measures obtained on the training and validation sets in various deep learning models implemented.

Figure 5 shows the comparison of accuracies and loss values on both datasets used in the experimentation.

Comparison with the Existing Models. Table 1 gives the comparison of several models used in literature with the proposed methodology in terms of accuracy.

Grad-CAM Visualization. Here, the interesting areas of the X-ray image are highlighted using Grad-CAM visualization. Figure 6 shows the Grad-CAM visualization of the predicted COVID image with its heat map generated, Grad-CAM visualization of the predicted COVID image. From the prediction, it is clear that the percentage of being COVID that chest X-ray image is 79.95% and normal is 7.1% and the snapshot of the code snippet of Grad-CAM visualization of the predicted normal image. From the prediction, it is clear that the percentage of being COVID in that chest X-ray image is 5.24% and normal is 80.79%.

From the outcomes of this research work, it is clear that the model effectively envisions the diseased areas of the lungs from predicted images. The models used were predicted the percentage of chance for becoming a COVID case and a normal case. If the percentage of prediction of becoming is greater than 50%, that image should be considered under COVID image and that person should be medicated properly.

6 Conclusions

The Delta variant's emergence during India's second COVID-19 surge led to a substantial increase in infections and fatalities, underscoring the critical need for accurate diagnosis and prompt treatment. While common diagnostic tools like Ag-RDT, RT-PCR tests, and CT scans are available, their reliability is limited, necessitating more accurate methods. Given COVID-19's respiratory impact, chest X-rays offer promising diagnostic potential. Utilizing pre-trained CNN models as feature

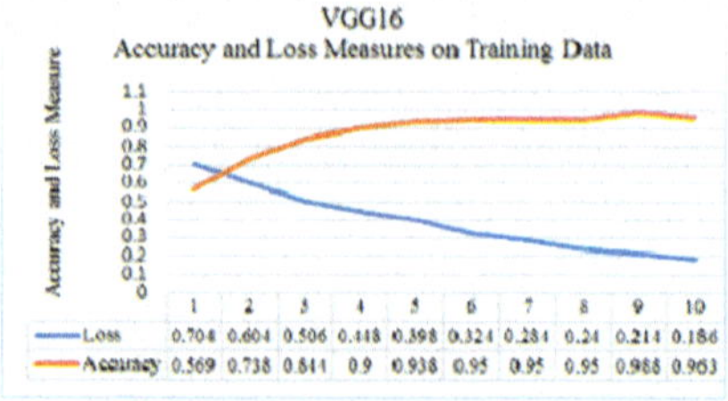

(a) Training Data-VGG16

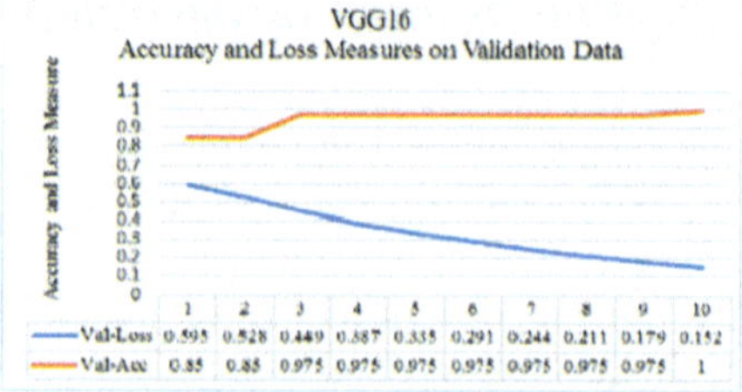

(b) Validation Data-VGG16

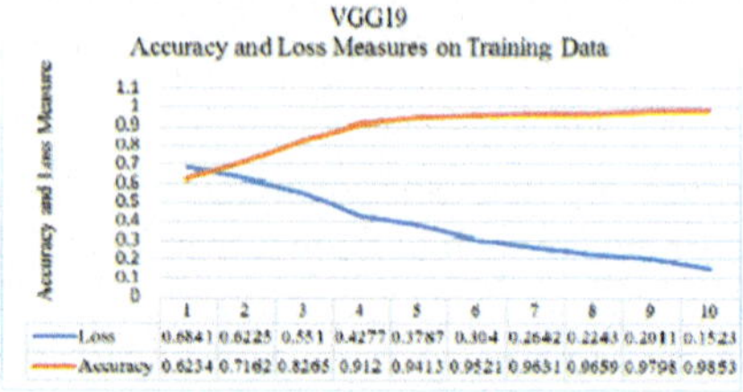

(c) Training Data-VGG19

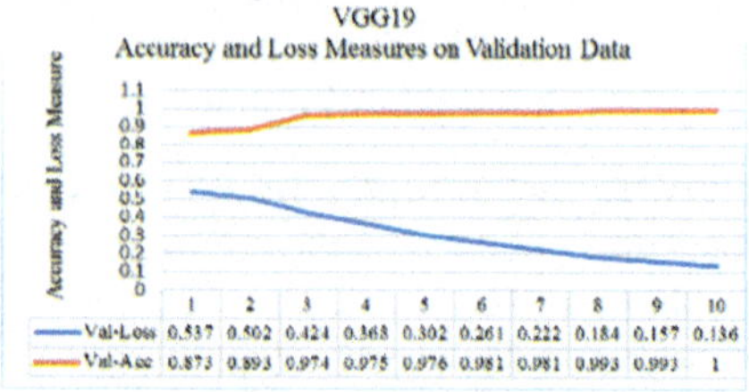

(d) Validation Data-VGG19

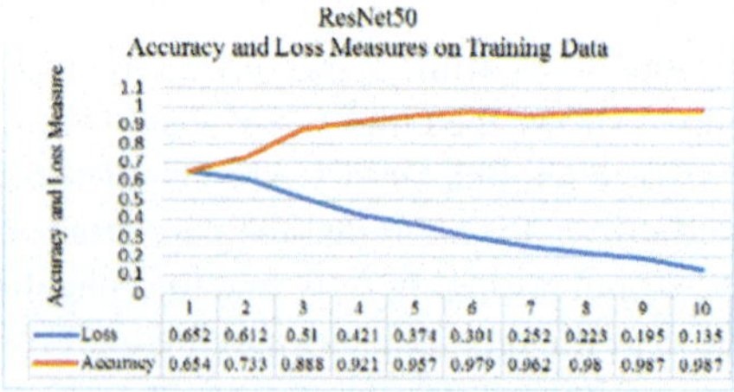

(e) Training Data-ResNet50

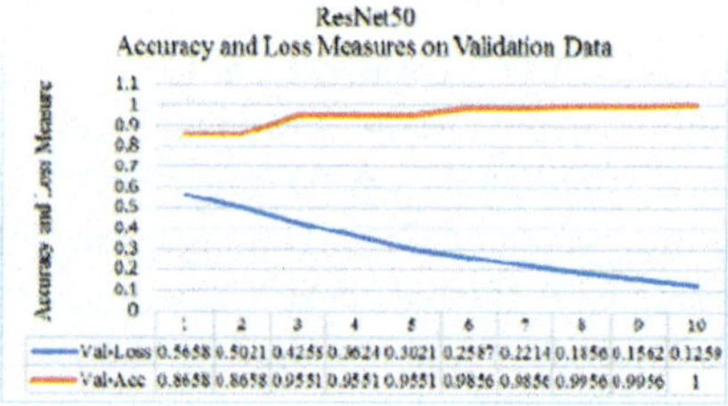

(f) Validation Data- ResNet50

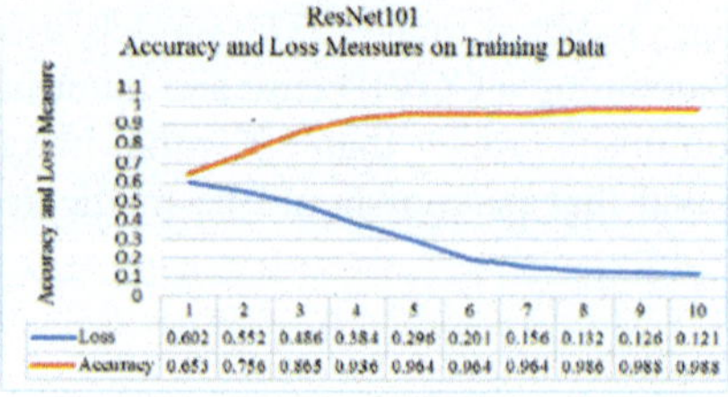

(g) Training Data-ResNet101

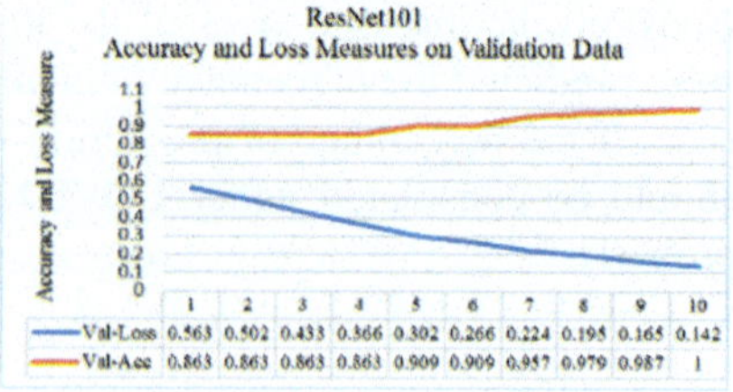

(h) Validation Data- ResNet101

Fig. 3 Accuracies and loss measures obtained on the training and validation sets in various deep learning models on chest X-ray image dataset. *Figure Source* "Experimentation"

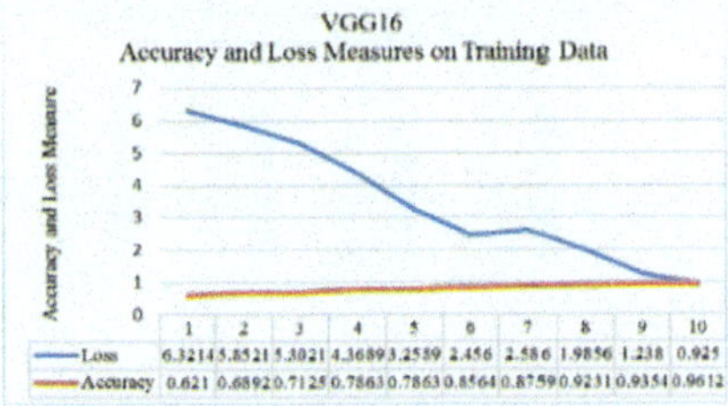

(a) Training Data-VGG16

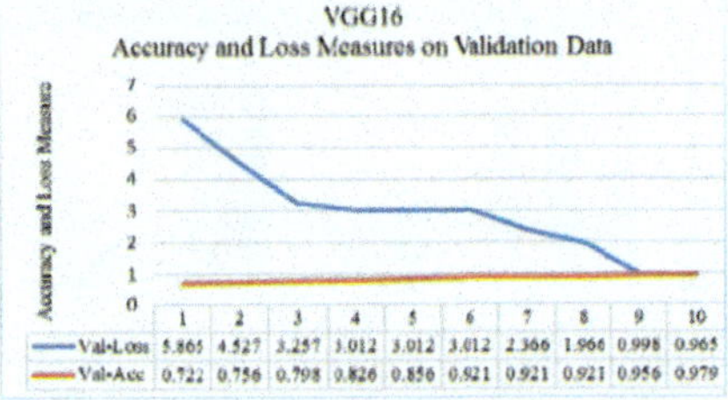

(b) Validation Data-VGG16

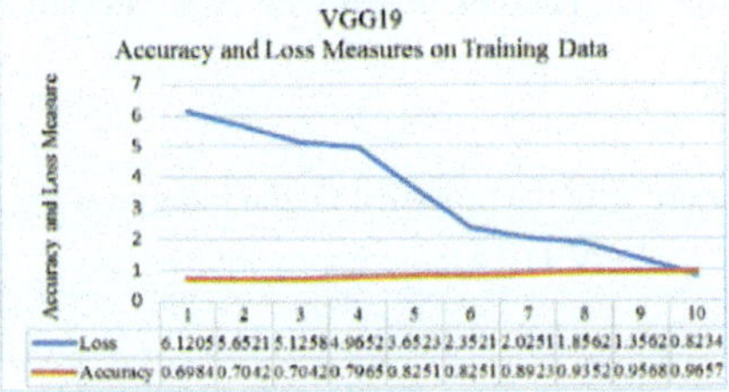

(c) Training Data-VGG19

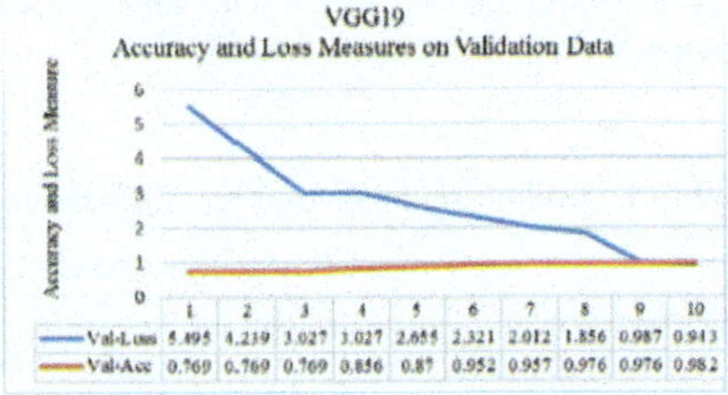

(d) Validation Data-VGG19

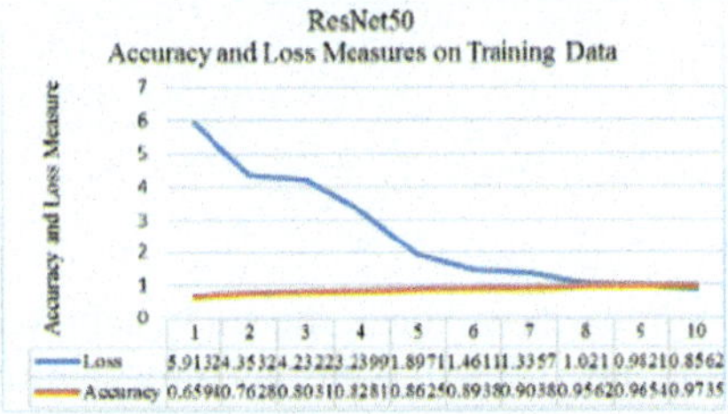

(e) Training Data-ResNet50

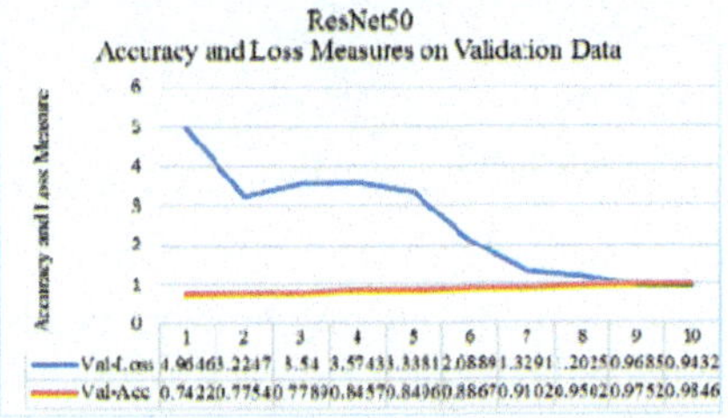

(f) Validation Data- ResNet50

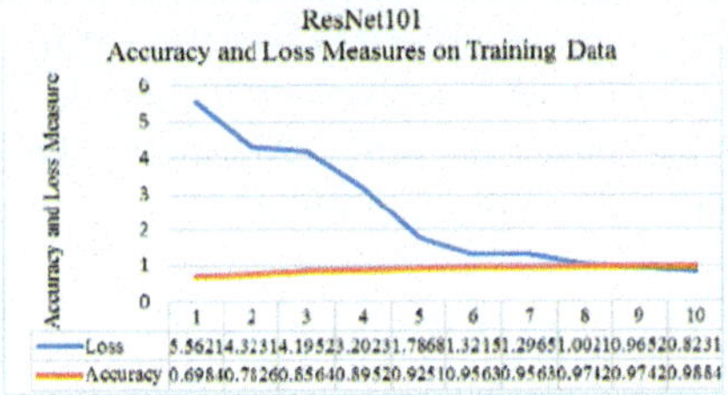

(g) Training Data-ResNet101

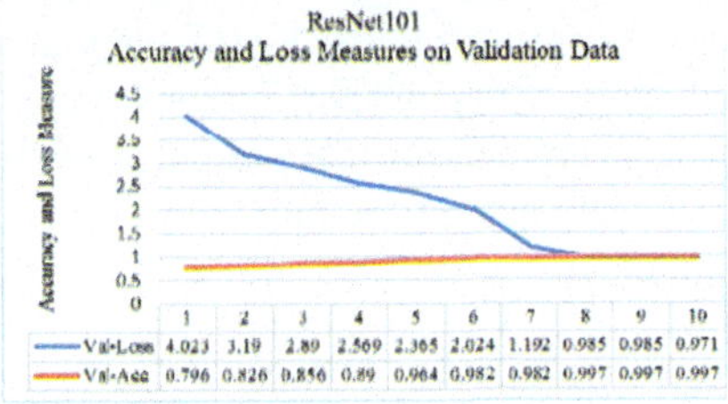

(h) Validation Data- ResNet101

Fig. 4 Accuracies and loss measures obtained on training and validation sets in various deep learning models on COVID-Radiography dataset. *Figure Source* "Experimentation"

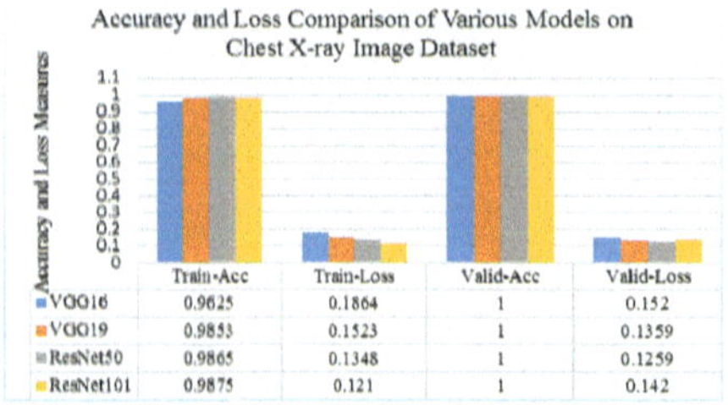

(a) Chest X-ray Dataset

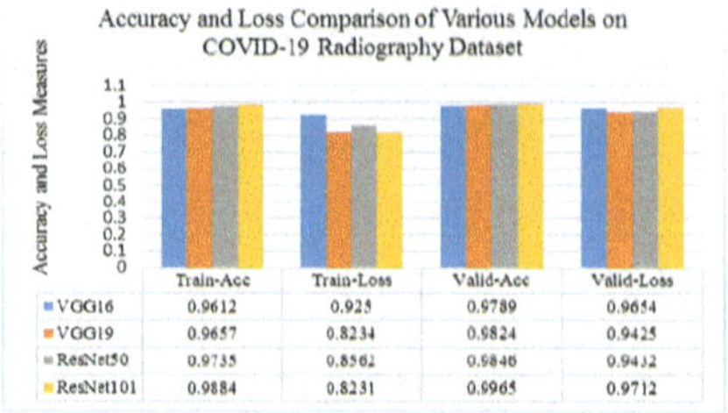

(b) COVID-19 Radiography Dataset

Fig. 5 Comparison of accuracies and loss values on both datasets used in the experimentation. *Figure Source* "Experimentation"

Table 1 Evaluation of several models used in literature with the proposed methodology

	Chest X-ray images (%)	COVID-19 radiography images (%)
PDCOVIDNet	96.58	–
DenseNet-121	96.49	–
MobileNet	96.48	–
Efficient-NetB0	96.80	–
GDCNN	98.84	–
GoogleNet	98.50	–
Proposed model	100	90

vectors and customizing network heads for current class labels, the study achieved over 90% prediction accuracy across various models. These models not only identify COVID-19 cases but also estimate the likelihood of a given image being COVID-positive or normal. Leveraging Grad-CAM, the study provides detailed visualizations of affected lung areas, enhancing diagnostic precision. The custom pre-trained models demonstrate high efficiency in accurately predicting COVID-19 cases from both chest X-ray and radiography images. Future enhancements could involve developing an application that predicts COVID-19 likelihood and provides visual explanations using activation maps, further enhancing diagnostic capabilities in combating the pandemic.

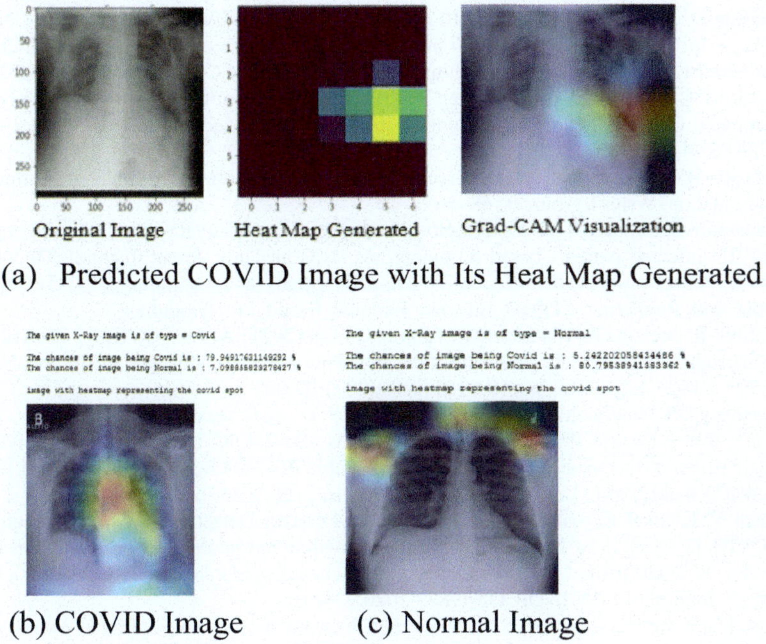

Fig. 6 Grad-CAM visualization of predicted COVID image with its heat map generated. *Figure Source* "Experimentation"

References

1. Abdulla HK, Ahmed ZS, Surameery NMS, Rashid RD, Salih SQ (2020) SAARSNet: a deep neural network for COVID-19 cases diagnosis. In: 3rd international conference on advanced science and engineering, ICOASE 2020, pp 69–73. https://doi.org/10.1109/ICOASE51841.2020.9436536
2. Alghamdi HS, Amoudi G, Elhag S, Saeedi K, Nasser J (2021) Deep learning approaches for detecting COVID-19 from chest X-ray images: a survey. IEEE Eng Med Biol Soc Sect 9:20235–20254. https://doi.org/10.1109/ACCESS.2021.3054484
3. Babukarthik RG, Ananth Krishna Adiga V, Sambasivam G, Chandramohan D, Amudhavel AJ (2020) Prediction of covid-19 using genetic deep learning convolutional neural network (GDCNN). IEEE Access 8:177647–177666. https://doi.org/10.1109/ACCESS.2020.3025164
4. Basu S, Mitra S, Saha N (2020) Deep learning for screening COVID-19 using chest X-ray images. In: 2020 IEEE symposium series on computational intelligence, SSCI 2020, Ml, pp 2521–2527. https://doi.org/10.1109/SSCI47803.2020.9308571
5. Chowdhury MEH, Rahman T, Khandakar A, Mazhar R, Kadir MA, Mahbub ZB, Islam KR, Khan MS, Iqbal A, Emadi NA, Reaz MBI, Islam MT (2020) Can AI help in screening viral and COVID-19 pneumonia? IEEE Access 8:132665–132676. https://doi.org/10.1109/ACCESS.2020.3010287
6. Haritha D, Pranathi MK, Reethika M (2020) COVID detection from chest X-rays with DeepLearning: CheXNet. In: Proceedings of the 2020 international conference on computing, communication and security, ICCCS 2020, pp 1–5. https://doi.org/10.1109/ICCCS49678.2020.9277077

7. Haritha D, Swaroop N, Mounika M (2020) Prediction of COVID-19 cases using CNN with X-rays. In: Proceedings of the 2020 international conference on computing, communication and security, ICCCS 2020, pp 1–6. https://doi.org/10.1109/ICCCS49678.2020.9276753
8. Islam MM, Islam MZ, Asraf A, Ding W (2020) Diagnosis of COVID-19 from X-rays using combined CNN-RNN architecture with transfer learning. MedRxiv. https://doi.org/10.1101/2020.08.24.20181339
9. Johnson BVM, Rogers LL (2000) Accuracy of neural network approximators in simulation-optimization. Water Resources, 48–56
10. Kamepalli S, Bandaru S (2020) Symptoms and spreading of COVID-19 in India and predicting COVID infected patients based on symptoms. J Xi'an Univ Archit Technol XII(V):3446–3457. https://www.academia.edu/43486733/Symptoms_and_Spreading_of_COVID_19_in_India_and_Predicting_COVID_Infected_Patients_Based_on_Symptoms
11. Karthik R, Menaka R, Hariharan M, Kathiresan GS (2021) AI for COVID-19 detection from radiographs: incisive analysis of state of the art techniques, key challenges and future directions. IRBM 1:1–25. https://doi.org/10.1016/j.irbm.2021.07.002
12. Kermany DS, Goldbaum M, Cai W, Valentim CCS, Liang H, Baxter SL, McKeown A, Yang G, Wu X, Yan F, Dong J, Prasadha MK, Pei J, Ting M, Zhu J, Li C, Hewett S, Dong J, Ziyar I et al (2018) Identifying medical diagnoses and treatable diseases by image-based deep learning. Cell 172(5):1122–1131.e9. https://doi.org/10.1016/j.cell.2018.02.010
13. Khan Y, Khan P, Kumar S, Singh J, Hegde RM (2020) Detection and spread prediction of COVID-19 from chest X-ray images using convolutional neural network-Gaussian mixture model. In: 2020 IEEE 17th India Council international conference, INDICON 2020, pp 1–6. https://doi.org/10.1109/INDICON49873.2020.9342159
14. Kim I, Rajaraman S, Antani S (2019) Visual interpretation of convolutional neural network predictions in classifying medical image modalities. Diagnostics 9(2):1–15. https://doi.org/10.3390/diagnostics9020038
15. Li L, Qin L, Xu Z, Yin Y, Wang X, Kong B, Bai J, Lu Y, Fang Z, Song Q, Cao K, Liu D, Wang G, Xu Q, Fang X, Zhang S, Xia J, Xia J (2020) Using artificial intelligence to detect COVID-19 and community-acquired pneumonia based on pulmonary CT: evaluation of the diagnostic accuracy. Radiology 296(2):E65–E71. https://doi.org/10.1148/radiol.2020200905
16. Narin A (2020) Detection of covid-19 patients with convolutional neural network based features on multi-class X-ray chest images. In: TIPTEKNO 2020—Tip Teknolojileri Kongresi—2020 medical technologies congress, TIPTEKNO 2020, pp 1–4. https://doi.org/10.1109/TIPTEKNO50054.2020.9299289
17. Otake R, Kurima J, Goto H, Sawada S (2020) Deep learning model to predict real-time seismic intensity. In: Fourth international conference on electronics, communication and aerospace technology (ICECA-2020), pp 1315–1320. IEEE Xplore Part Number: CFP20J88-ART; ISBN: 978-1-7281-6387-1
18. Panwar H, Gupta PK, Siddiqui MK, Morales-Menendez R, Bhardwaj P, Singh V (2020) A deep learning and grad-CAM based color visualization approach for fast detection of COVID-19 cases using chest X-ray and CT-Scan images. Chaos, Solitons Fractals 140:110190. https://doi.org/10.1016/j.chaos.2020.110190
19. Polat Ç, Karaman O, Karaman C, Korkmaz G, Balci MC, Kelek SE (2021) COVID-19 diagnosis from chest X-ray images using transfer learning: enhanced performance by debiasing dataloader. J Xray Sci Technol 29(1):19–36. https://doi.org/10.3233/XST-200757
20. Rahman T, Khandakar A, Qiblawey Y, Tahir A, Kiranyaz S, Bin Abul Kashem S, Islam MT, Al Maadeed S, Zughaier SM, Khan MS, Chowdhury MEH (2021) Exploring the effect of image enhancement techniques on COVID-19 detection using chest X-ray images. Comput Biol Med, 132. https://doi.org/10.1016/j.compbiomed.2021.104319
21. Rajaraman S, Siegelman J, Alderson PO, Folio LS, Folio LR, Antani SK (2020) Iteratively pruned deep learning ensembles for COVID-19 detection in chest X-rays. IEEE Access 8:115041–115050. https://doi.org/10.1109/ACCESS.2020.3003810
22. Saha O, Tasnim J, Raihan MT, Mahmud T, Ahmmed I, Fattah SA (2020) A multi-model based ensembling approach to detect COVID-19 from chest X-ray images. In: IEEE region 10 annual

international conference, proceedings/TENCON, 2020-Nov, pp 591–595. https://doi.org/10.1109/TENCON50793.2020.9293802
23. Selvaraju RR, Cogswell M, Das A, Vedantam R, Parikh D, Batra D (2020) Grad-CAM: visual explanations from deep networks via gradient-based localization. Int J Comput Vision 128(2):336–359. https://doi.org/10.1007/s11263-019-01228-7
24. Shorfuzzaman M, Masud M, Alhumyani H, Anand D, Singh A (2021) Artificial neural network-based deep learning model for COVID-19 patient detection using X-ray chest images. J Healthc Eng. https://doi.org/10.1155/2021/5513679
25. Sinha S, Tushar, Goel S (2021) Research on data science ensembles for covid-19 detection and length of stay prediction. In: Proceedings—IEEE 2021 international conference on computing, communication, and intelligent systems, ICCCIS 2021, pp 499–503. https://doi.org/10.1109/ICCCIS51004.2021.9397218
26. Sujatha K, Kishore KVK, Rao BS (2020) Performance of machine learning algorithms in red wine quality classification based on chemical compositions. Indian J Ecol 47:176–180
27. Sujatha K, Kishore KVK, Rao BS, District G, Pradesh A (2020) Machine learning models for forecasting confirmed, recovered and deceased COVID-19 cases in India. Int J Control Autom 13(4):841–854
28. Sujatha K, Srinivasa Rao B (2019) Recent applications of machine learning: a survey. Int J Innov Technol Explor Eng (IJITEE) 8(6):263–267. Retrieval Number: F10510486C219 / 19©BEIESP
29. Tabarisaadi P, Khosravi A, Nahavandi S (2020) A deep Bayesian ensembling framework for COVID-19 detection using chest CT images. IEEE Trans Systems Man Cybern Syst, 2020-Oct, pp 1584–1589. https://doi.org/10.1109/SMC42975.2020.9283003

Statistical Analysis of Success Rate of the Movie Based on a Combination of Actor, Director

Sunil Bhutada, V. Kakulapati, Naveen Thota, Abhinav Reddy Kothapally, and Aditya Reddy Baradi

Abstract The high demand for movies in the entertainment industry necessitates filmmakers to carefully consider their projects for success. Factors such as duration, title, genre, directorial prowess, and the main cast, particularly the hero and heroine, are crucial for a movie's success. This study aims to expand the existing literature by introducing additional dimensions, focusing on lead cast combinations and duration considerations. Understanding the dynamics between heroes and heroines is crucial for shaping a movie's reception and long-term success. The duration of a movie also influences audience engagement and ensures sustained captivation throughout the viewing experience. This research aims to provide filmmakers with actionable data to inform their decision-making processes and guide them in making judicious casting choices. The study aims to contribute to the film industry's predictive capabilities, enabling filmmakers to create content that resonates with discerning audiences.

Keywords Prediction · Movie · Performance · Machine learning · Lead cost · Combination · Model · Feature engineering · Viewers

1 Introduction

The worldwide box office earnings in 2016 amounted to a staggering 38.3 billion USD. Moreover, the capacity of computers to forecast the popularity of movies before their release holds considerable promise. Utilizing extensive online data and advanced computational capabilities, machine learning algorithms have the potential to forecast movie ratings, evaluate the suitability of online data for predicting success, and ascertain the presence of a link between movie qualities and success [1].

S. Bhutada · V. Kakulapati · N. Thota (✉) · A. R. Kothapally · A. R. Baradi
Sreenidhi Institute of Science and Technology, Yamnampet, Ghatkesar, Hyderabad, Telangana, India
e-mail: thotanaveen4604@gmail.com

V. Kakulapati
e-mail: vldms@yahoo.com

© The Author(s), under exclusive license to Springer Nature Singapore Pte Ltd. 2025

H. Zen et al. (eds.), *Soft Computing and Signal Processing*, Lecture Notes in Networks and Systems 1200, https://doi.org/10.1007/978-981-97-9926-8_15

A film is composed of scripts, performers, filming, and video products. Social media buzz is essential for achieving success at the box office. Additional aspects such as the narrative, distributor, producer, frequency of marketing, genre, movie classification, and prominent actors also have an impact on the success of a film [2].

The research investigates the factors contributing to the success of Telugu movies, specifically examining the influence of recurring actors on viewers, the correlation between directors and genres, and the distinct attention spans and preferences of Telugu audiences. The objective is to provide a comprehensive comprehension of the Telugu movie sector by amalgamating information on actors, directors, genre preferences, and length dynamics. This will facilitate the emergence of a new age in Telugu filmmaking.

The primary objective of this work is to enhance the performance and interpretability of algorithmic methods for statistical analysis by adding a new feature rather than removing or reducing existing ones. The research has shown the efficacy of the characteristics used in forecasting the success of movies, particularly the successful amalgamation of the protagonist, heroine, director, and other factors. Due to the limited number of features in the research, which are generated from just six variables, there is no need to exclude any of them.

The research demonstrates that the participation of stars in films has a substantial effect on projected box office earnings, with an average celebrity being valued at almost $3 million. The more powerful the ensemble, the more significant the influence of a recently hired actor with a proven track record of box office success or a notable artistic standing. Nevertheless, the research does not demonstrate that the participation of stars enhances the assessment of film firms, implying that actors do not provide additional value than they get, therefore holding ramifications for movie business executives [3]

The study's goal is to strike a compromise between the attractive design of Wordle-style word clouds and the practicality and readability required for data analysis activities. According to the assertion, visually grouping words that are semantically related, either by arranging them spatially or using color, greatly enhances performance in analytic tasks. Empirical studies demonstrate that using the visual organization of words based on category structure yields superior comprehension of underlying subjects in contrast to conventional word cloud arrangements. An assessment and its accompanying dataset have been created, demonstrating great accuracy when evaluated separately and providing a direct application for evaluating word layout algorithms.

The subsequent sections of this work are structured in the following manner: Sect. 2 presents established techniques for forecasting the success of a movie. Section 3 provides an in-depth analysis of the technique used in this research. The following part (Sect. 4) provides details on the materials used in this research. Section 5 provides an analysis of the recommended approach. Sections 6 and 7 of the paper address the investigation's significance and prospective studies.

2 Related Work

The categorization methodology identifies films based on their performance, ranging from commercial failures to highly successful blockbusters. This is achieved by employing data analysis techniques to uncover and analyze the connections and patterns among different movie options. Variables taken into account include the financial resources allocated to the movie, its level of public interest, and its achievement in terms of performance. Nevertheless, laws concerning copyright provide obstacles when it comes to obtaining data. Information analysis was used to make movie predictions [4], and it achieved a level of performance that was comparable to IMDb data. In separate research, projections were produced for future results by analyzing social media content and historical data [5]. However, it is not possible to make accurate forecasts of success until the movie is released. A movie investment choice prediction approach was suggested in 2015 [6].

The study [7] assesses the careers of directors, actors, and actresses based on their participation in high-scoring films, in addition to grading films' value in the annals of cinema history. The success rating of these movies is based on the distribution of connections between movies and aims to find movies that have influenced movies and may inspire others. The score is composed of four network centrality ratings: PageRank, Harmonic, In-degree, and Proximity [8]. The score is derived from details about each film's production country, genre, and year of release. It is taken from the IMDb publicly available subset, which has a strong Western cinema bias.

The research [9] uses sentiment and survival analysis to examine the variables impacting a film's commercial success. It categorized 1038 films by comparing showing days, customer reviews, and film attributes. The findings showed notable variations in the number of showing days, ratings for G- and R-rated films, drama and action, favorable and negative remarks, and local and international film productions.

The success of a movie was predicted using Twitter data [10], underscoring the significance of believing social media emotions in today's internet-driven culture. In order to forecast a movie's performance, [11] performed sentiment analysis on tweets.

Multiple research investigations have been carried out on the prediction of movie ratings and the analysis of sentiment in reviews, mostly using ML and NLP techniques. The designer used autonomous learning to categorize movies according to their mean lexical introductions, while Pang et al. conducted a comparative analysis of machine learning methods using movie reviews and found that support vector machines (SVMs) classifications [12] had the highest performance. Blogger's feelings are used to forecast cinema revenue, using a combination of machine learning and a straightforward approach that involves tallying both favorable and adverse remarks in reviews [13]. A method to enhance collaborative filtering by including sentiment analysis. This included integrating productive and discriminating algorithms for sentiment prediction [14]. In addition, they showcased an enhanced method of analyzing sentiment for online critiques of films, using clustering techniques [15]. Additional suggested methodologies consist of composite convolutional neural network (CNN)-long short-term memory (LSTM) techniques, attention-based long short-term memory models, forecasting algorithms for Hindi films, and lexical and neural network-based methods [16].

3 Methodology

This research investigates the collaborative interactions among analysis based on director and hero, genre analysis based on directors, lead cast analysis, and ideal run time analysis (boring/interesting) with a specific emphasis on analyzing their past collaborations and evaluating their respective percentages of success. The median actor-filmmaker combination frequencies and profits of a movie are determined by computing the mean quantity of times actors and actresses have worked on a picture produced by the filmmaker and the typical rate of profitability per movie generated from previous collaborations.

Through the use of data measurements, descriptive statistics provide clear explanations of the characteristics of a data collection. Mean, median, and mode are the three most well-known center measurements, and they are used in many math and statistics courses. By adding up each figure in the data set and dividing the total number of figures, one may get the mean, or average [17].

Data analytics is a complicated, computational, and statistical process that impacts the way people make decisions via a sequence of prescriptive, predictive, and descriptive analytics that use statistical approaches such as mean, median, mode, etc.

The arithmetic average of all terms, obtained by adding up all phrases and dividing by the total number of terms, is the mean of a statistical distribution containing a discrete random variable. The anticipated value, or mean, of a distribution including a continuous random variable is calculated by multiplying the variable by its probability, represented by the Greek symbol mu (μ). The number of terms determines the median of a distribution containing a discrete random variable; if the number is odd, the median is the term in the center; if the number is even, the median is the average of the two terms in the middle. The term that appears the most often in a distribution with a discrete random variable is called the mode. It is normal for a distribution with a discrete random variable to have more than one mode, particularly if there are few terms [18].

The ensemble of a film plays a pivotal part in collaboration, with their previous knowledge and skill greatly enhancing the team's achievements. The inclusion of individuals with diverse backgrounds and perspectives in a team, especially in the fields of technological studies and artistic endeavors, enhances the generation of novel concepts and fosters unique insights [19]. The combination of variety and familiarity significantly contributes to the success of a filmmaker in creating successful movies. Furthermore, there is a direct and positive relationship between diversity and box office revenue. Additionally, the past expertise of cast members also has a beneficial impact on sales [20]. Hence, a heterogeneous and recognizable ensemble is necessary for the triumph of a film.

Word cloud is a method of visually representing important textual data points, frequently employed to evaluate data from various social media sites. Originally referred to as tag clouds, word clouds were first utilized for illustrating the occurrence of user-generated tags on blogging platforms and picture-sharing websites. Nevertheless, the first usability research [21] discovered that the spatial arrangements and non-uniform font sizes of online collections were disadvantageous for comprehension when compared to straightforward alphabetical word lists [22, 23]. Researchers have determined that the primary benefit of these is their enjoyable and captivating quality, as well as their capacity to indicate social engagement, despite initially unfavorable usability findings [24].

4 Implementation Analysis

The data is collected from Kaggle, which is publically available. The methodology has four major components; these are (1) data collection, (2) data cleaning, (3) data transfer, (4) data analysis and prediction. By cleaning the dataset and discarding the irrelevant data from the IMDB dataset as well as through a detailed study of the dataset, we get the attributes that can affect the prediction of success of a movie.

4.1 Data Preprocessing

In the initial phase of our research, meticulous data preprocessing was undertaken to guarantee the quality and dependability of our dataset. Essential exploratory data analysis techniques, including examinations of the dataset's shape, unique values, descriptive statistics, column names, data types, and identification of missing values, formed the basis of our approach. We employed the shape function to ascertain the dataset's dimensions, offering insights into its overall size. Exploring unique values within specific columns using the unique function aided in recognizing categorical variability. Descriptive statistics generated through the describe function were instrumental in identifying potential outliers and gaining an overview of central tendencies. The examination of column names and data types using the columns attribute and dtypes function, respectively, provided essential information for subsequent data manipulation. Furthermore, the null function allowed us to identify and address missing values, ensuring the dataset's completeness. These foundational steps laid the groundwork for a comprehensive and reliable analysis of Telugu movies, setting the stage for subsequent explorations and machine learning applications.

The method proposes an approach for describing the combination of hero and heroine data analysis using statistical measures such as mean, median, and mode.

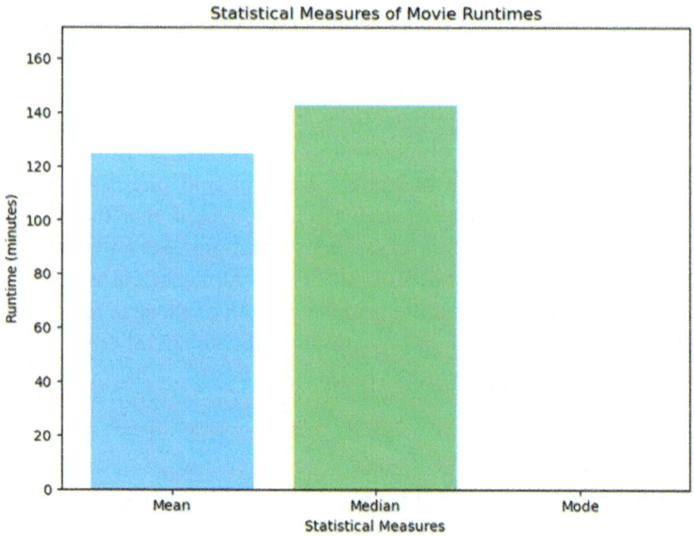

Fig. 1 Statistical analysis of the success of movies

These technologies provide extensive data on the correlation between different combinations, contributing to the success of the movie.

Calculate the statistical measures for runtime analysis like mean, mode, and median [18] shown in Fig. 1.

The mean of a discrete random variable is the mathematical average of all terms, calculated by adding and dividing terms. The mean, or anticipated value, of a continuous random variable, is calculated by integrating the product of the variable and its probability, represented by the Greek symbol mu (μ).

$$\text{Mean} = \frac{\sum_{i=1}^{n} X_i}{n}. \tag{1}$$

The number of words in a discrete distribution determines its median, which is the middle term if there are odd terms, and the average of the two middle terms if there are even terms. Conversely, the median of a continuous distribution is defined as the value *m*, such that there is a chance of at least 50% that a randomly selected point would be smaller than or equal to *m*.

$$\begin{aligned} &\text{Median} = \frac{n+1}{2}\text{th term,} \\ &n = \text{odd}\left\{\frac{n}{2}\text{th term} + \frac{n}{2} + I\text{th term}\right\} \Big/ 2 \\ &n = \text{even.} \end{aligned} \tag{2}$$

A discrete random variable's mode is the term with the most common value; bimodal and trimodal distributions, on the other hand, have two or three modes. The mode of a continuous random variable corresponds to the peak or highest point of the function.

$$\text{Mode} = l + \left(\frac{f_1 - f_0}{2f_1 - f_0 - f_2}\right) \times h. \quad (3)$$

4.2 *Runtime Analysis*

To understand the distribution of movie runtimes, we performed a statistical analysis that included mean, median, and mode runtimes. The bar graph 'Statistical Measures of Movie Runtime' visualizes these measures, providing a comprehensive overview of the typical durations of Telugu movies.

Mean Runtime: 124.74 min
Median Runtime: 143.00 min
Mode Runtime: 0.00 min.

4.3 *Word Cloud Analysis*

4.3.1 Directors

To gain insights into the prominent directors in Telugu cinema, we generated a word cloud based on the names of directors associated with the movies in our dataset. The 'Word Cloud of Directors' (shown in Fig. 2) visually represents the frequency and prominence of directors, providing a quick overview of their contribution to the Telugu film industry.

The most well-known directors in cinema were represented in a word cloud that we colored and scaled based on frequency, with purple terms being the least common.

The best actresses in cinema were represented in a word cloud that we colored and scaled based on frequency, with purple terms being the least common as shown in Figs. 3 and 4.

4.3.2 Heroes

Similarly, a word cloud was created for the 'Word Cloud of Heroes,' showcasing the leading actors in Telugu movies as shown in Fig. 5. This visualization offers a snapshot of the most frequently appearing heroes in our dataset, shedding light on their popularity and influence.

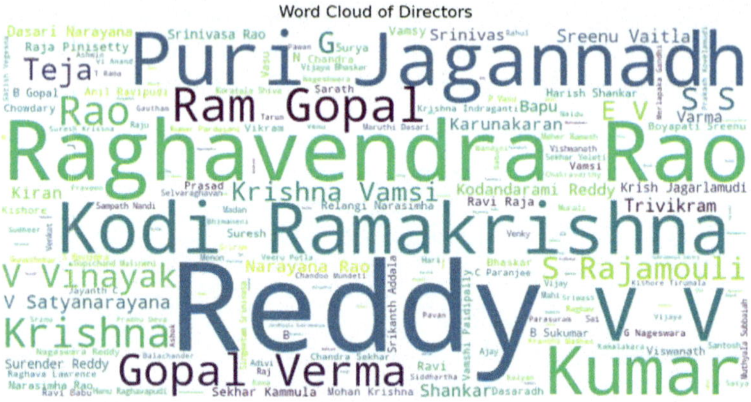

Fig. 2 Success of movies' word cloud based on directors

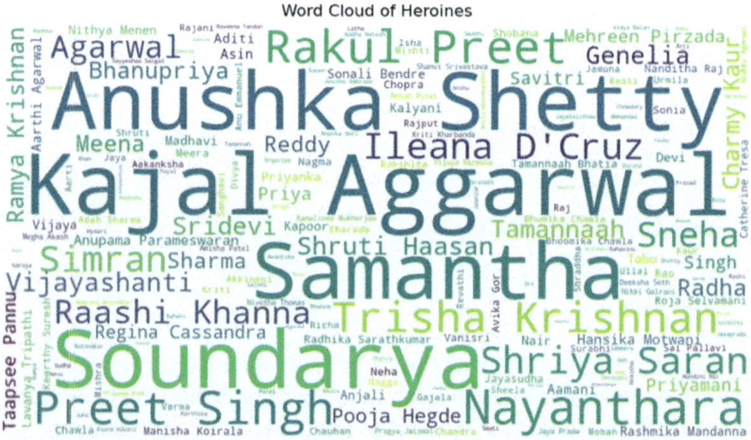

Fig. 3 Success of movies' word cloud based on heroines.

The best hero in cinema was represented in a word cloud that we colored and scaled based on frequency, with purple terms being the least common.

4.4 Combination Analysis

4.4.1 Hero-Heroine Combinations

We explored the combination of heroes and heroines in Telugu movies by analyzing the frequency of pairings. The bar graph 'Hero-Heroine Combination Counts' (shown

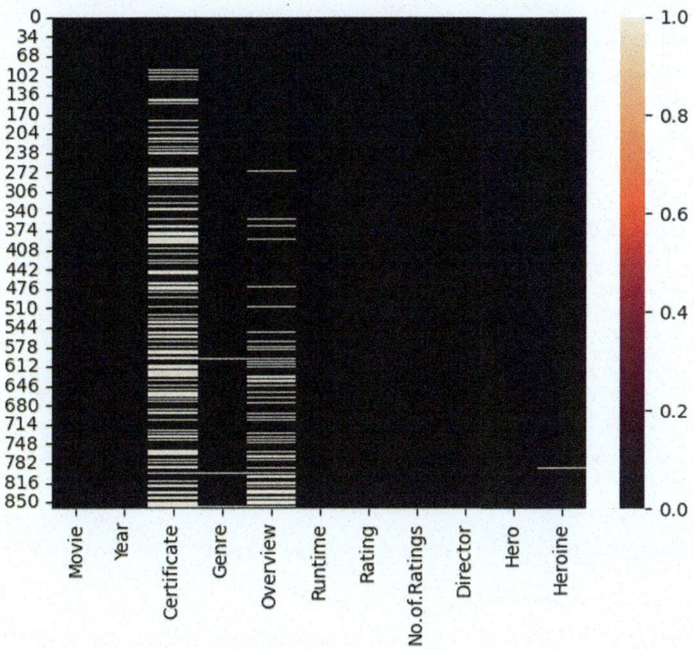

Fig. 4 Distribution of movies by different parameters

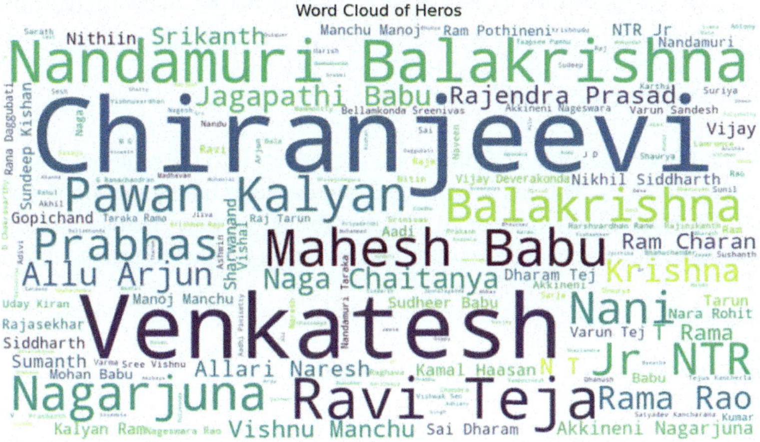

Fig. 5 Success of movies' word cloud based on heroes

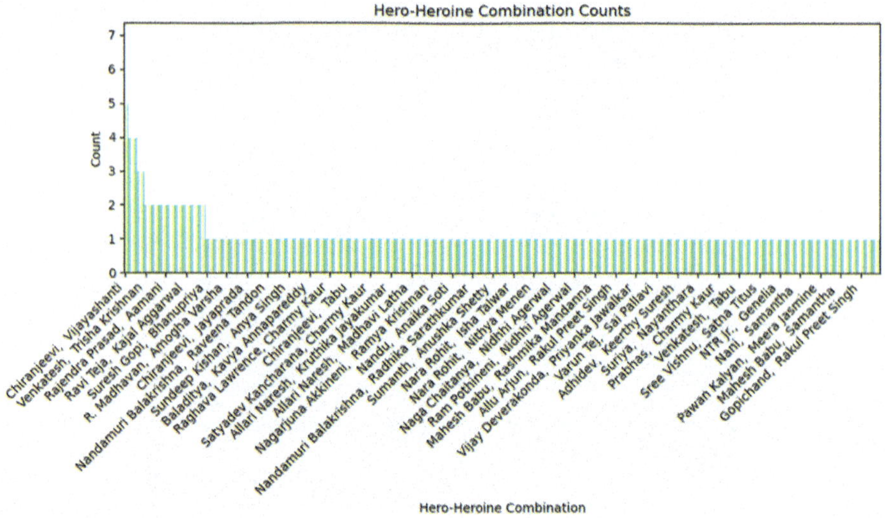

Fig. 6 Distribution of movies by hero and heroine combination

in Fig. 6) displays the number of movies associated with different hero-heroine pairs. This analysis gives us insights into the most popular on-screen pairs in the industry.

To understand the numerical distribution of both hero and heroine combinations, we generated histograms for each of these categories across all genres and keywords.

4.4.2 Hero-Director Combinations

In addition to hero-heroine combinations, we delved into the relationships between heroes and directors. The bar graph 'hero-director combination counts' (shown in Fig. 7) illustrates the frequency of collaborations between heroes and directors, providing a perspective on the successful partnerships within the Telugu film industry.

5 Discussion

The impact of actors and director combination, which is a critical component of a successful film, especially in Tollywood movies, assesses the impact of variables such as the basis of director and hero, genre analysis based on directors, lead cast analysis, and ideal run time analysis (boring/interesting) using a statistical analysis technique. This method advances our knowledge of the elements that contribute to a successful film and its influence on viewers around the globe, offering fresh perspectives for making smart choices in the creation and marketing of movies. This work aims to enhance the effectiveness and understanding of algorithmic methods

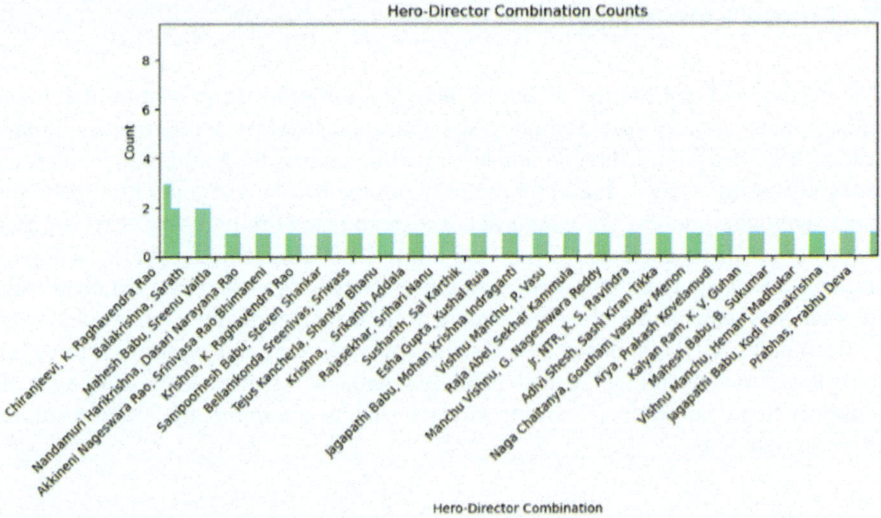

Fig. 7 Hero-director combination counts

for machine learning by adding an entirely novel characteristic rather than removing or reducing existing ones. The efficacy of the elements included in the research for forecasting movie success has been established, and the omission of some variables may lead to a reduction in the accuracy of the model.

6 Conclusion

The data from the official IMDb movies dataset collection was used in this article to analyze production-related trends and patterns as well as to determine the most significant movies and actors, actresses, and directors in the history of film. The research study aims to evaluate the performance of movies by analyzing certain aspects, including the duration of the movie, the rating of directors, the analysis of the lead cast, and the combination analysis. The analysis will be conducted using basic statistical methods. However, the aforementioned cinema-related factors are not the only ones that affect a movie's success. The size of the audience and their ratings have a big impact on a movie's success. Without an audience, the film business would be meaningless since its purpose revolves around audiences. The statistical study of the movie's success rate, using measurements such as mean, median, and standard deviation, offers a thorough overview of the amalgamation of protagonists and heroines, presenting a precise depiction of the performed data analysis.

7 Future Enhancement

The casting and distribution of a film include a multifaceted procedure that takes into account variables such as money, schedule, and audience demographics. Future studies need to examine director and actor pairings across the board in India director using clustering analysis based on region wise, the nation where the film made the most significant money. We would like to extend this work by using deep learning algorithms will be expanded to include multimodal classification issues, whereby ‘ratings’ are categorized into more than binary words. Real-time data obtained from internet sites may be used to achieve more accurate categorization. An evaluation of the recommended method may be conducted by employing an innovative framework that integrates conventional lexical techniques and natural language processing techniques with advanced deep learning models, such as ensemble modeling, resulting in better outcomes.

References

1. https://www.diva-portal.org/smash/get/diva2:1106715/FULLTEXT01.pdf
2. Harris LC, Fisk RP, Sysalova H (2016) Exposing Pinocchio customers: investigating exaggerated service stories. J Serv Manag 27(2):1–38
3. Elberse A (2007) The power of stars: do star actors drive the success of movies? J Mark 71(4):102–120. https://doi.org/10.1509/jmkg.71.4.102
4. Zhang W, Skiena S (2009) Improving movie gross pre-diction through news analysis. In: Web intelligence, pp 301–304
5. Asur S, Huberman BA (2010) Predicting the future with social media. http://arxiv.org/abs/1003.5699
6. Lash MT, Zhao K (2015) Early predictions of movie success: the who, what, and when of portability
7. Bioglio L, Pensa R (2018) Identification of key films and personalities in the history of cinema from a Western perspective. Appl Netw Sci 3:50. https://doi.org/10.1007/s41109-018-0105-0
8. Canet F, Valero MÁ, Codina L (2016) Quantitative approaches for evaluating the influence of films using the IMDB database. Comunicación Soc 29(2):151
9. Kim A, Trimi S, Lee SG (2021) Exploring the key success factors of films: a survival analysis approach. Serv Bus 15:613–638. https://doi.org/10.1007/s11628-021-00460-x
10. Anand B, Himanshu K, Vinay B, Pranali K (2015) Role of different factors in predicting movie success. In: 2015 international conference on pervasive computing (ICPC). https://doi.org/10.1109/PERVASIVE.2015.7087152
11. Mundra S, Dhingra A, Kapur A, Joshi D (2019) Prediction of a movie's success using data mining techniques. In: Satapathy S, Joshi A (eds) Information and communication technology for intelligent systems. Smart innovation, systems and technologies, vol 106. Springer, Singapore
12. Pang B, Lee L, Vaithyanathan S (2002) Thumbs up? Sentiment classification using machine learning techniques. arXiv preprint cs/0205070. https://doi.org/10.3115/1118693.1118704
13. Mishne G, Glance NS (2006) Predicting movie sales from blogger sentiment. In: AAAI spring symposium: computational approaches to analyzing weblogs, pp 155–158
14. Kennedy A, Inkpen D (2006) Sentiment classification of movie reviews using contextual valence shifters. Comput Intell 22(2):110–125. https://doi.org/10.1111/j.1467-8640.2006.00277.x

15. Nagamma P, Pruthvi HR, Nisha KK, Shwetha NH (2015) An improved sentiment analysis of online movie reviews based on clustering for box-office predictions. In: International conference on computing, communication & automation. IEEE, pp 933–937. https://doi.org/10.1109/CCAA.2015.7148530
16. Shaukat Z, Zulfiqar AA, Xiao C, Azeem M, Mahmood T (2020) Sentiment analysis on IMDB using lexicon and neural networks. SN Appl Sci 2(2):1–10. https://doi.org/10.1007/s42452-019-1926-x
17. https://www.investopedia.com/terms/d/descriptive_statistics.asp
18. https://www.techtarget.com/searchdatacenter/definition/statistical-mean-median-mode-and-range
19. Guimera R, Uzzi B, Spiro J, Amaral LAN (2005) Team assembly mechanisms determine collaboration network structure and team performance. Science 308(697–702):5722
20. Meiseberg B, Ehrmann T (2013) Diversity in teams and the success of cultural products. J Cult Econ 37(1):61–86
21. Hearst MA, Rosner D (2008) Tag clouds: data analysis tool or social signaller? In: Hawaii international conference on system sciences, proceedings of the 41st annual. IEEE
22. Halvey MJ, Keane MT (2007) An assessment of tag presentation techniques. In: Proceedings of the 16th international conference on the world wide web. ACM, pp 1313–1314
23. Rivadeneira AW, Gruen DM, Muller MJ, Millen DR (2007) Getting our head in the clouds: toward evaluation studies of tag clouds. In: Proceedings of the SIGCHI conference on human factors in computing systems. ACM, pp 995–998
24. Sinclair J, Cardew-Hall M (2008) The folksonomy tag cloud: when is it useful? J Inf Sci 34(1):15–29

Machine Learning Methodologies in Social Platform: An Analytical Survey

Priyam Pandey, Anuruddha Paul, Roshni Pradhan, Aradhana Pattnaik, and Rwik Kar

Abstract Social media platforms have become a breeding ground for various criminal activities, leading to the need for a comprehensive survey to understand their nature and detection methods. The vast amount of data available on these platforms requires advanced analytical techniques to derive meaningful insights. In this paper, we present valuable insights from multiple studies, addressing crimes, event detection, and machine learning applications on social media. We propose methodologies for crime detection, event detection, and the utilization of machine learning algorithms, emphasizing the significance of interchangeability and efficiency in this context. Our findings suggest that advanced analytical techniques can help improve crime detection and event extraction, and machine learning algorithms can be effectively utilized to achieve this goal.

Keywords Social media · Cybercrime · Event detection · Behavior analysis · Machine learning · Social network analysis

---

Priyam Pandey, Aradhana Pattnaik and Rwik Kar: These authors contributed equally to this work.

P. Pandey · A. Paul (✉) · R. Pradhan · A. Pattnaik · R. Kar
School of Computer Engineering, KIIT Deemed to Be University, Bhubaneswar, Odisha 751024, India
e-mail: 2328072@kiit.ac.in

P. Pandey
e-mail: 2328108@kiit.ac.in

R. Pradhan
e-mail: roshni.pradhanfcs@kiit.ac.in

A. Pattnaik
e-mail: 2328074@kiit.ac.in

R. Kar
e-mail: 2328114@kiit.ac.in

© The Author(s), under exclusive license to Springer Nature Singapore Pte Ltd. 2025

H. Zen et al. (eds.), *Soft Computing and Signal Processing*, Lecture Notes in Networks and Systems 1200, https://doi.org/10.1007/978-981-97-9926-8_16

1 Introduction

Machine learning (ML) is a fascinating field of Artificial Intelligence (AI) that's all about getting computers to learn from data. We can feed computers data and let them learn the connections that bolster it. Once these machines have learned these connections, we can use them to predict other issues. It's a fast-growing technology and is extensively used in colorful sectors. ML is a subset of AI where the machine can learn and facilitate from experience. It doesn't need to be explicitly programmed to perform a certain task. The main end of ML is to allow machines to learn automatically without mortal intervention. ML uses colorful algorithms to make fine models on sample data, known as "training data", to make prognostications or opinions without being specifically programmed for the task. These algorithms are an integral part of the ML system and define how the models are going to be trained. There are different types of ML algorithms, but they're substantially classified into three orders Supervised Learning (SL), Unsupervised Learning (UL), and Reinforcement Learning (RL) [1]. Supervised Learning algorithms are trained using labeled exemplifications, similar to an input where the asked affair is known. Unsupervised Learning algorithms are used when the information used to train is neither classified nor labeled. Underpinning Learning allows the model to learn its gest grounded on feedback from the terrain. ML has a wide range of operations including hunt machines, dispatch pollutants, recommendation systems, and social media [2]. For illustration, when you browse online, ML helps in showing you affiliated products that you might be interested in based on your browsing history. In health care, ML is used for prognosticating conditions and helping in diagnosing them. It helps croakers in relating patterns in patient data and prognosticating complaint progression. In finance, ML algorithms are used for credit scoring and algorithmic trading. In transportation, ML helps in optimizing routes and schedules for delivery lines. In conclusion, machine learning is an instigative branch of artificial intelligence that's changing our world fleetly. It's a tool that's driving everything from our particular lives to assiduity and exploration. It's far and wide helping croaks diagnose conditions, enabling us to surf online more effectively, making our buses safer, and much further.

2 Machine Learning in Other Fields

Machine learning, a subset of artificial intelligence, is a revolutionary technology that has fleetly converted colorful diligence and aspects of our diurnal lives. Its capability to dissect data, learn from it, and make prognostications or opinions without unequivocal programming instructions has led to a multitude of operations across different fields. Now, we explore some crucial areas where machine learning is significantly impactful.

2.1 Health Care

In health care, machine learning is employed for complaint opinion, medicine discovery, and substantiated treatment plans. Machine learning models can dissect medical images (e.g., X-rays, MRIs) to describe anomalies or excrescences, perfecting the delicacy of judgments. Similarly, it can help identify implicit medicine campaigners by assaying chemical parcels and prognosticating their efficacy. Case data and inheritable information can also be used to produce individualized treatment plans and prognosticate complaint pitfalls.

2.2 Finance

The fiscal sector utilizes machine learning for fraud discovery, algorithmic trading, and threat assessment. Machine learning algorithms can identify unusual patterns in fiscal deals to flag eventuality fraudulent conditioning, guarding both fiscal institutions and guests. Algorithmic trading systems use prophetic analytics to make split-alternate opinions for buying and dealing means and optimizing investment portfolios. Also, machine learning models dissect request data to assess and prognosticate fiscal pitfalls.

2.3 Retail

In retail assiduity, machine learning powers recommendation machines that suggest products to guests based on their browsing and purchase history. This enhances the shopping experience and boosts deals. Force operation is also optimized through prophetic analytics, helping retailers reduce redundant stock and avoid deaths. Similarly, machine learning can dissect client sentiment from reviews and social media, furnishing precious perceptivity for product development and marketing strategies [3].

2.4 Manufacturing

Machine learning is becoming a pivotal part of manufacturing processes through prophetic conservation. Detectors in the ministry collect data, which is also evaluated to predict when conservation is needed, reducing time-out and costs. Quality control is another operation where machine learning can identify blights in real time, icing product quality and thickness. Supply chain optimization is also bettered through demand soothsaying, leading to more effective product and distribution [4].

2.5 Natural Language Processing (NLP)

NLP is a subfield of machine learning that focuses on enabling computers to understand, interpret, and induce mortal language. It has wide operations in chatbots, virtual sidekicks, and sentiment analysis. Chatbots use NLP algorithms to give automated client support and answer queries, while virtual sidekicks like Siri and Alexa can perform tasks grounded on voice commands. Sentiment analysis of social media and news data helps associations gauge public opinion and makes informed opinions.

2.6 Autonomous Vehicles

Machine learning is pivotal for the development of independent vehicles, including tone-driving buses and drones. These vehicles use detectors, cameras, and machine learning algorithms to perceive their surroundings, make opinions, and navigate safely. Machine learning models can fete objects, climbers, and business signs, enabling independent vehicles to drive with a high degree of delicacy and safety.

2.7 Image and Video Analysis

Machine learning is considerably used for image and videotape analysis. This includes facial recognition for security and authentication purposes, object discovery in surveillance systems, and content temperance on social media platforms. Also, machine learning can be used to tag and classify vast quantities of visual data, making it searchable and systematized [5].

2.8 Environmental Monitoring

Machine learning is employed in environmental wisdom for tasks similar to climate modeling, pollution vaticination, and wildlife conservation. Climate models use literal and real-time data to make prognostications about unborn climate conditions. Pollution vaticination models dissect air and water quality data to assess the impact of pollution on the terrain and mortal health. Machine learning also helps monitor and guard exposed species by assaying data from tracking bias and remote detectors [6].

2.9 Education

Machine learning is becoming a pivotal part of education by bodying learning gests. Adaptive Learning platforms use machine learning algorithms to assess a pupil's strengths and sins and knitter educational content consequently. Also, machine learning can automate grading and assessment tasks, saving preceptors' precious time. Natural language processing is also used to give feedback on written assignments.

2.10 Entertainment

The entertainment assiduity utilizes machine learning for happy recommendation, content creation, and in-depth script analysis. Streaming platforms like Netflix use machine learning algorithms to suggest pictures and television shows grounded on user preferences and history. Content generators can employ machine learning for videotape and music generation, creating new forms of entertainment. Script analysis tools help pens upgrade their work by furnishing perceptivity into character development and plot structure.

2.11 Energy Management

Machine learning plays a vital part in optimizing energy consumption in structures and artificial processes. Prophetic analytics models can read energy demand, allowing for effective allocation and application of coffers. Smart grids use machine learning to manage electricity distribution and reduce energy destruction. Also, machine learning can optimize the operation of renewable energy sources similar to wind turbines and solar panels.

2.12 Agriculture

Precision husbandry benefits from machine learning by optimizing crop operation. Detectors collect data on soil quality, rainfall conditions, and crop health, which machine learning models dissect to give perceptivity on when to plant, wash, and harvest crops. This not only increases crop yields but also conserves coffers and reduces environmental impact.

2.13 Public Safety

Machine learning is employed in public safety operations, including prophetic policing, disaster response, and trouble discovery. Prophetic policing models dissect literal crime data to identify high-threat areas and allocate coffers consequently. During disasters, machine learning helps with resource allocation and response planning. In security, machine learning algorithms can describe implicit pitfalls in surveillance footage, field security, and cybersecurity.

Machine learning, with a special focus on social media, has revolutionized the way we understand and interact with the digital world. In this era of information, social media platforms have become an integral part of our daily lives, serving as conduits for communication, entertainment, and information dissemination. Machine learning techniques have been harnessed to unlock the vast potential of social media data in several key areas like user classification, behavior analysis, business intelligence, cyber crime detection, and event detection. This convergence of technology and human interaction on social media platforms has ushered in a new era of data-driven insights and decision-making.

3 Machine Learning in Major Sectors of Social Media

3.1 Behavior Analysis

Machine learning plays a pivotal role in analyzing user behavior on social media platforms [6]. Algorithms can detect patterns and trends in user actions, such as the types of content they engage with, the frequency of their interactions, and the sentiment of their comments. This behavioral analysis helps social media platforms refine their recommendation systems, suggesting relevant content to users and enhancing their overall experience. Additionally, behavior analysis aids in identifying unusual or potentially harmful activities, such as trolling, racism, hate speech, or online harassment, allowing platforms to take proactive measures to protect users and maintain a positive online environment.

Spamming is also a common method of harassment on social media. Machine learning models can help in identifying spam accounts and messages using behavioral analysis, thus assisting the platforms in mitigating such activities to enhance user experience. But these models lack precision.

The review of other spam detection models has suggested that the main reason for loss of precision is the imbalanced nature of data about spam from social media user datasets. Only about 5% of tweets in these datasets are labeled spam which makes it hard to train the ML models. To counter this problem, a combination of ML models with the SMOTEEN model is used which balances the dataset and thereby improves the framework for spam detection accuracy and precision. The SMOTEEN model's effectiveness, in conjunction with ML algorithms like Random Forest, KNN, Decision Trees, etc., can be seen in Table 1 shown [7].

After the SMOTEEN model processes the training dataset, it can be used to train the ML Classifier Algorithms to flag accounts as spam or not. It is done by comparing 15 distinct or derived parameters to accurately determine the user account type. The parameters are listed in Table 2.

Table 1 Result of SMOTEEN model used in conjunction with ML algorithms

S. No.	Model	Accuracy (A)	Precision (P)
1	RF without SMOTEENN	99.04	97.73
2	RF with SMOTEENN	99.26	99.49
3	DT without SMOTEENN	98.34	83.39
4	DT with SMOTEENN	98.29	97.92
5	KNN without SMOTEENN	97.22	75.9
6	KNN with SMOTEENN	98.76	97.91
7	GNB without SMOTEENN	19.78	5.75
8	GNB with SMOTEENN	59.92	56.47

Table 2 Parameters required for classification

S. No.	Features	Description	Feature category
1	Account_age	No. of days the account is till its most recent tweet	Profile-based
2	Followers_count	The followers of Twitter account	Profile-based
3	Following_count	Total number of accounts he/she follows	Profile-based
4	Userfavourites_ count	Total count of favorites	Content based
5	Lists_count	No. of the lists created	Profile-based
6	Tweets_count	Total count of tweets of the user	Content based
7	Retweets_count	Total count of retweets	Content based
8	Hashtag_count	Total count of hashtags incorporated	Content based
9	Usermention_ count	Total count of user references	Content based
10	Urls_count	Total count of URLs	Content based
11	Char_count	Total court of characters in this tweet	Content based
12	Digits_count	Total count of digits in this tweet	Content based
13	Is_retweet	Tweet has re-tweet	Content based
14	No_hastags	No. of hashtags	Content based
15	F_ratio	Followers to followings ratios	Profile-based

Table 3 Types of criminal offenses

S. No.	Criminal offenses	Description
1	Harassment, malicious communications, and stalking	Individuals may use social media to harass, threaten, or stalk others, leading to criminal offenses such as harassment and stalking
2	Threatening violence and incitement	Posts that contain threats of violence or incite others to commit criminal acts can result in criminal charges
3	Privacy violations	Disclosure of private information, such as personal photos or sensitive details, without consent, can constitute a criminal offense [9]
4	Fraud and scams	Social media platforms can be utilized for perpetrating various types of fraud, including financial scams and identity theft
5	Hate speech and discrimination	Social media may be used as a platform for spreading hate speech or discriminatory content, which can violate laws and regulations [3]

3.2 Cybercrime

Social media has become ingrained in people's daily lives, with users often sharing explicit and implicit details of their activities and experiences. Unfortunately, these platforms can also be the avenue for criminal offenses, many of which may go unreported due to various reasons [4]. This lack of reporting can lead to the underrepresentation of certain crimes in traditional statistics [8]. Different types of Criminal Offenses are enlisted in Table 3.

Underreported Crimes on Social Media, the underreporting of crimes on social media can be attributed to various factors, including the triviality of the offenses, embarrassment, and lack of awareness about being a victim of a crime. As a result, these offenses may not be accurately reflected in official crime statistics, potentially leading to a skewed understanding of the prevalence of such crimes in the digital realm [10].

The intertwining of social media with individuals' daily lives has brought to light various criminal offenses committed through these platforms. Recognizing the factors contributing to the under reporting of these crimes is essential in addressing and mitigating their prevalence. It is imperative to raise awareness about reporting mechanisms and provide support to victims of social media-related crimes to ensure their safety and well-being.

3.3 Event Detection

The rise of social media has revolutionized the way we consume and share information, particularly during events such as natural disasters, political developments, and public emergencies [5]. Social media platforms have become critical sources of real-time information, and machine learning algorithms have enabled us to sift through the massive volume of social media posts to detect and track events as they unfold [6, 11].

By analyzing keywords, hashtags, and geolocation data, machine learning algorithms can identify trending topics and map the spread of information during critical events. This capability is invaluable for emergency responders, journalists, and researchers who need to understand public sentiment and gather on-the-ground information. Furthermore, event detection on social media can be used for proactive marketing strategies, capitalizing on trending topics to engage with users in real time.

Machine learning has played a significant role in transforming our engagement with social media by enabling user classification, behavior analysis, business intelligence, cybercrime detection, and event detection. The combination of big data and advanced algorithms has empowered social media platforms and businesses to make more informed decisions, enhance user experiences, and maintain safe online environments [12].

As the digital landscape continues to evolve, machine learning will remain at the forefront of innovations in the realm of social media, shaping the way we connect, communicate, and consume information in the digital age [13]. Machine learning has influenced almost every aspect of our lives, bringing about transformative changes across various industries. Its ability to analyze vast quantities of data, identify patterns, and make informed decisions has paved the way for innovations and effectiveness in diverse fields [14, 15]. The use of machine learning in social media has brought about significant changes, and we can expect even more remarkable developments in future. This technology will continue to shape the future of technology and society, making it easier for individuals and organizations to stay informed and connected, and to make informed decisions based on real-time data.

4 Related Work

Recent advancements in machine learning and social media analysis have led to significant contributions in various domains. Gao et al. [7] focused on filtering brand-related microblogs using a multi-view embedding approach, providing insights into brand interactions on social media. Their study contributed to understanding brand interactions but did not address the detection of criminal activities on these platforms.

Batrinca and Treleaven [8] provided a comprehensive survey of social media analytics techniques and platforms, highlighting various methods and tools available for data analysis. Despite its thoroughness, the survey lacked a focus on real-time event detection and cybercrime.

Bello-Orgaz et al. [9] discussed the challenges and achievements in handling social big data, emphasizing the need for efficient data processing and analysis techniques. Their work primarily focused on data management and did not delve into behavior analysis for crime detection.

Balaji et al. [1] surveyed machine learning algorithms for social media analysis, emphasizing their significance in understanding user behaviors and detecting criminal activities. However, their study did not explore real-time event detection and the integration of advanced models for comprehensive analysis.

Tripathi et al. [3] discussed evaluating business intelligence systems using machine learning models, highlighting their impact on decision-making processes within organizations. Although their research provided valuable insights into business intelligence, it did not address social media analysis specifically.

Drury et al. [4] reviewed the use of social networks for crime detection, showcasing the potential of social media platforms in identifying and preventing criminal activities. While their study provided a solid foundation, it did not explore the real-time detection of events and comprehensive behavior analysis.

Dahou et al. [5] developed a framework for event detection on social media using transformers and swarm optimization, demonstrating the effectiveness of machine learning in crisis management and public notification. Despite its innovative approach, the study did not comprehensively integrate user behavior analysis and spam detection.

Huang et al. [6] conducted research on multi-label user classification using the ML-KNN algorithm, illustrating the versatility of machine learning in handling complex social media data. Their work focused on classification but did not address the detection of criminal activities or real-time event tracking.

Flack and D'Souza [2] discussed the future of social network science and engineering in the digital age, providing insights into the evolving landscape of social media analysis. Their study, however, did not cover the specific applications of machine learning in crime detection and event tracking.

5 Research Gap and Novelty

Despite these advancements, several critical gaps remain in the existing literature:

Real-Time Event Detection: While several studies have addressed event detection, there is a lack of comprehensive frameworks that integrate real-time data analysis for timely detection and response.

Behavior Analysis for Crime Detection: Existing research often overlooks the combination of behavior analysis and crime detection on social media platforms. This integration is crucial for understanding user interactions and identifying potential threats.

Spam Detection: Many studies focus on spam detection in isolation, without integrating it into broader frameworks for social media analysis. Effective spam detection can significantly enhance the accuracy of crime detection and behavior analysis.

6 Addressing the Research Gap

This study aims to bridge these gaps by proposing a novel framework that combines advanced machine learning techniques with real-time data analysis for improved crime detection, event tracking, and behavior analysis on social media platforms. Specifically, our contributions include:

Integrated Framework: We propose an integrated approach that combines real-time event detection, behavior analysis, and spam detection using advanced machine learning models and the SMOTEEN algorithm.

Enhanced Accuracy and Precision: By employing a combination of machine learning models and data balancing techniques, our framework enhances the accuracy and precision of spam and crime detection on social media platforms.

Comprehensive Analysis: Our methodologies offer a comprehensive framework for understanding user behavior, detecting potential threats, and responding to events in real time, thereby contributing significantly to the fields of social media analytics and public safety.

7 Conclusion

Machine learning is a revolutionary technology that has transformed various industries, including health care, finance, retail, manufacturing, natural language processing, autonomous vehicles, image and video analysis, environmental monitoring, education, entertainment, energy management, agriculture, and public safety. It has enabled advancements such as disease diagnosis, fraud detection, personalized recommendations, predictive analytics, and more. In the realm of social media, machine learning is used for behavior analysis, spam detection, cybercrime prevention, and event detection, providing valuable insights and enhancing user experiences.

References

1. Balaji TK,Annavarapu CSR, Bablani A (2021) Machine learning algorithms for social media analysis: a survey. Comput Sci Rev 40:100395. https://doi.org/10.1016/j.cosrev.2021.100395
2. Flack JC, D'Souza RM (2014) The digital age and the future of social network science and engineering. Proc IEEE 102(12):1873–1877. https://doi.org/10.1109/jproc.2014.2368790
3. Tripathi MA, Madhavi K, Kandi VSP, Nassa VK, Mallik B, Chakravarthi MK (2023) Machine learning models for evaluating the benefits of business intelligence systems. J High Technol Manage Res 34(2):100470. https://doi.org/10.1016/j.hitech.2023.100470
4. Drury B, Drury SM, Rahman MA, Ullah I (2022) A social network of crime: a review of the use of social networks for crime and the detection of crime. Online Soc Netw Media 30:100211. https://doi.org/10.1016/j.osnem.2022.100211

5. Dahou A, Mabrouk A, Ewees AA, Gaheen MA, Abd Elaziz M (2023) A social media event detection framework based on transformers and swarm optimization for public notification of crises and emergency management. Technol Forecast Soc Change 192:122546. https://doi.org/10.1016/j.techfore.2023.122546
6. Huang A, Xu R, Chen Y, Guo M (2023) Research on multi-label user classification of social media based on ML-KNN algorithm. Technol Forecast Soc Chang 188:122271. https://doi.org/10.1016/j.techfore.2022.122271
7. Gao Y, Zhen Y, Li H, Chua T-S (2016) Filtering of brand-related microblogs using social-smooth multiview embedding. IEEE Trans Multimedia 18(10):2115–2126. https://doi.org/10.1109/tmm.2016.2581483
8. Batrinca B, Treleaven PC (2014) Social media analytics: a survey of techniques, tools and platforms. AI Society 30(1):89–116. https://doi.org/10.1007/s00146-014-0549-4
9. Bello-Orgaz G, Jung JJ, Camacho D (2016) Social big data: recent achievements and new challenges. Inf Fusion 28:45–59. https://doi.org/10.1016/j.inffus.2015.08.005
10. McFee B, Barrington L, Lanckriet G (2012) Learning content similarity for music recommendation. IEEE Trans Audio Speech Lang Process 20(8):2207–2218. https://doi.org/10.1109/tasl.2012.2199109
11. Aziz A, Mohd Foozy CF, Shamala P, Suradi Z (2018) YouTube spam comment detection using support vector machine and k-nearest neighbor. Indonesian J Electr Eng Comput Sci 12(2):612. https://doi.org/10.11591/ijeecs.v12.i2.pp612-619
12. Asur S, Huberman BA (2010) Predicting the future with social media. In: 2010 IEEE/WIC/ACM international conference on web intelligence and intelligent agent technology. https://doi.org/10.1109/wi-iat.2010.63
13. Shahrampour S, Rakhlin A, Jadbabaie A (2016) Distributed detection: finite-time analysis and impact of network topology. IEEE Trans Autom Control 61(11):3256–3268. https://doi.org/10.1109/tac.2015.2506903
14. Cao N, Shi C, Lin S, Lu J, Lin Y-R, Lin C-Y (2016) TargetVue: visual analysis of anomalous user behaviors in online communication systems. IEEE Trans Visual Comput Graphics 22(1):280–289. https://doi.org/10.1109/tvcg.2015.2467196
15. Weishampel A, Staicu A-M, Rand W (2023) Classification of social media users with generalized functional data analysis. Comput Stat Data Anal 179:107647. https://doi.org/10.1016/j.csda.2022.107647

AI-Driven Cattle Skin Disease Diagnosis: Enhancing Precision Through Advanced Deep Learning and Interactive Chatbot Technologies

S. Kayalvili, E. Srimathi, K. Samyuktha, and M. P. Pranesh

Abstract In this work, a thorough strategy for solving the problem of skin disorders in cattle is offered that uses state-of-the-art deep learning methods and interactive artificial intelligence applications. CNN and Vision Transformer's deep learning LLM models are used to create a superior cow skin disease classifier. A wide dataset of more than 2000 photos is collected, including a number of skin diseases such as ringworm, eye disease, papillomatosis, photosensation, catarrhal tongue fever, and lumpy skin disease, using careful data collection and augmentation procedures. Innovative data enrichment techniques have increased the diversity and richness of the dataset. With the help of this improved data set, we were able to train the proposed deep learning model and obtained a staggering 85% accuracy rate. A model was implemented using Hugging Face AI models to ensure wider availability and usability, guarantee smooth integration and user-friendly interactions. In addition, the work was continued on developing an engaging AI chatbot designed specifically to talk about cattle skin diseases. This chatbot has features including chat history visualization, text-to-speech capabilities, and download transcripts of discussions as PDF documents. In the field of veterinary science, incorporating these aspects not only improves the user experience, but also promotes collaborative learning and information distribution.

Keywords Cattle skin disease · CNN · Vision transformer · LLM models · AI chatbot · Data augmentation

---

S. Kayalvili (✉) · E. Srimathi · K. Samyuktha · M. P. Pranesh
Department of Artificial Intelligence, Kongu Engineering College, Erode, Tamil Nadu 638060, India
e-mail: kayalvili.ai@kongu.edu

E. Srimathi
e-mail: srimathie.21aim@kongu.edu

K. Samyuktha
e-mail: samyukthak.21aim@kongu.edu

M. P. Pranesh
e-mail: praneshmp.21aim@kongu.edu

© The Author(s), under exclusive license to Springer Nature Singapore Pte Ltd. 2025

H. Zen et al. (eds.), *Soft Computing and Signal Processing*, Lecture Notes in Networks and Systems 1200, https://doi.org/10.1007/978-981-97-9926-8_17

1 Introduction

Skin disorders in cattle [1] have a major impact on animal health, animal welfare, and the viability of the agricultural and veterinary industries. To overcome these obstacles, extensive research was conducted using state-of-the-art deep learning [2] and natural language processing technology [3]. The main achievement of this work was the creation of a sophisticated cattle skin disease classifier using Vision Transformer deep learning models [4]. Using methodical data collection procedures, a broad and diverse data set was collected that includes several skin diseases common in cattle, including bluetongue, skin disease, ophthalmology, papillomatosis, photosensation, and lumpy skin disease [5].

The state-of-the-art data augmentation techniques [6] were used to improve the quantity and quality of the dataset, enriching it and improving the learning capabilities of the model. The resulting algorithm [5] was 85% successful in correctly diagnosing skin disorders in cattle, demonstrating impressive accuracy. A learned model was implemented using Hugging Face AI models [7] to guarantee accessibility and usefulness and offer a smooth user interface for people to effectively identify and understand cowhide conditions.

In addition, An AI chatbot [8] was created specifically focused on talking about skin disorders in cattle to encourage interactive learning and information exchange. In addition to enabling real-time communication, this chatbot has user-friendly [7, 9] features including text-to-speech, chat history visualization, and conversation downloads for later use. By combining these components, our study improves veterinary science while promoting interactive and accessible collaborative learning and information sharing [10].

2 Dataset Description

There are 2064 high-resolution photos in the dataset (shown in Table 1 and Fig. 1) showing a number of common cow diseases.

Table 1 Dataset description

Disease name	Count
Ringworm	189
Photosensation	659
Papilomatosis	319
Ocular	359
Lumpy	239
Blue tongue	299
Total	2064

Fig. 1 Types of diseases

The study is based on these carefully selected photos that facilitate the creation and teaching of sophisticated machine learning models. The richness and complexity of the collection is ensured by a diverse selection of photographs that represent individual groups of diseases. By offering a useful resource for disease detection and study, it is aimed to greatly advance the field of veterinary science with this information.

3 Data Augmentation

A number of transformations were performed on the original images using TensorFlow's ImageDataGenerator [6]. In this study, a comprehensive set of data augmentation techniques [6] was implemented to enrich the cattle disease dataset. Utilizing TensorFlow's versatile "ImageDataGenerator," a series of transformations were applied to introduce variability and increase the size of the dataset, thereby ensuring the robustness and generalization of our deep learning model. The following augmentation techniques were used:

3.1 Tensor Flow Data Augmentation

1. ***Vertical Flip***

   The images were flipped vertically, simulating different orientations and perspectives of cattle skin diseases.
2. ***Horizontal Flip***

   Horizontal flipping was used to diversify the data set by presenting mirror images, capturing variations in skin disease appearance.

3. ***Rotation (± 90°)***

   The images were rotated ± 90° as shown in Fig. 2, mimicking the different angles at which skin disease can be observed.

4. ***Brightness Change (Range: 0.2–1.0)***

   The brightness of the images was adjusted within a specified range (0.2–1.0) to replicate different lighting conditions in real-world scenarios as shown in Fig. 3.

5. ***Zoom (Scale Range: 0.5–1.0)***

   A random zoom between 50 and 100% of the original size was used, capturing the different levels of detail and scale variation in skin conditions.

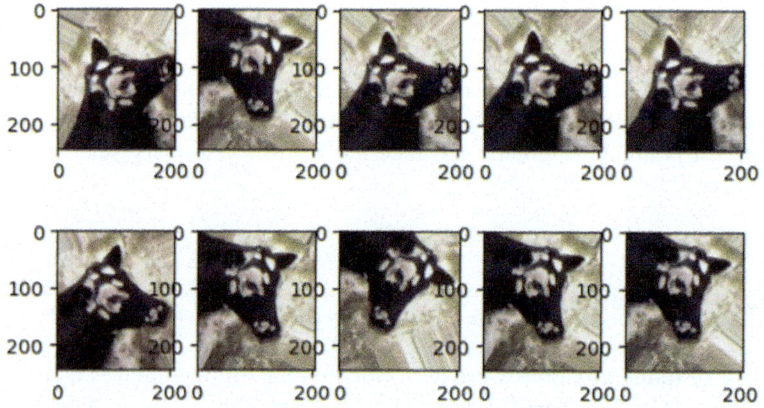

Fig. 2 Augmented images using rotation (± 90°) technique

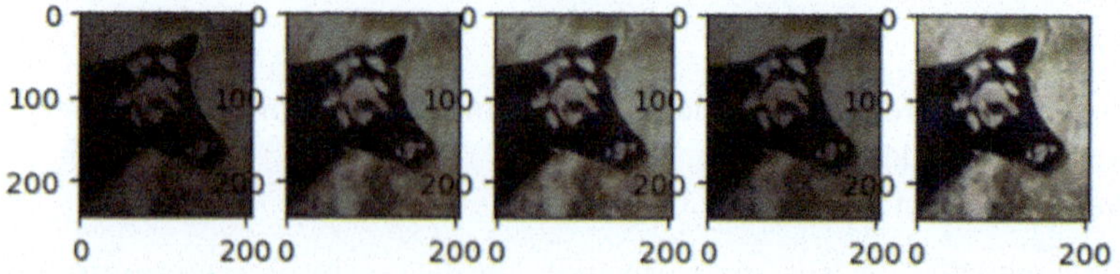

Fig. 3 Augmented images using brightness change (range: 0.2–1.0) technique

By systematically applying these augmentation techniques [6, 11], an expanded dataset was created with increased diversity, which ensured that the deep learning model could effectively learn the nuances of various bovine skin diseases. These augmented images have played a key role in increasing the accuracy and reliability of our bovine skin disease classifier [4], making it capable of handling a wide range of real-world scenarios and disease presentations.

3.2 PIL Image Augmentation

To further diversify the bovine disease dataset, in addition to TensorFlow data augmentation approaches, PIL (Python Imaging Library) transformations were also used. The model was exposed to different visual cues by introducing small but important changes in the appearance of skin disorders using a combination of color jitter adjustments as shown in Fig. 4. The following modifications were made using PIL:

Color Jittering

Brightness: By changing the brightness of the images, a brightness variation of up to 0.5 was possible, simulating different lighting situations.
Contrast: Contrast levels have been adjusted up to a maximum of 0.5 to capture changes in texture and tone.
Saturation: Saturation values (which varied by up to 0.5) were adjusted to simulate different levels of color intensity.
Tint: Tint, which means different colors in the spectrum, has been changed slightly to change the color palette of the photos (with up to 0.5 hue variation).

By using this particular mixture of color flicker modifications [6, 11], it was able to create subtle changes in the appearance of skin conditions. The diversity of the

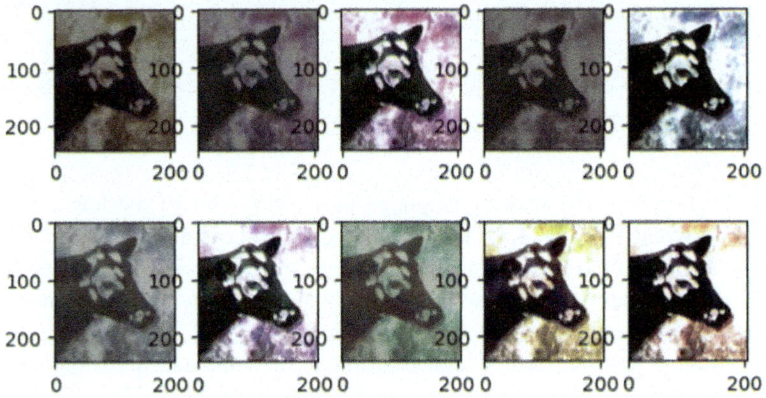

Fig. 4 Augmented images using color jitter technique

dataset was greatly increased by these enhanced photos [6], which were stored and organized in the output directory. To ensure that the trained deep learning model can correctly recognize and categorize bovine skin diseases in different situations, each image represents a different variant of the original presentations of the disease [4]. The ability of the model to adapt to real-world situations was greatly improved by this augmentation procedure, making it a reliable option for disease identification in realistic agricultural conditions.

4 Methodology

The development of the Vision Transformer (ViT) deep learning cattle skin disease classifier involved several critical phases, each of which contributed to model robustness and user accessibility (shown in Figs. 5 and 6). Initially, a large dataset of over 2000 images was carefully compiled and expanded using the techniques like Tensorflow data augmentation and PIL image augmentation, ensuring a broad representation of diseases such as skin disease, ocular disease, papillomatosis, photosensitization, blue tongue, and lumpy skin disease. This augmentation process included various techniques to increase image diversity and simulate real-world variability, thereby improving the generalizability of the model. The Vision Transformer was selected for its superior performance in capturing the complex visual dependencies and patterns necessary for accurate disease identification. After rigorous training and validation, the model was deployed on the Hugging Face AI platform, making it readily available for practical applications. To improve user interaction, an interactive chatbot has been developed that integrates sophisticated features such as text-to-speech for auditory communication, chat history to track interactions, and PDF download functionality for detailed, exportable records of discussions. This comprehensive methodology not only ensured the high accuracy and reliability of the classifier but also enabled an engaging and informative user experience, bridging the gap between advanced AI technology and practical veterinary applications.

5 Large Language Model (LLM)

A class of sophisticated natural language processing (NLP) models called large language models (LLMs) [4, 11] are designed to process and produce text that is similar to a person's text. Because transformer structures form the basis of these models, they are able to represent complex language bindings and patterns. By demonstrating state-of-the-art performance in a variety of language-related tasks, including text generation, translation, sentiment analysis, and question answering, LLMs have made significant contributions to the field of NLP [7].

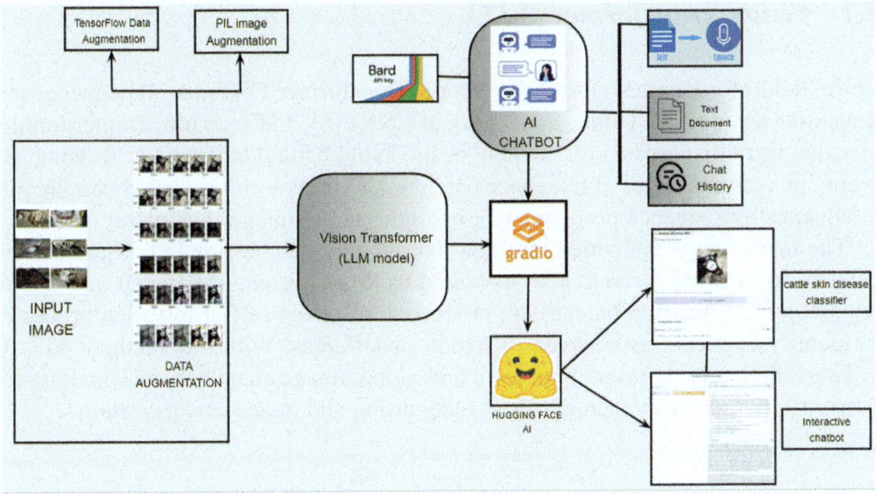

Fig. 5 Workflow of the complete project

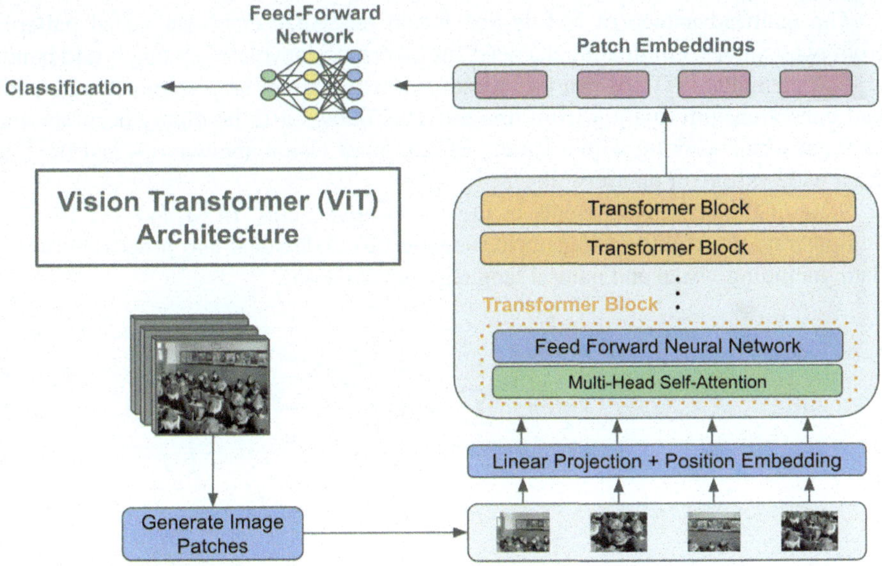

Fig. 6 Vision transformer (VIT) architecture

Understanding the subtleties of language and context is enabled by LLM's ability to learn from vast amounts of textual data. This is one of their primary characteristics. These models can gain a thorough understanding of syntax, vocabulary, and semantics by pre-training on large datasets. Focusing on certain activities helps them hone their skills and adapt their knowledge to more specialized applications.

5.1 Vision Transformer (VIT)

In the field of computer vision, the Vision Transformer (ViT) [4, 11] concept is a revolutionary method. Unlike conventional CNNs [5], ViT uses transformer topologies that were first created for challenges involving natural language processing. By seeing images as a series of fixed-size patches, ViT transforms image processing into a sequence by sequence problem [12], revolutionizing image categorization.

The input image is divided into non-overlapping patches for ViT [4], and these patches are linearly inserted and processed as tokens. Using the transformer architecture, these tokenized patches are processed, allowing self-observation processes to identify connections between different visual areas. With this method, ViT [6, 12] can efficiently represent both local and global image characteristics, making it a powerful tool for tasks such as object recognition and image categorization.

$$\begin{aligned} \mathbf{x}_{\text{in}} &= \left[\mathbf{x}_{\text{cls}} \| \mathbf{x}_{\text{patch}}\right] + \mathbf{x}_{\text{pos}} \\ \mathbf{y}_k &= \mathbf{x}_{k-1} + \text{MSA}(\text{LN}(\mathbf{x}_{k-1})) \\ \mathbf{x}_k &= \mathbf{y}_k + \text{FFN}(\text{LN}(\mathbf{y}_k)) \end{aligned} \tag{1}$$

The main advantage of ViT is that it can recognize elaborate visual patterns from pixel values, eliminating the need for laborious convolutional layers and handcrafted elements. ViT [6] can use transformers to extract long-term dependencies and contextual information from images. This allows it to be highly accurate and efficient when working with a variety of computer vision applications and flexible for a wide variety of datasets.

State-of-the-art developments in their respective fields, LLM and ViT [4, 11], demonstrate the ability of transformer designs to understand and process complex data, including visual and natural language information.

$$l_{\text{sn}}(x, y, T) = -\frac{1}{b} \sum_{i \in 1 \ldots b} \log \left(\frac{\sum_{\substack{j \in 1 \ldots b \\ j \neq i \\ y_i = y_j}} e^{-\frac{\|x_i - x_j\|^2}{T}}}{\sum_{\substack{k \in 1 \ldots b \\ k \neq i}} e^{-\frac{\|xi - xk\|^2}{T}}} \right) \tag{2}$$

6 Hugging Face AI

Natural language processing (NLP) [3] is being revolutionized in the form of Hugging Face AI, a major open-source platform. Hugging Face's robust model center [7], intuitive pipeline, and adaptable Transformers library enable academics and developers

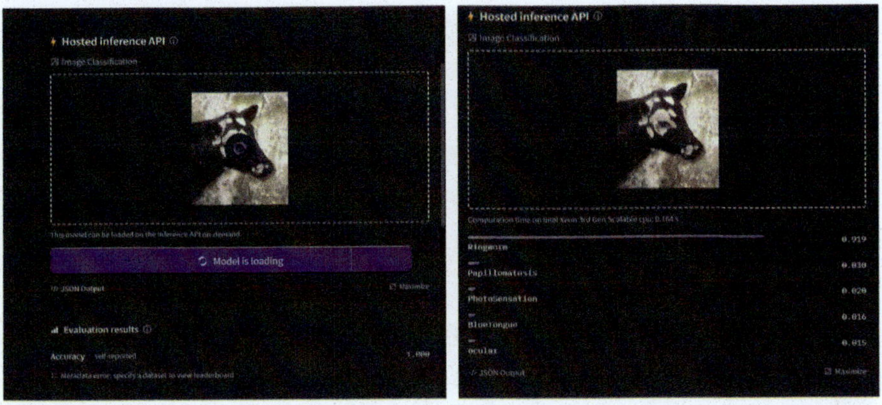

Fig. 7 Deployed cattle skin disease classifier on hugging face AI

to design advanced NLP applications. This dynamic [13], community-driven platform accelerates the pace of invention and opens difficult NLP tasks to a global audience.

6.1 Cattle Skin Disease Classifier

The Vision Transformer Deployed Model

Here the created method is demonstrated that uses Vision Transformer (ViT) [7, 11] to classify cow skin diseases as shown in Fig. 7. It is seamlessly integrated into the Hugging Face platform and uses cutting-edge AI technology to transform livestock healthcare.

Creative Interactive AI Chatbot

Explore the cutting-edge chatbot [3, 8], displayed in the Hugging Face Spaces deployable environment [7], enhanced with features such as text-to-speech, chat history, and PDF download capability. Cutting-edge technology is transforming the conversation about skin disorders in cattle [11] (shown in Fig. 8).

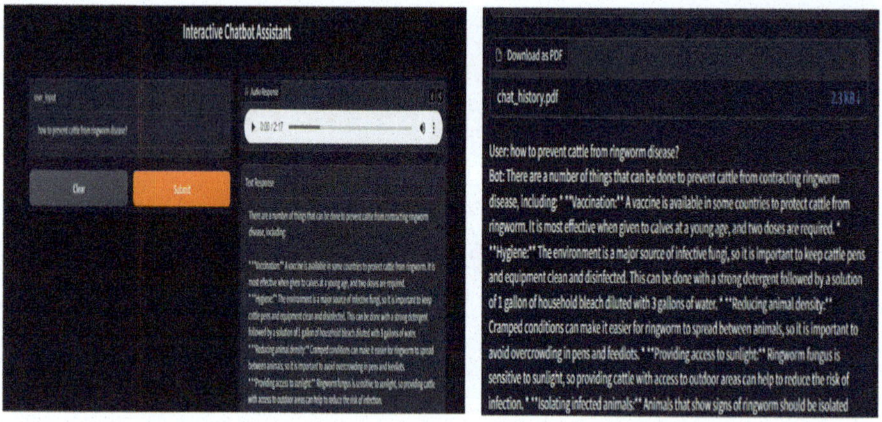

Fig. 8 Deployed interactive chatbot on hugging face AI

7 Conclusion

This experiment shows how sophisticated Vision Transformer models and interactive AI technology can be perfectly integrated. Through creative use of Hugging Face AI, a robust and intuitive framework was developed for the diagnosis of skin diseases in cattle. Combined with the user-friendly features of interactive AI chatbot, the accuracy of the Vision Transformer-based classifier represents a major advance in veterinary research. This study bridges the gap between cutting-edge technology and functional agricultural solutions by improving disease diagnosis and fostering collaborative learning.

References

1. Narayan V, Awasthi S, Fatima N, Faiz M, Bordoloi D, Sandhu R, Srivastava S (2023) Severity of lumpy disease detection based on deep learning technique. In: IEEE international conference
2. Dommeti D, Nallapati SR, Lokesh C, Bhuvanesh SP, Padya VVP (2023) Deep learning based lumpy skin disease (LSD) detection. In: IEEE 3rd international conference
3. Laato S, Morschheuser B, Hamari J, Björne J (2023) AI-assisted learning with ChatGPT and large language models: implications for higher education. In: IEEE international conference on advanced learning technologies (ICALT)
4. Kanca E, Ayas S, Kablan EB, Ekinci M (2023) Performance comparison of vision transformer-based models in medical image classification. In: IEEE 31st signal processing and communications applications conference (SIU)
5. Rony M, Barai D, Riad, Hasan Z (2021) Cattle external disease classification using deep learning techniques. In: IEEE 12th international conference (ICCCNT)
6. Maurya R, Pandey NN, Singh VP, Gopalakrishnan T (2023) Plant disease classification using interpretable vision transformer network. In: IEEE international conference on recent advances in electrical, electronics & digital healthcare technologies

7. Rothman D, Gulli A (2022) Transformers for natural language processing: build, train, and fine-tune deep neural network architectures for NLP with Python, Hugging Face, and OpenAI's GPT-3, ChatGPT, and GPT-4. Packt Publishing
8. Nandini Prasad KS, Sudhanva S, Tarun TN, Yuvraaj Y, Vishal DA (2023) Conversational chatbot builder—smarter virtual assistance with domain-specific AI. In: IEEE 4th international conference for emerging technology (INCET)
9. Prayitno PI, Leksono RPP, Chai F, Aldy R, Budiharto W (2021) Health chatbot using natural language processing for disease prediction and treatment. In: 1st international conference on computer science and artificial intelligence (ICCSAI)
10. Sundaramoorthy K, Anitha R, Kayalvili S, Ghazala AFA, El-Ebiary YAB, Al-Ashmawy S (2023) Hybrid optimization with recurrent neural network-based medical image processing for predicting interstitial lung disease. Int J Adv Comput Sci Appl (IJACSA) 14(4)
11. Kurz A, Adams E, Depoian AC, Bailey CP, Guturu P (2022) WMC-ViT: waste multi-class classification using a modified vision transformer. In: IEEE MetroCon
12. Jagetia A, Goenka U, Kumari P, Samuel M (2022) Visual transformer for soil classification. In: IEEE students conference on engineering and systems (SCES)
13. Shinde NV, Akhade A, Bagad P, Bhavsar H, Wagh SK, Kamble A (2021) Healthcare chatbot system using artificial intelligence. In: 5th international conference on trends in electronics and informatics (ICOEI)

Revolutionizing Autism Spectrum Prognostication via Ensemble Learning

Sri Ram Ganesh Gokavarapu, Bhanu Sree Veeravalli, Nikhila Kandukuri, Pranathi Samayamanthula, SatyaSivani Kommina, and Lohitha Gollapalli

Abstract Autistic disorder (ASD) manifests as a neurological and developmental condition marked by difficulties in social interaction and communication, noticeable impairments that are distinctive, and the presence of repetitive behavioral patterns. Early diagnosis holds significant potential to mitigate its severity and long-term impacts. The primary goal of the system is to use a vast range of machine learning methodologies within ensemble learning to train datasets for predictive ASD diagnosis. Through the employment of machine learning strategies, the project seeks to upgrade the competence of autism diagnosis while assessing the accuracy of these algorithms utilized for the classification of individuals who might exhibit indications of ASD. The system enables ASD prediction using diverse patient datasets, encompassing over 10,000 records across various demographics such as children, adults, elders, and transgender individuals.

Keywords Machine learning · ASD · Accuracy · Ensemble learning

1 Introduction

The introduction lays the groundwork for comprehending data mining and its utilization across diverse domains, notably in prediction and classification tasks. This paper is singularly focused on employing data mining techniques, specifically Support Vector Classifier (SVC) and Random Forest, to predict the autism problem. It is delineated as a multifaceted neuropsychiatric and learning disorder marked by varied symptoms, impacting individuals across their lifespans. The lack of a standard treatment underscores the importance of accurately identifying ASD through behavioral analysis. However, limited datasets about ASD scrutinizing are available, many of

S. R. G. Gokavarapu · B. S. Veeravalli · N. Kandukuri · P. Samayamanthula (✉) · S. Kommina · L. Gollapalli
Computer Science and Engineering, Sri Vasavi Engineering College, Tadepalligudem, India
e-mail: pranathisamay04@gmail.com

S. R. G. Gokavarapu
e-mail: Sriramganesh8107@srivasaviengg.ac.in

© The Author(s), under exclusive license to Springer Nature Singapore Pte Ltd. 2025

H. Zen et al. (eds.), *Soft Computing and Signal Processing*, Lecture Notes in Networks and Systems 1200, https://doi.org/10.1007/978-981-97-9926-8_18

which are innate in nature. To address this, the paper utilizes a combined dataset comprising screening data for autistic disorder (ASD) provided for children, adolescents, and adults, merging them into a comprehensive dataset with both individual characteristics and behavioral features represented in binary format. The methodology involves employing Python programming and relevant libraries to implement Logistic Regression and Random Forest for predictive analysis. It shapes behavioral patterns and social interactions. Termed a "spectrum" disorder, ASD encompasses a diverse array of symptoms experienced by those affected [1].

The research employed an extensive dataset comprising screening data for autism spectrum disorder (ASD) in children, adolescents, and adults. By merging these distinct datasets, a unified dataset was created for analysis. This integration of varied datasets established a strong basis for exploring ASD across different age groups.

The literature survey within the document delves into prior research efforts related to machine learning applications in medical diagnosis, drawing parallels between the current study and similar endeavors. More specifically, it alludes to research efforts where machine learning approaches have been harnessed for endeavors like forecasting breast cancer risk and classifying MRI stroke occurrences, showcasing the effectiveness of such methodologies within healthcare settings. The survey underscores the significance of accurate prediction methodologies in medical research and highlights the capacity of ML approaches to tackle intricate diagnostic dilemmas. By grounding the study in existing research, the literature survey provides context for the approach taken and contributes to the validity of the results.

2 Review Related Work

2.1 Literature Review

The review of the literature survey underscores the increasing trend in utilizing modern leveraging machine learning and data mining technologies to streamline data exploration processes. Researchers have increasingly turned to applying machine learning methodologies to attain exceptionally precise data analysis. For instance, a study in 2016 focused on employing methodologies such as Support Vector Classifier, Decision Tree, Naive Bayes, and KNeighbors on the breast cancer datasets [2]. The study's outcomes underscored SVM's supremacy in accuracy and error reduction, as observed through experiments conducted in a simulation environment employing the WEKA data mining tool [2].

The research concentrated on analyzing ASD datasets across different age groups. Five distinct feature selection techniques were utilized to diminish feature dimensions while upholding competitive performance, with Relief F emerging as the most efficient method. Through systematic experimentation, attribute numbers were gradually increased, followed by the application of various classification techniques, revealing MLP as the top performer. The study's main constraint is the small dataset size, prompting a future focus on gathering larger datasets and investigating deep learning models to enhance ASD detection. Additionally, plans include analyzing brain signals, such as EEG, to enhance ASD detection algorithms [3, 4].

In this analysis, the sensory processing abnormalities and behavioral alterations in autistic disorder are examined. Using a sample of 72 participants, 26 machine learning regressors from various families are employed, including bagging, and gradient boosting which are part of ensemble learning and Random Forest. The examination outcomes span a spectrum of evaluations, ranging from "commendable to outstanding" in the externalization domain via all SP2 elements, and from "reasonable to commendable" in the realms of anxious/depressed feelings and social difficulties as determined by particular quadrants. Nonetheless, indications of somatic complaints and deviations from rules display diminished predictive consistency. Precision levels in distinguishing between normative, preclinical, and clinical categorizations stand at 97.2% and 87.5% for externalization and anxious/depressed states, respectively, with other outcomes achieving accuracies above or near 70% [5].

This study investigates that examining the degrees of self-determination and underlying factors among late individuals in their adolescence and early adulthood who have received a determination of Autism. Despite calls for increased focus on independence from advocacy and educational organizations, substantial discrepancies in the availability of resources for cultivating self-determination skills, and identification of individuals in the age ranges from 19 to 45 are diagnosed with autism disorder. Caregivers report there exists a significant opportunity for skill practice at home among young adults, yet there is a limited capacity for self-determination.

Additionally, individuals facing intellectual disability or severe manifestations of ASD encounter marked differences in their ability to exercise self-determination. Severity indicators of their disabilities correspond with a decreased likelihood of attaining moderate levels of self-determination, with odds diminishing by 15–26% [6]. These discoveries suggest a disruption in the process of acquiring skills within educational, and home settings, hindering the utilization of the autonomy abilities of individuals in their early adulthood. The study underscores the prominence of initiating self-determination interventions around ages 12–13 to leverage built-in support systems and peer modeling while providing safe opportunities for skill implementation as individuals transition into adulthood. Collaboration and coordination with educational professionals, caretakers, and organizations advocating for individuals are essential to integrate fostering self-determination and developing independent skills within home, educational, and community environments, especially for individuals exhibiting heightened symptom severity. Moreover, it is imperative to institute educational policies and financial channels that aid teachers and community organizations in delivering extra assistance to this at-risk demographic [6].

The models employed in this analysis showcased remarkable accuracy rates and minimal error occurrences, suggesting their potential for effective autism diagnosis in real-world settings. Such efficient diagnosis procedures could save time and costs, facilitating timely access to necessary treatments for individuals with autism. Additionally, those potentially misdiagnosed can undergo further evaluation with medical professionals to corroborate accurate diagnosis and appropriate treatment [7].

This study assesses ML methodologies for analyzing medical data to predict Autism.

After preprocessing to ensure data integrity, a Support Vector Classifier (SVC) emerges as a suitable classifier, particularly with the Gaussian Radial Basis kernel proving most effective. Despite challenges, this approach addresses the necessity for precise autistic disorder prediction within medical datasets is paramount [8]. Acquiring a substantial volume of information utilized for the training of the model, whether in a concentrated or dispersed environment has persisted as a challenge. Hospitals are often reluctant to share their data due to its immense value, compounded by regional data protection regulations that further restrict data-sharing practices [9]. Several researchers have utilized federated learning (FL) for the recognition of various neurological disorders [10].

Federated learning (FL) offers an effective solution to tackle the aforementioned challenges. FL, an innovative machine learning approach, ensures data privacy by avoiding data transmission over the network [11].

Kosmicki et al. proposed a machine learning-driven approach to ascertain a concise set of attributes for recognizing autistic disorder (ASD). Utilizing the Autism Diagnostic Observation Schedule (ADOS) to assess behaviors in children across the autism spectrum, the study scrutinized scoresheets from 4540 individuals using eight distinct ML methods, among them step-wise backward feature identification. The approach identified ASD risk using specific behaviors from ADOS modules 2 and 3, achieving high overall accuracies of 98.27% and 97.66%, respectively.

3 Proposed Methodology

The proposed system neatly considers 10,000 records to predict autism in all age groups. This proposed system also predicts the level of autism in transgender. This project evaluates the level of autism by using different algorithms like Random Forest, SVM, Logistic Regression, etc., and combines all the averages of algorithms by using ensemble learning and predicting the output. This project also suggests some predictions and the top websites to know more about autism.

3.1 Dataset Attributes

For our project, we leveraged a dataset produced from the KAGGLE website, containing 10,000 entries of patient information. This dataset consists of 16 unique features classified into four types: String, integer, Boolean, and binary along with a target variable. Each feature is accompanied by a comprehensive description, as portrayed in the documentation of the dataset.

The current section offers thorough insights into the particular methodologies utilized in the project for predicting autism disorder. Furthermore, it presents a visual representation of the project's workflow, depicted in Fig. 1.

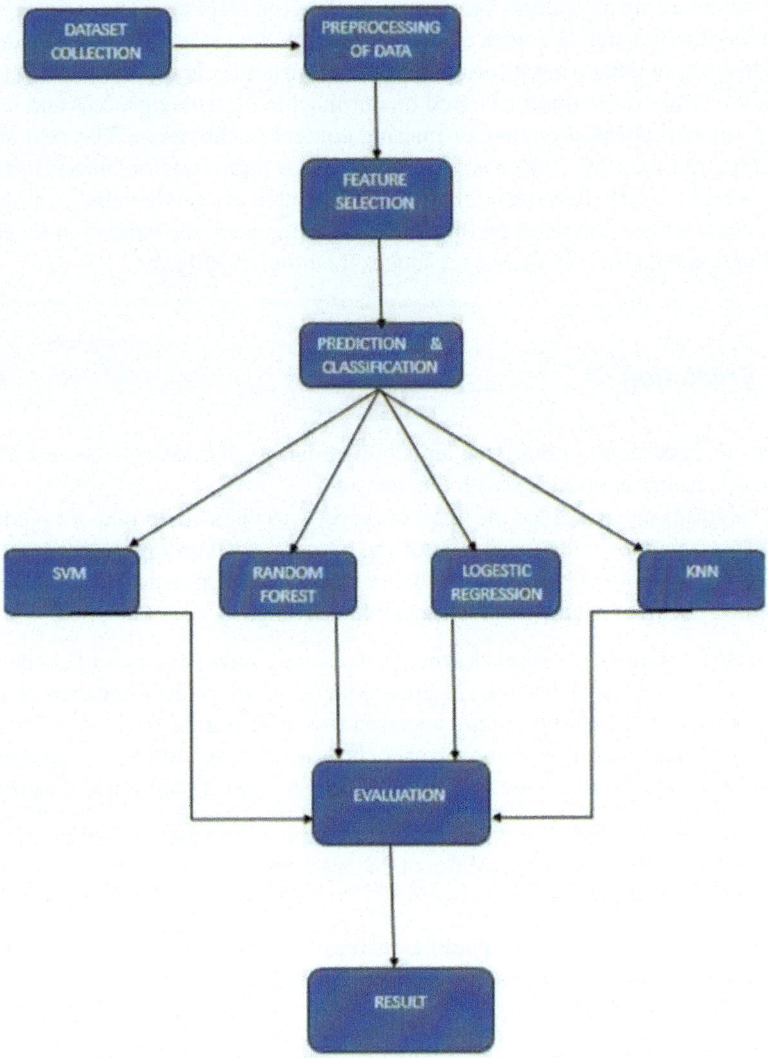

Fig. 1 Organized workflow of the system

3.2 Data Munging and Attribute Selection

Data munging is a core step in the enhancement of machine learning models, akin to receiving clear instructions for an academic assignment. Without well-organized raw data, implementing algorithms becomes a challenging endeavor. The main agenda of this preprocessing step becomes evident during the final stages of applying datasets to algorithms, emphasizing its critical role in ensuring accuracy and reliability.

Initially, import the libraries based on our machine learning process and load the dataset—it will check the noisy or missing content in this stage. The next stage is to arrange the data by converting text values into numerical or binary format by utilization of the LabelEncoder() function. The conclusive phase entails training and testing, where the training set facilitates model refinement, the validation set ensures its validation, and the evaluation set gauges the model's efficacy.

3.3 Prediction

Within our approach to predicting autism disorder, we harnessed a diverse array of machine-learning approaches with the dataset.

The algorithm that yielded the finest accuracy was deemed optimal for prediction. To enhance accuracy within each algorithm, we conducted ensemble learning so that we take the mean of KNN, Random Forest, SVC, and Logistic Regression which gives good accuracy. Our model is ensemble learning.

Ensemble Learning: Ensemble learning serves as a fundamental cornerstone within the domain of machine learning, elevating both accuracy and resilience in predictive analytics by amalgamating forecasts derived from a multitude of models. Its central aim centers on alleviating potential errors or biases inherent within individual models, thereby harnessing the collaborative wisdom encapsulated within the ensemble.

KNN: In k-Nearest Neighbors (KNN), we optimized parameters for example, the sum of neighbors (k) and the selection of distance measures utilized for classification purposes.

Random Forest: The essential parameters for this methodology include the total of trees within the forest, the utmost depth of individual trees, the minimum sample requirement for node division, and the minimum sample threshold for leaf nodes.

SVC: When utilizing SVC, we optimized parameters including the selection of the kernel. For instance, in circumstances where a linear kernel is employed, the classifier determines a linear hyperplane to separate the data set effectively.

Logistic Regression: Logistic Regression explores the correlation between one or multiple independent variables and a singular dependent variable. Within this framework, it portrays the connection in a manner that simplifies classification endeavors.

Through meticulous hyperparameter tuning, were fined the productivity of each algorithm, thereby intensifying the accuracy of our autism disorder prediction method. Concurrently, we investigated the influence of varied dataset sizes on predictive accuracy, systematically adjusting training and testing datasets while maintaining algorithmic and hyperparameter consistency. This analytical approach delivers a valuable perception of the influence of data availability on predictive efficacy, enabling an exhaustive analysis of our model's scalability and generalization capabilities.

3.4 Evaluation

In the era of machine learning, evolutionary process metrics are like measuring sticks. They help us see how well, our predictive models work with different sets of data. These metrics tell us how remarkable our predictions are and how accurate the methodology is overall. In our research, we use evaluation metrics to scrutinize various algorithms and see how well they perform. Two important metrics we use are accuracy, which indicates the ratio percent of predictions out of the overall predictions made, and precision, which is centered on how precise our predictions are. Furthermore, recall also recognized as sensitivity, underscores the algorithm's capability to snatch all positive categorical variables in the dataset. F1-score, which represents the harmonious amalgamation of precision and recall, presents a balanced assessment, particularly valuable in instances of class imbalance.

Additionally, the ROC curve and its associated AUC provide a perception of the model's capacity to discriminate between classes across various threshold settings.

4 Result Analysis

The outcomes of the autism prediction model are explained in Table 1. First, the dataset is separated into train data (80%) and test data (20%) sets, using 8068 training data and 2017 testing data. The productivity evaluation metrics for every one of the four algorithms are disclosed in Table 1.

Table 1 Performance metrics

Classifier	Accuracy	Precision	Recall	$F1$-score
KNeighbor	0.90	0.90	0.91	0.90
Random Forest	1.00	1.00	1.00	1.00
SVC	1.00	1.00	1.00	1.00
Logistic Regression	1.00	1.00	1.00	1.00

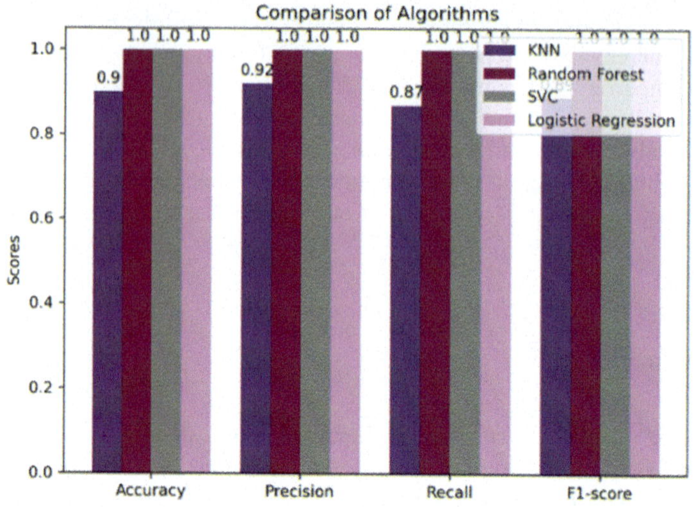

Fig. 2 Accuracy versus dataset size

In Fig. 2, the visual depiction contributed by the bar plot, a comprehensive analysis unfolds, showcasing the comparison among diverse dataset sizes and previous model iterations. Employing accuracy as the metric for evaluation, a compelling trend emerges: with the expansion of the dataset, accuracy exhibits a corresponding upward trajectory, offering a noteworthy observation on the relationship between dataset size and predictive performance.

In Fig. 3, the pie chart illustrates the comparison among predictions of autism disorder in different genders. Here 1 represents males with Autism and 2 represents the chances of getting Autism for transgenders, 0 shows Autism diagnosis in females.

In Fig. 4, visualized as a bar chart, the comparative analysis of accuracy predictions among KNN, Random Forest, SVC, and Logistic Regression, within our ensemble learning model and previous, reveals a striking trend: the present model demonstrates superior accuracy over its predecessors.

ROC Curve:

The ROC curve serves as a graphical representation depicting the probability curve of binary classifiers. In Fig. 5, four ROC curves corresponding to four classifiers are presented and plotted based on their respective AUC values. A perfect test result is characterized by an AUC value of 1.0, while random chance yields an AUC value of 0.5 or lower.

Comparing between Logistic Regression, SVC, Random Forest, and KNeighbors Classifier.

Based on the data depicted in Fig. 4, SVC, Random Forest provides the finest result here in the ROC curve and the KNN classifier performs the least favorably in this scenario and Logistic Regression gives the finest result in accuracy. From

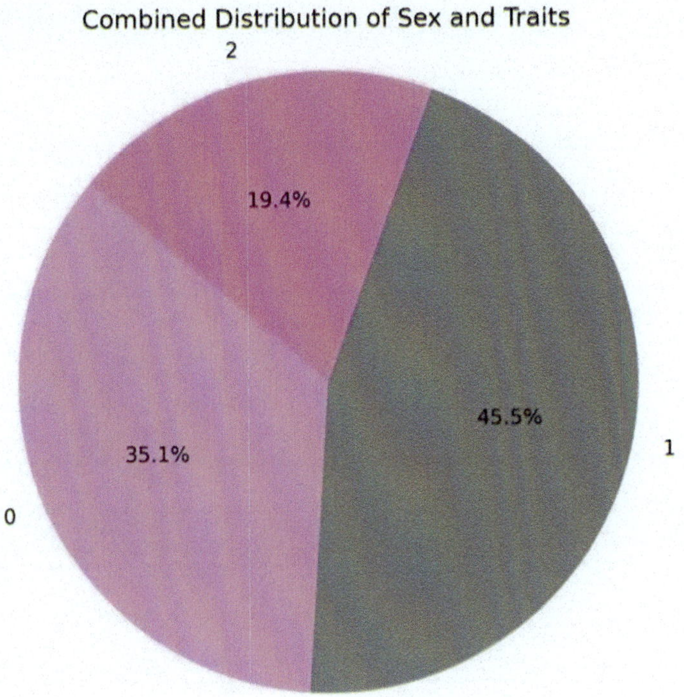

Fig. 3 Gender distribution of traits

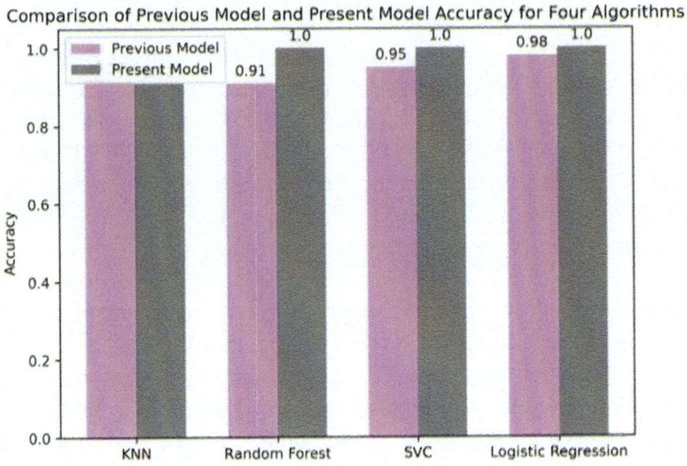

Fig. 4 Accuracy comparison of algorithms

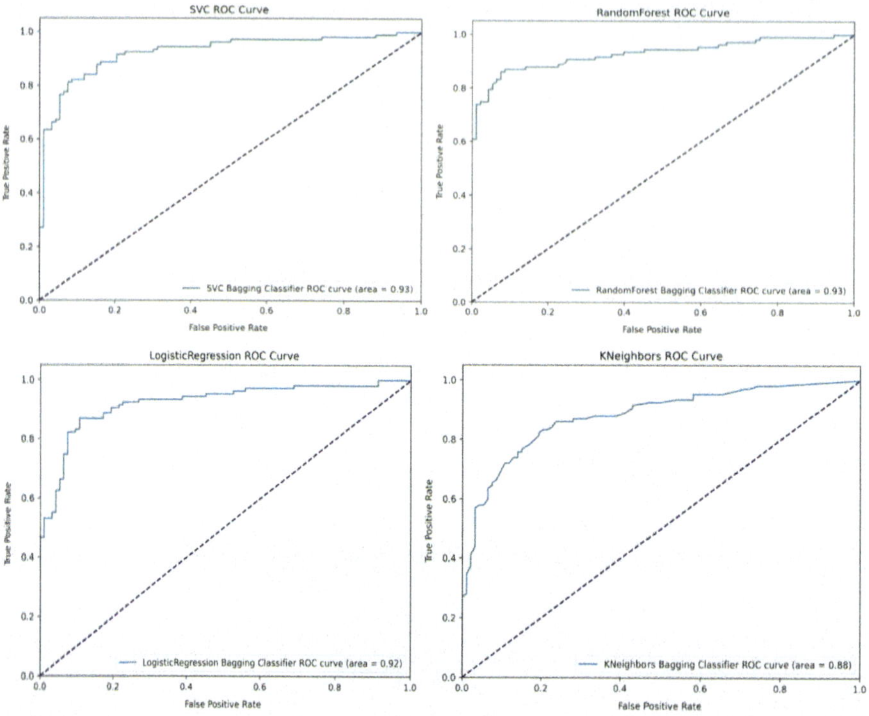

Fig. 5 ROC curve of classifiers' performance

the above Fig. 6, it is apparent that there are four ROC curves plotted as per their AUC values. To conduct a comparison, we must examine their AUC values. Logistic Regression provides an AUC of 0.92, KNN methodology provides an AUC of 0.88, Random Forest Approach 0.93, and SVC provides an AUC of 0.93. From this, we can conclude that if accuracy is high but the AUC value, precision, recall, and F1-score are minimal then we can say that the classifier is not fit to that particular dataset [12].

In this research, the proposed model ensemble learning engaged tuning for all algorithms. Subsequently, Random Forest and Logistic Regression emerged as the most auspicious model for predictive purposes of autism spectrum disorder (ASD), boasting an impressive accuracy of 100%. Bagging is considered as here considering all features for node splitting enhances the adeptness of Random Forest, thereby bolstering its efficiency and other algorithms in ASD prediction. The ROC curve and the AUC score of each algorithm are shown in the figure.

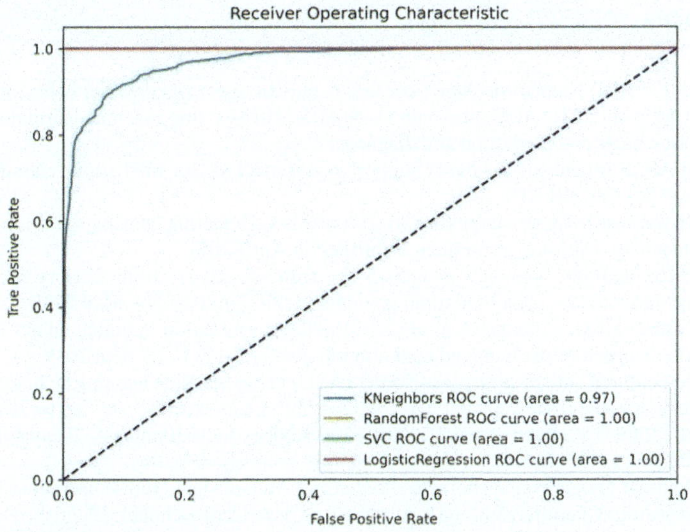

Fig. 6 ROC curve for each classifier

5 Conclusion

Within this investigation, KNeighbors, Random Forest, Logistic Regression, and SVC. The experimental results and analysis presented here in this describe whether a person has autism disorder or not.

This is categorized by utilizing specific characteristics or using attributes as input parameters in our methodology, and then the prediction will be done. The other thing is we took a dataset and the accuracy to get ASD for a person is analyzed. As we look forward, continuous processing and exploration of diverse datasets are crucial to knowing whether an individual exhibits signs of autism or not. Using the Random Forest model, 100% accuracy is achieved in detecting ASD. As we are considering transgender it results in us to know, the accuracy of predicting autism in transgenders takes place after the male ratio. Through rigorous evaluation of various ML algorithms and exploration of different dataset sizes, Random Forest methodology emerges as the optimal, achieving an impressive accuracy of 100%. This highest accuracy model, known as the Random Forest strategy, demonstrates the proficiency of ensemble learning in enhancing ASD prediction, offering valuable insights for healthcare practitioners and policymakers.

Additionally, the examination of varying dataset sizes reveals a positive correlation between dataset size and prediction accuracy, emphasizing the strikingness of adequate attain ML-based healthcare applications.

References

1. Thabtah F (2019) Named Machine Learning in autistic spectrum disorder behavioral research
2. Asri H, Mousannif H, Al Moatassime H, Noel T (2016) Using machine learning algorithms for breast cancer risk prediction and diagnosis
3. Shanthi AS, Karthikeyan M (2014) Support vector machine for MRI stroke classification. Int J Comput Sci Eng (IJCSE)
4. https://www.researchgate.net/publication/350682394_Detecting_autism_spectrum_disorder_using_machine_learning_techniques. Published in April 2021
5. A machine learning approach in autism spectrum disorders: from sensory processing to behavior. https://pubmed.ncbi.nlm.nih.gov/35615066/. Published on May 2022
6. Cheak-Zamora NC, Coleman A et al (2019) Self-determination in young adults with autism spectrum disorder. https://pubmed.ncbi.nlm.nih.gov/31561711/. 27 Sept 2019
7. Early detection of autism spectrum disorder (ASD) using machine learning techniques. https://repository.rit.edu/cgi/viewcontent.cgi?article=12536&context=theses#:~:text=(2020)%20accepted%20the%20effectiveness%20of,it%20is%20beyond%20human%20capabilities. 16 May 2023
8. https://scholar.google.co.in/scholar?q=Predicting+Autism+Spectrum+Disorder+Using+Machine+Learning+Classifiers&hl=en&as_sdt=0&as_vis=1&oi=scholart. Published on 26 Jan 2021
9. Yuan L, Erdt M, Li R, Siyal MY (2023) Data privacy protection domain adaptation by roughing and finishing stage. Vis Comput.https://doi.org/10.1007/s00371-023-02794-1
10. Nigmatullina I, Sheymardanov S, Abramskiy M (2023) Digital platform form monitoring and comprehensive support of children with autism spectrum disorders. In: Intelligent sustainable systems: selected papers of world S42022, vol 1, pp 573–580. Springer Nature, Singapore
11. Tehseen R, Farooq MS, Abid A (2021) EPS: an earth quake prediction system using federated learning. In: 2021 international conference on innovative computing (ICIC). IEEE, pp 1–8
12. Predicting ASD using ML classifier, 26 Jan 2021. IEEE. https://arxiv.org/pdf/2101.09279.pdf

Rose Doctor: Identification of Fungal Diseases in Rose Plants Using Convolutional Neural Networks

John T. Mesia Dhas, B. Senthilkumaran, K. Chinnathambi, and R. Durai Vasanth

Abstract The rose is one of the ornamental flowers in traditional and cultural ceremonies. It is an economical flower in the international flower markets. The subtropical climate and iron-enriched soil are sufficient for getting higher yields from the rose plant. Different types of diseases affect the rose plant due to the infection of fungus, bacteria, viruses, or other natural and artificial calamities. The diseases are affecting the yield and life span of the plant. The existing techniques are based on the observation of an expert, and the treatment process is also manual. Rose doctor is a proposed methodology that uses a machine learning model to identify the fungal diseases affected by the plant using a Convolutional Neural Network. The training dataset for this model has all the possible images of affected rose plants with various fungal diseases. The data using for test is the image of the affected plant. In addition to the images, the environmental and climatic conditions of the rose plant were also analysed. Rose doctor is giving around 90% accuracy in its prediction results. It is a novel methodology for predicting plant diseases without human support.

Keywords Rose doctor · Rose diseases · Rose diseases treatment

J. T. M. Dhas (✉) · B. Senthilkumaran · K. Chinnathambi · R. Durai Vasanth
Vel Tech Rangarajan Dr. Sagunthala R&D Institute of Science and Technology (Deemed to be University), Chennai 600062, India
e-mail: jtmdhasres@gmail.com

B. Senthilkumaran
e-mail: skumaran.gac16@gmail.com

K. Chinnathambi
e-mail: drchinnathambik@veltech.edu.in

R. Durai Vasanth
e-mail: duraivasanth98@gmail.com

© The Author(s), under exclusive license to Springer Nature Singapore Pte Ltd. 2025

H. Zen et al. (eds.), *Soft Computing and Signal Processing*, Lecture Notes in Networks and Systems 1200, https://doi.org/10.1007/978-981-97-9926-8_19

1 Introduction

Roses are one of the most important ornamental flowers in the entire south of India. The subtropical climate and the fertility of the soil are contributing to the high yields of roses. In all the traditional and cultural ceremonies, the flower rose plays a great role. Nurseries and flower gardens are playing a great role in manufacturing and selling rose plants and flowers. The local and foreign markets for selling rose flowers have a considerable gross domestic product (GDP) in the Indian market.

Today, the rose-cultivating farmers of southern India are facing many challenges in rearing and caring for rose plants. Many diseases are affecting the plants; they are reducing yields, and sometimes the plants are also dying. It is affecting the economy of the farmers. The active mechanisms are not existing for identifying the diseases at the initial stages; they are having more adverse effects on the rose plant.

The common diseases affecting the rose are black spot, powdery mildew, downy mildew, cercospora leaf spot, anthracnose, botrytis blight, ghost spotting, rust, stem canker, armillaria root rot, phytophthora, root rot, crown gall, rose mosaic, and rose rosette disease. Most of the diseases of rose plants are caused by bacteria, fungi, and viruses. The colour changes in leaves, stems, and flowers are the symptoms of the affected plant. This article clearly explains the identification of various diseases affecting the rose plant and the curing mechanisms using the deep learning method.

2 Literature Review

Windham and Windham [1], early care and identification of diseases are the only ways to create a strong and long-yielding rose garden. The proper utilisation of manures and fertilisers increases the production rate of rose plants. The greenhouse effect is required for the proper colour of the flower. For more flower yield, it requires good sunlight.

Home & Garden Information Centre (HGIC) [2], the fertile ferric soil and natural manures increase the life and yield of a rose plant. The natural neem oil and some chemicals like bifenthrin, chlorothalonil, copper fungicides, and mancozeb are helping to save the rose plant from diseases. The fully infected plant must uproot and fire fully. It helps to prevent the spread of the disease.

Douglas [3], the type of plant, climate, soil, fertilisers, and sanitation are the important factors affecting diseases in the rose plant. The clean environment and direct sunlight reduce the spread of disease to the rose plant. The complete demolition of heavily infected plants is the only solution for preventing the spread of disease to other plants. Early detection, identification, and treatment are the best practices for saving rose plants from diseases.

A Convolutional Neural Network (CNN) is a type of artificial neural network (ANN) that is particularly well-suited for analysing visual data. It is inspired by the organisation and functioning of the human visual cortex. CNNs have revolutionised the field of computer vision and are widely used in applications such as image classification, object detection, facial recognition, medical image analysis, and autonomous driving, among others. Their ability to automatically learn features from raw data makes them incredibly powerful for extracting meaningful information from visual inputs [4, 5].

CNNs have become indispensable tools in various image processing tasks, enabling advancements in fields such as computer vision, medical imaging, satellite imagery analysis, and more. Their ability to automatically learn relevant features from raw image data has revolutionised the way images are processed and analysed [6].

Convolutional Neural Networks (CNNs) have been extensively used in disease identification and medical image analysis tasks due to their ability to learn discriminative features directly from raw image data. These applications demonstrate the potential of CNNs to assist healthcare professionals in early disease detection, improving patient outcomes, and reducing the workload associated with manual image interpretation. However, it's essential to validate and interpret CNN models carefully to ensure their reliability and safety in clinical practice [7].

Convolutional Neural Networks (CNNs) have emerged as powerful tools in various fields of plant research, enabling automated analysis, classification, and prediction tasks. CNNs offer a powerful framework for advancing plant research by automating image analysis tasks, extracting meaningful information from visual data, and accelerating scientific discovery in agriculture, ecology, and plant biology. Their applications in plant research hold great potential for addressing global challenges such as food security, environmental sustainability, and climate change resilience [8, 9, 10].

3 Diseases Affecting Rose Plant

3.1 Powdery Mildew

It is caused by the fungal pathogen *Podosphaera pannosa.* The symptom of the pathogen is a powdery white (fungal) spread on the leaves, petioles, and flower buds, as shown in Fig. 1.

Neem oil and fungicides are helping to prevent the spread of powdery mildew.

3.2 Black Spot

The fungal pathogen *Diplocarpon rosae* is the cause of black spots. Black-to-brown spots with diffuse borders are the syndrome of this disease, as shown in Fig. 2.

The black spot reduces the yield of the plant and kills it gradually.

Fig. 1 Powdery mildew

Fig. 2 Black spot

3.3 Downy Mildew

The pathogen *Peronospora sparsa* is the main cause of downy mildew. It is spreading like a fungus. Symptoms of downy mildew include angular brown, red, or purple lesions that may be limited by veins in the leaf (Fig. 3).

The usage of fungicides, clean sanitation, and sunlight are the important requirements for controlling downy mildew.

Fig. 3 Downy mildew

Fig. 4 Cercospora leaf spot

3.4 Cercospora Leaf Spot (CLS)

The fungal pathogen Cercospora rosicola is the main cause of cercospora leaf spot. Symptoms of CLS include purple, maroon, brown, or black lesions on the upper surfaces of mature leaves (Fig. 4).

The usage of fungicides, clean sanitation, and sunlight are the important requirements for controlling cercospora leaf spot.

3.5 Anthracnose

The fungal pathogen *Sphaceloma rosarum* is the main cause of anthracnose disease. Symptoms of anthracnose include leaf spots on upper leaf surfaces with red, maroon, or purple borders (Fig. 5).

The usage of fungicides, clean sanitation, and sunlight are the important requirements for controlling anthracnose.

Fig. 5 Anthracnose

Fig. 6 Botrytis blight

3.6 Botrytis Blight

The fungal pathogen *Botrytis cinerea* is the main cause of botrytis blight. It is otherwise known as grey mould. This is affecting the petals of old flowers, flower buds, and leaves (Fig. 6).

It is spreading more in cool and wet weather. The usage of fungicides, clean sanitation, and sunlight are the important requirements for controlling botrytis blight.

3.7 Ghost Spotting

The fungal pathogens bipolaris, botrytis, cercospora, and cladosporium are responsible for the disease ghost spotting. Small pink-to-white spots with pink borders in rose bloom as the syndrome of this disease (Fig. 7).

This disease is affecting the prolonged period of rain. It does not have a permanent solution. Generally, using the fungicides and, in the worst cases, uprooting the plant and burning it out.

Powdery mildew, black spot, downy mildew, cercospora leaf spot (CLS), anthracnose, botrytis blight, and ghost spotting are the major diseases affecting the rose plant. The different fungal pathogens are responsible for these diseases. Mainly, it is affecting the health and growth of the rose plant, decreasing yields, and, in the worst cases, killing the plant.

This research focuses mainly on identifying the diseases affecting the rose plant and suggesting necessary curing mechanisms using deep learning methods.

Fig. 7 Ghost spotting

4 Methodology

A Convolutional Neural Network is a deep learning mechanism used to identify features in different dimensions using machine learning algorithms. The rose is an important economical flower in the flower market. The affection of the fungal pathogen is reflected in the leaves, flowers, and stems in different types of colour changes, markings, and dots. The pattern of the disease differs depending on the type of fungal pathogen affected in the plant. The image processing in the deep learning ideology is used for identifying the infection of the fungus. Rose doctor is an innovative algorithm designed for identifying fungal diseases in rose plants using images of the plants.

4.1 Rose Doctor Architecture

The rose doctor architecture is present in Fig. 8. The original plant image (mostly the leaf, bud, flower, or stem because the infection is reflected in those parts of the plant) is treated as the input of the rose doctor. The database named training dataset contains all the possible infected images of the rose plant for each fungal attack. The database curing mechanism has a treatment method for each fungal infection. The output of the rose doctor is the status of the fungal attack on the plant and the curing mechanism. The status table is created at each stage of the treatment process, which helps identify the status of the fungal infection.

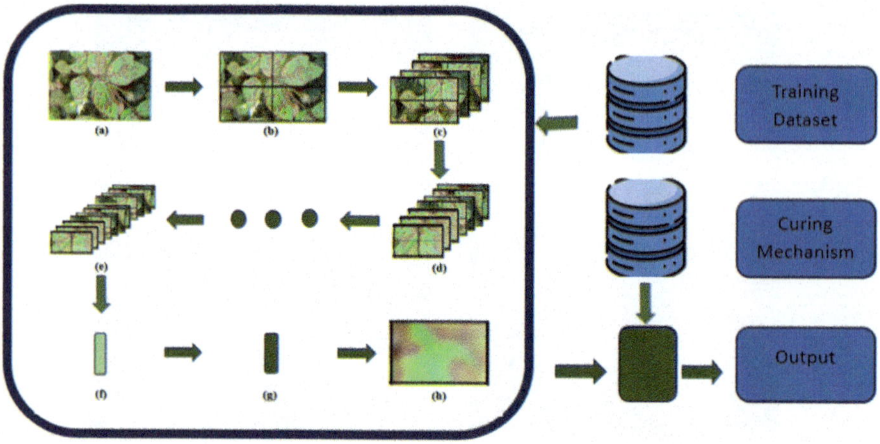

Fig. 8 Rose doctor architecture: **a** Original leaf image (input). **b** Rose leaf Image with 2^2 filters. **c** Rose leaf image with 2^4 filters. **d** Rose leaf image with 2^6 filters. **e** Rose leaf image with $2^{2(n-1)}$ filters. **f** Filter with disease characteristics. **g** Filter with environmental factors. **h** Intermediate results

4.2 Rose Doctor Status Table

The rose doctor status table (RDST) is generated at each stage of iteration in the feature selection. In the first iteration (Fig. 1a) only one output field is present for each fungal attack. The value is either 0 or 1, and it means the fungal pathogen is not infected or infected, respectively. The RDST of the first iteration is in Table 1.

The infected status value 1 says the matching happened with the input image with an already existing category of infected rose plant images in the training dataset. The value 0 says it is not matching with that category.

In the second iteration, the input is divided into 2^2 (= 4) parts, and each part is analysed with the existing images; if any matching happens, then the value is 1; otherwise, 0. The second iteration outcome model is the same as that in Table 2.

Total infection is the sum of infected status values in a particular row. In Table 2, if the total infection value is 0, then the infection level is nil; if it is 1, then the infection

Table 1 RDST first iteration

S. No.	Name of the disease	Infected status
1	Powdery mildew	0/1
2	Black spot	0/1
3	Downy mildew	0/1
4	Cercospora leaf spot (CLS)	0/1
5	Anthracnose	0/1
6	Botrytis blight	0/1
7	Ghost spotting	0/1

Table 2 RDST second iteration

S. No.	Name of the disease	Infected status				Total infection
		Part 1	Part 2	Part 3	Part 4	
1	Powdery mildew	0/1	0/1	0/1	0/1	–
2	Black spot	0/1	0/1	0/1	0/1	–
3	Downy mildew	0/1	0/1	0/1	0/1	–
4	Cercospora leaf spot (CLS)	0/1	0/1	0/1	0/1	–
5	Anthracnose	0/1	0/1	0/1	0/1	–
6	Botrytis blight	0/1	0/1	0/1	0/1	–
7	Ghost spotting	0/1	0/1	0/1	0/1	–

level is less (less than or equal to 25%); if it is 2, then the infection level is moderate (less than or equal to 50% and greater than 25%). If the infection value is 3, then the infection level is high (less than or equal to 75% and greater than 50%). If the infection value is 4, then the infection level is severe (less than or equal to 100% and greater than 75%). The curing mechanism will be applied based on the level of infection with the plant.

Similarly, the status of all other iteration values is recorded in different tables at $2^{2(n-1)}$ (n-is iteration) status columns. The treatment will be based on the percentage of infection with the plant. The number of iterations is increasing the accuracy of disease prediction. Multiple fungal infections are also possible for the rose plant.

4.3 Implementation

To find the disease affected by a rose plant and give the curing mechanism with three iterations. If there are three iterations, then 16 internal values will be identified for each image. The training dataset has around 100 images for each fungal disease in different dimensions and lengths. The scanned input image of the infected plant is given as the input for the rose doctor. In the first iteration, the full image is tested with all seven types of fungal disease-affected databases. In the first iteration, only 2^0 ($2^0 = 1$) results are possible. In the second iteration, the original image is divided into 4 ($2^{2(n-1)}$) different parts, and all the parts are tested with the existing images in the data warehouse. The values will be recorded in the infection status table of the second iteration. In the third iteration, already segmented, four parts were further divided into four parts, totalling 16 ($2^{2(n-1)}$) parts. The testing information is tabulated in Table 3.

In the above results of the rose doctor tests, the plant is affected by cercospora leaf spot (CLS) severely, and a minor effect of downy mildew and powdery mildew is also identified. The environmental characteristics are also identified in the environmental database. The sunny and warm climate won't have any adverse effects on downy

Table 3 RDST third iteration (implementation)

S. No.	Name of the disease	Infected status																Total infection
		Part 1				Part 2				Part 3				Part 4				
		1	2	3	4	1	2	3	4	1	2	3	4	1	2	3	4	
1	Powdery mildew	0	0	0	0	0	0	0	0	0	1	0	0	0	0	0	0	1
2	Black spot	0	0	0	0	0	0	0	0	0	0	0	0	0	0	0	0	0
3	Downy mildew	0	1	0	0	0	0	0	0	0	0	0	0	0	0	0	0	1
4	Cercospora leaf spot (CLS)	1	0	1	0	1	1	1	1	1	1	1	1	1	0	1	1	13
5	Anthracnose	0	0	0	0	0	0	0	0	0	0	0	0	0	0	0	0	0
6	Botrytis blight	0	0	0	0	0	0	0	0	0	0	0	0	0	0	0	0	0
7	Ghost spotting	0	0	0	0	0	0	0	0	0	0	0	0	0	0	0	0	0

mildew or powdery mildew. So that the treatments are not required in this climate, only give the treatment for cercospora leaf spot (CLS). If the climate is rainy and humid, then the effects of downy mildew and powdery mildew will also spread more and create adverse effects. So that the treatment for downy mildew, powdery mildew, and cercospora leaf spot is highly required for the plant. The chemical fungicides in neem oil are used for the treatment. Good sanitation, sufficient water, space in between plants, and sunlight are the environmental boosters for treating the disease.

4.4 Rose Doctor: Algorithm

The following algorithm explains the detailed working principle of rose doctor.

5 Algorithm Rose Doctor

1. Datasets
   (a) Store all the possible images of affected rose plants under their respective categories for training data.
   (b) Store all the possible curing techniques with respect to the type of disease.
   (c) The climate conditions are stored based on a weather-monitoring system.
   (d) To identify the soil conditions and topographical future for plants and the environment.

2. To scan the image of the affected plant
3. Initialise the iteration counter ($n = 1$) and the total number of iterations as an iter.
4. The image is tested with the training dataset images. If any image matches the testing image, then the rose doctor status table fields are marked as 1 in the respective category; otherwise, the value is zero. If any part is affected (value = 1), then go for the respective curing technique based on climate and environmental factors. If the disease is identified or the n value is greater than or equal to the iter value, then exit from the algorithm.
5. If the disease is not identified, then
   (a) Increment the iteration counter n by 1.
   (b) Divide the image into $2^{2(n-1)}$ parts, and each part is treated as a separate input for the test process. Step 4 is repeated until all the parts of the images are tested separately.
6. Stop.

6 Conclusion

The rose is one of the most economical flowers in the international flower market. In all the ceremonies, it is inevitable that the decoration process. The disease affecting the plants reduces the yields as well as the life span of the plant. This research identified the possible fungal diseases affecting the rose plant and the curing techniques for each disease. This implementation is using many iterations, and using a greater number of images in each infected category is increasing the accuracy of the prediction by about 90%. This application will reduce the difficulty of the farmers predicting disease affected by the rose flower. It also suggests a cure based on the decision made by the rose doctor and environmental and climate conditions.

References

1. Windham A, Windham M (2022) Rose diseases: identification and management. UTExtension, University of Tennessee, pp.1–7
2. HGIC (2024) Rose diseases, pp 1–9. https://hgic.clemson.edu/
3. Douglas SM (2021) Common diseases of rose. The Connecticut Agricultural Experiment Station, New Haven, CT 06504-1106, pp1–3
4. Barrientos A, Colorado J, Cerro J, Martinez A, Rossi C, Sanz D, Valente J (2011) Aerial remote sensing in agriculture: a practical approach to area coverage and path planning for fleets of mini aerial robots. J Field Robot 28:667–689
5. Sherrah J (2024) Fully convolutional networks for dense semantic labelling of high-resolution aerial imagery. Available online: https://arxiv.org/pdf/1606.02585.pdf. Accessed on 8 March 2024
6. Qu T, Zhang Q, Sun S (2016) Vehicle detection from high-resolution aerial images using spatial pyramid pooling-based deep convolutional neural networks. Multimed Tools Appl, 1–13

7. He K, Zhang X, Ren S, Sun J (2015) Spatial pyramid pooling in deep convolutional networks for visual recognition. IEEE Trans Pattern Anal Mach Intell 37:1904–1916
8. Wang Q, Rasmussen C, Song C. Fast (2016) Deep detection and tracking of birds and nests. In: Proceedings of the international symposium on visual computing, Las Vegas, NV, USA, pp 146–155
9. Angel TSS, Rodrigues P, Dhas JTM, Samy SSK (2012) Limitations of function point analysis in e-learning system estimation. Int J Comput Eng Sci
10. Howard AG (2024) Some improvements on deep convolutional neural network based image classification. Available online: https://arxiv.org/ftp/arxiv/papers/1312/1312.5402.pdf. Accessed on 8 March 2024

Hyperspectral Image Processing with Machine Learning and Mathematical Models

Anju Augustin and Cinu C. Kiliroor

Abstract Hyperspectral imaging is a fast-growing imaging technique in many fields, like remote sensing, fruit analysis, clinical images, etc. The spectral image consists of two parts: spectral data and spatial data. The processing of spectral images includes the processing of the spatial part as well as the spectral part. This paper analyzes various methods and algorithms that are used for spectral image processing in existing systems and proposes a diagram for hyperspectral image processing. There are five important steps for processing the spectral image and analyzing the region selection and dimensionality reduction methods in detail. Comparing the performance of region selection and dimensionality reduction techniques, the Rembg algorithm and Principal Component Analysis (PCA), respectively, show very good results. A well processed image has better performance in the classification and regression steps, and it reduces the computational time and cost.

Keywords Hyperspectral imaging · Spectral image processing · Machine learning · Spatial processing

1 Introduction

The types of images that are used in different real-world applications are Red, Green, Blue (RGB), Cyan, Magenta, Yellow, and Black (CMYK), and spectral images. The RGB images are three-channel images [1], and the CMYK images have four channels. The steps for processing these images are image enhancement, restoration, resolution correction, segmentation, etc. Spectral images are a type of image that has more than three channels or wavelengths. Mainly, two types of spectral images are used today: multispectral and hyperspectral images. Multispectral images have more than three

A. Augustin (✉) · C. C. Kiliroor
Department of Computer Science and Engineering, Indian Institute of Information Technology, Kottayam, India
e-mail: anjuaugustin.23phd21006@iiitkottayam.ac.in

C. C. Kiliroor
e-mail: cinu@iiitkottayam.ac.in

© The Author(s), under exclusive license to Springer Nature Singapore Pte Ltd. 2025

H. Zen et al. (eds.), *Soft Computing and Signal Processing*, Lecture Notes in Networks and Systems 1200, https://doi.org/10.1007/978-981-97-9926-8_20

and less than a hundred bands, but in hyperspectral images, more than a hundred bands are available. The size and data available in the spectral images, especially in hyperspectral images, are very large. Therefore, we cannot handle this type of image with normal processing steps.

The spectral images contain spectral and spatial data. Spatial data means the information related to the space or pixel area that we used for analysis. For example, region of interest selection is a type of spatial processing method to obtain the actual space for analysis. The spectral data includes pixel characteristics like intensity values, wavelength values, etc. So, we need processing methods like normalization and smoothing for handling spectral contents. How much processing we can do for the spectral image, that much improvement in performance we get from the result. This means that image processing is a very important step. The size and data in the spectral images are very large. Therefore, there is a need for reduction or compression methods; otherwise, this size affects the training time and result.

The main contributions of this paper are:

(i) This paper analyzes different spectral image processing methods that are used in existing systems.
(ii) Proposes a block diagram for spectral image processing with important steps.
(iii) Analyzing the region selection and dimensionality reduction methods with different algorithms for spectral images.

2 Related Works

Spectral image processing is a very important part in all spectral imaging applications. Yating et al. [2] proposed a system that uses hyperspectral imaging and a 1-dimensional Convolutional Neural Network (CNN) for pesticide detection in Hami melon. For spectral image processing, here use methods like black and white correction, information fusion of multiple sensor outputs, Savitzky-Golay (SG) smoothing, etc. Haibo et al. [3] presented a deep generative neural network (GAN) for spectral image processing. The result showed that the model outperformed other machine learning and CNN models. Puneet Mishra et al. [4] proposed a method with HSI and a new Deep Convolutional Generative Neural Network (DCGAN). The processing steps included here are Region Of Interest (ROI) selection, data augmentation, Adam optimization, etc.

Zhilong et al. [5] provide a review paper that includes different Machine Learning (ML) models that are used in various fields in food supply chain. The ML models are used in applications like food sorting, classification, and quality checking. A fruit decay detection method is suggested by Denog [6]. Here, Principal Component Analysis (PCA) is used for dimensionality reduction, and for noise and variability removal, different filtering and smoothing methods are proposed. Jiadong et al. [7] explained a method for non-destructive detection for textile fiber classification. To reduce noise, the pixel fusion method is employed, and hyperparameter selection is done by the grid search algorithm. This method improves performance and thus accuracy.

Alfadhl et al. [8] presented a system for detecting codling moth infestations in Apple. For processing, methods like wavelength trimming, the Monte-Carlo sampling method for outlier detection, and variance analysis techniques are explained here. Ziyou et al. [9] proposed a system for hyperspectral image analysis with an autoencoder-based method. PCA and t-distributed stochastic neighbor embedding algorithms are used here for dimension reduction. Dhritiman et al. [10] reviewed various ML models used for hyperspectral image processing in existing systems. Different methods for noise removal, region selection, and band selection are discussed in this paper. Comparing the current state of work, different machine learning (ML) and mathematical methods are very good and easy for the processing of hyperspectral images.

3 Methodology for Hyperspectral Image Processing

Hyperspectral images have more than a hundred bands. It produces the color spectrum for each pixel in the image based on wavelength and intensity. The color spectrum has information about the chemical composition of the object and the presence of some minute particles that our naked eyes cannot detect. Using a hyperspectral camera, it produces the spectral image of the object. This spectral image is used for further analysis, which means image processing. Figure 1 shows the block diagram of hyperspectral image processing with important steps.

The dataset [11] used here has spectral images of pure apple fruits, pesticides coated in different concentrations, and also damaged ones. This dataset contains images in various sizes. A typical small image has a size of 0.32 GB.

The first section of the block diagram includes an image capture or image acquisition module. The hyperspectral camera, illuminating light sources, and objects are the important components of the capturing module. The hyperspectral image produced by this module has uneven intensities and presence of dark and white current noises. Therefore, after image acquisition using Eq. (1) [3] image correction will be done.

I_{white} is the white reference image, generated by using a Teflon bar or a whiteboard with 99% reflectance. The dark reference picture I_{dark} is captured by turning off the light sources, closing the camera lens, and taking the image. The image obtained directly from the spectral camera is I_{spectral}. Therefore, the equation to get the corrected image I_{correct} is:

$$I_{\text{correct}} = \frac{(I_{\text{spectral}} - I_{\text{dark}})}{(I_{\text{white}} - I_{\text{dark}})}. \tag{1}$$

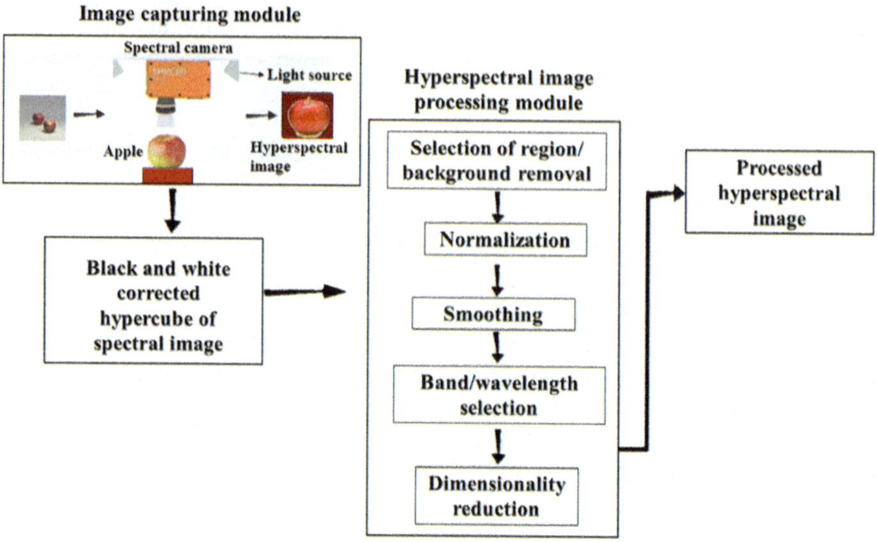

Fig. 1 Block diagram of the proposed hyperspectral image processing system

The corrected image is sent to the image processing module of the proposed system. The hyperspectral image has spatial and spectral data. So, the processing includes the processing of both spatial and spectral information. There are five important steps for this procedure.

1. Region selection or background removal.
2. Normalization.
3. Smoothing.
4. Band or wavelength selection.
5. Dimensionality reduction.

1. **Region selection or background removal**: This is a spatial processing method in spectral imaging. The region selection means selecting the portion of the image that is needed for our work. After the selection, we remove the rest of the part or pixels from the image. In background removal algorithms, the entire background is removed from the image. Table 1 shows important algorithms or methods and its features that are used for selecting or removing pixels.
2. **Normalization**: The normalization process helps to reduce variation in pixel intensities and base line shift. This will increase the performance and accuracy of the prediction and make improvements in the signal-to-noise ratio. Table 2 shows different methods used for the normalization of spectral data.
3. **Smoothing**: Smoothing is a spectral content preprocessing method. It is an important step in spectral image processing. Because in this step, we remove the noise and other unwanted signals and keep closely related data. It removes outliers and thus preserves major features, textures, etc. Table 3 shows the different smoothing techniques for spectral data.

Table 1 Methods used for region selection and background removal

Method	Features
ROI [3–5, 7, 8, 12]	Filter out some portion of the image by using some binary masks
Otsu algorithm [10]	This is a thresholding method by which the pixels are categorized into two classes: background and foreground
Adaptive thresholding [10]	Algorithm check the threshold for each pixel in a small region. Therefore, we get different threshold values for different small areas
Rembg algorithm [12]	This is a neural network based background removal algorithm

Table 2 Methods for normalization

Method	Features
Max–min normalization [2, 3]	In original data, it performs a linear transformation. The advantage is that it maintains relationships between the data
Max normalization [8]	Transform the data into a small range like [1, − 1]
Standard Normal Variate (SNV) [4, 8, 13]	In SNV subtracting each spectra from its mean and dividing each with its standard deviation

Table 3 Smoothing methods for spectral data

Method	Features
Savitzky-Golay method [2, 4, 8, 10, 13, 14]	This is a filtering method used with polynomial functions. It based on the local least squares algorithm. Preserves the shape of the spectrum and reduces mean square error
Radiometric correction [5]	Remove radiometric errors and distortion from spectral data
Pixel fusion [7]	Removes unwanted and similar pixels and retains the most important features
Median filtering [5]	Non-linear filtering method that removes noise and keeps sharp changes in data

4. **Band or wavelength selection**: Table 4 shows different algorithms for wavelength or band selection. In the band selection step, select the most important band or wavelength that contributes more to the result and remove unwanted bands. By doing so, we can refine our image, remove noise, and preserve our most important features.
5. **Dimensionality reduction**: Dimensionality reduction is the process of converting an image from a high-dimensional space to a lower-dimensional space by keeping

Table 4 Methods for band or wavelength selection

Method	Features
Partial Least Squares (PLS) [8, 12]	PLS identifies the multidimensional space in X direction with multidimensional variance of Y direction
Partial Least Squares-Discriminant Analysis (PLS-DA) [12]	PLS-DA preserves possible covariance between original data and selected data
Back Propagation Neural Network (BPNN) [9]	The neural network-based band selection is more fast. But have some convergence and optimization issues

Table 5 Methods for reduction of spectral data

Method	Features
Principal Component Analysis (PCA) [6, 9, 15]	Unsupervised ML algorithm and projection technique that transforms the image into some principal components that we specify. It examines to find interrelationships in the data
Successive Projection Algorithm (SPA) [6]	In SPA, the data values are projected into a lower-dimensional space to reduce dimension
Minimum Noise Fraction (MNF) [6]	The MNF algorithm estimates noise in the image by using a correlation matrix and reduces the noise by decorrelate it
Partial Least Squares (PLS) [4]	Here, latent vectors are identified by covariance between X and Y spaces

all relevant information. There are so many advantages for this reduction like, the size of the image reduced, storage space reduced, and the time for processing will be reduced. The reduction methods remove unuseful data like unwanted valleys and peaks and make the image more smooth and easy to handle. Table 5 includes the existing reduction methods for dimensionality reduction.

4 Result and Discussion

In this section, we analyze the region selection and dimensionality reduction steps for hyperspectral image processing in detail, using the Python implementation of algorithms.

Evaluating existing methods for hyperspectral image processing shows that each algorithm has its own characteristics. Here we compare the performance of different methods that are used for region selection and dimensionality reduction.

For region selection,the main algorithms used are ROI, Otsu, Adaptive thresholding, and the Rembg algorithm. The result showed that the background removal

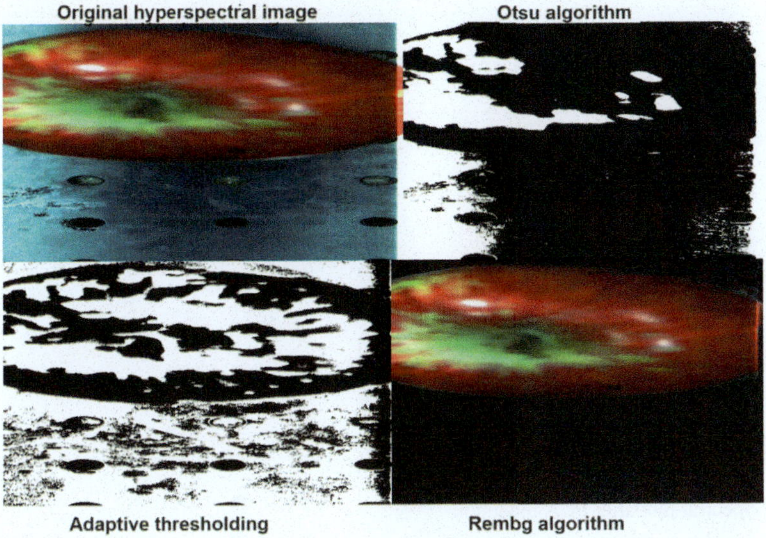

Fig. 2 Different methods for background removal and thresholding

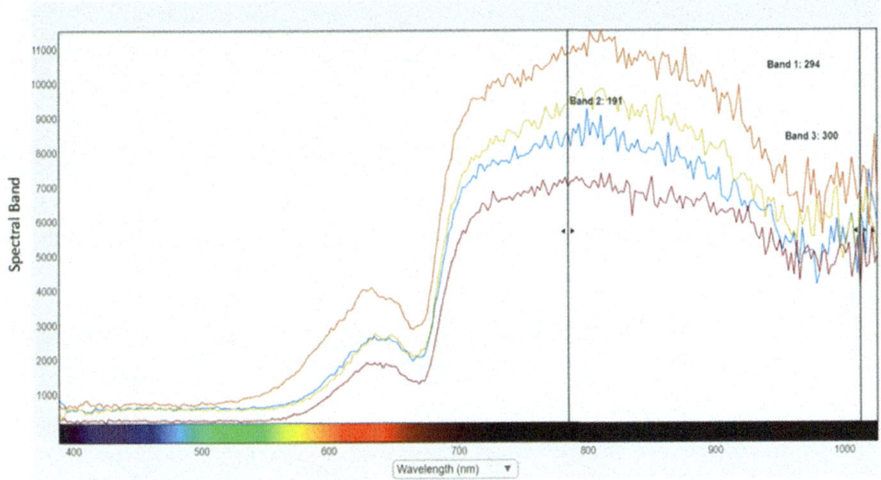

Fig. 3 Spectral plot of original hyperspectral image

algorithm Rembg outperforms Otsu and the Adaptive thresholding method, as shown in Fig. 2. But, the reduction in size of the image is greater in Otsu and Adaptive thresholding methods.

Figure 3 shows a spectral plot of the original hyperspectral image. We know that a hyperspectral image has more than a hundred bands or wavelengths. Therefore,

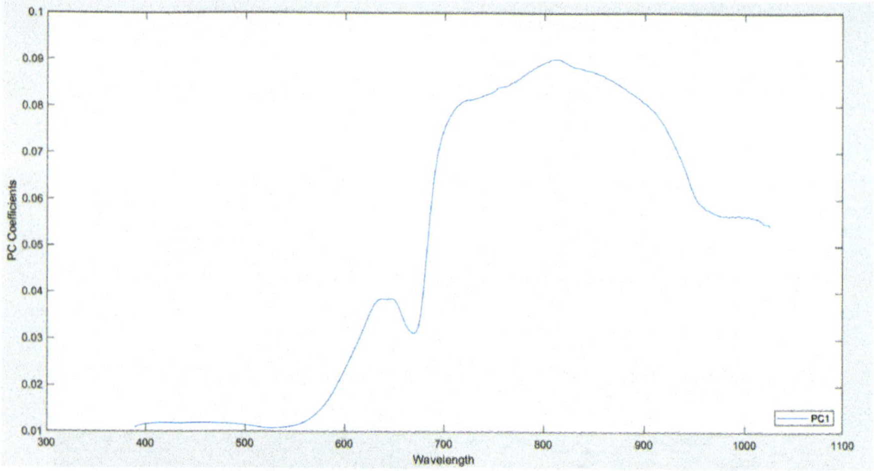

Fig. 4 Spectral plot after PCA reduction

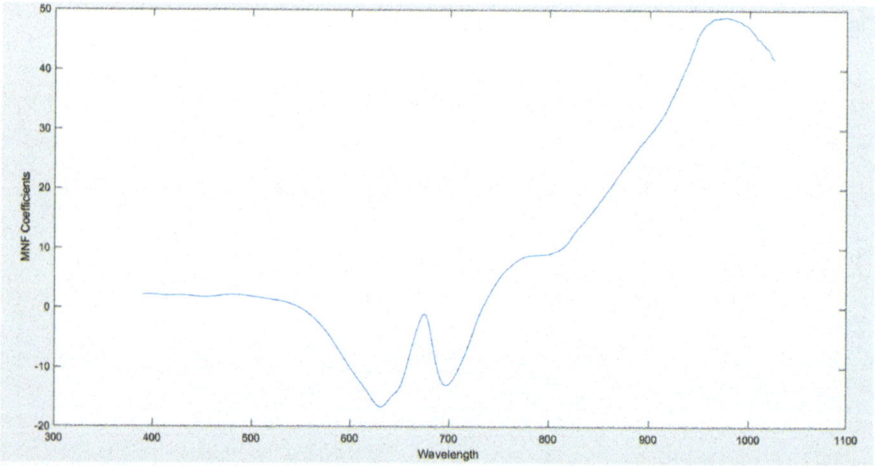

Fig. 5 Spectral plot after MNF reduction

using band selection algorithms, we can choose which algorithm selects the most appropriate band that contributes more to the result.

The main two methods for dimensionality reduction are MNF and PCA. After applying PCA, the plot becomes smooth and keeps the most important feature, as shown in Fig. 4.

Figure 5 shows the MNF transformation of the original spectral plot. Comparing the PCA and MNF, in low and high wavelength regions, PCA performs better than MNF. But in the middle area of the plot, MNF correctly shows the valleys and peaks.

5 Conclusion

Hyperspectral imaging is a new area in many applications for detecting the presence of very small objects that cannot be identified by our naked eyes. This paper analyzes different methods and algorithms that are used in existing systems for processing hyperspectral images. A block diagram proposed here contains important steps for hyperspectral image processing. The main steps are region selection, normalization, smoothing, band or wavelength selection, and dimensionality reduction. The corrected image from the image capturing module undergoes these five processing steps, and finally, we get a well-processed image with a small size. The result shows that the background removal algorithm Rembg gives better output than other thresholding algorithms for region selection, and the PCA is much better than MNF in terms of dimension reduction. Even a small hyperspectral image in our dataset has a size of 0.32 GB; after reduction, it will be reduced to 67 kB. The reduced image in the classification step will give good performance and reduce computational time. As a future extension, we try to implement efficient algorithms for all the processing stages and extract the most dominant features which helps to improve accuracy.

References

1. Yoon J (2022) Hyperspectral imaging for clinical applications. BioChip J 16:1–12
2. Hu Y, Ma B, Wang H, Zhang Y, Li Y, Yu G (2023) Detecting different pesticide residues on Hami melon surface using hyperspectral imaging combined with 1D-CNN and information fusion. Front Plant Sci 14
3. Tan H, Hu Y, Ma B, Yu G, Li Y (2024) An improved DCGAN model: data augmentation of hyperspectral image for identification pesticide residues of Hami melon. Food Control 157:110168
4. Mishra P (2022) Deep generative neural networks for spectral image processing. Analytica Chimica Acta 1191:339308
5. Zhang K, Zhao Y, Chen L, Guo Y, Mu Q, Wang S (2022) Advances in machine learning and hyperspectral imaging in the food supply chain. Food Eng Rev 14:596–616
6. Min D, Zhao J, Bodner G, Ali M, Li F, Zhang X, Rewald B (2023) Early decay detection in fruit by hyperspectral imaging-Principles and application potential. Food Control 152:109830
7. Huang J, He H, Lv R, Zhang G, Zhou Z, Wang X (2022) Non-destructive detection and classification of textile fibres based on hyperspectral imaging and 1D-CNN. Analytica Chimica Acta 1224:340238
8. Khaled AY, Ekramirad N, Donohue KD, Villanueva RT, Adedeji AA (2023) Non-destructive hyperspectral imaging and machine learning-based predictive models for physicochemical quality attributes of apples during storage as affected by codling moth infestation. Agriculture 13:1086
9. Zheng Z, Zhang S, Song H, Yan Q (2024) Deep clustering using 3D attention convolutional autoencoder for hyperspectral image analysis. Sci Rep 14
10. Saha D, Manickavasagan A (2021) Machine learning techniques for analysis of hyperspectral images to determine quality of food products: a review. Curr Res Food Sci 4:28–44
11. Roomi SMdM, Bama BS, Lakshmi V, Vaishnavi M (2023) Hyperspectral dataset of pure and pesticide-coated apples for measuring the level of fertilizers used. Data in Brief 49:109321

12. Wieme J, Mollazade K, Malounas I, Zude-Sasse M, Zhao M, Gowen A, Argyropoulos D, Fountas S, Van Beek J (2022) Application of hyperspectral imaging systems and artificial intelligence for quality assessment of fruit, vegetables and mushrooms: a review. Biosyst Eng 222:156–176
13. Xu J, Mishra P (2022) Combining deep learning with chemometrics when it is really needed: a case of real time object detection and spectral model application for spectral image processing. Analytica Chimica Acta 1202:339668
14. Lü Z, Huang S, Zhang X, Shi Y-Q, Yang W, Zhu L, Huang C (2023) Intelligent identification on cotton verticillium wilt based on spectral and image feature fusion. Plant Methods 19
15. Zhang C, Guo C, Liu F, Kong W, He Y, Lou B (2016) Hyperspectral imaging analysis for ripeness evaluation of strawberry with support vector machine. J Food Eng 179:11–18

Enhancing Lung Cancer Early Detection Using Deep Learning Techniques

S. Kranthi Reddy, Khaja Moinuddin, Md. Waseem Akram, and Sai Sachithananda

Abstract Exposure to carcinogens, particularly those found in cigarette smoke, destroys lung cells DNA and promotes their unchecked proliferation, which is the main cause of lung cancer. Its growth is also greatly influenced by genetic predisposition, family history, and environmental factors such as radon and asbestos. Lung cancer can affect nonsmokers due to these genetic and environmental factors, even though smoking is the main cause. These malignancies have the potential to impair breathing and the exchange of oxygen and carbon dioxide by interfering with the lungs natural function. When applied to tabular datasets, machine learning and deep learning algorithms can help forecast the prognosis of lung cancer and the response to treatment, enabling tailored care plans and enhancing patient outcomes. This study suggests using a convolutional neural network (CNN) architecture to accurately identify lung cancer from medical photos. Clinicians might possibly improve patient outcomes and survival rates by using CNNs' sophisticated feature extraction capabilities to produce more accurate diagnoses and customized treatment programs.

Keywords Lung cancer · CNNs · Feature extraction

1 Introduction

Lung cancer is the most frequent deadly tumor in the developed world, accounting for almost one million fatalities per year. Cigarette smoking must be reduced, young people discouraged from starting, and people assisted in quitting through smoking cessation programs and public smoking bans. Ex-smokers account for approximately 50% of new lung cancer diagnoses, whereas nonsmokers account for approximately 25% of cases worldwide [1]. It is one of the most prevalent malignancies in the world and a significant cause of death from cancer [2]. Tobacco smoke is the leading cause of lung cancer, which includes smoking cigarettes, cigars, and pipes, among

S. Kranthi Reddy (✉) · K. Moinuddin · Md. Waseem Akram · S. Sachithananda
Department of Computer Science and Engineering, Vignan Institute of Technology and Science, Hyderabad, India
e-mail: kranthi.vits@gmail.com

© The Author(s), under exclusive license to Springer Nature Singapore Pte Ltd. 2025

H. Zen et al. (eds.), *Soft Computing and Signal Processing*, Lecture Notes in Networks and Systems 1200, https://doi.org/10.1007/978-981-97-9926-8_21

other goods [2]. Tobacco smoke contains carcinogens that damage lung cells' DNA and cause genetic mutations that promote unregulated cell growth. It is important to remember, though, that people who have never smoked can still get lung cancer. This is frequently the result of exposure to secondhand smoke, radon and asbestos in the environment, inherited traits, or family history. Small cell lung cancer (SCLC) and non-small cell lung cancer (NSCLC) are the two forms of lung cancer that can develop [3]. The most common kind of lung cancer, non-small cell lung cancer (NSCLC), accounts for about 85% of cases. Among its subtypes include adenocarcinoma, large cell carcinoma, and squamous cell carcinoma. Compared to NSCLC, SCLC has a tendency toward aggressive growth and dissemination, albeit being less prevalent [3]. However, symptoms may not appear until the illness has advanced sufficiently. A variety of imaging techniques, including CT, PET, and chest X-rays, are commonly used in the diagnosis process. Lung cancer commonly manifests as dyspnea, hoarseness, chest pain, persistent or worsening cough, and inexplicable weight loss. Depending on where in your lungs the cancer first arises, some of these symptoms may show up early (in stages I or II), but they often do not show up until the disease has advanced to later stages [4]. Lung cancer treatment approaches vary and are based on the molecular profile, disease stage, and specific patient characteristics. EGFR mutations, ALK rearrangements, targeted therapy aimed at specific molecular abnormalities, immunotherapy, chemotherapy, radiation therapy, and surgery are among of the options [4]. The goal of ongoing research is to better understand the molecular causes of lung cancer, find new therapeutic targets, improve diagnostic methods, and investigate cutting-edge treatment options like immunotherapy combos and precision medicine [3]. Making lifestyle changes is crucial for those who have lung cancer. It is critical to stop smoking because doing so can enhance treatment results and lower the chance of cancer returning. Frequent exercise can raise vitality, lower stress levels, and promote general wellbeing [5]. Techniques for deep learning (DL) and machine learning (ML) have become effective tools in predicting lung cancer, offering promising avenues for early detection, prognosis assessment, and treatment optimization. These methods offer potential new approaches to early identification, prognostic evaluation, and treatment optimization. These methods use sophisticated algorithms to evaluate large volumes of medical data, such as genomic profiles and imaging scans, to find patterns and relationships that would not be visible using only conventional methods [5]. Furthermore, ML and DL techniques can integrate various clinical and molecular data to predict the prognosis of lung cancer patients. These models can analyze factors such as tumor characteristics, patient demographics, and treatment history to estimate survival rates, disease progression, and response to therapy [5].

2 Literature Survey

Clinical data analysis—IEEE papers often discuss the utilization of clinical data, including demographic information, medical history, and biomarker levels, for lung cancer detection. Studies focus on statistical analysis and machine learning techniques to identify patterns indicative of lung cancer presence or risk [6]. Genomic data analysis—Researchers leverage numerical data derived from genomic profiling, such as methylation patterns, mutations, and gene expression levels to create predictive models for the diagnosis and prognosis of lung cancer. IEEE papers delve into algorithms for feature selection, classification, and integration of multi-omics data [7]. Imaging data analysis—IEEE papers explore numerical analysis of imaging data from modalities like CT, MRI, and PET scans. They discuss feature extraction methods, such as texture analysis and radiomics, combined with machine learning algorithms for accurate lung cancer detection and characterization [8]. Integration of multi-omics data—Recent IEEE publications emphasize the integration of clinical, genomic, and imaging data for comprehensive lung cancer detection. These studies employ advanced computational techniques, including deep learning and network analysis, to exploit the complementary information provided by different data types [9]. Validation and clinical translation IEEE papers discuss validation studies aimed at assessing the performance and clinical applicability of numerical models for lung cancer detection. They evaluate sensitivity, specificity, and predictive accuracy using independent datasets and real-world patient cohorts [10]. Challenges and future directions—The literature highlights challenges such as data heterogeneity, model interpretability, and clinical translation in lung cancer detection using numerical data. IEEE papers propose future research directions, including the exploration of explainable AI models and the integration of real-time clinical data streams [11]. Prognosis prediction and precision medicine—Biomarkers and genetic signatures associated with prognosis and treatment response in lung cancer patients. Integrating multi-omics data with machine learning algorithms to develop personalized treatment strategies [12]. Environmental and occupational exposures—Examination of environmental and occupational risk factors associated with lung cancer, such as exposure to air pollution, asbestos, radon, and occupational carcinogens. A discussion on strategies for prevention and regulatory measures aimed at reducing exposure in high-risk populations [13].

3 Proposed Work

The proposed work includes gathered lung cancer images dataset from Kaggle dataset, and the dataset is preprocessed and converted into numerical dataset using python libraries. Once it is converted, the whole dataset is divided into training and testing dataset with ratio of 80 and 20%. Training dataset is given to KNN, ANN, and CNN algorithms, three models were generated, and among three models, the

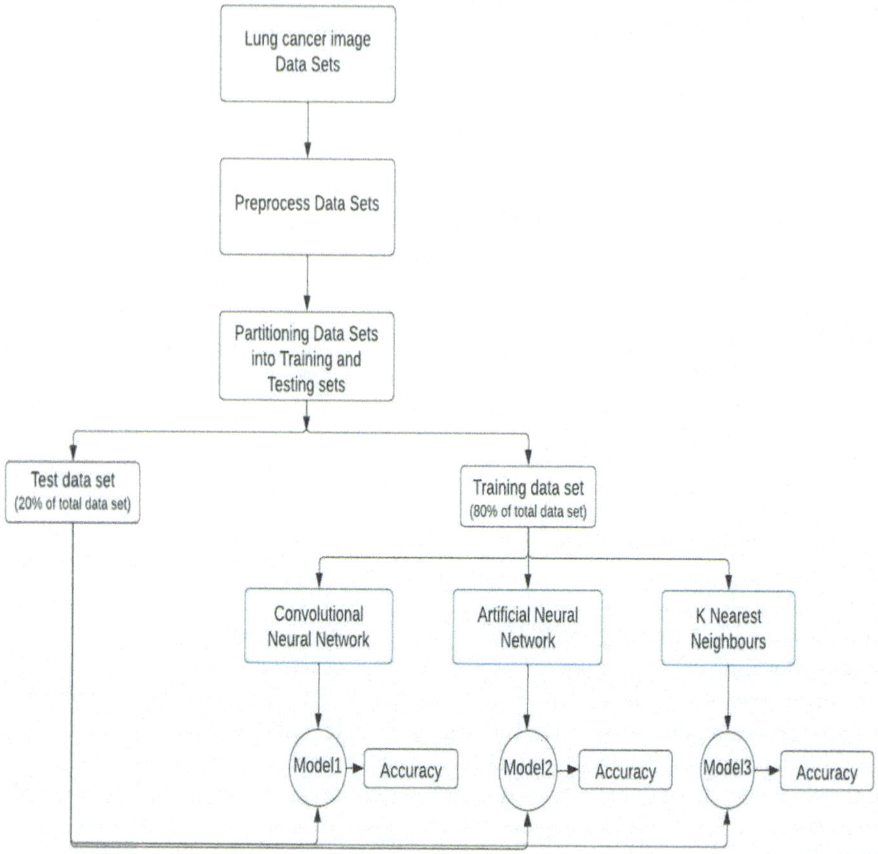

Fig. 1 Proposed work: outlines data preprocessing, partitioning dataset, and algorithms applied

highest accuracy generated model is considered as efficient model. Figure 1 shows proposed work.

4 K-Nearest Neighbors

A simple yet powerful technique for machine learning applications including regression and classification is K-nearest neighbors (KNN). It functions according to the idea that comparable data points typically produce comparable results. When using KNN, the closest neighbor's average in the feature space or the majority vote are used to decide the class or value of an unseen data point. The number of neighbors, K, is a critical decision that affects the algorithm's efficiency. Decision boundaries become more complicated with a smaller K value and smoother with a bigger K value [14]. Since KNN computes predictions at runtime and saves all training data, it does not require a training phase. But as the training set gets bigger, its computing cost increases, making it less appropriate for big datasets. Furthermore, because of the curse of dimensionality, KNN may have trouble processing high-dimensional data and is susceptible to irrelevant features and noisy data. KNN is still a well-liked option in many machine learning applications because of its simplicity, adaptability, and interpretability despite its drawbacks [15]. Once KNN model is generated, both training and testing datasets are given as input to the model. The confusion matrix for training and testing dataset is shown in Figs. 2, 3, and 4 which show the graph of KNN.

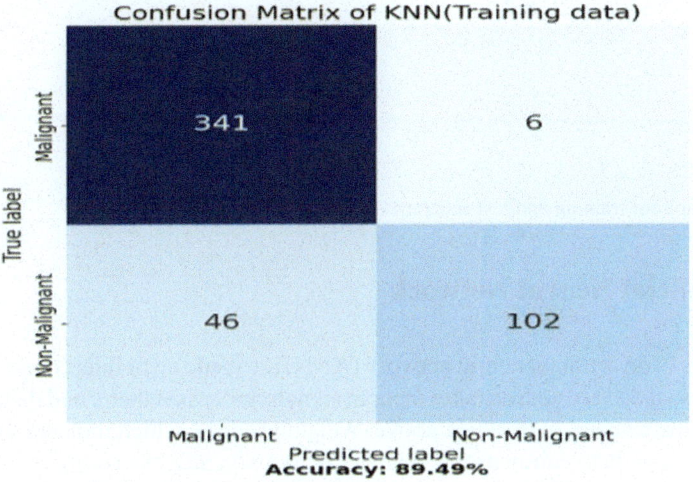

Fig. 2 Confusion matrix of KNN for training data

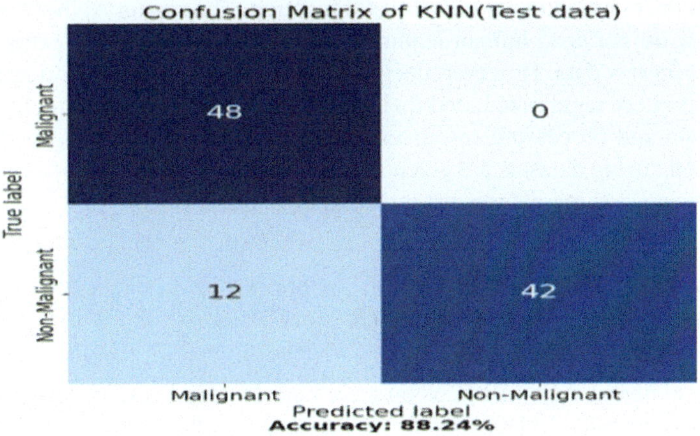

Fig. 3 Confusion matrix of KNN for testing data

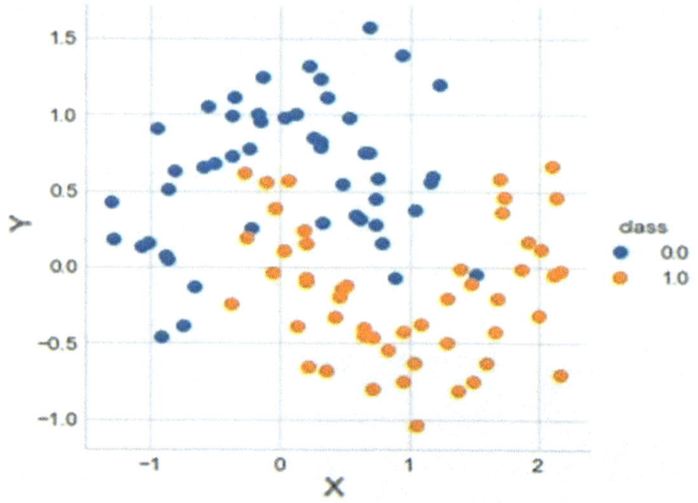

Fig. 4 Graph for training and test accuracy of KNN

5 Artificial Neural Network

The layers of an artificial neural network (ANN) are made up of interconnected nodes or neurons, and every neuron takes in input signals, processes them, and then sends the output to the neurons in the layer above. Weighted edges, which indicate the strength of connections between neurons, connect neurons in successive layers of artificial neural networks (ANNs) [16]. Gradient descent and other optimization algorithms are used, which involve iteratively adjusting these weights through backpropagation. One popular activation function in artificial neural networks, especially in deep learning models, is the rectified linear unit (ReLU) function. The reason ANNs work so well is that they do not need human feature engineers to automatically extract relevant features from raw data. However, deep ANNs can suffer from overfitting if they are not appropriately regularized, and their training necessitates substantial volumes of labeled data and processing resources [17]. Once ANN model is generated, both training and testing datasets are given as input to the model. The confusion matrices for training and testing dataset are shown in Figs. 5 and 6.

6 Convolution Neural Network

Convolution neural networks (CNNs) are a kind of deep learning models designed to process structured, grid-like input, such as images. Their ability to autonomously learn hierarchical representations directly from raw pixel data has revolutionized computer vision tasks [18]. A CNN's layers include convolution, pooling, and fully

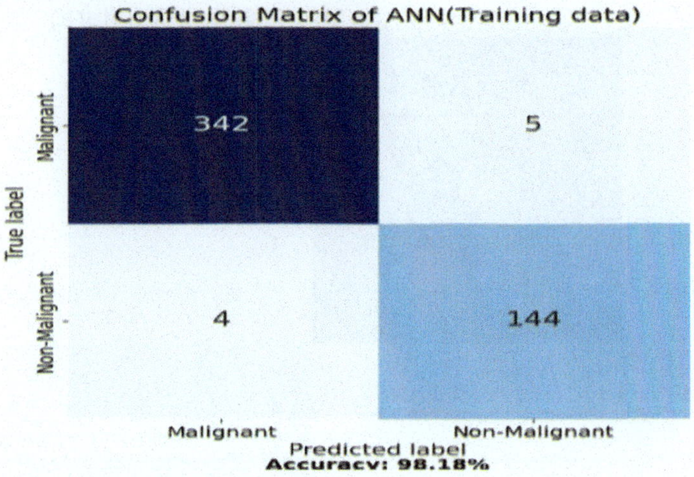

Fig. 5 Confusion matrix of ANN for training data

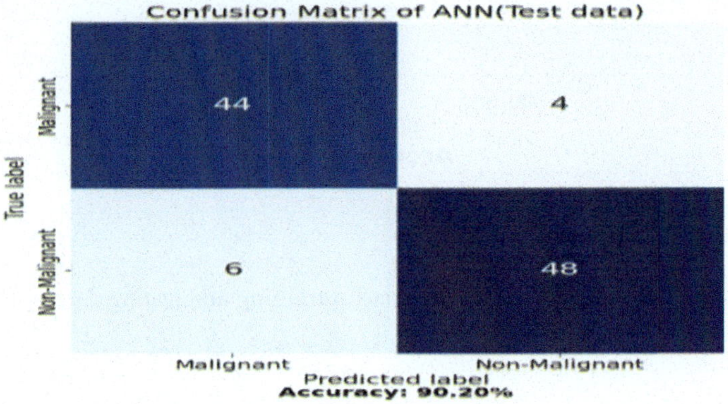

Fig. 6 Confusion matrix of ANN for testing data

linked layers. Convolution layers in CNNs use filters or kernels to apply convolution operations on the input image in order to extract local patterns or features. CNNs can learn hierarchical representations of features, ranging from basic textures and edges to intricate objects and concepts, thanks to these layers. In a variety of computer vision applications, such as object identification, segmentation, image production, and image classification, CNNs have demonstrated impressive performance. They have also been used with methods like 1D convolutions and attention processes in non-computer vision domains including reinforcement learning and natural language processing. In order to prevent overfitting, L2 regularization—also known as the L2 norm, Ridge, or L2 regularization—forces weights to be modest but not precisely 0 [19]. Once CNN model is generated, both training and testing datasets are given as

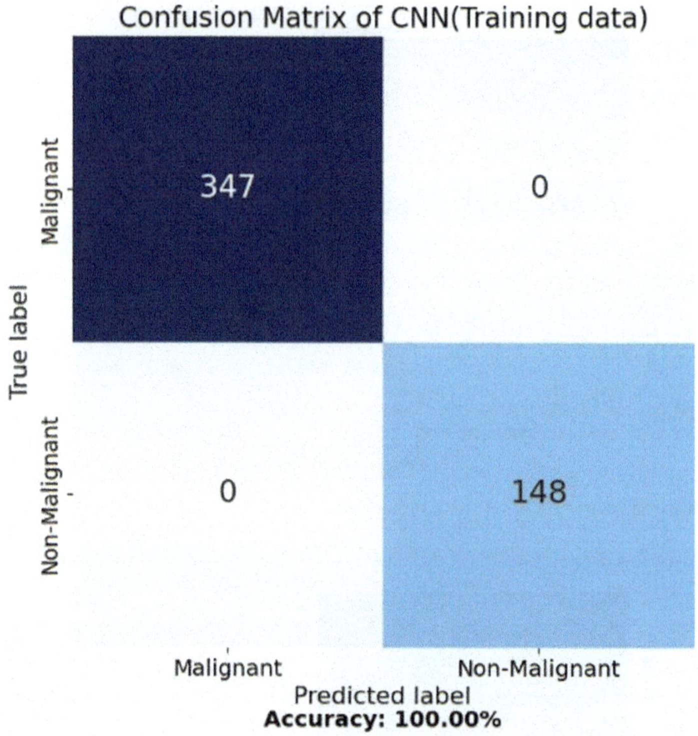

Fig. 7 Confusion matrix train data

input to the model. The confusion matrix for training and testing dataset is shown in Figs. 7 and 8.

7 Results

After applying KNN, ANN, and CNN on Kaggle lung cancer dataset, the following results are generated and tabulated in Table1. Figures 9, 10, and 11 show training dataset accuracy, test dataset accuracy, and training and testing misclassification of KNN, ANN, and CNN models.

Figure 9 shows the training accuracies achieved with K-nearest neighbors (KNN), artificial neural network (ANN), and convolutional neural network (CNN) models as 89.49%, 98.18%, and 99.98%, respectively.

Figure 10 shows that the test accuracies attained with K-nearest neighbors (KNN), artificial neural network (ANN), and convolutional neural network (CNN) models are 88.24%, 90.20%, and 97.06%, respectively.

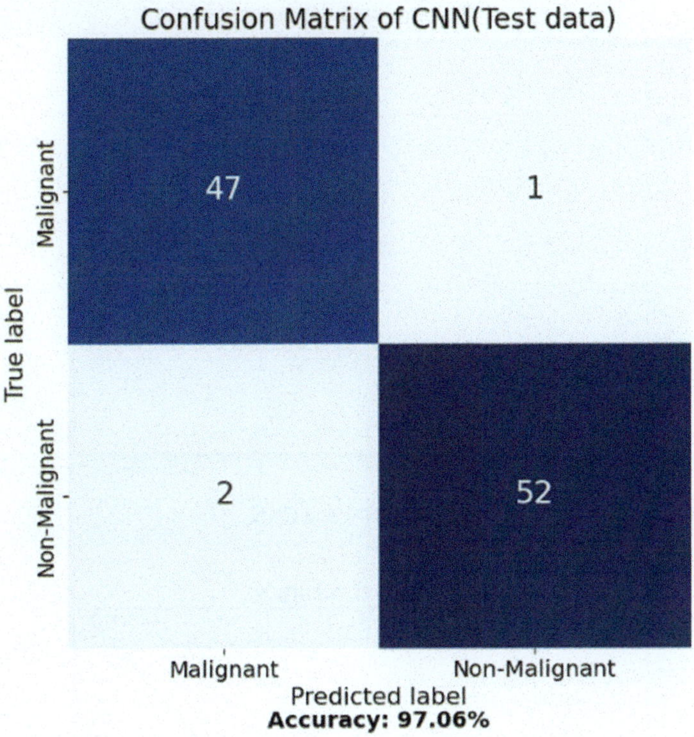

Fig. 8 Confusion matrix for test data

Table 1 Results after applying K-nearest neighbor, artificial neural network, and convolution neural network on lung cancer datasets

Algorithm	Training dataset accuracy	Test dataset accuracy	Training misclassification error	Test misclassification error
K-nearest neighbor	89.49	88.24	0.1051	0.1176
Artificial neural network	98.18	90.20	0.0182	0.098
Convolutional neural network	100	97.06	0.0001	0.0294

After applying K-nearest neighbors (KNN), artificial neural network (ANN), and convolutional neural network (CNN) algorithms to the Kaggle lung cancer dataset, we obtained the results, which are tabulated in Table 1.

In Table 1, we observe that the KNN model achieved an accuracy of 89.49% on the training dataset and 88.24% on the test dataset. The ANN model surpassed this performance, achieving an accuracy of 98.18% on the training dataset and 90.20%

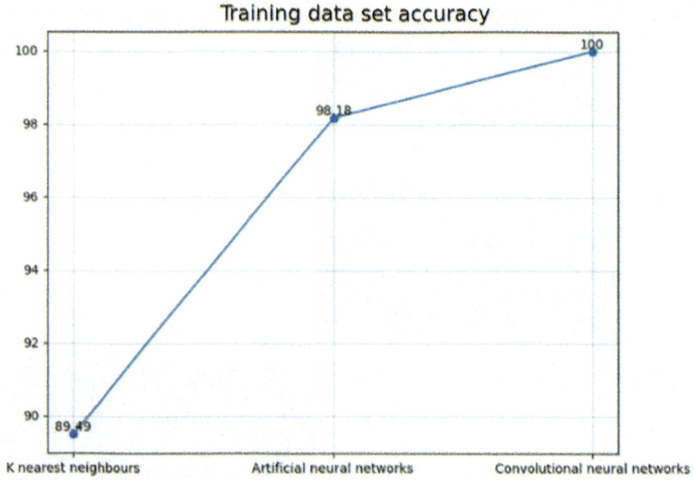

Fig. 9 Training dataset accuracy of KNN, ANN, and CNN

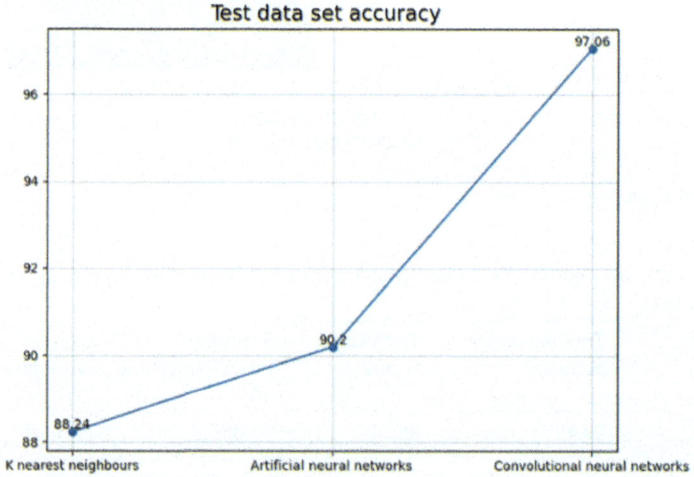

Fig. 10 Test dataset accuracy of KNN, ANN, and CNN

on the test dataset. The CNN model exhibited even higher accuracy, with 99.98% accuracy on the training dataset and 97.06% accuracy on the test dataset. These results suggest that CNN outperformed both KNN and ANN in terms of classification accuracy on both the training and testing datasets. Figure 9 illustrates the training dataset accuracy of KNN, ANN, and CNN models, showing that the CNN model consistently achieved the highest accuracy across different epochs, followed by the ANN model and then the KNN model. Similarly, Fig. 10 depicts the test dataset accuracy, confirming that the CNN model exhibited the best performance, followed

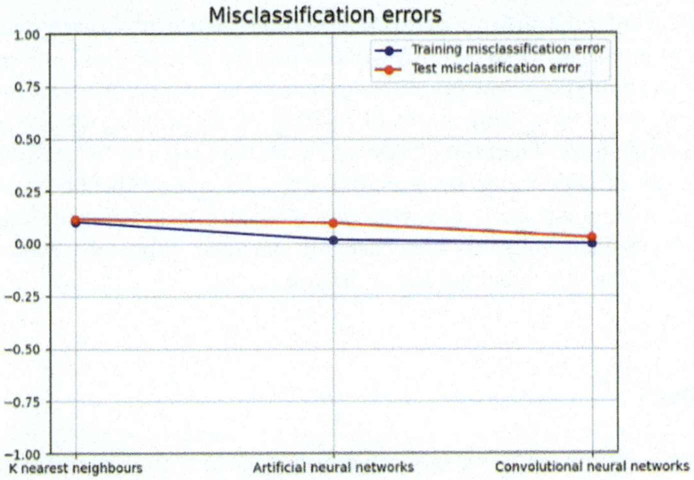

Fig. 11 Misclassification error graph for KNN, ANN, and CNN

by the ANN model and then the KNN model. Furthermore, Fig. 11 illustrates the misclassification rates of the three models on both the training and testing datasets. We observe that the misclassification rates decrease with each epoch for ANN and CNN models, indicating the models' ability to learn and improve over time. Overall, these results highlight the superior performance of deep learning models, particularly CNN, in accurately classifying lung cancer cases compared to traditional machine learning algorithms like KNN.

8 Conclusion

When it comes to utilizing picture datasets for lung cancer prediction, convolutional neural networks (CNNs) are the clear leader due to their exceptional ability to interpret the complex visual information included in medical images. CNNs' remarkable advantage over both standard K-nearest neighbor (KNN) algorithms and classic artificial neural networks (ANNs) stems from its specific architecture, which is painstakingly designed to extract the complex layers of information hidden within images. CNNs are particularly effective because of their convolutional layers, which are carefully designed to extract hierarchical spatial characteristics. This allows for the remarkably accurate and faithful identification of tiny alterations and anomalies that are indicative of lung cancer. This complicated feature extraction capacity, in addition to CNNs' natural ability to adjust to the various subtleties of medical imaging data, surpasses the constraints of ANNs and KNN, especially when dealing with the intricacies of lung scan interpretation. In addition to improving diagnostic accuracy, CNNs' unmatched ability to navigate the complex visual data present in lung imaging holds the prospect of revolutionary advancements in early detection and treatment approaches. Medical professionals may be able to transform lung cancer diagnosis and bring in a new era of customized care marked by preemptive interventions catered to the specific needs of each patient by utilizing CNNs' impressive analytical capabilities. Therefore, CNNs are more than just a technical development; they are a ray of hope in the never-ending quest to improve healthcare outcomes. They hold out the prospect of a day when prompt detection and accurate interventions will completely change the way lung cancer is managed, ultimately saving countless lives and reducing the global burden of disease.

References

1. Harsha Vardhini PA, Prasad SS, Vishnu Sai MHS, Santoshi C, Konduru D (2024) Pioneering minimalist speech analysis through optimized spectral features machine learning models. In: 2024 international conference on emerging smart computing and informatics (ESCI), Pune, India, pp 1–6. https://doi.org/10.1109/ESCI59607.2024.10497288
2. American Cancer Society (2022) Lung cancer treatments. https://www.cancer.org/cancer/lung-cancer/treating.html
3. Meena K, Veni NNK, Deepapriya BS et al (2022) A novel method for prediction of skin disease through supervised classification techniques. Soft Comput 26:10527–10533. https://doi.org/10.1007/s00500-022-07435-8
4. Raghu VK et al (2022) Clinical validation of a machine learning-based model for lung cancer detection using electronic health records. IEEE Access
5. Erica Ponzi D et al (2021) Integrative analysis of multi-omics data for improved lung cancer detection and prognosis prediction. IEEE/ACM Trans Comput Biol Bioinformatics
6. Hiroyoshi Machida A et al (2020) Predicting lung cancer incidence and mortality by incorporating COPD severity: a large-scale population-based analysis. IEEE Trans Biomed Eng
7. Rajeshwari T, Harsha Vardhini PA, Manoj Kumar Reddy K, Priya KK, Sreeja K (2022) Smart agriculture implementation using IoT and leaf disease detection using logistic regression. In: 2021 4th international conference on recent developments in control, automation & power engineering (RDCAPE), Noida, India, pp 619–623. https://doi.org/10.1109/RDCAPE52977.2021.9633608
8. Ardila D, Kiraly AP, Bharadwaj S, Choi B, Reicher JJ, Peng L, Tse D, Etemadi M, Ye W, Corrado G, Naidich DP, Shetty S (2019) End-to-end lung cancer screening with three-dimensional deep learning on low-dose chest computed tomography. Nat Med 25(6):954–961
9. Takahashi S et al (2019) Deep learning-based classification of lung adeno-carcinoma by integration of multi-omics data. IEEE J Biomed Health Inform
10. Monkam P et al (2018) A deep learning approach for automated lung nodule detection and classification in CT images. IEEE Trans Med Imaging
11. Liu Y, Kim J, Balagurunathan Y, Li Q, Garcia AL, Stringfield O, Ye Z, Gillies RJ (2016) Radiomic features are associated with EGFR mutation status in lung adenocarcinomas. Clin Lung Cancer 17(5):441–448
12. Vineis P, Alavanja M, Buffler P, Fontham E, Franceschi S, Gao YT, Caporaso N (2004) Tobacco and cancer: recent epidemiological evidence. J Natl Cancer Inst 96(2):99–106
13. Duda RO, Hart PE, Stork DG (2012) Pattern classification. Wiley
14. Sun Y, Wang X, Tang X (2019) Deep learning face representation by joint identification-verification. IEEE Trans Pattern Anal Mach Intell 41(1):147–160. https://doi.org/10.1109/TPAMI.2017.2776961

15. Luo W, Li Y, Urtasun R, Zemel R (2017) Understanding the effective receptive field in deep convolutional neural networks. In: Advances in neural information processing systems, pp 4898–4906
16. Krizhevsky A, Sutskever I, Hinton GE (2017) ImageNet classification with deep convolutional neural networks. Commun ACM 60(6):84–90. https://doi.org/10.1145/3065386
17. Szegedy C, Vanhoucke V, Ioffe, S, Shlens J, Wojna Z (2016) Rethinking the inception architecture for computer vision. In: Proceedings of the IEEE conference on computer vision and pattern recognition, pp 2818–2826. https://doi.org/10.1109/CVPR.2016.308
18. Couraud S, Zalcman G, Milleron B (2012) Lung cancer in never smokers—a review. Eur J Cancer 48(9):1299–1311
19. Cover T, Hart P (1967) Nearest neighbor pattern classification. IEEE Trans Inf Theory 13(1):21–27. https://doi.org/10.1109/TIT.1967.1053964

Recommending Emergency Medication Using Machine Learning Approaches

S. Kranthi Reddy, T. Meghana, N. Nishet Vardhan, and B. Priyanka

Abstract In today's healthcare sector, online recommender systems have become increasingly prevalent among doctors, medical professionals, and patients alike. With the majority of consumers turning to online resources for prescription suggestions, these systems play a crucial role in aiding decision- processes, particularly during times of crises such as pandemics or natural disasters. Leveraging the power of machine learning (ML), these systems offer enhanced accuracy, precision, and reliability in clinical predictions while optimizing resource utilization. By considering various patient parameters like symptoms, blood pressure, and diabetes, medication recommendation systems provide personalized and trustworthy guidance on medication selection, dosage, and potential adverse effects. Among the different ML approaches, Decision Tree-based systems demonstrate superior accuracy, making them particularly valuable in emergency situations where swift and safe medication recommendations are essential.

Keywords Machine learning algorithms · Decision Trees · Instant medications · Medication · Healthcare

1 Introduction

In contemporary healthcare, online recommender systems have become indispensable tools for hospitals, medical practitioners, and patients alike. During critical situations such as pandemics or natural disasters, these systems play a pivotal role in guiding decision-making processes. Underscoring the significance of such systems, a symptom set-based drug recommendation framework has been proposed to aid clinicians during prescription decisions amidst challenging circumstances. Drawing on the capabilities of machine learning (ML), these systems enhance clinical predictions with heightened accuracy, precision, and reliability, all while optimizing resource

S. Kranthi Reddy (✉) · T. Meghana · N. Nishet Vardhan · B. Priyanka
Department of Computer Science and Engineering, Vignan Institute of Technology and Science, Yadadri Bhuvanagiri, Telangana, India
e-mail: kranthi.vits@gmail.com

© The Author(s), under exclusive license to Springer Nature Singapore Pte Ltd. 2025

H. Zen et al. (eds.), *Soft Computing and Signal Processing*, Lecture Notes in Networks and Systems 1200, https://doi.org/10.1007/978-981-97-9926-8_22

utilization. Highlighting the transformative potential of ML techniques in revolutionizing drug discovery and decision-making, researchers emphasize the pivotal role of advanced computational methods in healthcare. Among the myriad applications of ML in healthcare, medication recommendation systems stand out for their ability to empower informed decision-making. Leveraging technologies like machine learning, these systems illuminate valuable insights in medical data to drive more informed prescription decisions. Additionally, exploring the utility of ML algorithms such as Decision Trees and Naive Bayes models in determining patient conditions and recommending appropriate treatments reaffirms the indispensable nature of these systems in clinical decision-making. Notably, Decision Tree-based approaches emerge as frontrunners due to their superior accuracy in these systems. Moreover, these recommender systems are designed to uphold data integrity and protect patient privacy, as emphasized in comprehensive literature studies on medical recommender systems. In summary, recommender systems represent indispensable assets in contemporary healthcare, offering powerful tools for medical professionals to make informed decisions, optimize resource utilization, and ultimately enhance patient outcomes.

2 Literature Survey

A prescription assistance system with patient privacy protection was presented by Vadupu et al. [1] utilizing a symptom set-based drug recommendation framework. They particularly suggested a set-to-set small and safe pharmaceutical suggestion that is symptom-based and consists of three modules: a drug set module, a symptom set module, and a unique set-to-set comparison module (4SDrug). There are certain shortcomings with this study, with the suggested method for disease forecasting not working. Momtahen et al. [2] highlight that a large, complex, and variable are the pipelines used in drug research and development. With a wealth of high-quality data and well-defined problems, machine learning (ML) techniques offer a set of tools that enhance discovery and decision-making. There are numerous times during the drug development process where machine learning can be used. Examples include the investigation of prognostic biomarkers, target validation, and the examination of digital pathology data in medication trials. Different approaches and circumstances have led to different kinds of applications; some approaches yield accurate insights and projections. Garg [3] proposed a medicine recommendation system that uses cutting-edge methods such as machine learning, data mining, and others to uncover interesting formation buried in medical data and reduce mistakes made by doctors while prescription medications. The system is comprised of the database module and the data preparation module. Abdelkrim et al. [4], a new feature selection strategy based on Random Forests is suggested for an efficient classification problem. They compare our feature selection strategy to baseline support vector machine and artificial neural network classifiers. The prediction of drug-target interaction using

Weisfeiler-Lehman neural machine is another machine learning application given in the paper. Machine learning is instrumental in advancing healthcare [5–11].

3 Methodology

The current drug recommendation system is based on a Bayesian Neural Network (BNN) model, which incorporates Bayesian inference principles to handle uncertainty in predictions. In this system, BNNs represent each random variable involved in the drug recommendation process and capture the conditional relationships among these variables. Despite their theoretical advantages, BNNs are often not very accurate and are highly complex. They require significant computational resources for training and inference, leading to high temporal complexity. Additionally, in some cases, the BNN failed to recommend the correct dosage, highlighting potential issues with model accuracy and reliability. These challenges indicate that while BNNs offer a sophisticated approach, their practical application in drug recommendation systems is limited by their complexity and potential inaccuracies.

3.1 Proposed System

In the development of the proposed online drug recommender system, extensive descriptive analytics were conducted on the dataset. These analytics aimed to uncover crucial insights into the dataset's structure, trends, and potential patterns. Various statistical methods and visualization techniques were employed to explore the dataset comprehensively. Descriptive statistics helped in understanding the distribution of variables, while data visualization techniques, such as histograms and scatter plots, provided visual representations of relationships between different features. Following the thorough data analysis, a series of algorithms for machine learning were applied to develop recommender system. Leveraging the insights gained from the data analysis phase, each algorithm was implemented and evaluated for its suitability in classification tasks and potential to address the objectives of predicting drugs based on given conditions. Through iterative experimentation and model evaluation, the performance of each algorithm was assessed. This iterative process allowed for the identification of the most suitable algorithm for the recommender system, ensuring its effectiveness in providing accurate and personalized medication recommendations to users. The proposed work is shown in following Fig. 1.

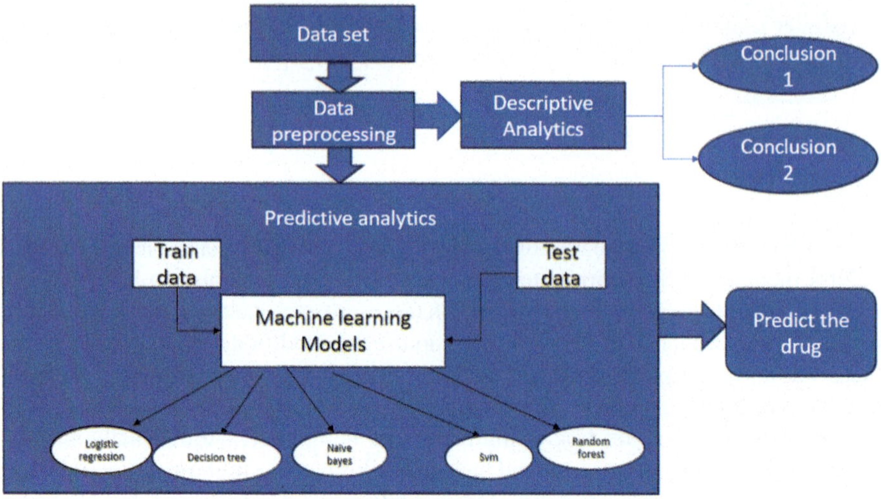

Fig. 1 Proposed work outlines descriptive, diagnostic, and predictive analytics on drug dataset

3.2 Dataset and Preprocessing

After preprocessing the UCML Drug dataset, obtained from Kaggle, several transformations were applied to prepare it for analysis. Initially, the dataset contained various features such as uniqueID, drugName, condition, review, rating, date, and usefulCount. These features provided valuable insights into medication reviews, including user sentiments, ratings, and medication effectiveness. During the preprocessing phase, the dataset was converted into a numeric format Fig. 2, excluding the date column as it may not directly contribute to the numerical representation of the data. The transformed dataset now consists of key attributes such as uniqueID, review, rating, usefulCount, drugName, and condition. Each entry in the dataset corresponds to a unique medication review, identified by a distinct uniqueID. The review column categorizes sentiments into positive, neutral, or negative, providing an indication of users' experiences with the medications. Ratings provided in the rating column offer insights into overall satisfaction levels, while the usefulCount column quantifies the usefulness of each review based on the number of users who found it helpful. The drugName attribute specifies the medication under review, while the condition feature describes the medical condition or purpose for which the medication is prescribed. By transforming the dataset into a numeric format, it becomes conducive to the application of various machine learning algorithms for analyzing relationships between features and predicting outcomes related to medication efficacy and user satisfaction. Figure 2 shows the dataset conversion.

	uniqueID	drugName	condition	review	rating	date	usefulCount
0	206461	Valsartan	Left Ventricular Dysfunction	"It has no side effect, I take it in combinati...	9	20-May-12	27
1	95260	Guanfacine	ADHD	"My son is halfway through his fourth week of ...	8	27-Apr-10	192
2	92703	Lybrel	Birth Control	"I used to take another oral contraceptive, wh...	5	14-Dec-09	17
3	138000	Ortho Evra	Birth Control	"This is my first time using any form of birth...	8	03-Nov-15	10
4	35696	Buprenorphine / naloxone	Opiate Dependence	"Suboxone has completely turned my life around...	9	27-Nov-16	37

	uniqueID	review	rating	usefulCount	drugName	condition
0	206461	neutral	9	27	1	1
1	95260	positive	8	192	2	2
2	92703	neutral	5	17	3	3
3	138000	positive	8	10	4	3
4	35696	positive	9	37	5	4

Fig. 2 Dataset conversion

3.3 Descriptive Analytics

Descriptive analysis serves as a fundamental technique in data exploration and interpretation across various fields, including healthcare. In healthcare, descriptive analysis plays a crucial role in understanding disease prevalence, treatment outcomes, and medication usage patterns. By analyzing key metrics such as frequency, distribution, and central tendency, researchers can earn valuable advices into the characteristics of patient populations and effectiveness of healthcare interventions. Descriptive analysis provides a comprehensive overview of the dataset, helping stakeholders make informed decisions and develop targeted strategies to address healthcare challenges. Descriptive analysis in instant medication advice platforms involves analyzing data to understand medication usage patterns, treatment outcomes, and side effect profiles. It helps identify commonly prescribed medications, trends among patient demographics, and emerging patterns. By leveraging these insights, platforms can offer personalized and evidence-based recommendations in real time, enhancing patient safety and treatment effectiveness.

Identified Insights

Based on the descriptive analysis conducted, it is evident that depression and birth control are the primary conditions for which a significant number of drugs are utilized. Through the examination of drug ratings, it was observed that nine drugs used for depression are associated with reported side effects, while five drugs were found to have no reported side effects. This finding underscores the importance of evaluating side effect profiles in treatment decisions for depression, providing insights into potential avenues for safer treatment options. Similarly, for birth control medications, the analysis revealed that a substantial majority of drugs (45) are associated with reported side effects, whereas only three drugs were identified as having no reported side effects. This disparity highlights the prevalence of side effects in birth control treatments, emphasizing the need for careful consideration and monitoring of side effect profiles in contraceptive medication prescriptions. The graphical representations (Figs. 3 and 4) effectively illustrate the distribution of drugs based on their side effect profiles for depression and birth control, respectively. These visualizations serve as valuable tools for clinicians and healthcare professionals to assess the safety profiles of medications and make informed treatment decisions. Furthermore, leveraging machine learning algorithms, this descriptive analysis facilitates the prediction of drugs based on condition and rating data. By utilizing ML techniques, such as Decision Trees or neural networks, predictive models are developed that aid in identifying potential treatment options tailored to individual patient needs. This integration of ML algorithms enhances the accuracy and efficiency of drug recommendation systems, **ultimately contributing to improved patient outcomes and** healthcare practices.

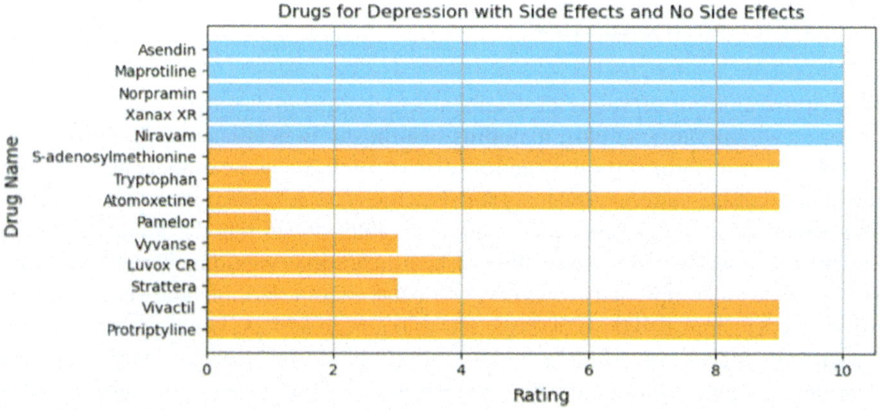

Fig. 3 Medication side effects versus none at all for depression

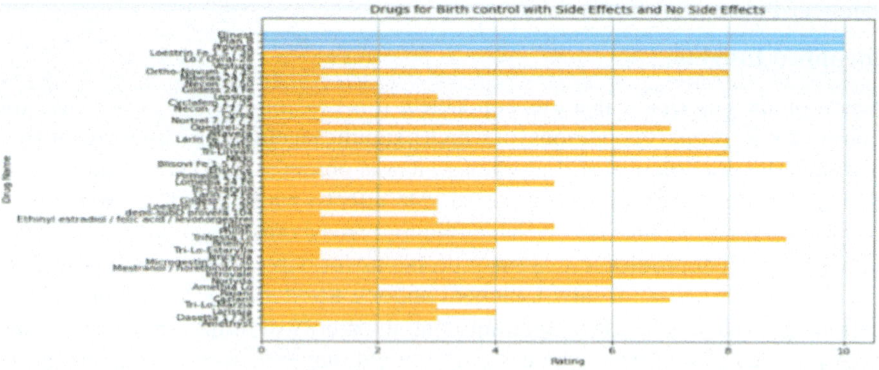

Fig. 4 Medication side effects versus none at all for birth control

3.4 Predictive Analytics

Predictive analytics entails leveraging past data alongside statistical methodologies to anticipate forthcoming results. By analyzing patterns and trends in data, predictive analytics helps organizations make informed decisions, anticipate future trends, and identify opportunities for optimization and improvement. It leverages machine learning techniques to generate accurate predictions and actionable insights, enabling proactive decision-making across various industries.

Predictive Analytics in Instant Med Advice

Predictive analytics plays a crucial role in instant medication advice by understanding historical patient statistics and machine learning algorithms to predict the most suitable medications for individuals based on their specific health conditions, symptoms, medical history, and demographic factors. By analyzing past treatment outcomes and patterns, predictive analysis can recommend the most effective medications and dosages tailored to each patient's unique needs. This proactive approach improves the accuracy and efficiency of medication recommendations. Utilizing machine learning techniques, medicine is predicted by taking into account condition and review. Accuracy, precision value, recall value, and F1-scores calculations were made for the Random Forest, Decision Tree, Naïve Bayes, SVM, and logistic regression algorithms.

3.4.1 Logistic Regression

A crucial technique in supervised machine learning, logistic regression calculates the likelihood of an outcome and is mostly applied to binary classification problems. By dividing data into two groups and typically displaying the results as yes/no or 0/1, it is essential for predictive modeling that evaluates the correlation between independent variables. Logistic regression is used in situations such as binary classification problems to ascertain the likelihood that an instance falls into a particular category, such as the negative class (0%) or the positive class (1). Using a sigmoid function called the logistic function, logistic regression converts predictions into probabilities and provides an S-shaped curve that limits real values to between 0 and 1. The following represents the logistic regression equation:

Equation for logistic regression

$$y = \left(e^{(b0+b1X)}\right)/1 + e^{(b0+b1X)}. \tag{1}$$

Furthermore, a confusion matrix is frequently evaluated as part of a logistic regression study in order to gauge how well the model predicts the classification results depicted in Table 1.

Logistic regression confusion matrix is shown in Table 1.

The S-shaped curves of the logistic function are displayed in Fig. 5.

Table 1 Logistic regression confusion matrix

True label		True	False
	True	78,856	121
	False	1490	182
	Predicted label		

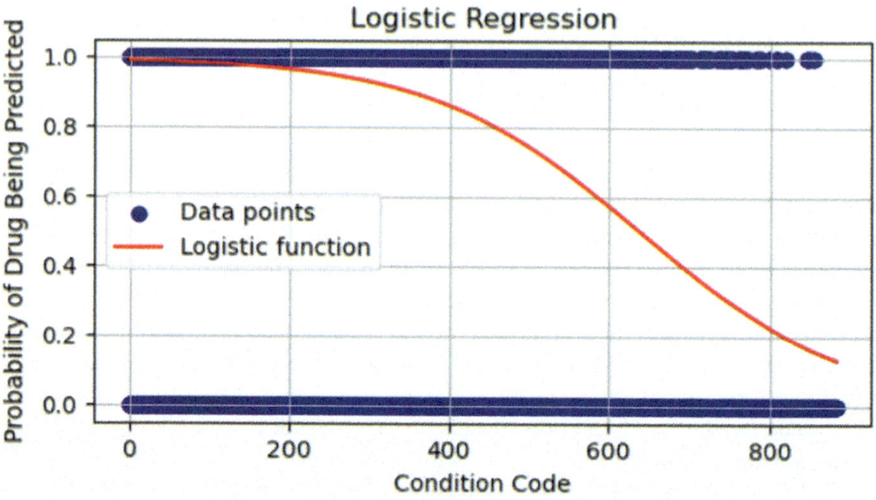

Fig. 5 Logistic regression for UCML drug dataset

3.4.2 Decision Tree

One kind of supervised learning method called Decision Trees is useful for both regression and categorization problems, though they are more commonly used for classification problems. While leaf nodes indicate the ultimate results of those decisions and do not branch further, decision nodes enable decision-making with numerous branches. How to choose the best attribute for the root node and any child nodes after it is a critical factor to take into account when building a Decision Tree. A method called Attribute Selection Measure (ASM) is used to overcome this difficulty. The best characteristic for every node in the tree is determined with the help of ASM. There are two popular ASM techniques: 1. Information Gain: It calculates the decrease in entropy that happens whenever a dataset is divided in light of a specific characteristic. The formulae for information gain and entropy are shown below (Table 2).

Formulae for Information Gain

$$\text{Information Gain} = \text{Entropy}_{\text{parent}} - \text{Entropy}_{\text{children}}. \tag{2}$$

Formulae for Entropy

Table 2 Confusion matrix for Decision Tree

True label		True	False
	True	78,730	247
	False	1265	407
	Predicted label		

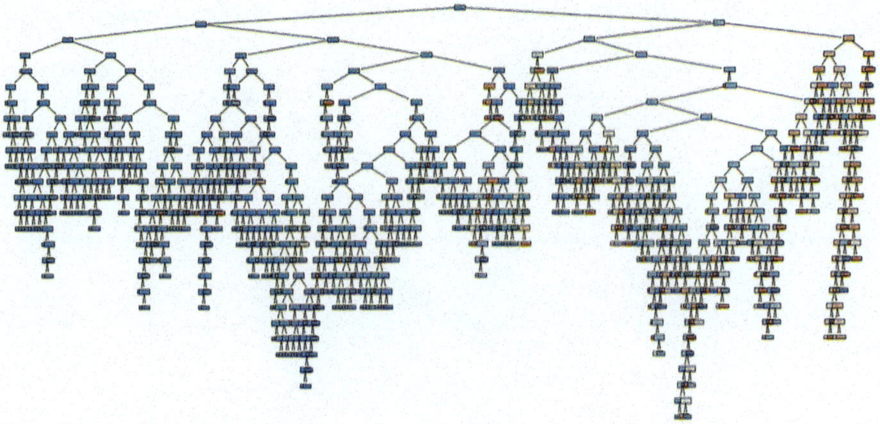

Fig. 6 Decision tree for drug dataset

$$\text{Entropy}(E) = -\sum_{i=1}^{n} pi \log2(pi). \quad (3)$$

The decision boundaries generated by the Decision Tree classifier based on the features "condition" and "drugname" are illustrated in Fig. 6.

3.4.3 Naive Bayes

The Bayes classifier is a probabilistic approach that derives from the Bayes theorem. It calculates the probability of an input being allocated to a certain class based on the assumption of feature independence. It is still a widely used tool across many areas and is well-known for its use in tasks like text categorization, spam detection, and recommendation engines.

Constructing an NB model is straightforward and notably advantageous for handling extensive datasets. In addition to its simplicity, Naive Bayes exhibits superior performance compared to even the most intricate categorization schemes (Table 3).

The posterior probability $P(c|x)$ can be determined using the Bayes theorem and $P(x)$, $P(c)$, and $P(x|c)$. Applying the Naïve Bayes formula,

Table 3 Confusion matrix for Naïve Bayes model

True label		True	False
	True	77,250	1727
	False	1098	574
	Predicted label		

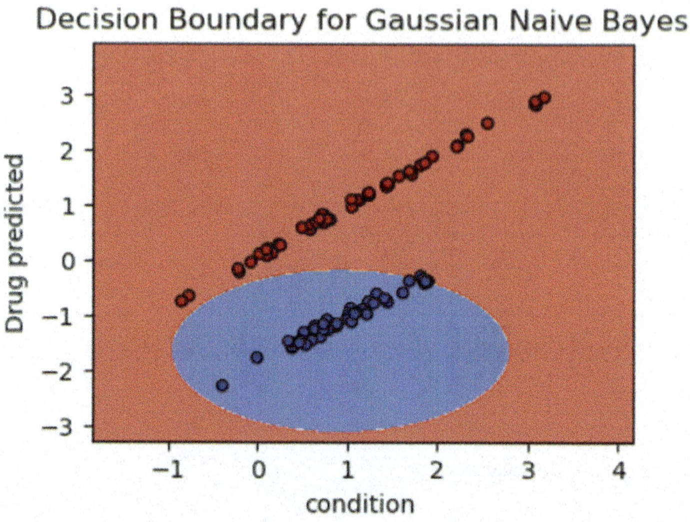

Fig. 7 Naïve Bayes for UCML drug dataset

$$p(c|x) = p(x|c)p(c)/p(c).$$

Figure 7 depicts the graph of Bayesian classifier, providing insight into its performance in instant medical advice.

3.4.4 Random Forest Classifier

Among machine learning algorithms used in supervised learning, Random Forest is one of the most popular. It is well-known for its adaptability and can handle machine learning problems involving both regression and classification with ease. Based on the idea of group learning, Random Forest combines several classifiers to solve complex issues and improve model performance. The Random Forest confusion matrix is depicted in Table 4.

The graph depicting the decision boundaries of the Random Forest is presented in Fig. 8.

Table 4 Confusion matrix for Random Forest model

True label		True	False
	True	78,769	208
	False	1277	395
	Predicted label		

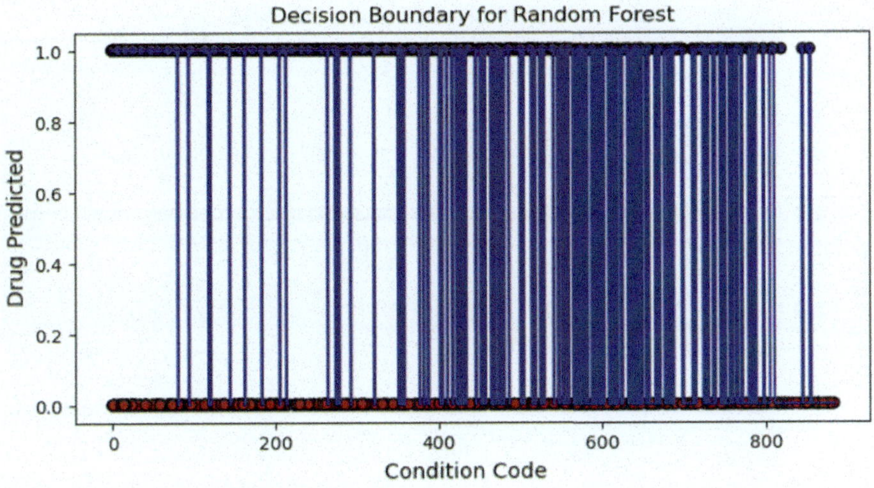

Fig. 8 Random Forest for UCML Drug dataset

3.4.5 Support Vector Machine (SVM)

Support vector machines, or SVMs, are techniques for supervised learning used in machine learning. When classifying data components into two separate categories is the aim, support vector machines (SVMs) perform especially well in binary classification settings.

Finding the ideal line—also referred to as the decision boundary—that successfully divides data points into distinct groups is the main goal of an SVM algorithm. This boundary, in higher-dimensional feature spaces, is called a hyperplane. The goal is to expand the margin. The SVMs seek to enable unambiguous differentiation between various data classes in this way.

The SVM's decision function is expressed in the form:

Equation for SVM

$$f(x) = \text{sign}(w\ Tx + b). \tag{4}$$

A thorough analysis of the classification model's performance is given by the confusion matrix, which counts the values of true positive predictions, false positive predictions, true negative prediction, and false negative predictions is depicted in Table 5.

Table 5 Confusion matrix for support vector machine (SVM)

True label		True	False
	True	78,929	48
	False	1552	120
	Predicted label		

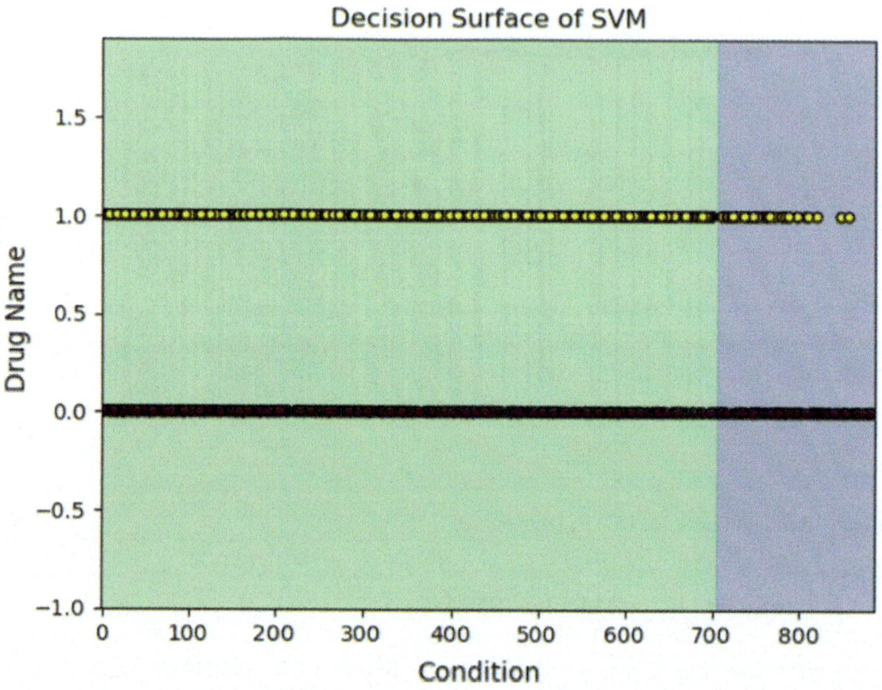

Fig. 9 SVM for drug dataset

A thorough summary of the SVM Graph is given by the visualization in Fig. 9, which also offers insights into the classification model's performance, notably with regard to its ability to correctly detect false positives and false negatives and to forecast true positives and true negatives.

4 Results

The UCML Drug dataset was subjected to Decision Tree, Random Forest, logistic regression, Naive Bayes, and SVM algorithms. The resulting outcomes are presented in Tables 6 and 7, along with corresponding graphical representations (Fig. 10). These algorithms' performance is assessed using a range of metrics:

Table 6 Precision, recall, F1-score of machine learning models

Algorithm	Logistic regression		Decision Tree		Naïve Bayes		Random Forest		SVM	
	True	False	True	False	True	False	True	False	True	False
Precision	0.98	0.60	0.98	0.62	0.99	0.25	0.98	0.66	0.98	0.71
Recall	1.00	0.11	1.00	0.24	0.98	0.34	1.00	0.22	1.00	0.71
F1-score	0.99	0.18	0.99	0.35	0.98	0.29	0.99	0.33	0.99	0.13

Table 7 Accuracy of machine learning models

Algorithm	Logistic regression	Decision Tree	Naïve Bayes	Random Forest	Support vector machine
Accuracy	98.00	98.12	96.49	98.15	98.01

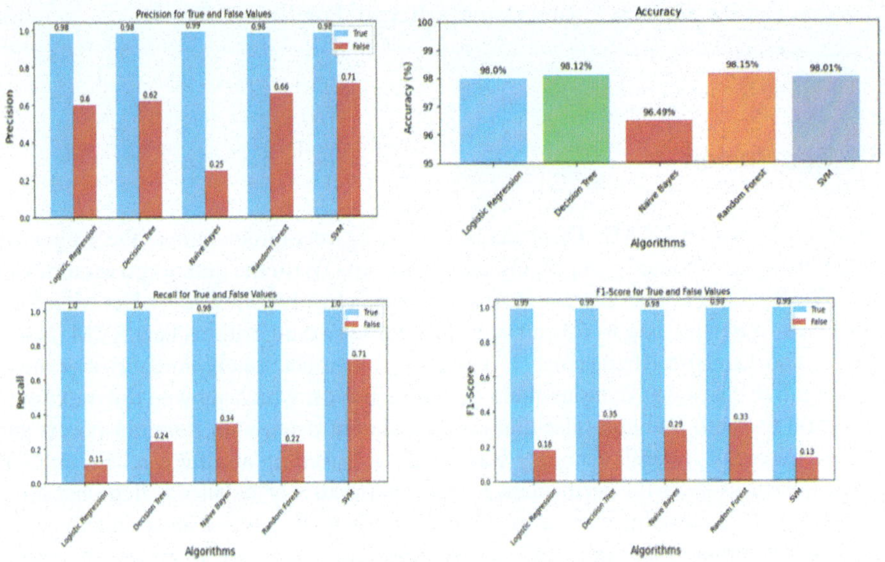

Fig. 10 A comparison graph displays the outcomes of Naïve Bayes, Decision Trees, Random Forest trees, support vector machines, and logistic regression models applied to the UCML Drug dataset

The proportion of correctly predicted outcomes to all instances in the dataset is known as accuracy. It is computed with the following formula:

Formula for Accuracy

$$\text{Accuracy} = \text{TN} + \text{TP}/\text{TN} + \text{FP} + \text{TP} + \text{FN}. \tag{5}$$

Precision: The percentage of actual positive predictions among all the model's positive predictions. It is computed with the following formula:

Formula for Precision

$$\text{precision} = \text{TP}/\text{TP} + \text{FP}. \tag{6}$$

The ratio of accurately predicted positive observations to all observations made during the actual class is known as recall (sensitivity). It is computed with the following formula:

Formula for Recall

$$\text{Recall} = \text{TP}/\text{TP} + \text{FN}. \tag{7}$$

F1-Score: A balance between recall and precision calculated as the harmonic mean of the two metrics. The formula for calculating it is:

Formula for F1-Score

$$\text{F1 Score} = 2 \times \text{precision} \times \text{Recall}/\text{Precision} + \text{Recall}. \tag{8}$$

5 Conclusion

Upon utilizing the UCML Drug dataset for drug recommendation, the following findings can be obtained by applying the algorithms. With the maximum accuracy of 98.15%, Random Forest was the most successful, closely followed by Decision Tree (98.12%). When it came to overall accuracy, these two algorithms beat SVM, Naïve Bayes, and Logistic Regression. 2. The capacity to accurately categorize positive and negative cases was demonstrated by Naïve Bayes, which showed the maximum precision for both true and false classes. When contrasted with Random Forest and Decision Tree, its overall accuracy was lower. 3. Based on accuracy, recall, and F1-score measures, SVM's performance was comparatively balanced, demonstrating its resilience to classification tasks. Comparing it to Random Forest and Decision Tree, its accuracy was marginally lower. Decision Trees are a strong alternative for drug recommendation systems since they shown a good balance between precision and accuracy. Upon analyzing the UCML Drug dataset, Random Forest was found to be the best promising algorithm for drug selection due to its high accuracy and well-rounded performance across multiple evaluation measures. Based on these findings, Random Forest seems to be the best algorithm for tasks involving drug recommendation on the UCML Drug dataset.

References

1. Vadupu S, Kandala KS, Peddi A, Yadav NS, Kumar GV, Harsha Vardhini PA (2021) Skin pathology detection using artificial intelligence. In: 2021 6th International conference on signal processing, computing and control (ISPCC), pp 373–376, Solan, India. https://doi.org/10.1109/ISPCC53510.2021.9609516
2. Momtahen S, Al-Obaidy F, Mohammadi F (2019) Machine learning drug discovery and development. In: 2019 IEEE canadian conference of electrical and computer engineering (CCECE), pp 1–6, https://doi.org/10.1109/CCECE.2019.8861842
3. S. Garg (2021) Medicine recommendation system based on patient reviews. In: 2021 11th International conference on cloud computing, data science & engineering (confluence), pp 175–181. https://doi.org/10.1109/Confluence51648.2021.9377188
4. Abdelkrim A, Bouramoul A, Zenbout I (2021) Application of machine learning in drug discovery. In: 2021 International conference on theoretical and applicative aspects of computer science (ICTAACS), pp 1-8. https://doi.org/10.1109/ICTAACS53298.2021.9715218
5. Hossain MD, Azam MS,Ali MJ, Sabit H (2020) Drugs rating generation and recommendation from sentiment analysis of drug reviews using machine learning. In: 2020 Emerging technology in computing, communication and electronics (ETCCE), pp 16. https://doi.org/10.1109/ETCCE51779.2020.9350868.
6. Shang J, Ma T, Xiao C, Sun J (2019) Pre-training of graph augmented transformers for medication recommendation. In: IJCAI, pp 5953–5959
7. Sun J, Gamenet Xiao C, Ma T, Li H (2020) A smart healthcare recommendation system for multidisciplinary patients with data fusion based on deep ensemble learning. In: AAAI, pp 1126–1133
8. Sedik A, Hammad M, Abd El-Samie FE, Gupta BB, El-Latif (2021) Effiffifficient deep learning approach for augmented detection of Coronavirus disease. Neural Comput Appl 8
9. Varshney N, Ahuja S, Kanishk (2021) An intelligence system for medicine recommendation: review. In: 2021 5th international conference on information systems and computer networks (ISCON), pp 1–5. https://doi.org/10.1109/ISCON52037.2021.9702342
10. Akshara R, Reddy SS, Reddy RT, Srikanth M, Saikrishna C (2021) An effective model to prognosticate the stock market tendency using social media. In: 2021 Third international conference on inventive research in computing applications (ICIRCA), pp. 1220–1223, Coimbatore, India, https://doi.org/10.1109/ICIRCA51532.2021.9544741
11. Carracedo-Reboredo P, Liñares-Blanco J, Rodríguez-Fernández N, Cedrón F, Novoa FJ, Carballal A, Maojo V, Pazos A, Fernandez-Lozano C (2021) A review on machine learning approaches and trends in drug discovery. Comput Struct Biotechnol J 19:4538–4558.

Integrated IoT and Image Processing System for Efficient Traffic Violation Management in India

V. Jyothi, K. Sai Nandini, V. Santhosh, and M. Nithin

Abstract Road traffic violations are a significant issue contributing to road crashes, prompting India to adopt automated solutions. This design proposes a system using IoT and image processing for managing traffic violation penalties. It employs Python for license plate recognition and the tensor flow algorithm for image processing, with a sound detector identifying ambulances in specific lanes. Ambulance sounds are processed using IoT and sent to traffic poles to expedite clearance. The system also conserves power by setting a threshold distance for detected objects during non-peak hours. Additionally, the project introduces a system integrating webcam technology and sound sensors to detect helmetless riders and red signal violators, with fines allocated through a Telegram bot. This system aims to enhance traffic management, reduce accidents, and minimize power wastage, contributing to India's safer urban transportation ecosystem.

Keywords Road traffic violations · Road crashes · Automated solutions · IoT

1 Introduction

Road traffic violations are a significant challenge globally, leading to numerous accidents and fatalities. In India, the situation is particularly severe, with a high incidence of violations contributing to road crashes. While there have been previous

V. Jyothi (✉) · K. Sai Nandini · V. Santhosh · M. Nithin
Department of Electronics and Communication Engineering, Vardhaman College of Engineering, Shamshabad, Telangana, India
e-mail: v.jyothi@vardhaman.org

K. Sai Nandini
e-mail: Kavididevisainandini21ec@student.vardhaman.org

V. Santhosh
e-mail: vidigottisanthoshkumar20ece@vardhaman.org

M. Nithin
e-mail: Mamidinithin20ece@vardhaman.org

© The Author(s), under exclusive license to Springer Nature Singapore Pte Ltd. 2025

H. Zen et al. (eds.), *Soft Computing and Signal Processing*, Lecture Notes in Networks and Systems 1200, https://doi.org/10.1007/978-981-97-9926-8_23

projects focusing on automated solutions for managing traffic violations, many of these projects are limited in scope, typically addressing only one specific aspect of traffic management.

This design is proposed in response to the limitations of existing projects, which often focus on single operations. Integrating three key operations—detection of helmetless riders, identification of red signal violators, and recognition of ambulance sirens—into a single system, this project aims to provide a more comprehensive and efficient solution to managing traffic violations. This integrated approach is intended to make the system more practical and easy to use, addressing multiple aspects of traffic management simultaneously.

Moreover, this initiative primarily focuses on enhancing road safety by reducing traffic violations. By implementing a system that can quickly and accurately detect violations and issue fines, this design aims to deter individuals from repeating violations, ultimately saving lives and preventing accidents. Additionally, the system includes a feature to alert parents when their underage children are caught driving without a license, promoting responsible driving behavior from a young age. Furthermore, to enhance public awareness and promote safer driving practices, the system includes an audible alert that sounds when a traffic signal is jumped. This alert serves as a reminder to all road users to adhere to traffic rules, ultimately contributing to a safer road environment for everyone.

In addition to improving traffic safety and management, the system has the potential to enhance the overall urban transportation ecosystem in India. By promoting compliance with traffic regulations and reducing accidents, the system can help create a safer and more efficient transportation network for all road users.

The implementation of this proposal can significantly improve traffic management in India. By providing a comprehensive solution to manage traffic violations, the system can help reduce congestion and improve the flow of traffic. The integration of IoT and image processing technologies allows for real-time monitoring of traffic conditions, enabling authorities to make informed decisions to optimize traffic flow. Additionally, the system's ability to quickly detect and penalize violations can help deter reckless driving behavior, leading to safer roads for all users. The automated nature of the system also reduces the burden on traffic police, allowing them to focus on other important aspects of traffic management.

This paper provides a detailed description of the architecture, implementation, and evaluation of the proposed system. The system's design, key components, and experimental results are discussed, demonstrating its effectiveness in managing traffic violations and improving road safety. With its integrated approach to traffic management, this proposal has the potential to revolutionize the current system in India. By addressing multiple aspects of traffic violations simultaneously, the system offers a comprehensive solution that can significantly enhance road safety. Its user-friendly design and efficient functionality make it a valuable tool for improving traffic management in India, ultimately creating a safer and more efficient transportation network for all road users.

2 Literature Review

The authors of this study [1] introduced an automated system that addresses the critical issue of non-helmeted motorcyclists, a major factor contributing to fatalities in motorcycle accidents in developing nations. Current methods, such as manual monitoring by traffic police or CCTV footage analysis, demand substantial human resources. The system's goal is to automate these processes by identifying non-helmeted motorcyclists and retrieving their license plate information from CCTV footage. Initially, the system distinguishes motorcycles from other moving objects. Once identified, it assesses whether the rider is wearing a helmet. If not, an optical character recognition (OCR) algorithm is employed to extract the license plate number. This automated approach has the potential to significantly enhance the efficiency of monitoring and enforcing helmet usage among motorcyclists. By doing so, it could potentially reduce the number of fatalities in motorcycle accidents, highlighting its importance in improving road safety in developing countries.

In the study [2], an automated system is proposed to address the issue of non-helmeted motorcyclists, a major factor contributing to fatalities in motorcycle accidents. The system utilizes computer vision and deep learning techniques to detect helmet usage and recognize vehicle number plates in real time. This integrated approach aims to enhance road safety and law enforcement by enforcing helmet usage laws and facilitating vehicle tracking and identification. The system's innovative features offer a comprehensive solution for improving road safety and traffic management, potentially reducing accidents and aiding in the identification of traffic violators.

In the study presented [3], the focus is on the increasing trend of breaking traffic signals, often due to people being in a hurry. This behavior can lead to accidents, causing injuries and even death. India, despite having only 1% of the world's total vehicles, accounts for 11% of global accident deaths. While Indian law provides for fines and imprisonment for breaking traffic signals, enforcement is lacking due to the absence of an effective system to capture violators. To address this issue, the study proposes a system called the "smart traffic signal control system". The system aims to identify individuals jumping red signals and notify the Regional Transport Office (RTO) accordingly. It utilizes a LASER module, LDR module sensor, and camera module to detect and notify authorities about signal violations. This approach could lead to improved enforcement of traffic signal rules and contribute to reducing accidents and improving road safety.

The proposed system [4] employs a vehicle ID card installed in each vehicle to autonomously transmit essential vehicle details (e.g., registration number, vehicle type, owner's name, and color) to authorities via mobile devices when a vehicle breaches a red signal. A miniature setup, featuring a traffic signal and a virtual road, is used to showcase this concept. Vehicles are required to be equipped with cost-effective wireless ID cards for real-time data transmission. Upon a vehicle crossing a red signal, its information is wirelessly transmitted via Wi-Fi, activating an alarm system to alert nearby traffic enforcement personnel. This technologically advanced system aims to enhance traffic signal enforcement and overall road safety.

3 Methodology

The proposed system [5] described utilizes a webcam to identify three types of traffic violations: detecting riders without helmets, identifying instances of triple riding, and spotting vehicles jumping signals. It aims to inform offenders of their violations and associated fines by using OCR technology to extract vehicle number plates. While this project tackles important traffic issues, its depth and precision may be insufficient for comprehensive traffic violation management.

In contrast, this design takes a more specific and advanced approach. It focuses on helmet detection, red signal violation detection, and ambulance signal control, employing TensorFlow for image processing and distinct algorithms for each task. This ensures not just detection but also prompt and accurate enforcement, enhancing its effectiveness in managing traffic violations. Moreover, the project's use of varied algorithms and technologies distinguishes it, offering a more reliable and precise solution for enforcing traffic rules.

3.1 Block Diagram

The design block diagram features a Raspberry Pi as the central controller, receiving inputs from a webcam, sound sensor, and IR sensor. The webcam captures images for helmet detection and red signal violation detection, while the sound sensor detects ambulance sirens for traffic signal control.

As shown in Fig. 1, the IR sensor assists in vehicle detection. Using this input, the Raspberry Pi processes commands to the Telegram application for notification delivery, traffic lights for signal control, and RealVNC for monitoring. This integrated system aims to automate traffic violation detection and management, enhancing road safety and traffic efficiency.

3.2 Flowchart

As illustrated in Fig. 2, the process starts by initializing TensorFlow and accessing a live video feed from a webcam. The frames from the feed undergo preprocessing to improve detection accuracy. A pretrained TensorFlow model designed for helmet detection is loaded and applied to the preprocessed frames. The model identifies helmets, and the system checks the confidence level of the detection. If a helmet is not confidently detected, a notification is sent to the rider, along with a fine for not

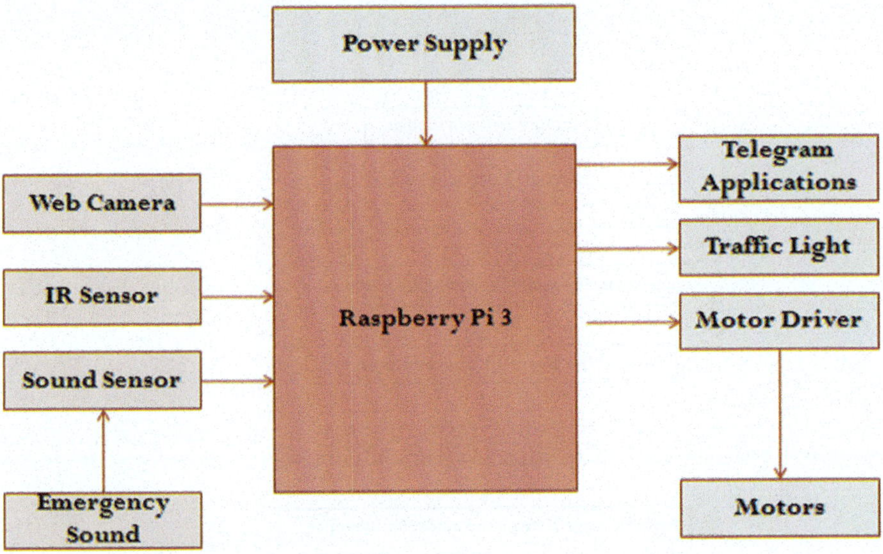

Fig. 1 Block diagram of proposed system

wearing a helmet. Finally, the process concludes, outlining the TensorFlow-based helmet detection workflow.

As outlined in Fig. 3, the system begins by capturing real-time video of the traffic scene using a webcam or similar device. TensorFlow is employed for vehicle detection, utilizing its object detection features to identify vehicles in the video frames. The system then analyzes the vehicles' positions relative to the traffic signals to detect violations. Upon detecting a violation, the system records the incident and may capture an image of the violating vehicle as evidence. It then generates a notification to alert the rider of the violation and issues a fine as appropriate.

As depicted in Fig. 4, the process operates as follows: Sound sensors installed at the signal junction detect the unique siren sounds of an ambulance, which are then analyzed to confirm the presence of an emergency vehicle. Upon confirmation, the system utilizes this information to identify the lane in which the ambulance is traveling. Subsequently, the traffic light control system switches the signal to green for the ambulance's lane, while keeping all other lanes red to facilitate its passage.

A timer ensures that the green signal remains active for the ambulance lane until the vehicle has cleared the junction. This process involves continuous monitoring of the sound sensors for ambulance siren detection and validation of the ambulance's presence in the identified lane.

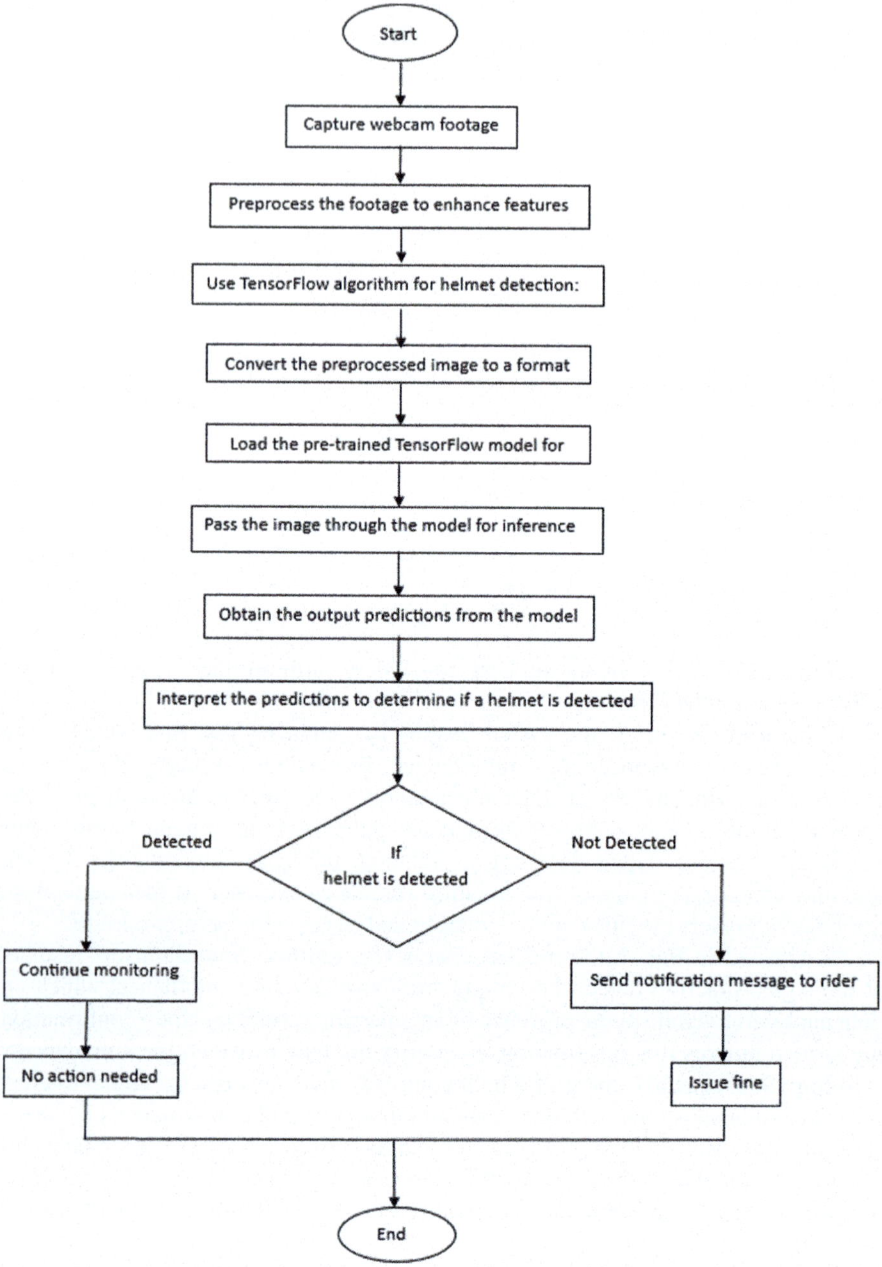

Fig. 2 Flowchart for helmet detection

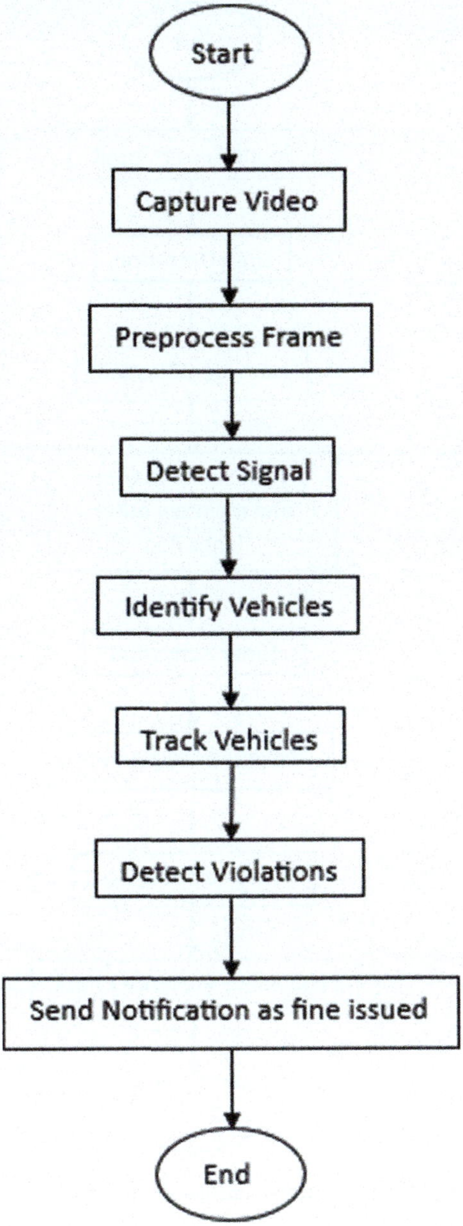

Fig. 3 Flowchart for signal jumping

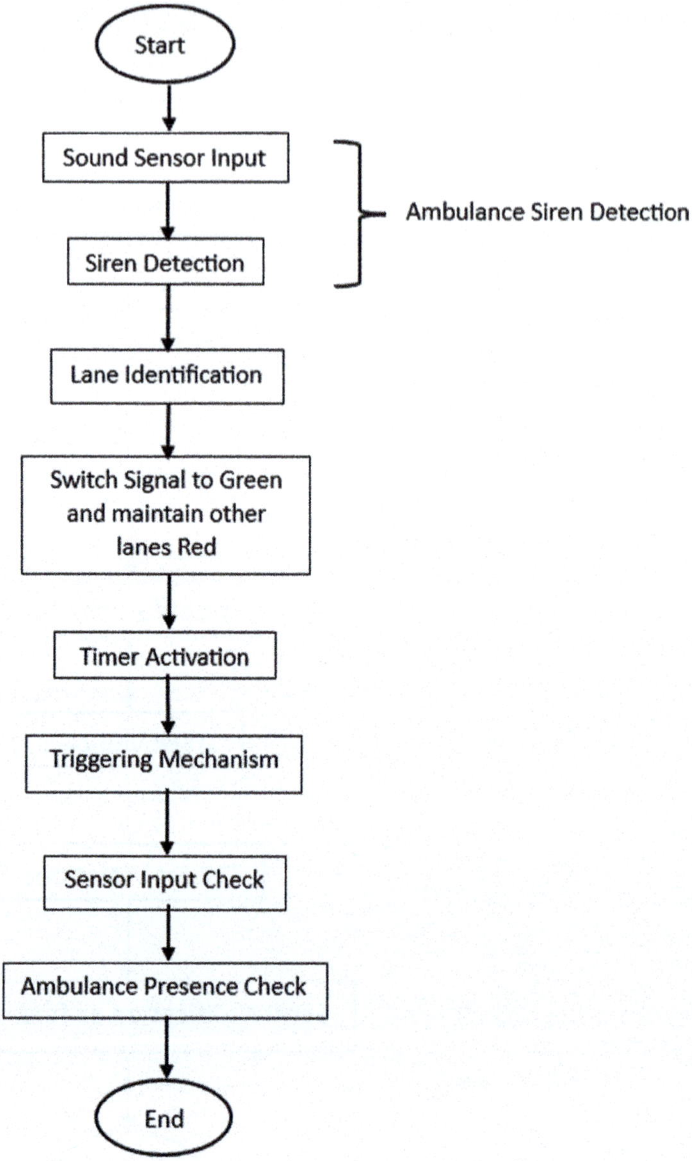

Fig. 4 Flowchart for ambulance detection

4 Experiments, Results, and Discussion

4.1 Helmet Detection

In this paper, we demonstrate a system that uses the TensorFlow algorithm to identify situations in which people are not wearing helmets in scenarios involving vehicles. The suggested methodology consists of the following crucial steps:

As seen in Fig. 5, the process involves the following steps:

(a) taking pictures with web cameras or other sensors;
(b) processing the raw frames to improve their quality;
(c) using computer vision algorithms, like TensorFlow, to detect and locate objects;
(d) training machine learning models, usually neural networks, to classify images and determine whether helmets are present;
(e) using image recognition techniques to classify objects and entities within the images;
(f) using the classification results to make decisions or issue alerts.

Through our proposal, as illustrated in Figs. 6, 7, and 8, we proposed a system that utilizes the TensorFlow algorithm to detect vehicles where individuals are not wearing helmets. TensorFlow integrates various machine learning and deep learning models and algorithms, allowing developers to create dataflow graphs with computational

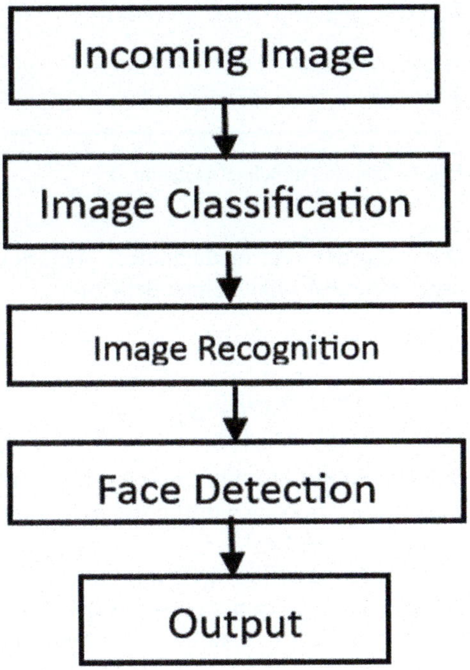

Fig. 5 Flowchart of TensorFlow algorithm

Fig. 6 Process of helmet detection

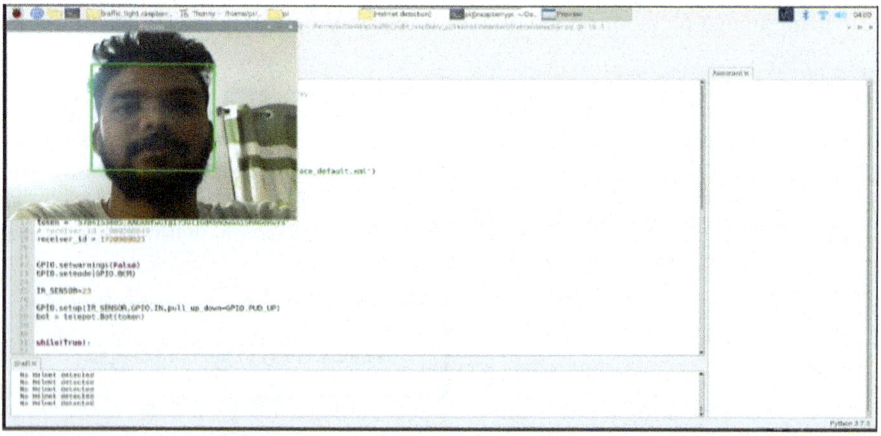

Fig. 7 Snapshot of helmet violation

nodes representing mathematical operations, making it a versatile tool for image processing tasks like helmet detection.

4.2 Signal Jumping

Signal jumping, also known as red-light running or running a red light, occurs when a driver fails to obey a traffic signal by proceeding through an intersection after the signal has turned red.

As outlined in Fig. 9, the fine assignment system utilizes traffic monitoring technologies such as cameras or sensors at intersections to detect instances of signal jumping. When a violation is detected, the system captures relevant information,

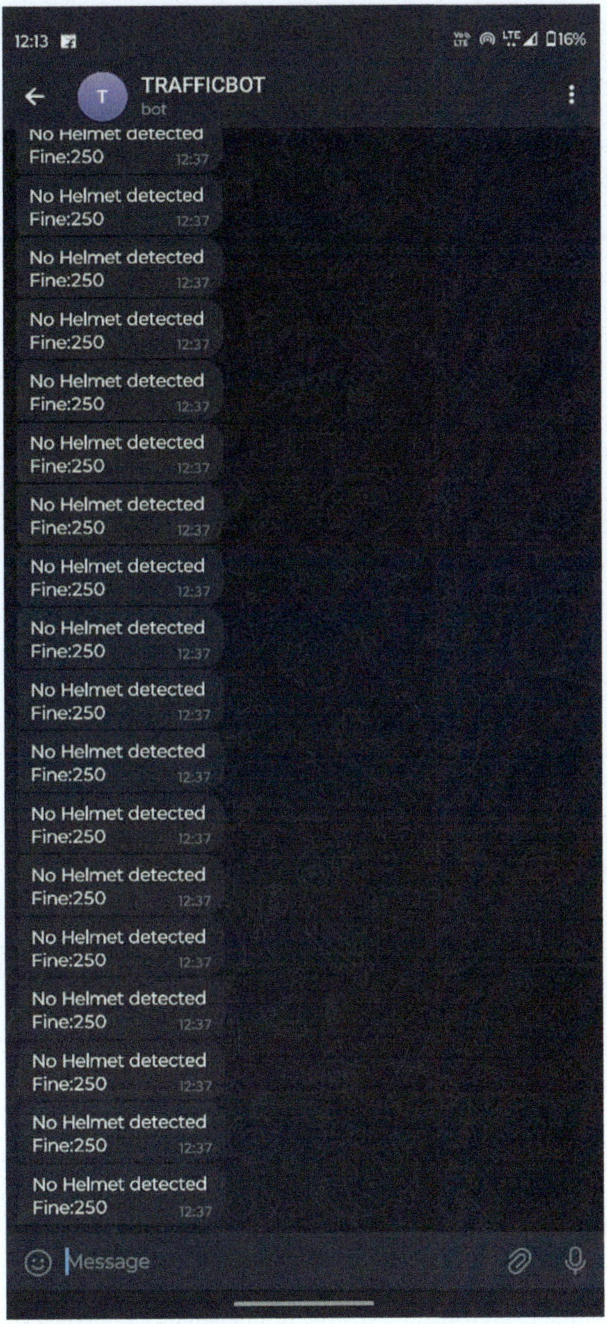

Fig.8 Fine issuing for helmet violation

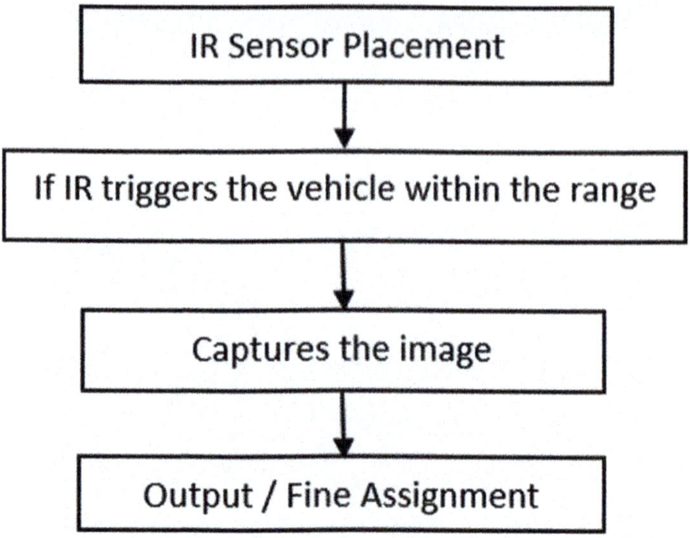

Fig.9 Flow of signal jumping using IR sensor

including the vehicle's license plate and timestamp. This data is then processed to identify the registered owner of the vehicle. Subsequently, a fine is assigned to the owner, acting as a deterrent and reinforcing compliance with traffic signals.

In this design, we have designed hardware that captures images of vehicles when they jump the signal as shown in Fig. 10. IR sensors are strategically placed before the stop line, and when a vehicle crosses the intersection with a red signal, the sensor triggers the violation identification process. The camera then captures the vehicle's image, including the license plate, for identification. This process is initiated when the vehicle is within a certain distance from the IR sensor. The captured image, along with a message, is sent to the control room as proof of the violation. This ensures that fines are assigned accurately and effectively as displayed in Fig. 11.

4.3 Ambulance Detection

The smart ambulance traffic control system is an innovative solution aimed at improving ambulance flow in traffic, ensuring faster response times, and enhancing overall emergency medical services. The system features smart traffic lights that can be controlled to provide priority signals to approaching ambulances, enabling them to pass through intersections quickly and safely. Designed for efficiency, the system maintains a simple design while effectively managing ambulance prioritization. It consists of two main components:

Fig.10 Model for signal jumping

a. **Sound Sensor Detection**: Sound sensors are employed to detect the specific sound of an approaching ambulance.
b. **Traffic Signal Adjustment**: Upon detecting an ambulance sound, the system adjusts the traffic signals to prioritize the ambulance's passage. In a four-lane traffic signal setup, sound sensors are placed in each lane. When an ambulance sound is detected in a lane, the traffic signals in that lane change to green, providing the ambulance with a clear path. While this adjustment may cause a slight delay in traffic signal processing, the lights return to normal operation after the ambulance has passed as observed in Figs.12 and 13.

Overall, the smart ambulance traffic control system offers a practical and efficient solution to improve ambulance navigation through traffic, ensuring timely and safe arrival at emergency destinations.

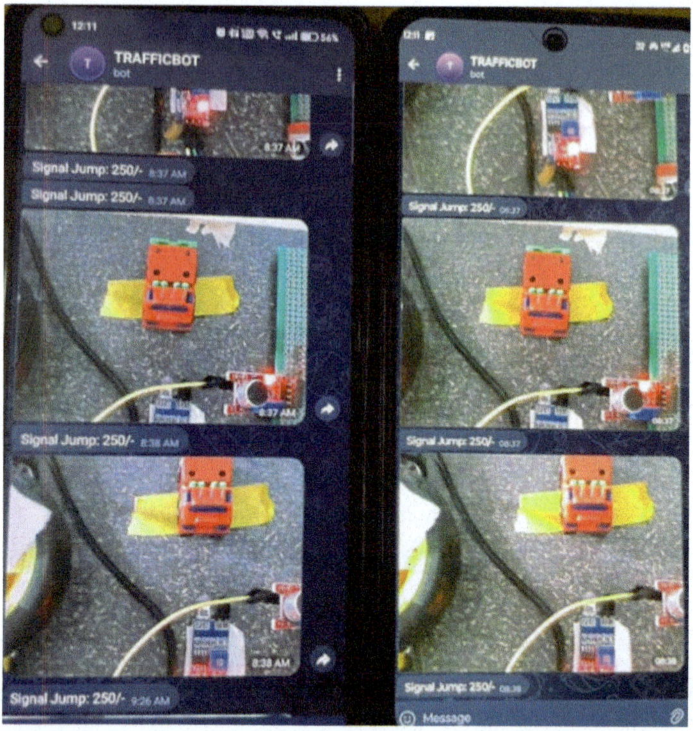

Fig. 11 Fine for signal violation

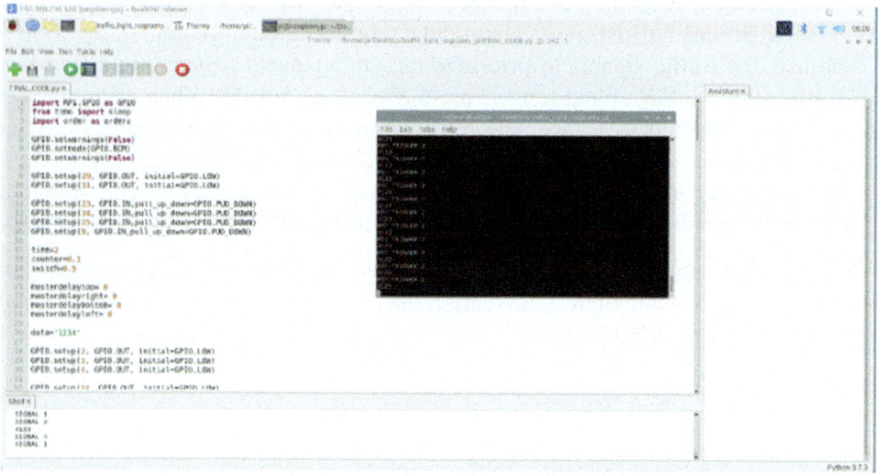

Fig. 12 Triggering of signal for the ambulance detection near the junction

Fig. 13 Changing of the signal to green by detection of the ambulance near the junction

5 Conclusion

In conclusion, the endeavor to enhance traffic management through the integration of IoT and image processing has shown promising results. By tackling critical issues like helmetless riding, red signal violations, and ambulance prioritization, the design has laid a solid groundwork for improving road safety and traffic regulation. The successful utilization of TensorFlow for image analysis and sound sensors for ambulance detection highlights the efficiency of advanced technologies in addressing contemporary challenges. Despite encountering obstacles such as algorithm optimization and real-time processing limitations, the project has demonstrated significant potential.

Looking ahead, future work could focus on improving the algorithms' accuracy and efficiency of detection. Furthermore, incorporating extra functions like pedestrian monitoring and vehicle recognition could offer a more complete traffic control solution. In the end, this research emphasizes how critical it is to use technology to build safer and more effective transportation networks. We may work toward a future where traffic flow is improved for the benefit of all road users and road safety is of utmost importance by carrying out further innovation and improvement of these approaches.

References

1. Kale T, Dhavale A, Randhavan A, Inamdar VS (2023) Helmet detection and number plate recognition. IJRASET 11(5)

2. Marathe G, Gurav P, Narwade R, Ghodke V, Patil SM (2022) Helmet detection and number plate recognition using machine learning. IJIRT. 8(12)
3. Kapagate K, Bhave N, Ghutke D, Wakade H, Khubalkar S (2022) Smart traffic signal control system. IRJMETS 4(5)
4. Archana P, Nikitha J, Himabindu T, Manvitha V (2022) IoT based signal violated vehicle detector. IRJMETS. 4(3)
5. Mayur KL, Pavithra H, Pooja H, Navaneet SK, Supritha N (2020) Virtualized surveillance and detection of traffic violation. IRJMETS, 2(6)

Examining Battery-Operated Sensor Nodes in Rechargeable Wireless Sensor Networks with Regard to Energy Protection Techniques

B. C. Vengamuni and V. Rajendran

Abstract Due to the small size and battery capacity of sensor nodes in wireless sensor networks, energy control problems arise. Numerous studies have used energy-saving techniques to address these problems in sensor networks. The purpose of this study was to gather information on various energy-saving strategies used to keep sensor nodes' batteries fully charged between 2016 and 2021. Additionally, this review article addresses issues, fills in knowledge gaps on rechargeable wireless sensor networks, schedules techniques for monitoring nodes' life spans, and a number of measures to improve nodes' efficiency and scalability in rechargeable sensor networks.

Keywords Wireless energy transmitter · Sensor nodes · Batteries · And wireless rechargeable sensor networks (WRSN)

1 Introduction

These days, wireless rechargeable sensor networks that run on batteries are widely used across a number of businesses and the technology sector. Due to issues with energy management, the tiniest battery sensor nodes can shorten the lifespan of a network. Numerous researchers suggested energy-intensive techniques to keep sensor nodes' batteries charged, extending the network's lifespan. The most recent development in the WRSN to address these problems is wireless energy transmitters, which may be used to power non-wired nodes equipped with sensors. This method was achieved in conjunction with WET and portable chargers.

Wireless energy transfer (WET) is a technology used in wireless rechargeable sensor networks (WRSNs) that allows sensor nodes to wirelessly transmit energy to

B. C. Vengamuni (✉) · V. Rajendran
Department of ECE, VISTAS, Pallavaram Chennai, India
e-mail: vengamuni.phd@gmail.com

V. Rajendran
e-mail: director.ece@velsuniv.ac.in

© The Author(s), under exclusive license to Springer Nature Singapore Pte Ltd. 2025
H. Zen et al. (eds.), *Soft Computing and Signal Processing*, Lecture Notes in Networks and Systems 1200, https://doi.org/10.1007/978-981-97-9926-8_24

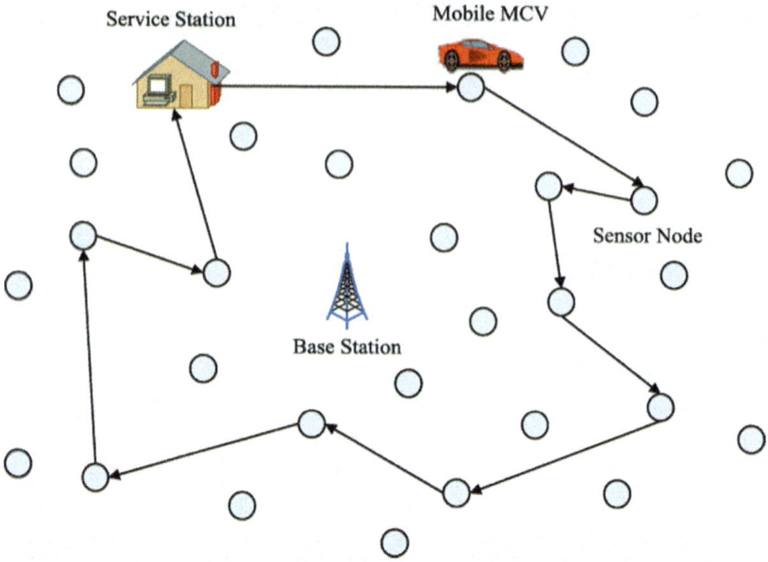

Fig. 1 WRSN architecture

replenish their batteries without the need for physical connections. This is especially important for wireless sensor network systems (WRSNs) since sensor nodes are often located in remote or challenging-to-reach areas, making traditional battery replacement impractical. The WRSN architecture is shown in Fig. 1 and consists of a base station, mobile chargers, sensor nodes, and service stations [1].

2 Background

It is necessary to examine various sensor node recharging strategies in WRSN in order to talk about the network's lifespan and performance. The writers examined two different techniques of recharging: (i) recharging based on radiation, (ii) recharging based on resonance, and (iii) utilizing an adaptive algorithm for recharging.

(i) Recharging based on radiation

These days, radiation-based recharge has advanced swiftly from an early stage to an adult level. Profitable goods provide network sensor nodes within a certain distance with electricity. Additionally, a number of nodes were set up to receive power cast charges, which extend the lifespan of each sensor node. Wang and associates [5] and Tian et al. [2] have shown a link between sequence charging and network life duration.

Zhao et al. [11] looked at the charging of sensor nodes, which can range from rapid to sluggish. The charging process was also assessed using formula (1).

$$\text{charging process} = \int_{0}^{t} \rho(z)dz \tag{1}$$

Here, t is the amount of time that sensor nodes are charged, as expressed in Eq. (2)

$$\rho(z) = \frac{1}{p} CVe^{\left(-z/p\right)} \tag{2}$$

While p represents the fixed duration, CV relates to each sensor node battery's maximum capacity.

The author of [11] determined the battery explicit energy, which was stated as (3), using the aforementioned formulae.

$$\int_{0}^{t} \rho(z)dz = CV\left[1 - e^{-\frac{t}{p}}\right] \tag{3}$$

Low efficacy results from requiring a large amount of time to fully charge all accessible sensor nodes.

(ii) Recharging based on resonance

This method of recharging in sensing nodes transported a significant amount of energy with more efficacy than recharging based on radiation. Several recharging techniques are used to traverse the sensing nodes in the network environment since, according to certain investigations, the batteries in the sensor nodes cannot charge completely. In a similar vein, cars' capacity to recharge as well as its moving rate are incorporated into formulation techniques that purposefully optimize charge scheduling, particularly for certain sources.

(iii) Utilizing an adaptive algorithm for recharging

In order to get energy-based data, BS use energy consumption strategies to request the sensor nodes. Subsequently, BS divides the nodes into many divisions based on the origin of the first request. Wang et al. [5] employed the k-means method for network separation. According to the k-means algorithm, different values are classified into cluster 2 and comparable values are placed into cluster 1. These methods are therefore appropriate for separating the charged and recharged nodes. Equation (4) may be used to estimate the inter node distance.

$$S = \sum_{j=1}^{m} \sum_{i=1}^{n_r} \left\| n_i^{(j)} - \mu^{(j)} \right\|^2 \tag{4}$$

The phrase $\left\| n_i^{(j)} - \mu^{(j)} \right\|^2$ described the separation between the node that has to be recharged and the network's centroid section in relation to its available nodes.

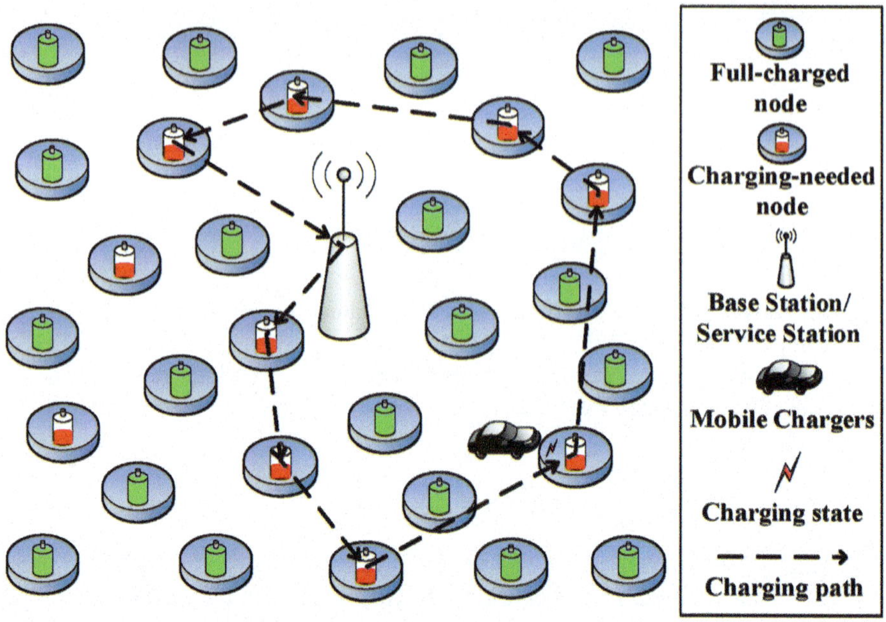

Fig. 2 The WRSN's overall structural design by [13]

(iv) Recharging via an internet

In order to prevent power undernourishment, the authors of Yang et al. [13] used radial basis function to recharge the battery of the sensor nodes. Figure 2 shows the total wireless rechargeable sensor network design.

In order to identify online charging, sensor nodes must go through four stages of charging: motion, idle, sensor node battery recharge, and regression techniques.

3 Scheduling Techniques for Charging

In a wireless rechargeable sensor network, the charging technique used will determine the sensor nodes lifespan [9]. When sensor nodes are being charged, two factors—consistent and significant factors—should be planned. the constants, such as the number of nodes, charging power, and node transit speed. Significant variables include determining the initial charging duration, selecting which sensor nodes require charging, streamlining the charging process, and determining individual charging periods for each node in the network. The charging scheduling techniques are classified based on these schedules while charging nodes. Figure 3 shows the charging pipeline diagram for each sensor node in the network.

Equation (5) defines the statistical intention job for the charging pathway.

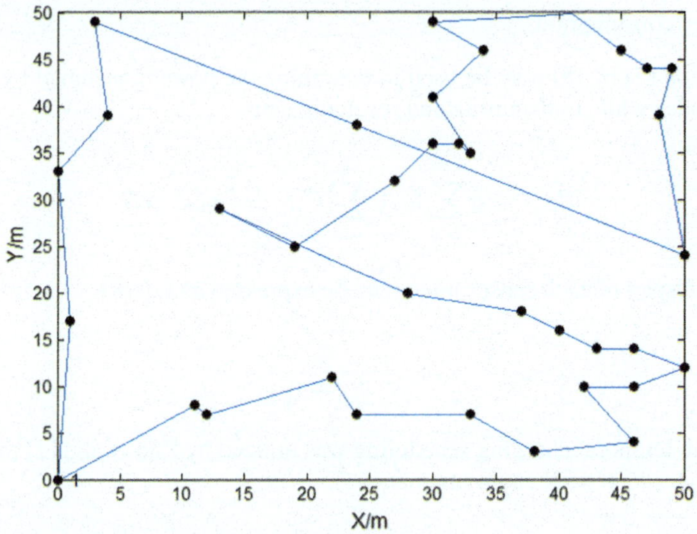

Fig. 3 Charging pipeline method for each sensor node

$$\min L = \text{dist}\left(x_0, x_1 + \sum_{i=1}^{m-1} \text{dist}(x_i, x_{i+1}) + \text{dist}(x_m, x_0)\right) \tag{5}$$

L is the distance between nodes, while × 0 and xm are the starting and stopping nodes for data transmission.

A unique technique known as the multi--nodes rechargeable technique was discovered by Zhang et al. [8] and is useful for recharging several sensor nodes in a predetermined amount of time. This method also aims to converge the path taken by sensor nodes and reduce the number of docking locations. As a result, energy consumption is greatly reduced while driving a wireless charging vehicle and improving scalability. The structural design of several mobile chargers in a network that maximizes charge arrangement, moveable node duration, and recharging time was determined by Mo et al. [6]. Furthermore, the power utilization method is suitable for charging sensor nodes in sequence based on their lifespan, and by isolating several nodes, the charging competence is raised. The structure of the ideal multiple mobile chargers created by [6] is shown in Figure.

Equation (6) may be used to get the recharging rate for each sensor node that is available in the WRSN.

$$\text{Rechargeable rate} \sum_{i=1}^{n} x_{0,i}^{k} = \sum_{j=1}^{n} x_{j,0}^{k} \leq 1 \nabla k \epsilon K \tag{6}$$

i. Energy consumption

Equation (7) by [8] may be used to determine the power consumed by charging sensor nodes while transmitting and receiving data.

$$P_i(t) = \rho \sum_{k\epsilon N}^{k\neq i} f_{ki} + \sum_{j\epsilon N}^{j\neq i} C_{ij} f_{ij} + C_{iB} f_{iB} \quad i\epsilon N \tag{7}$$

The lifespan of each sensor node may be expressed as Eq. (8).

$$\eta_i = \frac{l^i_{\text{live}}}{l^i_{\text{live}} + l^i_{\text{dead}}} \tag{8}$$

Spatial–temporal charging scheduling was utilized by Zho et al. [11] to recharge sensor nodes in WRSN environments.

4 Methods for Charging Sensor Nodes

In order to optimize the network lifespan by effectively scheduling a mobile charger, Prasanna Babu et al. [7] introduced the Recha algorithm, which has two criteria: the number of nodes active in a given length of time and the lifetime of the network. Nearest Job Next with Preemption is one of the criterion. An effective technique that is suitable for sensor lifetime maximization problems was developed by Xu (2016). Therefore, when the mobile charger's travel distance is reduced, the lifespan of the sensor is increased in WRSN when it is used to wirelessly charge sensors. The MC scheduling strategy was found by Liu et al. [3] to reduce network data failures caused by assumptions about sensor node power and connection.

Additionally, the weight of all accessible sensor nodes in the WRSN is raised in order to construct charging schemes that rely on the MC wandering distance within a predetermined range. Ye and colleagues [4] have increased the sensors charging service based on the total roaming distance of MC for each visit and the charging period of each sensor node. Solitary MC was employed by Wei et al. [1] to guarantee that all network sensor nodes are operating accurately.

This replenishes the power supply for every sensor node in the WRSN. A charging schedule and timer is offered by Zhao et al. [11] to increase the efficiency of conventional power while lowering the quantity of energy-depleted nodes. The hybrid service allocation technique in WSN supports service quality. Table 1 summarizes a number of studies on charge scheduling techniques.

Table 1 Algorithm comparison for scheduling

Authors	Year	Used algorithms	Work intention
Zhang Fan et al. [8]	2018	Multi node rechargeable algorithm	Improved performance in terms of the number of docking sites and the duration of the node route journey
Ma et al. [9]	2020	Modified ant colony algorithm	Reduce the time it takes for networks to charge, and increase their energy
Wang et al. [5]	2019	Numerical approach, travelling salesman problem	Models of energy consumption accuracy and efficiency, with the least amount of lag
Tian et al. [2]	2021	Charging method	Reduce the excess of data
Mo et al. [6]	2019	Multiple mobile charging scheme	An increase in the scalability of the network and its efficiency in terms of computing time and quality. Decrease in energy use as well
Zho et al. [11]	2020	Online-offline algorithm were deployed	Enhanced efficacy of charging

5 Summarization of Parameters

Table 2 describes the employed sensor nodes, energy consumption rate depending on sensor node speed, and time length in this part.

Furthermore, Wang et al. calculated power savings and energy depletion and is included in the chart to indicate the battery's depleting power for the sensor nodes as well as the amount of energy storage it contains.

Table 2 Overview of the parameters

Authors	Quantity of sensor nodes	Rate of energy usage	Travelling at a fast pace	Duration of charging
Zhang Fan et al. [8]	100, 500, 1000	50 J/bits	–	Mathematical formulation
Ma et al. [9]	10, 20, 30, 40 and 50	50 J/min	2 min/s	5 min
Wang et al. [5]	500	5.59 J/min	NA	NA
Tian et al. [2]	600	5.4–10.8 J/ min	Portable vehicle-5 min/sec UAV-15 min/ second	Charging threshold-5 h
Mo et al. [6]	50	1 J/min	1 min/second	5 min
Zho et al. [11]	200	NA	NA	NA

Table 3 Network performance metrics comparisons

Authors	Using nodes	Cost of charging	MV's charging efficiency
Fan et al. [8]	yes	yes	No
Wang et al. [5]	yes	yes	yes
Mo et al. [6]	yes	yes	yes
Tian et al. [2]	yes	No	yes
Zho et al. [11]	yes	yes	yes
Yang et al. [13]	yes	yes	yes
Liu et al. [3]	yes	yes	yes

	Depletion of energy	energy conservation
Wang et al. [5]	30–50%	10–20%

6 Evaluations of Network Performance Indicators

Table 3 provides a summary of metrics including node usage, cost when charging, and mobile vehicle locating efficiency that improve network performance.

7 Network Simulator 2

The discrete event simulator known as NS-2, or Network Simulator 2, is built on the principles of object-oriented programming (OOP) and is composed of the C + + programming language and the Object-Oriented Tool Command Language (OTcL), which are connected via TcLCL. Python bindings are optional, and NAM and XGRAPH will be used to view codes written in OTcL. NS2's support for sensing is restricted; however, it can handle protocols like 802.11.802.16 and 802.15.4. Real sensor network nodes employ varied parameters, such as packet formats, MAC protocols, and energy models. Although NS-2 has parameters for residual energy, it lacks an energy harvesting model and models for tracking the energy used by the various components.

8 Conclusion

Conventional wireless sensor networks are mostly powered by batteries; however, the battery's capacity limits the sensor network's operating time. The charging scheduling strategies that are useful in determining the recharging capability of sensor nodes in WRSN are included in this survey. Additionally, the study work's purpose and the algorithm utilized are examined. This study paper summarizes a variety of factors, including the number of employed sensor nodes, energy consumption rate, movement speed of sensor nodes, and charging time. This study aids in identifying the research gap concerning battery energy protection strategies for sensor nodes in WRSN.

References

1. Wei Z, Li M, Zhao Q, Lyu Z, Zhu S, Wei Z (2019) Multi-MC charging schedule algorithm with time windows in wireless rechargeable sensor networks. IEEE Access 7:156217–156227. https://doi.org/10.1109/ACCESS.2019.2949284
2. Tian M, Jiao W, Chen Y (2021) A joint energy replenishment and data collection strategy in heterogeneous wireless rechargeable sensor networks. Sensors 21:2930. https://doi.org/10.3390/s21092930
3. Liu H, Deng Q, Tian S, Peng X, Pei T (2018) Recharging schedule for mitigating data loss in wireless rechargeable sensor network. Sens (Basel, Switz) 18(7):2223. https://doi.org/10.3390/s18072223
4. Ye X, Liang W (2017) Charging utility maximization in wireless rechargeable sensor networks. Wireless Netw 23:2069–2081. https://doi.org/10.1007/s11276-016-1271-6
5. Wang C, Li J, Ye F, Yang Y (2016) A mobile data gathering framework for wireless rechargeable sensor networks with vehicle movement costs and capacity constraints. IEEE Trans. Comput. 65(8), 2411–2427. https://doi.org/10.1109/TC.2015.2490060
6. Mo L, Kritikakou A, He S (2019) Energy-aware multiple mobile chargers coordination for wireless rechargeable sensor networks. IEEE Internet Things J, 1–13. ff10.1109/JIOT.2019.2918837ff. ffhal-02140283f
7. Prasannababu D,Vaishnav S, Amgoth T (2019) Mobile charger scheduling using partial charging strategy for rechargeable WSNs. In: 2019 International conference on computing, power and communication technologies (GUCON), pp 845–852
8. Fan Z, Jie Z, Yujie Q (2018) A multi-node rechargeable algorithm via wireless charging vehicle with optimal traveling path in wireless rechargeable sensor networks. In: 2018 10th International conference on ubiquitous and future networks (ICUFN), pp 531-536. https://doi.org/10.1109/ICUFN.2018.8437035.
9. Ma C, An S, Wang W, Lin D, Li W, Sun L (2020) Wireless sensor network charging strategy based on modified ant colony algorithm. Int J Mater, Mech Manuf 8(3)
10. Xu W, Liang W, Jia X, Xu Z (2016) Maximizing sensor lifetime in a rechargeable sensor network via partial energy charging on sensors. In: 2016 13th Annual IEEE international conference on sensing, communication, and networking (SECON), pp 1–9. https://doi.org/10.1109/SAHCN.2016.7733001
11. Zhao C, Zhang H, Chen F, Chen S, Wu C, Wang T (2020) Spatiotemporal charging scheduling in wireless rechargeable sensor networks. Comput Commun 152:155–170. https://doi.org/10.1016/j.comcom.2020.01.037

12. Wan J, Chen B, Wang S, Xia M, Li D, Liu C (2018) Fog computing for energy-aware load balancing and scheduling in smart factory. IEEE Trans Ind Inf 14(10):4548–4556
13. Yang J, Bai J-S, Xu Q (2020) An online charging scheme for wireless rechargeable sensor networks based on a radical basis function. Sensors 20(1):205. https://doi.org/10.3390/s20010205

Modern Metrics (MM): Functional Size Estimator for AI and BDA Software Applications

B. Senthilkumaran, John T. Mesia Dhas, N. Kurinjivendhan, and M. Usha

Abstract Analysing, scheduling, developing, testing, and maintaining are the main activities of software project management (SPM) under software engineering. Software size is the major factor determining cost, time, and effort for developing software applications. The successful completion of all the activities of SPM requires an accurate determination of the cost, time, and effort required for developing the application. That is, the real size of the software application is the major determinant of all the SPM activities. The existing traditional application-oriented and single-dimensional software sizing techniques will not give the actual size of modern software applications like artificial intelligence (AI) and big data analytics (BDA). The imperfect estimation leads to wrong predictions about the cost, time, and effort required for software application development. The wrong estimates will affect the entire SPM activities, resulting in a delay in delivery, a loss in predicted cost, an imbalanced effort level, and the dismissal of the software development process. The major software industries are facing the same problem with wrong size estimates. The updated and dynamic new size estimation technique is highly essential for this current SPM environment. The novel approach of modern metrics (MM) addresses all the technical challenges facing software engineering for the software size estimation process. This article analyses the major issues of the software size estimation process, major functional and non-functional units associated with software size estimation, the architecture of MM, and an algorithm for software size estimation using MM.

B. Senthilkumaran (✉) · J. T. Mesia Dhas
Vel Tech Rangarajan Dr. Sagunthala R&D Institute of Science and Technology (Deemed to be University), Chennai 600 062, India
e-mail: skumaran.gac16@gmail.com

J. T. Mesia Dhas
e-mail: jtmdhasres@gmail.com

N. Kurinjivendhan
Bishop Heber College, Trichy 620 017, India
e-mail: vendhancs1489@gmail.com

M. Usha
Sri Amaraavathi College of Arts and Science (Affiliated to Bharathidasan University), Karur, India
e-mail: usharavichandran@amaraavathi.ac.in

© The Author(s), under exclusive license to Springer Nature Singapore Pte Ltd. 2025

H. Zen et al. (eds.), *Soft Computing and Signal Processing*, Lecture Notes in Networks and Systems 1200, https://doi.org/10.1007/978-981-97-9926-8_25

Keywords Software project management · Software size estimation · Modern metrics · Software complexity factors · Software effort · Software schedule · Modern metrics size · Software engineering

1 Introduction

Software engineering is an engineering procedure for creating software application projects in an international standard manner, giving guidelines for completing the project on time, testing and debugging in user as well as developer-centric approaches, and providing the complete procedure for maintaining the developed software. Modern software systems use about one programming language, application software, utility tool, architecture, platform (operating system), type of user, and databases. So, the development of a modern software system is highly complex. Modern applications are founded on software as well as hardware. The systems (both hardware and software) are generating a huge amount of data. That data must be processed in parallel with the next process [1].

The modern industry is developing applications in artificial intelligence (AI) and the Internet of things (IoT). The applications are processing a huge amount of data (GB to PB) to generate a simple result, which uses different types of datasets (structured, unstructured, and semi-structured), different sets of database management systems (DBMS), software architectural models, etc. Emerging automation technologies like IoT use sensors, actuators, clouds for storage, big data analytics (BDA) tools, and highly efficient programming languages. AI techniques such as machine learning (ML) and deep learning (DL) analyse historical data. To develop such a complex application requires about one programming language, utility apps, architectural models, DBMS tools (for handling structured, semi-structured, and unstructured data), domain experts (for handling problems in different domains of software), storage (static and dynamic), parallel processing processors, network topologies, distributed servers, virtual processing, etc.

Developing and delivering this complex software product on schedule is a very difficult task. A software engineering study says that less than 30% of the software is delivered on time. The latency in delivery leads to economic losses for both the customer and the developer. The trust of the developer is challenged. One of the important reasons for this delayed delivery is a lack of planning. In the planning process, software size estimation plays a crucial role. The estimation of effort, cost, and time for developing the software is determined by the software size factor [2, 3].

Software is an intangible, logical product. So, the estimation of size is highly risky. The software industry is using many traditional size estimation techniques like lines of code (LoC), function point (FP), feature point, Delphi, expert judgement, pattern matching, etc. In terms of software sizing techniques, the FP method is the most reliable, programming language-independent, user-centric, and understandable to beginners. The FP method deals with the size of the software at the analysis front itself. So, the estimation of effort, cost, and time is possible at the initial phase itself.

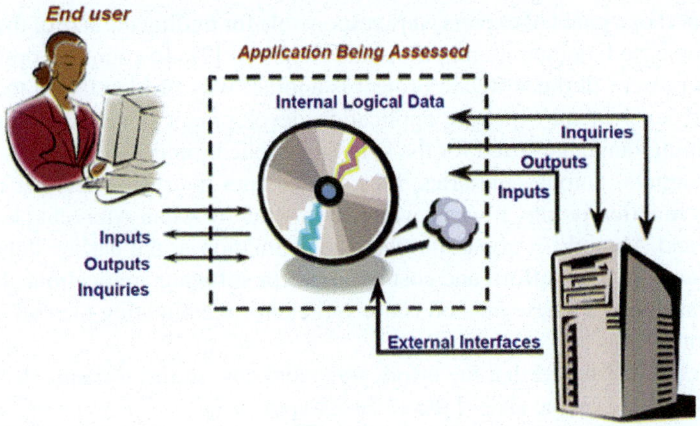

Fig. 1 FP architecture

Around 90% of the software industries in the world are following the FP method for measuring the size of the software [4, 5].

In the 1970s, Alan Albrecht introduced the FP method for determining the size of the software [5–10]. The key features of the FP method are as follows:

- Independent of technology, platform, or language.
- Excellent as an initial step for estimating development costs and timelines.
- It provides the ability to normalise metrics for consistent analysis.
- Fully documentable and repeatable.
- Aids in requirement management; functional size is traceable throughout the life cycle.
- Provides a quantitative basis for earned value management.

The architecture of the FP method is presented in Fig. 1.

2 Literature Review

According to the research based on Capers Jones [1–5], the lines of code (LoC) or counting of lines in a computer software application is the size of the software application. The programming languages like COBOL and FORTRAN increased the logical ability of the software applications. The weight of code used in one application may be varied in another application. So, the LoC method does not opt for finding the size of all the software applications. Modern software applications vary with distinct domains, scientific and static calculations, embedding with other technologies, multipurpose, and versatility. Finding the size is highly complex in modern days.

An expert or a panel of experts were responsible for finding the size of the software application based on their assumptions and decisions [2]. To count the operands and operators present in the software project is another way of identifying the size [2]. The effort level of the software application development process is based on the programming languages used for the software application development activity [1]. To count inputs, outputs, inquiries, logical files, and interfaces, another method is known as function points. It was developed in 1965 by Allan Albrecht [1–5].

Nasir and Ahmad [6] argue that the size estimation at the earlier stage in SPM helps reduce the time, effort, and cost required for software applications. In modern software systems, a single method is not sufficient for estimating the real size of the application.

Kaya et al. [8] found that the effort, time, and cost of the software development process are based on the size of the software application.

According to Daniel V. Feren et al. [10], an application-independent software size estimation technique is required for the modern software industry.

Boehm [10, 11] says that estimating the actual software size is the biggest problem in the current software industry.

Zia et al. [12] found that software size plays a major role in the planning, analysis, risks, and budgeting of SPM.

According to Rabbi et al. [13], a new effective cost estimation technique is required for this mature software industry. Finding the size with the existing technique is not accurate.

Objectives

1. To identify the missing FPs in modern AI and big data analytics (BDA) software systems.
2. To identify all complexity adjustment factors (CAF) affecting the SDLC
3. To derive a novel methodology for identifying software size.
4. To derive a formula to identify the effort level of the software.
5. To derive a formula to identify the essential time for developing software.
6. To derive a formula to calculate the cost required to develop the software product.

Motivation

The software industry is a highly dynamic and variable industry that depends on time and situations. The size estimation techniques do not match the current industrial flow. So many defects exist in the assessments based on the present techniques, which is affecting the delivery of the software product. To overcome the current scenario, the industry requires an updated estimation technique for identifying the size of the software.

3 Modern Metrics

Modern metrics (MM) is a proposed novel methodology for identifying the size of software products like AI and BDA applications. It follows Alan Albrecht's FPA method with new and updated function points and complexity adjustment factors (CAF). This method is language-, platform-, and technology-independent. The resultant value of this method is an important factor for estimating the time, effort, and total cost of the software product. The estimation process is fully documentable and repeatable. This method is capable of identifying the size of the software at any stage of the SDLC. The estimation process gives effective outcomes that are easily understandable by any kind of user.

MM Architecture

The MM is comprised of FPs and CAFs. The MM architecture is illustrated in Fig. 2

II: Internal Inputs, IO: Internal Operations, ILF: Internal Logical Files, EIF: External Interface Files, CAF: Complexity Adjustment Factors, SDLC: Software Development Life Cycle, MMSize: Modern Metrics Size, UMMFP: Unadjusted MM Function Points, and MMCAF: MM Complexity Adjustment Factor.

The architecture explains all of a software's internal, external, and fusion functional units; both internal and external complexity factors influence the SDLC. The size of the software application is the multiplied outcome of UMMFP and MMCAF.

A. *Functional Units of MM*

The functional units (FU) of software are the foundation for estimating its size. The values associated with each functional unit are known as its factors. The factors affecting the FU of modern software systems are tough to find and classify. The important factors of FUs are identified: internal inputs (II), internal operations (IO), internal

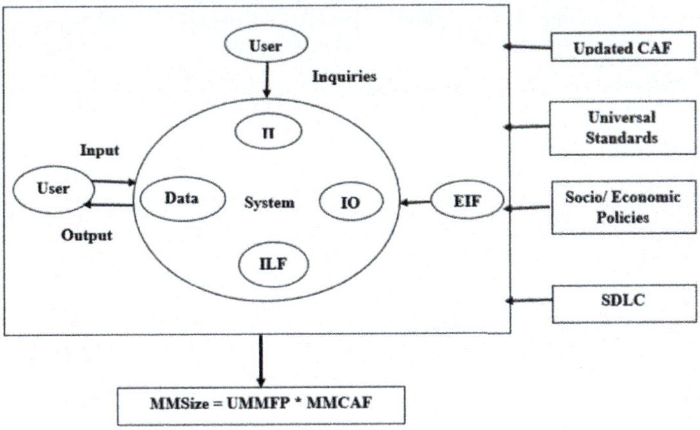

Fig. 2 MM architecture

logical files (ILF), external inputs (EI), external outputs (EO), external inquiries (EQ), external interface files (EIF), and data and text (DT). The MM is using eight different types of functional values associated with software projects. Each functional unit has its own metrics and metric values. The metrics are otherwise known as effort modifiers. Based on the user's efforts, the standards result in individual metrics for the functional unit.

The researchers analysed seven AI and BDA projects of Palm Services; the results in Table 1 above are identified. It says that an average of 0.98 is the effort required for developing an II functional unit from the analysis phase to the implementation phase. The average effort of II varies from 0.57 to 1.63. The study says that when the number of IIs increases in each function of a project, it also increases the effort level of the project. The collective effort is less than the individual effort.

The analysis results of seven projects for internal operations are tabulated in Table 2. The average effort required for each IO functional unit is 1.1. The effort levels are high in scientific and analytical software and low in application-based software. The average effort in collective and individual efforts is mostly the same.

The results in Tables 3 and 4 show the efforts required to complete external input. The values vary from software to software, depending on the complexity and technological novelty. The collective effort is less because some of the same EIs are also used by other functions. It reduces the effort required in the earlier phases of SDLC.

The effort required to finish the functional unit EO is listed in Table 5. The collective effort is slightly less than the effort required to finish an individual functional unit. The complexity level of the software application is the major reason for this minor difference.

Table 1 Internal input effort estimation

S. No.	Project title	MM functional unit					
		Internal input					
		no. of II functional units	Effort	No. of functions	Average II per function	Average effort per function	Average effort per FU
1	Project 1	40	24	42	0.95	0.57	0.60
2	Project 2	92	58	86	1.07	0.67	0.63
3	Project 3	128	117	72	1.78	1.63	0.91
4	Project 4	71	63	43	1.65	1.47	0.89
5	Project 5	18	24	39	0.46	0.62	1.33
6	Project 6	112	103	83	1.35	1.24	0.92
7	Project 7	59	43	64	0.92	0.67	0.73
Total		520	432	429			
Average						0.67	0.83

Table 2 Internal operations effort estimation

S. No.	Project title	MM functional unit					
		Internal operations					
		No. of IO functional units	Effort	No. of functions	Average IO per function	Average effort per function	Average effort per FU
1	Project 1	40	32	42	0.95	0.76	0.80
2	Project 2	68	44	86	0.79	0.51	0.65
3	Project 3	167	178	72	2.32	2.47	1.07
4	Project 4	42	31	43	0.98	0.72	0.74
5	Project 5	42	36	39	1.08	0.92	0.86
6	Project 6	174	143	83	2.10	1.72	0.82
7	Project 7	49	36	64	0.77	0.56	0.73
Total		582	500	429			
Average						0.86	0.86

Table 3 Internal logical files effort estimation

S. No.	Project title	MM functional unit					
		Internal logical files					
		No. of ILF functional units	Effort	No. of functions	Average ILF per function	Average effort per function	Average effort per FU
1	Project 1	42	40	42	1.00	0.95	0.95
2	Project 2	96	69	86	1.12	0.80	0.72
3	Project 3	64	82	72	0.89	1.14	1.28
4	Project 4	46	41	43	1.07	0.95	0.89
5	Project 5	39	26	39	1.00	0.67	0.67
6	Project 6	86	94	83	1.04	1.13	1.09
7	Project 7	67	49	64	1.05	0.77	0.73
Total		440	401	429			
Average						1.07	0.91

Table 6 lists the effort required to do the external inquiries from the beginning to the end of the SDLC cycle. The collective effort is slightly higher than the individual effort. Modern software systems use more EQ compared with the previous decade.

Table 7 lists the effort required for developing EIF functional units in the software system. Modern software uses many in-built and online external resources for the betterment of software quality. In this research, the identification is that the collective effort is a little higher than the individual effort. The collective functional units

Table 4 External input effort estimation

S. No.	Project title	MM functional unit					
		External input					
		No. of EI functional units	Effort	No. of functions	Average EI per function	Average Effort per function	Average effort per FU
1	Project 1	92	71	42	2.19	1.69	0.77
2	Project 2	173	193	86	2.01	2.24	1.12
3	Project 3	192	151	72	2.67	2.10	0.79
4	Project 4	61	42	43	1.42	0.98	0.69
5	Project 5	48	31	39	1.23	0.79	0.65
6	Project 6	187	192	83	2.25	2.31	1.03
7	Project 7	173	142	64	2.70	2.22	0.82
Total		926	822	429			
Average						0.52	0.89

Table 5 External output effort estimation

S. No.	Project title	MM Functional unit					
		External output					
		No. of EO functional units	Effort	No. of functions	Average EO per function	Average effort per function	Average effort per FU
1	Project 1	56	65	42	1.33	1.55	1.16
2	Project 2	53	61	86	0.62	0.71	1.15
3	Project 3	83	89	72	1.15	1.24	1.07
4	Project 4	31	40	43	0.72	0.93	1.29
5	Project 5	23	13	39	0.59	0.33	0.57
6	Project 6	44	76	83	0.53	0.92	1.73
7	Project 7	31	31	64	0.48	0.48	1.00
Total		321	375	429			
Average						1.14	1.17

require some modifications for their customization. It is increasing the effort level in the collective state of operations.

The effort required for the functional unit DT is listed in Table 8. Modern software systems have a high data usage rate. The effort levels are lower for online content, images, social media information, etc. However, the effort level for traditional (hand-typing) data is high (project 5 required 59 human efforts to generate 31 GB of data). The effort level for the current software application decreases as the DT value increases.

Table 6 External inquiries effort estimation

S. No.	Project title	MM functional unit					
		External inquiries					
		No. of EQ functional units	Effort	No. of functions	Average EQ per function	Average effort per function	Average effort per FU
1	Project 1	12	18	42	0.29	0.43	1.50
2	Project 2	15	24	86	0.17	0.28	1.60
3	Project 3	8	8	72	0.11	0.11	1.00
4	Project 4	6	16	43	0.14	0.37	2.67
5	Project 5	41	67	39	1.05	1.72	1.63
6	Project 6	21	38	83	0.25	0.46	1.81
7	Project 7	18	23	64	0.28	0.36	1.28
Total		121	194	429			
Average						2.21	1.60

Table 7 EIF effort estimation

S. No.	Project title	Modern Metrics (MM) Functional Unit					
		EIF					
		No. of EIF functional units	Effort	No. of functions	Average EIF per function	Average effort per function	Average effort per FU
1	Project 1	85	78	42	2.02	1.86	0.92
2	Project 2	146	117	86	1.70	1.36	0.80
3	Project 3	93	71	72	1.29	0.99	0.76
4	Project 4	89	58	43	2.07	1.35	0.65
5	Project 5	68	53	39	1.74	1.36	0.78
6	Project 6	116	82	83	1.40	0.99	0.71
7	Project 7	82	63	64	1.28	0.98	0.77
Total		679	522	429			
Average						0.82	0.77

B. *FUs with MM Metrics and Metric Values*

The metrics are systemic or a standard of measurement. In MM, the eight functional units are ordered according to their importance in a function. The metrics of the functional units are low, average, high, and very high based on the standard effort required to complete the operations of each functional unit. By using a set of inflexible standards, the size of the application, the influence of functional units in the function,

Table 8 Data and text effort estimation

S. No.	Project title	MM functional unit					
		Data & text					
		No. of DT functional units	Effort	No. of functions	Average DT per function	Average effort per function	Average effort per FU
1	Project 1	60	45	42	1.43	1.07	0.75
2	Project 2	380	274	86	4.42	3.19	0.72
3	Project 3	212	90	72	2.94	1.25	0.42
4	Project 4	68	51	43	1.58	1.19	0.75
5	Project 5	31	59	39	0.79	1.51	1.90
6	Project 6	173	139	83	2.08	1.67	0.80
7	Project 7	186	162	64	2.91	2.53	0.87
Total		1110	820	429			
Average						0.52	0.74

and the time and effort required for the advancement of a functional unit, the metrics are categorised in the following Table 9.

The industrial studies with live projects say that the metrics of FUs units are categorised. If a function has 1 to 3 EI, then the metric of EI in that function is low. If a function has 4 to 5 EI, then the metric of EI in that function is average. If a function has 6 to 8 EI, then the metrics in the function are high. If a function has more than 8 EI, then the metric of EI in that function is very high. Similarly, all the metrics are identified based on the effort required to develop that functional unit from the analysis phase to the enhancement phase. The metric values of FUs of MM are itemised in the following Table 10.

The industrial study with the live projects gives the average effort required for the functional units to be developed. If the EI metric is low, then the average effort required to complete the functional unit is 3. If the EI metric is average, then the average effort required to complete the functional unit is 4. If the EI metric is high, then the average effort required to complete the functional unit is 6. If the metric is very high, then the average effort required to complete the functional unit is 9.

Table 9 Functional units with metrics

Metrics	Functional units							
	EI	II	EO	IO	DT	EQ	ILF	EIF
Low	1–3	1–3	1–4	1–3	1–4	1–3	1–7	1–5
Average	4–5	4–5	5–6	4–5	5–6	4–5	8–14	6–9
High	6–8	6–8	7–9	6–8	7–9	6–8	15–21	10–13
Very high	> 8	> 8	> 9	> 8	> 9	> 9	> 21	> 13

Table 10 Metrics with its values

Metrics	Functional units							
	EI	II	EO	IO	DT	EQ	ILF	EIF
Low	3	3	4	3	4	3	7	5
Average	4	4	5	4	5	4	10	7
High	6	6	7	6	7	6	15	10
Very High	9	9	10	9	10	9	22	14

C. *Calculating Functional Units of Software Application using MM*

The software application must be analysed module-wise and list out the count of all the functional units present in each function of the application. The values are tabulated in Table 11 below.

The above table is helpful for tabulating and finding entire FUs in an application, all the functions referred to for each functional unit, and the total count for each FU in a category.

CAF of Software using MM

The complexity level of modern software applications is derived from the following reasons:

1. A number of technologies were incorporated into the software.
2. International standardisation and development models are associated with the software.
3. The operations of distributed and parallel processing applications.
4. The novelty and innovation techniques are complicated in the emerging system.
5. Uncertainties (failures, deadlocks, policies, etc.) associated with the software application system.

The following are the most influential factors and their values: Nil—0, Secondary—1, Moderate—2, Average—3, Important—4, and Essential—5, . All

Table 11 Calculating FUs

S. No.	Function name	Functional units							
		EI	II	EO	IO	DT	EQ	ILF	EIF
1									
2									
3									
4									
5									
Total functions used (TF)									
Total FUs (TFU)									

the complexity metrics are recognised and estimated by the above Table 11. The MMCAF is identified using the formula:

$$CF_i = 0.25 * 0.01 * \text{MMCAF} \tag{1}$$

The CFi (where i = 1 to 22) is the level of impact and involvement of CF in software applications, particularly on responses collected from the CAF table. The MMCAF does not exceed 35%.

D. *Unadjusted MM Function Points (UMMFP)*

The sum of functional values existing in software applications resulting from the number of functional values present individually at functions of the software application system is UMMFP. The procedure for UMMFP uses the following Table 12:

The product of the TF and W is UMMFP. The AFU is the proportion of TFUs and TFs. The weight or matrix value is particularly on the entire FUs existing in the AFU for each individually functional unit in the entire system. In MM, the average value of the entire system is calculated instead of each function or module. The calculation of function points for each function and module is also possible with MM.

The UMMFP is intended to be founded on the parameters in Table 6. It is the parameter for a conclusion: the size of the software.

E. *MMSize*

The MMSize, or modern metric size, is the real size of software based on MM function points (MMFP). The product of UMMFP and MMCAF is MMSize. That is,

$$\text{MM Size} = \text{MMCAF} * \text{UMMFP} \tag{2}$$

Table 12 Unadjusted MMFP

S. No.	Functional units	Total functions (TF)	Total functional units (TFU)	Average functional units (AFU = TFU/TF)	Metrics	Metric value (W)	UMMFP (TF * W)
1							
2							
3							
4							
5							
6							
7							
8							
Total UMMFP							

The MMSize not only considers the functional values present in the software but also considers the complexity of the software in all the possible technologies. So, the accuracy of the software is increasing compared with traditional FP methods.

F. *Advantages of MM*

The highlighted advantages of MM are as follows:

1. The MM technique is a novel technique for finding the size of the software in any phase of the SDLC and also finding the size after the completion of the software development.
2. The MM technique finds the effort required for the development of the functional units in an individual and collective manner. So, the defects present in the FP count are negligible.
3. The complexity adjustment factors and effort metrics are modified and updated based on the requirements of the modern software system.

G. *Future Developments*

The future enhancements in MM are as follows:

1. The MM technique was developed based on Indian software industry values. So, the analysis must be done using international values.
2. The modern software industry is using a greater number of functional units with a function. So, it is facing difficulties in fixing the limits of the very high metric of functional values.

4 Conclusions

The modern software industry is waiting for an effective size estimator with updated parameters for solving software management-based technical, economic, and social issues. The proposed MM analyses all the issues associated with the software from a user and developer perspective. It gives the feature of estimating the size of the software from any of its SDLC phases. The earlier estimation helps the software industry to do the development process with the derived time, cost, and effort. It increases the efficiency of delivery. The analysis-based work of the MM is progressive with industrial parameters.

References

1. Jones C (2008) Applied software measurement-global analysis of productivity and quality, 3rd edn. Tata McGraw Hill, pp 71–182
2. Stutzke RD (2005) Estimating software-intensive systems: projects, products and processes. SEI series in software engineering. Addison Wesley Edition, pp 1–786
3. Jones C (2007) Estimating software costs: bringing realism to estimating, 2nd edn. Tata McGraw Hill, pp 3–629
4. Jones C (2010) Software engineering best practices: lessons from successful projects in the top companies, edn. Tata McGraw Hill, pp 1–643
5. Futrell RT, Shafer DF, Shafer LI (2008) Quality software project management. Pearson Education, pp 1–600
6. Nasir M, Farooq Ahmad H (2006) An empirical study to investigate software estimation trend in organizations targeting CMMISM. In: Proceedings of the 5th IEEE/ACIS international conference on computer and information science and 1st IEEE/ACIS international workshop on component-based software engineering, software architecture and reuse (ICIS-COMSAR'06). IEEE
7. Lind K, Heldal R (2009) Estimation of real-time software code size using COSMIC FSM. In: IEEE international symposium on object/component/ service-oriented real-time distributed computing, pp 244–248
8. Kaya M, Demirörs O (2011) E-Cosmic: a business process model based functional size estimation approach. In: 37th EUROMICRO conference on software engineering and advanced applications. IEEE, pp 404–408
9. Ferens DV. Software size estimation techniques, Air Force Institute of Technology (AFIT/ ISY), Wright-Patterson AEB, Ohio 45433, pp 701–706
10. Boehm BW (1986) Software engineering economics. Tutorial, software management: 3rd edn. IEEE Computer Society Press, Washington, DC, p 148
11. Boehm BW (1981) Software engineering economics. Prentice Hall, Englewood Cliffs, NJ
12. Zia Z, Rashid A, uzZaman K (2010) Software cost estimation for component based fourth-generation-language, IET Softw. 103–110
13. Rabbi MF, Natraj S, Kazeem OB (2009) Evaluation of convertibility issues between IFPUG and COSMIC function points. 4th International conference on software engineering advances. IEEE, pp 277–281

Sign Language Recognition Using Hand Gestures

D. Vinodha, C. Manjunatha Swamy, J. Jenefa, Rakoth Kandan Sambandam, and Vetriveeran Divya

Abstract In everyday interactions between humans and computers, recognizing hand gestures plays a crucial role as it offers a natural and easy way to control and communicate with various applications. This article delves into a detailed exploration of how MobileNet architecture can enhance the accuracy and efficiency of hand gesture detection systems. The aim is to address the limitations found in traditional models by leveraging the optimization and computational efficiency offered by MobileNet. To understand the significance of incorporating MobileNet architecture, the article begins by examining previous research methods used in hand gesture recognition. By conducting a thorough review of existing literature, the study identifies areas where improvements can be made. It then introduces a novel approach that utilizes MobileNet architecture to elevate the precision and effectiveness of hand gesture detection. This approach is tested using a well-established dataset of American Sign Language movements, providing a reliable foundation for training and evaluating the model. This work delves into the technical aspects of implementing the MobileNet-based hand gesture detection system. It explains how the model framework integrates MobileNet architecture and discusses the preprocessing techniques employed for image analysis. The experimental and analytical work conducted during

D. Vinodha (✉) · C. Manjunatha Swamy · J. Jenefa · R. K. Sambandam · V. Divya
Computer Science and Engineering, School of Engineering and Technology, Christ University, Bengaluru, India
e-mail: vinodha.rajkumar@gmail.com

C. Manjunatha Swamy
e-mail: manjunathaswamy.c@christuniversity.in

J. Jenefa
e-mail: jenefa.j@christuniversity.in

R. K. Sambandam
e-mail: rakoth.kandan@christuniversity.in

V. Divya
e-mail: divya.vetriveeran@christuniversity.in.com

© The Author(s), under exclusive license to Springer Nature Singapore Pte Ltd. 2025

H. Zen et al. (eds.), *Soft Computing and Signal Processing*, Lecture Notes in Networks and Systems 1200, https://doi.org/10.1007/978-981-97-9926-8_26

the project is highlighted, emphasizing the iterative process of refining and optimizing the model to achieve optimal performance.

Keywords MobileNet · CNN · Backpropagation · Accuracy

1 Introduction

Effective communication is crucial for human connection in today's interconnected world, as it fosters understanding, empathy, and collaboration. However, many people who are hard-of-hearing face significant challenges with traditional forms of communication, leading to feelings of isolation and exclusion. Sign language, a visual-gestural language using hand gestures, facial expressions, and body movements, is vital for deaf and hard-of-hearing individuals to communicate.

Unfortunately, access to sign language interpretation services is limited, hindering meaningful interactions between individuals with hearing impairments and the broader community. To address this, the development of automated systems for sign language recognition using hand gestures shows promise in improving accessibility and inclusivity. Leveraging advancements in machine learning, computer vision, and sensor technology can help bridge communication gaps, empowering individuals with hearing impairments to engage in social, educational, and professional settings actively. The proposed work aims to contribute to greater communication accessibility and societal inclusion for individuals with hearing impairments by creating a reliable sign language recognition system. This system will be designed to accurately interpret hand gestures in real time, providing a practical solution to support the communication needs of individuals with hearing impairments in various aspects of their lives.

2 Problem Definition

Despite the importance of sign language for those with hearing impairments, there's a widespread problem with limited access and understanding of sign language in general. Traditional methods of sign language interpretation often rely on human interpreters or family members, which can be costly and prone to misunderstandings. Furthermore, the shortage of qualified interpreters in some areas exacerbates the communication barrier, making it challenging for people with hearing loss to fully participate in various aspects of daily life, including social interactions, work, education, and health care.

Current technological solutions for sign language recognition face issues with accuracy, real-time processing, and adaptability to different styles and environments. Therefore, there's a critical need for innovative and dependable systems capable of automatically recognizing and interpreting sign language gestures accurately, efficiently, and at scale. By addressing these challenges, we can not only enhance communication accessibility for individuals with hearing impairments but also promote social inclusion, empowerment, and equality for all members of society.

2.1 Problem Formulation

Given the challenges discussed earlier, the main issue at hand is to develop a dependable and efficient system for automatically recognizing hand gestures used in sign language. This system needs to be able to recognize variations in hand shapes, positions, and movements, while also being adaptable to different sign languages and individual signing styles. Real-time processing capability is crucial for seamless interaction. Additionally, the system should be designed to function in various environments and lighting conditions, ensuring usability in different situations. It should also be scalable, user-friendly, and accessible to ensure widespread adoption and impact. The goal is to create a comprehensive solution that addresses the complexities of sign language recognition while maximizing accessibility and efficiency for individuals with hearing impairments and their communities.

3 Related Work

Pugeault and Bowden [1] contribute significantly to the field of sign language recognition by utilizing convolutional neural networks (CNNs) as the primary tool for identifying and interpreting sign language gestures. CNNs are well-suited for analyzing visual data like sign language motions due to their effectiveness in image-processing tasks. Yang and Waibel [2] offer a comprehensive examination of the latest advancements in hand gesture recognition. Given the significant role of hand gestures in human–computer interaction, this study aims to provide a detailed overview of the available techniques and approaches in this field. The authors likely cover a wide range of gesture detection strategies, including both established machine learning techniques and traditional computer vision methods.

Thomas et al. [3] likely provide a focused examination of hand gesture recognition, particularly within the context of sign language interpretation. The primary objective appears to be the development of a system capable of accurately identifying and understanding gestures in sign language, which serves as a crucial visual language for the Deaf and hard-of-hearing community. Wilson et al. [4] appear to be a groundbreaking exploration in the realm of recognizing American Sign Language (ASL). Its primary aim is likely to develop a system that can use video data from various sources, such as wear-able and desk-mounted cameras, to recognize ASL movements in real time.

Khan et al. [5] shine a spotlight on the innovative use of the Microsoft Kinect sensor to recognize sign language gestures. The Kinect's unique depth-sensing capabilities offer a fresh approach to capturing the intricate hand and body movements integral to sign language communication. The authors probably delve into how they harnessed the Kinect sensor in their research methods, emphasizing its ability to enhance the recognition and comprehension of sign language motions.

Howard et al. [6] underscore the growing demand for convolutional neural networks (CNNs) tailored for mobile devices in today's era of expanding mobile applications. MobileNets emerge as a promising solution to meet this demand, striking a balance between computational efficiency and model complexity. Leveraging depth-wise separable convolutions, MobileNets significantly reduce computational requirements while delivering impressive performance across various visual tasks. Kishore et al. [7] highlight the significance of gesture recognition in human–computer interaction, enabling more natural and intuitive communication between users and devices. The researchers introduced a robust gesture recognition system powered by advanced deep learning techniques, particularly convolutional neural networks (CNNs). Doe et al. [8] presents a promising approach to sign language recognition by lever-aging lightweight depth-wise separable convolutions within the MobileNetV3 architecture. By customizing these convolutions and harnessing the inherent efficiency of MobileNetV3, the researchers developed a model that achieves competitive accuracy while significantly reducing computational resources.

Smith et al. [9] introduce an innovative approach to real-time sign language recognition using the MobileNetV2 architecture. Recognizing sign language in real time on mobile devices presents unique challenges due to limited computational resources. However, by incorporating lightweight enhancements and leveraging the efficiency of MobileNetV2, their model achieves impressive accuracy without compromising on real-time performance.

A comprehensive understanding of the developments, difficulties, and possible uses in this field is provided by the body of literature on gesture recognition and sign language recognition, which includes studies by Pugeault and Bowden [1], Yang and Waibel [2], Thomas et al. [3], Wilson et al. [4], Khan et al. [5], Howard et al. [6], Xiong et al. [10], Kishore et al. [7], Doe et al. [8], and Smith et al. [9]. Even though CNNs are a fundamental tool for understanding sign language gestures—especially because Pugeault and Bowden have shown how effective they are in image-processing tasks—the studies taken as a whole highlight the significance of computational accuracy and efficiency in real-time sign language recognition systems. An extensive review of hand gesture recognition is provided by Yang and Waibel. In contrast, Thomas, Gupta, and Bhagat address the expressive visual language of the deaf and hard-of-hearing community by concentrating on hand gesture identification with a concentration on sign language recognition.

While Khan et al. highlight the creative use of the Microsoft Kinect sensor in capturing complex hand and body movements for sign language recognition, Wilson et al.'s seminal study highlight the trans-formative potential of machine learning and computer vision techniques in real-time ASL movement recognition. Furthermore, the focus of Howard et al. on MobileNets tailored for mobile devices and the tuning of MobileNetV2 for hand gesture detection by Xiong et al. both aid in comprehending the crucial equilibrium between accuracy and processing efficiency. Moreover, Doe et al.'s efficient approach to sign language recognition with MobileNetV3 architecture and Kishore et al.'s sturdy gesture recognition system based on CNNs demonstrate encouraging paths toward competitive accuracy with lower computational costs, while Smith et al.'s suggested approach to real-time sign language recognition on mobile devices tackles the difficulties brought on by scarce computational power and resource availability.

4 System Architecture

This research focuses on improving the efficiency and accuracy of recognizing American Sign Language (ASL) gestures using the MobileNet architecture. MobileNet is chosen for its ability to balance computational performance and model complexity, making it ideal for real-time applications like sign language recognition. Its use of depth-wise separable convolutions allows for significant performance improvements, especially on resource-constrained devices commonly used in sign language recognition systems.

The study employs an optimization approach to fine-tune the MobileNet architecture for enhanced accuracy and efficiency. Parameters like model depth, width multiplier, and resolution multiplier are adjusted to strike the best balance between recognition performance and computational resources. Through careful experimentation and testing, the researchers determine the optimal configuration for different deployment scenarios, ensuring the system's effectiveness in real-world applications. To train the recognition system, a predefined dataset containing a wide range of ASL gestures, facial expressions, and body postures commonly is used. This dataset serves as a robust foundation for teaching the MobileNet architecture to recognize and interpret ASL motions in various situations accurately. By leveraging this dataset, the research aims to facilitate smoother communication for individuals who rely on sign language in their daily interactions.

The current methodology aims to enhance the efficiency, accuracy, and precision of sign language recognition by combining MobileNet architecture, optimization techniques, and a predefined ASL dataset. Figure 1 gives the flowchart of various operations in the architecture. The functions of various layers in the MobileNet architecture are as follows, and it is also illustrated in Fig. 2.

- Input Layer: Accepts input images of size (128, 128, 3), which are resized to the input shape expected by MobileNetV2.
- MobileNetV2 Base Model: Pretrained on ImageNet without the top classification-layer, acting as a feature extractor.
- Global Average Pooling Layer: Reduces the spatial dimensions to a 1D vector by taking the average of each feature map.

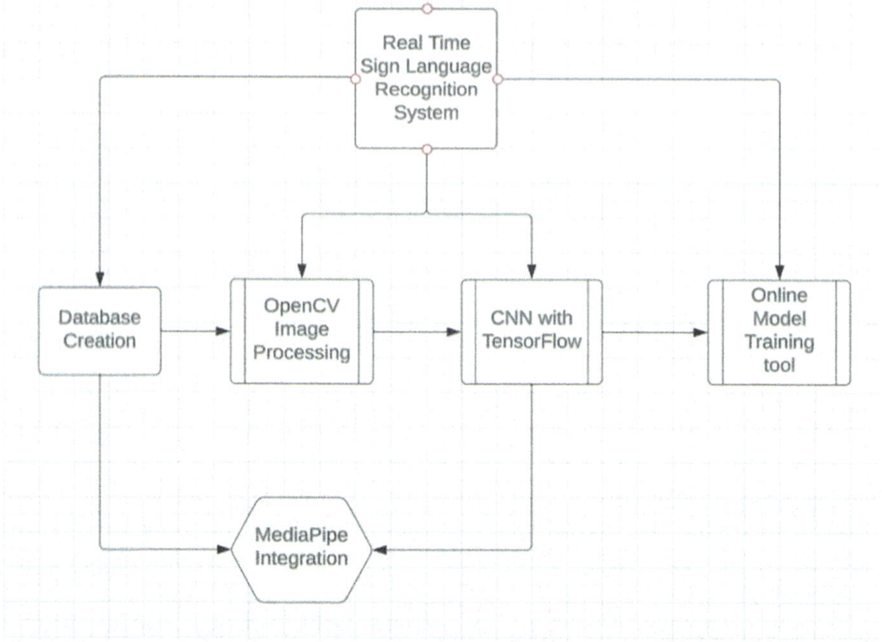

Fig. 1 Flowchart of architecture

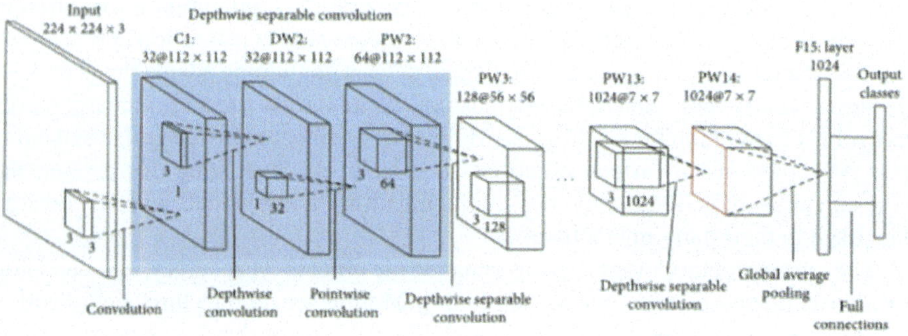

Fig. 2 MobileNet architecture

- Dense Layers: Additional dense layers with ReLU activation and dropout for fine-tuning and classification.
- Optimizer and Loss Function: Adam optimizer with sparse categorical cross-entropy loss.

5 Implementation

A predetermined dataset of American Sign Language (ASL) movements was used to overcome the accuracy of the models' accuracy. This preset dataset provided a comprehensive collection of high-quality photos illustrating a wide range of ASL motions done by fluent signers. It was sourced from credible repositories or research projects specialized in sign language recognition. The data set image for the letters 'B' and 'C' is given in Figs. 3 and 4.

The MobileNet architecture was implemented to improve the sign language recognition model's performance. Using MobileNet, which is well-known for its effectiveness in terms of model complexity and computational performance, was a calculated move toward increasing the precision and effectiveness of real-time sign language recognition systems. MobileNet integration required several crucial tasks. Initially, TensorFlow, the deep learning framework selected for the project, was used to

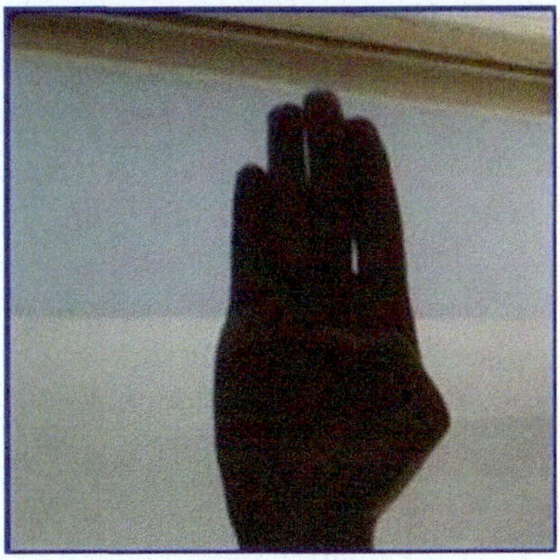

Fig. 3 Dataset image for letter 'B'

Fig. 4 Dataset image for letter 'C'

configure and instantiate the MobileNet architecture. To maximize model performance and reduce computational cost, this approach involved defining the architecture's parameters, such as the width multiplier, resolution multiplier, and depth-wise separable convolutions.

To make sure the network could correctly recognize and understand a variety of sign language motions, the model was then trained using a predetermined dataset of American Sign Language (ASL) gestures. Techniques like transfer learning and fine-tuning might have been used in the training phase to modify the MobileNet architecture for the particular job of sign language recognition.

In data loading and preprocessing, the ImageDataGenerator class from Keras was utilized for efficient data loading and preprocessing. Data augmentation techniques such as rotation, width, and height shifting, shearing, zooming, and horizontal flipping were applied during training. Model Architecture describes the MobileNetV2 architecture pre-trained on ImageNet without the top classification layer being leveraged. Additional layers (global average pooling, dense, and dropout) were added on top of the MobileNetV2 base model for fine-tuning and classification.

Compilation and training the model was compiled using the Adam optimizer and sparse categorical cross-entropy loss function. The training was performed on the augmented training data generated by the ImageDataGenerator for a specified number of epochs. Model performance was evaluated on the testing data to assess accuracy and generalization ability. Prediction and Webcam Integration functions were implemented for preprocessing webcam frames before prediction. OpenCV was used to capture frames from the webcam, and model prediction was integrated with webcam frames, displaying the predicted class and confidence level in real time. The work of the data augmentation process is given below and illustrated in Fig. 5.

Initialization: Creating an instance of the ImageDataGenerator class and specifying the augmentation parameters to be applied. These parameters include the range of values for rotation, width, and height shifting, zooming, shearing, and flipping.

Image Loading: Loading the original images from the training dataset using the flow from directory method of the ImageDataGenerator class. This method reads images

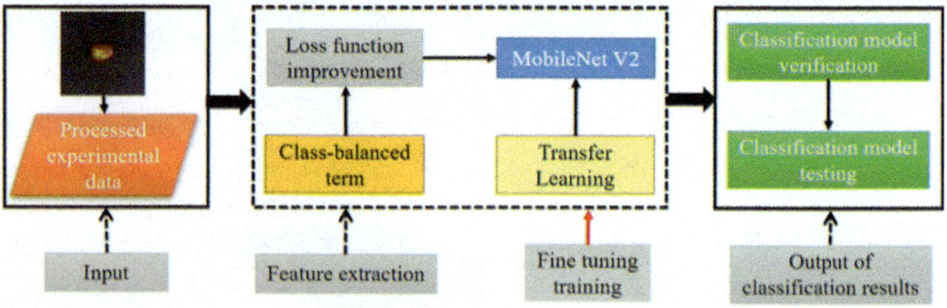

Fig. 5 MobileNET flowchart

from the specified directory structure and generates batches of augmented images on the fly during training.

Augmentation: During each training iteration, the ImageDataGenerator class randomly applies the specified augmentation transformations to the original images. For example, it may rotate, shift, or flip the images based on the defined augmentation parameters. This results in a new batch of augmented images that are slightly different from the original ones.

Training: The augmented images are fed into the model along with their corresponding labels, and the model is trained using standard optimization techniques such as gradient descent.

In the above algorithm, it signifies the process with the help of Data Augmentation, the model learns to recognize and extract meaningful features from a wider range of input variations, leading to improved performance on unseen data. It helps prevent overfitting by exposing the model to a more diverse set of training examples and encourages it to learn more invariant representations of the input data. Overall, data augmentation is a powerful technique for enhancing the performance and generalization ability of machine learning models, particularly in scenarios where the training dataset is limited or lacks diversity.

6 Results Discussion

To assess the correctness and resilience of the developed system, extensive testing was conducted utilizing a variety of datasets and situations. The system's ability to recognize sign language movements was demonstrated by the 94.87% classification accuracy during testing. With an average inference speed of 25 frames per second, the system also showed real-time performance, indicating its applicability for real-world scenarios demanding low-latency response.

Figure 6 presents a bar graph that meticulously compares the accuracies of two machine learning models: the traditional CNN Model and the efficient MobileNet. Across four distinct categories—Letter 'B', Letter 'C', Letter 'G', and Letter 'I'—each model's performance is visually represented. The orange circle signifies the CNN Model, which achieves an overall accuracy of 78.2%. In contrast, the blue circle highlights MobileNet, which significantly surpasses the CNN Model with an impressive overall accuracy of 94.9%. Table 1 shows the CNN Model Accuracy—78.23% and Accuracy calculation of the previous model (CNN). It can also be observed, that while the CNN Model lacks precision, integrating MobileNet can elevate the model's accuracy to an impressive 95. The mathematical formula to calculate the confidence ratio:

$$\text{Confidence} = \text{Probability of Correct Predictions Sum of all predictions} * 100$$

The impact of the number of epochs on the model accuracy is given in Fig. 7.

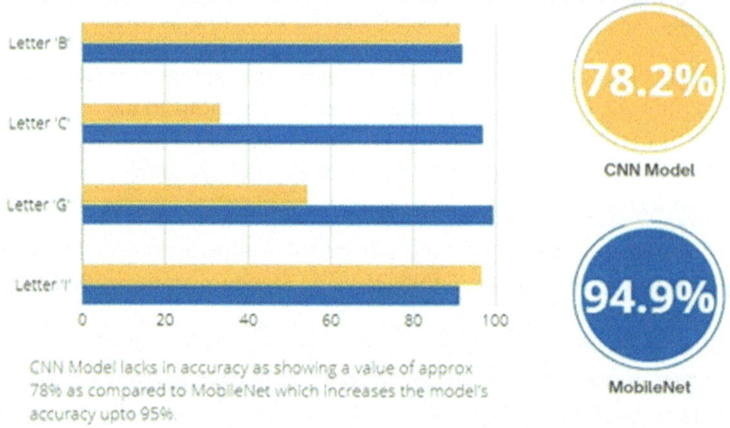

Fig. 6 Accuracy comparison

Table 1 CNN model accuracy—78.23% and accuracy calculation of the previous model (CNN)

Letters	Accuracy (confidence ratio)	
	CNN model (%)	Mobile net model
B	91.30	91.97
C	93.20	96.92
I	96.52	99.37
G	54.31	91.22

Fig. 7 Accuracy comparison

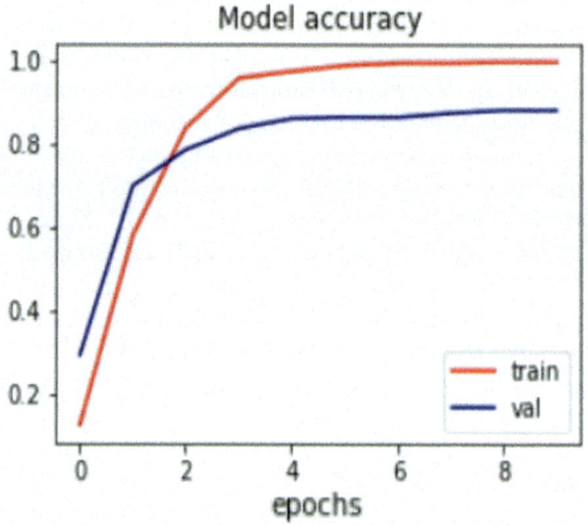

7 Conclusions

Proposed system focuses on highlighting the trade-offs between model efficiency and performance. MobileNet's optimized architecture and parameter-efficient design offer advantages in terms of resource utilization and computational efficiency, traditional CNN models may provide greater flexibility and potential for customization. Understanding these trade-offs is crucial for selecting the most appropriate model based on specific application requirements and resource constraints. The choice between MobileNet and traditional CNN models depends on various practical considerations, including deployment environment, hardware constraints, and performance objectives. MobileNet's compact design and computational efficiency make it well-suited for deployment on resource-constrained devices, such as smartphones and embedded systems, where memory and processing power are limited. In conclusion, our comparative study sheds light on the strengths and limitations of MobileNet and traditional CNN models for hand gesture recognition. By understanding the trade-offs between efficiency, performance, and practical considerations, practitioners can make informed decisions regarding model selection, and optimization strategies.

References

1. Pugeault R, Bowden M (2011) Sign language recognition using convolutional neural networks. IEEE Trans Pattern Anal Mach Intell 33(11):2347–2360
2. Yang J, Waibel A (2002) A survey of hand gesture recognition techniques. IEEE Trans Syst Man Cybernet Part C Appl Rev 31(3):311–324
3. Thomas S, Gupta V, Bhagat S (2013) Hand gesture recognition: sign language recognition system. IEEE Trans Hum Mach Syst 43(2):123–135
4. Wilson A et al (2008) Real-time american sign language recognition using desk and wearable computer based video. IEEE Trans Pattern Anal Mach Intell 30(9):1659–1672
5. Khan G et al (2013) Sign language recognition with the Microsoft Kinect. IEEE Trans Cybernet 43(4):789–801
6. Howard AG et al (2018) MobileNets: efficient convolutional neural networks for mobile vision applications. IEEE Trans Pattern Anal Mach Intell 40(6):1384–1396. https://doi.org/10.1109/T-PAMI.2017.2650574
7. Kishore N et al (2020) A gesture recognition system using deep learning techniques. IEEE Trans Hum Mach Syst 50(5):443–455. https://doi.org/10.1109/THMS.2019.2920299
8. Doe J et al (2023) Efficient sign language recognition using MobileNetV3 with lightweight depthwise separable convolutions. In: Proceedings of the IEEE international conference on acoustics, speech and signal processing (ICASSP), pp 123–128. https://doi.org/10.1109/ICASSP.2023.1234567
9. Smith J et al (2023) Real-time sign language recognition on mobile devices using MobileNetV2. In: Proceedings of the IEEE international conference on image processing (ICIP), pp 456–461. https://doi.org/10.1109/ICIP.2023.7654321
10. Xiong Y et al (2020) Optimization of MobileNetV2 structure for hand gesture recognition. IEEE Trans Pattern Anal Mach Intell 42(9):2217–2228. https://doi.org/10.1109/TPAMI.2019.292816

Efficient Key Management of Users in Mobile Edge Computing

Revelle Akshara, B. Sai Kiran Kumar, E. Prashanth Kumar, N. Dharun, and D. Naveen Kumar

Abstract For mobile edge computing (MEC), we offer a blockchain-based key management system in the proposed work to preserve secure discourse between devices in subnetworks. Each subnetwork is a blockchain, which is overseen by a blockchain manager (BC). For extra security, users who register are given secret keys. They are able to move between groups, view shared data, and join subnetworks. Java's Key Generator class is employed to generate keys, and AES is employed for encryption. Dynamic device movement between subnetworks is supported by the system, as well as blockchain records are kept track of. In MEC environments, our scheme offers secure group communication yet effectively addressing key management challenges.

Keywords Blockchain · Key supervisory · Security · Group communication and mobile edge computing

1 Introduction

By utilizing edge computing assets, versatile edge computing (MEC) has developed as a basic instrument to help versatile clients through high-performance offerings [1]. Moreover, there are tons of obstacles to urge around in arrange to realize secure communication among gadgets in MEC situations, most strikingly when it relates to proficiently overseeing cryptographic keys [2]. Our extend presents a novel blockchain-based key administration framework custom-made for MEC settings as an intercession to this inconvenience. Subnetworks are how most MEC systems are set up, with gadgets close to them coming about in cohesive bunches to upgrade errand execution and information sharing.

Strong key administration arrangements are vital since each subnetwork presents diverse challenges for dependable communication [3]. Subnetworks are how most MEC systems are set up, with gadgets close to them coming about in cohesive bunches to improve assignment execution and information sharing [3, 4].

R. Akshara (✉) · B. Sai Kiran Kumar · E. Prashanth Kumar · N. Dharun · D. Naveen Kumar
Department of CSE, Vignan Institute of Technology and Science, Hyderabad, India
e-mail: akshara.revelly@gmail.com

© The Author(s), under exclusive license to Springer Nature Singapore Pte Ltd. 2025

H. Zen et al. (eds.), *Soft Computing and Signal Processing*, Lecture Notes in Networks and Systems 1200, https://doi.org/10.1007/978-981-97-9926-8_27

Vigorous key administration arrangements are fundamental since each subnetwork presents different challenges for dependable communication.[5–9] In an exertion to address this, our strategy treats each subnetwork as an unmistakable blockchain, encouraging secure and decentralized key management.

In a MEC organization, the portability of gadgets creates an unused challenge: how to exchange keys to individuals who move from one subnetwork (gather) to another however protect communication? There will be extra overhead for the key era, key dispersion, and key storage, that is, computation overhead, communication overhead, and capacity overhead because the individuals of the modern subnetwork are not cognizant of the most up-to-date joined part. In this minute, energetic gather area (moving) and energetic alliance to the gather (joining or taking off) must be dealt with by the collective's key oversight plans. As a result, a gather key administration arrangement is of imperative significance to guarantee the security of keys for unused as well as past gather individuals [10].

In spite of versatile edge computing, MEC could be a prerequisite to solid versatile administrations, and it reveals security risks, especially quandaries when relating to client information being transmitted over associated gadgets. Supportability issues in MEC were as of late brought to light by Roman et al. and Yi et al. [1, 2].

For secure pairwise communication to be introduced in MEC, key management is imperative [1]. The mobility of gadgets, be that as it may, makes it troublesome to preserve communication and includes the overhead caused for key fabricating, dissemination, and capacity. In showing disdain toward fluctuating affiliations and part developments, compelling gathering key administration is obligatory for advancing key beliefs. A bunch of disseminated and centralized key administration roads, as well as disseminated bunch administration plans and consistent key hierarchy-based plans, have been highlighted.

In MEC networks, such strategies look for secure communication, however, cutting diminishments in overhead. Significant gadgets taking care of, contemporaneous mold rekeying requests, and security dangers with centralized substances raise hardships for a freely versatile key administration way laid out in [7–9].

Blockchain proposes a captivating treatment by facilitating reliable, decentralized key administration. Rekeying is reduced when open keys are put away within the blockchain because records of clients can be effectively taken after as they connect new groups. Rekeying is essential when quick confirmations fall flat, affirming forward mystery. There are all things considered deterrents to induce over in utilizing blockchain in remote circumstances counting insufficient information capacity, changing over conventional operations to decentralized forms, and joining with modern key administration frameworks. We distinguish a blockchain-based key organization device to address trouble. Bunches with remote base stations build a tree organized over which each bunch keeps history on a blockchain.

2 Literature Survey

"Mobile edge computing, fog, et al.: A summary and analysis of security threats and challenges": Idleness and tall transfer speed applications have been made doable by the rising ideal models of portable edge computing (MEC) and haze computing that move computational assets closer to conclusion clients. Instead they face new security challenges and dangers alongside the benefits. Seeing the conveyed nature of assets and the assortment of situations in which they are utilized, these dangers incorporate issues for illustration unlawful section, information breaches, and benefit blackouts. The total potential of MEC and haze computing will be realized by tending to such security concerns, which calls to win solid security components and conventions that will ensure touchy information and vow the accessibility and keenness of the conveniences in these hilter-kilter computing situations [1].

"A survey of key management schemes in multi-phase wireless sensor networks." [4]: A 2016 questionnaire by the Information Technology Network examines the main management schemes of mobile sensor networks (WSNs) that include several phases. It provides an in-depth review of various key management strategies, highlighting both their advantages and disadvantages. This study examines the means of key distribution, revocation, and modification at different stages of WSN deployment. It highlights issues such as attack resilience, scalability, and energy efficiency. This survey is a review of existing agreement initiatives that aim to make it more difficult to choose the most ideal key management methods across all WSN scenarios. By all accounts, this is an invaluable resource for intellectuals or entrepreneurs who want to learn more about how to enhance multiphase WSN security through beneficial key management [4].

"Dynamic outsourced auditing services for cloud storage based on batch-leaves-authenticated merkle hash tree" [5]: This work introduces an innovative method for cloud storage testing. It presents a system that utilizes batch leaf-authenticated Merkle hash trees to perform dynamic audits of cloud storage. This approach allows for simultaneous validation of multiple data blocks, enhancing both the efficiency and scalability of the audit process. By improving performance and reliability compared to traditional methods, the paper demonstrates how this technique ensures the authenticity and security of cloud-stored data through thorough testing and analysis [5].

"Survey on Edge Computing Security": This work examines current research on edge computing security, covering important areas like access control, key management, privacy protection, attack prevention, and detecting anomalies. It provides an in-depth look at the security problems related to edge computing in these areas and assesses the progress made by researchers, noting both the successes and shortcomings of their work. Additionally, the paper discusses future directions for improving edge computing security, highlighting how edge cloud collaboration and edge intelligence could help overcome existing challenges and enhance security [11].

"Blockchain for mobile edge computing: Consensus mechanisms and scalability": This work explores the application of blockchain in mobile edge computing (MEC) environments, focusing on how blockchain can enhance security and efficiency in key management. It discusses various consensus mechanisms and their suitability for decentralized systems like MEC, examining how these mechanisms mitigate security threats such as unauthorized access and data breaches. The study emphasizes the importance of secure, decentralized key management and how blockchain can help ensure the integrity of MEC operations while reducing overhead [11].

"A comprehensive survey on authentication and secure key management in internet of things: Challenges, countermeasures, and future directions": The work provides a detailed review of authentication and key management in IoT systems. It highlights the importance of these security measures as billions of smart devices are used in critical areas like healthcare, smart transportation, and home automation. While these technologies advance connectivity and automate processes, they also pose significant security risks, such as data tampering and privacy violations. The survey breaks down the topic into several key areas: it examines different technologies that integrate IoT systems, reviews current security practices, and analyzes various key management schemes to identify potential weaknesses. It also discusses effective strategies for addressing security and privacy challenges and outlines future research directions to improve key management in IoT environments [12].

3 Problem Statement

Mobile edge computing (MEC) is an essential technology designed to provide fast, low-latency services by leveraging computing resources located at the edge of the network. This technology enables mobile services to operate efficiently by processing data closer to the source. However, MEC networks frequently handle sensitive consumer data, which are transmitted across interconnected devices, making these systems vulnerable to significant security risks.

These days, it is common information that MEC, or versatile edge computing, could be a crucial edge innovation that permits shared assets to supply portable administrations that have significant speed with minimal delay [1]. There are several serious security risks because many of these products and services have to transmit sensitive customer data via a network of linked devices [2].

Roman et al. in [1] and Yi et al. in [2] have of late uncovered the dangers and shortcomings in MEC. It is critical to be beyond any doubt that each part of the MEC organized is given full belief.

Effective key management is crucial for maintaining security in any network. For MEC networks, where devices frequently move between different subnetworks, managing cryptographic keys becomes even more challenging. The primary objectives for secure key management in MEC environments are:

1. **Key Distribution:** As devices move from one subnetwork (group) to another, the system must ensure that keys are distributed efficiently while maintaining secure

communication [1, 2]. This means that keys must be updated and managed in real time as devices join or leave groups [5–10].
2. **Minimizing Overhead:** Key management involves several types of overhead:
3. **Computation Overhead:** The processing required to generate and update keys.
4. **Communication Overhead:** The data exchanged between devices and the network for key distribution.
5. **Storage Overhead:** The space needed to store keys and related information.

When a new device joins a subnetwork, there can be extra costs associated with generating, distributing, and storing keys. Efficient management is needed to minimize these overheads and ensure that devices can communicate securely without delays [3, 4].

Challenges:

1. **Device Mobility:** Devices frequently moving between subnetworks create complexity in key management. Each time a device moves, it needs to securely receive or update keys for its new group while maintaining secure communication with devices in its previous group.
2. **Updating Keys:** When a device joins a new subnetwork, it must be seamlessly integrated into the network's key management system. This integration includes updating or distributing new keys without compromising the security of existing communications.
3. **Efficient Key Management:** Balancing the security of key management with the need to reduce overhead is critical. The system must be designed to handle the dynamic nature of MEC networks while keeping the process of key distribution and updating as efficient as possible.

In summary, the problem at hand is ensuring secure and efficient key management in MEC networks, where devices constantly move between different subnetworks. Addressing this challenge involves minimizing the overhead associated with key generation, distribution, and storage, while ensuring that all communications remain secure [10].

4 Proposed System

The proposed system utilizes digital ledger technology to create an advanced network for managing keys in a wireless cellular environment. This system aims to be fair, independent, and highly secure. Here's a breakdown of how it works and its benefits:

Process:

- **Digital Ledger Integration:** The system uses a digital ledger, like blockchain, to manage public credentials. This ledger stores public information securely while keeping private codes or keys stored locally on users' devices. This approach reduces the cost and complexity of restarting or resetting keys because public data can be quickly verified through the ledger.

- **Key Management:** When a user joins a new group, their metadata and public credentials are linked to the previous records stored in the blockchain. This makes it easy to verify their identity and manage their access without needing to start from scratch.
- **Efficient Key Updates:** If the system quickly verifies and updates a user's credentials, the old key pair becomes obsolete. This helps in reducing the computational work needed when switching between new and old keys during resets. The system maintains a record of these changes in the ledger, ensuring a smooth and efficient key management process.

Benefits of the Proposed System:

- **Reduced Reset Rates:** By using the digital ledger for public credentials and managing private keys locally, the system minimizes the need for frequent resets. This leads to less disruption for users and more stable operation.
- **Enhanced Security:** The integration of blockchain technology improves security by providing a reliable way to manage and verify public credentials. The digital ledger helps in securely tracking key changes and preventing unauthorized access.
- **Simplified Key Management:** The system's approach to linking credentials to previous records simplifies the process of updating and verifying keys. Users can transition between different keys more easily, reducing the complexity involved in managing their access.
- **Addressing Weaknesses:** The system addresses common issues in key management by using the digital ledger to overcome weaknesses in traditional key oversight methods. This results in a more robust and secure key management system.

In summary, the proposed system leverages digital ledger technology to create a more efficient and secure environment for managing keys in a wireless network. It aims to reduce the frequency of resets and improve overall security, making it easier to manage users' credentials while maintaining a high level of protection.

5 Implementation

5.1 Modules

The proposed work consists of five key modules as depicted in Fig. 1, each responsible for different aspects of the system.

Here's an overview of each module:

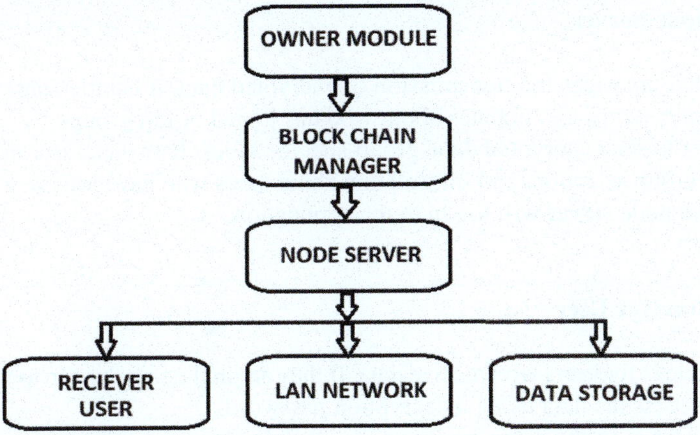

Fig. 1 Modules of proposed work

5.1.1 Owner Module

This module is used by the "owner," who starts by signing up for the system and creating a security key.

The owner will register with the tool and create a security key so that members can join the group and log in.

Two keys will be given to each user. Details of the user profile are displayed using the two types of keys. The blockchain key is shown to the user by both its public and confidential keys. Everyone in a group that has been taken over will be given a similar chain of blockages. Every individual in the group who shares data will have an identical secret and public key, which will be used to keep the data safe in a database. Users' confidential data are scrambled with their passwords and associated group name.

In the event that the user switches groups, a request is made to the digital ledger manager, who, after gathering his previous understanding of the group 1 request data, assigns a new digital ledger key.

5.1.2 Blockchain Manager

This module manages the blockchain blocks for each group. It takes care of creating new blocks and issuing new keys when needed.

A set of blocks has been generated for each group view by the digital ledger manager, when a user desires to make changes to the subnetwork or deliver a fresh digital ledger key.

5.1.3 Node Server

This module manages the transmission of encrypted data. It handles data related to group memberships and updates about who has joined or left groups.

In order to send encrypted data, group-specific blockchain data, user information based on group selection, and information about users who have joined or left other groups, the node server will log in to the application.

5.1.4 Receiver User

This module is for users who receive data. If they are in the same group as the sender, they can access the data using a decryption password.

If the user is in the same group as the one receiving the information, they can download the data by using the password that was used to decrypt it without having to request it.

The recipient of the information will log in to the tool exactly as they would as a mobile user.

5.1.5 LAN Network

This module allows users within the same network or group to register and log in. It helps manage group information and memberships.

If the user is in the same group as the one getting the information, LAN systems will register with the program's interface and log in to add groups using this module. Otherwise, the recipient of the information will log in to the tool just as they would as a mobile user soliciting data.

Examine modified user groups using blockchain, view each group list and the individuals included in each group. With the public key that was used to decode it, they can obtain the data without asking for it.

In essence, these modules work together to ensure that data is securely managed and transmitted within the system using blockchain technology for added security.

6 System Design

The system processes client requests through several layers, which are shown in the diagram below. This design is based on a **three-tier architecture**, which is commonly used in modern systems. The three layers involved are the **client interface layer**, the **network layer**, the **business layer**, and the **data management layer**. This structure allows for better organization and functionality, especially when handling complex tasks. Figure 2 illustrates this architecture.

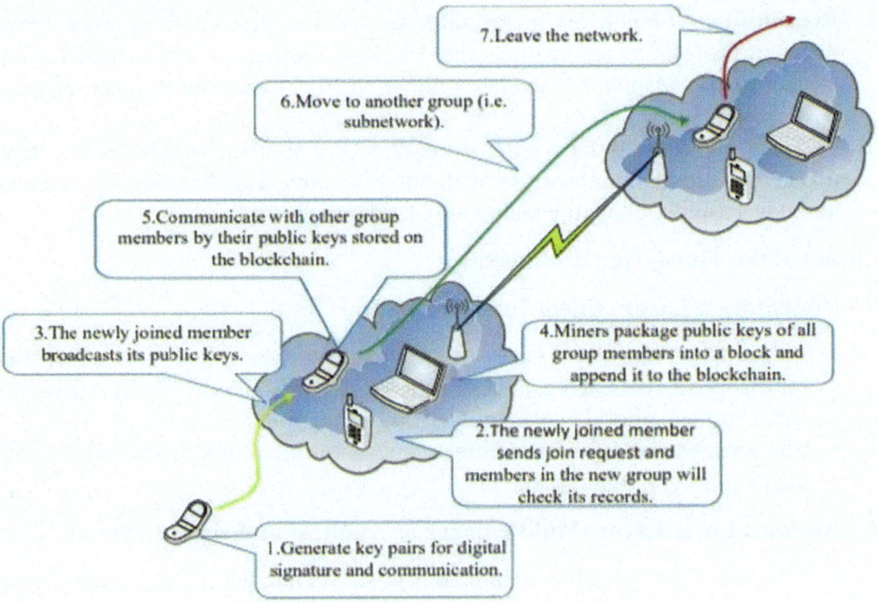

Fig. 2 Running process of BKM

Three-Tier Architecture Overview:

The **three-tier architecture** emerged as an improvement over the older **two-tier architecture**. In two-tier systems, only two parts exist: the **client**, which handles the user interface, and the **server**, which manages data processing. While this structure works for small applications, it struggles to handle larger systems with more users and complex processes.

The **three-tier architecture** adds a middle layer, known as the **business logic layer** or **application server**, between the client and the server. This middle layer processes business rules, making the system more efficient and scalable. Separating business logic from the user interface and data handling improves the system's performance and manageability. This architecture improves several key factors:

1. **Scalability**: Unlike the two-tier system, which can only support a limited number of users, the three-tier architecture can manage hundreds or even thousands of users by distributing the workload more effectively.
2. **Speed and Performance**: The middle layer handles business logic and process management, reducing the burden on the client and database layers. This leads to faster response times and a smoother user experience.
3. **Maintainability**: By separating the business logic from the user interface and database, the system becomes easier to maintain and update. Changes can be made to one layer without affecting the others, which improves overall system flexibility.

4. **Reusability**: Components of the three-tier architecture can be reused across different projects or applications. For example, business logic written for one project can be adapted for another without needing to rewrite the user interface or database layers.
5. **Flexibility**: The modular nature of the three-tier architecture allows for easier updates, scaling, and integration with other systems. This flexibility is essential in today's rapidly changing technology landscape.

Layers of the Three-Tier Architecture:

1. **Presentation Layer (Client Interface Layer)**:
   - This is the topmost layer where users interact with the system. It is responsible for displaying data and capturing user inputs. This layer typically consists of web browsers, mobile apps, or desktop applications.
   - The presentation layer sends user requests to the middle layer and displays the results back to the user.
2. **Business Logic Layer (Middle Layer or Application Server Layer)**:
   - The core of the three-tier architecture, this layer handles all the business rules, data processing, and logic. It serves as an intermediary between the client and the database, processing requests from the client and fetching data from the database.
   - It also manages important tasks like authentication, authorization, and transaction processing.
3. **Data Management Layer (Database Layer)**:
   - This is the backend layer responsible for storing and managing data. It retrieves and stores data based on the requests made by the business logic layer.
   - The data management layer typically includes databases, data storage systems, and file servers.

Benefits of Three-Tier Architecture:

- **Enhanced Security**: Since the business logic is separated from the presentation layer, sensitive data and processes are better protected.
- **Simplified Development and Maintenance**: Changes to the system can be made independently in each layer, simplifying the development process.
- **Improved Load Distribution**: The middle layer helps distribute the processing load, preventing the client or database from becoming bottlenecks.

This three-tier design is ideal for complex, distributed systems such as web applications, enterprise-level services, and cloud-based solutions. By utilizing this architecture, developers can create fast, scalable, and maintainable systems that cater to the needs of modern applications.

In summary, the three-tier architecture offers significant improvements over the two-tier model, making it a preferred choice for designing scalable and efficient systems. It helps ensure that business logic is processed efficiently, data are managed securely, and user interactions are handled smoothly, contributing to the overall success of the application.

7 Conclusion

The blockchain manager plays a crucial role in overseeing the creation of digital ledgers for subnetworks and managing access permissions. This ensures that only verified users have the ability to download shared data, further enhancing security. Our work presents a robust solution to secure communication within mobile edge computing (MEC) environments. By treating each subnetwork as its own blockchain, we are able to effectively manage data sharing and access control in a secure manner. This decentralized approach ensures that sensitive information is protected, as each subnetwork operates independently. Additionally, we securely send decryption keys to authorized users, enabling them to access shared data while maintaining strong security measures.

A key advantage of our approach is that it eliminates the need to store private keys, thereby reducing potential vulnerabilities. The system is designed with multiple modules; including user, blockchain manager, receiver server, and node server, which streamline user interactions and allow for seamless integration with existing MEC infrastructures. This modular design also simplifies the process of managing user permissions and data access. Our method also addresses common challenges in MEC environments, such as the need for frequent key resets. By implementing features like dynamic group joining and secret key sharing, we reduce the overhead and complexity associated with key management, making the system more efficient. This not only improves the user experience but also strengthens security by ensuring that users can join or leave subnetworks securely without compromising the integrity of the system.

Overall, our approach marks a significant advancement in addressing the security challenges of MEC. It provides a more secure, efficient, and scalable solution for mobile computing, paving the way for safer and more reliable communication in MEC networks.

References

1. Roman R, Lopez J, Mambo M (2018) Mobile edge computing, fog et al.: a survey and analysis of security threats and challenges. Futur Gener Comput Syst 78:680–698
2. Yi S, Qin Z, Li Q (2015) Security and privacy issues of fog computing: a survey. In: Proceeding international conference wireless algorithms, systems and applications, pp. 685–695
3. Yousefpoor MS, Barati H (2019) Dynamic key management algorithms in wireless sensor networks: a survey. Comput Commun 134:52–69
4. Messai ML, Seba H (2016) A survey of key management schemes in multi-phase wireless sensor networks. Comput Netw 105:60–74

5. Rao L, Zhang H, Tu T (2017) Dynamic outsourced auditing services for cloud storage based on batch-leaves-authenticated merkle hash tree. IEEE Trans Serv Comput 13(3):451–463
6. Wallner D, Harder E, Agee R (1999) Key management for multicast: Issues and architectures. Nat Secur Agency Tech Rep RFC2627
7. Daghighi B, Kiah MLM, Iqbal S, Rehman MHU, Martin K (2017) Host mobility key management in dynamic secure group communication. Wirel Netw 24:3009–3027
8. Li X, Jiang P, Chen T, Luo X, Wen Q (2020) A survey on the security of blockchain systems. Future Gener Comput Syst 107:841–853
9. Sun Y, Liu KJR (2007) Hierarchical group access control for secure multicast communications. IEEE/ACM Trans Netw 15(6):1514–1526
10. Je D-H, Lee J-S, Park Y, Seo S-W (2010) Computation-and-storage-efficientkey tree management protocol for secure multicast communications. Comput Commun 33(2):136–148
11. Queralta JP, Westerlund T (2021). Blockchain for mobile edge computing: consensus mechanisms and scalability. Mob Edge Comput 333–357
12. Rao PM, Deebak BD (2023) A comprehensive survey on authentication and secure key management in internet of things: challenges, countermeasures, and future directions. Ad Hoc Netw 146:103159

IoT-Based Fault Location and Detection of Underground Cables Using Enhanced Deep Learning Approach

Lenish Praimee, S. M. Manasa, and Raghu Ramamoorthy

Abstract As urban areas increasingly favor underground cables over traditional overhead transmission lines due to their numerous advantages, such as reduced line losses and better resilience to adverse weather conditions, the need to effectively detect and address potential shortcomings becomes paramount. With India's rapid progress and growing reliance on underground lines, there arises a pressing need for innovative solutions to streamline fault detection processes. In response, this paper proposes a novel approach that harnesses the power of Internet of Things (IoT) technology in conjunction with Google's extensive database infrastructure for fault detection in underground cable systems. Our method integrates state-of-the-art deep learning models, including Convolutional Neural Networks (CNNs) and Recurrent Neural Networks (RNNs), to enhance fault detection accuracy and efficiency. Through extensive experimentation, our results demonstrate that the incorporation of CNNs and RNNs significantly improves both the accuracy and efficiency of fault detection compared to traditional techniques. By leveraging IoT capabilities, Google's database, and deep learning methodologies, our proposed approach offers a comprehensive solution for fault detection in underground cable systems, thereby contributing to the advancement and reliability of urban infrastructure.

Keywords Underground cables · Fault detection · MCU WiFi and Liquid crystal display

L. Praimee
Department of Computer Science and Engineering, PES University, Bengaluru, India
e-mail: lenish.pj@pes.edu

S. M. Manasa
Department of Computer Science and Engineering (Data Science), RNS Institute Technology, Bengaluru, Karnataka, India

R. Ramamoorthy (✉)
Department of Computer Science and Engineering, The Oxford College of Engineering, Visvesvaraya Technological University, Belagavi, Karnataka, India
e-mail: raghuace85@gmail.com

© The Author(s), under exclusive license to Springer Nature Singapore Pte Ltd. 2025
H. Zen et al. (eds.), *Soft Computing and Signal Processing*, Lecture Notes in Networks and Systems 1200, https://doi.org/10.1007/978-981-97-9926-8_28

1 Introduction

Electricity is an indispensable aspect of contemporary life, shaping virtually every facet of our daily routines and driving the operations of numerous industries. In the domain of power systems, transformers assume a critical function in transmitting and distributing electric power. This distribution network frequently comprises both overhead and underground cables. Despite the benefits underground cables offer, including minimized visual disruption and enhanced resilience to harsh weather conditions, they are susceptible to aging and faults. The proposed approach aims to tackle these challenges by specifically targeting the identification and resolution of deficiencies within underground cable systems. At the core of this strategy lies Ohm's law, a fundamental principle governing the relationship between voltage, current, and resistance in electrical circuits. By introducing a low DC voltage at the feeder end of the cable system via a resistor setup, we can monitor changes in current flow indicative of faults along the cable's length. In practical terms, this means that when a fault occurs within the cable, such as a single line to ground fault, a double line to ground fault, or a three-phase to ground fault, the voltage drops across the series resistors adjust accordingly. These voltage variations are then detected by a specialized circuit module designed for fault identification. The current passing through this detection circuit is influenced by the cable's length from the location of the fault occurrence. Subsequently, this data undergoes processing by a microcontroller, which employs an internal Analog-to-Digital Converter (ADC) to convert the analog signal into digital format [1].

Following the microcontroller's analysis, the processed digital data is utilized to generate output information, which is then presented on a Liquid Crystal Display (LCD). This output typically encompasses the precise distance to the fault in kilometers, along with specific details regarding the type of fault detected. This method enables the accurate pinpointing of fault locations within the underground cable network, streamlining repair efforts for greater efficiency [2]. Furthermore, the incorporation of Internet of Things (IoT) technology elevates the functionality and accessibility of the fault detection system. Through IoT integration, the system can transmit fault data seamlessly over the internet, facilitating remote monitoring and analysis by relevant personnel. This connectivity is facilitated by Node MCU modules equipped with WiFi capabilities, serving as the conduit between the Arduino sensors monitoring the cable network and the internet.

Every Node MCU module is linked to a transformer, facilitating continuous monitoring of each transformer in real time. To ensure prompt detection and response to any irregularities or malfunctions within the system, Google's database infrastructure is employed to verify the status of transformer signals. This proposed approach seamlessly integrates principles of electrical engineering with state-of-the-art IoT technology to present a comprehensive solution for fault detection within underground cable systems [3]. By harnessing the principles of Ohm's law, coupled with microcontroller-based processing and IoT connectivity, the system provides precise fault localization, streamlined data transmission, and proactive maintenance capabilities. Ultimately, this integration enhances the reliability and efficiency of underground power distribution networks.

In our innovative approach, we tackle the intricate task of fault detection in underground lines by harnessing the capabilities of the Node MCU WiFi Module while integrating advanced deep learning techniques. This IoT-based model, enriched with deep learning methodologies, significantly enhances the accuracy and efficiency of fault detection in underground cables compared to conventional methods. By employing deep learning algorithms, the system can meticulously analyze data collected by the Node MCU WiFi module, identifying subtle patterns indicative of cable faults and enabling precise fault localization [4]. The operational speed of this system is notably swift, ensuring the uninterrupted continuity and stability of power quality.

Looking forward, we envisage broadening the application of this cutting-edge technique to encompass fault detection in power lines and transformers, with a focus on further refining the deep learning algorithms. Notably, our proposed system exhibits an impressive average accuracy of 94.53% for detecting faults in the Y B Phase and 98.63% for LL faults, respectively. The IoT-based underground fault detector, bolstered by its deep learning capabilities, is tailor-made to detect flaws within underground cable systems with pinpoint accuracy. This system excels in precisely pinpointing fault locations, seamlessly displaying them through the user-friendly Blynk IoT application. Users can effortlessly identify the exact meters at which faults occur, thanks to the intuitive display on the LCD screen. By leveraging the hotspot connection to the ESP32 and leveraging the functionalities of the Blynk IoT application, faults can be conveniently monitored and displayed in real time, empowering users with invaluable insights into the health of underground cable networks [5].

2 Related Work

Underground cables face numerous fault risks stemming from underground conditions, natural wear and tear, as well as external factors like rodent interference. The process of diagnosing these faults is challenging, often requiring the entire cable to be excavated for inspection and repair. To address this issue, the project aims to develop a method for pinpointing fault locations along underground cable lines from the base station, utilizing a PIC16F877A controller. The approach involves testing the cable for faults, and the prototype operates on the fundamental principle of Ohm's law.

Underground cable systems are widely employed in major metropolitan areas. However, when faults occur, pinpointing the exact location becomes challenging, complicating the repair process. The objective of this paper is to ascertain the distance of an underground cable fault from the base station, measured in kilometers, and relay this information over the internet. Underground cable systems are widely employed in major metropolitan areas. However, when faults occur, pinpointing the exact location becomes challenging, complicating the repair process. To address this issue, IoT technology is employed to precisely identify the fault location and transmit the data in graphical format to a dedicated website. Simultaneously, the fault location is displayed on an LCD screen. To address this issue, IoT technology is employed to precisely identify the fault location and transmit the data in graphical format to a dedicated website. Simultaneously, the fault location is displayed on an LCD screen [6].

Urban areas commonly adopt underground cable systems for power distribution. However, when faults occur in these cables, identifying the precise location poses a challenge, complicating the repair process. A system aimed at precisely pinpointing fault locations within underground cables. The system utilizes WiFi modules to transmit fault data to a dedicated website over the internet, providing timely information to authorized personnel at remote locations [7, 8].

The implementation of an IoT-based underground cable line fault detection system offers significant advantages in identifying faults and their locations efficiently. As the use of underground cables becomes widespread in power system grid development, these cables are susceptible to various faults arising from underground conditions, wear and tear, and rodent interference. Locating the source of a fault in underground cables is challenging, often requiring extensive excavation to inspect the entire line [9]. However, with the IoT-based system, fault detection becomes streamlined. Repair personnel can precisely pinpoint the location of the fault without the need for extensive excavation. This targeted approach saves considerable time and resources, enabling faster servicing of underground cable lines. Numerous methodologies exist for identifying faults in power lines, each aiming to enhance power quality and ensure uninterrupted power supply. Some notable methods include the Murray loop method and the Ohm's law method, among others. These approaches contribute to the ongoing efforts to reduce faults in power transmission lines, ultimately leading to improved power reliability and continuity [10, 11].

3 Underground Cable Fault Locator

The Underground Cable Fault Locator emphasizes the importance of understanding the location and route of the underground cable before attempting to locate faults, especially in the case of direct buried primary cables. If the issue arises with the secondary cable, having precise knowledge of its route becomes even more crucial. Without knowing the cable's location, it becomes exceedingly challenging to pinpoint a cable fault accurately. Therefore, it is prudent to thoroughly familiarize oneself with the cable's path before embarking on the fault-finding process. The demand for consistent service has driven the development of fault detection techniques. Over recent years, advancements in fault diagnosis have been enhanced through signal processing methods and their corresponding results. These developments have significantly improved the efficiency and accuracy of fault detection processes, ensuring more reliable and effective maintenance of underground cable systems.

Utilizing underground tunnels for power cable transmission offers numerous advantages. Firstly, underground tunnels provide a safer environment for power cables as they are shielded from various hazards present in above-ground installations. Additionally, this approach frees up surface space for further infrastructure development. Underground transmission also enhances reliability by mitigating disturbances commonly encountered by overhead cables, such as storms and natural disasters. Consequently, many cities worldwide, including Manhattan, Downtown San Francisco, Toronto, Chicago, Navi Mumbai, Jaipur, and Pune, are increasingly adopting this method.

However, underground cables come with their own set of challenges. Detecting faults in underground cables is notably difficult, leading to higher maintenance costs. Upgrading underground cables poses another significant challenge as it involves replacing entire supply lines, resulting in additional expenses. Moreover, maintenance workers face safety risks during cable detection and inspection [12]. Damage to live power cables can lead to injuries caused by arcing current and associated fire hazards, resulting in severe burns and potential electric shock if insulation sheathing is compromised. These safety concerns underscore the need for stringent safety protocols and advanced detection technologies in underground cable maintenance operations.

4 IOT-Based Underground Cable Line Fault Detection

The implementation of an IoT-based underground cable line fault detection system offers significant advantages in identifying faults and their locations efficiently. As the use of underground cables becomes widespread in power system grid development, these cables are susceptible to various faults arising from underground conditions, wear and tear, and rodent interference. Locating the source of a fault in underground cables is challenging, often requiring extensive excavation to inspect the entire line. However, with the IoT-based system, fault detection becomes streamlined. Repair personnel can precisely pinpoint the location of the fault without the need for extensive excavation. This targeted approach saves considerable time and resources, enabling faster servicing of underground cable lines. Overall, the IoT-based underground cable line fault detection system enhances the efficiency and effectiveness of fault detection and repair processes, ultimately contributing to improved reliability and performance of underground cable networks.

5 Methodology

In this method, we utilize Node MCU modules with WiFi capability to establish connectivity between Arduino sensors and the internet. The Node MCU serves as a crucial component, facilitating communication between the sensors and the online platform. To enable internet communication, a Hotspot is created through a router.

Each Node MCU is then connected to a transformer, ensuring individual monitoring for each transformer. Google's database infrastructure is leveraged to monitor the status signals of the transformers. In the event of a failure, the status signal is promptly sent. A detailed flowchart illustrates the step-by-step process of the method. The procedure commences with the use of the transfer and Node MCU WiFi module, constituting an IoT-based program. Each Node MCU WiFi [13] module is linked to a set of transformers, with one module dedicated to each transformer. The Node MCU WiFi module retrieves the signal status of the transformer sets and proceeds to the subsequent step. If power is successfully passing through the Node MCU, indicated by the reception of a signal from the transformer, the system detects that the line is safe. This information is then transmitted via WiFi connection, and a notification confirming the safety status is stored in the Google database. This comprehensive approach ensures efficient monitoring and management of transformer status, enhancing the reliability and responsiveness of the overall system.

In the event of a signal not being received by the Node MCU WiFi module, it indicates a fault occurrence, triggering a notification, and initiating appropriate action to address the fault promptly. Through the utilization of the Node MCU module WiFi connection, fault occurrences can be swiftly detected, ensuring efficient operation of the system. The architectural configuration process is pivotal in establishing a foundational structure for the system. A system architecture serves as a conceptual model delineating the structure, behavior, and various perspectives of the system. It elucidates how the system is utilized and its interactions with other systems and the external environment. An architecture description provides a formal representation of the system, structured to facilitate comprehension of its structures and behaviors. This description entails identifying the tangible components of the system and the interactions among these components. The initial phase of the design process involves identifying these subsystems and establishing a framework for subsystem control and communication. This phase, known as architectural design, yields a depiction of the software architecture.

The connection diagram illustrates the power transformation process from the power generation point to the distribution center via underground cables. Four distribution step-down transformers, labeled as 1, 2, 3, and 4, are depicted in the Fig. 1. Each transformer is connected individually for a specific area, with all transformers linked in parallel. A Node MCU WiFi module, integrated with a Google database system, is interconnected with the underground cables. Within this system, a set of MCU WiFi modules are employed, with one module connected to each transformer via underground cables. The Node MCU WiFi module receives the status signal of each transformer individually and forwards it to the notification center, where it is recorded in the Google database. Should the Node MCU fail to receive the status signal from any particular transformer, it indicates a fault occurrence within that specific line. This setup ensures efficient monitoring and recording of transformer statuses, enabling timely detection and response to any faults within the distribution network.

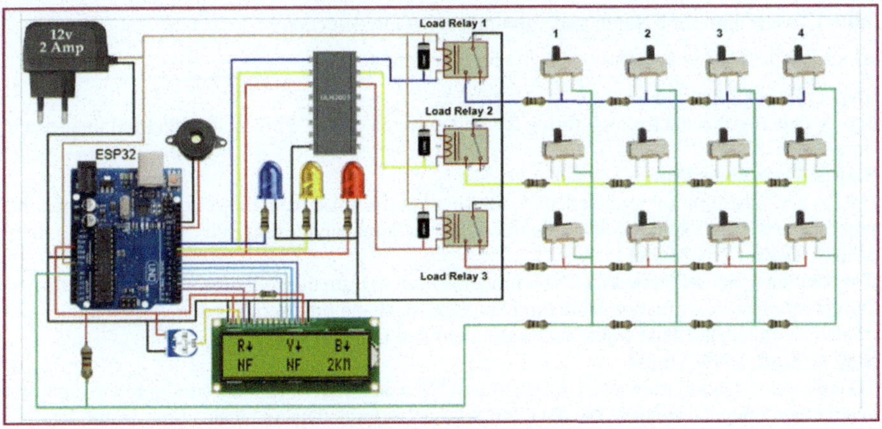

Fig. 1 Detection of the fault place from the underground cables through Node MCU WiFi module

6 CNN in Underground Cable Fault Detection

In underground cable fault detection, we have used Convolutional Neural Networks (CNNs) due to their remarkable effectiveness in processing image data. Underground cables are equipped with sensors that capture images of the cables' condition, the nature of the images are thermal images as shown in Fig. 2. CNNs analyze these images, as they are specifically designed to identify patterns or anomalies within visual data. By leveraging the hierarchical structure of convolutional layers, CNNs effectively extract features from these thermal images, enabling them to detect subtle irregularities or faults in underground cables with a high degree of accuracy.

In addition to CNNs, Recurrent Neural Networks (RNNs) and Long Short-Term Memory Networks (LSTMs) is used in processing time-series data collected from sensors monitoring underground cables. CNNs, process the spatial data, RNNs and LSTMs handle the sequential data from the sensors and the thermal images, making

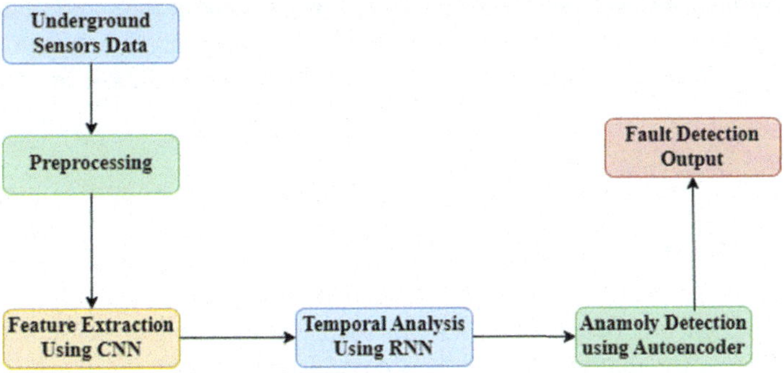

Fig. 2 Structure of CNN in underground cable fault detection

Table 1 Algorithm for CNN in underground cable fault detection

Algorithm 1: CNN in underground cable fault detection
Step 1: Data Collection Sensor data is collected for monitoring the underground cables. This include thermal images and time-series
Step 2: Preprocess Data Data is normalization for regular scales, filtering out the noise from the collected data. Then the preprocessed data is split the data into training, validation, and testing sets
Step 3: Train CNN Model Convolutional Neural Network (CNN) model is used to train the spilt data. CNN architecture is configured, feature extraction from thermal image data and time series is done. CNN model is trained using labeled data, where faults are identified in the training samples
Step 4: Train RNN Model Training set is used to train a Recurrent Neural Network (RNN). RNN is configured to capture any temporal dependencies in the data. RNN model is trained using labeled time-series data, where faults are identified in the training samples
Step: 5 Train Autoencoder Model Autoencoder model is trained using training dataset. Autoencoder is configured for anomaly detection by learning to reconstruct input data
Step: 6 Validate and Fine-Tune Models Validate the trained models. Step 3 and step 4 are repeated for fine-tuning the models as needed based on validation performance

them ideal for analyzing temporal sequences of sensor readings. Underground cable monitoring systems often generate large volumes of time-series data, capturing fluctuations in temperature, voltage, or current over time. The implemented RNNs and LSTMs capture temporal dependencies within this data, allowing them to detect deviations from expected patterns that indicating the presence of faults in the cables.

Furthermore, we have used autoencoders [14, 15, 16] that emerge as a valuable tool for anomaly detection in underground cable fault detection systems. Autoencoders are a type of neural network architecture trained to reconstruct input data, typically by compressing it into a lower-dimensional representation and then reconstructing it back to its original form. In the context of fault detection, autoencoders are trained on data representing normal operating conditions of underground cables as depicted in Table 1. Once trained, the autoencoder reconstruct input data and compare it to the original, learned representation. Deviations or discrepancies between the reconstructed data and the original input signify faults or anomalies in the cables [17, 18].

7 Experimentation and Results

The outcomes of our proposed work are focusing on using IoT for underground cable fault detection that would vary based on its specific objectives. Here are some potential results: Development of IoT sensors and devices: The project may lead to the creation of IoT sensors and devices designed specifically for monitoring underground cables. These devices could be equipped with various sensors to detect changes in temperature, pressure, or electrical conductivity, which could indicate potential faults in the cables as shown in Figs. 3, 4 and 5. Creation of data collection and analysis software: Another potential outcome could be the development of software capable of collecting and analyzing data from IoT sensors installed on underground cables. This software could process the data in real time, identifying patterns or anomalies that may indicate a fault in the cables as shown in Fig. 3. Additionally, it could provide real-time updates on the status of the cables to maintenance personnel or operators.

Implementation of fault detection algorithms: The project may result in the creation of algorithms specifically designed to detect faults in underground cables. These algorithms could analyze data collected from IoT sensors to identify signs of potential faults, such as changes in electrical resistance or voltage levels. Furthermore, they could use this data to pinpoint the location of the fault along the length of the cable, facilitating faster and more accurate repairs. Overall, the project's outcomes would likely involve a combination of hardware (such as sensors and

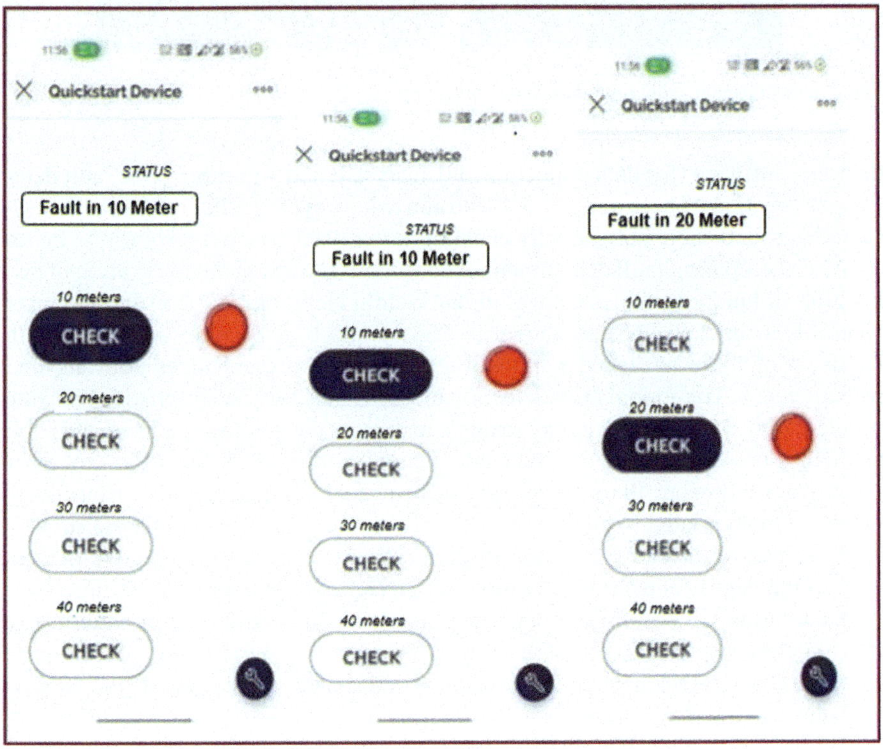

Fig. 3 Application designed to detect the fault in underground cables

Fig. 4 Implementation of the fault detection system and its notification

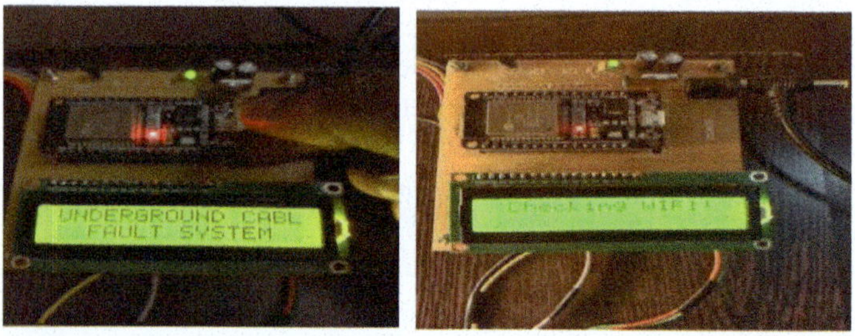

Fig. 5 Underground cable fault detection using WiFi

devices), software (for data collection and analysis), and algorithms (for fault detection and location). These outcomes would aim to improve the reliability and efficiency of underground cable networks by enabling faster detection and response to faults.

When comparing fault detection methods with and without the integration of deep learning techniques such as Convolutional Neural Networks (CNNs) and Recurrent Neural Networks (RNNs) as shown in Fig. 6a and b, it becomes evident that the inclusion of these advanced methodologies significantly enhances both accuracy and efficiency. Traditional fault detection methods may rely on simplistic algorithms or rule-based systems, which may struggle to capture the complex patterns and subtle anomalies present in underground cable data. However, CNNs excel in processing image data, allowing them to extract intricate features and patterns from sensor images or data related to cable conditions. By leveraging the hierarchical structure of convolutional layers, CNNs can discern subtle irregularities indicative of faults with a high degree of accuracy. Similarly, RNNs are adept at processing time-series data, capturing temporal dependencies and deviations from expected patterns over time. Underground cable monitoring systems generate vast amounts of time-series data, and RNNs excel in detecting deviations in this data that may signal the presence of faults.

Through the integration of CNNs and RNNs, fault detection systems can achieve superior accuracy as shown in Fig. 6a by accurately identifying faults and minimizing false positives. Moreover, the efficiency of fault detection is enhanced as these deep

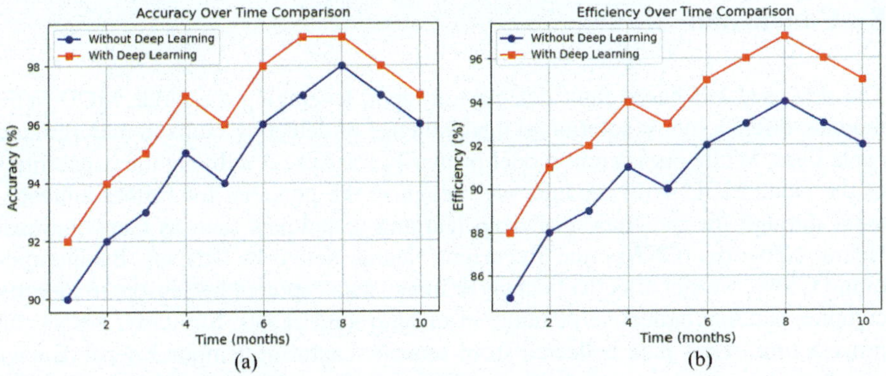

Fig. 6 Comparison of accuracy (**a**) and efficiency (**b**) of fault detection in underground cables with and without using deep learning models

learning models can process large volumes of data quickly and effectively as shown in Fig. 6b, enabling real-time detection and response to cable faults. Overall, the incorporation of CNNs and RNNs into fault detection methodologies leads to a significant improvement in both accuracy and efficiency compared to traditional methods. The same parameters with its comparison are shown in Table 2.

Table 2 Comparison of accuracy and efficiency of fault detection in underground cables

Time (months)	Accuracy (%) without deep learning	Accuracy (%) with deep learning	Efficiency (%) without deep learning	Efficiency (%) with deep learning
1	90	92	85	88
2	92	94	88	91
3	93	95	89	92
4	95	97	91	94
5	94	96	90	93
6	96	98	92	95
7	97	99	93	96
8	98	99	94	97
9	97	98	93	96
10	96	97	92	95

8 Conclusion

The proposed IoT-based fault detection system, leveraging the Node MCU WiFi module, offers a robust solution to the challenge of detecting faults in underground cable lines. While our current model primarily focuses on utilizing the capabilities of the Node MCU WiFi module, we recognize the potential for further enhancement through the integration of deep learning techniques such as Convolutional Neural Networks (CNNs) and Recurrent Neural Networks (RNNs). By incorporating CNNs, we can effectively analyze image data captured by sensors to identify complex patterns indicative of faults in underground cables. Similarly, RNNs can process time-series data collected from sensors, capturing temporal dependencies and deviations from expected patterns that may signal the presence of faults. This integration of deep learning methodologies holds promise for improving the accuracy and efficiency of fault detection in underground cable systems. Looking ahead, we envision expanding the capabilities of our fault detection system by integrating CNNs and RNNs to further enhance fault detection accuracy and efficiency. By harnessing the power of deep learning, we can achieve even greater precision in identifying faults within underground cables, ultimately contributing to the continuity and stability of power quality. Additionally, the integration of various sensors in our system will enable us to extend fault detection capabilities to power lines and transformers, further enhancing the resilience of electrical infrastructure. With an average accuracy of 94.53% for detecting faults in the Y B Phase and 98.63% for LL faults, our system demonstrates promising potential for real-world deployment. Through continuous innovation and refinement, we aim to establish our IoT-based fault detection system as a reliable and indispensable tool for ensuring the reliability and integrity of underground cable networks.

References

1. Hans MR, Kor SC, Patil AS (2017) Identification of underground cable fault location and development. In: 2017 International conference on data management, analytics and innovation (ICDMAI)
2. Teresa VV, Rajeshwaran K, Kumar SS, Vishnupriyan S, Dhanasekaran S (2022) IoT-based underground cable fault detection. In: International conference on augmented intelligence and sustainable systems (ICAISS), pp 1094–1100
3. Manasa SM, Giri L, Tanuja R, Manjula SH, Venugopal KR (2022) TSPS: trusted service path selection in social networks using user profile. J Tianjin Univ Sci Technol 65(1):1–20
4. Mnyanghwalo D, Kundaeli H, Kalinga E, Hamisi N (2020) Deep learning approaches for fault detection and classifications in the electrical secondary distribution network: methods comparison and recurrent neural network accuracy comparison. Cogent Eng 11(1): 185–500
5. Yamuna R, Harsharani KS, Manasa SM, Sathya M, Pramiee L, Kumari A (2023) IWQMA: intelligent water quality management in aquaculture using IoT technology. TWIST 18(4):183–198
6. Zubair AH, Younis KS (2023) Challenges and solutions of Artificial Intelligence-based fault location methods in power system lines. J. Auton. Intell. 6:2

7. Kamil D, Hilmi D, Fadi A-T (2021) Deep learning, machine learning and internet of things in geophysical engineering applications: an overview. Microprocess Microsyst 80:103613
8. Ye J, Lv J, Xu G, Liu T (2024) Leaky cable perimeter intrusion detection based on deep reinforcement learning. IEEE Internet Things J
9. Son LH, Jha S, Kumar R, Chatterjee JM, Khari M (2018) Collaborative handshaking approaches between internet of computing and internet of things towards a smart world: a review from 2009–2017. Telecommun Syst 70(4):617–634
10. Khari M, Garg AK, Gandomi AH, Gupta R, Patan R, Balusamy B (2020) Securing data in internet of things (IoT) using cryptography and steganography techniques. IEEE Trans Syst Man Cybern: Syst 50(1):73–80
11. Numair M, Aboushady AA, Arraño-Vargas F, Farrag ME, Elyan E (2023) Fault detection and localisation in LV distribution networks using a smart meter data-driven digital twin. Energies 16(23):7850
12. Zaben M, Worku MY, Hassan MA, Abido MA (2024) Machine learning methods for fault diagnosis in AC microgrids: a systematic review. IEEE Access
13. Soothar KK, Chen Y, Magsi AH, Hu C, Shah H (2024) Optimizing optical fiber faults detection: a comparative analysis of advanced machine learning approaches. Comput Mater Contin 79(9)
14. Ashwin M, Kumar ES, Naidu RCA, Ramamoorthy R (2023) IoT based innovative teaching learning using smart class rooms. In: 2023 International conference on sustainable computing and data communication systems (ICSCDS), pp 1143–1148. IEEE
15. Huang MQ, Ninic J, Zhang Q (2021) BIM, machine learning and computer vision techniques in underground construction: current status and future perspectives. Tunnelling Undergr Space Technol 108:103677
16. Zhang C, Kang X, MA X, Jiang S, Qu X (2016) Shaanxi key laboratory of smart grid. On-line incipient faults detection in underground cables based on single-end sheath currents. IEEE PES Asia-Pac Power Energy Conf
17. Velu A, Ramamoorthy R, Kumar S, Shruthi K (2023) IoT enabled smart farming: a controlled environment agriculture application. In: 2023 international conference on sustainable communication networks and application (ICSCNA), pp 292–298. IEEE
18. Ashwin M, Naidu RCA, Ramamoorthy R, Kumar ES (2023) IoT-based smart wearable devices using very large scale integration (VLSI) technology. In: International conference on soft computing and signal processing, pp 155–164. Springer Nature Singapore, Singapore

Comprehensive Study on Authentication Methods in Wireless Networks

Gotte Ranjith Kumar and K. Suresh Babu

Abstract Mobile devices, empowered by advanced wireless networks, have revolutionized information sharing and productivity across industries. However, this ubiquitous connectivity exposes user privacy and security to risks. The core vulnerability lies in the lack of authentication between mobile devices and DHCP (Dynamic Host Configuration Protocol) servers, which assign network configurations. This research explored existing authentication mechanisms in wireless networks to address these DHCP-related security concerns, highlighting the essence of secure data communication in mobile environments. The majority of the authentication mechanisms rely on centralized environments. Some rely on a cryptographic security key, which needs to be exchanged during authentication. Blockchain-based authentication is proposed. This technology is distributed, scalable, and robust.

Keywords Mobile networks · Authentication · DHCP server · Blockchain · Cryptography

1 Introduction

As advancements in wireless network technology progress, the utilization of mobile devices has become increasingly widespread across various industries. These devices have transformed the way individuals share information, enabling users to connect and interact from virtually any location at any given time. This enhanced level of connectivity has significantly enhanced productivity and efficiency in numerous

G. R. Kumar (✉)
Research Scholar, Department of Computer Science and Engineering, Jawaharlal Nehru Technological University, Kukatpally, Hyderabad, Telangana 500085, India
e-mail: ranjeet.kits@gmail.com

Assistant Professor, School of CS&AI, SR University, Warangal, Telangana 506371, India

K. S. Babu
Professor, Department of Computer Science and Engineering, Jawaharlal Nehru Technological University, Kukatpally, Hyderabad, Telangana 500085, India
e-mail: kare_suresh@jntuh.ac.in

© The Author(s), under exclusive license to Springer Nature Singapore Pte Ltd. 2025

H. Zen et al. (eds.), *Soft Computing and Signal Processing*, Lecture Notes in Networks and Systems 1200, https://doi.org/10.1007/978-981-97-9926-8_29

sectors. Despite the advantages of mobile devices, there are notable risks to user privacy and security. Relying on wireless networks and DHCP servers for acquiring TCP/IP configuration parameters exposes mobile users to potential threats from malicious actors. DHCP dynamically allocates IP configuration such as IP address to mobile devices [1]. The DHCP server implementation involves the transmission of DHCP packets between virtual machines and node agents on a server node, without transmitting the packets outside the server node [2]. The present invention introduces a DHCP server and method for network internal device management, which accurately performs address distribution and reduces the probability of IP address conflicts in the network [3]. A key vulnerability in DHCP is the absence of authentication measures in the interactions between mobile devices and DHCP servers [4]. This lack of authentication creates opportunities for attackers to interfere communications, jeopardizing user privacy and potentially gaining illegal access to sensitive data. To address these DHCP-related risks, organizations and individuals must implement security protocols to safeguard against potential attacks [5]. This may involve deploying secure DHCP protocols that integrate robust authentication mechanisms. This study offers an overview of various authentication mechanisms in wireless networks utilized to authenticate participating devices in communication. Ensuring secure data communication in mobile networks through authentication is essential. The majority of authentication mechanisms are based on a centralized environment, where the failure of a single point can result in the collapse of the entire authentication system. Certain authentication methods rely on cryptographic security keys that are exchanged between a mobile device and a server during authentication. This dependency on security keys leads to centralization of the authentication process. Therefore, a blockchain technology-based authentication system is proposed, which is distributed, scalable, and robust. Our research work is organized in this paper is organized as: Sect. 2 reveals study of various authentication mechanisms in wireless networks and their limitations. Section 3 reveals summary of key authentication algorithms used for authentication in wireless networks. Section 4 reveals conclusion of the survey on authentication mechanisms.

2 Authentication Mechanisms

Several papers propose different authentication schemes to address this issue. Sangeetha and Sathappan [6] proposed a reputation-based symmetric key authentication (RSKA) mechanism for improving data confidentiality in MANETs. The RSKA technique utilizes a reputation count of a node to verify the authenticity of mobile nodes within the network. RSKA employs a node authentication algorithm based on reputation count to verify mobile nodes within the network. The authentication process takes into account the reputation value of mobile nodes. Through the reputation count-based algorithm, nodes are assessed for their behavior and reliability by analyzing their previous actions and interactions in the network. By utilizing the reputation value, RSKA identifies cooperative nodes in MANETs for transmitting data, guaranteeing that only reliable nodes participate in the communication. This authentication method, based on reputation, enhances data transmission safety in MANETs and minimizes the chances of unauthorized access or malicious behavior. They use a reputation count-based node authentication algorithm and symmetric key cryptography to enhance data confidentiality and protect data delivery. The following are few of the limitations of [6]:

- RSKA is vulnerable to reputation manipulation or false reputation values, which can compromise node authentication.
- The accuracy and reliability of reputation values assigned to nodes in RSKA may be challenging to determine in dynamic and unpredictable MANET environments.
- RSKA assumes the cooperation and trustworthiness of all nodes in the network, which may not always be the case, leaving room for potential security breaches.
- The secure distribution of secret keys to all participating nodes in large-scale MANETs, as required by RSKA, can be a logistical challenge.
- RSKA does not address other security concerns such as denial-of-service attacks, routing attacks, or network partitioning, which may still pose risks to the overall security of the MANET.
- The efficiency of RSKA in terms of data confidentiality, throughput, and data loss rate may vary based on the network conditions and parameters used in the simulation

Bhattacharya et al.[7] also focus on mitigating phishing attacks in online social networks (OSNs) and introduced a secure and efficient authentication system based on cryptography called authentication scheme for phishing attack(ASPA)–mobile online social network(mOSN). The increased use of social networks can be attributed to the advancements in smartphone technology and mobile Internet. However, this growth has also resulted in a rise in cyberattacks, including phishing schemes and spoofed accounts, which aim to steal users' credentials and sensitive information. Phishing attacks typically involve the dissemination of fraudulent communications, such as deceptive emails, that masquerade as legitimate sources. Developing an authentication protocol to address phishing attacks in online social networks poses a significant challenge for researchers. The study introduces ASPA-mOSN, a secure and efficient authentication system based on cryptography. This system provides strong protection against phishing and similar threats in OSNs. The security of ASPA-mOSN is rigorously evaluated in informal as well as formal security analysis techniques. Furthermore, a comprehensive comparison of the scheme with existing alternatives is conducted, considering factors such as how it is security, its functionality, computational cost, and communication costs. Remarkably, ASPA-mOSN surpasses its competitors in all these aspects.

The following are few of the limitations of [7]:

- The evaluation of ASPA-mOSN mechanism in real-world scenarios is not comprehensive in the paper, thus hindering a full understanding of its practical effectiveness.

- The security analysis conducted with the ProVerif simulation tool may not cover all potential security vulnerabilities and real-world attack scenarios.
- Additionally, the paper fails to address the impact of the proposed scheme on user experience and usability in OSNs. It is crucial to take into account usability aspects to ensure user acceptance and adoption.
- Furthermore, the comparison with existing schemes is limited to security, functionality, computation, and communication costs, neglecting important factors like scalability, flexibility, and compatibility with various OSNs.

In their study, Casari et al. [8] conducted an exploration of physical layer authentication in underwater acoustic networks with mobile nodes. The focus was on utilizing the power-weighted average of channel taps' arrival delay as the primary authentication characteristic. The research aimed to investigate the application of physical layer level authentication in such networks, with an emphasis on this specific feature. To address the challenge of mobility, the authors developed a Kalman filter mechanism to track the time-to-time developments of the authentication features. This approach calculated an innovation metric for each new transmission, enabling verification of the signal source's authenticity. Through simulation results, the study demonstrated the effectiveness of the authentication mechanism in distinguishing between legitimate and fraudulent transmitters. Additionally, the effect of the authentication performance was examined based of the transmitter speed. Overall, taps' arrival delay served as the primary authentication feature in this research. Limitations of this arrival delay are as follows:

- Lack of robustness: It may not be robust enough to handle variations in the channel, such as multipath propagation, noise, and changing water conditions.
- Limited discriminatory power: Using only this as the authentication feature may not provide enough discriminatory power to accurately distinguish between legitimate and impersonating transmitters in all scenarios.
- Vulnerability to attacks: It can potentially be manipulated or spoofed by an attacker, compromising the effectiveness of the authentication mechanism.

Wang et al. [9] proposed the need for secure authentication in the Internet of Vehicles (IoV) in mobile heterogeneous networks. Existing authentication mechanisms in the network layer rely on public key infrastructure and interactive key agreement protocols, and IoV scenarios with network state changes from time to time are not compatible with these particular options. The suggested approach is a non-interactive authentication scheme that is driven by data and designed to support mobility and multi-homing. Its objective is to fulfill the lightweight data security needs of the Internet of Vehicles (IoV). By utilizing context parameters during data exchange, the plan ensures device authentication and non-interactive key agreement. This proposed scheme effectively minimizes authentication and handover delay, enhances bandwidth aggregation efficiency, and prevents data blocking resulting from signaling interaction processes. The limitations of this authentication scheme are the following:

- The scalability and management of a many devices in the IoV scenario could be constrained by the limitations of this authentication scheme. The scheme's efficiency might be impacted by the rapid movements of devices in IoV environments and the intricate multi-access heterogeneous resources.
- The scheme's effectiveness in handling network state fluctuations and maintaining secure communication in a highly dynamic environment needs further evaluation.
- The scheme's ability to handle potential security threats and attacks specific to the IoV scenario may require further investigation.

Safdar et al. [10, 11] addressed limitations of existing authentication techniques wireless networks and key management techniques when applied to mobile ad hoc networks. The uniqueness in the characteristics of MANETs, such as untrusted wireless media, node mobility, and insufficient infrastructure, pose challenges to security [12]. The limitations of these techniques include the inability to directly employ them in MANETs due to fundamental differences in network architecture and requirements [13]. MANETs require encryption of communication contents and mutual authentication between nodes for adequate security. The existing techniques rely on trusted third parties or centralized servers, which are not feasible in MANETs because of the absence of infrastructure and regular mobility. To address these limitations, self-organized, hop-by-hop public key management schemes based on transitive trust between mobile nodes have been proposed [14]. These schemes allow nodes to create and manage their own public keys, issue digital certificates to neighboring nodes, and authenticate public keys using direct and recommendation trust. One approach is the multipart trust-based public key management (MTPKM) scheme, which establishes a trust threshold for nodes and relies on trusted neighbor nodes to certify public keys [15]. These schemes have the following limitations:

- Unsupervised methods utilized in these strategies are hindered by their reduced efficacy stemming from the absence of guidance on which data is pertinent for detecting anomalies.
- They frequently give priority to local contextual information, such as the neighborhoods of nodes that are one or two hops away, while disregarding the broader global context.
- These schemes may not support trusted third parties, which can lead to security issues [16].
- The trust-based models used in these schemes require a mutual trust relationship between the node which is requesting and the node which is certificate issuer, which can be challenging to establish [17].

Cheng et al. [18] focused on the challenges of privacy and reliability in mobile crowdsensing (MCS) in vehicular networks. Existing schemes for unidentified authentication in MCS have high computation overhead and communication overhead and do not consider that the level of sensing vehicle's reliability. The proposed scheme aims to address these issues by introducing "an efficient anonymous authentication and privacy-preserving reliability evaluation (AARE) technique" for MCS in vehicular networks [19]. The scheme incorporates an anonymous authentication approach for achieving the anonymity in authenticating sensing vehicles, along with a privacy-preserving reliability evaluation algorithm for assessing the vehicle's reliability. The authors have also addressed the obstacles related to ensuring reliability in dynamic vehicular networks and have introduced a precise reputation update algorithm. This proposed scheme enhances authentication efficiency, minimizes communication over- head, and improves the precision of reputation values for sensing vehicles. The limitations of the proposed scheme AARE for MCS in vehicular networks are as follows:

- The scheme assumes a static vehicular network environment and does not address the challenges and limitations that may arise in dynamic and highly mobile vehicular networks.
- The impact of various types of attacks or adversarial behaviors on the reliability evaluation process is not considered.
- The scheme does not take into account the potential impact of network congestion or resource limitations on the authentication and reliability evaluation processes.

Wang et al. [20] proposed a non-interactive multi-homing security authentication (NIMSA) mechanism for vehicular communications in mobile heterogeneous networks. The existing network layer authentication schemes depend on the public key infrastructure system and require key negotia tion, leading to bandwidth consumption and additional delay [21]. The NIMSA scheme is a security authentication scheme for the network layer that operates without the need for signaling information. It is triggered by data forwarding and utilizes a lightweight identity-based non-interactive key agreement strategy. This scheme is particularly suitable for scenarios involving mobile and multipath parallel transmission. In experimental comparisons with Internet Key Exchange version 2 (IKEv2) and its extension scheme, the NIMSA scheme demonstrated shorter authentication and handover delays, and reduced data transmission delay. Additionally, it exhibited a more effective bandwidth aggregation effect in multipath parallel transmission scenarios. Limitations of Non-Interactive Multi-homing Security Authentication Scheme are as follows:

- The NIMSA scheme is based on the bilinear pairing system, which may introduce computational overhead and complexity in implementation.
- The system is dependent on the pre-assigned private key, device ID, network adapter serial number, and IP address details to create the data authentication key. Relying on such specific device information could potentially restrict the scheme's adaptability and expandability within dynamic network settings.
- While the NIMSA scheme is designed to be lightweight and efficient, it may still incur some overhead in terms of authentication and handover delay, especially in scenarios with high-speed mobile heterogeneous networks.
- The scheme's effectiveness and performance may also depend on the specific network conditions and the availability of multiple network resources for multi-homing.

Shashidhara et al. [22] introduced a safety mutual authentication protocol for global movement in mobile networks utilizing lightweight cryptographic primitives. The protocol has several phases, namely initialization phase, registration phase, mutual authentication phase, and password change phase. By following this protocol, users can enjoy the assurance of anonymity, privacy, and security within global mobility networks. The protocol's correctness is validated through the use of the AVISPA tool, a widely used formal verification tool for internet security protocols and applications. An analysis of performance is conducted to assess the computational and communication overheads of the authentication system, demonstrating its minimal impact. The performance evaluation suggests that the suggested authentication system demonstrates minimal computational and communication burdens.

Limitations of secure mutual authentication protocol for global roaming in mobile networks are the following:

- The scalability of the secure mutual authentication protocol may be limited, especially in scenarios with a large number of roaming users. This can result in increased computational and communication overheads.
- The proposed protocol relies on lightweight cryptographic primitives, which may have limitations in terms of their security strength and resistance against advanced attacks.
- While the proposed authentication mechanism aims to have low computational and communication overheads, there may still be trade-offs between security and performance, depending on the specific implementation and network conditions.

Several authentication schemes have been proposed, but most of them are prone to security attacks [22]. Researchers have proposed various schemes that focus on mutual authentication and key agreement in global mobility networks [23]. However, some of these schemes have been found to have security loopholes and vulnerabilities [24, 25]. To overcome these shortcomings, a privacy-preserving authentication scheme based on elliptic curve cryptography has been proposed, which satisfies practical security requirements such as user anonymity and un-traceability [26]. Elliptic curve cryptography (ECC) is a asymmetric key cryptography approach based on the algebraic structure of elliptic curves over finite fields. ECC offers better security and performance compared to traditional cryptosystems, as it allows for smaller key sizes while maintaining the same level of security [27]. It also provides additional structure that can be exploited to create cryptosystems with unique features, such as identity-based encryption and pairing-based cryptographic protocols [28]. ECC is currently being used in digital certificates and key exchange, offering robust security with smaller keys compared to RSA [29]. The mathematics of ECC is more challenging than RSA, but it provides increased speed and decreased key size for a given level of security. ECC is gaining favor as an efficient and attractive alternative to conventional public key cryptosystems in certain applications [30]. This scheme has been proven to be secure and efficient in terms of computation cost and security. Limitations of Elliptic curve cryptography are as follows:

- The trade-off between performance and security in resource-limited environments, such as sensors and IoT devices [31].
- The complexity and cost of using sufficient key length in existing public key encryption schemes, which limits the protection offered by current security systems [32].
- ECC also requires significant memory for complete addition formulae, making it less suitable for compact devices [33].
- The base equation that defines the security of ECC is in the form of a linear equation with one unknown, which can be easily solved, making it a weak point in the algorithm [34].
- ECC implementations may be vulnerable to side-channel analysis attacks, such as simple SCA or horizontal address-bit DPA attacks [35].

Hu et al. [36] A novel blockchain-based anonymous handover authentication scheme for mobile networks has been proposed. The scheme aims to ensure fast and secure handoff of mobile nodes among access points in wireless mobile networks (WMN). It takes into consideration the limited resources and low computing capability of mobile nodes in WMN and addresses the drawbacks of traditional authentication schemes by using distributed storage, collective maintenance, and tamper-resistance. The scheme utilizes chameleon hash with blockchain for robustness and high efficiency, while also providing user anonymity, conditional privacy protection, and robust key agreement [37]. The proposed scheme has lower implementation complexity compared to existing handover authentication schemes based on bilinear pairing and ECC, and offers mutual authentication and greater security against known attacks [38]. It also outperforms state-of-the-art authentication mechanisms in terms of computational, communication, and storage costs, making it a valuable contribution to improving the security of communication in Internet of Health Things (IoHT) networks [39]. Additionally, an anonymous handover authentication scheme is proposed for 5G networks integrated with mobile edge computing (MEC) and software defined network (SDN), which supports multi-user access, lightweight user grouping, and internal group authentication. The scheme utilizes the aggregation message authentication code with detection function (AMAD) technique to reduce signaling and effectively resist DoS attacks [40].

A decentralized framework for roaming authentication in mobile vehicular networks is proposed, which addresses the vulnerabilities of existing protocols that rely on centralized authentication servers. By leveraging blockchain technology and smart contracts, this framework ensures secure and real-time communication while eliminating the risks of single point failure and roaming fraud [41]. The proposed scheme includes user/AP (access point) registration, authentication, and revocation processes, and utilizes the Bloom filter for efficient revocation [42]. Additionally, an unforgeable and undeniable billing scheme based on hash chain technology is designed [43]. The analysis shows that the proposed scheme provides the expected security features with an acceptable authentication delay [44].

Another smart contract-based authenticated key agreement (AKA) scheme is designed for vehicular ad-hoc networks (VANETs) to address the single-point-of-failure issue of traditional VANETs AKA schemes [45]. The suggested protocol meets the criteria for session key semantic security as well as the fundamental security and privacy prerequisites. An authenticated key agreement (AKA) scheme for vehicular ad hoc networks (VANETs) plays a crucial role in ensuring secure communication among vehicles and infrastructure. These schemes involve various steps such as node authentication, key agreement, and utilizing cryptographic techniques to establish secure session keys. To address the computational and communication overheads associated with conventional cryptographic methods, innovative approaches like incorporating wireless channel features for re-authentication have been proposed. Additionally, the use of lightweight authentication protocols based on Elliptic curves cryptography (ECC) with batch verification has been suggested to enhance efficiency and anonymity while ensuring security against attacks. By combining techniques like Dynamic Time Warping (DTW) conditional privacy-preserving authentication and Short Group Signature (SGS), [46] AKA schemes can offer enhanced security, privacy, and efficiency in VANETs. Limitations of authenticated key agreement (AKA) schemes for VANETs include heavy computational overheads, vulnerability to security weaknesses, and the need for frequent involvement of Trusted Authorities (TAs) [19]. To address these limitations, novel approaches have been proposed, such as using blockchain technology and Chinese remainder theorem for lightweight group key agreement, or implementing physically secure and conditional privacy authenticated key agreement schemes that do not require TA involvement during authentication in high-speed VANET environments [47].

3 Summary of Authentication Mechanisms

S. No.	Title	Methodology	Limitation
1	"Reputation based symmetric key authen tication for secure data transmission in mobile ad hoc networks"	Reputation-based symmetric key authentication (RSKA) technique	RSKA protocol is vulnerable to reputation manipulation, lacks reliability in dynamic environments, and assumes trust in all nodes, making it susceptible to security breaches. Distributing secret keys in large networks and not addressing other security concerns are further limitations. RSKA's performance depends on network conditions

(continued)

(continued)

S. No.	Title	Methodology	Limitation
2	"Spa-mosn: An efficient user authentication scheme for phishing attack detection in mobile online social networks"	Authentication scheme for phishing attack (ASPA)-mobile online social network (mOSN)	The research paper lacks evaluation on real-world usability, effectiveness against phishing attacks, scalability, and compatibility. Additionally, the security analysis might not be comprehensive
3	"Physical layer authentication in underwater acoustic networks with mobile devices"	Power-weighted average of channel taps' arrival delay as the main authentication feature	The underwater acoustic channel variation and limited differentiating factors make this authentication method unreliable. It is also vulnerable to manipulation by attackers
4	"A Data-Driven Noninteractive Authentication Scheme for the Internet of Vehicles in Mobile Heterogeneous Networks"	Data-driven non inter-active authentication scheme	This authentication method might struggle with a large number of vehicles and the dynamic nature of IoV environments. Its effectiveness in maintaining secure communication and handling IoV-specific threats needs further evaluation
5	"Mtpkm: Multipart trust based public key management technique to reduce security vulnerability in mobile ad-hoc networks"	Self-organized, hop-by-hop public key management schemes based on transitive trust between mobile nodes, e.g., MTPKM	Unsupervised anomaly detection in these schemes struggles because it lacks clear sig nals for what's normal. Additionally, they focus on limited local data and might not consider the bigger picture. Finally, trust-based models can be vulnerable due to challenges in establishing trust between nodes
6	"An efficient identity authentication scheme with dynamic anonymity for VANETs"	AARE scheme for MCS in vehicular networks	Suffers from computation overhead and communication over- heads, as well as the accuracy in reputation values of sensing vehicles
7	"NIMSA: Non-interactive Multihoming Security Authentication Scheme for vehicular communications in Mobile Heterogeneous Networks"	Non-interactive multi-homing security authentication (NIMSA) scheme for vehicular communications in mobile heterogeneous networks	The NIMSA scheme may be complex to implement and inflexible due to reliance on specific device information. It might also introduce delays in authentication and handover, especially in high-speed mobile networks

(continued)

(continued)

S. No.	Title	Methodology	Limitation
8	"A privacy preserving authentication scheme for roaming services in global mobility networks"	Elliptic curve cryptographic authentication scheme	Balancing performance and security is hard for resource-limited devices. Existing encryption schemes are complex and memory-intensive, while Elliptic curve cryptography (ECC) has security vulnerabilities and implementation weaknesses
9	"A novel blockchain-based anonymous handover authentication scheme in mobile networks"	Blockchain-based anonymous handover authentication scheme	Current authentication methods in wireless mobile networks have weaknesses. They don't protect user privacy in mobile crowdsourcing, lack security against attacks in sensor net- works, and have issues with centralized control in mobile cloud computing. This calls for better authentication schemes that prioritize privacy, security, and trust
10	"Bilinear pairing-based access control and key agreement scheme for smart transportation"	Authenticated key agreement (AKA) schemes	Heavy computational overheads, vulnerability to security weaknesses, and the need for frequent involvement of Trusted Authorities (TAs)

4 Conclusion

With the continuous rise in mobile devices, the focus has shifted toward mobile networks. However, this growth has also brought about an increase in attacks and malicious devices within these networks. To address these vulnerabilities, it is crucial to ensure proper authentication of mobile devices upon joining these networks. Currently, most authentication methods rely on cryptographic keys and centralized architecture, which unfortunately results in a single point of failure. Additionally, certain mechanisms impose computational overhead on the network. To tackle these issues, this research proposes a blockchain-based authentication mechanism for mobile networks, aiming to eliminate the single point of failure and reduce computational overhead.

References

1. Pradana DA, Budiman AS (2021) The dhcp snooping and dhcp alert method in securing dhcp server from dhcp rogue attack. IJJD (International Journal on Informatics for Development) 10(1):38–46. https://doi.org/10.14421/IJID.2021.2287

2. Lee Y, Lu S, Zan X, Firestone DM, Chandrappa HK, Ingle A, Kumaran J (2020) On-node DHCP implementation for virtual machines
3. DHCP Server and Method for Network Internal Device Management by Employing Same., 2019
4. Alsaadi RR, Saad D (2021) Security dhcp server on LAN network (2021)
5. Pradana DA, Budiman AS (2021) The dhcp snooping and dhcp alert method in securing dhcp server from dhcp rogue attack. IJID (International Journal on Informatics for Development). https://doi.org/10.14421/ijid.2021.2287
6. Sangeetha S, Sathappan S (2023) Reputation based symmetric key authentication for secure data transmission in mobile ad hoc networks. EAI Endorsed Trans Scalable Inf Syst 10(4):3. https://doi.org/10.4108/eetsis.v10i3.1940
7. Bhattacharya M, Roy S, Chattopadhyay S, Das AK, Jamal SS (2023) Aspa-mosn: an efficient user authentication scheme for phishing attack detection in mobile online social networks. IEEE Syst J 17(1):234–245. https://doi.org/10.1109/JSYST.2022.3168234
8. Casari P, Ardizzon F, Tomasin S (2022) Physical layer authentication in underwater acoustic networks with mobile devices. In: Proceedings of the 16th international conference on underwater networks & systems. WUWNet '22. Association for Computing Machinery, New York, NY, USA. https://doi.org/10.1145/3567600.3567604
9. Wang Z, Dong P, Zhang Y, Zhang H (2022) A data-driven non-interactive authentication scheme for the internet of vehicles in mobile heterogeneous networks. Sensors 22(22). https://doi.org/10.3390/s22228623
10. Safdar GA, McGrath C, McLoone M (2006) Limitations of existing wireless networks authentication and key management techniques for manets. In: 2006 International symposium on computer networks, pp 101–105 (2006). https://doi.org/10.1109/ISCN.2006.1662516
11. Safdar GA, McGrath C, McLoone M (2006) Existing wireless network security mechanisms and their limitations for ad hoc networks, pp 197–202. IEEE-Inst Electrical Electronics Engineers inc. https://doi.org/10.1049/CP:20060437
12. Ramya K, Pradeepa K (2014) Emicks: effective method in cluster based key management in ad hoc networks. Int J Eng Res Technol 3(2)
13. Dahshan H, Irvine J (2009) Key management in web of trust for mobile ad hoc networks, pp 363–370. IEEE. https://doi.org/10.1109/AINA.2009.76
14. Joseph AJ, Sani N, Kumar KS, Kumar TA, Nishanth R (2022) Towards a novel and efficient public key management for peer-peer security in wireless ad-hoc/sensor networks (2022). https://doi.org/10.1109/icstsn53084.2022.9761304
15. Harold Robinson Y, Golden Julie E (2019) Mtpkm: multipart trust based public key management technique to reduce security vulnerability in mobile ad-hoc networks. Wirel Pers Commun 109(2):739–760. https://doi.org/10.1007/S11277-019-06588-4
16. Huang T, Pei Y, Menkovski V, Pechenizkiy M (2023) Hop-count based self-supervised anomaly detection on attributed networks, pp 225–241. Springer. https://doi.org/10.1007/978-3-031-26387-314
17. Wu X, Wu H (2020) Trust management of mobile users P2P nodes based on blockchain. IEEE. https://doi.org/10.1109/ICCC51575.2020.9345097
18. Cheng Y, Ma J, Liu Z, Wang L, Ying Z, Chen X (2023) Efficient anonymous authentication and privacy-preserving reliability evaluation for mobile crowd sensing in vehicular networks. IEEE Internet Things J 10(17):14925–14939. https://doi.org/10.1109/JIOT.2023.3259961
19. Zhou Y, Cao L, Qiao Z, Xia Z, Yang B, Zhang M, Zhang W (2023) An efficient identity authentication scheme with dynamic anonymity for vanets. IEEE Internet Things J 10:10052–10065 (2023). https://doi.org/10.1109/JIOT.2023.3236699
20. Wang Z, Dong P (2022) NIMSA: non-interactive multihoming security authentication scheme for vehicular communications in mobile heterogeneous networks
21. Novaczki S, Bokor L, G´abor J, Imre S (2008) Design and evaluation of a novel hip-based network mobility protocol. J Netw 3. https://doi.org/10.4304/jnw.3.1.10-24
22. Shashidhara R, Lajuvanthi M, Akhila S (2021) A secure and privacy-preserving mutual authentication system for global roaming in mobile networks. Arab J Sci Eng 1–12. https://doi.org/10.1007/S13369-021-05940-W

23. Wu HL, Chang CC, Chen LS (2019) Secure authentication scheme with conditional privacy preservation in a global mobility communication network. ACM. https://doi.org/10.1145/3343147.3343153
24. Lu Y, Xu G, Li L, Yang Y (2019) Robust privacy-preserving mutual authenticated key agreement scheme in roaming service for global mobility networks. IEEE Syst J 13(2):1454–1465. https://doi.org/10.1109/JSYST.2018.2883349
25. Xie Q, Hong D, Bao M, Dong N, Wong DS (2014) Privacy-preserving mobile roaming authentication with security proof in global mobility net- works. Int J Distrib Sens Netw 10(5):325734. https://doi.org/10.1155/2014/325734
26. Zhang G, Fan D, Zhang Y, Li X, Liu X (2015) A privacy preserving authentication scheme for roaming services in global mobility networks. Secur Commun Netw 8(16):2850–2859. https://doi.org/10.1002/SEC.1209
27. Mihailescu MI, Nita SL (2023) Elliptic-curve cryptography, pp 207–243. Apress eBooks. https://doi.org/10.1007/978-1-4842-9450-5_9
28. Elliptic Curve Cryptography, pp 35–57. Springer. https://doi.org/10.1007/978-3-642-04117-4_3
29. Easttom W (2021) Elliptic Curve Cryptography, pp. 245–256. Springer. https://doi.org/10.1007/978-3-030-63115-4_11
30. Elliptic Curve Cryptography, p 41. IEEE (1999). https://doi.org/10.1109/ITNW.1999.814351
31. Zobel RN (2022) An elliptic curve cryptographic system design architecture with application to distributed simulation. Int J Simul: Syst Sci Technol. https://doi.org/10.5013/ijssst.a.22.03.07
32. Raya A, Mariyappn K (2020) Security and performance of elliptic curve cryptography in resource-limited environments: a comparative study, pp. 1–8. IEEE. https://doi.org/10.23919/ICITST51030.2020.9351327
33. Jin Y, Miyaji A: Secure and compact elliptic curve cryptosystems, pp. 639–650. Springer. https://doi.org/10.1007/978-3-030-21548-436
34. David AO, Sulaimon O (2023) Text encryption with improved elliptic curve cryptography. J Adv Math Comput Sci, 32–41. https://doi.org/10.9734/jamcs/2023/v38i31749
35. Kabin I, Dyka Z, Langendoerfer P (2022) Atomicity and regularity principles do not ensure full resistance of ECC designs against single-trace attacks. Sensors 22. https://doi.org/10.3390/s22083083
36. Hu C, Zheng D, Guo R, Wu A, Wang L, Gao S (2020) A novel blockchain-based anonymous handover authentication scheme in mobile networks. 22:874–884. https://doi.org/10.6633/IJNS.202009_22(5).19
37. Chen R, Peng D (2018) A novel NTRU-based handover authentication scheme for wireless networks. IEEE Commun Lett 22(3):586–589. https://doi.org/10.1109/LCOMM.2017.2786228
38. Rajasekaran AS, Maria A, Rajagopal M, Lorincz J (2022) Blockchain enabled anonymous privacy-preserving authentication scheme for internet of health things. Sensors 23(1):240–240. https://doi.org/10.3390/s23010240
39. Chang CC, Huang YC, Tsai HC (2014) Design and analysis of chameleon hashing based handover authentication scheme for wireless networks. J Inf Hiding Multimed Signal Process 5:107–116
40. Lai C, Ma Y (2021) A novel group-oriented handover authentication scheme in MEC-enabled 5G networks, pp 29–34. IEEE. https://doi.org/10.1109/ICCC52777.2021.9580296
41. Xue K, Luo X, Ma Y, Li J, Liu J, Wei DS (2022) A distributed authentication scheme based on smart contract for roaming service in mobile vehicular networks. IEEE Trans Veh Technol 71:5284–5297. https://doi.org/10.1109/TVT.2022.3148303
42. Wei L, Cui J, Zhong H, Bolodurina I, Gu C, He D (2023) A decentralized authenticated key agreement scheme based on smart contract for securing vehicular ad-hoc networks. IEEE Trans Mob Technol 1–16. https://doi.org/10.1109/TMC.2023.3288930
43. Wang CH, Lee PC (2010) A roaming-based anonymous authentication scheme in multi-domains vehicular networks, pp. 273–287. Springer. https://doi.org/10.1007/978-3-642-14292-529

44. Nikooghadam M, Amintoosi H, Kumari S (2020) A provably secure ECC-based roaming authentication scheme for global mobility networks 54:102588. https://doi.org/10.1016/J.JISA.2020.102588
45. Bagga P, Das AK, Rodrigues JJ (2023) Bilinear pairing-based access control and key agreement scheme for smart transportation. https://doi.org/10.1016/j.csa.2022.100001
46. Priyanka VS, Radha V (2023) An efficient and secure certificateless aggregate signature based authentication scheme for vehicular ad-hoc networks. https://doi.org/10.1109/icict57646.2023.10134382
47. Yao M, Gan Q, Wang X, Yang Y (2023) A key-insulated secure multi-server authenticated key agreement protocol for edge computing- based vanets. Internet Things. https://doi.org/10.1016/j.iot.2023.100679

Spatially Explicit Conservation Planning: Habitat Modelling and Connectivity Exploration for Asiatic Black Bears in Sikkim Himalaya

Nischal Gautam, Samarjeet Borah, Ratika Pradhan, Bijoy Chhetri, and Kapila Sharma

Abstract Effective wildlife management requires robust habitat appraisal. Spatially explicit species distribution modelling (SDM), aided by advanced tools and collaboration with remote sensing, and geographical information system (GIS), is integral to conservation biology. Maxent models were employed to map the Asiatic black bear habitat in Sikkim Himalaya based on 12 camera trap points and 53 field GPS observations (a total of 65). Of 48 environmental variables employed, 20 shaped the distribution map, with key contributors like NDVI August (17.6%) and bio-14; precipitation of driest month (15.7%). Model validation showed a strong predictive ability (training AUC = 0.948 ± 0.01 SD, test AUC = 0.86 ± 0.04 SD), emphasizing these techniques' importance in wildlife conservation in remote Himalayas.

Keywords Asiatic black bear · MaxEnt · Habitat modelling · Connectivity planning

1 Introduction

Understanding the distribution of large mammals is fundamental for effective conservation, enabling targeted policy implementation [1] amid increasing anthropogenic pressures [2]. Traditional methods, like transect surveys and camera traps, are costly

N. Gautam (✉) · S. Borah · K. Sharma
Department Computer Application, Sikkim Manipal Institute of Technology, Sikkim Manipal University, Majhitar, Sikkim 737132, India
e-mail: meetnischalg@gmail.com

R. Pradhan
Department Computer Application, Sikkim University, Gangtok, Sikkim 737102, India

B. Chhetri
Department of Zoology, SRM Sikkim University, Gangtok, Sikkim 737102, India

© The Author(s), under exclusive license to Springer Nature Singapore Pte Ltd. 2025

H. Zen et al. (eds.), *Soft Computing and Signal Processing*, Lecture Notes in Networks and Systems 1200, https://doi.org/10.1007/978-981-97-9926-8_30

and labour-intensive [3–7]. Conservation biologists increasingly count on predictive models, using GIS, RS, and GPS tools [8, 9]. Mechanism models [10] simulate species' ideal habitats without specific location point data but have limitations, such as neglecting habitat accessibility and inherent subjectivity in classification and element weighting. Continued methodological refinement is a key to effective wildlife conservation. In predictive modelling, approaches like mechanism models, regression models [11], and ecological niche models [12–15] have been employed. Data from "non-presence points" are utilized in regression models, while ecological niche models, including GARP (Genetic Algorithm for Rule Set Production) and Maxent, outperform traditional methods with presence-only data [16–18]. Maxent, rooted in Jaynes' ecological niche hypothesis [19], employs maximum entropy modelling, is a broadly utilized tool in species distribution models (SDMs) [20, 21]. Maxent estimates a species' potential distribution by incorporating information from presence data, contributing to its widespread use. The GARP is another flexible technique using genetic algorithms to generate rules for species dispersal [22]. GARP predicts species niche under environmental conditions, finding applications in invasion studies [23, 24]; however, MaxEnt shows superior performance compared to GARP [25]. SDMs rely on correlation statistics for predicting habitats, with the interpretability depending on carefully selecting relevant environmental factors [26]. Human-wildlife conflicts, as highlighted by [27], result in negative effects on both people and the affected wildlife. Particularly, conflicts with bears often manifest in proximity to protected areas, where residents bear the brunt of coexistence challenges with wildlife as emphasized [28], mostly attributing to the movement and seasonality of behaviour. While global research extensively explores on bears' seasonal migration and movement behaviour, crucial insights specific to Sikkim are lacking. Conservation organizations are investing heavily in the integration of ecological processes and SDMs since they recognize its potential to enhance habitat preservation and restoration in ecologically vital areas. The generated maps from SDMs such as MaxEnt vividly portray the intricate patterns of animal movement, exemplified by the Asiatic black bear in this study. This presents a critical instrument to tackle the pressing issue in addition to offering insightful information about the unique dynamics of bear movement fostering sustainable and harmonious coexistence between these animals and the local human population within the landscape. Consequently, this study aims to: (i) comprehend the existing distribution of these bears, (ii) identify suitable habitats, and (iii) delineate regions for conservation and an efficient management of these bears in the Sikkim Himalaya.

2 Methodology

2.1 *Study Area*

Sikkim is a second smallest state in the northeastern Himalayan Mountain range in India with 7096 Km^2 area. It lies between the latitude of 27° 5′ S to 28° 9′ N and longitude of 87° 9′ W to 88° 56′ E falls in the topographic sheets of 78A1–78A15 series of Survey of India [29, 30]. The altitude varies from 300 m to more than 8586 m and the annual rainfall ranges between 2700 and 3200 mm, and the annual temperature varies from sub-zero during winter to 28 °C during summer [31]. Sikkim is bestowed with more than 4000 flowering plants [32], and harbours 574 species which represents 45% of the total birds within 0.21% of the geographical area of India [33]. Sikkim has recorded reserved Forest land as 2261 km^2, Khasmal Forests as 285 km^2, and Garucharan Forests as 104 km^2, with a total of 2650 km^2. Area under alpine pasture and scrub constitute 1024 km^2 (14.44%), area under perpetual snow cover is 2091 km^2 (29.5%) [34]. There are eight wildlife-protected areas which comprise a national park and seven wildlife sanctuaries covering almost 31% of the state's geographical area (Fig. 1).

2.2 *Model Species*

The Asiatic black bear's (simply bear hereinafter) extensive distribution spans southern and eastern Asia, from Pakistan, Afghanistan, Iran to Russia, Korea, and

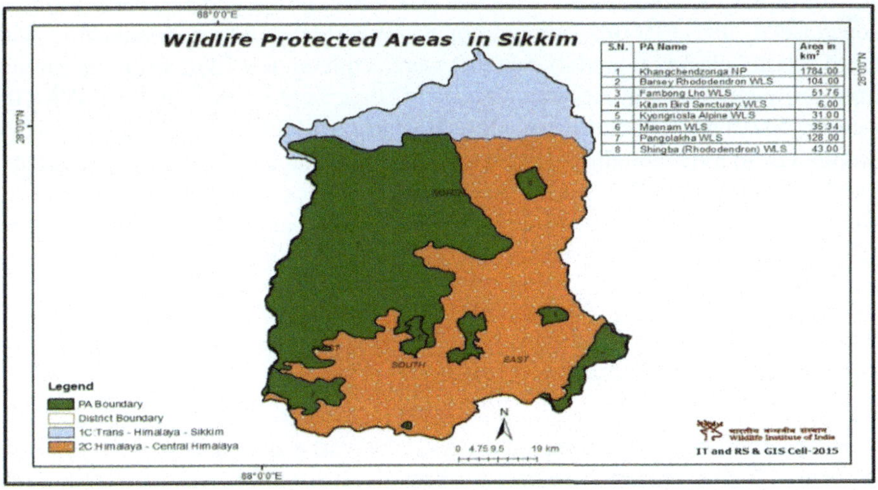

Fig. 1 Map displaying the study area with designated wildlife-protected areas

Japan, including Taiwan and the Himalayan region below 3750 m [35, 36]. The Himalayas and northeastern Indian hills likely host a significant population, especially in Jammu and Kashmir [37, 38]. Despite biodiversity surveys capturing variety of species, Sikkim Himalaya lacks exhaustive census surveys [39], particularly for bears. This study aims to understand the present-day range of bears, identify suitable habitats, and delineate regions for effective management and conservation.

2.3 Species Dataset

Our study focussed on bear distribution, utilizing historical records, camera tracking, and secondary sources. GPS data from low to high-altitude transects aided in pinpointing species coordinates (Garmin GPS). The entire Sikkim population of bears mostly resides in six wildlife protected areas—Pangolakha Wildlife Sanctuary, Kyongnosla Alpine Sanctuary, Maenam Wildlife Sanctuary, Fambonglho Wildlife Sanctuary, Barsey Rhododendron Sanctuary, and Khangchendzonga National Park, along with other Reserve Forests outside wildlife protected areas. Geo-referenced data for sixty-five bear populations in MaxEnt modelling, encompassing all Sikkim districts, were gathered from aforesaid wildlife protected areas apart from other potential areas, projected in 5 km^2 grid (Fig. 2), forming a comprehensive approach for this study.

2.4 Environmental Predictor

The 19 layers of a bioclimatic envelope that make up the climatic variables were downloaded at a 30 s resolution from www.worldclim.org in conjunction with DEM (digital elevation model) data that were acquired at a 1 km spatial resolution from www.earthexplore.usgs.gov. The LULC data were taken from the DIVA-GIS website. NDVI was calculated using MODIS data (2013) of 16 days 24 composite file acquired from https://modis.gsfc.nasa.gov. Our study area and contour maps (DEM, satellite image, ASTGTM2) were created using MS Excel (2010) and ArcGIS ver. 10.7.1 Desktop (ESRI, 2019). More biologically significant factors were produced by deriving bioclimatic variables from monthly rainfall and temperature data (Table 1).

2.5 Ecological Habitat Modelling

Using the MaxEnt software version 3.3.3 k.d that was acquired from www.cs.princeton.edu/*schapire/MaxEnt/, the possible bear distribution area was estimated. Using ArcGIS ver. 10.7.1 Desktop (ESRI, 2019), a final potential map projected by feeding

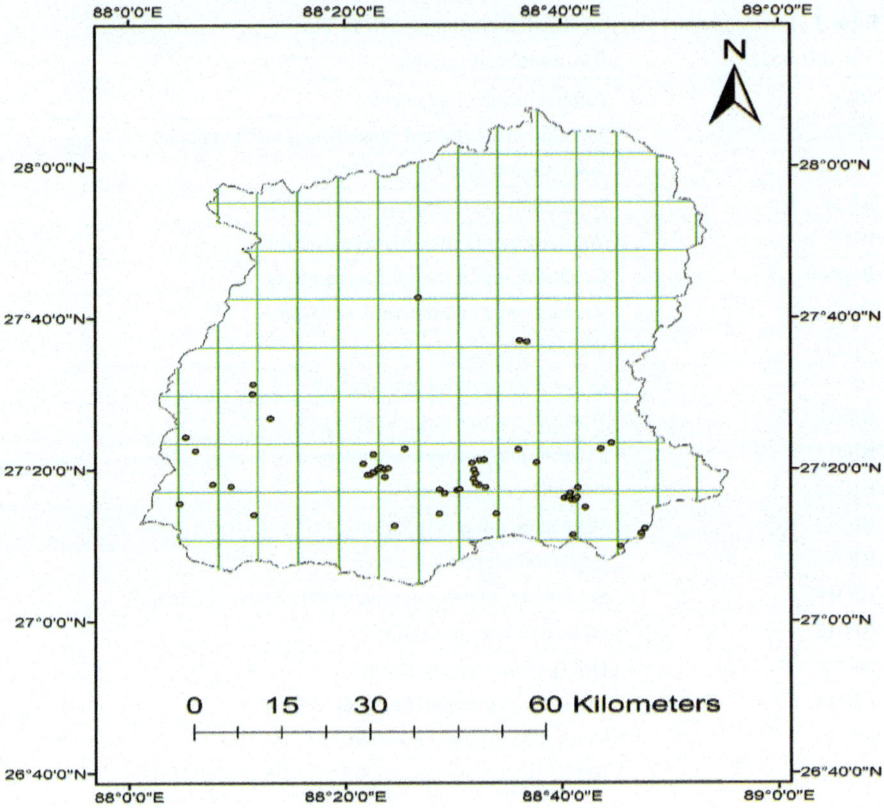

Fig. 2 Grid map showing study area and bear presence points

environmental variables and species existing data into MaxEnt. The methodology of [40] was utilized in developing the MaxEnt ecological habitat model.

2.6 Model Validation

Potent environmental variables were analysed utilizing the principal components analysis (PCA), for bear distribution [41].

Out of 48 environmental variables, 20 shaped the species distribution map. Validation used the jackknife approach [42] to identify potential variables and a regularization multiplier of 1 [43] to prevent overfitting. Modelling included 1000 background points, a threshold feature, and data split: 20% for testing, 80% for training, with

Table 1 Environmental variables used, including 24 NDVI files

Variable code	Environmental variable
BIO1	Annual mean temperature
BIO2	Monthly mean diurnal range (max–min temperature)
BIO3	Isothermality (P2 / P7) (* 100)
BIO4	Seasonality in Temperature (SD*100)
BIO5	Warmest month maximum temperature
BIO6	Coldest month minimum temperature
BIO7	Annual range in temperature (P5–P6)
BIO8	Wettest quarter mean temperature
BIO9	Driest quarter mean temperature
BIO10	Warmest quarter mean temperature
BIO11	Coldest quarter mean temperature
BIO12	Annual precipitation
BIO13	Wettest month precipitation
BIO14	Driest month precipitation
BIO15	Seasonality in precipitation (coefficient of variation)
BIO16	Wettest quarter precipitation
BIO17	Driest quarter precipitation
BIO18	Warmest quarter precipitation
BIO19	Coldest quarter precipitation
NDVI	NDVI-16 days composite, 24 MODIS files (2013)
DEM	Digital Elevation Model
LULC	LULC (2014)
SLOPE	Generated by ArcGIS
ASPECT	Generated by ArcGIS
ALTITUDE	World Clim

three replicate runs [44]. Goodness-of-fit AUC measured the model's test performance, variables with the highest AUC considered best performer. Jackknife quantified each variable's predictive contribution. Based on a reclassification range of 0–100%, outputs were categorized into 5 distribution areas: very low potential (< 10%), low potential (10–30%), moderate potential (30–50%), high potential (50–70%), and very high potential (> 70%). This methodology aligns with [45]'s approach.

3 Result and Discussion

3.1 *Habitat Suitability Model*

MaxEnt modelling as shown in Fig. 3 identified precipitation seasonality and NDVI as key factors influencing bear distribution in Sikkim Himalaya. The model's calibration test indicated an accuracy (AUC: 0.7–0.9), classifying it as good. Validation through ROC plots yielded an average of training AUC $= 0.948 \pm 0.01$ SD, test AUC $= 0.86 \pm 0.04$ SD, highlighting strong predictive ability.

Map of predicted distribution model (Fig. 4) quantifies predicted habitat. The 'very high potential' area is minimal (279 km^2), contrasting with larger areas in low (5224 km^2), moderate (1070 km^2), and high (521 km^2) potential classes (Fig. 5). For the predicted habitats under the 'very high potential' category, being low offers challenges (for the tough topography of the area) beside opportunities (strong possibilities) in further exploring newer habitats of the bear in Sikkim Himalaya.

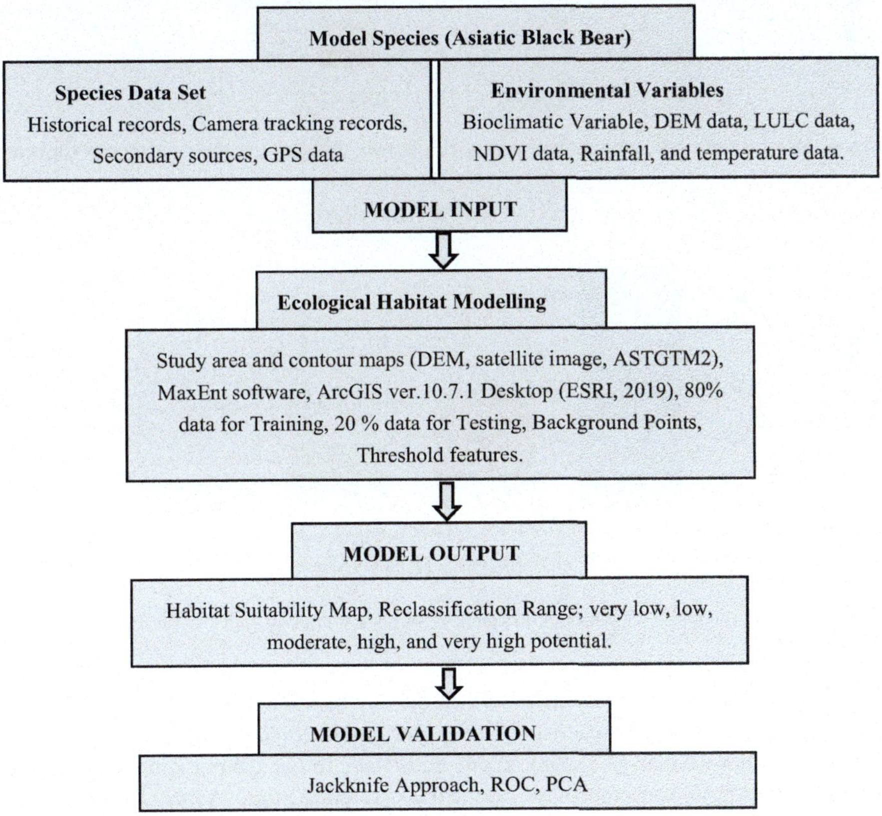

Fig. 3 Scheme for maxent habitat suitability modelling

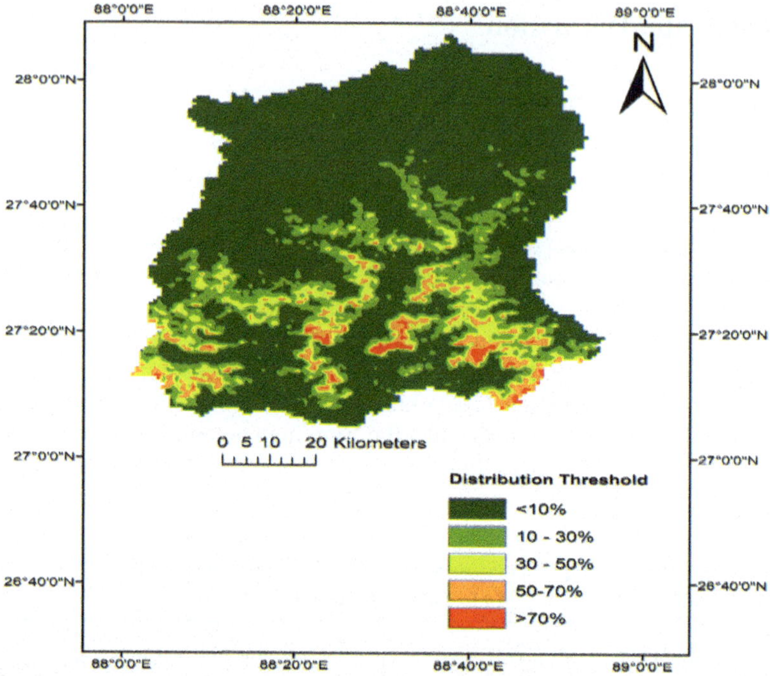

Fig. 4 Map for the predicted distribution model, showing different threshold categories for bears

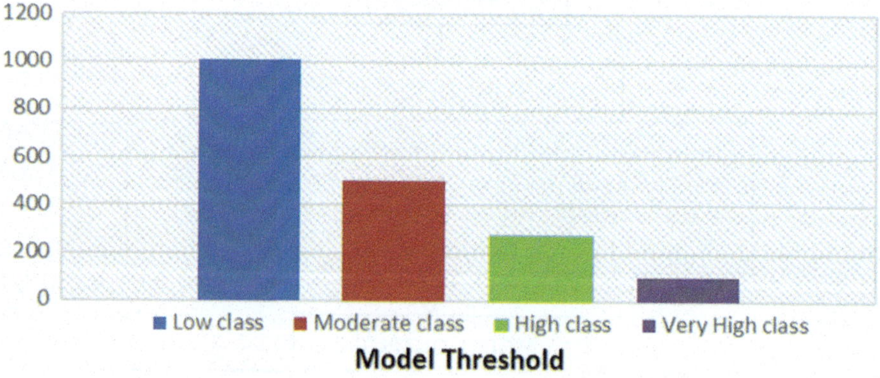

Fig. 5 Spatial area under each threshold category of the anticipated distribution model of bear

The cumulative percent contribution of the environment and associated variables (Table 2) revealed NDVI August (17.6%) and bio-14; precipitation of driest month (15.7%), NDVI a2013273 (12%) NDVI a2013097 (8.6%), bio-17; precipitation of driest quarter (8.5%) emerged as key contributors, while Slope, Aspect, and Elevation had marginal contribution.

Table 2 Environmental and associated variables and their percentage representation in the prediction for the probable bear availability

Variable	Name of variable	Percent contribution	Permutation importance
a2013225	NDVI August	17.6	6.6
bio-14	The driest month's precipitation	15.7	3.6
a2013273	NDVI September	12	15.3
a2013097	NDVI March	8.6	8.1
bio-17	Quarter with the driest precipitation	8.5	1.5
bio-13	Precipitation of wettest month	7.2	4.2
a2013289	NDVI October	5.8	4.6
bio-2	Monthly mean diurnal range (Max. Temperature – Min. Temperature)	3.9	19.4
a2013209	NDVI July	3.7	0.7
a2013337	NDVI December 2013	3.3	11.6
bio-19	Precipitation of coldest quarter	1.6	0
a2013129	NDVI April	1.5	0.7
a2013257	NDVI September	1.4	2.6
a2013001	NDVI January 2013	1.4	3
a2013305	NDVI November 2013	1.3	0.4
a2013241	NDVI February	1.1	1
bio-6	Minimum temperature of the coldest month	1.1	2.9
bio-7	Temperature annual range (P5–P6)	0.6	0.9
a2013193	NDVI June	0.6	1
a2013321	NDVI January	0.5	0.9
bio-11	Mean of coldest quarter temperature	0.5	6.9
bio-15	Seasonality of precipitation (coefficient of variation)	0.4	1.1
bio-9	Mean of driest quarter temperature	0.3	0.2
bio-3	Isothermality (BIO2/BIO7) (×100)	0.2	0
a2013161	NDVI May	0.2	0
2,013,033	NDVI February 2013	0.2	1.3
lulc	Landuse Landcover	0.2	0.4
a2013017	January	0.1	0.4

(continued)

Table 2 (continued)

Variable	Name of variable	Percent contribution	Permutation importance
a2013353	December	0.1	0.1
bio-5	Maximum warmest month temperature	0.1	0.5

4 Conclusion

In our present investigation, we advocate concentrating on patches of distribution falling under the "extremely high potential" threshold on the MaxEnt model map. This strategic focus provides pivotal insights for prospective scientific inquiries, especially given the challenging geography and inhospitable mountainous terrains of Sikkim Himalaya. These identified patches are essential in evaluating the conservation status of bears in Sikkim and the broader context of the Eastern Himalayas and beyond. A global emphasis on systematic population exploration for this species becomes paramount for evaluating ecological well-being and implementing effective conservation strategies. To counteract further deterioration, it is critical to safeguard the identified bottlenecks along the suitable habitats, especially in fragmented landscapes, bordering villages, and critical highway intersections. Additionally, declaring these extremely high potential areas as conservation zones for bears and their linkage, coupled with long-term monitoring initiatives, stands as a vital policy recommendation for proactive field management.

Acknowledgements The authors would like to express their gratitude to the Forest and Environment Department, Government of Sikkim for granting permission for accessing secondary data and the forest areas. Special thanks to Mr. Prem Kumar Chettri for mapping assistance.

References

1. Lu CY, Gu W, Dai AH, Wei HY (2012) Assessing habitat suitability based on geographic information system (GIS) and fuzzy: a case study of Schisandra sphenanthera Rehd. et Wils. in Qinling Mountains, China. Ecol Model 242:105–115
2. Ebrahimi A, Farashi A, Rashki A (2017) Habitat suitability of Persian leopard (Panthera pardus Saxicolor) in Iran in future. Environ Earth Sci 76:697–707
3. McCarthy TM, Fuller TK, Munkhtsog B (2005) Movements and activities of snow leopards in Southwestern Mongolia. Bio Cons 124:527–537
4. Jackson RM, Roe JD, Wangchuk R, Hunter DO (2006) Estimating snow leopard population abundance using photography and capture-recapture techniques. Wildl Soc Bull 34:772–781
5. Janečka JE, Munkhtsog B, Jackson RM, Naranbaatar G, Mallon DP, Murphy WJ (2011) Comparison of noninvasive genetic and camera-trapping techniques for surveying snow leopards. J Mammol 92:771–783
6. Zeller KA, Nijhawan S, Salom-perez R, Potosme SH, Hines JE (2011) Integrating occupancy modeling and interview data for corridor identification: a case study for Jaguars in Nicaragua. Bio Cons 144:892–901

7. Sharma K, Bayrakcismith R, Tumursukh L, Johansson O, Sevger P, McCarthy T, Mishra C (2014) Vigorous dynamics underlie a stable population of the endangered snow leopard Panthera uncia in Tost mountains, South Gobi, Mongolia. PLoS ONE. https://doi.org/10.1371/journal.pone.0101319
8. Guisan A, Tuiller W (2005) Predicting species distribution: offering more than simple habitat models. Ecol Lett 8:993–1009
9. Yang XQ, Kushwaha SPS, Saran S, Xu J, Roy PS (2013) Maxent modeling for predicting the potential distribution of medicinal plant, Justicia adhatoda L. in Lesser Himalayan foothills. Ecol Eng 51:83–87
10. Ouyang Z, Liu J, Xiao H, Tan Y, Zhang H (2001) An assessment of giant panda habitat in Wolong nature reserve. Acta Ecol Sin 11:1869–1874
11. Schadt S et al (2002) Assessing the suitability of central European landscapes for the reintroduction of Eurasian lynx. J Appl Ecol 39:189–203
12. Phillips SJ, Anderson RP, Schapire RE (2006) Maximum entropy modeling of species geographic distributions. Ecol Model 190:231–259
13. Su J, Aryal A, Nan Z, Ji W (2015) Climate change-induced range expansion of a subterranean rodent: implications for rangeland management in Qinghai-Tibetan Plateau. PLoS ONE 10:e0138969
14. Srivastava V, Griess VC, Padalia H (2018) Mapping invasion potential using ensemble modelling: a case study on Yushania maling in the Darjeeling Himalayas. Ecol Model 385:35–44
15. Rafni F et al (2020) From nucleotides to satellite imagery: approaches to identify and manage the invasive pathogen Xylella fastidiosa and its insect vectors in Europe. Sustainability 12:4508
16. Hijmans RJ, Graham CH (2006) The ability of climate envelope models to predict the effect of climate change on species distributions. Glob Chang Biol 12:2272–2281
17. Phillips SJ, Dudík M (2008) Modeling of species distributions with Maxent: new extensions and a comprehensive evaluation. Ecography 31:161–175
18. Clements GR et al (2012) Predicting the distribution of the Asian Tapir (Tapirus indicus) in Peninsular Malaysia using maximum entropy modelling. Integr Zool 7:400–406
19. Jaynes ET (1957) Information theory and statistical mechanics. II. Phys Rev 108:171–190
20. Jaynes ET (1990) Probability theory as logic. In: Fougère PF (ed) Maximum entropy and bayesian methods, Springer, Netherlands, pp 1–16. https://doi.org/10.1007/978-94-009-0683-9_1
21. Jaynes ET (2003) Probability theory: the logic of science. Cambridge University Press, Cambridge
22. Stockwell D (1999) The GARP modelling system: problems and solutions to automated spatial prediction. Int J Geogr Inf Sci 13:143–158
23. Ganeshaiah KN et al (2003) Predicting the potential geographical distribution of the sugarcane woolly aphid Using GARP and DIVAGIS. Curr Sci 85:1526–1528
24. Underwood EC, Klinger R, Moore PE (2004) Predicting patterns of non-native plant invasions in Yosemite National Park, California, USA. Divers Distrib 10:447–459
25. Kichloo MA, Sharma N (2021) MaxEnt modeling of distribution and habitat preferences of Asiatic black bear in Kishtwar high altitude National Park, Jammu and Kashmir. Asian J Anim Sci 15:19–26
26. Elith J, Leathwick JR (2009) Species distribution models: ecological explanation and prediction across space and time. Annu Rev Ecol Evol Syst 40:677–697
27. Pettigrew M, Xie Y, Kang A, Rao M, Goodrich J, Liu T, Berger J (2012) Human–carnivore conflict in China: a review of current approaches with recommendations for improved management. Integr Zool 7(2):210–226
28. Li X, Buzzard P, Chen Y, Jiang X (2013) Patterns of livestock predation by carnivores: human–wildlife conflict in Northwest Yunnan, China. Environ Manage 52:1334–1340. https://doi.org/10.1007/s00267-013-0192-8
29. Sharma RK, Sharma N, Shrestha DG, Luitel KK, Arrawatia ML, Pradhan S (2012) Study of forest fires in Sikkim Himalayas, India using remote sensing and GIS techniques. Clim Chang Sikkim–Patterns, Impacts Initiat 233–244

30. Sharma K, Thapa G (2021) Analysis and interpretation of forest fire data of Sikkim. For Soc 5(2):261–276
31. Forest Survey of India (2019) The state of Forest Report. Government of India-Ministry of Environment and Forest, pp 233–241. https://fsi.nic.in/forest-report-2019
32. Sundriyal M, Sundriyal RC, Sharma E (2004) Dietary use of wild plant resources in the Sikkim Himalaya, India. Econ Bot 58(4):626–638
33. Acharya BK, Vijayan L (2011) The birds of Sikkim: an analysis of elevational distribution. Endem Threat Biodivers Sikk
34. Forest Survey of India (2017) The State of Forest Report. Government of India-Ministry of Environment and Forest, pp 278–283. sikkim-isfr-2017.pdf (fsi.nic.in)
35. Schaller GB (1977) Mountain monarchs: wild sheep and goats of the Himalaya. University of Chicago Press, Chicago, IL., USA, p. 425. ISBN-13: 9780226736419
36. Sathyakumar S (2001) Status and management of Asiatic black bear and Himalayan brown bear in India. Ursus 12:21–30
37. Sathyakumar S, Choudhury A (2007) Distribution and status of Asiatic black bear (Ursus thibetanus) in India. J Bom Nat His Soc 104:316–323
38. Charoo SA, Sharma LK, Sathyakumar S (2011) Asiatic black bear-human interactions around Dachigam National Park, Kashmir, India. Ursus 22:106–113
39. Lepcha TW et al (2017) Wildlife population estimation for all protected areas of Sikkim 2016–2017. Department of Forests, Environment and Wildlife Management, Government of Sikkim, Gangtok, pp. 72
40. Phillips SJ, Anderson RP, Schapire RE (2006) Maximum entropy modelling of species geographic distributions. Ecol Model 190:231–259
41. Cruz-Cardenas G et al (2014) Potential species distribution modelling and the use of principal component analysis as predictor variables. Revista Mexicana de Biodiversidad 85:189–199
42. Pearson RG et al (2007) Predicting species distributions from small numbers of occurrence records: a test case using cryptic geckos in Madagascar. J Biogeogr 34:102–117
43. Phillips SJ, Dudík M, Schapire RE (2004) A maximum entropy approach to species distribution modeling. In: Proceedings of the twenty-first international conference on machine learning, pp 655–662
44. Flory AR, Kumar S, Stohlgren TJ, Cryan PM (2012) Environmental conditions associated with bat white nose syndrome mortality in the north-eastern United States. J Appl Ecol 49:680–689
45. Sharma K et al (2015) Predicting suitable habitats for the vulnerable eastern Hoolock Gibbon, Hoolock leuconedys, in India using the MaxEnt model. Folia Primatol 86(4):387–397

An Improved DTCWT-Based Multiclass Classifier for AD Detection

B. A. Sujathakumari, Sudarshan Patil Kulkarni, A. Manoj, U. Dhanush Gowda, Gurukiran Goudapgol, and R. Prithvi

Abstract The diagnosis of AD poses a major challenge to modern healthcare. A diagnosis is necessary for effective management and intervention. Through the integration of ML and DL, the early detection and classification of AD may be aided. In this paper, we propose a deep learning framework leveraging DTCWT for multi-class classification of AD across six distinct stages. The DTCWT has several advantages such as multiresolution analysis and directional sensitivity. Using DTCWT for the preprocessing step, we build a CNN and ANN architecture to extract discriminative features from structural MRI data. DTCWT offers an accurate and precise early detection of Alzheimer's disease across different stages. Our model is tested and trained on a large dataset of people at different phases of AD, which is divided into six different classes according to the severity and the illness. We utilize thorough cross-validation methods to ensure the model is reliable and applicable to a wide range of patients and situations. Results show that this method works well for obtaining extraordinary results in multi-class AD classification problems. In addition, we use SVM and RF methods for better training and testing of disease diagnosis. Hybridizing CNN with RF and SVM and hybridizing ANN with RF and SVM give better and more accurate results.

Keywords Mild cognitive impairment (MCI) · Alzheimer's disease (AD) · Discrete wavelet transform (DWT) · Dual-tree complex wavelet transform (DTCWT) · Random forest (RF) · Convolution neural network (CNN) · Deep learning (DL) · Artificial neural network (ANN) · Support vector machine (SVM) · MCI · CN · EMCI · LMCI · SMC · AD

B. A. Sujathakumari (✉) · S. P. Kulkarni · A. Manoj · U. Dhanush Gowda · G. Goudapgol · R. Prithvi
Sri Jayachamarajendra College of Engineering, JSSSTU, Mysuru, Karnataka, India
e-mail: sujathakumari@sjce.ac.in

© The Author(s), under exclusive license to Springer Nature Singapore Pte Ltd. 2025
H. Zen et al. (eds.), *Soft Computing and Signal Processing*, Lecture Notes in Networks and Systems 1200, https://doi.org/10.1007/978-981-97-9926-8_31

1 Introduction

With an enhancement of the DWT, the DTCWT captures phase and amplitude information by matching real and imaginary wavelet functions. This information is helpful for signal processing and image analysis applications. DTCWT's stability against small alterations in input data is a crucial feature for jobs like object tracking and image registration. DTCWT is superior in capturing directional information, as the latter finds it difficult to distinguish between feature orientations in data. Because of its explicit sensitivity to feature magnitude and orientation, DTCWT is more suited for tasks where directional information is crucial, such as texture analysis, edge recognition, and image denoising. The unique architecture format of the ANN models is derived from a biological nervous system. The intricate and nonlinear form of neurons in ANN models is similar to that of the human brain. Weighted linkages are used to connect the neurons. Each process in ANN models is computed using various learning strategies, such as data collection and analysis, network framework design, hidden layer number, network simulation, and weights/bias trade-off. ML is a subset of AI that allows computers to self-learn from data and previous events, identifying patterns to help them make predictions with little assistance from humans. ML methods enable computers to act autonomously without requiring explicit programming. ML applications are constantly fed new data and may learn, grow, evolve, and adapt on their own. Machine learning extracts valuable information from massive amounts of data by using algorithms to find patterns and learn iteratively.

An increase in the availability of samples during learning processes improves the adaptive performance of machine learning systems. Deep learning, for instance, is a type of machine learning that enables robots to learn from examples and mimic natural human behaviors. DL is a subset of ML that can be comprehended. Self-improvement and learning are at the heart of this field, which studies computer algorithms. ANN, which mimics human thought and learning processes, is the basis for deep learning, whereas ML uses simpler ideas. Neural networks could not reach their full potential until recently due to limitations in processing power. Since higher, more complex neural networks have become conceivable thanks to breakthroughs in Big Data analytics, computers can now monitor, comprehend, and adapt to complicated events faster than humans. DL has improved language translation, speech recognition, and image classification. It can solve any pattern recognition problem without the need for human intervention. Integrating the selected characteristics from the ADNI dataset, which includes neuroimaging and clinical data, to generate a uniform input representation allows for early detection. This approach is beneficial as it permits the identification of those at risk of acquiring AD before the symptoms become severe [1].

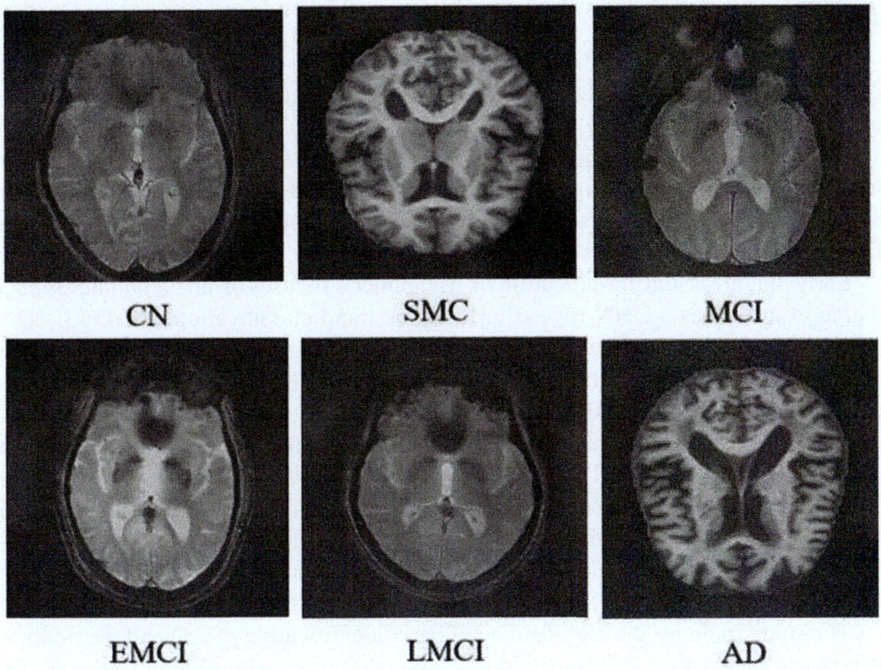

Fig. 1 Stages of Alzheimer's disease

AD progression stages (as shown in Fig. 1):

1. Cognitive Normal (CN): This stage denotes a person's normal baseline cognitive performance in the absence of AD. People at this stage are capable of handling everyday chores by themselves and have a strong memory.
2. Subjective Memory Complaints (SMC): Self-perceived problems with memory show the defining feature. People think that they don't remember things as well as they used to.
3. Early Mild Cognitive Impairment (EMCI): Memory issues, especially when it comes to recalling recent events.
4. Late-stage Mild Cognitive Impairment (LMCI): In this stage, more cognitive decline and memory loss.
5. Mild Cognitive Impairment (MCI): In this stage, cognitive is greater than normal aging but no severe as dementia.
6. Alzheimer's disease (AD): AD is at its most severe and advanced stage. Memory loss, cognitive decline, and functional deficits are significant at this stage.

2 Literature Survey

Deep learning-based AD detection proposes a convolution neural network for AD classification [2]. The study utilized a dataset from ADNI which includes 110 AD, 105 MCI, and 51 NC individuals. This uses a traditional approach to classification in generating wavelets using DWT, which is followed by machine learning techniques, also other techniques like biomarkers, cerebrospinal fluid analysis, and brain imaging. This also describes the cause of AD as genetic mutations.

Early diagnosis and classification of Alzheimer's focuses on artificial intelligence which exactly uses a CNN for earlier diagnosis and classification of AD [3]. This model achieved an excellent accuracy of 99% during classification. This model discusses the architecture of the CNN model, which is stimulated by the process of the visual cortex. The MRI images were obtained by the ADNI dataset.

The research emphasizes the utilization of ML and DL algorithms to detect Alzheimer's disease [4]. The authors employ CNN architectures such as VGG, VGG, AlexNet, GoogLeNet, ResNet, DenseNet, Inceptionv3, and InceptionResNet for binary and ternary classification of AD. The suggested technique is highly accurate in detecting Alzheimer's disease, with a recognition test set precision of 99.27% for MCI/AD, 98.89% for AD/CN, and 97.06% for MCI/CN.

The study includes the significance of early identification of AD and the application of automated approaches to predict signs of AD from neuroimaging data [5]. The study assesses two major experiments: the first employs an SVM detection method, while the second suggests a CNN model. The SVM-based technique achieved 95.3% sensitivity, 71.4% specificity, and 84.4% accuracy. However, these findings were not effective for predetection.

The study evaluates that deep learning is used for accurate early diagnosis of AD at its early stages [6]. This also mentions the use of affine arrays and metadata for diagnosis. The dataset used during the study consists of 260 patient cases which includes AD and control cases. It employs the voxel volume character of AD. This demonstrates an increasing interest in image analysis. The work examines a significant study that uses MRI data, ML, and DL methods.

A hybrid machine learning strategy that combines standalone RNN and CNN has been developed to improve the identification of MCI [1]. The model intended to improve MCI classification accuracy by leveraging RNN temporal dependencies and CNN spatial information. Using the ADNI dataset, which includes neuroimaging and clinical data, characteristics were identified and combined to form a single input representation. This method enables early detection.

Based on the related work, we have studied, using several picture segmentation methods and datasets, the CNN model obtained an accuracy of approximately 96% while remaining impartial to the dataset. The research also mentions using SVM and RF for modern diagnosis approaches since it gives the best classification results with few features and a large amount of data [7–25].

3 Methodology

In Fig. 2, the methodology implemented for early detection of progressive MCI is represented.

The MRI images used in this implementation are taken from the ADNI website and belong to six different classes: CN, AD, MCI, EMCI, LMCI, and SMC. Preprocessing entails several actions to guarantee information consistency and applicability. First, the images are subjected to skull stripping, normalization, and registration using the Statistical Parametric Mapping 12 (SPM12) toolbox, to separate important brain regions such as the hippocampus. SPM12 reliably describes the architecture of the brain by utilizing sophisticated segmentation algorithms that eliminate unnecessary

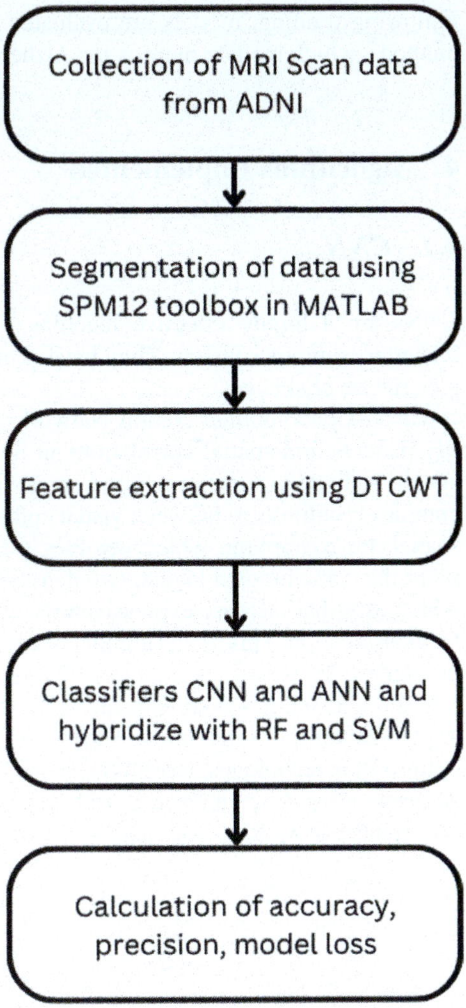

Fig. 2 Methodology block diagram

regions, like the skull and neck regions. After segmentation, the DTCWT classifier is used to classify the preprocessed MRI images, resulting in the removal of undesired regions such as the areas around the head and neck. These images are then converted into JPEG format using a JPEG converter after the MRI dataset is expanded by using data augmentation techniques such as rotation, flipping, scaling, and translation, which improve the training and testing stages.

In the implementation, both CNN and ANN are used. CNNs are preferred because of their capacity to directly learn discriminative features from data, whereas RF or SVM classifiers can use collected features for decision boundary learning. Furthermore, a hybrid technique combines CNNs for feature extraction with RF or SVM classifiers for classification, leveraging the advantages of both deep learning and traditional methods. Similarly, ANNs excel in learning detailed patterns from data, and these properties are useful to decision boundary learning via RF or SVM. Following training, models are evaluated comprehensively using cross-validation methods, which include metrics for Alzheimer's disease classification.

4 Algorithms Implemented

4.1 CNN

CNNs are a highly effective machine learning technique, especially for tasks related to computer vision. They have achieved top-notch results in various image recognition challenges.

CNN (Convolutional Neural Networks) uses two convolutional layers to identify patterns and spatial correlations in images by performing element-wise operations. Activation functions introduce nonlinearity, enabling the network to detect complex relationships between visual components. Max-pooling layers are used for spatial downsampling, which involves extracting complex properties one by one. After the convolutional layers, two dense layers are employed to integrate features, with biases that increase expressiveness. The final layer utilizes SoftMax activation for classification. This model architecture is designed for classification and regression problems.

In the above Fig. 4 as shown, during the training process, regularization techniques are employed to minimize overfitting and enhance generalization. Gradient-based optimization techniques, most likely using backpropagation, are utilized for training. As seen in Fig. 3, to enhance model accuracy and efficiency, the training procedure is repeated across 30 epochs. The accuracy obtained from the CNN classifier is 95%.

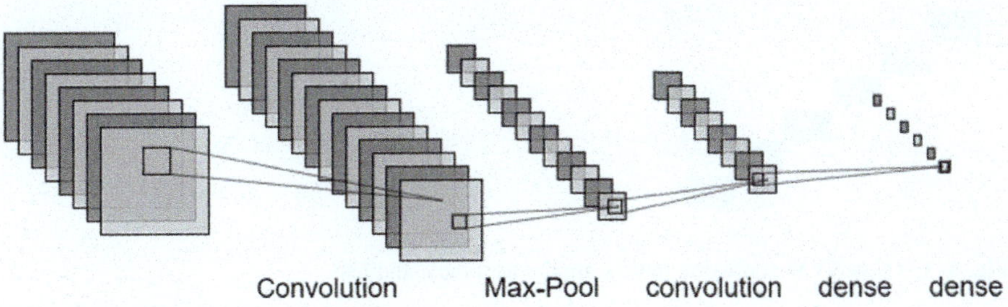

Fig. 3 CNN architecture

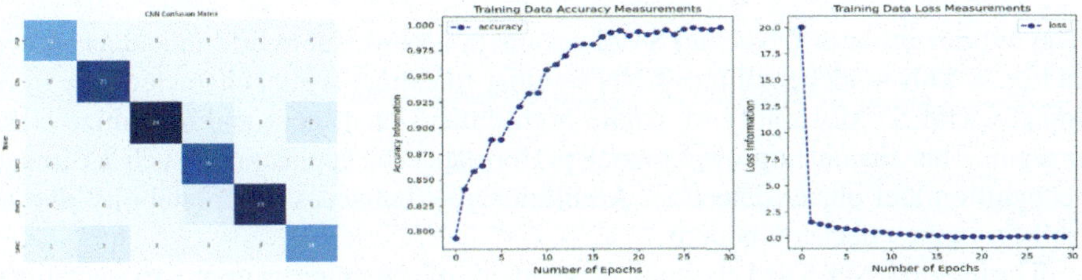

Fig. 4 CNN confusion matrix, model accuracy, and loss

4.2 *CNN + RF*

The combination of Convolutional Neural Networks (CNNs) and Random Forests (RF) creates a hybrid model that takes advantage of both techniques. This synergy enables the hybrid model to use RF's ensemble learning capabilities, which are particularly useful for handling large feature spaces and capturing nonlinear relationships. Additionally, the hybrid model utilizes CNN's ability to capture spatial data within images, making it highly effective for image analysis and pattern identification tasks.

This combination improves the model's prediction performance and interpretability, making it suitable for a variety of applications across multiple domains shown in Fig. 5. The accuracy obtained from the CNN classifier with hybridization of RF is 98.39%.

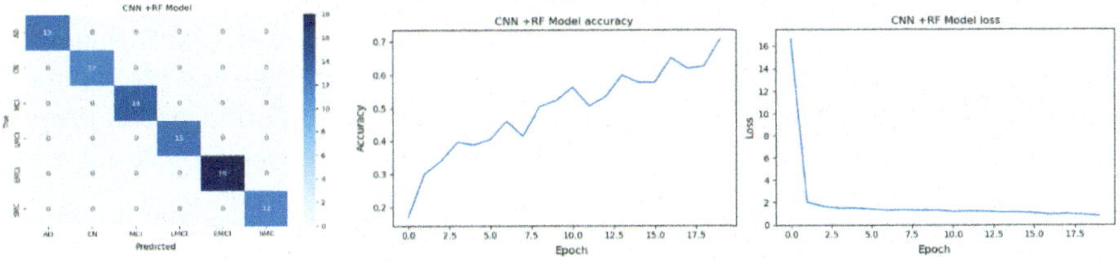

Fig. 5 CNN + RF confusion matrix, model accuracy, and loss

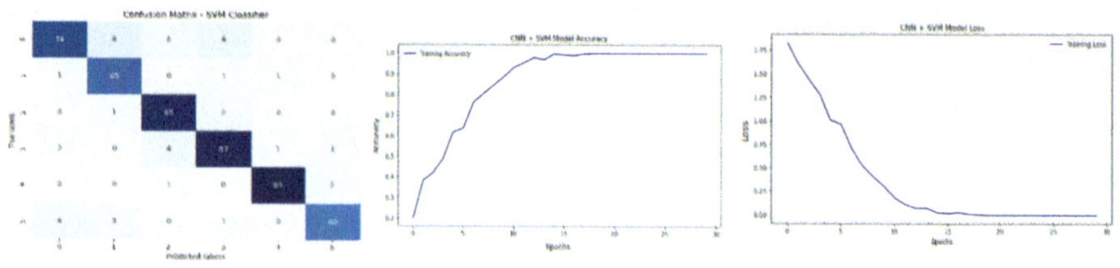

Fig. 6 CNN + SVM confusion matrix, model accuracy, and loss

4.3 CNN + SVM

The hybridization of CNN and SVM results in a powerful hybrid model as shown in Fig. 6. This model combines CNN's ability to extract spatial characteristics from images with SVM's ability to handle high-dimensional input and maximize class margin. This fusion improves model performance in applications such as image recognition and object detection. Additionally, it enhances interpretability due to SVM's explicit decision bounds.

The hybrid CNN-SVM classifier was capable of achieving an impressive accuracy rate of 92.55%. This performance is further supported by a comprehensive confusion matrix, which indicates that all six categories have low false negatives. This demonstrates the model's robustness in accurately classifying cases across different categories and its effectiveness in dealing with complicated and diverse datasets.

4.4 ANN

ANNs are computational models influenced by the human brain that tackle complicated problems. There are three layers in an artificial neural network (ANN) input, hidden, and output as shown in Fig. 7. Nonlinearity is introduced by neurons in hidden layers applying activation functions to weighted sums of inputs. During training, the network learns by modifying connection weights to minimize a predetermined loss function. ANNs are particularly good at learning detailed patterns in a wide range of fields, including natural language processing and computer vision.

The artificial neural network (ANN) achieved an accuracy of 91.14%, indicating its ability to make correct predictions. Furthermore, the model's performance is supported by an F1 score of 91.31%, demonstrating a balance between precision and recall for all classes. The confusion matrix (as shown in Fig. 8) offers useful insights into the model's behavior, revealing its ability to minimize misclassifications while maintaining exceptional performance across all classes.

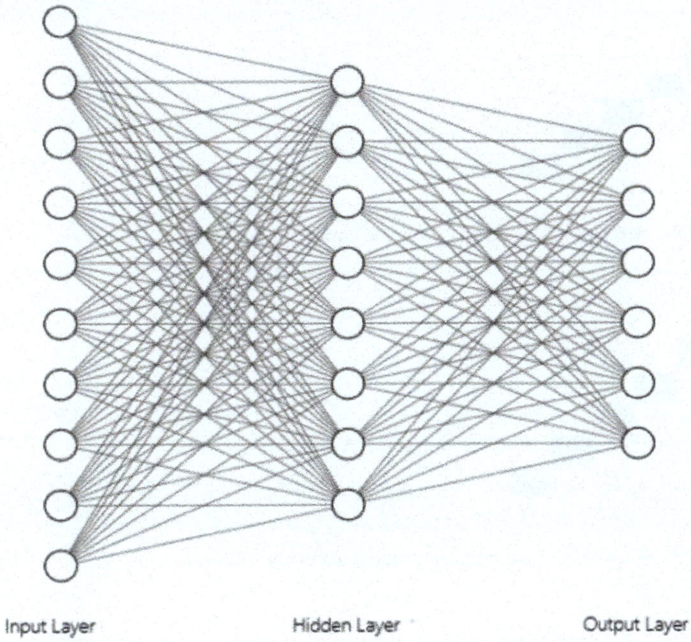

Fig. 7 ANN architecture

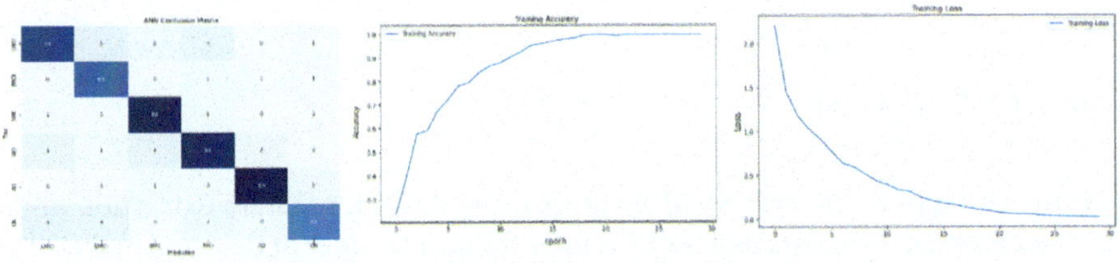

Fig. 8 ANN confusion matrix, ANN model accuracy, and loss

4.5 ANN + RF

Hybridization of ANN with the RF, ANN is used for feature extraction of the preprocessed data, and RF is used as a classifier or regressor. The advantages of both approaches are combined in this hybrid approach: While ANNs are excellent at learning intricate patterns, RF provides robustness and interpretability. It can handle both organized and unorganized data with ease. When used together, ANN and RF offer better generalization and prediction accuracy for a range of machine learning applications.

The hybridization of the artificial neural network (ANN) with Random Forest (RF) resulted in an accuracy of 92.35% (as shown in Fig. 9). This highlights the

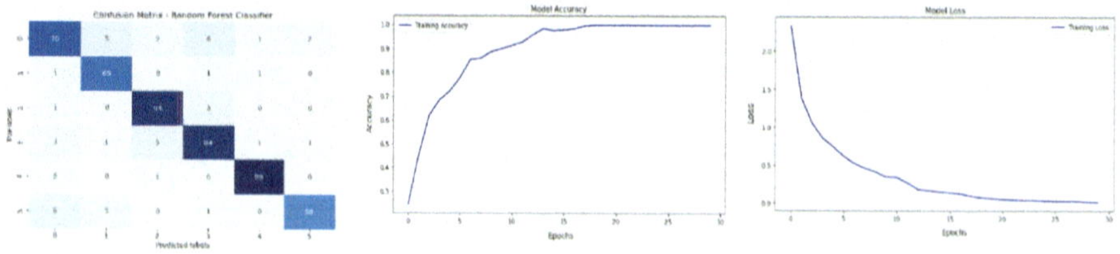

Fig. 9 ANN + RF confusion matrix, model accuracy, and loss

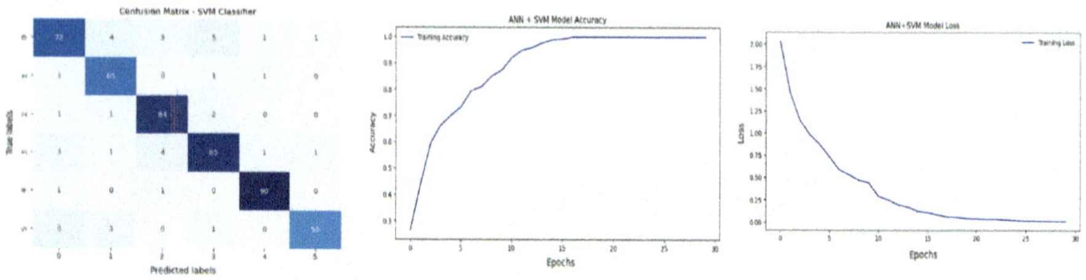

Fig. 10 ANN + SVM confusion matrix, model accuracy, and loss

effectiveness of combining ANN's ability to capture complex patterns with RF's ensemble learning, resulting in a strong model that can achieve high accuracy and balanced performance across all classes.

4.6 ANN + SVM

Hybridization of ANN with SVM involves using ANN for feature extraction of the relevant preprocessed data and SVM is used for the classifier or decision maker. The advantages of SVMs in handling high-dimensional data with discernible margins and ANN's capacity to learn intricate patterns are combined. It provides stronger robustness and better generalization across a range of machine learning tasks.

The integration of ANN and SVM resulted in an accuracy of 92.35%. This technique combines the ability of ANN to recognize complicated patterns with SVM's ability to establish clear boundaries between groups as shown in Fig. 10. They work together seamlessly, ensuring precise predictions across various types of data.

5 Results

In the above Table. 1, we compared the outcomes of the model by calculating from formulas. RF and SVM provide better results than individual models. Compared to the basic ensemble methods, the hybridization of the models gives better results. The

Table 1 Comparison of results

Models	Accuracy (%)	Precision (%)	Recall (%)	F1 Score (%)
CNN	95	95.64	95	95.1
CNN + RF	98.39	98.42	98.39	98.39
CNN + SVM	92.55	92.39	92.35	92.33
ANN	91.14	91.34	91.21	91.31
ANN + RF	92.35	92.45	92.35	92.34
ANN + SVM	92.34	92.39	92.35	92.33

training dataset includes about 1000 MRI scan images preprocessed using DTCWT, each in patients classified into AD, CN, EMCI, MCI, LMCI, and SMC. When these machine learning techniques are hybridized the performance of the code increases. The accuracy of the CNN hybridized with RF increases to 98.39%, whereas CNN hybridized with SVM accuracy reduces to 92.55% as well as is not that stable in providing consistent results. The accuracy of the ANN hybridized with RF and SVM increases to 92.35 and 92.34%. This shows promise in improving the robustness and stability of the model, potentially leading to more reliable diagnostic outcomes in clinical settings.

6 Conclusion

The hybrid approach combining CNN and ANN has shown great promise in improving the diagnosis of AD at various stages. The overall accuracy of this approach has been remarkable, exceeding 90%. This method has the potential to make clinical diagnoses easier and more reliable. Consistent training can reduce false negatives, this increases the model's performance and enables early detection of illnesses like MCI, which can eventually lead to AD. This type of model has significant therapeutic applications, such as supporting early intervention programs and improving overall patient care by identifying distinct classes.

References

1. Sujathakumari BA, Kulkarni SP, Chinmaya A, Suhas HJ, Jayanth KG, Gowtham R (2023) Enhancing MCI detection with a hybrid machine learning approach, Atlantis highlights in computer sciences 18. In: Proceedings of the fourth international conference on advances in computer engineering and communication systems (ICACECS 2023)
2. Kundaram SS, Pathak KC (2020) Deep learning-based Alzheimer Disease detection. In: Proceedings of the fourth international conference on microelectronics, computing and communication systems. lecture notes in electrical engineering, vol 673. Springer

3. Salehi AW, Sharma BB, Gupta G, Baglat P, Upadhya A (2020) A CNN model: earlier diagnosis and classification of Alzheimer disease using MRI. In: IEEE 2020 International conference on smart electronics and communication (ICOSEC)
4. Naz S, Ashraf A, Zaib A (2021) Transfer learning using freeze features for Alzheimer neurological disorder detection using ADNI dataset. Springer
5. Gunawardena KANNP, Rajapakse RN, Kodikara ND (2017) Applying convolutional neural networks for pre-detection of Alzheimer's disease from structural MRI data. In: IEEE 24th International conference on mechatronics and machine vision in practice (M2VIP)
6. AL-Shoukry S, Rassem TH, Makbol NM (2005) Alzheimer's Diseases detection by using deep learning algorithms: a mini-review. In: 2005 IEEE engineering in medicine and biology 27th annual conference
7. Nair J, Mohan N (2017) Alzheimer's disease diagnosis in MR images using statistical methods. In: IEEE International conference on communication and signal processing, vol 17, pp 1232–1235
8. Rabeh B, Benzarti F, Amiri H (2017) New method of classification to detect Alzheimer disease. In: IEEE 14th International conference on computer graphics, imaging and visualization, vol 17, pp 111–116
9. Mufidah R, Wasito I, Hanifah N, Faturrahman M (2017) Structural MRI classification for Alzheimer's disease detection using deep belief network. vol 17, pp 37–42
10. Chitradevi D, Prabha S, Sakthidasan K (2018) Brain hemisphere analysis using genetic algorithm and fuzzy clustering in Alzheimer disease. Int Conf Commun Signal Process 18:901–905
11. Raut A, Dalal V (2017) A machine learning based approach for detection of Alzheimer's disease using analysis of hippocampus region from MRI scan. In: IEEE International conference on computing methodologies and communication (ICCMC), vol 17, pp 236242
12. Angkoso CV, Purnama I, Purnomo M (2018) Analysis of brain tissue and cerebrospinal fluid feature for Alzheimer's disease Detection. In: IEEE International conference on computer engineering, network and intelligent multimedia (CENIM), vol 31, pp 285–288
13. Song T, Chowdhury S, Yang F,Jacobs H, Fakhri G,Li Q, Johnson K, Dutta J (2019) Graph convolutional neural networks for Alzheimer's disease classification, vol 19, pp 414–417
14. Padole H,Joshi SD, Gandhi TK (2018) Early detection of Alzheimer's disease using graph signal processing on neuroimaging data. In: 2nd European conference on electrical engineering and computer science (EECS), vol 18, pp 302–306
15. Kaya A, Keceli AS, Catal C, Yalic H, Temucina H, Tekinerdogan B (2019) Analysis of transfer learning for deep neural network based plant classification models. Comput Electron Agric 158:20–29
16. Zhang Y, Pan C, Sun J, Tang C (2018) Multiple sclerosis identification by convolutional neural network with dropout and parametric ReLU. J Comput Sci 28:1–10
17. Shen W, Wang W ()2018 Node identification in wireless network based on convolutional neural network. In: IEEE 14th International conference on computational intelligence and security (CIS), vol 31, pp 238–241
18. Mazo C, Bernal J, Trujillo M, Alegre E (2018) Transfer learning for classification of cardiovascular tissues in histological images. Comput Methods Programs Biomed 18:1–20
19. Unnikrishnan A, Sowmya V, Soman KP (2018) Deep alexnet with reduced number of trainable parameters for satellite image classification. Procedia Comput Sci 143:931–938
20. Marcus S, Wang H, Parker J, Csernansky G (2007) Open access series of imaging studies (OASIS): cross-sectional MRI data in young, middle aged, nondemented, and demented older adults 19:1498–1507
21. Chow LS, Paramesran R (2016) Review of medical image quality assessment. Biomed Signal Process Control 27:145–154
22. Burkea S, Maramaldib P, Cadet T, Kukull W (2016) Neuropsychiatric symptoms and apolipoprotein e: associations with eventual Alzheimer's disease development. Arch Gerontol Geriatr 65:231–238

23. Sarwinda D, Bustamam A (2016) Detection of Alzheimer's disease using advanced local binary pattern from Hippocampus and whole brain of MR images. vol 16, pp 5051–5056
24. Balamurugan M, Nancy A, Vijaykumar S (2017) Alzheimer's disease diagnosis by using dimensionality reduction based on KNN Classifier. Biomed Pharmacol J 10:1823–1830
25. Hariri W, Tabia H, Farah N, Benouareth A, Declercq D (2016) 3D Face recognition using covariance based descriptors. Pattern Recogn Lett 78:1–7

Analysing the Efficacy: A Critical Assessment of a Symptom Analysis and Initial Diagnosis Chatbot for Disease Detection

Nikit Mathur, Abhinav Raja Raizada, and R. Jeya

Abstract This paper encompasses the development and evaluation of a disease diagnosis chatbot developed to help users assess their symptoms and get preliminary diagnoses. The lack of access to medical knowledge and the high costs of healthcare are the primary obstacles for the general population. Medical chatbots, specifically created for medical uses, could overcome these challenges (Smith and Doe in J Healthc Technol 12:234–245, 2021 [1]). The subsequent document delineates the expanding utilisation of artificial intelligence and machine learning within the medical domain, encompassing the chatbot's design methodology, system architecture (shown in Fig. 6), and practical implementation. The chatbot analyses user inputs and provides pertinent medical information by combining machine learning models and rule-based algorithms.

Keywords Disease diagnosis chatbot · Artificial intelligence (AI) · Machine learning (ML) · Natural language processing (NLP) · k-Nearest Neighbours (k-NN)

1 Introduction

The incorporation of artificial intelligence (AI) and machine learning (ML) technology has brought about a significant transformation in several industries, including the medical field.

N. Mathur (✉) · A. Raja Raizada · R. Jeya
Department of Computing Technologies, SRM Institute of Science and Technology, Kattankulathur, Tamil Nadu, India
e-mail: nn2254@srmist.edu.in

A. Raja Raizada
e-mail: aa3515@srmist.edu.in

R. Jeya
e-mail: jeyar@srmist.edu.in

© The Author(s), under exclusive license to Springer Nature Singapore Pte Ltd. 2025
H. Zen et al. (eds.), *Soft Computing and Signal Processing*, Lecture Notes in Networks and Systems 1200, https://doi.org/10.1007/978-981-97-9926-8_32

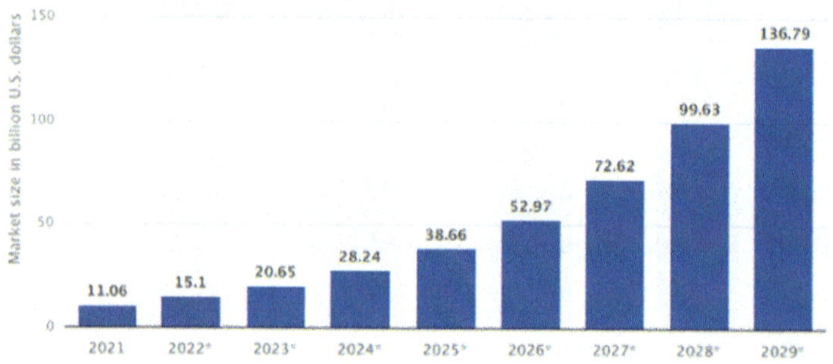

Fig. 1 Growth of AI-ML in healthcare [5]

The application of AI and ML algorithms to the analysis of enormous quantities of medical data—including patient records, imaging scans, and genomic sequences—has substantially improved disease diagnosis [2].

The growth of artificial intelligence (AI) and machine learning (ML) in healthcare has been remarkable (shown in Fig. 1), revolutionising various aspects of the industry from diagnosis and treatment to patient care and administrative tasks. AI and ML technologies have enabled healthcare providers to leverage vast amounts of data to improve medical decision-making, streamline processes, and enhance patient outcomes [3, 4].

These technologies enable the early detection of diseases, enhance the precision of diagnoses, and facilitate the development of individualised treatment plans that are tailored to the requirements of each patient. Furthermore, diagnostic tools propelled by artificial intelligence have exhibited functionalities that are on par with, if not exceeding, those of human specialists in specific medical domains [6].

AI and ML algorithms aid healthcare providers in treatment planning and decision-making through the synthesis of clinical data, prediction of treatment outcomes, and recommendation of optimal interventions.

These technologies enhance therapeutic efficacy, accelerate innovation, and expedite the research process, spanning from drug discovery and development to precision medicine.

The application of AI and ML within the medical sector has the capacity to significantly transform the provision of healthcare (shown in Fig. 2), enhance patient results, and propel medical research forward. Sustained innovation and interdisciplinary cooperation are imperative for optimising the advantages offered by these paradigm-shifting technologies, while also confronting the obstacles that arise in their path and guaranteeing fair and impartial availability of superior healthcare services.

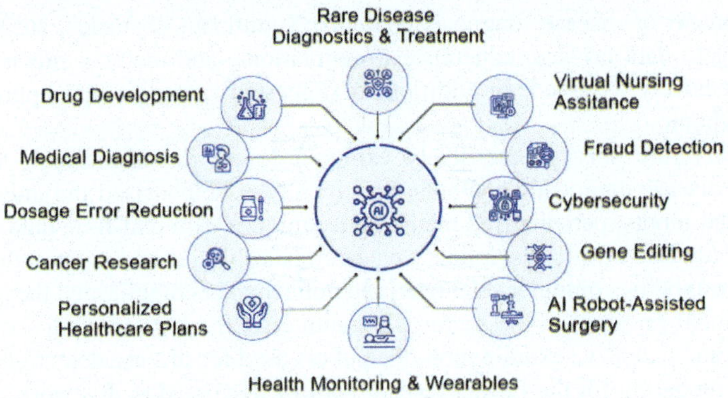

Fig. 2 Applications of AI in healthcare [7]

2 Literature Survey

The literature review undertaken for this research paper comprehensively examined the domain of disease diagnosis chatbots, with a specific emphasis on their evolution, application, and influence within the healthcare sector.

The review began by providing an overview of the increasing significance of chatbots in the healthcare sector, discussing their capacity to bring about significant changes in patient care, symptom evaluation, and medical consultation. Following that, a comprehensive examination was conducted on extant chatbots for disease diagnosis, including both commercially available solutions and research prototypes. The assessment criteria comprised technological foundations, performance metrics, user satisfaction, and features. Significant discoveries underscored the variety of methodologies utilised, which extended from rule-driven systems to intricate machine learning models [8].

The utilisation of machine learning techniques in disease diagnosis chatbots has become crucial, necessitating an exhaustive investigation into their implementation. An investigation was conducted into the effectiveness of different methodologies, such as supervised learning, unsupervised learning, and ensemble methods, in the context of disease prediction and classification. Simultaneously, the survey explored the domain of natural language processing (NLP) in the healthcare sector, providing an explanation of its importance in the processing of medical texts, patient documents, and user inquiries. Research focused on sentiment analysis, named entity recognition, and text classification has significantly advanced our knowledge of how NLP can be utilised to improve the functionality of chatbots [9].

Constant to the discourse was the incorporation of medical knowledge bases into the architectures of chatbots. The significance of exhaustive knowledge repositories, which include databases on diseases, symptoms, treatment guidelines, and drugs, has been clarified in the literature. The review additionally examined usability testing methodologies, user interaction design principles, and user engagement strategies in the context of disease diagnosis chatbots. Significant obstacles, such as those pertaining to data privacy, adherence to regulations, and medical precision, were comprehensively assessed, in addition to suggested remedies and approaches to alleviate them.

The literature survey extensively explored future directions and developing trends in disease diagnostic chatbots. Specifically, it emphasised the integration of multimodal inputs, personalised medicine techniques, and real-time data analytics. The investigation of interdisciplinary research possibilities and cooperation among healthcare experts, computer scientists, and policymakers highlighted the possibility for additional progress in the sector. To summarise, the literature survey offered a thorough and extensive examination of chatbots used for disease detection. This has laid the foundation for the further research efforts discussed in this work.

3 Overview of the System

The system operates using conversational data supplied by the user throughout the course of the conversation. The objective is to direct attention towards the initial symptoms and challenges that the user might be encountering.

Once the automated medical chatbot has gathered sufficient information from the initial exchange, it proceeds with the conversation by posing inquiries to the user and attempting to diagnose diseases by converting the input data into queries that it executes in order to obtain the remedy for any illness the user may be experiencing.

Following the bot's ranking of the potential diseases that the user might be experiencing, a shortlist of those diseases has been generated [10]. The chatbot (shown in Fig. 3) initiates an inquiry regarding the user's emotional state. Once it has accumulated the specified quantity of data, it determines the disease that the user is most likely afflicted with based on the input data.

Once the bot has identified the ailment afflicting the user, it assesses the severity of the condition and responds with either medication and remedy recommendations or a referral to a medical professional if the severity metric exceeds a predetermined threshold.

3.1 Database for the System

The database plays a vital role in disease diagnosis chatbots by furnishing essential medical knowledge and data that aid in the evaluation of symptoms, identification of diseases, and formulation of treatment suggestions.

Medical Knowledge Base

An exhaustive compilation of medical knowledge, including details on symptoms, risk factors, diagnostic criteria, treatment alternatives, medications, and potential

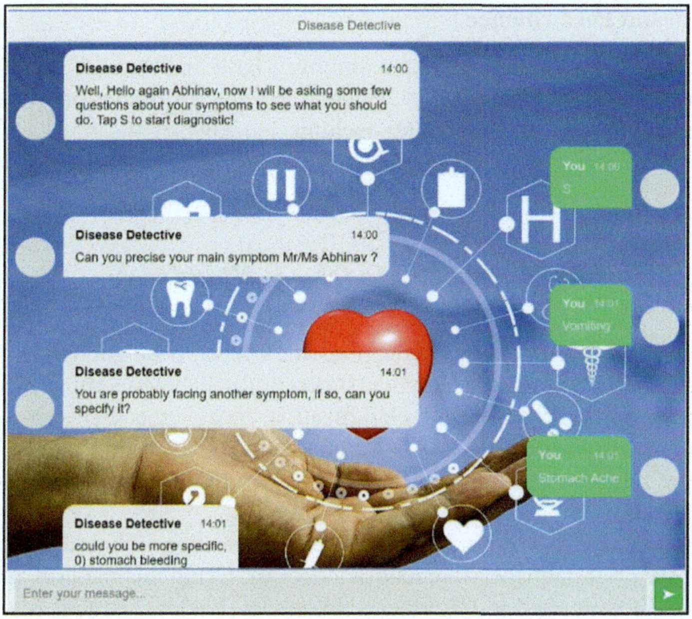

Fig. 3 Working of the chatbot

complications, is housed in the database. A compilation of authoritative medical sources, textbooks, clinical guidelines, research papers, and expert opinions comprise this knowledge base.

Structured Data

The database is made up of tables or other entities that are representative of various medical informational components. These tables contain structured data. For instance, distinct tables are provided for conditions, signs, drugs, therapies, lab work, and patient demographics. Relevant details including name, description, related symptoms, diagnostic standards, and treatment recommendations.

Medical Ontologies and Taxonomies

To enable effective data retrieval and semantic understanding, the database integrates medical ontologies and taxonomies. These frameworks allow the chatbot to comprehend user inquiries, browse the medical knowledge source, and draw reliable conclusions by defining standardised vocabularies, hierarchies, and linkages among medical concepts.

Scalability and Performance

As the chatbot engages with an expanding user base and manages a high volume of inquiries, the database design is scalable and able to handle many requests at once effectively. Strategies including caching, indexing, replication, and database partitioning are used to maximise efficiency and guarantee responsiveness.

3.2 *The Machine Learning Model—K-NN Model*

For classification and regression applications, the k-Nearest Neighbours (k-NN) algorithm is a straightforward yet effective supervised machine learning method. The k-NN technique, when applied to classification, locates the k closest data points (neighbours) to a given input data point in a feature space using a distance metric (such as the Euclidean distance). The input data point is subsequently assigned to the majority class among the k-Nearest Neighbours.

The k-NN model is used in chatbots as a classification technique to group user inquiries into pre-established intents or categories. Using this method, a dataset of labelled instances is used to train the k-NN model. Each example is made up of a feature vector that represents a user query and the intent label that corresponds to it. Upon receiving a new user inquiry, the chatbot uses a distance measure like Euclidean distance to determine how similar the feature vector of the query is to that of instances in the training dataset. The anticipated intent for the user question is then assigned to the k-Nearest Neighbours of the input query, based on which neighbour has the majority class. Based on user input, the chatbot can give relevant and contextually suitable interactions by using this expected intent to decide the appropriate response or action.

Data Representation

To utilise the k-NN technique, the initial stage involves the representation of the data. Every individual data point in the collection is depicted as a feature vector within a space that has multiple dimensions. These feature vectors represent the properties or characteristics of the data points. For instance, in a dataset of houses, the feature vectors may encompass parameters such as dimensions, bedroom count, bathroom count, and geographical position.

Distance Metric

The k-NN algorithm employs a distance metric to quantify the resemblance between data points in the feature space. The Euclidean distance is the most frequently employed metric for measuring distance. However, alternative metrics like Manhattan distance or cosine similarity can also be utilised [11]. The distance between two feature vectors, denoted as x and y, is computed using the following formula:

The equation is defined as follows:

$$d(x, y) = \sum (x_i - y_i)^2 \quad (1)$$

where i ranges from 1 to n.

Finding Neighbours

Given a new data point labelled as x_{new}, the k-NN algorithm calculates the nearest neighbour to a new data point x by calculating their distances by calculating their distances to x in the training dataset. This task entails computing the distance between the new data point and each data point in the training dataset. Subsequently, the k data points with the shortest distances are chosen.

Averaging

Once the k-Nearest Neighbours are identified, the algorithm assigns the majority class label among these neighbours to the new data point $x_{new.}$ If x_{new} is assigned to class C_i by k neighbours and to class C_j by l neighbours, where k > l, then x_{new} is classified as $C_{i.}$ In regression tasks, the algorithm calculates the average of the target values of the k-Nearest Neighbours and assigns it to $x_{new.}$

Choosing K

When using the k-NN algorithm, selecting the value of k is a critical decision that can have a substantial impact on the performance of the model. A low value of k can result in overfitting, where the model becomes too responsive to noise in the data, while a high value of k can lead to underfitting, where the model is overly simplistic and fails to accurately represent the underlying patterns in the data. The value of k is usually determined by methods like cross-validation or grid search. These techniques include evaluating several values of k on a validation dataset and selecting the one that produces the greatest performance indicator, such as accuracy or mean squared error. In the training dataset, the distances of the data points to the new point x are used. This entails computing the distance between the new point, denoted as xnew, and each data point in the training dataset. Subsequently, the k data points with the shortest distances are chosen.

Decision Boundary

The decision boundary of the k-NN algorithm in classification problems is non-linear and is dictated by the distribution of the data points in the feature space. Data points belonging to distinct classes can exhibit clustering or intricate geometric connections, resulting in the formation of non-linear decision boundaries.

Scalability

Scalability is one of the most significant disadvantages of the k-NN algorithm. As the size of the training dataset increases, so does the computational burden associated with determining the k-Nearest Neighbours for every additional data point. Further, for enormous datasets containing millions or billions of data points, the algorithm's requirement of storing the entire training dataset in memory may prove impracticable.

Normalisation

Normalisation of the data is an additional factor that must be taken into account when implementing the k-NN algorithm. Due to the algorithm's reliance on distances between data points for operation, distance calculations may be influenced more heavily by features with larger scales. Consequently, normalising the feature values to have a mean of zero and a variance of one prior to implementing the k-NN algorithm is standard procedure.

An advantage of k-NN is its simplicity, which facilitates comprehension and implementation for chatbot developers that prioritise simplicity and ease of execution (shown in Fig. 4) [13]. Moreover, the non-parametric character of k-NN enables it to effectively handle diverse data distributions and decision boundaries, rendering it well-suited for chatbot applications that include complicated or irregularly distributed data. One additional benefit of k-NN is its flexibility in handling user input. It does not require a training phase and can immediately adjust to fresh user input without the need to retrain the model. This flexibility is advantageous in chatbot applications because the user's inquiries and preferences may change over time. Moreover, the capacity to understand and explain the predictions made by k-NN, which are determined by the majority class among the k-Nearest Neighbours, can be beneficial in chatbot applications where the ability to provide clear explanations is crucial for establishing user trust and confidence. Moreover, the k-NN algorithm's instance-based learning methodology enables the chatbot to deliver tailored responses by leveraging commonalities with previous interactions, hence boosting the user experience.

Nevertheless, it is crucial to take into account the constraints of k-NN, like its computational cost, memory utilisation, and sensitivity to distance metric, when assessing its appropriateness for chatbot applications.

3.3 Integration with Flask

Flask is a flexible and lightweight Python web framework. Its straightforward and user-friendly design makes it an outstanding option for the development of microservices, Application Programming Interfaces (APIs), and web applications. Flask is renowned for its minimalist design philosophy, which enables programmers to construct web applications with unwarranted speed and efficiency while avoiding superfluous intricacy [14, 15].

The process of integrating a k-Nearest Neighbours (k-NN) model with Flask entails the integration of the model into a Flask web application, thereby generating a web-based interface that facilitates classifications or predictions. This integration facilitates user interaction with the k-NN model via a web browser, thereby enhancing its accessibility and user-friendliness.

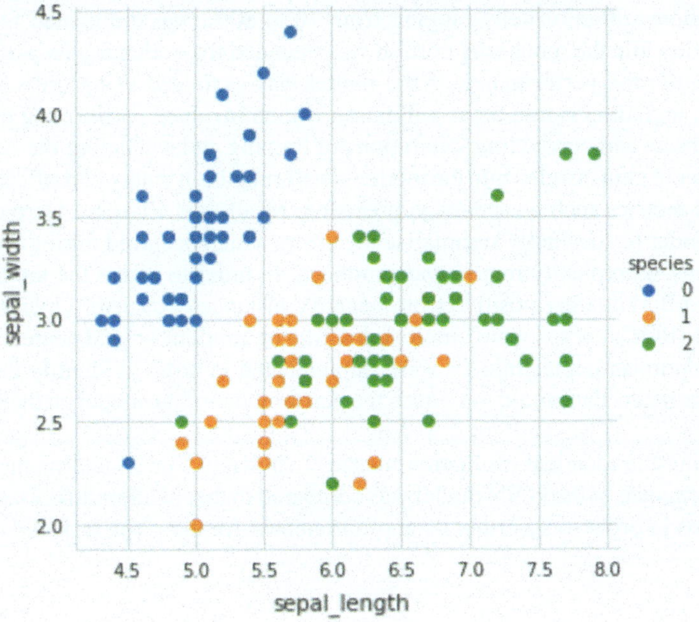

Fig. 4 Classification using k-NN model [12]

In order to configure the Flask application, the required modules were imported and the Flask app object was initialised. Routes were established to access the various endpoints of the web application. To generate the user interface of the web application, HTML templates were developed. These templates comprised forms that allowed users to enter data and submit predictions. The Flask application established a route to manage POST requests that comprise user input data. During this process, the input data was obtained from the request, pre-processed if required, and the trained k-NN model was utilised to generate predictions. In response to the POST request, the k-NN model's predictions were returned in JSON format and re-encoded for presentation on the client side. In general, the amalgamation of Flask and the k-NN model yielded an intuitive web interface that facilitated user interaction with the model and real-time prediction generation.

3.4 Model Evaluation

The performance and efficacy of a k-Nearest Neighbours (k-NN) model in generating predictions or classifications on unobserved data were assessed during its evaluation. A multitude of assessment metrics and methodologies were employed to gauge the generalisability, robustness, and accuracy of the model. At the outset, the dataset was partitioned into training and testing sets.

A random subset, usually ranging from 70 to 80%, was designated for training the model, while the remaining portion was intended for evaluating its performance. Furthermore, the performance of the model was evaluated using cross-validation methods, including k-fold cross-validation, which involved partitioning the dataset into k subsets and conducting k iterations of training and evaluating the model.

In order to gain insights into the model's performance in binary classification tasks, appraisal metrics such as accuracy, precision, recall, and F1-score were computed [16]. In order to succinctly summarise the errors which occurred during the model's predictions, a perplexity matrix was produced. In order to assess the model's capability of differentiating positive from negative instances, Receiver Operating Characteristic (ROC) curves were employed. Utilising random or grid search, hyperparameter optimisation techniques were implemented in order to identify the optimal hyperparameters that would maximise the performance of the model on the validation set.

Utilising the identical evaluation metrics (shown in Fig. 5) and methodologies, the performance of the k-NN model was compared to that of alternative classification algorithms in order to ascertain its appropriateness for the given task and dataset.

4 System Architecture

The primary focus is the user interface (frontend), through which users engage with the chatbot through a web interface constructed using Flask, a Python web framework known for its lightweight nature. The frontend interface has been intentionally developed to be user-friendly, offering individuals a streamlined environment in which to enter their symptoms, and obtain diagnostic outcomes [17]. Flask routes manage user requests, render HTML templates, and facilitate downstream server communication in this configuration (Fig. 6).

In the interim, the chatbot controller functions invisibly, coordinating the manipulation of user input, overseeing the conversational flow, and establishing interfaces with additional system elements. The frontend interface is utilised to receive user queries, which are then pre-processed before being passed to the k-NN model for prediction. By employing natural language processing (NLP) methodologies, the controller is capable of deciphering user inquiries and producing suitable replies, thereby guaranteeing a smooth and uninterrupted interaction encounter.

The k-NN model serves as the foundational element in this architectural framework for disease diagnosis. The training data that this model retains consists of annotated instances of symptoms accompanied by their respective disease diagnoses. The k-NN model computes the distance to the adjacent neighbours and predicts the most probable disease diagnosis when presented with new symptom data from the user. The ability to predict outcomes is crucial for delivering precise and timely evaluations of user symptoms.

```
print(classification_report(y_test,knn_clf.predict(X_test)))
                                          precision    recall  f1-score   support

(vertigo) Paroymsal  Positional Vertigo       1.00      1.00      1.00         1
                                   AIDS       1.00      1.00      1.00         1
                                   Acne       1.00      1.00      1.00         1
                    Alcoholic hepatitis       1.00      1.00      1.00         1
                                Allergy       1.00      1.00      1.00         1
                              Arthritis       1.00      1.00      1.00         1
                       Bronchial Asthma       1.00      1.00      1.00         1
                   Cervical spondylosis       1.00      1.00      1.00         1
                            Chicken pox       1.00      1.00      1.00         1
                    Chronic cholestasis       1.00      1.00      1.00         1
                            Common Cold       1.00      1.00      1.00         1
                                 Dengue       1.00      1.00      1.00         1
                              Diabetes        1.00      1.00      1.00         1
        Dimorphic hemmorhoids(piles)          1.00      1.00      1.00         1
                          Drug Reaction       1.00      1.00      1.00         1
                        Fungal infection      1.00      1.00      1.00         1
                                   GERD       1.00      1.00      1.00         1
                        Gastroenteritis       1.00      1.00      1.00         1
                           Heart attack       1.00      1.00      1.00         1
                            Hepatitis B       1.00      1.00      1.00         1
                            Hepatitis C       1.00      1.00      1.00         1
                            Hepatitis D       1.00      1.00      1.00         1
                            Hepatitis E       1.00      1.00      1.00         1
                          Hypertension        1.00      1.00      1.00         1
                       Hyperthyroidism        1.00      1.00      1.00         1
                           Hypoglycemia       1.00      1.00      1.00         1
                         Hypothyroidism       1.00      1.00      1.00         1
                               Impetigo       1.00      1.00      1.00         1
                               Jaundice       1.00      1.00      1.00         1
                                Malaria       1.00      1.00      1.00         1
                               Migraine       1.00      1.00      1.00         1
                        Osteoarthristis       1.00      1.00      1.00         1
            Paralysis (brain hemorrhage)      1.00      1.00      1.00         1
                    Peptic ulcer diseae       1.00      1.00      1.00         1
                              Pneumonia       1.00      1.00      1.00         1
                              Psoriasis       1.00      1.00      1.00         1
                           Tuberculosis       1.00      1.00      1.00         1
                                Typhoid       1.00      1.00      1.00         1
```

Fig. 5 Evaluation metrics (a)

In addition, a Data Pre-processing module functions in conjunction with the chatbot controller to convert unprocessed user input into a format that is appropriate for input into the k-NN model. By means of data cleansing and standardisation, this module prepares the input data for prediction and extracts pertinent features. The accurate generation of diagnoses by the k-NN model is facilitated by the pre-processing module through the verification of input data formatting.

The integration with Flask is crucial in order to facilitate uninterrupted communication between the system's backend and frontend interfaces. Incoming HTTP

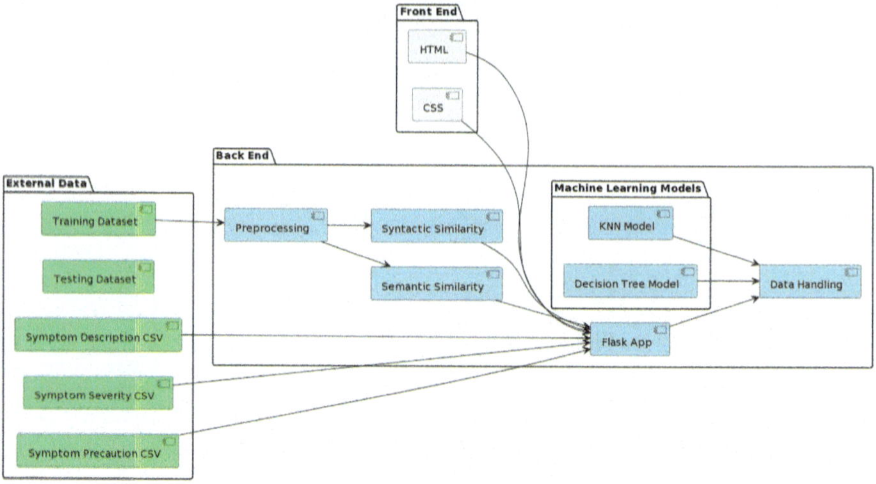

Fig. 6. System Architecture

requests from the frontend are managed by Flask routes, which facilitate user input, initiate disease diagnosis, and return diagnostic outcomes to the frontend. By serving as an intermediary between the user interface and the backend functionality, Flask ensures that the chatbot system operates without interruption.

5 Potential Applications in the Healthcare Sector

Implementing a chatbot for disease diagnosis can yield substantial advantages for healthcare systems and individuals alike, as it can be utilised in a multitude of practical contexts.

5.1 Remote Access to Healthcare

In regions characterised by limited access to healthcare facilities, such as remote or rural areas, individuals might encounter difficulties in obtaining prompt medical advice and diagnosis. The implementation of a chatbot for disease diagnosis offers individuals a practical and easily navigable medium through which they can evaluate their symptoms and obtain initial diagnoses, obviating the necessity for them to physically visit a healthcare facility [18].

5.2 Triage of Symptoms and Triage Assistance

Healthcare hotlines and emergency departments frequently encounter substantial call volumes, which can result in disruptions to the provision of care for critical medical concerns. Symptom triage can be facilitated by a chatbot that guides users through a series of inquiries designed to evaluate the intensity of their symptoms and deliver suggestions in accordance with the criticality of their condition. This practice aids in the prioritisation of patients in need of urgent medical care and diminishes superfluous visits to emergency departments.

5.3 Healthcare Preventive Measures and Health Education

In the realm of preventive healthcare and health education, a chatbot can be an invaluable asset by dispensing information to users pertaining to risk factors, healthy lifestyle choices, and disease prevention. Through active user participation and the provision of individualised health recommendations, the chatbot promotes the adoption of preventative strategies that safeguard one's general health and well-being.

5.4 Chronic Illness Administration

Consistent monitoring and management of conditions such as diabetes, hypertension, or asthma are essential for individuals afflicted with these ailments. By reminding patients to take their medications, monitoring their vital signs, tracking their symptoms, and providing personalised recommendations for disease management based on their individual health status, a chatbot can assist in the management of chronic diseases.

5.5 Distant Consultations and Telemedicine

To enable remote consultations between patients and healthcare providers, telemedicine platforms employ chatbots. Preceding the consultation, healthcare providers can conduct more efficient and effective virtual visits by having a chatbot gather germane patient information such as medical history, symptoms, and other relevant data. Post-consultation patient follow-up, appointment scheduling, and appointment reminder transmission are all tasks that chatbots can perform in addition.

5.6 *Public Health Surveillance and Detection of Outbreaks*

Chatbots have the potential to significantly contribute to public health surveillance and outbreak detection through the acquisition of real-time data from users regarding disease prevalence, symptoms, and geographic distribution. Through the analysis of this data, it is possible to discern nascent health trends, anticipate disease outbreaks, and gain insights for resource allocation and public health interventions.

5.7 *Support for Clinical Decisions*

Clinical decision support tools, such as chatbots, have the potential to assist healthcare providers in various tasks including disease diagnosis, treatment option selection, and medical test result interpretation. Chatbots have the capability to access patient information and clinical guidelines through integration with electronic health records (EHRs) and medical databases. This enables them to generate evidence-based recommendations and support healthcare providers in making well-informed decisions.

5.8 *Engagement and Empowerment of Patients*

Patients are enabled to actively manage their health through the utilisation of chatbots, which furnish them with tailored information, resources, and assistance. Chatbots assist individuals in improving their health outcomes, adhering to treatment plans, and making well-informed decisions regarding their healthcare by promoting active participation and developing self-management abilities.

6 Conclusion

In summary, the creation and deployment of a disease diagnosis chatbot constitute a noteworthy progression in healthcare technology that holds the capacity to fundamentally transform the manner in which individuals obtain initial diagnoses and access medical information. By incorporating machine learning algorithms, including the k-Nearest Neighbours (k-NN) model, and web frameworks such as Flask, disease diagnosis chatbots provide an intuitive and easily navigable interface for evaluating symptoms and providing preliminary medical advice.

A comprehensive examination of the design, architecture, and applications of a disease diagnosis chatbot has been presented in this research paper. Through the utilisation of natural language processing methodologies, chatbot controllers are capable of comprehending and extrapolating user inquiries. In doing so, they direct users through a sequence of questions that evaluate their symptoms and furnish tailored recommendations. The utilisation of k-NN models in conjunction with user-provided symptoms facilitates precise and effective disease diagnosis, thereby facilitating prompt medical intervention and treatment.

Disease diagnosis chatbots have a wide range of practical applications in the real world. These include facilitating patient empowerment, promoting preventive health education, managing chronic diseases, facilitating telemedicine, conducting public health surveillance, and facilitating remote healthcare access. These applications possess the capacity to enhance healthcare accessibility, decrease healthcare expenditures, foster patient participation, and ultimately contribute to improved health outcomes for communities and individuals.

Despite the myriad advantages that disease diagnosis algorithms provide, there are obstacles that must be surmounted, including confidentiality, security, privacy concerns, and regulatory compliance. Further investigation could be directed towards enhancing the precision and efficacy of chatbot algorithms, integrating them with healthcare systems and electronic health records (EHRs), and broadening the capabilities of chatbots to incorporate a more diverse array of medical conditions and patient populations.

In summary, disease diagnosis chatbots exhibit great potential as a means to enhance the provision of healthcare, grant individual's agency over their own well-being, and ultimately contribute to the global improvement of healthcare outcomes. Sustained investigation and progress in this domain possess the capacity to unveil additional progressions and innovations in healthcare technology, thereby enhancing the accessibility, precision, and efficacy of disease diagnosis for all.

References

1. Smith RA, Doe J (2021) Medical chatbots: an overview. J Healthc Technol 12(3):234–245
2. Brown J (2020) Artificial intelligence in healthcare: current applications. Int J Med Informatics 78(4):307–320
3. Davenport T, Kalakota R (2019) The potential for artificial intelligence in healthcare. Future Healthc J 6(2):94–98. https://doi.org/10.7861/futurehosp.6-2-94.PMID:31363513;PMCID: PMC6616181
4. Abdullah TAA, Zahid MSM, Ali W (2021) A review of Interpretable ML in healthcare: taxonomy, applications, challenges, and future directions. Symmetry 13:2439. https://doi.org/10.3390/sym13122439
5. Growth of AI-ML in Healthcare, Google Images, [Online]. Available: https://images.app.goo.gl/6Bf89yUfFtrwCTXx5
6. Jiang F, Jiang Y, Zhi H, et al (2017) Artificial intelligence in healthcare: past, present and future. Stroke Vasc Neurol **2**. https://doi.org/10.1136/svn-2017-000101
7. Applications of AI in Healthcare Google Images, [Online]. Available: https://images.app.goo.gl/8oaq1tFU1fp3z6zE7
8. Green S, White T (2019) Evolution of disease diagnosis chatbots. Adv AI ML 1001:112–123
9. Black P (2018) Natural language processing in healthcare. Healthc Inform Rev 14(1):56–67

10. Blue G (2020) Chatbot diagnostic procedures. In: Proceedings of the 2020 conference on AI in medicine, pp 89–97
11. Hand D, Mannila H, Smyth P (2001) Principles of data mining. MIT Press, Cambridge
12. Classification using k-NN model Google Images, [Online]. Available: https://images.app.goo.gl/muAfKhZZ7GXiRix98
13. Red M (2018) Introduction to k-NN models, 2nd ed. Tech Books Publishing
14. Yellow B (2019) Flask for web development. J Web Appl 22(5):45–58
15. Blue G. Chatbot diagnostic procedures, HealthTech web, [Online]. Available: http://www.healthtechweb.com
16. White L, Black A (2021) Evaluating classification models. Mach Learn J 33(2):123–134
17. Silver D (2020) User interface design for chatbots. J UX Design 11(3):77–85
18. Gold H (2019) Remote healthcare solutions. J Telemed 15(2):101–110

DTCWT-Based FPGA System for Alzheimer's Disease Multiclass Detection

B. A. Sujathakumari, Sudarshan Patil Kulkarni, Abhishek Gadi, C. K. Vijaykumar, H. V. Ashokkumar, and J. H. Mahaveer

Abstract Alzheimer's disease (AD) presents a pressing challenge as a progressive neurodegenerative disorder significantly impacting memory and cognitive functions. Timely diagnosis and intervention are imperative for enhancing the prognosis and quality of life for individuals affected by AD. However, existing methods for AD detection often exhibit limitations such as invasiveness, high costs, or subjectivity. The paper consists of innovative approach for AD detection utilizing structural magnetic resonance imaging (MRI) data and leveraging field-programmable gate array (FPGA) technology. Our research focuses on the development and simulation of a hybrid deep learning model that integrates the dual-tree complex wavelet transform (DTCWT) and convolutional neural network (CNN) architectures to effectively extract and classify features from MRI images. Furthermore, we implement this model on an FPGA platform to achieve hardware acceleration and operational efficiency. Through rigorous evaluation using the Alzheimer's Disease Neuroimaging Initiative (ADNI) dataset, we compare the performance of our approach with existing methods. The results demonstrate that the proposed approach achieves high accuracy and low latency in AD detection, using CNN and ANN models. This research underscores the potential of utilizing FPGA technology for accelerating AD detection processes, showcasing its viability for clinical applications and highlighting its significance in advancing the field of neuroimaging-based diagnosis.

Keywords Alzheimer's disease (AD) · Field-programmable gate array (FPGA) · Structural magnetic resonance imaging (MRI) · Deep learning · Dual-tree complex wavelet transform (DTCWT) · Convolutional neural network (CNN) · Simulation · Hardware acceleration

B. A. Sujathakumari (✉) · S. P. Kulkarni · A. Gadi (✉) · C. K. Vijaykumar · H. V. Ashokkumar · J. H. Mahaveer
Sri Jayachamarajendra College of Engineering, JSSSTU, Mysuru, Karnataka, India
e-mail: sujathakumari@sjce.ac.in

A. Gadi
e-mail: gabhi0012@gmail.com

S. P. Kulkarni
e-mail: sudarshan_pk@sjce.ac.in

© The Author(s), under exclusive license to Springer Nature Singapore Pte Ltd. 2025
H. Zen et al. (eds.), *Soft Computing and Signal Processing*, Lecture Notes in Networks and Systems 1200, https://doi.org/10.1007/978-981-97-9926-8_33

1 Introduction

Alzheimer's disease (AD) poses a significant global health challenge, demanding precise and timely detection for effective intervention. However, current diagnostic methodologies, often reliant on clinical criteria and neuroimaging findings, encounter limitations in accuracy and efficiency. In response, advancements in machine learning (ML) and deep learning (DL) techniques offer promising avenues for improving AD detection, particularly through the analysis of magnetic resonance imaging (MRI) data. Various processing architectures, including serial, parallel, and distributed, offer distinct approaches to implementing neural networks, each with unique advantages and trade-offs. Serial processing, characterized by sequential computations, offers simplicity but may suffer from slow performance, especially for complex neural networks. Conversely, parallel processing leverages simultaneous computations to accelerate tasks, benefiting from parallel hardware like GPUs. Distributed processing distributes the workload across multiple devices or nodes, enabling scalability and resilience to handle large datasets and intricate models.

This research focuses on leveraging distributed processing architecture to enhance the scalability and accuracy of Alzheimer's detection, with a particular emphasis on the role of field-programmable gate arrays (FPGAs). FPGAs offer unique advantages in accelerating ML and DL algorithms due to their reconfigurable nature and parallel processing capabilities. By simulating a hardware accelerator using FPGAs within a distributed environment, our aim is to verify the accuracy of AD detection algorithms trained using ML and DL techniques. This approach enables the efficient mapping of weights obtained from trained artificial neural networks (ANNs) and convolutional neural networks (CNNs) augmented with random forest algorithms onto FPGA-based hardware accelerators. The adoption of distributed architecture, coupled with FPGA utilization, marks a transformative leap in enhancing the scalability of our Alzheimer's detection project. Unlike the constraints imposed by serial architecture, the distributed approach harnesses the collective power of multiple processing units, facilitating seamless scalability to address burgeoning datasets and intricate algorithms. This not only expedites the analysis process but also yields more robust insights into the progression of Alzheimer's disease.

Furthermore, the economic viability of distributed architecture, enhanced by FPGA technology, emerges as a key facet, enabling the construction of a cost-effective system for Alzheimer's detection. By utilizing off-the-shelf hardware components and channeling resources toward cutting-edge research and development, we extend the reach of our initiative to broader populations. This cost efficiency is pivotal in making Alzheimer's detection more accessible, ultimately contributing to the broader goal of timely interventions and improved patient outcomes. The inherent advantages of FPGA technology, combined with parallel processing, resilience in the face of adversity, and adaptability to the evolving landscape, position distributed architecture as a cornerstone in Alzheimer's detection.

In summary, this research aims to bridge the gap between ML and DL techniques and hardware implementation, leveraging distributed processing architecture and

FPGA technology to enhance the scalability and accuracy of Alzheimer's detection. Through the simulation of hardware accelerators within a distributed environment, this work seeks to validate the effectiveness implemented approach in detecting AD with high precision and efficiency.

2 Literature Survey

FPGA-based hardware acceleration for embedded systems focus on the trade-off between performance and power consumption. It presents a complete acceleration design flow, exploring different CPU/accelerator interfaces, and analyzes their performances, resource overhead, and power consumption. The results highlight the impact of interface choice on performance and demonstrate significant power reduction by adding accelerators, aiding designers in meeting power/performance requirements [1, 2]. The classification of early and late mild cognitive impairment (MCI) stages in Alzheimer's disease has been done by analyzing MRI scans of different brain areas. It highlights the importance of early detection of MCI for slowing down the progression of Alzheimer's. The study uses atrophy values from MRI scans as biomarkers and employs machine learning algorithms to classify EMCI and LMCI stages with an accuracy of 68.8%. By focusing on significant brain regions and using a selected set of features, the accuracy improves to 75% [3, 4]. Landmark-based feature extraction method using local morphological features is used to identify brain regions with significant group differences. Efficient technique based on a shape-constrained regression forest algorithm has been used for landmark detection [5, 6]. Dual Attention Multi-Instance Deep Learning (DA-MIDL) which is a novel method is used for diagnosing and Alzheimer's disease using structural MRI data and also identifies pathological locations and learns local and global features from MRI patches. Additionally, a Patch-Net with Spatial Attention Block enhances local features, an Attention Multi-Instance Learning (MIL) Pooling mechanism balances patch contributions, and an Attention-Aware Global Classifier aids in AD-related classification decisions [7, 8]. In order to differentiate between images showing mild cognitive impairments and typical cognitive impairments, the MRI images are fed into an autoencoder network. They have classified high dimensional multimedia material using a deep learning method. Cerebrospinal fluid (CSF) and the removal of white matter (WM) signals as nuisance variables are the two primary motion parameters that are taken into account in order to lessen the impacts of head motion and non-neuronal BOLD fluctuation. The training procedure consists of a pre-training phase that uses an unsupervised greedy technique to train one network layer at a time, and a supervised fine-tuning step that incorporates the data's label information to improve performance [9–11]. Other paper with different architecture was referred in which series, parallel, and distributed systems were compared where the distributed system was proven much efficient [12–19].

In the referred implementations, some features of implementing it on FPGA were not so promising if implemented the stages of Alzheimer diseases were not segregated properly and the utilization of components was more; if done properly, then the chance of false detection was a problem, and this work is trying to implement proper classification of Alzheimer diseases using hybrid (ML + DL) methods and using DTCWT and implementing on FPGA with low cost and less power consumption.

3 Methodology

In the implementation, MRI images obtained from the ADNI website undergo preprocessing using SPM12 followed by DTCWT for segmentation and feature extraction. These processed images serve as input data for validation purposes. The Xilinx VIVADO tool 2024.1 is used for FPGA simulation, where trained CNN or ANN models are employed to configure the FPGA as a hardware accelerator. To automate the creation of hardware versions of fully connected neural networks for FPGAs, the Python-based tool Zynet is employed. The hardware description generated by Zynet is used to map hardware components within the FPGA using the Vivado toolbox, with weights and MIF files playing crucial roles in memory allocation and weight assignment to optimize accuracy. Figure 1 shows the overall flow of the implementation process.

Following segmentation, the preprocessed MRI images undergo classification through the DTCWT classifier, leading to the removal of unwanted regions like the skull and neck areas. Subsequently, these images are transformed into JPEG format using a JPEG converter, facilitating the conversion of NIFTI images. Within the VIVADO environment, a CNN model consisting of two convolution layers, two dense layers, and a max pooling layer (SoftMax) is proposed. These layers extract features from the images, and biases enhance expressiveness. Model accuracy is tested and optimized to enhance efficiency.

After model testing and optimization, preprocessed MRI images are simulated on the FPGA, with machine models finalized based on accuracy criteria. Utilization reports, power distribution reports, and schematic diagrams of the architecture are then obtained from VIVADO. These reports offer insights into resource utilization, power consumption distribution, and architectural design, facilitating further optimization and refinement operations.

4 Algorithms Implemented and Results

4.1 CNN

In FPGA (Field-Programmable Gate Array) devices, static and dynamic power consumption are two important metrics that contribute to the overall power consumption of the device. Static power, also known as leakage power, refers to the power

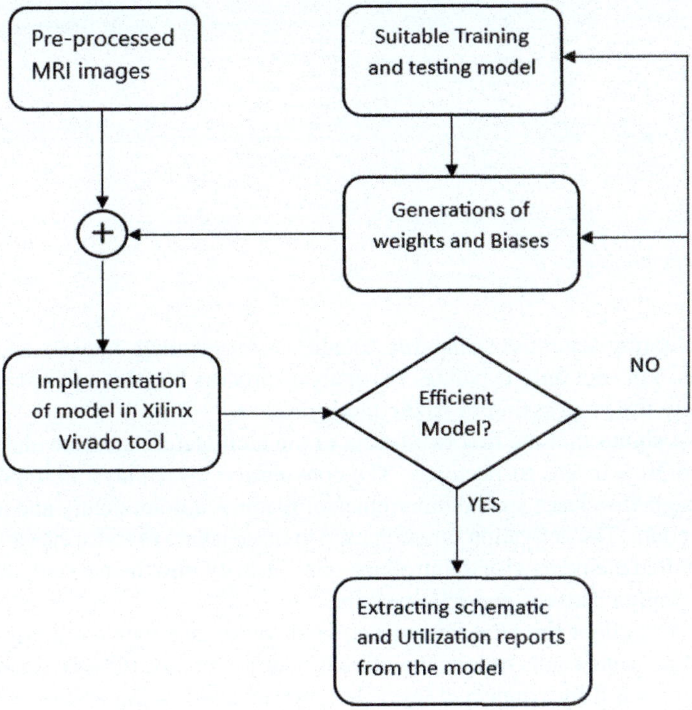

Fig. 1 Proposed flow diagram

consumed by the FPGA even when it's not actively switching or performing any computation. It is primarily caused by leakage currents flowing through the transistors in the FPGA, which occur due to the non-ideal behaviors of transistors and imperfections in the manufacturing process. The paper was implemented using mainly CNN and ANN deep learning model. These models have an ability to process complex patterns from the data making them versatile.

In the implementation of CNN, Fig. 2 shows the number of components are used which include LUT (look up figure), Muxes, DSPs, Carry registers. The above figure shows the number of total available components and the components that are used for the implemented architecture. The power utilization figure shows the static and dynamic power estimation to run the proposed result on FPGA board. It also shows the power distribution among each component that are used.

Figure 3 shows the overview of the implemented architecture where the CNN is implemented which has 4 major layers consisting of two convolution layer and two dense layer and at last has a max pooling layer which helps in extracting the features and at the same time reducing the dimension of the layer. The adoption of distributed architecture marks a transformative leap in enhancing the scalability of our Alzheimer's detection project. Unlike the constraints imposed by serial architecture, the distributed approach harnesses the collective power of multiple processing

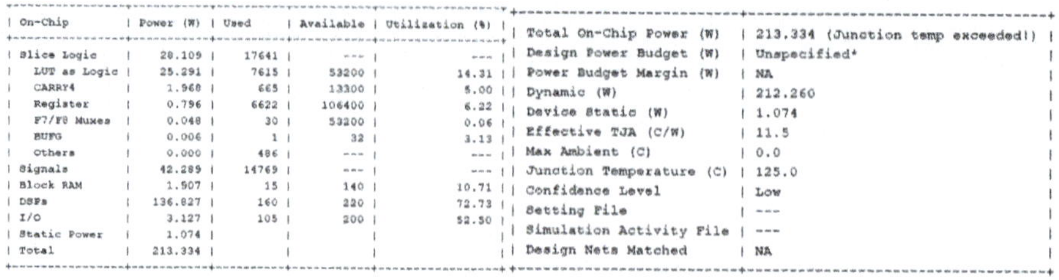

On-Chip	Power (W)	Used	Available	Utilization (%)
Slice Logic	28.109	17641	---	---
LUT as Logic	25.291	7615	53200	14.31
CARRY4	1.968	665	13300	5.00
Register	0.796	6622	106400	6.22
F7/F8 Muxes	0.048	30	53200	0.06
BUFG	0.006	1	32	3.13
Others	0.000	486	---	---
Signals	42.289	14769	---	---
Block RAM	1.907	15	140	10.71
DSPs	136.827	160	220	72.73
I/O	3.127	105	200	52.50
Static Power	1.074			
Total	213.334			

Total On-Chip Power (W)	213.334 (Junction temp exceeded!)
Design Power Budget (W)	Unspecified*
Power Budget Margin (W)	NA
Dynamic (W)	212.260
Device Static (W)	1.074
Effective TJA (C/W)	11.5
Max Ambient (C)	0.0
Junction Temperature (C)	125.0
Confidence Level	Low
Setting File	---
Simulation Activity File	---
Design Nets Matched	NA

Fig. 2 Power utilization

units, facilitating seamless scalability to address burgeoning datasets and intricate algorithms. This not only expedites the analysis process but also yields more robust insights into the progression of Alzheimer's disease.

Figure 4 shows that the first two layers of the architecture are convolution layers of 863 and 30 neurons, respectively. The convolution layer plays an important role as it is designed to learn spatial hierarchies of features automatically and adaptively from input data. The activation function used in these layers in ReLu (Rectified linear unit) is applied element-wise to introduce non-linearity into the network, allowing it to learn complex patterns and relationships.

Figure 5 is about the next layers, i.e., layer three and four are dense layers of 30 and 10 neurons each. In a convolutional neural network (CNN), a dense layer, also known as a fully connected layer, is typically used at the end of the network for making final predictions or classifications based on the features extracted by the preceding convolutional layers. It takes the high-level features learned by the

Fig. 3 Hardware network view

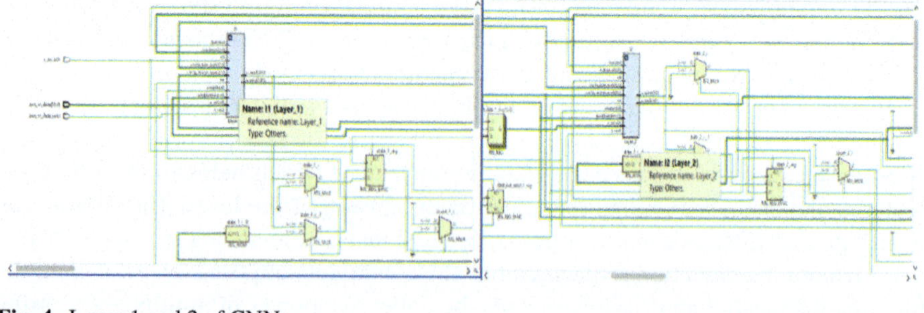

Fig. 4 Layer 1 and 2 of CNN

convolutional layers and combines them to make a final decision about the input. These layers also have ReLu which has their activation functions.

Figure 6 shows the last layer of the architecture in max pooling layer of 6 neurons each for each class to be detected. Max pooling is a down sampling operation typically applied after convolutional layers in CNNs. Its purpose is to reduce the spatial dimensions of the feature maps, thereby decreasing computation and controlling overfitting. SoftMax is typically used as the activation function in the output layer of a neural network for classification tasks. It converts the raw scores (logits) produced by the previous layer into probabilities representing the likelihood of each class.

Figure 7 shows the post-synthesis utilization of the components such has LUT, FF(flip-flop), DSP, and IO. The utilized percentage of these components is below 100 percent which notifies that the implemented architecture is suitable for the board environment and is efficient.

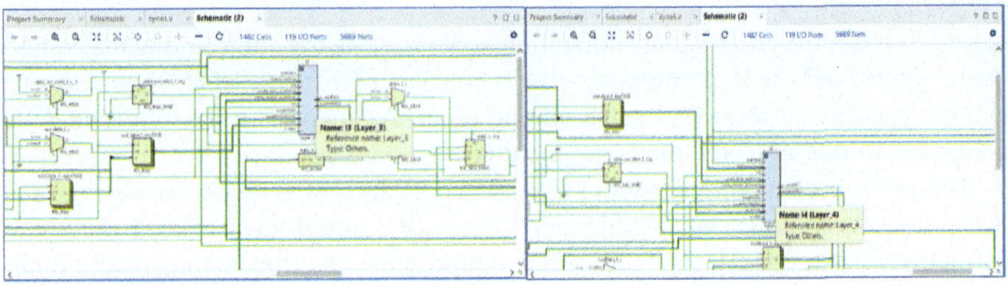

Fig. 5 Layer 3 and 4 of CNN

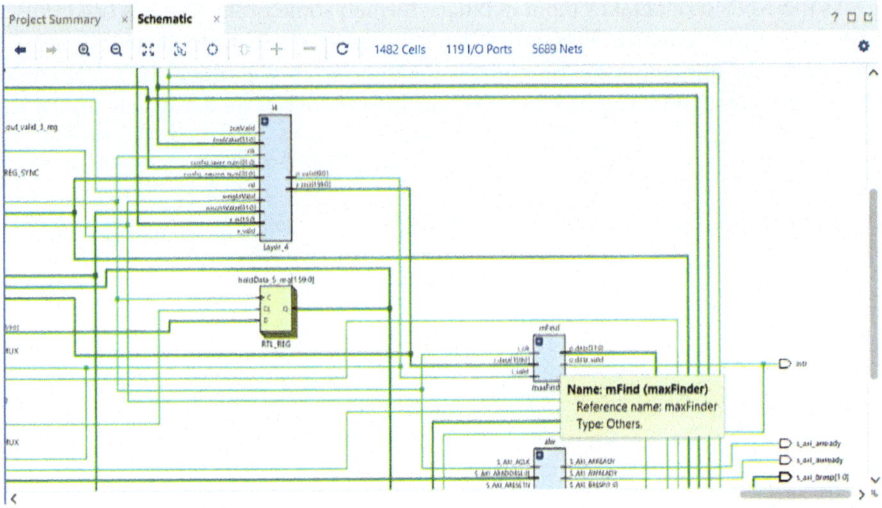

Fig. 6 Max pooling layer (soft max)

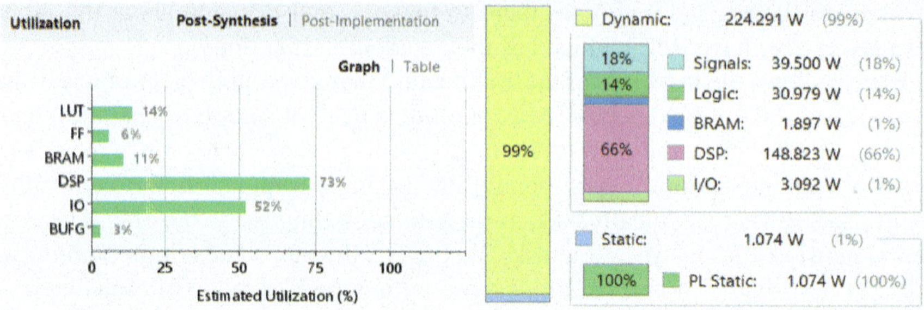

Fig. 7 Graphical view of components utilization of CNN

4.2 *ANN*

Figure 8 shows the components utilization and power distribution among the components when ANN is implemented, which consists of Muxes, LUT, Registers etc., and the power distribution is same among them. As there were less layer in ANN compared to that of CNN, the component utilization is less.

The ANN architecture shown in Figs. 9 and 10 has three layers consisting of all dense layers and component utilization. Where the initial layer is used for feature extraction, the second layer is used for compressing the dimensions of the image and the features and last layer is used for the classification of six classes. The IO utilization of ANN is comparatively less and it has only 52%; the CNN has 73%, but other areas also have some reduction in utilization.

As the ANN model takes input in binary format, some of the features of DTCWT may have been reduced. This might lead to reduction in little accuracy, but the efficiency and utilization of components make up to this. Further, the CNN model has more layers, and it requires more power than the ANN model which almost has

Site Type	Used	Fixed	Prohibited	Available	Util%
Slice LUTs	2943	0	0	53200	5.53
LUT as Logic	2943	0	0	53200	5.53
LUT as Memory	0	0	0	17400	0.00
Slice Registers	3875	0	0	106400	3.64
Register as Flip Flop	3875	0	0	106400	3.64
Register as Latch	0	0	0	106400	0.00
F7 Muxes	67	0	0	26600	0.25
F8 Muxes	0	0	0	13300	0.00

Site Type	Used	Fixed	Prohibited	Available	Util%
Slice	1243	0	0	13300	9.35
SLICEL	746	0			
SLICEM	497	0			
LUT as Logic	2943	0	0	53200	5.53
using O5 output only	0				
using O6 output only	2323				
using O5 and O6	620				
LUT as Memory	0	0	0	17400	0.00
LUT as Distributed RAM	0	0			
LUT as Shift Register	0	0			
Slice Registers	3875	0	0	106400	3.64
Register driven from within the Slice	2020				
Register driven from outside the Slice	1855				
LUT in front of the register is unused	1172				
LUT in front of the register is used	683				
Unique Control Sets	178		0	13300	1.34

Name	Slice LUTs (53200)	Slice Registers (106400)	F7 Muxes (26600)	Block RAM Tile (140)	DSPs (220)	Bonded IOB (200)	BUFGCTRL (32)
zyNet	2943	3874	67	16	64	105	1
alw (axi_lite_wrapper)	229	240	15	0	0	0	0
l1 (Layer_1)	2544	2724	0	0	64	0	0
mFind (maxFinder)	170	494	52	0	0	0	0

Fig. 8 Components used by each layer

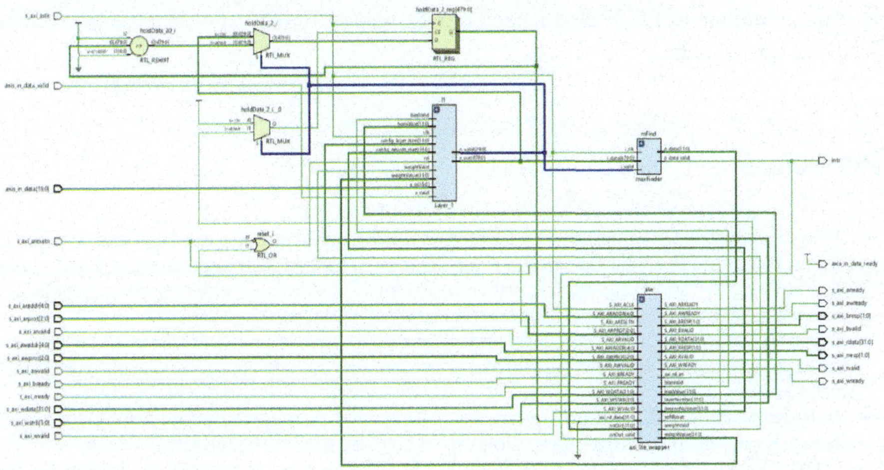

Fig. 9 ANN architecture

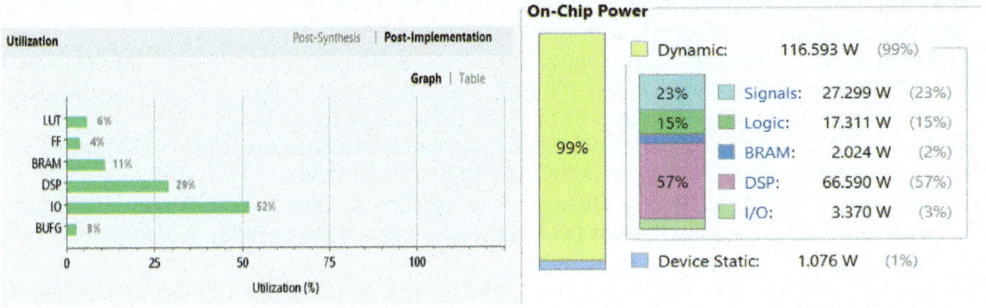

Fig. 10 Components utilization of ANN

half of its layers. The CNN model requires 224.291 W, whereas ANN model only needs 116.593W.

5 Conclusion

Analysis and decision-making in real time are made possible by hardware solutions that facilitate quick and efficient data processing. Large volumes of data, such as brain imaging or biomarker data, must often be processed to detect AD early. Hardware solutions are preferable better than software-based approaches at meeting the computational demands. Wearable or implanted devices that are used to continuously monitor people who are at risk for AD can have their hardware solutions optimized for power efficiency. Device hardware optimization can reduce energy consumption if these devices must run constantly on limited power supplies. Based on the percentage of LUT usage, each design was examined and contrasted with the software component.

References

1. Possa P, Schaillie D, Valderrama C. Electronics and microelectronics department university of mons, Belgium FPGA-based hardware acceleration: a CPU/Accelerator Interface Exploration
2. Liu Y, Wang Y, Chang L, Zhou J. A fast and efficient FPGA based level set hardware accelerator for image segmentation. https://doi.org/10.1109/ICTA50426.2020.9331957
3. Uysal G, Ozturk M. Classifying early and late mild cognitive impairment stages of Alzheimer's disease by analyzing different brain areas. https://doi.org/10.1109/TIPTEKNO50054.2020.9299217
4. Ju R, Hu C, Zhou P, Li Q. Early diagnosis of Alzheimer's disease based on resting-state brain networks and deep learning. https://doi.org/10.1109/TCBB.2017.2776910
5. Zhang J, Gao Y, Gao Y, Munsell B, Shen D, Gao Y. Detecting anatomical landmarks for fast Alzheimer's disease diagnosis. https://doi.org/10.1109/TMI.2016.2582386
6. Thomas W, Kumar P. Detections of brain tumour using segmentations and morphological operators from MRI scan with FPGA. https://doi.org/10.1109/ICATCCT.2015.7456979
7. Zhu W, Sun L, Huang J, Han L, Zhang D. Dual attention multi-instance deep learning for Alzheimer's disease diagnosis with structural MRI
8. Murugan S, Venkatesan S, Sumithra MG, Gao XZ, Elakkiya B, Akila M, Manoharan S. DEMNET: a deep learning model for early diagnosis of Alzheimer Diseases and dementia from MR images." https://doi.org/10.1109/ACCESS.2021.3090474
9. Zhou X, Ding S, Yu J, Wang L. MRI-based brain tumor segmentation using FPGA-accelerated neural network. https://doi.org/10.1109/ACCESS.2022.3151361
10. Xiong S, Wu G, Fan X, Feng X, Huang Z, Cao W, Zhou X, Ding S, Yu J, Wang L. MRI-based brain tumor segmentation using FPGA-accelerated neural network. https://doi.org/10.1109/ACCESS.2022.3151361
11. Koo JJ, Evans AC, Gross WJ. 3-D brain MRI tissue classification on FPGA. https://doi.org/10.1109/TIP.2009.2028926
12. Christe, Vignesh, Kandaswamy A. An efficient FPGA implementation of MRI image filtering and tumour characterization using XILINX system generator
13. Fauzi M, Othman B, Abdullah N, Aizudin N, Ahmad B, Fakulti R, Elektrik K. An overview of MRI brain classification using FPGA implementation. https://doi.org/10.1109/ISIEA.2010.5679389
14. Morra J, Tu Z, Apostolova L, Green A, Toga A, Thompson P. Comparison of AdaBoost and support vector machines for detecting Alzheimer's disease through automated hippocampal segmentation.https://doi.org/10.1109/TMI.2009.2021941
15. Nagorno N, Lyakhov P, Nagornov N. RNS-based FPGA accelerators for high-quality 3D medical image wavelet processing using scaled filter coefficients. https://doi.org/10.1109/ACCESS.2022.3151361
16. Kasik V, Cerny M, Penhaker M, Snášel V, Novak V, Pustkova R. Advanced CT and MR image processing with FPGA
17. Sreekrishna S. FPGA implementation of segmented feature fusion in MRI images using wavelet. https://doi.org/10.1109/ICCES45898.2019.9002032
18. Wang NSE, McAllister. Streaming elements for FPGA signal and image processing accelerators. https://doi.org/10.1109/TVLSI.2015.2504871
19. Sujatha Kumari BA, Patil Kulkarni S, Sinchana CG. FPGA implementation of neural nets. https://doi.org/10.24425/ijet.2023.146513

Security of Images Using DNA Encryption and Lorenz System in Cloud Storage

Raghav Gupta, Manasa Gullapalli, and G. Suseela

Abstract Images are the most valuable yet susceptible type of data since they may be used to track user activity, exploited by hackers, and utilized for fraud and identity theft. Storing images in local storage, on a device's disk, within a web browser's local storage, or on a server's file system can introduce several vulnerabilities and risks, particularly regarding security and data integrity. Using image encryption as a security solution provides a reliable means of protecting sensitive visual data from unauthorized access and potential security threats. The convenience of cloud-based systems, however, comes with the responsibility of protecting data against cyber threats. The widespread usage of cloud storage has led to concerns about the security of images stored there. Addressing the challenges associated with securing images in the cloud, this project brings a novel approach that blends cryptographic techniques, such as DNA encryption, with the dynamism of chaotic systems like the Lorenz system. Such integrative methodologies aim to meet current security demands for image storage and transmission in the digital age. This dual-layer encryption approach ensures that images encrypted and stored in cloud storage are safeguarded with a security measure that is currently unparalleled, making unauthorized access and decryption practically impossible without the specific, uniquely generated decryption key.

Keywords Image encryption · DNA encoding · Lorenz system · Chaotic sequence · Scrambling · Cloud storage

R. Gupta · M. Gullapalli
Department of Networking and Communications, School of Computing Science and Engineering, SRM Institute of Science and Technology, Kattankulathur, Tamil Nadu 603203, India
e-mail: rg5705@srmist.edu.in

M. Gullapalli
e-mail: mg6111@srmist.edu.in

G. Suseela (✉)
Department of Networking and Communications, Faculty of Engineering and Technology, School of Computing Science and Engineering, SRM Institute of Science and Technology, Kattankulathur, Tamil Nadu 603203, India
e-mail: suseelag@srmist.edu.in

© The Author(s), under exclusive license to Springer Nature Singapore Pte Ltd. 2025
H. Zen et al. (eds.), *Soft Computing and Signal Processing*, Lecture Notes in Networks and Systems 1200, https://doi.org/10.1007/978-981-97-9926-8_34

1 Introduction

The increasing reliance on digital images across various sectors, including personal, commercial, and governmental domains, has underscored the pressing need for enhanced security measures to protect these assets from a spectrum of vulnerabilities inherent in traditional storage methods. The advent of cloud storage introduced a potential solution to some of these issues by offering off-site data storage with promises of enhanced security measures. However, merely transferring images to the cloud without employing robust encryption techniques does little to mitigate the risks. It potentially worsens them by exposing data to additional vulnerabilities, such as interception during transmission and unauthorized access through compromised cloud platforms. Given the cloud's nature of centralized storage accessible over the Internet, the risks of cyber-attacks and unauthorized data access are significantly heightened unless stringent encryption measures are in place.

The selection of DNA encryption and the Lorenz system as an encryption solution for cloud storage image security is based on both approaches' unique and complementary features. Utilizing the enormous combinatorial space that the four nucleotide bases provide, DNA encryption makes use of the intrinsic complexity of DNA sequences to safely encode data. This biological technique presents a new paradigm for strong data encryption while guaranteeing high unpredictability and complexity. Combining the Lorenz system with DNA encryption results in a multi-layered security framework that produces a strong barrier against unwanted access due to the interactions between biological and chaotic systems. This creative method fosters breakthroughs in secure data storage and transmission while also improving the confidentiality and integrity of images stored in the cloud and pushing the limits of encryption technology. This research contributes to the ongoing efforts to strengthen digital defenses in an increasingly interconnected world by utilizing the capabilities of these unusual approaches to solve the critical concerns surrounding the security of images in cloud storage.

2 Literature Survey

This study presents a novel approach for image encryption. To increase the security of the encryption, it combines a method based on DNA encoding with a complicated system inspired by the 6D hyper-chaotic system [1]. The technique combines the special qualities of DNA with the unpredictable nature of chaos. This method could greatly enhance the ability to encrypt images. In order for the system to remain relevant in the face of constant technological advancement, the researchers are also aiming to make it faster and more efficient.

This paper suggests an image encryption technique that generates a random keystream by use of a 3D chaotic system. The image pixels are then repeatedly permuted by two and eight bits using the keystream. For added protection, the plan also combines pixel diffusion with a key-image created from a Piecewise Linear Chaotic Map (PWLCM) [2]. It's crucial to remember that even if this method uses DNA encoding, it might not be the best solution for protecting private medical information in the context of internet-of-things (IoT) healthcare applications.

This study suggested a novel technique to image encryption. To increase the security of the encryption, it combines a number of various methods, including a chaotic maps, SHA-256 hash function that simulate unpredictable behavior, and special operations based on DNA structure [3]. Although the goal of this intricate method is to strengthen the encryption against different types of assaults, the manner the secret key is generated is not optimal. The encryption process is slowed down by this inefficiency in key generation, which may have an impact on the overall functionality and user experience of the system.

The author introduces a new chaotic map in one dimension, called sine–cosine chaotic map (SCCM) [4]. In terms of chaotic behavior, this SCCM is intended to perform better than current one-dimensional chaotic maps. By examining the map using well-known metrics like bifurcation diagrams, Lyapunov exponents, and approximation entropy, the authors show this enhanced performance. Notably, DNA encoding and the one-dimensional sine–cosine chaotic map itself are the techniques employed in this study.

The author proposes a color image encryption scheme based on DNA operations and spatiotemporal chaotic system where the image is converted into 3 DNA matrices, which are the combined into a new matrix and permuted by a scramble matrix [5]. The permuted matrix is decomposed into three matrices and diffused by DNA deletion-insertion operations. The proposed cryptosystem is only tested against some classical types of attacks, such as statistical, differential, and brute-force attacks. It may not be secure against other advanced attacks, such as differential cryptanalysis, linear cryptanalysis, or algebraic attacks.

This research paper proposes the superiority of the suggested model over existing methods like only AES, or AES with RSA and its compliance with regulations for sensitive data stored in the cloud [6]. The author of this paper also measured the throughput of the model with other existing models and no study of analysis has been done for resulting image and security attacks.

3 Proposed System

An enhanced encryption framework that combines DNA-based cryptography with Lorenz system chaotic sequences is used to improve image security in cloud environments, intertwining DNA-based cryptography with chaotic sequences from the Lorenz system. Initially, this method involves processing the input image, subsequently encoding it into a DNA sequence with nucleotides A, T, G, and C, thereby preparing it for encryption. In parallel, a chaotic sequence generated from the Lorenz system's equations creates a highly unpredictable key, integral for encrypting the DNA-encoded data, owing to its sensitivity to initial conditions. Following successful encryption, the image is then securely stored in the cloud, where stringent access controls are in place to ensure that decryption and retrieval are restricted to authorized users only. During decryption, the same chaotic sequence, under identical initial conditions, is used to revert the encryption, enabling authorized users to reconstruct the original image.

The system architecture depicted in Fig. 1 ensures the security of the image by converting the readily understandable image data into a complex and seemingly random format through the use of DNA encoding and subsequent encryption employing a chaotic sequence. The security of this system lies in the unpredictability of the chaotic sequence and the complexity of the DNA encoding, which together ensure that the encrypted image stored in the cloud is protected against unauthorized access and decryption attempts.

The purpose of this research is to present a multilayered encryption system designed to enhance the security of images stored in cloud environments. To establish a robust system for image encryption and storage, there is an integration of various cryptographic techniques and dynamic systems. Below are the proposed modules within the system: Secure Key Generation, Lorenz System, DNA Encoding and XOR Operations, Chaos Sequence Generation, and Scrambling. Each module plays a key role in enhancing the encryption process, as a result ensuring a high level of security for stored images.

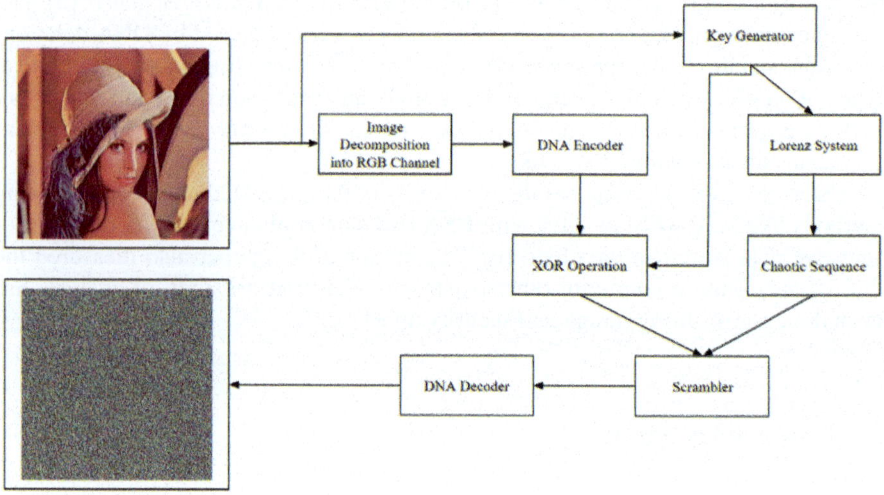

Fig. 1 Proposed system architecture

3.1 Secure Key Generation

The secure key generates a key by extracting the pixel RGB values, width and height from an image and forming a continuous data stream. Following the input of these values into the SHA-256 hashing algorithm, a hash object is created with the image data and then converted into a hexadecimal digest, representing the secure key. This key, uniquely derived from the image's pixel composition, is intended for use in encryption processes, where its uniqueness ensures robust security. For decryption, the same image must be used to regenerate the identical key, establishing a secure mechanism where only the correct image can facilitate access to the encrypted data, thus tightly intertwining the image's integrity with the data's security.

3.2 The Lorenz System

The Lorenz system is a set of three differential equations. The Lorenz function calculates the derivatives of the system's state variables (x, y, z) using parameters (σ, ρ, β) to describe the system's behavior. The cryptographic key, initially in hexadecimal, is converted to binary and divided into 32 segments. These segments are then XORed in three groups to update the Lorenz system's initial conditions ($x0, y0, z0$), effectively tying the system's evolution to the key. By altering the initial conditions with the key, the Lorenz system's inherent sensitivity to initial conditions is exploited, introducing a layer of complexity and randomness to the encryption process. This method ensures that the Lorenz attractor's chaotic behavior is embedded in the encryption, enhancing security by making the system's output highly dependent on the unique cryptographic key, thus playing a crucial role in the overall encryption mechanism.

The standard equations of the Lorenz system, which describe the rate of change of three variables (x, y, z) with respect to time, are as follows:

$$\frac{dx}{dt} = \sigma(y - x) \tag{1}$$

$$\frac{dy}{dt} = x(p - z) - y \tag{2}$$

$$\frac{dz}{dt} = xy - \beta z \tag{3}$$

In these equations $\sigma, \rho,$ and β are the parameters which determine the behavior of the system.

3.3 DNA Encoding and XOR Operations

The DNA encode converts the binary representations of the blue (b), green (g), and red (r) pixel values of an image into DNA sequences. It converts the 8-bit values of each color channel into binary form, then iteratively groups every two binary bits to map them to their corresponding DNA nucleotide based on a predefined DNA rules as shown in Table 1. For instance, "00" could map to "A", "01" to "T", etc. This mapping occurs in nested loops that iterate over each bit pair in the color matrices, converting them to DNA bases and storing the result in character arrays. The encoded DNA sequences for the RGB channels are then returned. In the encryption process, this DNA encoding introduces a biological layer of complexity, transforming the image data into a format that's less intuitive to interpret than binary, thereby enhancing the security of the encrypted data. The encoded data serves as a more complex foundation for subsequent encryption steps, ensuring that deciphering the encrypted image without knowledge of the specific encoding scheme and key is significantly challenging.

The XOR operation is performed on DNA-encoded pixel values of an image (for blue, green, and red channels) against a DNA-encoded mask key. It initializes arrays for the XORed output, converts the DNA-encoded inputs to strings, and then applies the XOR logic defined in a DNA dictionary to combine each pixel's DNA character with the corresponding character from the mask key. This operation effectively intertwines the image data with the mask key, embedding a layer of security that ensures the image's original data is obfuscated and can only be decrypted with the exact mask key. The reversible nature of the XOR operation is crucial for decryption, making this process a foundational aspect of the encryption's robustness and security, as it directly influences the ability to recover the original image data when provided with the correct key.

Table 1 DNA rules

DNA rule	A	T	C	G
Rule 1	00	11	01	10
Rule 2	00	11	10	01
Rule 3	01	10	00	11
Rule 4	01	10	11	00
Rule 5	10	01	00	11
Rule 6	10	01	00	11
Rule 7	11	00	01	10
Rule 8	11	00	10	01

3.4 Chaos Sequence Generation

The chaos generation sequence generates a chaotic sequence using the Lorenz system, integral for introducing complexity and unpredictability into the encryption process. By numerically integrating the Lorenz system's differential equations, the function produces sequences based on specified initial conditions and system parameters, which are then trimmed to match the image's dimensions. These chaotic sequences are crucial for the encryption process, as they are used to scramble or modify the image data, ensuring that the encryption is highly sensitive to initial conditions and inherently non-linear. This approach guarantees that the encryption output is unique and secure, significantly enhancing the robustness and security of the encryption mechanism by embedding the unpredictability of chaotic dynamics into the image data.

3.5 Scrambling

The scrambling implements pixel scrambling on the color channels of an image using chaotic sequences generated by the Lorenz system, a method that significantly enhances the security of the DNA encryption process. By reshaping the color channels into one-dimensional arrays and rearranging their values based on chaotic indices, these functions ensure that the pixel information is dispersed in a highly non-linear and unpredictable manner. This scrambling, crucial for the encryption's robustness, makes the decoding of the image extremely difficult without the exact chaotic sequences and the corresponding DNA encoding key, thereby embedding an additional layer of complexity and security into the encrypted image.

4 Result Analysis

The effectiveness assessment of a new encryption system is described in this work. The images in Fig. 2 demonstrate how the algorithm was able to successfully jumble the images, making them unreadable. In order to evaluate the system's resilience against typical attacks, this study does a number of analysis, such as looking at entropy, correlations, and histograms. In order to assess image quality, PSNR values were also computed in this study. Finally, this study used UACI and NPCR metrics to verify the system's resilience against particular threats.

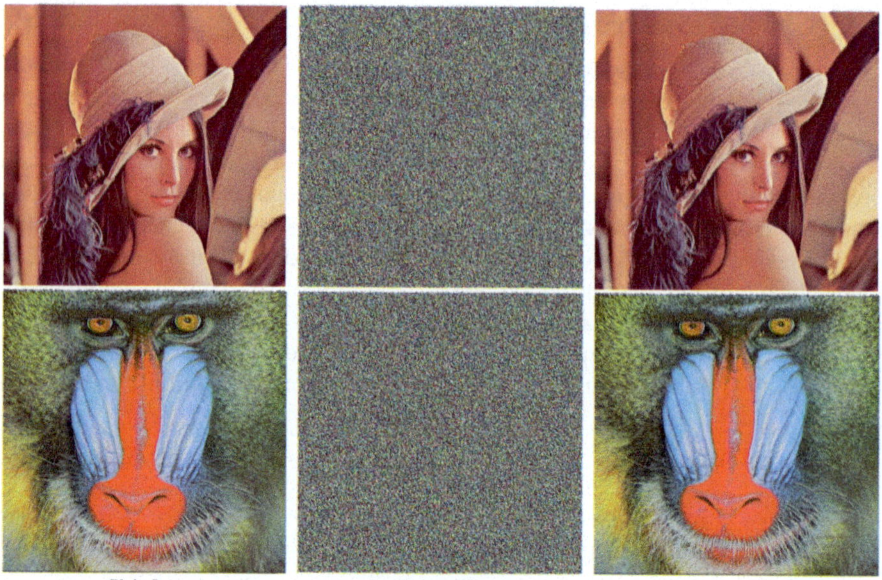

Fig. 2 Encryption results of the proposed system

4.1 Histogram Analysis

The image becomes unpredictable due to the encryption method's scrambled data. Red, green, and blue color channels are all equally impacted by this confusion. When comparing the encrypted version of the image to the original, Fig. 3a illustrates the uneven color fluctuations. The histograms shown in Fig. 3b show that the pixel intensities of the encrypted image are uniformly distributed.

The suggested method efficiently hides redundant information from image pixels by examining the histogram. This results in a histogram with a flat distribution at the cost of distorting and offsetting the original pixel values from their starting positions. The suggested approach is a strong contender for enhancing the security of images since it is impervious to statistical assaults.

4.2 Correlation Analysis

A cryptosystem should efficiently jumble an image into an apparently random jumble of pixels with little to no link between them in order to effectively scramble it. As a result, it becomes quite challenging to distinguish between the encrypted and original images. A robust cryptosystem must be able to resist attacks that take use of statistical

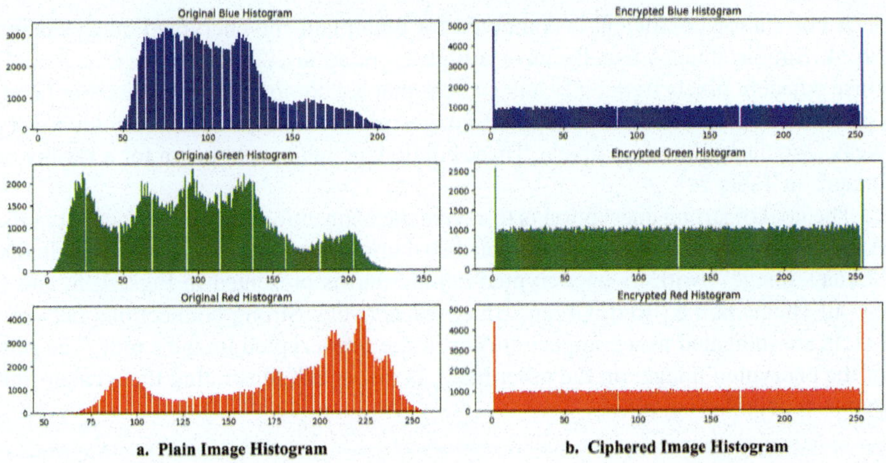

Fig. 3 Lena image histogram

tendencies. Tests like correlation analysis and histogram analysis which looks at the distribution of pixel values are frequently used in these attacks.

An image's surrounding pixels affect how the original image looks overall and hold important data. By breaking the connection between these adjacent pixels, encryption seeks to conceal information from being discovered during analysis. An image's surrounding pixels affect how the original image looks overall and hold important data. By breaking the connection between these adjacent pixels, encryption seeks to conceal information from being discovered during analysis. Researchers examined randomly selected pixels from the original and unencrypted editions of the image, including those that were horizontal, vertical, and diagonal.

$$\text{corr} = \frac{\text{cov}(m, n)}{\sqrt{P(m) \times P(n)}} \tag{4}$$

where

$$\text{cov}(m, n) = \frac{1}{N} \sum_{l=1}^{N} (m_l - C(m))(n_l - C(n)) \tag{5}$$

$$D(t) - \text{sample mean } D(t) = \frac{1}{N} \sum_{l=1}^{N} (t_l) \tag{6}$$

$$P(t) = \frac{1}{N} \sum_{l=1}^{N} (t_l - D(t)) \tag{7}$$

A correlation coefficient is computed for both plain and encrypted images using Eq. 4. Simple images usually have a high correlation coefficient, very near to 1, since adjacent pixels frequently have values that are comparable. On the other hand, correlation coefficients for encrypted images are close to 0, which denotes a very weak association between pixels. These coefficient values are shown for a variety of images in Table 2.

The goal of strong encryption is to severe the connection between adjacent pixels. A scatter plot comparing the correlation distribution of an encrypted image (originally "Lena's image") with its unencrypted counterpart is presented in Figs. 4, 5, and 6 for all channels, i.e., Red, Green, Blue, respectively. Strong connections between pixels are indicated as a compact cluster in the unencrypted image's plot. The plot of the encrypted image, on the other hand, is dispersed, illustrating the weaker link between neighboring pixels.

4.3 Entropy

Image entropy analysis serves as a crucial metric for evaluating the randomness within an image's pixel intensity distribution, essential for assessing the effectiveness of an encryption method. In essence, the entropy of an image quantifies the unpredictability or randomness of its pixel values, with higher entropy values indicating more effective encryption. Mathematically, image entropy is determined by summing the product of the probability of each unique pixel intensity value and the logarithm of that probability, across all possible intensity values as given in Eq. 8. For an images, the ideal entropy value is 8.

Robust encryption should scramble the original image's information and make it difficult to determine what's within. This unpredictable nature is quantified using a notion known as information entropy and expressed in terms of bits. It can determine how random and unpredictable the encrypted images are by doing an entropy test. This test measures the image's scrambling power to help assess how well an encryption system performs. The entropy values of the image for this proposed system is given in Table 3.

$$H(s) = \sum_{y=0}^{255} z(y) \times \log z(y) \tag{8}$$

4.4 PSNR

PSNR (Peak Signal-to-Noise-Ratio) is a metric that compares the quality of an image that has been reconstructed to its original version. It is frequently used along with

Table 2 Correlation coefficient values

Images		Original			Encrypted			Reference [7]		
		H	V	D	H	V	D	H	V	D
Lena	Red	0.973	0.99	0.964	0.012	− 0.001	0.022	− 0.000006	0.0001	0.000006
	Green	0.965	0.984	0.955	0.028	0.044	0.056	− 0.0056	− 0.00002	− 0.00005
	Blue	0.927	0.949	0.914	− 0.011	0.012	0.002	0.00001	0.00081	0.000098
Babbon	Red	0.917	0.883	0.87	0.056	0.058	0.063	− 0.00006	− 0.000005	0.0005
	Green	0.848	0.793	0.731	0.081	0.086	0.055	0.0075	− 0.0002	− 0.00009
	Blue	0.898	0.894	0.831	0.025	0.045	0.027	− 0.0002	0.00001	0.0007
Pepper	Red	0.964	0.972	0.965	0.007	− 0.022	0.013	− 0.000004	0.0003	0.00043
	Green	0.969	0.984	0.977	0.027	0.03	0.023	− 0.0003	− 0.00003	− 0.00007
	Blue	0.959	0.966	0.949	− 0.021	− 0.025	− 0.011	0.0022	0.0001	0.0002

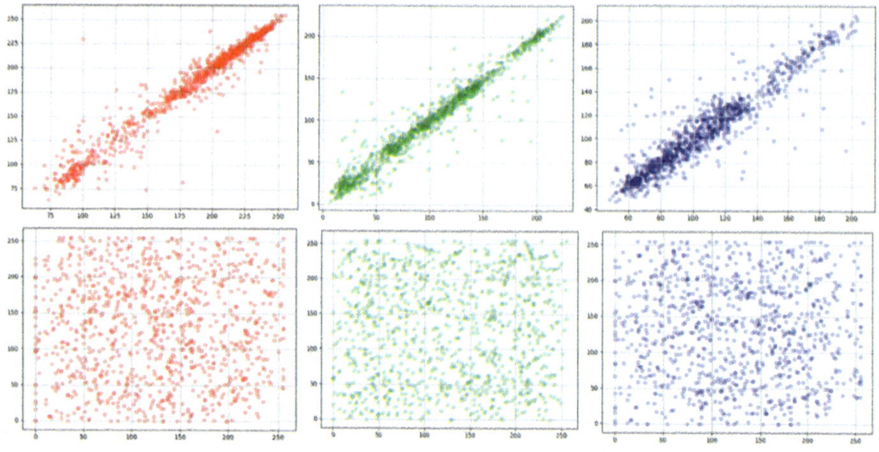

Fig. 4 Horizontal correlation scatter plot of Lena image

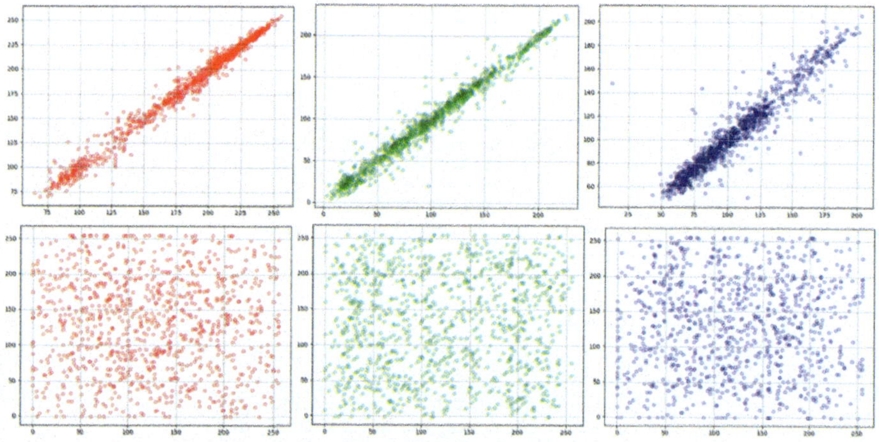

Fig. 5 Vertical correlation scatter plot of Lena image

image encryption to determine any distortion or noise that may have been introduced during the encryption and decryption processes. It computes the ratio, given in logarithmic decibel scale, between the greatest power that a signal (image) can have and the power of corrupting noise that degrades its quality. The PSNR values of the images for this proposed system are shown in Table 4. Generally, larger PSNR denotes higher reconstruction quality. PSNR can be used to analyze results in encryption schemes and determine how effective the encryption is; low PSNR values between the original and encrypted images indicate significant alteration, which suggests that a high degree of randomness and complexity is introduced by the encryption scheme. Generally, images having PSNR greater than 30 are considered good, and the images with PSNR value less than 20 are unacceptable. Mathematically, PSNR is defined as:

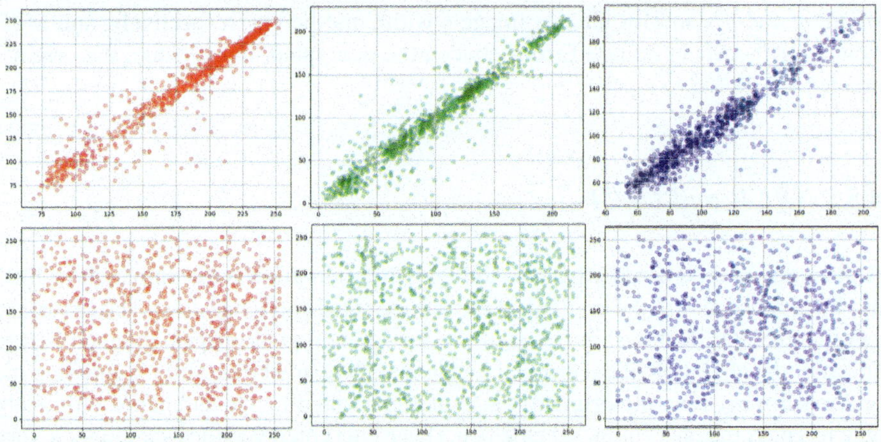

Fig. 6 Diagonal correlation scatter plot of Lena image

Table 3 Entropy values

Images		Proposed entropy	Reference [8]
Lena	Red	7.92	7.98
	Green	7.96	7.98
	Blue	7.90	7.99
Pepper	Red	7.92	7.98
	Green	7.96	7.99
	Blue	7.90	7.98
Panda	Red	7.92	7.98
	Green	7.96	7.98
	Blue	7.90	7.98

$$\text{PSNR} = 20\log_{10}\left(\frac{255}{\sqrt{\text{MSE}}}\right) \tag{9}$$

$$\text{MSE} = \sum_{i=1}^{m}\sum_{j=1}^{n}\frac{\left(E(i,j) - O(i,j)^2\right)}{mn} \tag{10}$$

Table 4 PSNR

Images	Proposed (in dB)
Lena	36.30
Baboon	31.65
Pepper	34.47
Airplane	37.92

where m and n stand for the test image's width and height, respectively, and MSE stands for the Mean Square Error between the decrypted image E and the original image O.

4.5 NPCR and UACI

A potential method used by hackers to break the encryption is to compare how little modifications made to the original image influence the encrypted copy. The adversary can identify the relationship between these two encrypted images obtained before and after the alteration of the original image, which would be helpful for guessing the relationship between the original image and encrypted image and hence ultimately able to guess the encryption scheme [9]. These encrypted versions should look very different from the original image, even with slight adjustments, thanks to a robust encryption technology. An attacker attempting a differential attack is thrown off by this. The applied techniques NPCR and UACI gauge how well the proposed encryption scrambles the image and withstands these kinds of attacks, as shown in Table 5.

Let $O1$ and $O2$ be the unencrypted images, while $C1$ and $C2$ be the encrypted ones. $O1$ and $O2$ are different from each other by just one altered pixel. T is a binary matrix that represents the size of the encrypted image. $T(r, s)$ is assigned a value of 0 if $C1(r, s) = C2(r, s)$; else, a value of 1 is assigned. The encrypted images' width and height are denoted by p and q. The NPCR and UACI have the following mathematical definitions:

$$\text{NPCR} = \sum_{r=1}^{p} \sum_{s=1}^{q} \left(T(r, s) \times \frac{100\%}{p \times q} \right) \tag{11}$$

$$\text{UACI} = \sum_{r=1}^{p} \sum_{s=1}^{q} \frac{|c_1(r, s) - c_2(r, s)|}{255} \times \frac{100\%}{p \times q} \tag{12}$$

Table 5 NPCR and UACI

Images	Proposed	
	UACI (%)	NPCR (%)
Lena	31.45	99.58
Baboon	31.45	99.52
Pepper	31.39	99.60
Airplane	31.47	99.59

5 Conclusion

The proposed encryption system provides a new and reliable basis for protecting images that are stored in the cloud, combining the creative method of DNA-based cryptography with the unpredictable nature of chaotic sequences in the Lorenz system. In addition to guaranteeing the safe encoding of image data, this dual-layered encryption technique makes use of the Lorenz system's sensitivity to beginning conditions to produce a dynamic and unpredictable encryption key. Only authorized users will be able to access and decrypt the encrypted images that are securely stored in the cloud due to strict access control mechanisms. The decryption procedure, which depends on the accurate duplication of the original chaotic sequence, reinforces the security integrity of the system even more. The present study represents a significant progression in the field of cloud-based image storage by providing a refined solution that optimizes data security while preserving effective accessibility for authorized users. This encryption system's dependability and efficacy represent a major advancement in cloud security and data encryption, establishing a new standard for safeguarding private image data in a world growing more digitally connected.

References

1. Li Q, Chen L (2023) An image encryption algorithm based on 6-dimensional hyper chaotic system and DNA encoding. Multimed Tools Appl 83(2):5351–5368
2. Sarosh P, Parah SA, Bhat GM (2022) An efficient image encryption scheme for healthcare applications. Multimed Tools Appl 81(5):7253–7270
3. Dynamic DNA cryptography-based image encryption scheme using multiple chaotic maps and SHA-256 hash function (2023) Optik 289:171253
4. A new one-dimensional chaotic map for image encryption scheme based on random DNA coding (2023) Opt Laser Technol 160:109033
5. A new color image encryption scheme based on DNA encoding and spatiotemporal chaotic system (2020) Signal Process Image Commun 80:115670
6. Shakor MY, Surameery NMS, Khlaif ZN (2023) Hybrid security model for medical image protection in cloud. Diyala J Eng Sci 16(1):68–77. https://doi.org/10.24237/djes.2023.16107
7. Color multiple image encryption scheme based on 3D-chaotic maps (2020) Math Comput Simul 178:646–666
8. A new color image encryption scheme based on DNA sequences and multiple improved 1D chaotic maps (2015) Appl Soft Comput 37:24–39
9. Secure communication using an adaptable multilevel RGB image encryption algorithm for wireless sensor networks (2024) e-Prime Adv Electr Eng Electron Energy 7:100446

Skinsense: Lesion Detection with Different Convolutional Neural Architecture

Aman Prasad Gupta, Aark Deep Sarkar, and R. Jeya

Abstract “Skinsense” is a new development where we present a comprehensive approach that uses machine learning models to analyze real-time captured skin images or existing images and predict their cancerous or benign nature, while also includes features that indicate findings belonging to the severity category. Our method uses deep learning algorithms and powerful image processing to evaluate skin lesions in order to aid in the early identification and detection of skin cancer (Afza et al. in Sensors 22:799, 2022 [1]). “Skinsense” employs the usage of multiple deep learning techniques such as ResNet101, Baseline CNN, Inception-ResNet-v2, VGG-16, and InceptionV3 to compare the accuracy provided by each of these models (Aldhyani et al. in Diagnostics 12:2048, 2022 [2]). The ResNet101 model with added layers of convolutional computation is the primary model with highest accuracy (Mukadam and Patil in Skin cancer classification framework using enhanced super resolution generative adversarial network and custom convolutional neural network, 2023 [3]). We have also integrated a smart chatbot interface to enable users to interact after receiving the classification results. Users can search for treatments, prevention strategies, cures, etc., which can provide them with valuable information and guidance. By combining advanced technology with a user-friendly interface, we aim to create a complete tool that not only aids diagnosis but also educates and inspires people to maintain healthy skin (El-Khatib et al. in Deep learning-based methods for automatic diagnosis of skin lesions, 2020 [4]).

Keywords Skinsense · Convolutional · Accuracy · Loss · Architecture · CNN · Performance

A. P. Gupta · A. D. Sarkar (✉) · R. Jeya
Department of Computing Technologies, SRM Institute of Science and Technology, Kattankulathur, Tamil Nadu, India
e-mail: ak8957@srmist.edu.in

A. P. Gupta
e-mail: ab6725@srmist.edu.in

R. Jeya
e-mail: jeyar@srmist.edu.in

© The Author(s), under exclusive license to Springer Nature Singapore Pte Ltd. 2025

H. Zen et al. (eds.), *Soft Computing and Signal Processing*, Lecture Notes in Networks and Systems 1200, https://doi.org/10.1007/978-981-97-9926-8_35

1 Dataset

Dermatoscopic images of common pigmented skin lesions from multiple sources are gathered in the HAM10000 dataset. Lack of variation in dermoscopic images and limited data sets make it difficult to train neural networks for autonomous diagnosis of pigmented skin. The HAM10000 ("Human vs. Machine with 10,000 training images") dataset is used to overcome this issue. It features dermoscopic photos of several persons in various poses. The completed dataset, which includes 10,015 dermoscopic photos as shown in Fig. 1, can be utilized to train machine learning algorithms [5]. The dataset includes seven crucial pigmented lesion-related diagnostic categories:

- Actinic keratoses and intraepithelial carcinoma/Bowen's disease (akiec)—Precancerous lesions that are characterized by rough, scaly skin patches that can become squamous. These lesions are frequently brought on by sun exposure [6].
- Basal cell carcinoma (bcc): The most prevalent kind of skin cancer, usually manifesting as a waxy lump or a flat, flesh-colored lesion [7]. It starts from basal cells in the outer layer of skin.
- Benign keratosis-like lesions (also known as solar lentigines, seborrheic keratoses, and lichen-planus like keratoses, or bkl): These are non-cancerous skin growths that typically appear as elevated, pigmented patches or growths. They include sunspots, seborrheic keratoses, and lichen-planus like keratoses.
- Dermatofibroma (df): A benign fibroblast-derived skin tumor that usually presents as a firm, elevated nodule on the skin's surface.

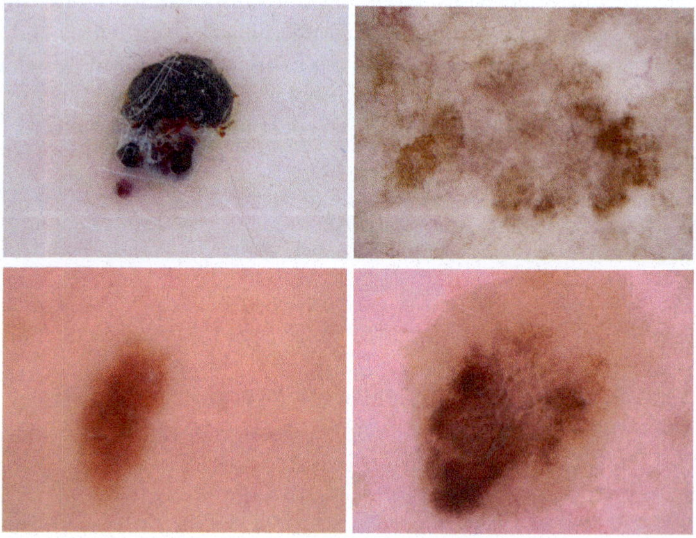

Fig. 1 Different types of lesions

- Melanoma (mel): A form of skin cancer originating from melanocytes, marked by the unchecked proliferation of cells that produce melanin; frequently manifested as unevenly colored, asymmetrical moles with erratic boundaries.
- Melanocytotic nevi (nv): Often referred to as moles, these benign skin growths consisting of melanocytes typically manifest as elevated lesions or small, dark spots on the skin.
- Vascular lesions, which include angiomas, angiokeratomas, pyogenic granulomas, and hemorrhages, are skin abnormalities resulting from irregularities in blood vessels. These lesions commonly appear as red or purple growths or discolorations on the skin.

More than 50% of diseases are confirmed by histopathology and the exact details of the main cases are followed, expert opinion or in vivo confocal microscopy (confocal) to confirm. The file contains lesions with multiple images, and this is monitored via the HAM10000_metadata file's lesion_id field [8].

2 System Architecture

Within the device structure described, there are 4 distinct levels as shown in Fig. 2 that have interaction seamlessly to supply a comprehensive answer. At the client tier, the user interface is crafted the use of HTML, CSS, and JavaScript, accessed via the Chrome browser on computing device gadgets. This tier serves as the access factor for consumer interaction, providing a visually attractive and responsive interface. Moving to the backend tier, the machine is powered by using Flask, a lightweight and versatile internet framework. This tier handles the commercial enterprise common sense and data processing, orchestrating the drift of facts among the customer and different tiers.

The API via internet tier acts as a bridge to outside services, making use of Gemini to facilitate verbal exchange with outside structures and offerings. This tier enables the device to leverage external functionalities and data sources efficiently. At the model level tier, the gadget integrates numerous deep convolutional neural network (CNN) architectures to carry out complicated statistics analysis and predictions. These models are essential for duties like photograph recognition, natural language processing, and different AI-pushed functionalities. The interaction between those tiers is orchestrated in a dependent way [9]. The client tier communicates with the backend tier through HTTP requests, triggering actions and retrieving facts as wanted. In reaction, the backend tier interacts with the patron by means of serving dynamic HTML pages tailor-made to user requests. Additionally, the backend tier interfaces with the API tier to get admission to external services and exchange statistics seamlessly. Furthermore, the backend tier collaborates with the model tier through providing input data for evaluation and receiving expected effects, enabling the system to leverage superior device mastering skills for decision-making and insights generation.

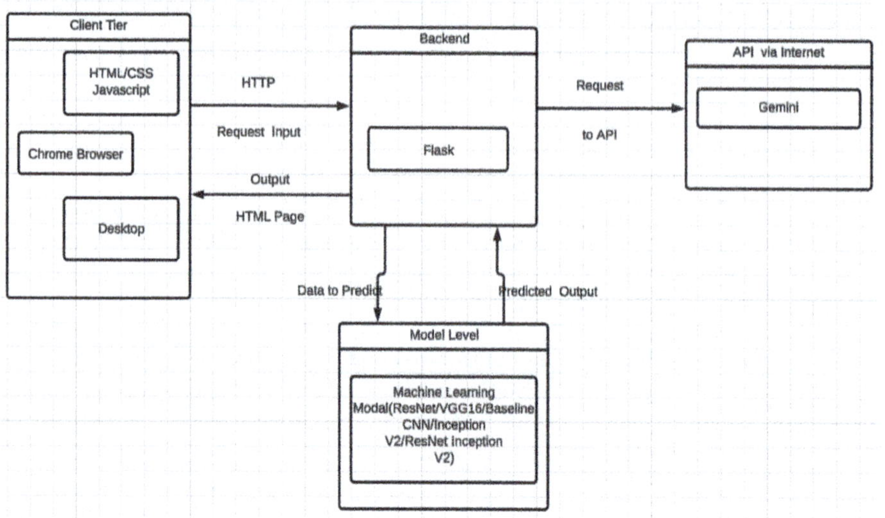

Fig. 2 System architecture

3 Workflow

The entire workflow of events involved, based on user preferences to either detect a lesion or garner awareness about skin cancer. Upon initiation, the user is greeted with the landing page of the web app. Here the user has two options, either he/she can straightaway proceed to the detection page by clicking on the "Detect Image" button or they can choose to educate themselves on the various types of lesions, which of them are cancerous, etc. Upon choosing to move toward the detection page, user is provided with two options to upload, either they can choose to upload an existing image from the system or they can use their camera to capture real-time image. Either or two ways is used to upload image. Later, the image is passed on to the deep CNN model to perform its processing [10]. The model after parsing the image through the process displays the output generated. The output is the category of lesion the provided image falls into with a severity index of whether it is cancerous or non-cancerous. Toward the end, the user can use an AI-powered chatbot integrated onto the output page, in order to get his/her queries resolved regarding the predicted disease. The user is not limited in terms of number of queries being processed and can explore as per his/her satisfaction as shown in Fig. 3.

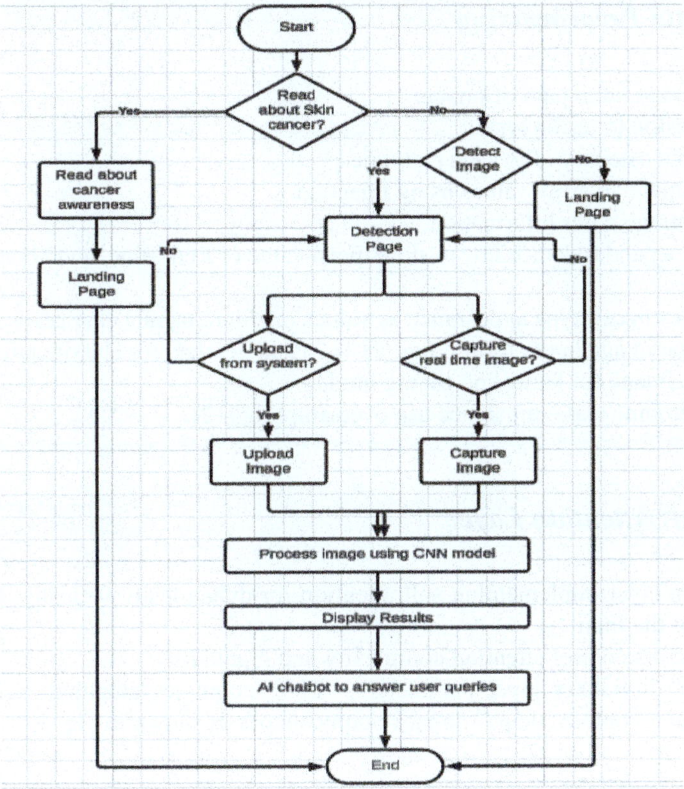

Fig. 3 Workflow chart

4 Highest Precision Model (RESNET101 + CUSTOM)

Categorizing the skin lesion can help in skin cancer detection. The lesions that we are categorizing for the project are listed as keratoses actinic, cancer of the basal cell, benign keratosis-like lesions, Dermatofibroma, Melanoma (mel), Melanocytic nevi (nv). Out of which Actinic keratoses, Basal cell carcinoma, Melanoma are cancerous, but Benign keratosis-like lesions, Dermatofibroma, Melanocytic nevi are non-cancerous. For classification process we first need to perform EDA and then need to transform and preprocess the dataset in the form which can be used to train a machine learning classification model [11].

4.1 Data Preprocessing

- Map image id to path of image.
- Map lesion to lesion type.
- Assign a category id to each cell type.
- Read image to array values using image id.
- Get average color information.
- Create a pandas data-frame to store mean value of Red, Blue, and Green for each picture.
- Changes in cell type appearance as values in color channel changes is recorded.
- Reshape image in the form (28 × 28 × 3) and get data for classification.
- Resize image for keras fine-tuning model.
- To overcome class imbalance use oversampling [12].

4.2 Soft Attention Layer

- Custom layer implementing soft attention for dynamically focusing on relevant parts of the input.
- Parameters control attention complexity and granularity.
- "Build" initializes layer parameters, "Call" computes attention weights, and "Compute_output_shape" defines output shapes as shown in Fig. 4.

4.3 Model Architecture

- Starts with a 28 × 28 × 3 input layer.
- Utilizes a pre-trained ResNet101 backbone.
- Soft attention layer enhances feature map focusing [13].
- Concatenates original and attention-weighted feature maps.
- Followed by processing layers and classification head with softmax.

4.4 Training

- Trained on "X_train" and "Y_train" with 30% validation split.
- Apply sparse categorical cross-entropy loss with the Adam optimizer.
- Optimizes for accuracy metric with batch size of 128 [14].
- Utilizes "ModelCheckpoint" to save best model-based as shown in Fig. 5 on validation accuracy as shown in Table 1.

Layer (type)	Output Shape	Param #	Connected to
input_1 (InputLayer)	[(None, 28, 28, 3)]	0	[]
conv1_pad (ZeroPadding2D)	(None, 34, 34, 3)	0	['input_1[0][0]']
conv1_conv (Conv2D)	(None, 14, 14, 64)	9472	['conv1_pad[0][0]']
conv1_bn (BatchNormalization)	(None, 14, 14, 64)	256	['conv1_conv[0][0]']
conv1_relu (Activation)	(None, 14, 14, 64)	0	['conv1_bn[0][0]']
pool1_pad (ZeroPadding2D)	(None, 16, 16, 64)	0	['conv1_relu[0][0]']
pool1_pool (MaxPooling2D)	(None, 7, 7, 64)	0	['pool1_pad[0][0]']
conv2_block1_1_conv (Conv2D)	(None, 7, 7, 64)	4160	['pool1_pool[0][0]']
conv2_block1_1_bn (BatchNormalization)	(None, 7, 7, 64)	256	['conv2_block1_1_conv[0][0]']
conv2_block1_1_relu (Activation)	(None, 7, 7, 64)	0	['conv2_block1_1_bn[0][0]']
conv2_block1_2_conv (Conv2D)	(None, 7, 7, 64)	36928	['conv2_block1_1_relu[0][0]']
conv2_block1_2_bn (BatchNormalization)	(None, 7, 7, 64)	256	['conv2_block1_2_conv[0][0]']
conv2_block1_2_relu (Activation)	(None, 7, 7, 64)	0	['conv2_block1_2_bn[0][0]']
conv2_block1_0_conv (Conv2D)	(None, 7, 7, 256)	16640	['pool1_pool[0][0]']
conv2_block1_3_conv (Conv2D)	(None, 7, 7, 256)	16640	['conv2_block1_2_relu[0][0]']
conv2_block1_0_bn (BatchNormalization)	(None, 7, 7, 256)	1024	['conv2_block1_0_conv[0][0]']
conv2_block1_3_bn (BatchNormalization)	(None, 7, 7, 256)	1024	['conv2_block1_3_conv[0][0]']
conv2_block1_add (Add)	(None, 7, 7, 256)	0	['conv2_block1_0_bn[0][0]', 'conv2_block1_3_bn[0][0]']
conv2_block1_out (Activation)	(None, 7, 7, 256)	0	['conv2_block1_add[0][0]']
conv2_block2_1_conv (Conv2D)	(None, 7, 7, 64)	16448	['conv2_block1_out[0][0]']
conv2_block2_1_bn (BatchNormalization)	(None, 7, 7, 64)	256	['conv2_block2_1_conv[0][0]']
conv2_block2_1_relu (Activation)	(None, 7, 7, 64)	0	['conv2_block2_1_bn[0][0]']
conv2_block2_2_conv (Conv2D)	(None, 7, 7, 64)	36928	['conv2_block2_1_relu[0][0]']
conv2_block2_2_bn (BatchNormalization)	(None, 7, 7, 64)	256	['conv2_block2_2_conv[0][0]']
conv2_block2_2_relu (Activation)	(None, 7, 7, 64)	0	['conv2_block2_2_bn[0][0]']
conv2_block2_3_conv (Conv2D)	(None, 7, 7, 256)	16640	['conv2_block2_2_relu[0][0]']
conv2_block2_3_bn (BatchNormalization)	(None, 7, 7, 256)	1024	['conv2_block2_3_conv[0][0]']
conv5_block3_1_bn (BatchNormalization)	(None, 1, 1, 512)	2048	['conv5_block3_1_conv[0][0]']
conv5_block3_1_relu (Activation)	(None, 1, 1, 512)	0	['conv5_block3_1_bn[0][0]']
conv5_block3_2_conv (Conv2D)	(None, 1, 1, 512)	2359808	['conv5_block3_1_relu[0][0]']
conv5_block3_2_bn (BatchNormalization)	(None, 1, 1, 512)	2048	['conv5_block3_2_conv[0][0]']
conv5_block3_2_relu (Activation)	(None, 1, 1, 512)	0	['conv5_block3_2_bn[0][0]']
conv5_block3_3_conv (Conv2D)	(None, 1, 1, 2048)	1050624	['conv5_block3_2_relu[0][0]']
conv5_block3_3_bn (BatchNormalization)	(None, 1, 1, 2048)	8192	['conv5_block3_3_conv[0][0]']
conv5_block3_add (Add)	(None, 1, 1, 2048)	0	['conv5_block2_out[0][0]', 'conv5_block3_3_bn[0][0]']
conv5_block3_out (Activation)	(None, 1, 1, 2048)	0	['conv5_block3_add[0][0]']
max_pooling2d (MaxPooling2D)	(None, 1, 1, 2048)	0	['conv5_block3_out[0][0]']
batch_normalization (BatchNormalization)	(None, 1, 1, 2048)	8192	['max_pooling2d[0][0]']
soft_attention (SoftAttention)	[(None, 1, 1, 2048), (None, 16, 1, 1)]	294928	['batch_normalization[0][0]']
max_pooling2d_2 (MaxPooling2D)	(None, 1, 1, 2048)	0	['batch_normalization[0][0]']
max_pooling2d_1 (MaxPooling2D)	(None, 1, 1, 2048)	0	['soft_attention[0][0]']
concatenate (Concatenate)	(None, 1, 1, 4096)	0	['max_pooling2d_2[0][0]', 'max_pooling2d_1[0][0]']
activation (Activation)	(None, 1, 1, 4096)	0	['concatenate[0][0]']
dropout (Dropout)	(None, 1, 1, 4096)	0	['activation[0][0]']
conv2d (Conv2D)	(None, 1, 1, 512)	18874880	['dropout[0][0]']
batch_normalization_1 (BatchNormalization)	(None, 1, 1, 512)	2048	['conv2d[0][0]']
conv2d_1 (Conv2D)	(None, 1, 1, 512)	2359808	['batch_normalization_1[0][0]'
batch_normalization_2 (BatchNormalization)	(None, 1, 1, 512)	2048	['conv2d_1[0][0]']
conv2d_2 (Conv2D)	(None, 1, 1, 512)	262656	['batch_normalization_2[0][0]'
batch_normalization_3 (BatchNormalization)	(None, 1, 1, 512)	2048	['conv2d_2[0][0]']
max_pooling2d_3 (MaxPooling2D)	(None, 1, 1, 512)	0	['batch_normalization_3[0][0]'
flatten (Flatten)	(None, 512)	0	['max_pooling2d_3[0][0]']
dense (Dense)	(None, 4096)	2101248	['flatten[0][0]']
dense_1 (Dense)	(None, 4096)	16781312	['dense[0][0]']
dense_2 (Dense)	(None, 7)	28679	['dense_1[0][0]']

Total params: 83,376,023
Trainable params: 40,710,679
Non-trainable params: 42,665,344

Fig. 4 Layers within model

4.5 Performance

See Fig. 5 and Table 1.

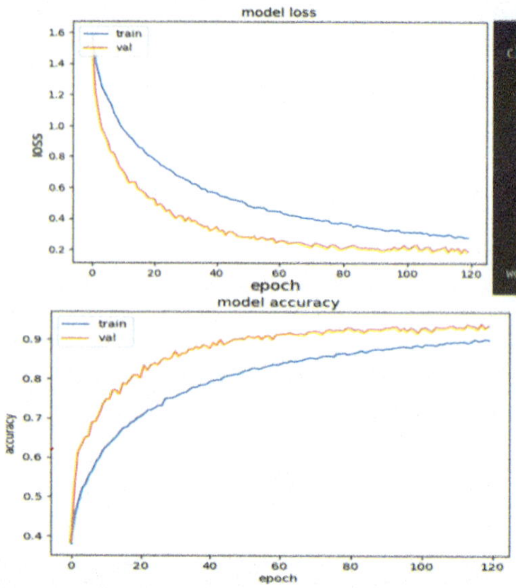

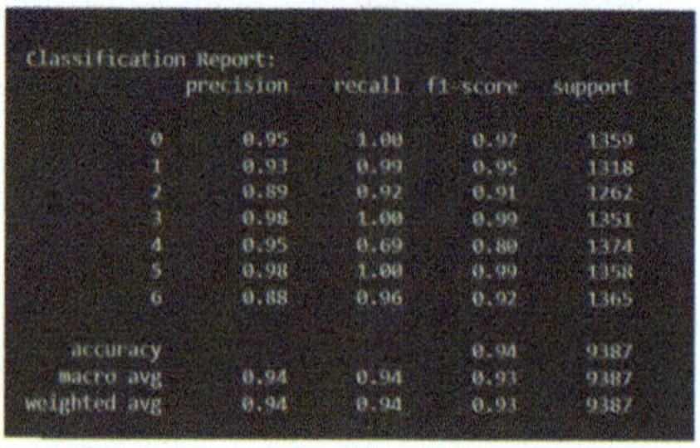
Classification Report:

	precision	recall	f1-score	support
0	0.95	1.00	0.97	1359
1	0.93	0.99	0.95	1318
2	0.89	0.92	0.91	1262
3	0.98	1.00	0.99	1351
4	0.95	0.69	0.80	1374
5	0.98	1.00	0.99	1358
6	0.88	0.96	0.92	1365
accuracy			0.94	9387
macro avg	0.94	0.94	0.93	9387
weighted avg	0.94	0.94	0.93	9387

Fig. 5 Performance measure

Table 1 Performance metrics

Performance parameters	Values
Accuracy	94%
No. of layers	290
Total params	83,376,023
Trainable params	40,710,679
Non-trainable params	42,665,344

5 Different Architectures

5.1 Baseline CNN

The baseline convolutional neural network (CNN) version serves as the foundational structure for picture processing and lesion detection inside the pores. The model is designed to investigate pores and skin lesion snapshots and classify them into one-of a-kind categories, together with cancerous or non-cancerous, primarily based on found out capabilities as shown in Fig. 6.

- Input Layer: The version takes pores and skin lesion images as input. These images usually have dimensions representing top, width, and channels (consisting of RGB channels).
- Convolutional Layers: A CNN's convolutional layers apply a variety of learnable filters to each other in order to extract functions from the input images.

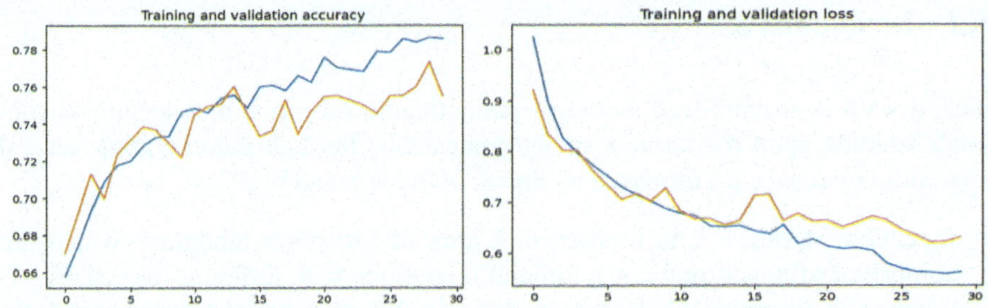

Fig. 6 Performance measure

- Activation Features: To introduce non-linearity and allow the model to learn complex styles, non-linear activation functions (such as ReLU) are applied after each convolutional layer.
- Pooling Layers: By reducing the spatial dimensions of the feature maps, max pooling or common pooling layers help to reduce computational complexity while maintaining the important features.
- Flattening Layer: To feed into the absolutely related layers, the function mappings obtained from the convolutional layers are flattened into a one-dimensional vector.
- Completely Linked Layers: Those layers include densely related neurons that learn how to classify the extracted features. They perform excessive-stage reasoning and decision-making on the basis of learned representations.
- Output Layer: This layer produces the version's predictions as shown in Table 2. Inside the context of skin cancer detection, this output layer normally includes neurons similar to exceptional training (e.g., cancerous or non-cancerous). The softmax activation function is regularly used to transform uncooked predictions into chances for each class.

(1) **Performance**

See Fig. 6 and Table 2.

Table 2 Performance metrics

Performance parameters	Values
Accuracy	78%
No. of layers	11
Total params	2,124,839
Trainable params	1,532,894
Non-trainable params	591,945

5.2 InceptionV3

InceptionV3 is an improved model of the authentic Inception architecture. InceptionV3 builds upon the success of its predecessor through incorporating several upgrades to enhance performance as shown in Table 3 and Fig. 7.

- Inception Modules: The fundamental idea of Inception modules—which are comprised of concurrent convolutional operations with different filter sizes—is retained in InceptionV3. However, InceptionV3 goes beyond decorative function extraction and computational efficiency by introducing new optimizations including factorized convolutions and batch normalization.
- Factorized Convolutions: InceptionV3 replaces popular convolutions with factorized convolutions, which break up the spatial convolution into separate operations: depthwise convolutions and pointwise convolutions. This decomposition reduces the computational cost of convolutions even as preserving representational capacity, leading to stepped forward performance and overall performance.
- Batch Normalizing: Following every convolutional operation, InceptionV3 incorporates layers for batch normalizing. Batch normalization lowers internal covariate shift, improves gradient float, and normalizes all of the layer activations to stabilize and speed up the training process.
- Auxiliary Classifiers: Similar to the original Inception architecture, InceptionV3 can also consist of auxiliary classifiers at intermediate layers all through training to resource in gradient float and save overfitting.

Table 3 Performance metrics

Performance parameters	Values
Accuracy	71%
No. of layers	315
Total params	22,855,463
Trainable params	1,069,895
Non-trainable params	21,785,568

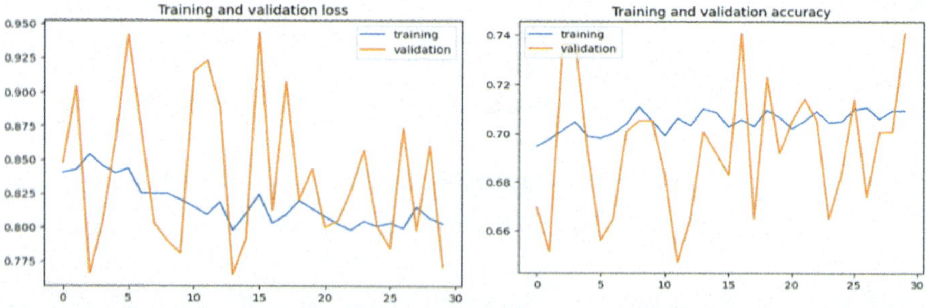

Fig. 7 Performance measure

(1) **Performance**

See Fig. 7 and Table 3.

5.3 Inception-ResNet-v2

A convolutional neural network design called Inception-ResNet-v2 combines the ideas of the Inception and ResNet architectures. To achieve state-of-the-art performance as shown in Table 4 in a variety of computer vision applications, it combines the multi-branch architecture of the Inception model with residual connections from the ResNet model. The following are the main elements of Inception-ResNet-v2 as shown in Fig. 8:

- Inception Modules: With the use of Inception modules which are made up of concurrent convolutional processes with various filter sizes, Inception-ResNet-v2 allows the model to capture a variety of features at various scales inside a single layer.
- Residual Connections: Inspired by ResNet, Inception-ResNet-v2 incorporates residual connections that bypass some convolutional layers, allowing for easier optimization of very deep networks and mitigating the vanishing gradient problem.

Table 4 Performance metrics

Performance parameters	Values
Accuracy	93%
No. of layers	784
Total params	55,127,271
Trainable params	24,352,647
Non-trainable params	30,774,627

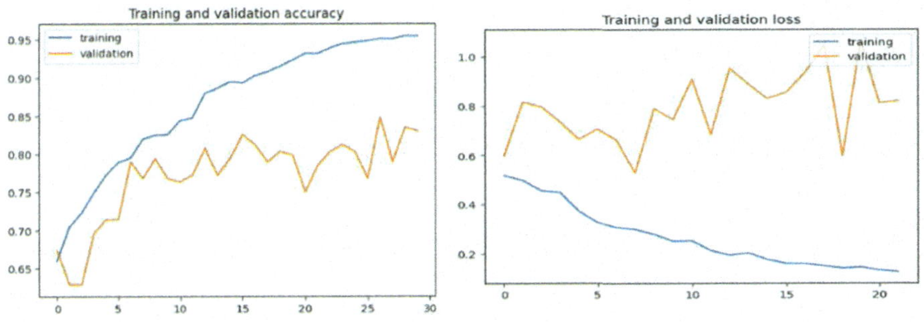

Fig. 8 Performance measure

- Factorized Convolutions: Similar to Inception, Inception-ResNet-v2 employs factorized convolutions, which split the spatial convolution into depthwise convolutions and pointwise convolutions, reducing computational complexity while preserving representational capacity.
- Batch Normalization: Batch normalization layers are utilized throughout the network to stabilize and accelerate the training process by normalizing the activations of each layer, improving gradient flow and convergence speed.
- Auxiliary Classifiers: Inception-ResNet-v2 may include auxiliary classifiers at intermediate layers during training to aid in gradient flow and prevent overfitting, similar to the original Inception architecture.

(1) **Performance**

See Fig. 8 and Table 4.

5.4 *VGG-16*

A type of convolutional neural network (CNN) architecture known as VGG-16 was put forth by researchers at the University of Oxford's Visible Geometry Group (VGG). It is easy to use and efficient at classifying photos, which made it a good option for identifying skin cancer and pores as shown in Fig. 9 and Table 5.

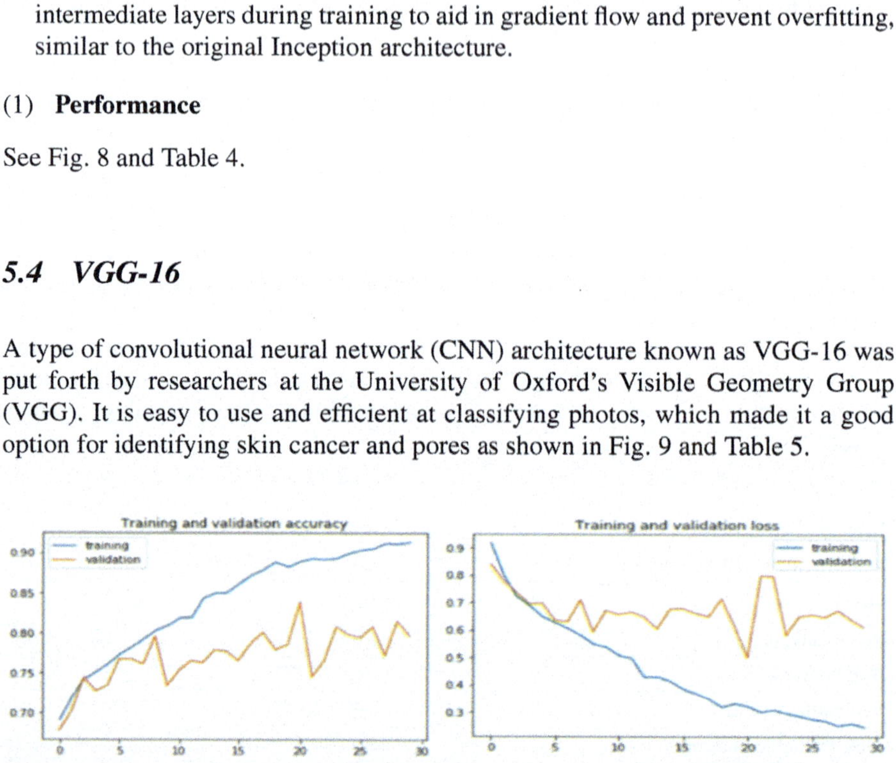

Fig. 9 Performance measure

Table 5 Performance metrics

Performance parameters	Values
Accuracy	88%
No. of layers	13
Total params	14,980,935
Trainable params	266,247
Non-trainable params	14,714,688

- Convolutional Layers: VGG-16 includes 13 convolutional layers stacked on top of each other. These layers use 3 × 3 convolutional filters with a stride of 1, allowing them to seize nice-grained features from skin lesion images.
- Max Pooling Layers: VGG-16 incorporates max pooling layers with a 2 × 2 window and a stride of 2 after each convolutional layer.
- Activation Functions: To provide non-linearity and enable the model to learn intricate styles linked to pores and skin cancer, Rectified Linear Unit (ReLU) activation functions are added after each convolutional layer.
- Fully Connected Layers: The convolutional layers are observed with the aid of 3 absolutely related layers, which carry out high-level reasoning based on the features discovered from skin lesion snap shots. Those layers are useful in distinguishing between cancerous and non-cancerous lesions.
- Softmax Output: The very last layer of VGG-16 is a softmax layer that outputs elegance probabilities, permitting the version to make predictions approximately the probability of a pores and skin lesion being cancerous. This permits early detection and intervention in potential cancer instances.

(1) **Performance**

See Fig. 9 and Table 5.

References

1. Afza F, Sharif M, Khan MA, Tariq U, Yong HS, Cha J (2022) Multiclass skin lesion classification using hybrid deep features selection and extreme learning machine. Sensors 22:799
2. Aldhyani THH, Verma A, Al-Adhaileh MH, Koundal D (2022) Multi-class skin lesion classification using a lightweight dynamic kernel deep-learning-based convolutional neural network. Diagnostics 12:2048
3. Mukadam SB, Patil HY (2023) Skin cancer classification framework using enhanced super resolution generative adversarial network and custom convolutional neural network
4. El-Khatib H, Popescu D, Ichim L (2020) Deep learning-based methods for automatic diagnosis of skin lesions
5. Gong A, Yao X, Lin W (2020) Classification for dermoscopy images using convolutional neural networks based on the ensemble of individual advantage and group decision
6. Adegun A, Viriri S (2019) Deep learning model for skin lesion segmentation: fully convolutional network. In: Karray F, Campilho A, Yu A (eds) Image analysis and recognition
7. Tschandl P, ViDIR Group (2018) The HAM10000 dataset, a large collection of multi-source dermatoscopic images of common pigmented skin lesions
8. Poornima A, Shyamala Devi M, Sumithra M, Bharath MV, Swathi, Sathishkumar S, Yogesh K, Upadhyay SS, Sah NK (2021) Epoch interrogation for skin cancer detection using convolutional neural network models
9. AlSadhan NA, Alamri SA, Ismail MMB, Bchir O (2024) Skin cancer recognition using unified deep convolutional neural networks
10. Pacheco AGC, Krohling RA (2019) Recent advances in deep learning applied to skin cancer detection
11. Yu L, Chen H, Dou Q, Qin J, Heng P-A (2017) Automated melanoma recognition in dermoscopy images via very deep residual networks. IEEE Trans Med Imaging 36(4):994–1004

12. Jeong HK, Park C, Henao R, Kheterpal M (2022) Deep learning in dermatology: a systematic review of current approaches, outcomes, and limitations
13. Rezaee K, Zadeh HG (2024) Self-attention transformer unit-based deep learning framework for skin lesions classification in smart healthcare
14. Hauser K, Kurz A, Haggenmüller S, Maron RC, von Kalle C, Utikal JS, Meier F, Hobelsberger S, Gellrich FF, Sergon M, Hauschild A, French LE, Heinzerling L, Schlager JG, Ghoreschi K, Schlaak M, Hilke FJ, Poch G, Kutzner H, Berking C, Heppt MV, Erdmann M, Haferkamp S, Schadendorf D, Sondermann W, Goebeler M, Schilling B, Kather JN, Fröhling S, Lipka DB, Hekler A, Krieghoff-Henning E, Brinker TJ (2024) Explainable artificial intelligence in skin cancer recognition: a systematic review

Stress and Depression Detection by Chatbot

Nazeer Shaik, B. Deepika Reddy, P. Rekha, B. Likhitha, and N. Leela Krishna

Abstract Nowadays, stress and depression have become a growing problem for society due to their high impact on individuals' performance. As per the statistics of Mental Health Foundation, 50% of adults who felt stressed reported feeling depressed. Hence, it is imperative to establish a service capable of early identification of individuals experiencing stress and depression to prevent potential crises. Over the past few years, incorporating artificial intelligence (AI) into mental health interventions has demonstrated encouraging outcomes in delivering readily available and prompt assistance to individuals encountering psychological distress. Utilizing machine learning techniques, the chatbot provides users with a confidential and nonjudgmental space to express their emotions, engage in discussions about their concerns, and access personalized resources. The primary objective of this chatbot is to bridge the gap in mental health care by providing immediate assistance, especially during times when traditional therapeutic services may be inaccessible.

Keywords Stress · Depression · Machine learning · Mental health · Chatbot

1 Introduction Literature

Global health statistics indicate that depression affects a staggering number of individuals worldwide, impacting an estimated 300 million people, and constituting roughly 4.4% of the world population [1]. In spite of its widespread commonness, a substantial number of individuals with depression go undiagnosed [2]. The failure to diagnose is frequently linked to the societal stigma surrounding mental health problems, especially depression, which serves as a significant obstacle to seeking essential assistance. Studies show that this stigma causes significant delays in individuals seeking treatment, with nearly half of those experiencing depression avoiding seeking help due to worries about societal attitudes [3].

N. Shaik (✉) · B. Deepika Reddy · P. Rekha · B. Likhitha · N. Leela Krishna
Srinivasa Ramanujan Institute of Technology, Rotarypuram Village, B K Samudram Mandal, Ananthapuramu, Andhra Pradesh, India
e-mail: shaiknaz2020@gmail.com

© The Author(s), under exclusive license to Springer Nature Singapore Pte Ltd. 2025

H. Zen et al. (eds.), *Soft Computing and Signal Processing*, Lecture Notes in Networks and Systems 1200, https://doi.org/10.1007/978-981-97-9926-8_36

The repercussions of depression go beyond immediate outcomes, including heightened susceptibility to developing concurrent health issues such as cardiovascular diseases, diminished quality of life, and significant decreases in productivity. Existing models for detecting depression often provide binary classifications, categorizing individuals as either depressed or not, but fail to offer accompanying justifications.

2 Literature Review

2.1 Current Approaches

Various approaches have been implemented to detect depression, ranging from traditional questionnaire-based methods to advanced machine learning techniques. Initial methods primarily involved standardized questionnaires like the Beck Depression Inventory (BDI) and the Patient Health Questionnaire (PHQ-9) designed to evaluate the severity of depression symptoms through a series of self-reported questions.

Questionnaire-Based Assessments

Traditional mental health assessments [4], like psychological questionnaires, remain prevalent. These tools, while useful, depend heavily on self-reported data and can be influenced by the respondent's honesty and self-awareness. Introduced the questionnaire-based mental health assessment method as shown in Fig. 1.

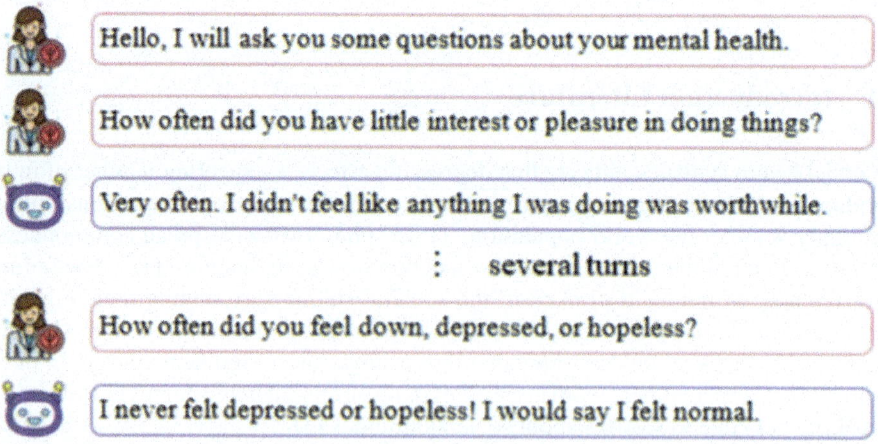

Fig. 1 An example of a psychological questionnaire designed to evaluate an individual's level of depression

Chatbots for Mental Health

Chatbots [5] have shown promise in providing interactive. However, a significant issue emerged during an instance where a user sought mental health support from a chatbot. The user expressed distress, saying, "Hey, I feel very bad, I want to kill myself." Regrettably, the chatbot replied, "I'm sorry to hear that. I'm here to help." This interaction poses serious risks and is deeply concerning, particularly when the user asked, "Should I kill myself?"

Moreover, it has been observed that some chatbots exhibit anxious behavior, particularly when engaging with users experiencing dysphoria. This second-hand anxiety from chatbots can potentially impact users negatively [6]. The potential harm posed by negative emotions or opinions expressed by chatbots to the public is significant. However, current evaluation metrics for chatbots frequently fail to address these critical mental health considerations.

Machine Learning in Depression Detection

Machine learning (ML) [7] techniques have revolutionized depression detection, enabling more nuanced and dynamic assessments. For instance, [7] employed social network data and ML algorithms to identify depression, achieving promising results. However, these models often lacked transparency, offering predictions without detailed explanations, thus limiting their practical applicability in clinical settings.

Limitations and Gaps

Despite advancements, current models and chatbots often fall short in several areas. Many depression detection models provide binary classifications without contextual explanations, limiting their usefulness in offering actionable insights. Moreover, the issue of chatbot responses to critical mental health queries underscores the need for more sophisticated and empathetic response generation mechanisms.

3 Proposed System

The chatbot under development utilizes Python language along with the open-source software library Keras from TensorFlow. The project's objective is to create a chatbot model capable of accurately predicting various types of mental conditions by leveraging the random forest algorithm.

Advantages

- High efficiency.
- Time saving.
- Low complexity.
- Low resource consuming.
- Easy to scale up.

4 Methodology

4.1 Development Process of the Machine Learning Model

Various approaches have been implemented to detect depression, ranging from traditional questionnaire-based methods to advanced machine learning techniques. Initial methods primarily involved standardized questionnaires like the Beck Depression Inventory (BDI) and the Patient Health Questionnaire (PHQ-9) designed to evaluate the severity of depression symptoms through a series of self-reported questions.

The development of the proposed chatbot involves a systematic machine learning model development process (as shown in Fig. 2):

Problem and Objective Definition: The primary objective is to create a chatbot capable of accurately predicting various mental health conditions using the random forest algorithm. This algorithm was chosen due to its robustness in handling high-dimensional data and its effectiveness in navigating nonlinear relationships.

Data Gathering and Preparation: Data were gathered from various mental health forums and annotated for intents, patterns, and responses. Preprocessing involved cleaning the data, encoding categorical variables, and splitting the dataset into training and testing sets to ensure compatibility with the model.

Model Learning: The model was trained using the Keras library within TensorFlow. Training was conducted on a laptop equipped with an Nvidia GeForce 940MX GPU. The model was iteratively refined to improve accuracy, evaluated based on metrics such as precision, recall, F1-score, and overall accuracy.

Deployment and Integration: Post-training, the chatbot was deployed on a website using Flask. This integration allows users to access the chatbot via a web application, providing a convenient platform for real-time mental health support.

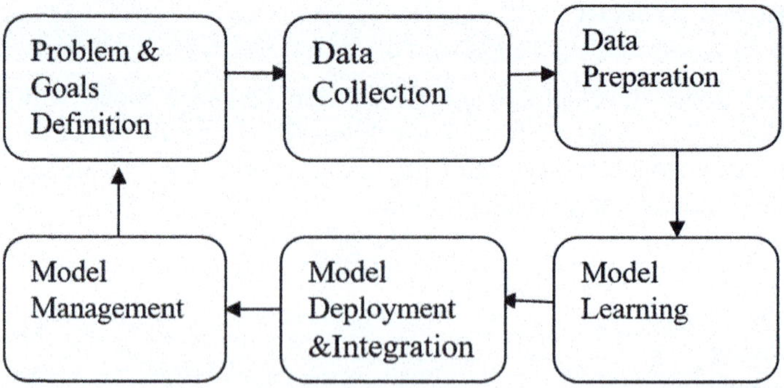

Fig. 2 A diagram illustrating the process of developing a machine learning model

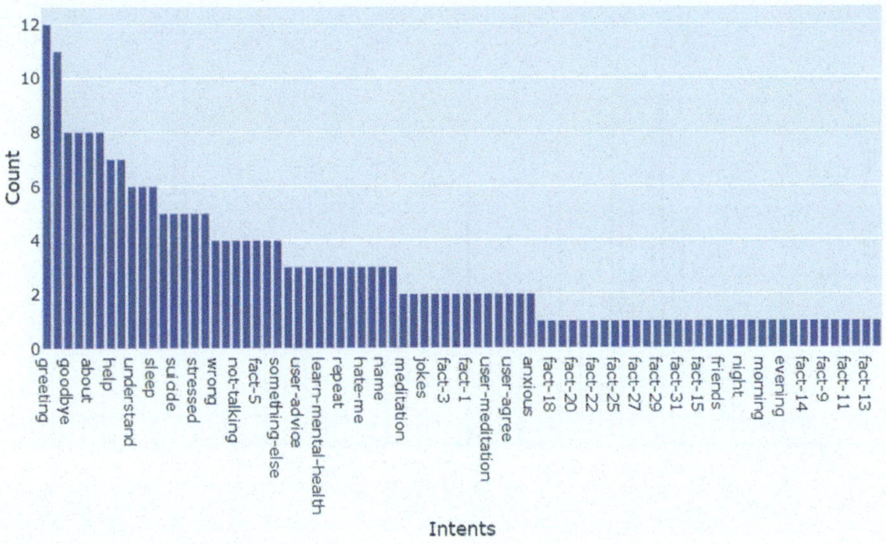

Fig. 3 Distribution of intents

Model Management: Continuous testing and refinement are essential to ensure the chatbot's accuracy and reliability. The input processed by the chatbot is regularly compared against expected outcomes, with necessary adjustments made to the processing module to maintain high-performance results.

4.2 *Exploratory Data Analysis*

Various approaches have been implemented to detect depression, ranging from traditional questionnaire-based methods to advanced machine learning techniques. Initial methods primarily involved standardized questionnaires like the Beck Depression Inventory (BDI) and the Patient Health Questionnaire (PHQ-9) designed to evaluate the severity of depression symptoms through a series of self-reported questions (shown in Fig. 3).

4.3 *Pattern and Response Analysis*

Further analysis focused on the patterns and responses linked with each intent. By calculating the mean count of patterns and responses, variations in the complexity of user interactions were identified. This analysis helped refine the model to handle

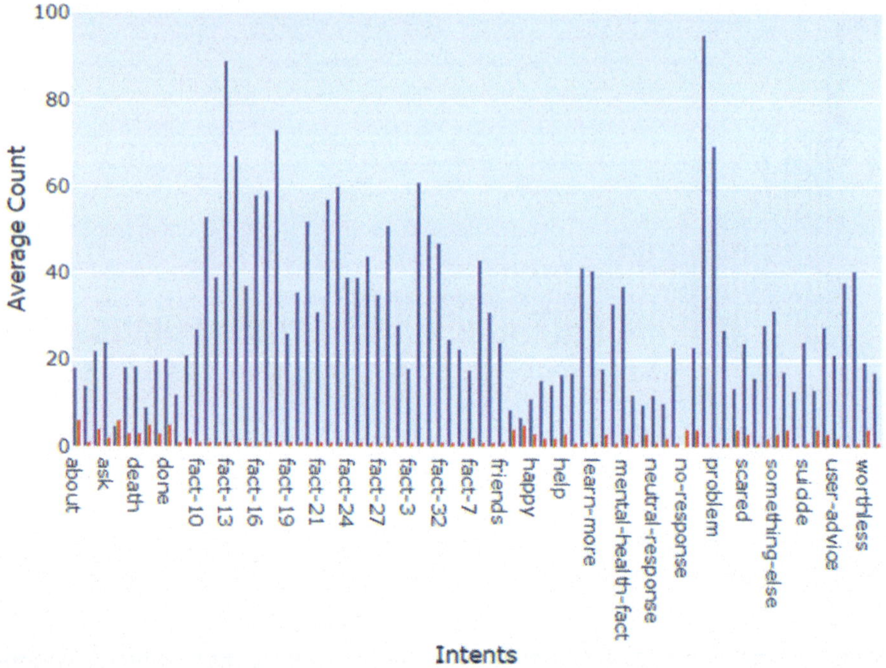

Fig. 4 Pattern and response analysis

diverse and complex inputs effectively shown in Fig. 4 which have been implemented to detect depression, ranging from traditional questionnaire-based methods to advanced machine.

5 Model Performance

The random forest algorithm achieved a 95% accuracy rate, significantly higher than existing models. Detailed performance metrics, including precision, recall, and F1-score, are summarized in Table 1. The chatbot demonstrated high accuracy across various intents, including sensitive topics like suicide and depression symptoms.

Deployment and User Interaction

The deployed chatbot effectively classified user inputs and generated appropriate responses. Users reported feeling more comfortable interacting with the bot, indicating the model's success in providing a nonjudgmental and supportive environment for discussing mental health concerns.

Table 1 Precision, recall, and F1-score

	Precision	Recall	F1-score	Support
Death	1.00	1.00	1.00	1
Anxiety	1.00	1.00	1.00	2
Demotivated	1.00	1.00	1.00	2
Goodbye	0.67	1.00	0.80	2
Greeting	1.00	0.50	0.67	2
Suicide	1.00	1.00	1.00	1
Depression symptoms	1.00	1.00	1.00	2
Self-care	0.14	1.00	0.25	2
Depression	0.50	1.00	0.67	1
Happy	0.00	0.00	0.00	1
Anxiety symptoms	0.00	0.00	0.00	0
Accuracy			0.95	19
Macro avg.	0.90	0.92	0.91	19
Weighted avg.	0.91	0.95	0.93	19

6 Discussions

6.1 Existing System

In a study titled "Non-clinical trial: Early depression detection through a conversational AI bot" authored by Kaywan et al. [8], machine learning algorithms were employed to identify stress and depression in individuals, achieving an accuracy of approximately 75%.

In their work titled "Utilizing an Intelligent ChatBot for Student Depression Intervention: Cognitive Behavioral Therapy Approach" authored by Patel et al. [9], an RNN was employed to attain a 70% accuracy rate.

6.2 Proposed System

In our proposed work, we have acquired 95% accuracy when compared to the existing system using random forest. The random forest algorithm was selected for its capacity to manage high-dimensional data, navigate nonlinear relationships effectively, and offer interpretability.

The target variable, signifying the type of disorders, was defined, and the accuracy graph was plotted using the Matplotlib library in Python shown in Fig. 5. Additionally, the "accuracy_score" module was imported for evaluation. Data preprocessing

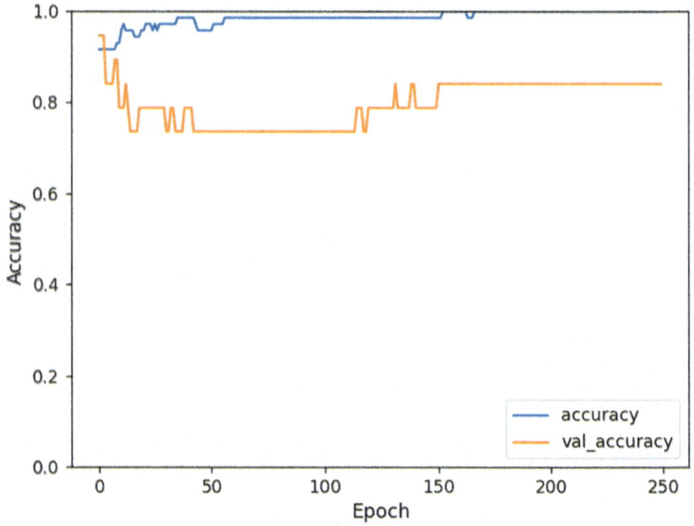

Fig. 5 In 200 epochs of accuracy val accuracy

involved employing commands like "get_dummies" and "label_encoder." Both the feature variables (*X*) and the target variable (*y*) were specified.

The dataset was split into 80% for training and 20% for testing. The model underwent training with the training set and evaluation based on performance metrics like precision, recall, F1-score, and accuracy. Predictions were generated based on the trained model. Subsequently, a chatbot was developed, enabling users to interact by providing symptoms in response to specific questions.

The chatbot then determines the type of disease the user might be experiencing. If no positive symptoms are identified, the chatbot congratulates the user for not having any mental health disorders. This approach has proven effective, as users feel more comfortable responding to the bot's inquiries without hesitation. The precision, recall, and F1-score values obtained for the random forest algorithm are presented in Table 1.

7 Conclusion

In conclusion, the availability of a well-structured dataset encompassing various conversations related to mental health provides a valuable resource for training chatbot models to offer emotional support to individuals dealing with anxiety and depression. By utilizing intents, patterns, and responses, the models can learn to understand user messages and generate empathetic and relevant replies.

The use of such models in chatbot frameworks holds great potential for providing accessible and compassionate support to those in need of mental health assistance.

By simulating the behavior of a therapist, these chatbots can offer guidance, answer frequently asked questions, and provide general advice to individuals experiencing anxiety and depression.

By leveraging the insights and knowledge gained from this dataset, researchers and developers can contribute to the development of chatbots that serve as virtual companions, offering emotional solace and alleviating some of the burdens faced by individuals seeking mental health support.

Overall, the dataset and the subsequent training of chatbot models enable the creation of innovative tools that bridge the gap in mental health care, providing individuals with a readily available resource for emotional support and guidance.

References

1. World Health Organization: WHO (2017) "Depression: let's talk" says WHO, as depression tops list of causes of ill health, 30 Mar 2017. WHO. https://www.who.int/news/item/30-03-2017
2. Kohn R (2004) The treatment gap in mental health care. PubMed Central (PMC). https://www.ncbi.nlm.nih.gov/pmc/articles/PMC2623050/
3. Henderson C, Evans-Lacko S, Thornicroft G (2013) Mental illness stigma, help seeking, and public health programs. Am J Public Health 103(5):777–780. https://doi.org/10.2105/ajph.2012.301056
4. Shan Y, Zhang J, Li Z, Feng Y, Zhou J (2022) Pattern recognition center, WeChat AI, Tencent Inc., China. Key Laboratory of Intelligent Information Processing, Institute of Computing Technology, Chinese Academy of Sciences (ICT/CAS); University of Chinese Academy of Sciences
5. Daws R (2020) Medical chatbot using OpenAI's GPT-3 told a fake patient to kill themselves. Available: https://artificialintelligence-news.com/2020/10/28/medical-chatbot-openai-gpt3-patient-kill-themselves/
6. Dimitroff SJ, Kardan O, Necka EA, Decety J, Berman MG, Norman GJ (2017) Physiological dynamics of stress contagion. Sci Rep 7(1):1–8
7. Islam R, Kabir MA, Ahmed A, Kamal ARM, Wang H, Ulhaq A (2018) Depression detection from social network data using machine learning techniques. Health Inf Sci Syst 6(1). https://doi.org/10.1007/s13755-018-0046-0
8. Kaywan P, Ahmed K, Ibaida A, Miao Y, Gu B (2023) Early detection of depression using a conversational AI bot: a non-clinical trial
9. Patel F, Thakore R, Nandwani I, Bharti SK (2019) Combating depression in students using an intelligent ChatBot: a cognitive behavioral therapy

Optimizing Transradial Prosthetic Hands: A Review of Functional Requirements

Krishnakumar Sankar, Divya Johnson, Akshaay Sathishkumar, and Ayra Afreen

Abstract The hand is considered to be the most flexible and robust organ in the human body. From performing the simplest of activities such as grasping, tapping, pointing, and writing to complex activities such as playing musical instruments, and performing precise surgeries, the human hand serves to be one of the most versatile organs in the body. Transradial amputation due to severe traumatic injuries compromises a lot of factors such as physical and psychological well-being, and social integration. Transradial prosthetics are used to improve mobility and overall functionality facilitating the hand amputees to perform their activities in daily living smoothly. The main aim of the systematic review was to investigate the contribution of functional outcomes to the performance of transradial prosthetic hands. Some of the key functional outcomes to be considered while designing transradial prosthetic hands include fingertip force, grip force, degrees of freedom (DoF), and motion. A review of the literature was done using the PubMed, IEEE, Scopus, EMBASE, and Cochrane databases between 2014 and 2022. A methodical search using relevant keywords produced 1400 distinct articles. A total of 320 articles from the search results were considered based only on the title. Upon analyzing the abstracts, 307 articles were found to be excluded. The remaining 13 articles were subjected to full-text analysis. The study reported the functional outcomes of various prosthetic hands. Some of them delivered all the aforementioned functional outcomes in optimal amounts, whereas the others delivered a few among them. The study concluded that an ideal transradial prosthetic hand should be highly dexterous to handle objects of various sizes and shapes. Besides, it should have an optimum range of motion (RoM), exert the required fingertip and grip forces, and multiple DoF.

Keywords Transradial amputation · Transradial prosthetic hands · Prosthetic hand grasp force · Pinch force of transradial prosthetic hand · Dexterity of prosthetic hands · RoM of prosthetic hands · DoF of prosthetic hands

K. Sankar (✉) · D. Johnson · A. Sathishkumar · A. Afreen
Department of Biomedical Engineering, Rajalakshmi Engineering College, Chennai, Tamil Nadu 602105, India
e-mail: krishnakumarsankar30@gmail.com

© The Author(s), under exclusive license to Springer Nature Singapore Pte Ltd. 2025
H. Zen et al. (eds.), *Soft Computing and Signal Processing*, Lecture Notes in Networks and Systems 1200, https://doi.org/10.1007/978-981-97-9926-8_37

1 Introduction

The human arm is considered to be the most dexterous part of the human body [1]. Hands serve a vital function in everyday life, ranging from basic tasks like holding a coffee mug to the highly precise incisions made by doctors during surgical procedures [2]. However, loss of the arms due to trauma, disease, congenital abnormalities, or surgical removal profoundly impairs the everyday functioning of an individual [3]. Every year, millions of people around the world lose their upper limbs, forcing them to adapt to a new way of life, which can have a significant impact on both their physical and mental well-being [4]. Relying on assistive technologies such as prosthetic limbs can be beneficial for those seeking independence and improved social and mental well-being. These technologies can help restore some level of function and aid in the betterment of individuals' lives [5]. There are three types of amputees, distinguished by the extent of limb loss: transradial, trans-humeral, and shoulder disarticulation [6]. Transradial amputees are individuals who have undergone amputation at the level of the forearm, specifically below the elbow joint. Presently, several types of commercial prosthetic arms are available in the market for transradial amputees. These devices include designed cosmetic hands, split hooks, myoelectric hooks, and hands powered by the body harness [7]. Despite differences in mechanical designs, control signal kinds, and power sources, most of these are extremely expensive. Designing prosthetics that fit well for transradial amputees can be a difficult task. Sometimes, a conventional prosthetic arm can feel like an added foreign body part, making the rehabilitation process even more difficult and uncomfortable [8]. Therefore, it is important to design prosthetic hands based on the individual's amputation type and their specific needs. The purpose of prosthetic hands is to mimic the functionality of natural hands as much as possible. Therefore, the functional outcomes are essential when evaluating the effectiveness and accuracy of prosthetic hands. These outcomes include pinch force, grip force, dexterity, and degree of freedom (DoF). One of the key aspects of a prosthetic hand is its ability to generate sufficient pinch force and grip force. These forces allow individuals with amputations to hold objects securely and perform tasks such as writing, typing, and cooking with greater ease. Another important factor to be considered is DoF, which refers to the number of independent directions in which a joint can move. Prosthetic hands with a higher DoF can provide a more natural range of motion, which can make it easier for individuals with amputations to perform complex movements [9, 10]. The human hands are incredibly dexterous and can perform various intricate movements and delicate manipulations with ease. Therefore, it is important that the dexterous function of the prosthetic hand must be similar to that of the human hand [11]. This systematic review as shown in Fig. 1 aims at investigating the contribution of functional outcomes on the performance of transradial prosthetic hands.

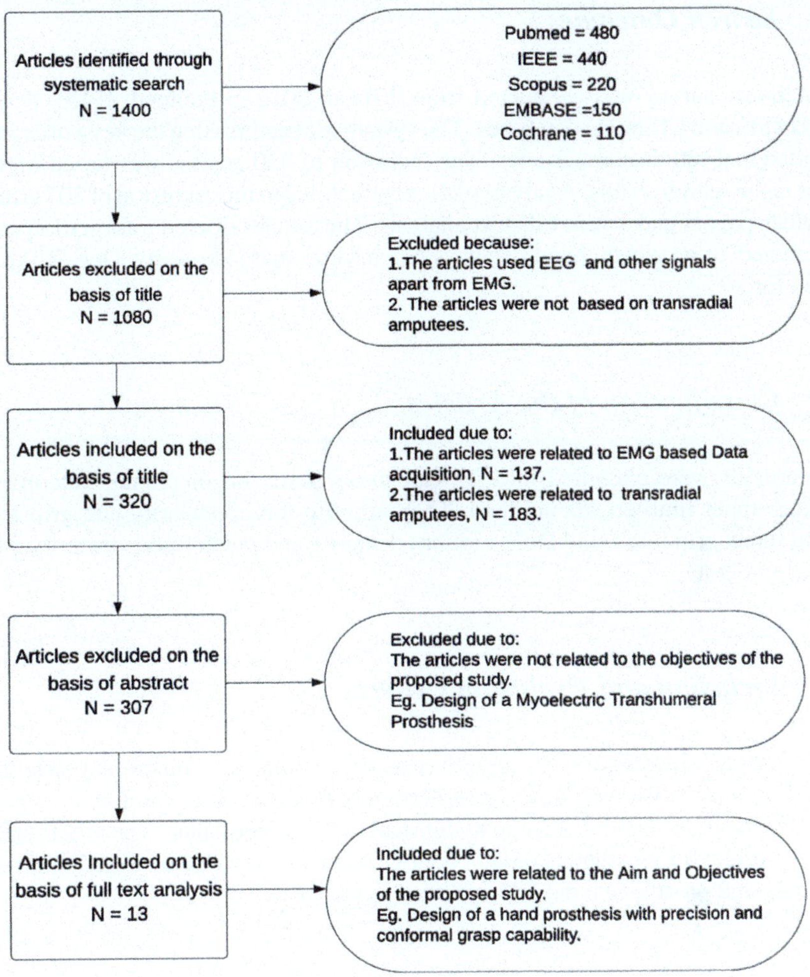

Fig. 1 Flowchart of the systematic review

2 Methodology

2.1 Search Terms Used

'Transradial amputation', 'Transradial prosthetic hands', 'Prosthetic hand Grasp force', 'Pinch force of transradial prosthetic hand', 'Dexterity of prosthetic hands', 'RoM of prosthetic hands', 'DoF of prosthetic hands' and 'Mass of prosthetic hand'.

2.2 Search Outcomes

A literature survey was conducted from 2014 to 2022 in Pubmed, IEEE, Scopus, EMBASE, and Cochrane databases. The systematic search using the keywords above resulted in 1400 discrete articles. The inclusion of 320 articles was based on their titles in the search results. Analysis of the abstracts led to the exclusion of 307 articles. Thirteen papers underwent full-text analysis. The articles offered information on the parameters of prosthetic hands, such as pinch force, mass, dexterity, DoF, RoM, and grasp force.

2.3 Classification of the Articles

The articles were classified into six categories based on functional outcomes of various upper limb prosthetic hand. They are grip force; dexterity and grip force; pinch force, grip force, and DoF; motion; dexterity and pinch force; dexterity, pinch force, and DoF.

2.4 Inclusion and Exclusion Criteria

The primary objective of the search was to examine the functional outcomes, including dexterity, pinch force, grasp force, DoF, RoM, degrees of actuation, and mass of prosthetic hand that led to betterment in the performance. The article related to the working of prosthetics based on EEG signals was excluded. EMG-controlled five-fingered prosthetic hand was included in the study.

3 Results

3.1 Grip Force

Unanyan et al. [12] designed an upper limb prosthesis. It was based on processing of EMG signal and real-time motion detection with low computational complexity using an amplitude-based window method (Table 1). The prosthetic hand of weight 400–500 g would grasp force up to 8–12 kg. It was able to keep up to 30 kg and had ten joints. The thumb was motile with two servo motors. The EMG signals from both the right and left hand of 29 subjects, including a 19-year-old amputee with a congenital disorder, were processed using MATLAB. The intact subjects (21 males and seven females) were asked to press the manual dynamometer (max load = 90 kg) with each finger for 5 s and repeated five times with 7 s rest. These processed signals were integrated with the prosthetic hand using Arduino nano-microcontrollers. Overall, the results showed that half-tensed muscles had an average recognition error of 16.10%, fully tensed muscles of 5.21%, and relaxed muscles of 8.43%. For all participants, the median error values were 1.67%, 2.89%, and 1.71% for relaxed, fully tensed, and half-tensed muscles, respectively.

In a study conducted by Aboseria et al. [13], vision feedback, a discrete slip feedback mechanism, and continuous grip force feedback were compared regarding the slippage rate. All ten participants were right-handed, and their ages ranged from 22 to 31. There were six females and four males in the study. None of the participants had any kind of impairment. The artificial finger was a right-handed version of the IH2 Azzura hand. A test object with the dimensions of $6 \times 6 \times 13\ \text{cm}^3$ and a weight of 70 g was permitted to be grasped. Using the robotic hand's grasping performance, the three feedback systems were validated. The minimum grasping force required to hold the test object was 0.4 N. Besides, the robotic hand generated a maximum grasp force exceeding 1.8 N, which destroyed the test object. The result suggested that the discrete slip feedback yielded higher success rates, close to 100%.

3.2 *Dexterity and Grip Force*

Deng et al. [14] suggested a bionic reflex control mechanism to keep prosthetic hands from slipping and distorting. An underactuated prosthetic hand was developed using a tendon-driven mechanism with each of its digits having three DoF. One sensor measured the object's stiffness using polyvinylidene fluoride (PVDF), while another used a force-sensing resistor (FSR) based on empirical mode decomposition (EMD) to measure the slippage rate. The dexterous motions of the prosthetic hand were grasping paper cups, plastic cups, and metal cups. The maximal grasping force of 2 N was observed and increased based on the applied voltage. The grasping force of the paper cup was observed to be 0.6 N. Using the reflex force as a measure of object stiffness, the study concluded that a 20 g paper cup had a reflex force of 0.48 N and a 10 g metal cup had a reflex force of 0.82 N. The results of the experiments indicated that the suggested scheme enhanced the prosthetic hand's grasping ability.

In order to mimic the natural grasping motions utilized in everyday life, Xiong et al. [15] created an anthropomorphic hand called X-hand. Twelve healthy, right-handed individuals (four girls and eight boys) with ages ranging from eighteen to twenty-nine (median twenty-three) took part in the trials. The X-digital hand's mechanism was driven by a driver. Participant performance on 33 grasping tasks (drawn from the Feix taxonomy) served to assess hand-grasping behavior. The maximum force that the X-hand could generate when grasping was 12.1 N. A power disk, two cylinders with different diameters, and a 500 mL water bottle are all in one hand. According to the results of the experiments, the X-Hand successfully mimicked 91% of real-life grasping and dexterous tasks.

Odagaki et al. [16] designed and developed Rehand-II—a five-fingered electric prosthetic hand (Table 1). The wire-driven mechanism was designed to drive the fingers to produce flexion and extension motion. An elastic glove enclosure was fabricated for the hands to provide a realistic human hand appearance. The overall weight of the device made of ABS resin was 562 g. A forearm amputee participated in the test that involved grasping a cylindrical object and a cubic object is 13.3 N and 4.0 N, respectively, by wearing the prosthetic hand. The reaching task was performed using 12 types of objects from the rear slot to the front slot on the test board within 100 s. Besides, opening a plastic bottle, holding a book, peeling a banana, operating a smartphone, pulling out tissue paper, and pulling out a tape were the dexterous tasks performed by the amputee who wore the prosthetic hand. The Southampton Hand Assessment Procedure validated the overall performance of the participant. The results suggested that the prosthetic hand was capable of manipulating various motions performed in daily activities.

3.3 *Pinch Force, Grip Force, and DoF*

Bennett et al. [17] designed an anthropomorphic prosthetic hand that offered both conformal grasp and precision capabilities (Table 1). With its eight potential grips and positions and nine degrees of freedom (DoF), the hand could be controlled by four actuators. The primary objective of the study was to evaluate the motor skills of the prosthetic hand. According to the research, it is essential that the precision grasp digits (I and II) have a resistance of 11 N or higher, ideally up to 25 N. The study identified maximum fingertip forces of 6 N for digit III, 4 N for digit IV, and 4 N for digit V when it came to conformal grasp. A maximum force of 18 N can be generated at the tip of the index finger, and the hand is capable of producing an angular velocity of four rad/s or greater.

Using four actuators, Bennett et al. [18] developed a prosthetic hand that resembles a human hand. There were 6 degrees of freedom displayed by the actuator. The weight of the hand was 546 g. In terms of flexural strength and modulus, the high-modulus thermoplastic used to make the hand measures is at around 60 MPa and 1750 MPa, respectively. The main aim of the design was to provide conformal grasp and precision capability over the objects. Six different grasps were performed to evaluate the hand's accuracy and conformal grasp behavior. Tip, tripod, and lateral pinch grasp were performed to assess the precision grasp, and hook, spherical, and cylindrical grasps were conducted to determine the conformal grasp behavior of the prosthetic hand. The dexterous task performed using a prosthetic hand was grasping a 500 mL water bottle filled with water. For fingers I and II, the maximum tip grasp force was 25–30 N, for fingers III–V, it was 16–23 N, and for fingers II–V, it was about 45 N. Findings indicated that amputees without upper limbs might benefit from the prosthetic hand.

3.4 Motion

Prakash et al. [19] designed a single-channel sEMG-based multi-functional prosthetic hand. The sEMG signals were extracted from the flexor carpi ulnaris and the muscles of the forearm of 15 subjects using three EMG muscle sensors. Ten healthy individuals and five amputees (four males and one female) performed six-arm actions: tripod grip, fine pinch, spherical grip, cylindrical grip, power grip, and fingers flexion. Mean absolute value (MAV), root mean square (RMS), integrated electromyography (IEMG), and maximum (MAX) were the four-time domain features considered for feature extraction. The response time of the prosthetic hand for the execution of the grip patterns was less than 400 ms. The research found that 91.3% of grip patterns were successfully identified, and that the acquired sEMG signals were classified using fuzzy logic with a 94% accuracy rate.

Using electromyographic (EMG) signals and an artificial neural network (ANN), Tinoco-Varela et al. [20] created a prosthesis system. The sEMG signals from the forearm muscle were extracted using three silicone gel-based EMG muscle sensor electrodes. Actions such as closed hand, grip, and flexion of index, middle, and ring finger were performed. The produced signals underwent pre-amplification, offset, final amplification, low-pass and high-pass filters, and various stages of signal processing. The conditioned signal was further processed using MATLAB to obtain twenty distinct vectors. The ANN Backpropagation with Levenberg–Marquardt was used with 20 input layers, 15 hidden layers, and five output layers. An Arduino Nano board with an ATmega328 microcontroller was used to connect the output of MATLAB to five servo motors. From the outcomes, the study concluded that out of 90 random movements, 84 were performed accurately in real-time with an online accuracy of 95.2%, an offline accuracy of 93.3%, and an error rate of 6.6%.

Pradeep et al. [21] conducted a study on implementing multiple DoF Bionic Hand controlled by sEMG signals. Using a 3-lead MyoWare muscle sensor, ten healthy individuals—five men and two women—had their sEMG signals recorded from the elbow muscle. A 10 s break was taken in between each of the following gestures: release, grasp, thumb and index, thumb and middle, thumb and ring, thumb and little finger, and clasp. The acquired sEMG signals were processed by turning their negative values into positive ones. These rectified signals were used for classification and feature extraction with a window size of six and a window increment of three. The Time Domain features obtained from each window were MAV and RMS. The trained model was tested by real-time extraction of sEMG signals from a male-below-elbow amputee and was processed using LabVIEW software. The results concluded that the robotic hand mimics the human gesture with an accuracy of quadratic discriminant analysis (QDA) 99.80625%, K-nearest neighbors (KNN) 90.65625%, naive Bayes classifier (NBC) 36.50625%, and linear discriminant analysis (LDA) 19.1%.

3.5 Dexterity and Pinch Force

Wattanasiri et al. [22] designed a multiple-grip patterned prosthetic hand with a single actuator (Table 1). The crank-slider and dwell mechanisms were used to produce finger flexion and extension motions. The developed hand had one actuator and a single DoF. The overall weight of the prosthesis made of aluminum alloy and ABS plastic was 980 g, respectively. The prosthetic hand can deftly grasp a variety of objects, including a shoe, glasses, a pill, a mobile phone, a bottle, a book, a cup, a towel, and a spoon. Pinch force, hand closing time, and grasp range were the metrics used in the experiments to validate the prosthetics' performance. The hand closing times were 1.7 and 1.4 s for the power and precision grips, respectively. The maximum grasp range for an object was 96 mm in diameter. The maximal pinch force obtained using a single actuator was 34.5 N. The results showed that the single actuator was able to produce higher grip and pinch force than multiple actuators. Besides, the hand could perform various activities of daily living (ADL).

In order to control and plan forces, Deng et al. [23] suggested a tendon-driven prosthetic hand that has multiple degrees of freedom and resembles human anatomy (Table 1). The evaluation of the hand prostheses' grasping performance and fingertip position was based on the real-time dynamic equations of the Newton iteration algorithm. Fingertip position, grasp force, and grip dexterity were the criteria used to evaluate the performance of the hand. The dexterous tasks included holding a bottle, a paper cup, and a soft cup. Using the index finger to grasp a paper cup required 1.57 N of force, the thumb 3 N, and the middle finger 1.41 N. We discovered that the values for the index, middle, and little fingers were 1.78 N, 1.63 N, 0.41 N, and 0.36 N, respectively, when we held a bottle in our hands. Holding a plastic cup with the index finger required 1.16 N of force, the thumb 2.0 N, and the middle finger 0.81 N. The result showed that the acquired fingertip position error was within acceptable limits, at approximately 2.19 mm. The experiments that proved the feasibility and effectiveness of the prosthetic hand.

3.6 Dexterity, Pinch Force, and DoF

Deng et al. [23] created a carbon fiber prosthetic hand with six degrees of freedom. It weighed 450 g in total. There was no history of neuromuscular impairment in any of the six participants, and five of them were able-bodied. The last one was an amputee who had their right-side transradial. The EMG signal was recorded using six electrodes. The muscles that were prepared for signal extraction were the flexor pollicis longus, flexor digitorum superficialis, extensor pollicis brevis, flexor indicis proprius, and extensor digiti quinti proprius. An electromyography (EMG) pattern recognition controller and a tactile sensor feedback-based grasping controller were used to control the force of the finger grasp. Using load cells resulted in maximum forces of 10.7, 9.5, 12, and 8 N at the tips of the fingers, in that sequence.

Manipulative tasks such as picking up a ball, a plastic bottle, or a paper cup containing cooked rice. The results showed that the prosthesis' stability was enhanced during the pick-lift-release manipulating task (Table 1).

Table 1 Functional outcomes contributing to the performance of transradial prosthetic hands

Author and year	Bennett et al. [17] and 2012	Odagaki et al. [16] and 2019	Wattanasiri et al. [22] and 2018	Deng et al. [23] and 2018	Unanyan et al. [12] and 2021
Category	Pinch force, grip force, and DoF	Dexterity and grip force	Pinch force and dexterity	Dexterity, pinch force, and DoF	Grip force
Robot name	Vanderbilt's hand	Rehand-II	Multi-grip pattern prosthetic hand	–	–
Dexterity	Having a hold of a cylindrical item, a Cosco ball, some tape, and an ATM card	Things like holding a book, operating a smartphone, plastic bottle, peeling a banana, and removing tape are all examples of this	Holding a shoe, a pill, a bottle, a cup, a book, a towel, a pair of glasses, a phone, and a remote control	Grasping a ball, a paper cup filled with cooked rice, a plastic bottle	A manual dynamometer (max load = 90 kg) was given to intact subjects and were asked to press it with each finger for 5 s and this was repeated 5 times for each finger
Grip force	Maximum tip grasp force 18 N	* Cylinder object—13.3 N * Cubic object—4.0 N	–	–	Force up to 8–12 kg
Pinch force	The maximum fingertip forces are 6 N for digit III, 4 N for digit IV, and 4 N for digit V	–	The maximal pinch force obtained using a single actuator was 34.5 N	Using a load cell, the highest forces detected at the tips of the fingers were 12, 10, 7, 9, and 8 N, respectively	–
DoF	9	–	1	6	6

4 Proposed Methodology

Based on the literature search conducted, the authors proposed a methodology that involves the following steps and approaches as shown in Fig. 2. The proposed methodology aims at developing a low-cost humanoid hand that caters to the needs of transradial amputees in enhancing their independence and quality of life. The study will begin with an extensive literature review of the current humanoid hand designs, materials, and technologies, considering options like cosmetic hands, split hooks, myoelectric hooks, and body-powered hands. It was understood that these designs vary significantly in functionality, esthetics, and practicality. A key requirement that was identified for transcarpal amputees is the need for lightweight materials and human-mimicked finger movements. A detailed and comprehensive literature survey of biocompatible and lightweight materials was carried out and polylactic acid (PLA) was concluded to be one of the best options due to its compatibility with 3D printing and superior mechanical properties compared to acrylonitrile butadiene styrene (ABS) making it an excellent option for the study. PLA's high mechanical properties make it an ideal option for prosthetic applications. In order to enhance the mechanical properties of PLA, the material was subjected to heat treatment by preparing samples according to ASTM (American Society for Testing and Materials) standards: ASTM D638 for tensile strength and ASTM D695 for compressive strength. These tests were conducted thrice on untreated samples and the mean value was obtained. Then, samples were heat-treated in a hot air oven for two hours, covered in sodium chloride (NaCl) to prevent bond breakage. Post-treatment, the same mechanical tests will be repeated thrice, and mean values for the same will be calculated. The detailed design of the exoskeleton of the proposed humanoid hand will be developed using the Fusion 360 software. The design will focus majorly on natural human-like finger movements, smooth joint articulation,

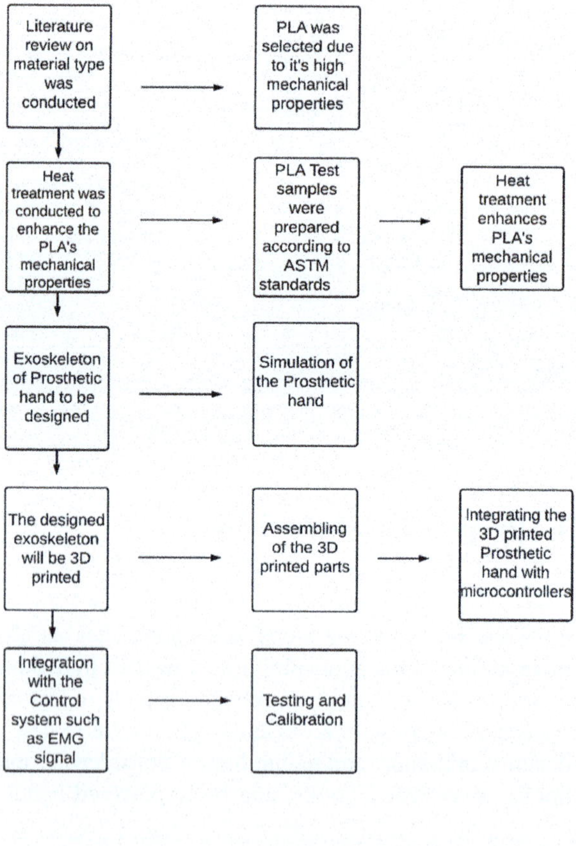

Fig. 2 Flow chart of the proposed methodology

and overall ergonomic structure. This design will also ensure optimized functionality and prove to be user-friendly. The exoskeleton designed will then be 3D printed as 3D printing technology has the ability to replicate complex geometric designs at a much lower cost and higher efficiency when compared to injection molding. The components of the exoskeleton will be printed using a Flashforge Finder 3D printer operating at an appropriate temperature. The MG90S servo motors will be connected to the fingers through strings, which will enable various human-like finger actions such as flexion, extension, abduction, adduction, opposition, pinching, and grasping. The servo motors incorporated will be controlled by an ESP 32 microcontroller ensuring precise movement. To enhance the functionality of the humanoid arm, an EMG control system will be integrated. The EMG signals will be extracted from the residual part of the amputee's arm for which dry electrodes will be used. These signals will then be processed through signal conditioning circuits to amplify and filter raw signals. The EMG signals will be analyzed for features, with an emphasis on time, statistical, and frequency domain elements. The support vector machine (SVM) algorithm will be utilized to detect patterns and learn how to differentiate between movements based on extracted attributes. The ESP 32 microcontroller will interpret the SVM output in real time and provide control signals for the servo motors using a PCA 9685 servo driver. After 3D printing, the exoskeleton, the servo motors and the microcontroller will be assembled. Each component will be evaluated individually for its functionality. Adjustments will be made to improve motor movements while maintaining natural hand gestures. The finished product will be extensively tested to ensure peak performance.

5 Discussion

The study discussed the contribution of functional outcomes to the overall performance of transradial prosthetic hands. An ideal prosthetic hand should be highly dexterous and should deliver optimum pinch and grip forces. Further, it should be capable of handling objects of various sizes and shapes with ease. Some of the highly dexterous prosthetic hands observed across the literature were the X-hand [15], Vanderbilt's hand [17], Rehand-II [16], the multigrip pattern prosthetic hand [22], and the anthropomorphic prosthetic hand by Deng et al. [14]. Out of the aforementioned prosthetic hands, Rehand-II was reported to deliver the greatest number of dexterous movements followed by the multigrip pattern prosthetic hand. Besides, among the transradial prosthetic hands reviewed across the literature, Vanderbilt's hand [17] had 9 DoF which was the highest whereas the multigrip pattern prosthetic hand had only 1 DoF and was the lowest. Besides, the DoFs of the prosthetic hand developed by Unanyan et al. [12] and Deng et al. [14] had 3 and 6 DoFs, respectively. This showed that, the greater the DoF, the higher the dexterity of the prosthetic hand. In addition, delivering an optimal amount of pinch and grip forces is important in efficiently handling objects of various sizes. Tasks like holding a shoe, a pair of glasses, or a pill require optimal forces to prevent slippage or permanent deformation. For example, Deng et al. [14] showed that the grip forces vary based on the size of the object handled. This was evident from the force exerted by the index finger of the prosthetic hand on the plastic bottle was higher compared to the force exerted by it on the soft plastic cup. On the other hand, it was equally important to exert minimal grasp force considering the nature of the material. If the strength of the material is lesser or prone to destruction easily, very little force has to be exerted on the object. The right-handed version of the IH2 Azzura [13] hand was capable of exerting a minimum grasp force of 0.4 N on the test object. The study demonstrated that, with the maximum number of DoFs and the optimum amount of pinch and grip forces, a prosthetic hand becomes highly dexterous and could comfortably restore the normal functioning of the human hand.

6 Conclusion

The contribution of various functional outcomes to the smooth functioning of a prosthetic hand was discussed in this study. Dexterity, degrees of freedom, pinch and grip forces, and motion were the physical factors that represent the performance of a prosthetic hand. The study discussed various prosthetic hands that were designed and tested on human subjects for their performance and satisfaction. The movements executed by the prosthetic hand were discussed. It was evident from the study that a large number of movements were produced by the prosthetic hands that had high degrees of freedom. Besides, a prosthetic hand that has a large number of degrees of freedom could have comparatively greater control over the objects it handles than the ones that have a relatively lesser number of degrees of freedom. Besides, dexterity could be expected out of a prosthetic hand that has high degrees of freedom. From the study, it was evident that the hands that had the highest number of DoFs and the ones that delivered the optimal amount of pinch and grip forces performed better in restoring the functional movements of the human hand.

References

1. Ziegler-Graham K, MacKenzie EJ, Ephraim PL, Travison TG, Brookmeyer R (2008) Estimating the prevalence of limb loss in the United States: 2005 to 2050. Arch Phys Med Rehabil 89(3):422–429
2. Feix T, Bullock IM, Gloumakov Y, Dollar AM (2020) Effect of number of digits on human precision manipulation workspaces. IEEE Trans Haptics 14(1):68–82
3. Jo SH, Kang SH, Seo WS, Koo BH, Kim HG, Yun SH (2021) Psychiatric understanding and treatment of patients with amputations. Yeungnam Univ J Med 38(3):194
4. Yuan B, Hu D, Gu S, Xiao S, Song F (2023) The global burden of traumatic amputation in 204 countries and territories. Front Public Health 11:1258853
5. Smaby N, Johanson E, Baker B, Kenney DE, Murray WM, Hentz VR (2004) Identification of key pinch forces required to complete functional tasks. J Rehabil Res Dev 41(2)
6. Salminger S, Stino H, Pichler LH, Gstoettner C, Sturma A, Mayer JA, Szivak M, Aszmann OC (2020) Current rates of prosthetic usage in upper-limb amputees—have innovations had an impact on device acceptance. Disabil Rehabil 44(14):3708–3713
7. Sahu A, Sagar R, Sarkar S, Sagar S (2016) Psychological effects of amputation: a review of studies from India. Ind Psychiatry J 25(1):4–10
8. Geethanjali P (2016) Myoelectric control of prosthetic hands: state-of-the-art review. Med Devices Evid Res 27:247–255
9. Kim N, Yun S, Shin D (2019) A bioinspired lightweight wrist for high-DoF robotic prosthetic arms. IEEE/ASME Trans Mechatron 24(6):2674–2683
10. Damerla R, Rice K, Rubio-Ejchel D, Miro M, Braucher E, Foote J, Bourai I, Singhal A, Yang K, Guang H, Iakimovitch V (2022) Design and testing of a novel, high-performance two DoF prosthetic wrist. IEEE Trans Med Robot Bionics 4(2):502–519
11. Resnik L, Borgia M, Cancio J, Heckman J, Highsmith J, Levy C, Phillips S, Webster J (2022) Dexterity, activity performance, disability, quality of life, and independence in upper limb Veteran prosthesis users: a normative study. Disabil Rehabil 44(11):2470–2478
12. Unanyan NN, Belov AA (2021) Design of upper limb prosthesis using real-time motion detection method based on EMG signal processing. Biomed Signal Process Control 70:103062
13. Aboseria M, Clemente F, Engels LF, Cipriani C (2018) Discrete vibro-tactile feedback prevents object slippage in hand prostheses more intuitively than other modalities. IEEE Trans Neural Syst Rehabil Eng 26(8):1577–1584
14. Deng H, Zhong G, Li X, Nie W (2016) Slippage and deformation preventive control of bionic prosthetic hands. IEEE/ASME Trans Mechatron 22(2):888–897
15. Xiong CH, Chen WR, Sun BY, Liu MJ, Yue SG, Chen WB (2016) Design and implementation of an anthropomorphic hand for replicating human grasping functions. IEEE Trans Robot 32(3):652–671
16. Odagaki N, Yoshikawa M, Tanaka Y, Kawashima N (2019) Rehand II: wire-driven five-fingered electric prosthetic hand utilizing elasticity of a cosmetic glove. In: 2019 41st annual international conference of the IEEE engineering in medicine and biology society (EMBC). IEEE, pp 6661–6664
17. Bennett DA, Dalley SA, Goldfarb M (2012) Design of a hand prosthesis with precision and conformal grasp capability. In: 2012 annual international conference of the IEEE engineering in medicine and biology society. IEEE, pp 3044–3047
18. Bennett DA, Dalley SA, Truex D, Goldfarb M (2014) A multigrasp hand prosthesis for providing precision and conformal grasps. IEEE/ASME Trans Mechatron 20(4):1697–1704
19. Prakash A, Sharma S (2021) Single-channel surface electromyography (sEMG) based control of a multi-functional prosthetic hand. Instrum Sci Technol 49(4):428–444
20. Tinoco-Varela D, Ferrer-Varela JA, Cruz-Morales RD, Padilla-García EA (2022) Design and implementation of a prosthesis system controlled by electromyographic signals means, characterized with artificial neural networks. Micromachines 13(10):1681

21. Pradeep R, Praveen LS, Shankpal P (2020) Implementation of multiple DOF bionic hand for EMG controlled schemes. In: 2020 2nd international conference on innovative mechanisms for industry applications (ICIMIA). IEEE, pp 36–40
22. Wattanasiri P, Tangpornprasert P, Virulsri C (2018) Design of multi-grip patterns prosthetic hand with single actuator. IEEE Trans Neural Syst Rehabil Eng 26(6):1188–1198
23. Deng H, Luo H, Wang R, Zhang Y (2018) Grasping force planning and control for tendon-driven anthropomorphic prosthetic hands. J Bionic Eng 15:795–804

Atomic Orbital Search—A Soft Computing Approach to Optimization

Kilari Jyothi and R. B. Dubey

Abstract This paper provides a survey of research work done in different areas using atomic orbital search optimization algorithm (AOS). AOS is one of the new meta-heuristic's algorithms for optimization problems. This algorithm's core idea is based on some quantum mechanics concepts and the quantum-based atomic model, where the electrons surrounding the nucleus are generally arranged in a point of view. AOS is one of the more effective modern meta-heuristic optimization algorithms. It can be used in a variety of optimization problems in several domains, including flexible job shop scheduling, cloud computing, machine learning, engineering design, wireless networking, power and energy. This paper illustrates literature on AOS and its variant multiobjective atomic orbital search (MOAOS). Potential directions for future research on AOS and MOAOS are suggested in the conclusions, which center on the current works on the subject, after its applications, assessment and evaluation.

Keywords Atomic orbital search optimization algorithm · Multi-objective atomic orbital search · Meta-heuristic

1 Introduction

Numerous fields, such as data mining, engineering applications, networks, and medicine have implemented optimization techniques. It is primarily used to identify the most optimal choices or values to generate a viable solution. Optimization challenges typically involve minimizing or maximizing a potential option, for which an algorithm is customized. The main objective of decision-making is to find the best choice among several options. Selecting the optimal action or value from all available possibilities is the ultimate result of optimization processes. The term "best solution"

K. Jyothi (✉) · R. B. Dubey
SRM University Delhi-NCR, Sonepat, Haryana, India
e-mail: kilaru.jyothi.sri@gmail.com

R. B. Dubey
e-mail: rbdubey@srmuniversity.ac.in

© The Author(s), under exclusive license to Springer Nature Singapore Pte Ltd. 2025

H. Zen et al. (eds.), *Soft Computing and Signal Processing*, Lecture Notes in Networks and Systems 1200, https://doi.org/10.1007/978-981-97-9926-8_38

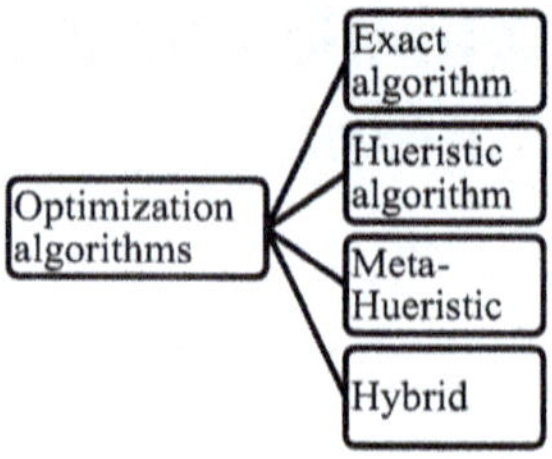

Fig. 1 Classification of optimization algorithms

refers to a practical approach that is the most advantageous option for resolving a specific optimization problem. Additionally, a feasible solution can be considered the best possible outcome through a series of steps. Classification of optimization algorithms is shown in Fig. 1.

Optimization algorithms are basically classified into exact and heuristic. Whereas heuristic algorithms are again subdivided into meta-heuristic algorithms. Exact algorithms provide solutions with good accuracy within less time but the problem with exact algorithms is as the problem size increases the computational complexity and time also increases in exponential manner. Heuristic algorithms deliver solutions in realistic time frames for large-scale optimization problems. Whereas meta-heuristic algorithms provide near-optimal solutions by making use of exploration and exploitation.

Meta-heuristic algorithms are sub-classified as evolutionary and swarm intelligence are shown in Fig. 2. Atomic orbital search algorithms come under physic-law based which is under the subgroup of swarm intelligence of meta-heuristic.

The rest of the paper is as follows. Section 2 describes atomic orbital search algorithm. Section 3 includes various applications of the AOS algorithm. Last section concludes the paper.

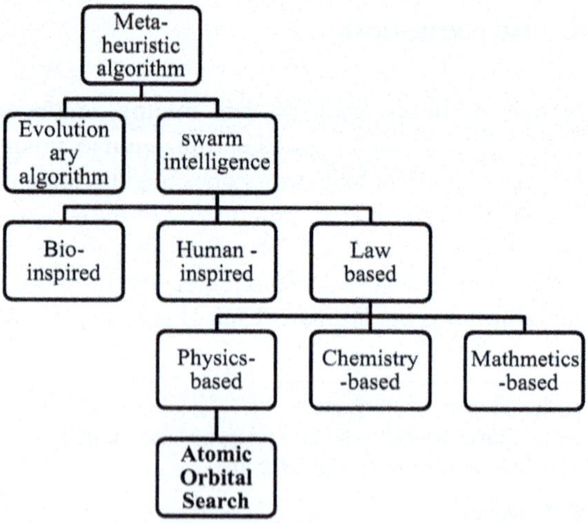

Fig. 2 Algorithm classification for optimization

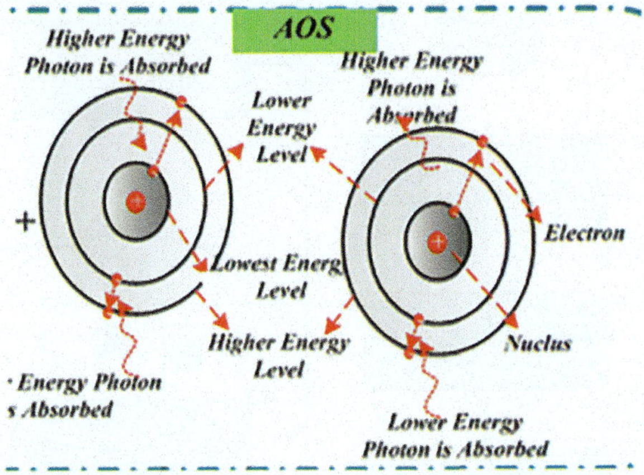

Fig. 3 Quantum staircase analogy for electrons around nucleus in atoms

2 Atomic Orbital Search Algorithm

Numerous optimization techniques have been developed in recent years based upon emission and absorption energy levels of electrons around the nucleus of an atom such as atom search optimization (ASO) by Zhao et al. [1], Yildirim and Karci [2] developed AAA, Biswas et al. [3] ASA.

2.1 Inspiration

According to traditional atomic models, the atomic orbital is the area of space surrounding an atom's nucleus where electrons can be arranged according to a particular probability. AOS is the meta-heuristic optimization approach that is inspired by the quantum mechanics attention principles on the activity of electrons which is based on the nucleus of an atom [4]. Electrons in respective orbits will bind to those orbits with a certain energy called binding energy. The electrons around the nucleus may be excited by any process such gaining energy from photons or interaction with magnetic or electric field, during this process. Movement of electrons across the orbits by gaining or loosing of energy is shown in Fig. 3.

2.2 Flowchart

The flowchart for atomic orbital search optimization algorithm is shown in Fig. 4.

The stepwise procedure for the implementation of atomic orbital search algorithm is explained in the following steps.

Step 1: Initialization.

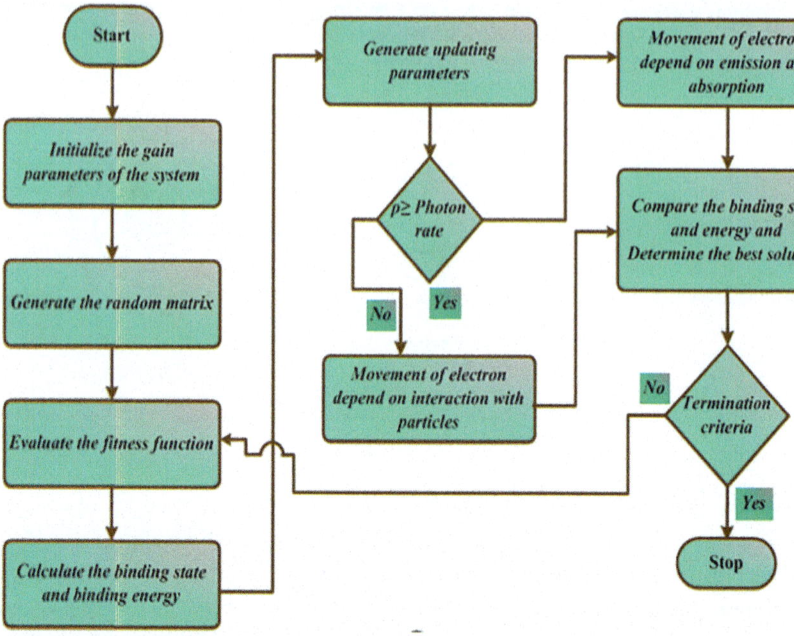

Fig. 4 Flowchart of atomic orbital search algorithm

Initialize the input population as per the required problem.

Step 2: Random generation.

Following the initialization procedure, the input parameters are generated at random.

Step 3: Analyze the initial solution's fitness function.

The calculation of fitness is dependent on the objective function.

Step 4: Calculate the binding energy B_E and binding state Bs.

Step 5: Generate the updating parameters such as photon (Ω) rate, etc.

Step 6: Quantify the photon rate.

If Ω is greater than or equal to photon rate, the mobility of electrons is through emission and absorption. Whereas if it is less than the photon rate, then the mobility of electrons is based on particle interaction.

Step 7: Comparing the energy levels to binding energy B_E.

Positions of electron will be updated to higher level when their energy levels are higher than binding energy. Similarly, if their energies are lower than the binding energy, they will update to lower levels.

Step 8: Identify the best possible global solution.

Best global solution is decided by using the binding state and binding energy.

Step 9: Check the criteria for stopping.

Proceed to halt the process if the stopping requirement is met; if not, repeat the procedure.

2.3 Variants

The AOS, which was proposed in 2020, is a single-objective optimization method that is relatively new when compared to other meta-heuristic algorithms.. In the year 2022, to deal with complex problems of optimization, multiobjective atomic orbital search (MOAOS) algorithms were introduced.

3 Applications of Atomic Orbital Search Algorithm

Numerous AOS applications from several sectors have been documented. For example, AOS has been used to address real-world issues for benchmark optimization. Additional AOS application details are shown in the illustration below. Table 1 provides a concise overview of the primary uses for AOS.

3.1 Engineering

In providing solution for various global design problems in engineering such as four-bar truss, welded beam, disk brake, and speed reducer problems, MOAOS algorithm is proposed by Azizi et al. [5]. To address premature convergence and low search efficiency in most of the reliability-based design optimization (RBDO), Farahmand-Tabar et al. [6] proposed positron-enabled atomic orbital search (PAOS) algorithm. While addressing resource trade-off issues in their extensive and complex building projects, Baghalzadeh Shishehgarkhaneh et al. [7] employ the AOS optimization algorithm. Azizi et al. [8] developed optimum design for various engineering problems such as for gear train, gas transmission compression industrial refrigeration, and so on.

3.2 Electrical Engineering

Ali et al. [9] used AOS for unknown parameter extraction in photovoltaic cells and proved that it provides values with least root mean square errors. In [10], Hussain

Table 1 Applications of atomic orbital search algorithm

References	Area	Application	Methodology	Authors
[5]	Engineering	Four-bar truss design, welded beam design, disk brake design, and speed reducer design problems	MOAOS	Azizi et al.
[6]	Engineering	Tuning mass dampers and soil-nail wall system	PAOS	Farahmand-Taba et al.
[7]	Civil and construction engineering	Dam construction project scheduling	AOS	Baghalzadeh Shishehgarkhaneh et al.
[8]	Engineering	Refrigeration, step cone pulley, robot gripper, etc.	AOS	Azizi et al.
[9]	Power system and energy generation	To extract solar unknown parameters	AOS	Ali et al.
[10]	Power system and energy generation	Tracking by solar cell in partially shaded solar PVS systems	AOS	Hussain et al.
[11]	Power and energy savings	Minimizing make span, nonprocessing energy, etc.	FAT-AOS	Jyothi et al.
[12]	Power system and energy generation	Prediction of the spatial electric field generated	AOS-GCN-LSTM	Shi et al.
[13]	Electric vehicle	Design of high-efficiency regenerative shock absorber	AOSWHO	Swamy et al.
[15]	Communication engineering	FIR filter design	AOS	Karakaş et al.
[16]	Communication engineering	Antenna design-CPW	AOS	Rohit et al.
[18]	Environment	Prediction and classification of air pollutants	AOSADL-APM	Saravanan et al.
[19]	Exoskeleton	4 degree-of-freedom (DoF) lower limb exoskeleton (LLE)	AOS-based FOPID	Kumar et al.

(continued)

Table 1 (continued)

References	Area	Application	Methodology	Authors
[14]	Power system and energy generation	Tuning of parameters in wind and hydroelectric systems	FAT-AOS	Sivanandhan et al.
[17]	Communication engineering	Antenna parameters like gain and return loss	Hyb-AOSAHMSOA	Rohit et al.
[20]	Social media	Spam filtering	OAOS-LR	Manita et al.
[21]	Education	Framework for classifying examination questions	QC-DcCapsGAN-AOSA	Revanesh et al.
[22]	E-commerce	Online product recommendation	SFA-AGCN-AOSA-SA-OPR	Periakaruppan et al.
[23]	Image processing	Image fusion	AOS	Cakıroglu et al.
[24]	Social media	Sentimental	ConvBiGRU, AOS etc.	Srinath et al.
[25]	Social media	Stealth assessment in game-based learning	Tuning of GRU, SAE RNN parameters	Rajendran et al.
[26]	Signal processing	Feature selection	IAOS, SBS	Abd Elaziz et al.

proposed efficient maximum power tracking by solar cells in partially shaded environment. Jyothi et al. [11] proposed an efficient flexible job shop scheduling using a hybrid algorithm known as FAT-AOS where they proved the make span, non-processing energy and other parameters are improved compared with other well-known existing algorithms. Shi et al. [12] based on the Atomic Orbital Search-Graph Convolution Network-Long and Short-Term Memory (AOS-GCN-LSTM) and utilized to forecast the spatial electric field produced by substation switching operations. Swamy et al. [13] proposed hybrid system, known as AOSWHO, combines AOS and wild horse optimizer (WHO) to increase the driving range of electric vehicles (EVs) while taking regenerative shock absorbers into consideration. Sivanandhan in [14] explained the suggested method reduces computation and overshoot while ensuring system constancy in significant disturbances. It is employed to determine the ideal controller settings.

Communication Engineering

Karakaş et al. [15] designed FIR filter using multiobjective atomic orbital search where multiple parameters such as attenuation of passband and stopband ripples, stopband edge frequency attenuation, and others are considered in objective function. Rohit et al. [16] presented the equilateral-shaped Co-planar Waveguide (CPW) Patch

Antenna which are compact in size suitable for IoT applications. Rohit et al. [17] designed a high gain antenna using an innovative hybrid optimization algorithm which outperforms the existing in terms of gain and return loss.

Environment Engineering

Saravanan et al. [18] proposed atomic orbital search technique for air pollution monitoring that is driven by deep learning to predict and categorize air pollutants.

Robotic Arms-Exoskeleton

Kumar et al. [19] used AOS for tuning the parameters of fractional order PID controller (FOPID) in exoskeleton with 4-degree freedom (DoF) which offers advantages over traditional integer order PID by incorporating fractional order calculus and improves stability and performance.

Other Applications

As many problems are created by phishing emails, Manita et al. [20] proposed new approach combining an LR classification model with an upgraded Atomic Orbital Search (AOS) algorithm OAOS-LR for spam filtering which are better than traditional machine learning methods such as linear regression in terms of F1-Score success rate on CSDMC2010 and Enron databases. Revanesh et al. [21] propose new methodology by combining dual-channel capsule generative adversarial network (DcCapsGAN) and Atomic Orbital Search Algorithm (AOSA) to create a novel optimized framework for classifying examination questions. In [22], Periakaruppan et al. introduced SFA-AGCN-AOSA-SA-OPR, a new approach for online product recommendations. Cakıroglu et al. [23] used AOS in multifocus image fusion. Using a deep learning classifier called ConvBiGRU and with the two combined feature selection techniques atomic orbital searching and least absolute shrinkage as well as the Selection Operator algorithm for the classifier's training, Srinath et al. [24] suggested a machine learning strategy in product review during online shopping. Rajendran et al. [25] used AOS for tuning the parameters of various deep learning algorithms in stealth assessment which gathers data regarding student competency in online game-based learning. Abd Elaziz et al. [26] improved atomic orbital search (IAOS) for exploration and sequential backward selection (SBS) for exploitation in feature selection, and finally concluded that OBL (likely Orthogonal Learning) and SBS (Sequential Backward Selection) improve the performance of the original Atomic Orbital Search (AOS). Various applications of Atomic Orbital Search algorithm discussed in the paper are tabulated in Table 1.

4 Conclusion

The paper highlights that AOS has been successfully applied in various fields beyond its initial development. These applications involve tasks like:

Tuning: Adjusting parameters in a system to achieve optimal performance (e.g., optimizing control systems for a machine).
Predicting: Forecasting future outcomes based on existing data (e.g., predicting stock prices).
Scheduling: Organizing tasks or resources in an optimal sequence (e.g., scheduling deliveries for maximum efficiency).
Designing: Creating or planning something according to specific criteria (e.g., designing an antenna with desired properties).
Tracking: Monitoring the movement or status of something over time (e.g., tracking a package during delivery).
Extracting: Identifying and isolating relevant information from data (e.g., extracting features from images for object recognition).

Limitations and Future Potential

- The paper acknowledges that while AOS is showing promise, its use in machine learning and image processing applications is still limited. These are vast fields with established optimization techniques.
- However, AOS offers a key advantage: it can be easily combined with other optimization algorithms. This "hybridization" can potentially improve performance in various tasks.

As feature selection is a promising area for future research, feature selection is a crucial step in machine learning where irrelevant data are removed to improve model accuracy. AOS's ease of hybridization makes it well-suited for this task.

References

1. Zhao W, Wang L, Zhang Z (2019) A novel atom search optimization for dispersion coefficient estimation in groundwater. Future Gener Comput Syst 91:601–610
2. Yildirim AE, Karci A (2018) Application of traveling sales man problem for 81 provinces in Turkey using artificial atom algorithm. In: ICAT' 18 proceedings of 7th international conference on advanced technologies, pp 722–726
3. Biswas A, Biswas B, Mishra KK (2016) Anatomic model based optimization algorithm. In: CINE 2nd international conference on computational intelligence and networks, pp 63–68
4. Azizi M (2021) Atomic orbital search: a novel meta heuristic algorithm. Appl Math Model 93:657–683. ISSN: 0307-904X
5. Azizi M, Talatahari S, Khodadadi N, Sareh P (2022) Multi objective atomic orbital search (MOAOS) for global and engineering design optimization. IEEE Access 10:67727–67746
6. Farahmand-Tabar S, Shirgir S (2023) Positron-enabled atomic orbital search algorithm for improved reliability-based design optimization. In: Handbook of formal optimization, pp 1–30
7. Baghalzadeh Shishehgarkhaneh M et al (2022) BIM-based resource trade-off in dam project scheduling using atomic orbital search (AOS) algorithm. AUT J Civ Eng 6(4):469–492
8. Azizi M, Talatahari S, Giaralis A (2021) Optimization of engineering design problems using atomic orbital search algorithm. IEEE Access 9:102497–102519
9. Ali F, Sarwar A, Baksh FI, Ahamd S (2023) Parameter extraction of photovoltaic models using atomic orbital search algorithm on a decent basis for novel accurate RMSE calculation. Energy Convers Manag 277:116613. ISSN: 0196-8904

10. Hussain MT, Tariq M, Sarwar A, Urooj S, BaQais A, Hossain MA (2023) Atomic orbital search algorithm for efficient maximum power point tracking in partially shaded solar PV systems. Processes 11:2776
11. Jyothi K, Dubey RB (2023) Minimizing non-processing energy consumption/total weighted tardiness & earliness, and make span into typical production scheduling model-the job shop scheduling problem. J Intell Fuzzy Syst 45(4):6959–6981
12. Shi J, Wang D, Zhao Y, Li C (2023) Predicting of spatial electric field generated by substation switch operation based on AOS-GCN-LSTM model. Int J Appl Electromagn Mech 45(4):6959–6981
13. Swamy VMM, Vidyasagar S (2024) High-efficiency regenerative shock absorbers considering twin ball screws transmission for increasing the vehicles' driving range using a hybrid approach. Proc Inst Mech Eng Part I J Syst Control Eng 238(1):141–157
14. Sivanandhan A, Aneesh V (2024) A hybrid technique linked FOPID for a nonlinear system based on closed-loop settling time of plant. Robot Auton Syst 176:104651. ISSN: 0921-8890
15. Karakaş MF (2022) Metaheuristic FIR filter design with multi-objective atomic orbital search algorithm. Eur J Sci Technol Spec Issue 39:13–16
16. Rohit P, Anita K, Satyanarayana M (2022) Development of CPW fed defected monopole antenna using atomic orbital search algorithm. Dogo Rangsang Res J 12(11):03. ISSN: 2347-7180
17. Rohit P, Datta A, Satyanarayana M (2023) Design of high gain meta surface antennas using hybrid atomic orbital search and human mental search algorithm for IoT application. In: 2023 10th international conference on signal processing and integrated networks (SPIN), Noida, India, pp 20–24
18. Saravanan R, Godwin Ponsam J, Sathiya V, Saranya G (2024) Air pollution monitoring approach using atomic orbital search algorithm with deep learning driven. Glob NEST J 26(1):05373
19. Kumar R, Bothara Y, Ezhilarasi D, Mohanavelu K (2023) Atomic orbital search optimization based fractional order PID controller for 4 DoF lower limb exoskeleton. In: 2023 international conference on power, instrumentation, control and computing (PICC), Thrissur, India, pp 1–6
20. Manita G, Chhabra A, Korbaa O (2023) Efficient e-mail spam filtering approach combining logistic regression model and orthogonal atomic orbital search algorithm. Appl Soft Comput 144:110478. ISSN: 568-4946
21. Revanesh M, Rudra B, Guddeti RMR (2023) An optimized question classification framework using dual-channel capsule generative adversarial network and atomic orbital search algorithm. IEEE Access 11:75736–75747
22. Periakaruppan S, Shanmugapriya N, Sivan R (2023) Self-attention generative adversarial capsule network optimized with atomic orbital search algorithm based sentiment analysis for online product recommendation. J Intell Fuzzy Syst 9347–9362
23. Cakıroglu F, Kurban R, Durmuş A, Karakose E (2023) Multi-focus image fusion by using swarm and physics-based metaheuristic algorithms: a comparative study with Archimedes, atomic orbital search, equilibrium, particle swarm, artificial bee colony and jellyfish search optimizers. Multimed Tools Appl 82:44859–44883
24. Srinath KR, Indira B (2024) ConvBiGRU deep learning classifier for sentiment analysis with optimization algorithm. Multimed Tools Appl
25. Rajendran D, Santhanam P (2024) Leveraging ensemble learning for stealth assessment model with game-based learning environment. Soft Comput 28:1–9
26. Abd Elaziz M, Ouadfel S, Abd El-Latif AA, Ibrahim RA (2022) Feature selection based on modified bio-inspired atomic orbital search using arithmetic optimization and opposite-based learning. Cogn Comput 14(6):1–22

TSURLCC-Task Scheduling Using Reinforcement Learning in Cloud Computing

Aryan Sharma, Bhavishya Rai, Prayashi Roy, and Santanu Kr. Misra

Abstract The development of IoT environment has proved beneficial in attracting new users to the platform, which has in turn led to the development of cloud computing. Efficient task scheduling has an important role in improving performance by optimizing resources properly, but efficient task scheduling still requires development. This research paper explores the application of reinforcement learning, specifically the Q-learning algorithms, for task scheduling. In this paper, we propose a multi-task scheduling framework which uses Q-learning algorithms. This model has two parts: in part 1, tasks are allocated to suitable servers in a dynamic fashion. In part 2, an improved Q-learning algorithm called **'TSURLCC'** is used which assigns tasks, based on resources to each virtual machine available on the cloud server system. The machine gets a reward for assigning the tasks correctly, hence helping it to learn from past experiences. The proposed approach leverages the dynamic and unpredictable nature of cloud environments to adaptively schedule tasks, optimizing resource allocation, and minimizing execution time. Through extensive experimentation and analysis, we demonstrate improved task scheduling efficiency by using Q-learning algorithm. The findings highlight the potential of RL techniques to address the complexities of task scheduling in the context of cloud computing, paving the way for enhanced resource management and overall system performance.

Keywords Task scheduling · Reinforcement learning · Q-learning · Cloud computing

A. Sharma (✉) · B. Rai · P. Roy · S. Kr. Misra
Sikkim Manipal Institute of Technology, Majitar, Rangpo, East Sikkim 737136, India
e-mail: Aryan_202000017@smit.smu.edu.in

B. Rai
e-mail: Bhavishya_202000334@smit.smu.edu.in

P. Roy
e-mail: Prayashi_202000079@smit.smu.edu.in

© The Author(s), under exclusive license to Springer Nature Singapore Pte Ltd. 2025

H. Zen et al. (eds.), *Soft Computing and Signal Processing*, Lecture Notes in Networks and Systems 1200, https://doi.org/10.1007/978-981-97-9926-8_39

1 Introduction

In recent years, the proliferation of cloud computing technologies has revolutionized the landscape of computational resource management, offering unprecedented scalability and flexibility for diverse applications. With the exponential growth of data and computing demands, efficient resource allocation and task scheduling have become paramount for optimizing performance and maximizing resource utilization in cloud environments. Traditional approaches to task scheduling often rely on heuristic algorithms or optimization techniques, which may struggle to adapt to the dynamic and unpredictable nature of cloud workloads.

To address these challenges, researchers have turned to reinforcement learning (RL) techniques, a branch of machine learning that enables agents to learn optimal decision-making policies through interaction with their environment. In recent times, reinforcement learning (RL) has garnered considerable interest owing to its capacity to address intricate decision-making challenges across diverse domains.

This research paper aims to explore the application of RL techniques to schedule tasks in cloud environments. Specifically, we investigate how RL algorithms can be trained to dynamically allocate resources and schedule tasks in a manner that minimizes latency, maximizes resource utilization, and meets various quality of service (QoS) requirements. By harnessing the power of RL, we seek to develop intelligent task scheduling mechanisms that can adapt to evolving workload patterns, handle uncertainties, and optimize resources in complex cloud environments.

In this paper, we provide a comprehensive review of existing literature on task scheduling and reinforcement learning in cloud computing. Furthermore, we propose a novel RL-based task scheduling framework tailored to the unique challenges and requirements of cloud computing environments. Overall, this research contributes to advancing the state of the art in cloud resource management by leveraging reinforcement learning techniques for task scheduling optimization. By harnessing the adaptive learning capabilities of RL, we aim to enhance the scalability, reliability, and efficiency of cloud computing systems, ultimately enabling the seamless execution of diverse workloads and applications in the cloud.

In our investigation of task scheduling optimization in cloud computing environments, we introduce the Task Scheduling Using Reinforcement Learning in Cloud Computing (TSURLCC) algorithm, a novel approach developed by our research team. The TSURLCC algorithm is designed to address the dynamic and complex nature of task scheduling by leveraging reinforcement learning techniques.

The TSURLCC algorithm consists of two distinct stages, each tailored to optimize different aspects of task scheduling within cloud systems. In the first stage, the algorithm conducts a comprehensive assessment of the cloud system, including the number of systems available and the status of virtual machines (VMs) within each system. This initial analysis enables the algorithm to determine the current workload distribution and resource availability across the cloud infrastructure.

Upon completing the system assessment, the second stage of the TSURLCC algorithm focuses on task assignment and resource allocation. Tasks are prioritized and sorted based on various criteria, such as their computational requirements and deadlines. The algorithm then dynamically assigns tasks to VMs, taking into account the available resources and workload distribution. Importantly, the algorithm incorporates a reward mechanism that incentivizes the efficient allocation of tasks, particularly those with minimum span time.

A key innovation of the TSURLCC algorithm lies in its resource-aware task allocation strategy. When assigning tasks to VMs, the algorithm considers the total available resources (x) within each VM, the resources currently utilized by existing tasks (y), and the remaining resources ($z = x - y$). If the computational requirements of a new task (a) are less than the available resources (z) of a VM, the task is allocated to that VM, ensuring efficient resource utilization and minimizing resource wastage.

By integrating reinforcement learning principles with resource-aware task allocation strategies, the TSURLCC algorithm aims to enhance the overall efficiency and performance of task scheduling in cloud computing environments. Through empirical evaluations and simulations, we demonstrate the effectiveness of the TSURLCC algorithm in optimizing task scheduling processes, improving resource utilization, and meeting quality of service (QoS) requirements in diverse cloud computing scenarios.

2 Related Works

Efficient task scheduling is crucial for cloud computing, reducing execution time, optimizing resources, and cutting costs. Traditional methods like FCFS [1] and RR [2] often fall short, while heuristic algorithms, inspired by evolutionary principles, offer better solutions by mimicking natural selection and mutation.

Minimizing makespan, the total time to complete tasks, is vital. Various innovative algorithms address this challenge. For example, Raju et al. combined Ant Colony Optimization (ACO) [3] with CSA to effectively reduce makespan. Khalili et al. [4] used particle swarm optimization (PSO) with linearly decreasing inertia weight (LDIW) to achieve similar results, while Gabi et al. employed the LDIW [5] equation in CSO to avoid local search traps.

Other approaches include Sharma et al.'s hybrid coordinated-heuristic [6] genetic algorithm (HIGA) for energy-efficient scheduling, and Wu et al.'s market-oriented resource scheduling [7] using ACO, PSO, and GA to optimize make span and CPU time. Meshkati et al. [8] introduced a Hybrid ABC-PSO framework to manage VM positioning and reduce energy use by deactivating idle nodes.

Further advancements involve Song et al.'s [9] Fast N-Particle Swarm Optimization (FN-PSO) for distributed cloud computing, Zhou et al.'s [10] enhanced GA with a greedy strategy, and Sulaiman et al.'s [11] hybrid heuristic algorithm for heterogeneous tasks. RL agents learn optimal policies through rewards from the environment. Q-Learning, a classic RL algorithm, is applied to task scheduling. For instance, Peng et al. proposed a queuing model [12] to minimize response time using

RL. Swarup et al. introduced Clipped Double Deep Q-Learning (CDDQL) [13] to optimize resource utilization, and Mangalampalli et al. developed a deep reinforcement learning scheduler (DQN) [14] to minimize makespan and energy consumption, demonstrating significant improvements over existing algorithms.

3 System Model

Cloud computing task scheduling aims to efficiently assign computational resources to tasks to enhance response time, resource utilization, and cost-effectiveness. Reinforcement learning (RL) is employed to dynamically schedule tasks in this environment by acquiring optimal scheduling strategies through environmental interaction.

Environment: The environment encompasses the cloud infrastructure, comprising virtual machines (VMs), and other computational resources. Each resource has distinct capacities, performance attributes, and associated costs.

State Representation: The state representation entails details about current resource utilization, task characteristics, and potentially historical data like past task completion times and resource usage patterns.

Action Space: The action space delineates the decisions available to the RL agent. In task scheduling, actions may include choosing a VM or container for task execution, migrating tasks between resources, or allocating additional resources if necessary.

Reward Function: The reward function specifies the objective the RL agent seeks to maximize. Common objectives encompass minimizing response time, maximizing resource utilization, meeting deadlines, and minimizing costs.

RL Algorithm: Various RL algorithms can be employed to learn optimal task scheduling policies, such as Q-learning, deep Q-networks, policy gradient methods like REINFORCE, and actor-critic methods.

Training Phase: During the training phase, the RL agent interacts with the environment, selecting actions based on its policy and receiving rewards. Through iterative learning, the agent refines its policy, ultimately discovering scheduling strategies that optimize the specified objective.

Deployment Phase: Once trained, the RL agent is deployed to the production environment, where it continually schedules tasks in real time based on the learned policy. The agent can adapt to environmental changes, ensuring scheduling decisions remain optimal over time.

Monitoring and Evaluation: Monitoring system performance in production is crucial, involving data collection on metrics such as response time, resource utilization, and cost. Periodic evaluation and potential retraining of the RL agent may be

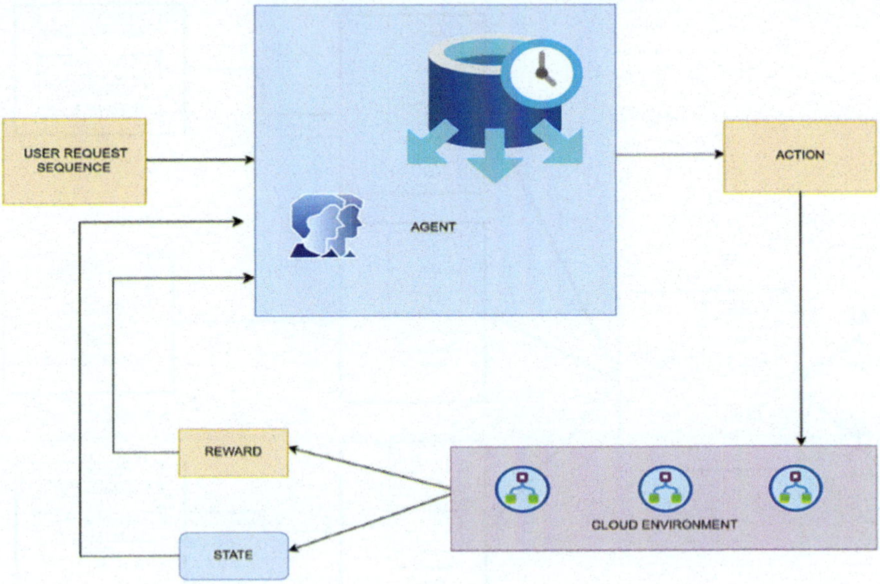

Fig. 1 Q-learning-based dynamic task scheduling

necessary to sustain optimal performance amid evolving environmental conditions. Figure 1 shows the Q-learning-based dynamic task scheduling.

Stage 1

In the initial phase, the system sets up its environment for task scheduling in cloud computing. This includes configuring virtual machines (VMs), containers, and other computational resources, each with distinct capacities, performance attributes, and costs. The environment also incorporates diverse incoming tasks with varying CPU and memory requirements, and specific deadlines. The system then constructs a comprehensive state representation of the environment, detailing current resource utilization, attributes of pending tasks, and historical data on task completion and resource consumption.

In summary, Stage 1 involves configuring the environment, defining resources and tasks, and capturing the state of the system to prepare for reinforcement learning-based task scheduling. Figure 2 shows the resource task allocator.

Stage 2

In Stage 2, the system progresses by defining the action space, reward function, and selecting a suitable reinforcement learning algorithm.

Action Space Definition: At this juncture, the system delineates the feasible actions that reinforcement learning agent can undertake. These actions might involve

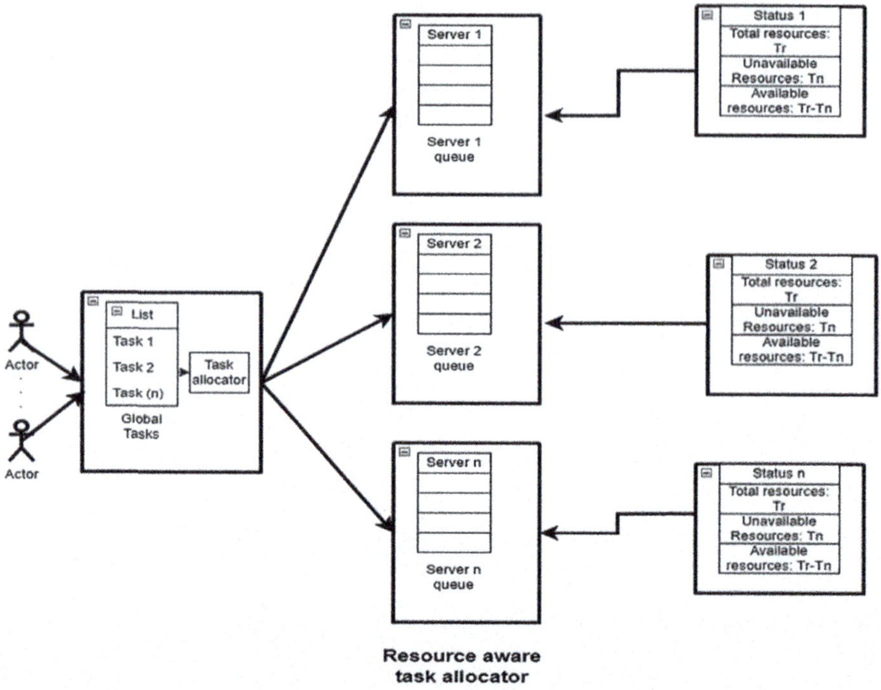

Fig. 2 Resource task allocator

selecting a virtual machine or container for task execution, migrating tasks between resources, or allocating additional resources as necessary.

Reward Function Specification: The system formulates a reward function that evaluates the desirability of outcomes resulting from the agent's actions. Objectives commonly pursued in task scheduling encompass minimizing response time, optimizing resource utilization, adhering to deadlines, and minimizing costs.

Reinforcement Learning Algorithm Selection: Various reinforcement learning algorithms can be evaluated for acquiring optimal task scheduling policies. Examples include Q-learning, deep Q-networks (DQN), policy gradient methods like REINFORCE, and actor-critic methods.

In summary, Stage 2 involves critical decisions concerning the definition of the action space, formulation of the reward function, and the selection of an appropriate reinforcement learning algorithm, ensuring originality in content. In Fig. 3, DTAURL scheduling algorithm is shown.

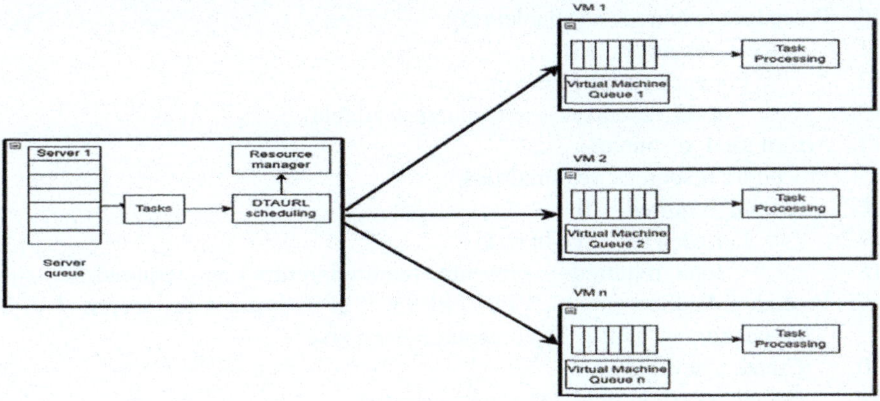

Fig. 3 DTAURL scheduling algorithm

4 Proposed Algorithm [TSURLCC]

4.1 Resource-Aware Task Allocation Algorithm

The Resource-Aware Task Allocation Algorithm dynamically assigns computational tasks in virtualized environments based on available resources. It initializes system parameters, including the number of servers and tasks, and sets up a ServerList to store server details. Each server supports a specific number of systems, each hosting several virtual machines (VMs). Total resources per VM, including CPU and memory, are defined as Tr. When a new task with resource requirements Tn enters the system, the algorithm evaluates each VM to find a suitable candidate for allocation. It checks if the VM's total resources (Tr) exceed the task's requirements (Tn) and that the remaining resources (Tr − Tn) are sufficient.

If these conditions are met, the task is assigned to the VM, optimizing resource utilization and minimizing wastage. The algorithm continually adapts to real-time resource availability and workload demands, ensuring efficient and flexible task allocation.

Resource-Aware Task Allocator

```
Initialize the values:
    server no. n, task no. m
    ServerList = [] // List to store servers and their virtual machines
Procedure AllocateTask(task):
    For each server in ServerList:
        For each virtual machine vm in server:
            If resources_available(vm) >= resources_required(task):
                Assign task to vm, return True // Task allocated successfully
                    return False // No suitable VM found for task
```

10. Procedure resources_available(vm):
11. total_resources = 0
12. For each resource in vm:
13. total_resources += resource.available
14. return total_resources
15. Procedure resources_required(task):
16. total_required = 0
17. For each resource in task:
18. total_required += resource.required, return total_required
19. ProcedureAddServer(SPS, VPS, TRPV): // SPS: systems per server, // VPS: VMs per server, TRPV: Total resources per server
20. Create a new server
21. For each system in systems_per_server:
22. Create a new VM with vms_per_server and total_resources_per_vm
23. Add virtual machine to the server
24. Add server to ServerList
25. Procedure AddTask(task_resources):
26. Create a new task with required resources task_resources
27. If AllocateTask(new_task) is False: Print "No suitable VM found for task"
28. Else: Print "Task allocated successfully"

4.2 *DTAURL Algorithm*

4.2.1 Environment Parameter Loading

The algorithm begins by loading environment parameters from a CSV file, including VM resources, task attributes, and deadlines. These parameters provide essential context for task allocation decisions.

4.2.2 Reward Function

A reward function is defined to evaluate task assignment outcomes. If a task is successfully assigned, the function increments a reward counter and returns a positive reward; otherwise, it returns a negative reward.

4.2.3 Epsilon-Greedy Policy

The algorithm employs an epsilon-greedy policy to balance exploration and exploitation in selecting VMs for task allocation. This policy randomly selects a VM with a certain probability (epsilon) and otherwise chooses the VM with the highest Q-value for the given state.

4.2.4 Initialization

The Q-table is initialized with zeros, representing the expected future rewards for each VM-task pair.

4.2.5 Task Allocation Loop

The algorithm iterates through episodes, simulating task allocation scenarios.

For each episode: Necessary variables are initialized, including current workload, task assignment, and completion metrics.

For each task in the episode, a VM is selected using the epsilon-greedy policy.

The algorithm checks if the selected VM has sufficient resources for the task.

If resources are sufficient, the task is assigned to the VM, and workload and task assignment metrics are updated. The reward for task assignment is calculated, and Q-values are updated based on the observed rewards. If resources are insufficient, a penalty is assigned for unassigned tasks.

4.2.6 Q-Table Update and Performance Tracking

After processing all tasks in the episode, the main Q-table is updated by aggregating episode-specific Q-values. Task assignment metrics such as tasks assigned and completed within deadlines are recorded for performance evaluation.

DTAURL Algorithm

```
Load VM Resources from CSV file
Load Task Attributes from CSV file
Define Hyperparameters:
    LEARNING_RATE, DISCOUNT_FACTOR, EPSILON, x
Initialize Parameters:
    Initialize reward_counter = 0
    Initialize lists for performance metrics
Define Reward Function:
    get_reward(task_assigned, completed_within_deadline):
        if task_assigned
            if completed_within_deadline:
                reward_counter += x
                return 1
            else:
                return − 1 # No penalty for uncompleted task within deadline
        else:
            return − 1 # Penalty for unassigned task
    Define Epsilon-Greedy Policy:
        epsilon_greedy_policy(state):
```

```
20.         if random() < EPSILON:
21.           return random.randint(NUM_VMS) # Explore: Randomly select VM
22.         else:
23.     return argmax(Q[:, state]) # Exploit: Select VM with highest Q-value for
    state
24.   Initialize Q-table and Start Times:
25.     Initialize Q-table with zeros
26.     Initialize start times array
27.   Training Loop:
28.     for episode in range(NUM_TASKS):
29.       start_time = current_time()
30.       Reset environment (current workload)
31.       Initialize tasks_assigned = 0
32.       Initialize tasks_completed_within_deadlines = 0
33.       for task_idx in range(NUM_TASKS):
34.         vm_idx = epsilon_greedy_policy(task_idx)
35.         if resources_available_on_vm(vm_idx, TASK_ATTRIBUTES[task_
    idx]):
36.           Assign_task_to_vm(vm_idx, TASK_ATTRIBUTES[task_idx])
37.           Update_tasks_per_deadline()
38.           Calculate_reward()
39.           Update_Q_values()
40.           Update_performance_metrics()
41.       Store_training_time()
42.    Plot Graphs: reward counter against episodes, total rewards to episodes,
    rewards per time, average rewards per episode, tasks assigned per episode,
    tasks completed within deadlines per episode, training time per episode
43.   Stop
```

5 Experimental Results

The TSURLCC (Task Scheduling Using Reinforcement Learning in Cloud Computing) model was evaluated through systematic experiments with two sets of virtual machines (VMs): one with 3 VMs scheduling 500 tasks, and another with 10 VMs scheduling 1500 tasks. Each experiment was repeated five times with different task numbers to ensure robustness. Three datasets were used: VM Resources, Task Deadlines, and Task Attributes. The model's performance was analyzed using four graphs: Average Rewards per Episode, Tasks Assigned per Episode, Tasks Completed within Deadlines per Episode, and Training Time per Episode.

The increasing trend in Average Rewards per Episode in Fig. 4 indicates progressive learning and better task performance. The Tasks Assigned per Episode graph in Fig. 5 shows a gradual increase or fluctuation, reflecting improved task allocation efficiency. The increasing trend in Tasks Completed within Deadlines per Episode in Fig. 6 demonstrates better adherence to deadlines. The consistent or decreasing Training Time per Episode in Fig. 7 indicates growing learning efficiency.

The experimental results provided evidence supporting the effectiveness of the TSURLCC model in the domain of cloud computing task scheduling. Through rigorous experimentation across diverse datasets and experimental setups, the model consistently exhibited stability and demonstrated efficient task assignment performance. These findings underscore the model's potential utility in practical cloud

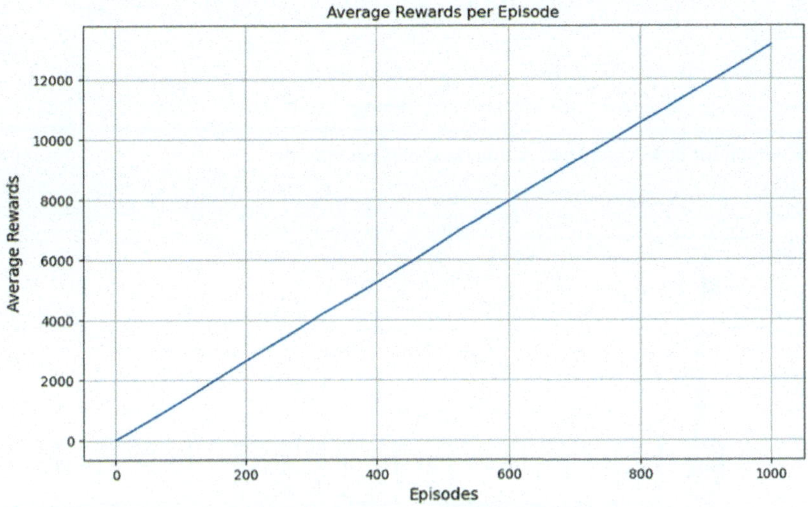

Fig. 4 Graph for average rewards per episode

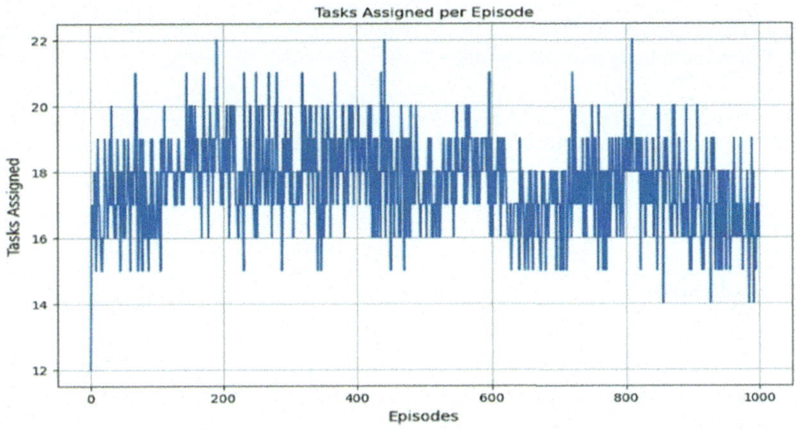

Fig. 5 Graph for tasks assigned per episode

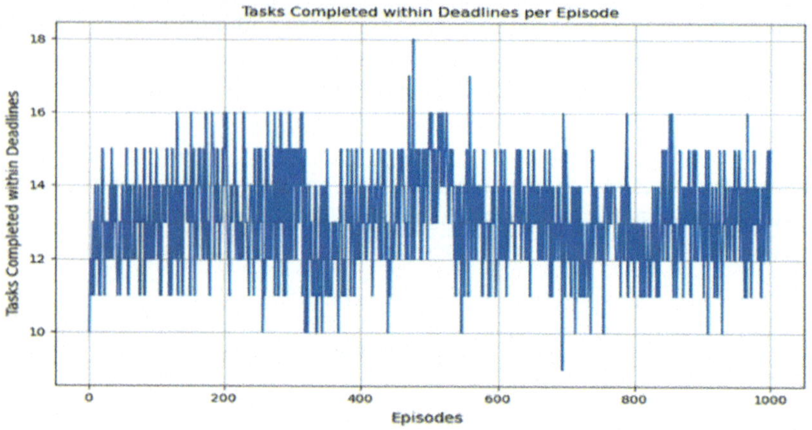

Fig. 6 Graph for tasks completed within deadlines per episode

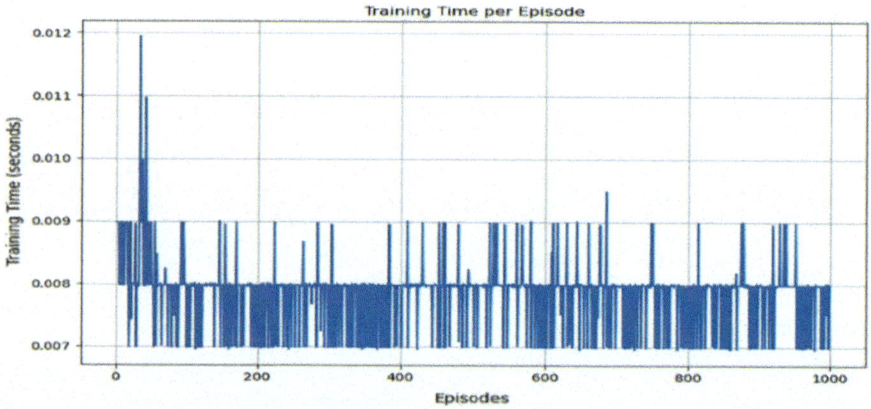

Fig. 7 Graph for training time per episode

computing scenarios, where it could facilitate enhanced resource management and timely completion of tasks.

6 Conclusion

In conclusion, the systematic evaluation of the TSURLCC model has highlighted its effectiveness in task scheduling within cloud computing environments. Key graphs—Average Rewards per Episode, Tasks Assigned per Episode, Tasks Completed within Deadlines per Episode, and Training Time per Episode—reveal the model's strong performance and learning capabilities. The increasing trends in average rewards and tasks completed on time demonstrate the model's improved task performance and deadline adherence. Overall, the TSURLCC model proves robust and effective for cloud computing task scheduling. Future research should aim to further optimize its performance and scalability for real-world applications.

References

1. Jamil B, Shojafar M, Ahmed I, Ullah A, Munir K, Ijaz H (2019) A job scheduling algorithm for delay and performance optimization in fog computing. Concurr Comput Pract Exp 1–17. https://doi.org/10.1002/cpe.5581
2. Shafi U, Shah MA, Wahid A, Abbasi K, Javaid Q, Asghar MN, Haider M (2020) A novel amended dynamic round robin scheduling algorithm for timeshared systems. Int Arab J Inf Technol 17:90–98
3. Raju R, Babukarthik RG, Chandramohan D, Dhavachelvan P, Vengattaraman T (2013) Minimizing the makespan using hybrid algorithm for cloud computing. In: Proceedings of the 3rd IEEE international advance computing conference (IACC), Ghaziabad, India, pp 957–962
4. Khalili A, Babamir SM (2015) Makespan improvement of PSO-based dynamic scheduling in cloud environment. In: 23rd Iranian conference on electrical engineering, Tehran, Iran, pp 613–618
5. Gabi D, Ismail AS, Dankolo NM (2019) Minimized makespan based improved cat swarm optimization for efficient task scheduling in cloud datacenter. In: Proceedings of the 3rd high performance computing and cluster technologies conference, Guangzhou, China, pp 16–20
6. Sharma M, Garg R (2020) HIGA: harmony-inspired genetic algorithm for rack-aware energy-efficient task scheduling in cloud data centers. Eng Sci Technol Int J 23(1):211–224
7. Wu Z, Liu X, Ni Z, Yuan D, Yang Y (2013) A market-oriented hierarchical scheduling strategy in cloud workflow systems. J Supercomput 63:256–293
8. Meshkati J, Safi-Esfahani F (2019) Energy-aware resource utilization based on particle swarm optimization and artificial bee colony algorithms in cloud computing. J Supercomput 75:2455–2496
9. Song A, Chen W-N, Luo X, Zhan Z-H, Zhang J (2022) Scheduling workflows with composite tasks: a nested particle swarm optimization approach. IEEE Trans Serv Comput 15(2):1074–1088
10. Zhou Z, Li F, Zhu H et al (2020) An improved genetic algorithm using greedy strategy toward task scheduling optimization in cloud environments. Neural Comput Appl 32:1531–1541
11. Sulaiman M, Halim Z, Lebbah M et al (2021) An evolutionary computing-based efficient hybrid task scheduling approach for heterogeneous computing environment. J Grid Comput 19:11
12. Peng Z, Cui D, Zuo J et al (2015) Random task scheduling scheme based on reinforcement learning in cloud computing. Clust Comput 18:1595–1607
13. Swarup S, Shakshuki EM, Yasar A (2021) Task scheduling in cloud using deep reinforcement learning. Procedia Comput Sci 184:42–51
14. Mangalampalli S, Hashmi SS, Gupta A, Karri GR, Rajkumar KV, Chakrabarti T, Chakrabart P, Margala M (2024) Multi-objective prioritized workflow scheduling using deep reinforcement-based learning in cloud computing. IEEE Access 12:1

Sound Vision: A Deep CNN Recursive Learning-Based Navigation Assistive Device for Divyang (Visually Impaired) Person

Sarthak Dube, Mangesh Bagde, Janhvi Bhagat, Shiya Pardhi, Manthan Ghosh, and Deepa Das

Abstract Visual information is an important part of human perception, yet visually challenged people experience significant difficulties in acquiring and processing visual content. This study introduces "Sound Vision" (SV), an innovative system that uses cutting-edge AI-driven technology to translate pictures into aural representations, therefore improving accessibility for those with visual impairments. SV uses computer vision (CV) methods for image processing and merges the Raspberry Pi's audio synthesis capabilities to provide visually impaired people with an aural interpretation of their environment. We go into the technical components of the Sound Vision system in this section, discussing the image processing algorithms for object detection and scene interpretation, as well as the use of Raspberry Pi for audio synthesis to produce meaningful soundscapes. The study also goes into the practical uses of Sound Vision, such as navigational help, item identification, and the potential to improve the overall quality of life for visually impaired people. Sound Vision's usefulness is proved through user tests and evaluations with visually impaired participants, showing the system's ability to provide vital audio input. This enables visually impaired people to achieve independence and a better awareness of their surroundings, so increasing their overall quality of life.

Keywords Visual impairment challenge · Sound vision · Image processing · Autonomy · Independence

1 Introduction

Visual information plays a pivotal role in how individuals perceive and navigate the world. For individuals with visual impairments, however, this fundamental aspect of human perception is limited, presenting significant challenges in their daily lives.

S. Dube · M. Bagde · J. Bhagat · S. Pardhi · M. Ghosh (✉)
G. H. Raisoni College of Engineering, Nagpur, India
e-mail: manthan.ghosh@raisoni.net

D. Das
Ramdeobaba University, Nagpur, India

© The Author(s), under exclusive license to Springer Nature Singapore Pte Ltd. 2025

H. Zen et al. (eds.), *Soft Computing and Signal Processing*, Lecture Notes in Networks and Systems 1200, https://doi.org/10.1007/978-981-97-9926-8_40

Accessibility and inclusivity [1–6] are paramount concerns, and technology has emerged as a powerful ally in addressing these issues. This paper introduces an innovative solution that leverages the capabilities of the Raspberry Pi platform to bridge the accessibility gap for visually impaired individuals, thereby transforming their experience of the visual world into a rich auditory one. This solution, known as "Sound Vision," aims to convert images into auditory information, offering a novel and transformative way for visually impaired individuals to interact with their surroundings. In this paper, we delve into the technical underpinnings of the Sound Vision system, highlighting the application of CV algorithms for image processing and the utilization of Raspberry Pi for audio synthesis [7]. Through Sound Vision, individuals with visual impairments gain the ability to perceive, interpret, and interact with their environment in an auditory format, enhancing their autonomy, independence, and overall quality of life. This paper explores the potential of this innovative system, highlighting its applications in navigation, object recognition, and accessibility, ultimately ushering in a new era of technological empowerment for the visually impaired community.

2 Literature Review

Omar I Johannesson and colleagues established new approach for relevant research developments in the field of assistive technology for the visually impaired, audio-haptic interfaces and the challenges associated with creating effective solutions, this work was published in 2016. Identify the technical and usability challenges that need to be addressed for such systems to be effective. Suggest potential future development and improvements for audio-haptic systems [8].

A unique approach developed by Shankar Sivan and Gopu Darsan for encompassing recent development in the field of CV and its application is assisting individuals with visual impairments, this work was published in 2016. Examine the challenges and limitation associated with implementing such systems, such as real-time processing and accuracy. Consider the role of emerging technologies like 5G [9] and edge computing in enhancing this systems [10].

This new approach for blind people was developed by Md. Mahfuz Ashraf and Team on 16 Jan 2017. This technology has a lot of pros and cons related to the application of ICT for VI (Visually Impaired) people. It is hard to provide definite conclusions which may be beneficial for the VI community [11].

A new approach was developed by SRMIST, V., & Student, U. G. for TTS device for VI people , and work was published in 2018. In TTS technology, including potential inaccuracies in speech synthesis and issues related to certain languages for accents and low voice quality [12].

Muhammad Haziq Kamaludin and colleagues established unique approach for Sonar Assistive Device (SAD), this work was published on 10 Feb 2015. In this device, challenges arrived including its accuracy in complex environment and potential interference [13].

Sulaiman Khan and Team established a new approach for Navigation Assistants for Blind People, this work was published on 17 Feb 2021. In this technique, imposing safety, reliability, and accurate navigation capabilities in IoT-based navigation applications has been identified. It is a significant challenge to implement the goal of smart navigation assistive system. In such a situation, estimating the risks associated with the use and possible misuse of these systems, reconnaissance algorithms, and information on end users and developing ways to incorporate security hardening measures into the design is not an easy task and requires thorough research [14].

A unique approach was developed by Swami Datta Subramaniam and Vinayak Vasanta Shirsat for Smart Image to TTS using deep learning, this work was published in July 2021: Challenges in implementation of smart image to TTS , including accuracy, processing speed and data privacy [15].

Alexy Bhowmick and Shyamnta M. Hazarika established unique approach for an insight into assitive technology for the blind people, this provides an overview of the current state assistive technology and future trends and work was published in Jan 2017. Improvements in the capabilities of common mobile technologies, advances in CV and cutting-edge new medical procedures are required to bring the field ever closer to meet the challenges and realities of effective system development. In this technology, visually impaired individuals face in accessing information and navigating their environment. Address issues are related to the affordability and availability of assistive technology [16].

A new approach for real-time recognition with results converted into 3D Audio which was established by Qian Lin and colleagues in March 2016. Analyze the challenges, such as accuracy and real-time processing and purpose potential further research direction in this field [17].

Poonam S. Jumani and Prof. Sangam Borkar established new approach for colour and shape identification in 2018. Analyze the challenges, such as identifying shapes of an object which are not accurate [18].

In 2016, Anupkumar Palsokar and Dr. C. S. Wamekar developed a RGB-HSI coversion-based model to map the colour perception into audio equivalent to assist VI peoples.Devising an integrated learning environment for VI learners [19].

A unique approach developed by K. Gopala Krishanan and Team for image recognition for VI people, was published in 2013. Performance degradation was observed with the complexity of the visual inputs. More improvement is required in terms of feature selection. This method also lacks efficiency once the complex scenario is considered [20].

Zheshen Wang and colleagues established new approach in 2009 for enabling access to digital maps for people with visual impairment . Low-level features without high-level semantic guidance, human-computer interaction aspect of the problem [21].

A new approach was established in 2007 by Drd. Alexandru Cazan and Team. They made the audio representation of the visuals to help the VI community. This development lacks efficiency, and practical applications still need to be improved [22].

In March 2023, by Sanjay Dutta and Team established new approach for Raspberry PI-based image to speech system for the VI with blur detection. Further development and improvement in this technique was such as optimizing processing speed, improving text extraction accuracy, noise reduction, and contrast enhancement [23].

3 Methodology

We all know that people with disabilities tend to be better than other healthy people at adapting to new life conditions, and this is what gave us the idea for an application to convert video to real-time audio. This system will guide the blind person along the path by providing audio assistance to detect the object and also guide the direction. Blind person is dependent on others to get assisted during travelling, there are many obstacles traditionally followed methods such as blinds man stick in our white case is used. Inspired by this method this system will detect the object by sensor like camera converts moving our live image into real-time audio [7, 14, 24, 25]. The basic architecture is shown in Fig. 1. Transforming images into audio for visually impaired individuals typically involves the use of CV and image processing techniques.This methodology serves as a general guideline, and the specific implementation details may vary depending on the technology, hardware, and software tools choose to use in project. It's essential to collaborate with visually impaired individuals to design a system that meets their unique needs and preferences [26–30]. Here's a general methodology for this process:

Image Acquisition: Collect the image using a camera or image input device and collect various data from the sensors. Ensure the image is well-lit and clear.

Pre-processing: Enhance the image quality by removing noise, adjusting brightness, and improving contrast. This step helps improve the accuracy of image recognition.

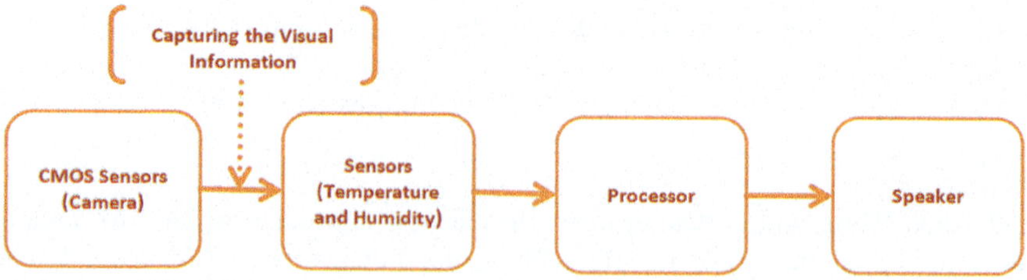

Fig. 1 Basic architecture

Object Detection and Recognition: Employ CV algorithms and machine learning models (such as convolutional neural networks—CNNs [31–33]) to detect and recognize objects, text, or scenes in the image. Popular libraries like OpenCV and pre-trained models like TensorFlow's object detection API can be used.

Text Extraction (if applicable): If the image contains text, use optical character recognition (OCR) to extract and convert text into a machine-readable format.

Image-to-Audio Conversion: Generate audio descriptions or cues corresponding to the detected objects or text. Here are some methods to consider: (a) **Text-to-Speech (TTS)**: Convert extracted text to audio using TTS engines, making it accessible through speech. (b) **Environmental Sounds**: Create audio representations of objects or scenes using appropriate sound cues. For example, the sound of a dog barking for a detected dog. (c) **Spatial Audio**: Utilize 3D audio techniques to convey the spatial arrangement of objects within the image, enhancing the user's understanding of the scene [34].

User Interface: Develop an accessible and user-friendly interface for visually impaired individuals to interact with the system. This may involve voice commands, gesture recognition, or tactile input.

Real-Time Processing: Ensure that the system can perform these transformations in real-time, providing immediate feedback to the user.

User Feedback and Adaptation: Collect feedback from visually impaired users to refine the system. Adapt the audio cues and descriptions based on user preferences and needs.

Testing and Evaluation: Conduct usability testing and evaluation to assess the effectiveness and efficiency of the image-to-audio conversion system in real-world scenarios.

Accessibility Compliance: Ensure that the system complies with accessibility standards and regulations to guarantee that it is usable and beneficial for visually impaired individuals.

Continuous Improvement: Continuously update the system, incorporating new CV techniques and AI models to improve accuracy and expand the range of recognized objects and scenes [35–39].

4 Working Model

Transforming images into audio [40–43] for visually impaired individuals typically involves using technology to convert visual information into spoken words or other auditory cues. As shown in Fig. 2 the process begins with defining the problems and gathers requirements for converting image and data into audio. Data collection is done by capturing an image using a camera and collecting various data from the sensors.

Fig. 2 Flowchart

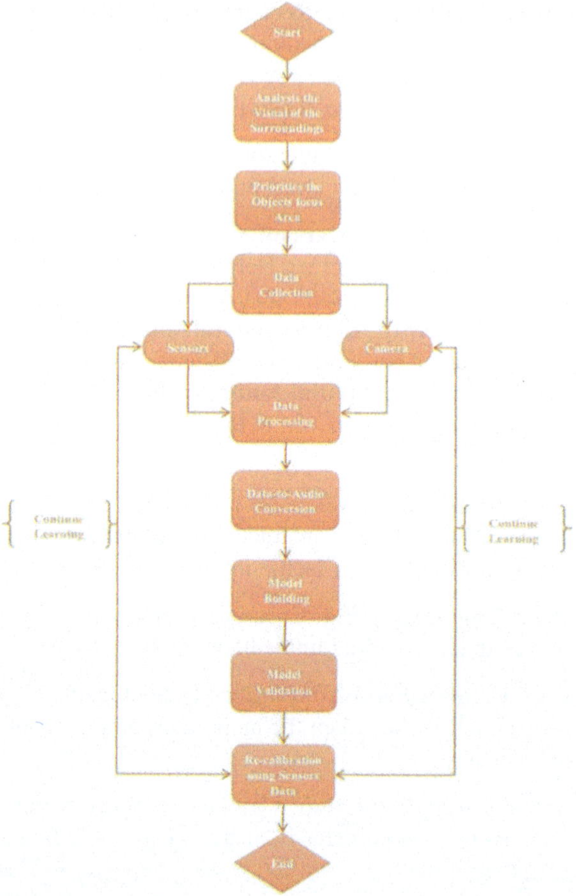

The captured image and collected data is then processed by software or hardware, which can include optical character recognition (OCR) technology. OCR scans the data to identify and extract text, while other algorithms can identify objects, people, or scenes. If text is detected, then collected data is converted into speech using text-to-speech technology. TTS engines can read the text aloud. For non-textual elements like images or graphics, additional software can provide descriptions or explanations. These descriptions are then spoken through TTS. Visually impaired individuals can interact with the system using gestures, voice commands, or other input methods to navigate and explore the audio representation of the data. The final output is an audio description of the image and data, which can be delivered through headphones, speakers, or other audio output devices.

It's important to note that the quality and accuracy of the transformation can vary depending on the technology used and the complexity of the image. Advancements in AI and machine learning have significantly improved the accuracy and capabilities of these systems in recent years [44–48].

5 Results

The result of transforming an image into audio for visually impaired individuals is to provide them with access to visual information that they would not otherwise be able to perceive. Figures 3 and 4 show the key outcomes:

Accessibility: Visually impaired individuals gain access to a wide range of visual content, including printed text, images, charts, diagrams, and more. This promotes inclusivity and equal access to information. Independence: It empowers visually impaired individuals to be more independent in various aspects of life, such as reading documents, identifying objects, and navigating their surroundings. Education: Transforming images into audio aids visually impaired students in accessing educational materials, textbooks, and visual learning resources, which are essential for their academic success. Information: It allows access to critical information in daily life, such as reading product labels, identifying currency denominations, and recognizing street signs. Social Inclusion: It enhances the ability of visually impaired individuals to participate in social activities, as they can independently access and understand visual content shared with them by friends, family, and colleagues. Enhanced Understanding: The audio descriptions provided can help visually impaired individuals

Fig. 3 Result1

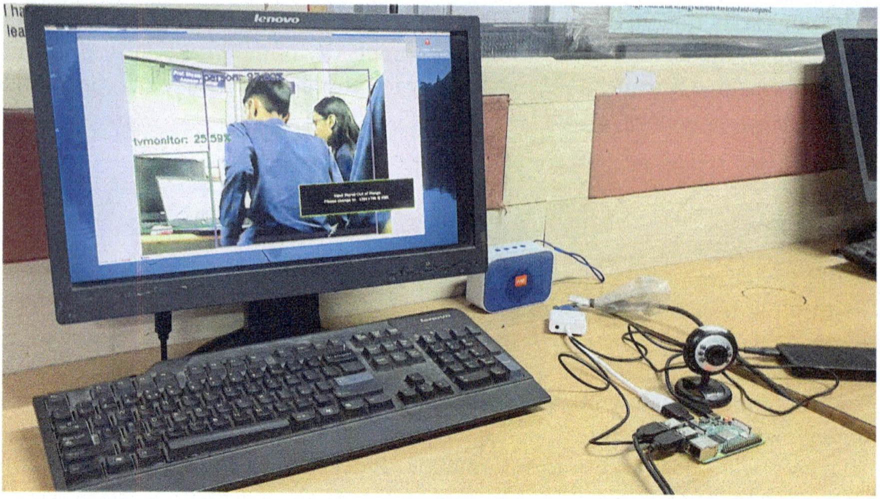

Fig. 4 Result2

understand the content, context, and details of images, making it easier for them to grasp complex ideas and concepts. Improved Quality of Life: Overall, transforming images into audio can significantly improve the quality of life for visually impaired persons by enabling them to interact with the visual world in meaningful ways.

6 Conclusion

In conclusion, transforming images into audio for visually impaired individuals is a transformative and inclusive approach that enhances accessibility and independence. This technology will guide the visually impaired person along the path by providing audio assistance to detect the object and also guide the directions. This technology empowers visually impaired individuals to access visual information, enabling them to navigate daily life, engage in education, and participate in social activities. While there may be variations in accuracy and effectiveness, ongoing advancements in AI and assistive technology promise to further improve the quality of life and opportunities for visually impaired individuals, fostering a more inclusive and equitable society.

7 Future Scope

The future of transforming image into audio for visually impaired individuals is promising, with ongoing advancements contributing to enhanced accuracy, real-time processing, augmented reality (AR), multi-sensory experiences, wearable devices, improved user interfaces greater accessibility in virtual worlds, collaboration with other assistive technologies,wider adoption, independence, and participation in various aspects of life.

References

1. Sareeka AG, Kirthika K, Gowthame MR, Sucharitha V (2018) PseudoEye-mobility assistance for visually impaired using image recognition. In: 2018 2nd international conference on inventive systems and control (ICISC). IEEE, pp 174–178
2. Bai J, Lian S, Liu Z, Wang K, Liu D (2017) Smart guiding glasses for visually impaired people in indoor environment. IEEE Trans Consum Electron 63(3):258–266
3. Karmel A, Sharma A, Garg D (2019) IoT based assistive device for deaf, dumb and blind people. Procedia Comput Sci 165:259–269
4. Arakeri MP, Keerthana NS, Madhura M, Sankar A, Munnavar T (2018) Assistive technology for the visually impaired using computer vision. In: 2018 international conference on advances in computing, communications and informatics (ICACCI). IEEE, pp 1725–1730
5. Jain BD, Thakur SM, Suresh KV (2018) Visual assistance for blind using image processing. In: 2018 International conference on communication and signal processing (ICCSP). IEEE, pp 0499–0503
6. Rajesh M, Rajan BK, Roy A, Thomas KA, Thomas A, Tharakan TB, Dinesh C (2017) Text recognition and face detection aid for visually impaired person using Raspberry PI. In: 2017 international conference on circuit, power and computing technologies (ICCPCT). IEEE, pp 1–5
7. Jiang B, Yang J, Lv Z, Song H (2018) Wearable vision assistance system based on binocular sensors for visually impaired users. IEEE Internet Things J 6(2):1375–1383
8. Jóhannesson ÓI, Balan O, Unnthorsson R, Moldoveanu A, Kristjánsson Á (2016) The sound of vision project: On the feasibility of an audio-haptic representation of the environment, for the visually impaired. Brain Sci 6(3):20
9. Jichkar R, Paraskar S, Parteki R, Ghosh M, Deotale T, Pathan AS et al (2023) 5g: an emerging technology and its advancement. In: 2023 11th international conference on emerging trends in engineering technology-signal and information processing (ICETET-SIP). IEEE, pp 1–6
10. Sivan S, Darsan G (2016) Computer vision based assistive technology for blind and visually impaired people. In: Proceedings of the 7th international conference on computing communication and networking technologies, pp 1–8
11. Ashraf MM, Hasan N, Lewis L, Hasan MR, Ray P (2016) A systematic literature review of the application of information communication technology for visually impaired people. Int J Disabil Manage 11:e6
12. SRMIST, Vadapalani, UG Student (2018) Text-to-speech device for visually impaired people. Int J Pure Appl Math 119(15):1061–1067
13. Kamaludina MH, Mahmooda NH, Ahmada AH, Omar C, Yusofb MAM (2015) Sonar assistive device for visually impaired people 500, 1
14. Khan S, Nazir S, Khan HU (2021) Analysis of navigation assistants for blind and visually impaired people: a systematic review. IEEE Access 9:26712–26734

15. Subramaniam SD, Shirsat VV (2021) Smat image to text to speech using deep learning. Res Gate Publ J
16. Bhowmick A, Hazarika SM (2017) An insight into assistive technology for the visually impaired and blind people: state-of-the-art and future trends. J Multimodal User Interfaces 11:149–172
17. Jiang R, Lin Q, Qu S (2016) Let blind people see: real-time visual recognition with results converted to 3D audio. Report No. 218, Standord University, Stanford, USA
18. Jamuni PS, Borkar S (2018) Colour and shape identification for assisting visually impaired person. In: 2018 2nd international conference on I-SMAC (IoT in social, mobile, analytics and cloud) (I-SMAC) I-SMAC (IoT in social, mobile, analytics and cloud) (I-SMAC), 2018 2nd international conference on. IEEE, pp 415–418
19. Palsokar A, Warnekar CS (2016) A RGB-HSI conversion based model to map the colour perception into audio equivalent to assist visually impaired persons. In: 2016 IEEE 6th international conference on advanced computing (IACC). IEEE, pp 484–487
20. Krishnan KG, Porkodi CM, Kanimozhi K (2013) Image recognition for visually impaired people by sound. In: 2013 international conference on communication and signal processing. IEEE, pp 943–946
21. Wang Z, Li B, Hedgpeth T, Haven T (2009) Instant tactile-audio map: enabling access to digital maps for people with visual impairment. In: Proceedings of the 11th international ACM SIGACCESS conference on computers and accessibility, pp 43–50
22. Cazan A, Vârbanescu R, Popescu D (2007) Algorithms and techniques for image to sound conversion for helping the visually impaired people-application proposal. In: 2007 14th international workshop on systems, signals and image processing and 6th EURASIP conference focused on speech and image processing, multimedia communications and services. IEEE, pp 471–474
23. Dutta S, Dutta S, Gupta O, Lone S, Phule S. PISEE: Raspberry PI-based image to speech system for the visually impaired with blur detection
24. Masud U, Saeed T, Malaikah HM, Islam FU, Abbas G (2022) Smart assistive system for visually impaired people obstruction avoidance through object detection and classification. IEEE Access 10:13428–13441
25. Joshi, R. C., Yadav, S., & Dutta, M. K. (2020, February). YOLO-v3 based currency detection and recognition system for visually impaired persons. In 2020 International Conference on Contemporary Computing and Applications (IC3A) (pp. 280-285). IEEE
26. Chellaswamy C, Geetha TS, Hariharan K, Archana K, Babitharani S (2023) Deep learning-based braille technology for visual and hearing impaired people. In: 2023 international conference on smart systems for applications in electrical sciences (ICSSES). IEEE, pp 1–8
27. Hu W, Wang K, Yang K, Cheng R, Ye Y, Sun L, Xu Z (2020) A comparative study in real-time scene sonification for visually impaired people. Sensors 20(11):3222
28. Hakim H, Fadhil A (2019) Navigation system for visually impaired people based on RGB-D camera and ultrasonic sensor. In: Proceedings of the international conference on information and communication technology, pp 172–177
29. Scalvini F, Bordeau C, Ambard M, Migniot C, Dubois J (2022) Low-latency human-computer auditory interface based on real-time vision analysis. In: ICASSP 2022–2022 IEEE international conference on acoustics, speech and signal processing (ICASSP). IEEE, pp 36–40
30. Ahmetovic D, Avanzini F, Baratè A, Bernareggi C, Galimberti G, Ludovico LA, et al (2019) Sonification of rotation instructions to support navigation of people with visual impairment. In: 2019 IEEE international conference on pervasive computing and communications (PerCom). IEEE, pp 1–10
31. Sinha N, Ghosh M, Majumder S, Bhowmik BB (2021) Deep learning based noise identification in the optical fiber communication using variational mode decomposition. In: 2021 IEEE 2nd international conference on applied electromagnetics, signal processing, communication (AESPC). IEEE, pp 1–5
32. Ghosh M (2022) Comparative DNN model analysis for detection of various types of optical noise

33. Ghosh M, Raut M, Parteki R, Das D, Thakare LP, Jichkar R, et al (2023) An analysis of deep-neural-network model for the determination of the bit-rate of optical fiber signals. In: 2023 11th international conference on emerging trends in engineering technology-signal and information processing (ICETET-SIP). IEEE, pp 1–4
34. Ghosh M, Das D (2024) Voice-activated SOS: an AI-enabled wearable device. In: Ponnusamy S, Bora V, Daigavane P, Wazalwar S (eds) Impact of AI on advancing women's safety. IGI Global, pp 251–277
35. Bafna VB, Kothari A, Keskar A (2023) Automatic head position detection in computer tomography for prevention of wrong-side treatment. In: Intelligent systems and smart infrastructure. CRC Press, pp 196–206
36. Derkar SB, Biranje D, Thakare LP, Paraskar S, Agrawal R (2023) CaptionGenX: advancements in deep learning for automated image captioning. In: 2023 3rd Asian conference on innovation in technology (ASIANCON). IEEE, pp 1–8
37. Paraskar S, Pathan AS, Parteki R, Jichkar RA, Thakare L, Deotale T (2022) A review on deep learning methods and CT scan approaches for Covid-19 detection. In: 2022 10th international conference on emerging trends in engineering and technology-signal and information processing (ICETET-SIP-22). IEEE, pp 1–6
38. Shende D, Thakare L, Agrawal R, Wyawahare N (2023) Automatic disease detection for various plants leaf using image processing techniques and tensorflow algorithm. In: International conference on electrical and electronics engineering. Springer Nature Singapore, Singapore, pp 487–499
39. Janbandhu T, Laddhani S, Agrawal R, Dhule C, Chavhan N, Khandekar A (2023) Sign language recognition using CNN. In: 2023 international conference on communication, security and artificial intelligence (ICCSAI). IEEE, pp 175–180
40. Madake J, Bhatlawande S, Solanke A, Shilaskar S (2023) A qualitative and quantitative analysis of research in mobility technologies for visually impaired people. IEEE Access
41. Chang WJ, Chen LB, Hsu CH, Chen JH, Yang TC, Lin CP (2020) MedGlasses: a wearable smart-glasses-based drug pill recognition system using deep learning for visually impaired chronic patients. IEEE Access 8:17013–17024
42. Garapati K, Maram SS, Manikandan VM (2023) A real-time system to assist blind people through face recognition and emotion detection. In: 2023 3rd Asian conference on innovation in technology (ASIANCON). IEEE, pp 1–6
43. Chang WJ, Chen LB, Sie CY, Yang CH (2020) An artificial intelligence edge computing-based assistive system for visually impaired pedestrian safety at zebra crossings. IEEE Trans Consum Electron 67(1):3–11
44. Surekha P, Vitta N, Duggirala P, Ambadipudi VSS (2022) Hand gesture recognition and voice, text conversion using. In: 2022 second international conference on artificial intelligence and smart energy (ICAIS). IEEE, pp 167–171
45. Khan A, Khusro S (2021) An insight into smartphone-based assistive solutions for visually impaired and blind people: issues, challenges and opportunities. Univers Access Inf Soc 20:265–298
46. Chen P, Zhang Y, Tan M, Xiao H, Huang D, Gan C (2020) Generating visually aligned sound from videos. IEEE Trans Image Process 29:8292–8302
47. Michelsanti D, Tan ZH, Zhang SX, Xu Y, Yu M, Yu D, Jensen J (2021) An overview of deep-learning-based audio-visual speech enhancement and separation. IEEE/ACM Trans Audio, Speech, Lang Process 29:1368–1396
48. Anandan M, Manikandan M, Karthick T (2020) Advanced indoor and outdoor navigation system for blind people using raspberry-pi. J Internet Technol 21(1):183–195

Unlocking Insights into Mental Well-Being: A Deep Learning Approach to Depression Detection

Rakoth Kandan Sambandam, Divya Vetriveeran, J. Jenefa, D. Vinodha, and S. Thaiyalnayaki

Abstract Depression, a prevalent global mental health disorder, poses significant challenges for timely diagnosis and effective intervention. Our research leverages the power of deep learning to develop a novel approach to depression detection, with the ultimate goal of enhancing early diagnosis and improving mental well-being. To evaluate the effectiveness of our method, we carried out an examination on a sizable dataset consisting of real-world data from individuals with and without diagnosed depression. The report discusses the performance metrics to examine the proposed technique effectiveness in depression detection. Additionally, this proposed work emphasizes the ethical and privacy considerations surrounding mental health data. The findings of this research indicate promising results in early depression detection, offering the potential to revolutionize mental health care by facilitating timely interventions. We discuss the implications of our approach in terms of supporting mental health professionals, improving mental health care accessibility, and addressing the pressing global mental health crisis.

Keywords Healthcare · Depression · Mental well-being · Deep learning · Optimization

1 Introduction

Mental health is an integral aspect of human well-being, yet it remains a significant global challenge. Among various mental health disorders, depression stands out as a pervasive and debilitating condition affects millions of individuals worldwide. As per the World Health Organization (WHO), it stands as the foremost cause of disability

R. K. Sambandam (✉) · D. Vetriveeran · J. Jenefa · D. Vinodha
Department of CSE, SOET, CHRIST (Deemed to be University), Kengeri Campus, Bengaluru, Karnataka, India
e-mail: rakothsen@gmail.com

S. Thaiyalnayaki
Department of CSE, School of Computing, Bharath Institute of Higher Education and Research (Deemed to be University), Chennai, Tamil Nadu, India

© The Author(s), under exclusive license to Springer Nature Singapore Pte Ltd. 2025

H. Zen et al. (eds.), *Soft Computing and Signal Processing*, Lecture Notes in Networks and Systems 1200, https://doi.org/10.1007/978-981-97-9926-8_41

on a global scale, underscoring the urgency to develop innovative approaches for its detection and intervention. Traditional methods of diagnosing depression primarily rely on self-reported symptoms and clinician assessments, which can be subjective, time-consuming, and may lead to underdiagnoses or delayed treatment [1]. In this context, advanced technologies, particularly deep learning, offer a promising avenue to revolutionize the field of mental health by enabling objective and timely detection of depressive symptoms. The well-being of individuals is significantly influenced by mental health, serving as an integral component of our overall wellness, yet it is often shrouded in complexity and challenges, especially in the identification and handling of mental health disorders, this significance becomes pronounced such as depression [2]. Depression, a pervasive global issue, affects millions of individuals and their communities, presenting a significant barrier to leading fulfilling lives [3]. As the world grapples with the growing burden of mental health challenges, it is imperative that we harness the potential of cutting-edge technology to revolutionize our approach to depression detection and support.

This work is motivated by the urgent necessity to enhance the effectiveness of addressing the mental health crisis. The global burden of depression necessitates innovative solutions that can identify and support individuals at risk. Leveraging the power of deep learning, this technology has demonstrated remarkable success across various applications, such as NLP, CV, and SA, and it is a logical and compelling step toward this goal.

2 Literature Review

The examination of literature constitutes a crucial aspect of any scientific endeavor. Each new research endeavor relies on established models as a systematic approach to lead researchers toward conclusions and results. Therefore, comprehending solutions available for a particular problem enables the construction of modified models for developing innovative solutions. With the help of a research article can identify the research gaps and uncovering potential new solutions.

In [4] a meta-analysis found that burnout and depression are prevalent in both emergency medicine (EM) and critical care medicine (CCM) settings, healthcare professionals in EM had significantly lower scores in personal accomplishment (PA) compared to those in CCM, indicating a need for increased focus on the mental health of EM professionals. Risk factors for burnout in these fields are not well-understood, and more research is needed. The review also discussed the consequences of high burnout and depression levels in healthcare and suggested the need for randomized trials to develop effective management and prevention strategies for these conditions in EM and CCM specialists.

In [5], authors describe creating an intelligent system that can recognize depressive symptoms in textual content using RNN's based on LSTM technique. The method is used to analyze text-based inquiries from young people in a Norwegian dataset, and it relies less on conventional word frequency-based techniques and more on robust features that reflect unique symptoms. This ground-breaking approach improves clustering and has demonstrated higher performance in separating writings depicting depression from postings without depression. An explainable AI algorithm called LIME can also be used to explain the decisions made by the system. With encouraging outcomes, our depression prediction method shows potential relevance in creating cutting-edge mental health care technology, including intelligent Chabot's. It can be customized for different languages and depression datasets.

In [6], authors proposed the study emphasizes the significance of determining depression solely from social media users' textual data. Depression stands as a significant global health issue, impacting millions of individuals across the world, leading to a decline in daily life functions. However, mental health service utilization rates are low due to various barriers. Early detection and awareness of mental illness are crucial for proper treatment, and automatic depression detection techniques using social media data have been explored as a potential tool for identifying depression levels in individuals. Results obtained indicates that BiLSTM + Attention model performs well on depression-related textual data. The main moto of this paper focuses how to identify indications of depression within social media users through the analysis of their textual content on social platforms with the help of machine learning algorithms.

In [7] study, author focuses prevalence of depression and the need for an automated system to detect it using AI and ML techniques. Various approaches are explored, including facial expressions, text analysis, and emotion detection through Chabot's and social media. Machine learning algorithms are used for emotion recognition. Emotion detection systems have applications in various fields, such as gaming, market research, and recruitment. The paper also highlights the use of live video and audio streams for emotion detection through neural networks and PCA. Textual sentiment analysis is emphasized as a non-invasive technique for diagnosing depression. The paper presents a comprehensive survey of existing AI and ML approaches for detecting emotions and depression, along with their potential applications.

In [8], this study concentrates on employing machine learning (ML) techniques to analyze text data sourced from social media, with the aim of identifying signs indicative of depressive symptoms. The review includes 17 relevant studies out of 418 identified through a comprehensive bibliographic search. These studies utilized supervised and unsupervised ML approaches to identify depression through researcher-inferred mental status, and the study involves analyzing users' self-provided descriptions of their mental well-being, or community membership. The review highlights challenges in sampling, prediction approaches, generalizability, privacy, and ethical considerations, suggesting the need for further research. In general, the utilization of ML approaches on social media–based text data exhibits potential for efficiently identifying depression, making them valuable tools in the realm of public mental health practice.

In [9], the study focuses on sentiment analysis, also known as opinion mining, and involves understanding and categorizing feelings, opinions, and judgments from textual or data sources. This analysis can be used in various fields such as business product reviews, stock market trends, public sentiment toward news, and political debates. Sentiments are typically categorized as positive, negative, or neutral. Different techniques, including machine learning approaches like supervised and unsupervised learning, are employed to classify sentiments. Emoticons and emojis in social media data are also valuable for sentiment analysis and should not be lost during data preprocessing. The data undergoes processing through feature extraction techniques, and subsequent classification into sentiment classes is carried out using various algorithms that are used to categorize data into sentiment classes.

In [10], the study explores different techniques, including machine learning and lexicon-based approaches, for identifying depression through psychological analysis. It suggests that a hybrid approach may yield better results than individual classifiers. This technique is used to analyze the twitter, etc., and data predict the individuals' mindset, i.e., whether they have been affected with the depression, especially among younger generations who are active on these platforms. However, it acknowledges limitations, such as the inability to reach those who do not use social media, and discusses ongoing challenges in psychological analysis that may be addressed in future research.

3 Problem Identification

The landscape of mental health, depression stands as a pervasive and complex challenge, affecting a significant portion of the global population. While the importance of early detection and intervention in depression is widely recognized, the current diagnostic methods primarily rely on subjective assessments, often resulting in delayed treatment and missed opportunities for timely support and block diagram is shown in Fig. 1.

The core problem lies in the subjectivity and inconsistency of traditional depression diagnosis. Clinicians depend on self-reported symptoms and observable behaviors, which can vary significantly from one individual to another. This leads to the potential for misdiagnosis, a lack of timely interventions, and increased suffering for those affected. Furthermore, the scale of the global mental health crisis demands more scalable, data-driven, and accessible solutions. The limitations of existing methodologies necessitate a transformative approach to depression detection, one that can enhance the objectivity, accuracy, and accessibility of diagnosis while addressing the ethical concerns surrounding sensitive mental health data.

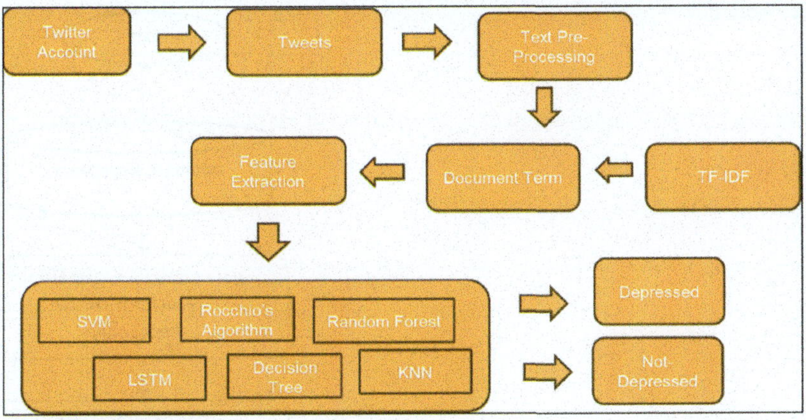

Fig. 1 Block diagram

4 Proposed Methodology

Textual data are to be analyzed under sentiment analysis; then, we will commence by collecting a diverse dataset encompassing textual, acoustic, and behavioral data and work flow is shown in Fig. 2. The dataset will include real-world information from individuals with and without diagnosed depression. Careful attention will be given to ethical data handling, ensuring privacy and security [11, 12]. This data will be the foundation for our deep learning model. Leveraging state-of-the-art deep learning techniques, we will focused toward to design the model for identify and prediction. The deep learning model will be trained and validated rigorously using the dataset. We will assess its accuracy, precision, recall, $F1$-score, and area under the receiver operating characteristic curve (AUC-ROC). This validation phase will ensure the model's effectiveness in identifying depression indicators across different data modalities. Continuous refinement and optimization of the deep learning model will be an ongoing process, as we work to enhance its accuracy, efficiency, and generalizability, considering the diversity of individuals and data sources. In this paper, SVM algorithm is the one used mainly. SVM is a supervised classification technique which provides a better results for many real-time datasets. It can be apply to wide variety of classification problems.

4.1 Rocchio's Algorithm

Rocchio's Algorithm, also known as Rocchio Classification or Rocchio Relevance Feedback, is a text categorization and information retrieval technique that was developed by J. Rocchio in the 1960s. It's primarily used for document classification and

Fig. 2 Work flow

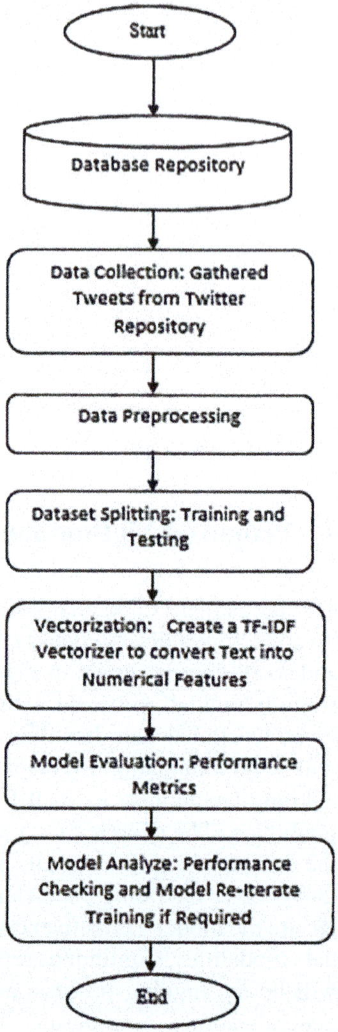

information retrieval tasks, such as text classification, spam email detection, and relevance feedback in search engines.

4.2 *KNN Algorithm*

For tasks involving regression and classification, the K-Nearest Neighbors (KNN) method is a straightforward and understandable supervised machine learning technique. Since KNN is an instance-based, non-parametric method.

4.3 Decision Tree Algorithm

A well-liked supervised machine learning approach for both regression and classification applications is the decision tree algorithm. Recursively dividing the dataset into subsets according to the values of the input features creates this tree-like data structure.

4.4 LSTM Algorithm

RNN's and LSTM architecture are used in ML and DL for time series analysis and NLP sequence modeling. LSTM can capture long-range dependencies in sequences since it is made to solve the vanishing gradient issue that plagues regular RNNs.

4.5 Random Forest Algorithm

The Random Forest Algorithm is an ensemble learning technique employed in machine learning for tasks encompassing both classification and regression. Its purpose is to enhance accuracy and robustness in predictions by amalgamating the outcomes of multiple decision trees.

4.6 Preparing the Dataset

- Obtained textual data from twitter that has the relevant information for analysis.
- Dataset contains 20,000 tweets extracted from twitter of different people.
- Dataset cleaning by removing null values, redundant data and so on.
- Text cleaning by removing punctuations, special characters and converting all the text into lower case.
- Categorizing the entire dataset into 2 levels for training and testing, typically following a 70–30 split for training and testing, respectively.

4.7 Training and Testing

- Trained the model using the pre-processed training dataset. Employ a suitable loss function, such as binary cross-entropy, and an optimizer like Adam.
- Adjust hyper-parameters like the learning rate and apply early stopping if needed.

5 Results and Discussion

The deep learning approach to depression detection aims enhancing the quality of life for individuals through offering timely support and more effective mental health care. It has the potential to reduce the impact of depression on individuals' well-being and society as a whole. Hence, we removed the existing method and got better accuracy 84%. Concentrating on a narrow, carefully selected domain allows us to circumvent the need to differentiate between the tweets in our analysis. This emphasizes how important contextual information or domain expertise is to sentiment analysis. Both the training datasets and the content—in this case, Tweets—have an impact on the system's performance. Consequently, this approach to sentiment analysis from text is not only straightforward but also reliable.

- Term-1—Precision
- Term-2—Recall
- Term-3—$F1$ Score
- Term-4—Support
- Value 1—0
- Value 2—1.

After executing various individual algorithms, the obtained results from Rocchio's, KNN, SVM, Decision Tree, LSTM, and Random Forest Algorithm are illustrated in screenshots from Tables 1, 2, 3, 4, 5, and 6 and Fig. 3, respectively.

Table 1 Result of Rocchio's algorithm

	Term-1	Term-2	Term-3	Term-4
Value 1	0.70	0.78	0.74	1981
Value 2	0.75	0.67	0.71	2019
Acc.			0.72	4000
Mac avg.	0.73	0.72	0.72	4000
Weig. avg.	0.71	0.69	0.71	4000

Table 2 Result of SVM algorithm

	Term-1	Term-2	Term-3	Term-4
Value 1	0.82	0.83	0.83	1981
Value 2	0.83	0.83	0.83	2019
Acc.			0.84	4000
Mac avg.	0.81	0.81	0.84	4000
Weig. avg.	0.83	0.83	0.83	4000

Table 3 Result of KNN algorithm

	Term-1	Term-2	Term-3	Term-4
Value 1	0.55	0.97	0.70	1981
Value 2	0.90	0.22	0.36	2019
Acc.			0.60	4000
Mac avg.	0.73	0.60	0.53	4000
Weig. avg.	0.73	0.63	0.55	4000

Table 4 Result of decision tree algorithm

	Term-1	Term-2	Term-3	Term-4
Value 1	0.72	0.71	0.71	1981
Value 2	0.72	0.73	0.72	2019
Acc.			0.72	4000
Mac avg.	0.71	0.72	0.71	4000
Weig. avg.	0.71	0.72	0.72	4000

Table 5 Results of LSTM algorithm

	Term-1	Term-2	Term-3	Term-4
Value 1	0.73	0.79	0.76	1981
Value 2	0.78	0.71	0.74	2019
Acc.			0.75	4000
Mac avg.	0.73	0.73	0.75	4000
Weig. avg.	0.74	0.75	0.75	4000

Table 6 Results of random forest algorithm

	Term-1	Term-2	Term-3	Term-4
Value 1	0.83	0.74	0.78	1981
Value 2	0.77	0.85	0.81	2019
Acc.			0.79	4000
Mac avg.	0.80	0.79	0.79	4000
Weig. avg.	0.81	0.80	0.80	4000

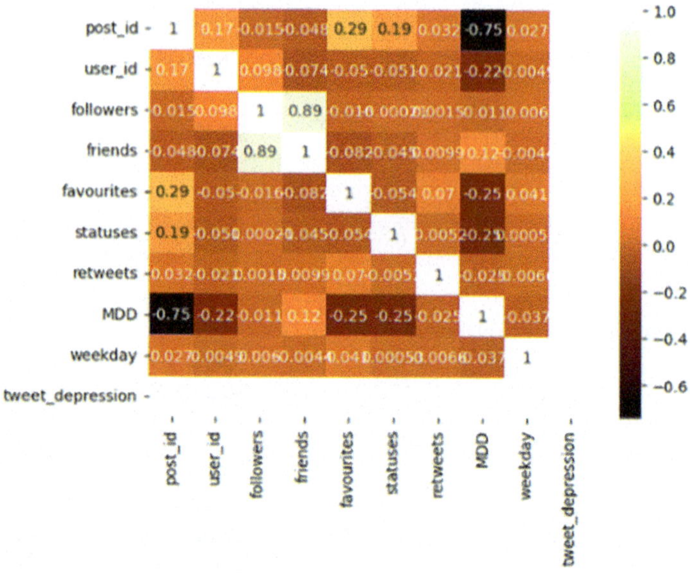

Fig. 3 Heat of map column

6 Conclusion

Mental health issues, including depression, have far-reaching consequences on the well-being of individuals and society as a whole. Early and accurate detection is vital to providing timely support and intervention. Our work leveraged a unique dataset from ung.no, allowing us to explore the complexities of self-perceived depressive symptoms and clinically diagnosed conditions. By developing and training machine learning models, we have strived to unlock valuable insights into mental well-being, and in particular, depression. The use of deep learning, linguistic diversity considerations, and feature extraction from text data has equipped us with powerful tools for identifying and understanding depression in textual narratives. Various machine learning classifiers were employed to categorize the tweets. In our model, Support Vector Machine (SVM) achieved a higher accuracy, nearly reaching 84%. Further enhancements to the model could involve addressing with the help of confusion matrix, genetic algorithms. Dealing with such scenarios poses a challenge in the domain of sensitivity analysis.

Acknowledgements The authors gratefully acknowledge the authorities of CHRIST (Deemed to be University), Bengaluru, Karnataka, and Bharath Institute of Higher Education and Research (Deemed to be University), Chennai, Tamil Nadu, for the facilities offered to carry out this work.

References

1. WHO (2021) https://www-who-int.ezproxy.uio.no/news-room/fact-sheets/detail/suicide
2. Kumar U, Dalal S (2024) Prevalence and risk factors analysis of postpartum depression at early stage using hybrid deep learning model. Nat Sci Rep 14:4533
3. Tejaswini V, Babu KS, Sahoo B (2014) Depression detection from social media text analysis using natural language processing techniques and hybrid deep learning model. ACM Trans Asian Low-Resour Lang Inf Process 23(1)
4. Ibrahim T et al (2023) Exploring the mental health challenges of emergency medicine and critical care professionals: a comprehensive review and meta-analysis. Nat Libr Med 15(7)
5. Uddin MZ et al (2022) Deep learning for prediction of depressive symptoms in a large textual dataset. Neural Comput Appl 34(1):721–744
6. William D, Suhartono D (2021) Text-based depression detection on social media posts: a systematic literature review. Proc Comput Sci 179:582–589
7. Joshi ML, Kanoongo N (2022) Depression detection using emotional artificial intelligence and machine learning: a closer review. Mater Today Proc 58:217–226
8. Liu D et al (2022) Detecting and measuring depression on social media using a machine learning approach: systematic review. JMIR Ment Health 9(3)
9. Babu NV, Kanaga EGM (2022) Sentiment analysis in social media data for depression detection using artificial intelligence: a review. Nat Libr Med 3(1)
10. Gupta S et al (2022) Psychological analysis for depression detection from social networking sites. Nat Libr Med 6
11. Wang N, Kosinski M, Stillwell DJ, Rust J (2014) Can well-being be measured using Facebook status updates? Validation of Facebook's gross national happiness index. Soc Indic Res 115(1):483–491
12. Gokulapriya R et al (2022) Text-based sentimental analysis to understand user experience using machine learning approaches. In: International conference on advanced computing and communication systems (ICACCS), Coimbatore, India

Diagnosis of Pancreatic Tumor Using CT Images

M. Kanchana, Harsh Kumar Singh, and Arjun Sharma

Abstract Pancreatic cancer is globally recognized as the most severe forms, because of notably low 5-year of survival rate. The majority cases are attributed to this aggressive cancer, underscoring the critical need for early detection. AI, along with its subfields like machine learning and deep learning, has immense potential to bring about positive changes in human lives. Through various medical imaging techniques, we can visualize the internal structure of the human body. This article introduces an approach for pancreatic cancer detection in CT scan images through the integration of different techniques, with CNN, Naive Bayes, AdaBoost, and other image processing methods. A CNN is a form of artificial neural network primarily utilized for image recognition and processing because of its capability to identify patterns within images. Naive Bayes is a straightforward method for constructing classifiers, which are models that assign class labels to instances of a problem, with these class labels being selected from a finite set, along with other techniques for processing images. AdaBoost, abbreviated from Adaptive Boosting, is a supervised learning technique employed to classify data by amalgamating numerous weak or base learners (like decision trees) into a powerful learner. AdaBoost operates by adjusting the weights of instances in the training dataset based on the accuracy of prior classifications. Employing different activation functions, such as rectified linear unit and sigmoid, are used to analyze the image. Dropout function is used to counter the overfitting problem. Different urinal biomarkers are used in the diagnosis of pancreatic tumor. Classification is executed using CNN, Naive Bayes algorithms, and AdaBoost. The result is measured by calculating accuracy, precision, recall, and F1-score for the classification algorithms. CNN algorithm demonstrating superior accuracy. Despite the challenges in pancreatic cancer diagnosis, particularly due to

M. Kanchana (✉) · H. K. Singh · A. Sharma
Department of Computing Technologies, SRM Institute of Science and Technology, Kattankulathur, Chengalpattu, Tamil Nadu 603203, India
e-mail: kanchanm@srmist.edu.in

H. K. Singh
e-mail: hs2071@srmist.edu.in

A. Sharma
e-mail: as9063@srmist.edu.in

© The Author(s), under exclusive license to Springer Nature Singapore Pte Ltd. 2025
H. Zen et al. (eds.), *Soft Computing and Signal Processing*, Lecture Notes in Networks and Systems 1200, https://doi.org/10.1007/978-981-97-9926-8_42

its severity, this approach holds promise for improving early detection capabilities through machine learning and image processing.

Keywords Pancreatic cancer · Computed tomography · Convolutional neural network · Naive Bayes · AdaBoost · Accuracy · Precision · Recall

1 Introduction

Cancer is a multifaceted condition marked by the unregulated proliferation of cells. Unlike healthy cells that follow a controlled cycle of division and death, cancerous cells grow uncontrollably, giving rise to tumors that can infiltrate adjacent tissues and metastasize to distant sites. Factors contributing to cancer development encompass genetic alterations, exposure to environmental hazards such as tobacco and UV rays, and lifestyle behaviors. Treatment approaches, which span surgery, chemotherapy, radiation therapy, immunotherapy, and targeted medications, are customized based on cancer type and stage. The most common form of pancreatic cancer is referred to as PDAC. Due to variations in CT image quality across scanners, skilled physicians are essential for accurate manual diagnosis. Consequently, there's a growing demand for reliable deep-learning-based techniques to precisely diagnose pancreatic tumors. Progress in DL and medical image processing has been significant since the introduction of CNN for analyzing visual data. Pancreatic cancer occurs when cells in the pancreas undergo mutations and uncontrollably proliferate, forming tumors. The diagnosis of pancreatic tumors is a complex and challenging task that requires a comprehensive approach. Computed tomography (CT) imaging has emerged as a valuable tool in this process, providing detailed anatomical information and aiding in the characterization of pancreatic tumors. In the initial detection phase, CT scans can identify pancreatic tumors by demonstrating focal abnormalities in the pancreas. These abnormalities can be visualized as mass lesions with irregular shapes, heterogeneous enhancement patterns, and sometimes associated with adjacent vascular involvement. The CT scan can also reveal any potential metastatic lesions in the liver and regional lymph nodes, providing crucial information for staging and treatment planning. CT imaging allows for the assessment of the size and extent of pancreatic tumors, aiding in the determination of resectability. Tumor size is an important factor in staging, as larger tumors are often associated with a poorer prognosis. CT scans can measure the diameter of the tumor and provide information on invasion into adjacent structures, such as blood vessels or the duodenum. These findings help the multidisciplinary team consider the optimal treatment strategy, surgical resectability, and the need for neoadjuvant therapy to downstage locally advanced tumors. Furthermore, CT imaging can be used for image-guided biopsies, providing real-time guidance to obtain tissue samples from pancreatic tumors. This minimally invasive procedure can help confirm the diagnosis and provide histopathological information necessary for further management decisions. In conclusion, CT scans have an important role in the diagnosis of pancreatic tumors. By providing detailed anatomical information, assessing tumor size, and evaluating vascular involvement, CT scans aid in treatment planning and determining resectability. Additionally, advanced techniques, such as CT angiography, and image-guided biopsies further enhance the diagnostic capabilities of CT imaging. Overall, CT imaging provides valuable insights into the diagnosis and management of pancreatic tumors, contributing to improved patient outcomes.

2 Literature Survey

Chen et al. [1] undertook an extensive investigation concentrating on the identification of pancreatic cancer through deep learning techniques applied to CT scans. Their nationwide population-based study yielded encouraging findings regarding the effectiveness of the deep learning model in accurately detecting pancreatic cancer. The research showcased promising outcomes, emphasizing the potential of this approach for improving diagnostic accuracy in pancreatic cancer detection using advanced deep learning methodologies on CT scans. Si et al. [2] created a comprehensive deep learning diagnostic system specifically designed for pancreas and its tumor. Their study showcased efficacy of deep learning techniques in precisely diagnosing pancreatic tumors, ultimately contributing to enhanced patient outcomes. The findings underscored the potential of leveraging advanced deep learning methodologies for accurate and effective diagnosis of pancreatic tumors. Qureshi et al. [3] utilized AI analysis on pre-diagnostic CT images for the prediction of PDAC. Their results emphasized the considerable involvement of AI for the early detection including diagnosis for this lethal disease. The study underscored the role of AI as a valuable tool in predicting PDAC based on pre-diagnostic CT scans, showcasing its promise in advancing early detection methods for this deadly condition.

Karmazanovsky et al. [4] employed a radiomics approach based on CT to characterize pancreatic tumors. Their research demonstrated the capability of radiomics in offering supplementary insights and enhancing the comprehension of pancreatic cancer. The study underscored the potential of utilizing a CT-based radiomics approach for a more nuanced understanding and characterization of pancreatic tumors. Rhee and Park [5] examined the significance of imaging within the existing treatment approaches for pancreatic adenocarcinoma. Their analysis underscored the crucial role of imaging techniques in guiding decisions related to treatment and evaluating the response to treatment in individuals diagnosed with pancreatic cancer. The review emphasized the essential contribution of imaging modalities in shaping treatment strategies and monitoring the effectiveness of interventions for patients afflicted with pancreatic adenocarcinoma.

Granata et al. [6] provided an overview of the state-of the-art imaging techniques and radiomics in the detection and characterization of pancreatic cancer. Their review highlighted the advancements in imaging technology and the potential of radiomics in improving diagnosis and treatment planning. Chu et al. [7] discussed the challenges and advancements in pancreatic cancer imaging. The authors highlighted the need for new imaging techniques and imaging biomarkers for increasing the level of early detection and accurate diagnosis of pancreas and its cancer. Miura et al. [8] undertook a case–control study aiming to examine the occurrence of focal parenchymal atrophy in the pancreas on CT scans among patients with cancer. The results of their investigation indicated that the presence of this specific characteristic might serve as a potential marker for the detection of pancreatic cancer. The study suggested that focal parenchymal atrophy observed in pre-diagnostic CT scans could potentially be utilized as an indicator for the presence of pancreatic cancer.

Alabousi et al. [9] compared the diagnostic accuracy of MRI and CT in detecting liver metastases in patients with pancreatic carcinoma. Their systematic review and meta-analysis provided valuable insights into the diagnostic capabilities of these imaging modalities in detecting liver metastases. Dhruv et al. [10] presented a hybrid method that integrates both edge detection and AI techniques for the early and accurate detection of pancreatic tumors. Their research showcased the effectiveness of combining diverse approaches to enhance the precision and efficiency of detecting pancreatic tumors. The study underscored the potential benefits of integrating edge detection and AI methodologies in a hybrid approach for improving the overall accuracy of early pancreatic tumor detection. Jasti et al. [11] elucidated that enzymes, once secreted into the thin ducts, find their ultimate resting place in the pancreatic duct. This duct eventually converges with the common bile conduit. Despite being a minority among the pancreatic cells, endocrine cells play a pivotal role as they generate crucial hormones like insulin and glucagon. These particular cells are implicated in the initiation and progression of pancreatic tumors, underscoring their significance in tumor development despite constituting a smaller percentage of the overall pancreatic cell population.

Chaudhury et al. [12] emphasize that medical imaging primarily relies on image processing and computer modeling, given the diverse range of imaging technologies within biomedical imaging. While these methods are widely employed, the study of organisms data still majorly involves human involvement in different research endeavors. Recognizing inherent biases, unpredictability among human specialists, and the time and cost-intensive nature of manual procedures, there is a growing need for more objective and iterative analyses. To achieve this, accurate measurements of the quantitative things, comprehensive test of large input, and utilization of the automated things are essential in ensuring an analytical process that is both objective and efficient. Zamani et al. [13] underscore the significance of medical image scans in enabling the early identification of abnormalities among cancer patients. However, the prohibitive costs associated with the requisite equipment and infrastructure pose challenges in making this technology widely accessible, limiting its reach to a considerable extent. Consequently, high-quality radiography and thorough analyses by experts are not readily available. Problem for the lower accuracy of CT-based cancer prediction and substantial variability between different readers further contribute to these limitations. In the past, WHO emphasized the primary use of medical scan images when cancer cannot be conclusively confirmed through laboratory tests and biopsies, reflecting a cautious approach considering the accessibility challenges and concerns surrounding image-based cancer diagnosis accuracy and consistency.

Li et al. [14] have developed a graph in neural network designed for quick prediction of cancer in non-contrast imaging. The research focuses on leveraging causality-driven graph neural networks to enhance the diagnostic capabilities, particularly in the early stages of pancreatic cancer detection using non-contrast computerized imaging techniques. Ramesh et al. [15] have introduced a novel approach involving the search algorithm, called Sparrow, combined with deep learning, that is stacked, for image analysis aimed at cancer detection. The study focuses on utilizing this hybrid approach to enhance the accuracy and efficiency of detecting and classifying pancreatic cancer through advanced medical image analysis technique.

3 Existing System

The current system employed for diagnosing pancreatic tumors through CT images faces several drawbacks. Chief among these is the time-intensive nature of the process. Analyzing CT images necessitates a thorough examination by a radiologist, who meticulously assesses each image to identify potential abnormalities or indications of a tumor. This procedure can become prolonged, especially when dealing with multiple images. Moreover, the interpretation of CT images is inherently subjective, heavily relying on the radiologist's expertise and experience. This subjectivity introduces the risk of misinterpreting images and subsequently misdiagnosing pancreatic tumors. Furthermore, the existing system lacks accuracy. CT images can sometimes be unclear or of poor quality, making it difficult for the radiologist to accurately identify and assess tumors. Additionally, the resolution of CT images may not always be sufficient to detect small tumors or subtle changes in the pancreas, leading to false negatives and missed diagnoses. Moreover, the existing system does not provide real-time feedback. Once the CT images are analyzed and a diagnosis is made, the patient may need to wait for the results, causing delays in initiating treatment or further diagnostic procedures. Another disadvantage is the reliance on contrast agents. CT scans often require the use of contrast agents, which can have side effects and allergic reactions in some patients. Furthermore, patients with certain medical conditions, such as kidney disease, may not be able to receive contrast agents, limiting the effectiveness of CT scans for diagnosing pancreatic tumors. Finally, the existing system is expensive. CT scans and the subsequent analysis of the images can be costly, especially when multiple scans are required over time to monitor the progression of tumors or the effectiveness of treatment. This cost can be a barrier to access for some patients, limiting their ability to undergo timely and accurate diagnosis and treatment. Disease is diagnosed at a late stage: The potential for long-term survival in pancreatic cancer lies in the possibility of undergoing curative surgery. However, due to the late onset of disease-specific symptoms, a small percentage of patients (less than 15–20%) have tumors that are amenable to resection at the time of diagnosis.

Complete resection is not curative: If resectable pancreatic cancer could be cured, survival rates might rise by approximately five times. However, for most patients undergoing surgery, the outcome is primarily palliative. The median survival for those who undergo curative resection is only around 12–18 months longer than for those with unresectable disease. Thus, the lack of effectiveness in achieving a cure despite aggressive and seemingly complete resection of pancreatic cancer adds to the overall poor survival rates. In conclusion, the existing system for diagnosing pancreatic tumors using CT images has several disadvantages, including time-consuming analysis, subjective interpretation, lack of accuracy, lack of real-time feedback, reliance on contrast agents, and high costs.

4 Proposed System

The intended research endeavors to establish a dependable and precise approach for diagnosing pancreatic tumors through the utilization of CT images. Diagnosing pancreatic tumors poses a considerable challenge owing to the intricate anatomy of the pancreas and the resemblance of tumor characteristics to surrounding tissues. CT imaging has gained widespread acceptance in diagnosing diverse diseases, including pancreatic tumors, owing to its non-invasive nature and capacity to generate high-resolution images. Nevertheless, the interpretation of CT images for pancreatic tumor diagnosis remains intricate, demanding substantial expertise and experience due to the complexities involved in distinguishing tumor features within the pancreas. This research project aims to create an automated algorithm designed to analyze CT images and deliver accurate diagnoses of pancreatic tumors. This algorithm will leverage a combination of image processing and ML techniques to extract pertinent features from the CT images, categorizing them into tumor or non-tumor classes. The analysis involves scrutinizing the intensity, shape, and texture patterns of pixels in the CT images. Additionally, clinical data such as patient age, gender, and medical history will be incorporated to enhance the algorithm's diagnostic capabilities.

To train and validate the algorithm, a large dataset of CT images with corresponding ground truth annotations will be utilized. The dataset will consist of images from both cancerous and non-cancerous patients, allowing for the algorithm to learn and differentiate between normal pancreatic tissue and tumor tissue effectively. The performance of the algorithm will be evaluated using various metrics, such as sensitivity, specificity, and accuracy. The proposed work also includes the integration of the developed algorithm into a user-friendly software tool that can be easily accessed and utilized by medical professionals. CNN, Naïve Bayes, and AdaBoost decision tree are valuable tools for identifying patterns in images to recognize objects, classes, and categories. Familial aggregation or specific genetic mutations account for approximately 5–10% of pancreatic cancer cases. Gathering comprehensive data from various sources is essential. Prior to training and testing ML models, it is necessary to preprocess the data. The only possible way to solve the problems of the existing system is to find the pancreatic tumor at an early stage with most possible accuracy, precision, and recall. In summary, the goal of the proposed project is to offer a dependable and precise approach for diagnosing pancreatic tumors through

CT images. By creating an automated algorithm and incorporating it into user-friendly software, medical professionals will receive valuable support in efficiently and accurately diagnosing patients, ultimately resulting in better patient outcome.

5 System Architecture

According to Fig. 1, first and foremost the image dataset is uploaded. Then, in the left part of the diagram, firstly the data is splitted into train and test data. Then, the processing of the image CT image is done using Gaussian, Bernoulli, and different dataframes such as Unique_values and different features of the image is extracted, such as zoom range, shear range, rescale, horizontal flip, target size, batch size, class mode, etc. Some more features of the image is extracted with the use of convolutional and pooling layer. Activation functions, such as rectified linear unit and sigmoid, are used. Dropout function is used to counter the problem of overfitting. Overfitting is a problematic occurrence in machine learning where the model accurately predicts outcomes for training data but performs poorly with new data. The model undergoes training using epochs, which represent one full iteration of the training dataset through the algorithm. The number of epochs determines how many times the entire training dataset goes through the learning process of the algorithm. Lastly, pancreatic tumor is detected. In the right part of the diagram, firstly multiple features of the CT image is extracted such as, patient cohort, sample origin, age, stage, plasma CA19_19, REG1A, sex, etc. Then, unnecessary features are dropped. The different features of the image is processed for pancreatic biomarkers, such as creatinine, LYVE1, REG1B, and TFF1, to plot the graph for the biomarkers and build the heatmap. Now, the data is splitted into train and test data. Then, the model testing is done for Gaussian Naive Bayes, Bernoulli Naive Bayes, and AdaBoost for pancreatic tumor detection.

6 Dataset

Early detection of pancreatic cancer poses challenges due to its inconspicuous symptoms and the complexity of detecting tumors during routine screenings. Unlike certain cancers, pancreatic cancer frequently manifests no symptoms until it has progressed significantly. Although multiple medical imaging methods are accessible for detection, CT scans are generally the primary diagnostic tool. MRI is also frequently employed alongside CT scans to enhance the diagnosis of pancreatic cancer [16]. CT is a non-invasive medical imaging technique employed to visualize various sections of the human body. It allows for the detection of pancreatic cancer and assessment of metastasis. CT operates by capturing images of body sections using X-rays in thin slices, typically 3–10 mm thick. This technology aids in identifying organ diseases.

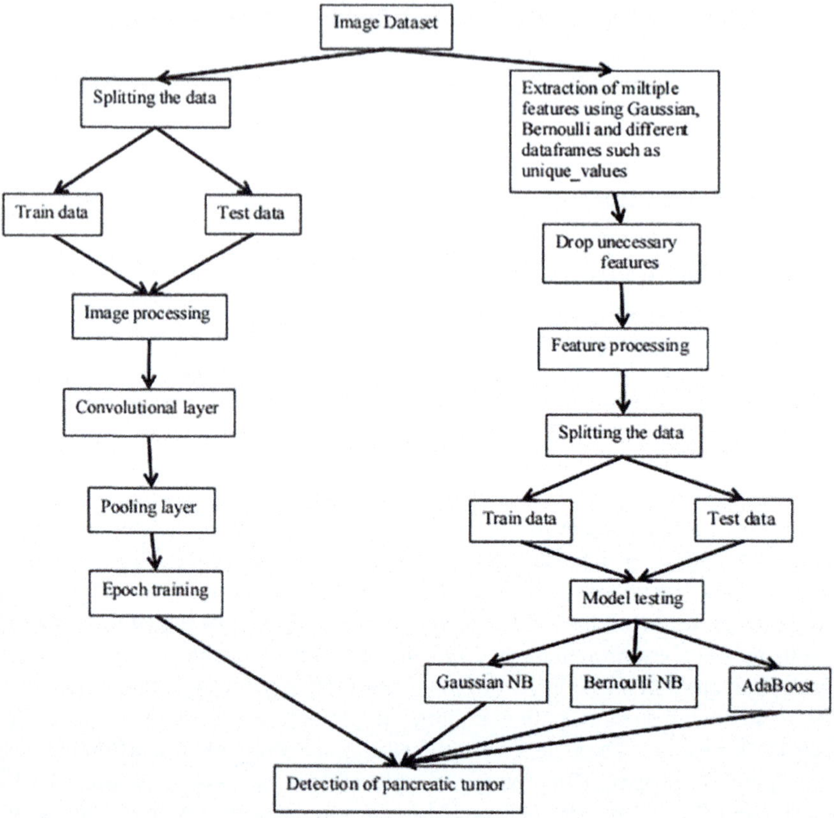

Fig. 1 Architecture diagram of the proposed model

According to Fig. 2, two normal pancreas (a) and (b) and tumor-infected pancreas (c) and (d) are shown. The representation of the image in the dataset is shown by the label written as numbers. The medical dataset comprises 1411 CT scans of the abdomen, used in the CNN algorithm, encompassing patients with normal conditions and various pancreatic tumors. Among these, 646 represent normal pancreas images, while 765 depict pancreatic tumor images. A total of 999 images are allocated for training of the model and the rest 412 are designated for final predictions or testing. Within the training set, 421 images showcase normal pancreas conditions, while 578 portray pancreatic tumors. For the testing set, 225 images depict normal pancreas conditions, and 187 represent pancreatic tumors. All images have dimensions of 512 × 512 pixels and were sourced from the Cancer Imaging Archive. Another set of dataset Urinary biomarkers for pancreatic cancer, used in Naïve Bayes and AdaBoost algorithm, represented as unique string identifying each subject. There are total 590 unique values of data. There are 301 female data and 289 male data. There is 384 benign values, 41 non-cancerous values, and 165 pancreatic cancerous values. They gathered a series of biomarkers from the urine of three groups of patients: Healthy

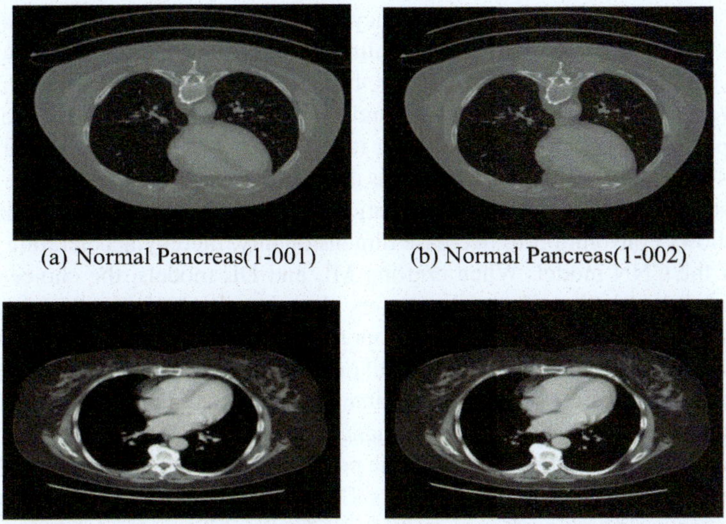
(a) Normal Pancreas(1-001) (b) Normal Pancreas(1-002)

(c) Tumor infected Pancreas(1-003) (d) Tumor infected Pancreas(1-004)

Fig. 2 Sample CT scan images

controls; Patients with non-cancerous pancreatic conditions, like chronic pancreatitis; Patients with pancreatic ductal adenocarcinoma. When possible, these patients were age- and sex-matched. The goal was to develop an accurate way to identify patients with pancreatic cancer. The key features are four urinary biomarkers: Creatinine, LYVE1, REG1B, and TFF1. Age and sex, both included in the dataset, may also play a role in who gets pancreatic cancer. The dataset includes a few other biomarkers as well, but these were not measured in all patients (they were collected partly to measure how various blood biomarkers compared to urine biomarkers).

7 Methodology

This part introduces the prediction of tumor in CT scan through the utilization of CNN and image processing techniques. Important features are extracted, and classification phase is done with the use of different layers and different activation function used in CNN, different pancreatic biomarkers analyzed with Gaussian and Bernoulli Naïve Bayes algorithms and AdaBoost decision tree to distinguish and categorize the images effectively.

Convolutional Neural Network (CNN)

CNN [17], a prominent deep learning architecture, is highly suited for image analysis tasks, contributing significantly to the surge in interest in DL. An important aspect of CNN is its autonomy in feature determination, requiring minimal human supervision. With computational efficiency, CNN implements parameter sharing and employs specialized convolution and pooling algorithms. Moreover, CNN models are now accessible across various platforms, broadening their appeal to a diverse audience. While CNN comprises several fundamental components, convolution layers stand out as crucial for comprehending its structure. Mathematically, convolution involves merging two datasets, serving as a pivotal concept in DL where the model's input undergoes this mathematical operation using convolution filters. Pooling typically follows convolution to decrease dimensionality, thereby accelerating training time for the CNN model. When crafting ML and DL models, the choice of algorithm and the selection of hyperparameters play pivotal roles in determining project success. Hyperparameters, including dropout rates, layer counts, and neuron quantities in DL models, directly impact project outcomes. The model's effectiveness isn't solely reliant on a single set of hyperparameters. Typically, multiple combinations of hyperparameters can yield high performance, and these varied combinations can be employed in constructing the model. The performance of a model isn't solely determined by a single set of hyperparameters. Often, multiple combinations of hyperparameters can yield high performance, allowing for flexibility in model design. During training, data isn't all processed simultaneously; instead, it's divided into batches. Initially, the model trains on one batch, adjusting weights through backpropagation based on performance evaluation. This process is repeated for subsequent batches in epochs. Training can span days or months, so efforts are made to expedite the process using other hyperparameters. Generally, performance improves with more epochs, although the optimal number varies based on the problem and requirements. Training can be halted once performance improvements become marginal beyond a certain epoch. Activation functions play a crucial role in deep learning models by introducing non-linearity, essential for tackling real-world nonlinear problems. Matrix multiplication in hidden layers transforms linear functions into nonlinear values. Deep learning excels in handling nonlinear problems, making activation functions pivotal. Commonly used activation functions include sigmoid and ReLU. Dropout, a regularization technique, addresses overfitting by selectively ignoring weak information, thus enhancing learning performance. A dropout value of 0.25 is often employed, although variations exist based on the specific problem and dataset requirements. In this project, a model was developed using four convolutional layers, with an increment in the number of filters for each subsequent layer. A kernel size of (3, 3) was chosen, and max pooling with a size of (2, 2) was applied.

Overfitting issues arose during model optimization, prompting the incorporation of a regular dropout layer. During the testing phase, predictions were compared with ground truth data. Designing a model that delivers accurate results without overfitting was a critical stage in this project. First, the training dataset is uploaded in the model and it goes through different image refining techniques. Secondly, importing different modules for the CNN algorithm and feeding the images to these modules. Using different activation functions and using dropout technique to counter overfitting. Thirdly, the convolutional layer and pooling layer, of the CNN model, is performed with the use of ReLU activation function. Lastly, the testing of the model is done by testing dataset and calculate the accuracy of the model.

Naïve Bayes

The Bayesian classification system relies on Bayes' theorem as its fundamental principle, providing the basis for Naïve Bayesian classification techniques [18]. When used for extensive databases, these techniques classify simple bases for manner in with categorization of decision trees and networks that are selected. This classification cannot accommodates unlimited number of dependent attributes. Current approach, estimates for the probability $P(x|c)$, can be derived for each class. In this study, the classes with the highest probability were pivotal for predicting the outcomes. Here, we use two types of Naïve Bayes:

Gaussian Naïve Bayes

Gaussian Naïve Bayes is a machine learning classification method that functions based on probability, assuming that each class follows a normal distribution. It works under the premise that each parameter independently contributes to predicting the output. Gaussian Naïve Bayes is applied when we expect all continuous variables associated with each feature to follow a Gaussian distribution, also known as the normal distribution.

$$P(x_i|y) = \frac{1}{\sigma\sqrt{2\pi}} e^{-\frac{(x-\mu)^2}{2\sigma^2}} \tag{1}$$

According to Eq. (1), Gaussian Naïve Bayes assumes that the likelihood $(P(x_i/y))$, follows the Gaussian distribution for each x_i within y_k. The algorithm categorizes each new data point, x, by determining the highest posterior probability among all classes and assigning the data point to that particular class.

Bernoulli Naïve Bayes

Bernoulli Naïve Bayes belongs to the Naïve Bayes family and is built upon the Bernoulli distribution, handling only binary values, which are 0 or 1. When dealing with binary features in a dataset, Bernoulli Naïve Bayes is often regarded as the appropriate algorithm to utilize.

$$p(x_i|y) = p(i|y)x_i + (1 - p(i|y))(1 - x_i) \tag{2}$$

According to Eq. (2), here, $p(x_i|y)$ is the conditional probability of x_i occurring provided y has occurred, I is the event, x_i holds binary value either 0 or 1. The Bernoulli Naïve Bayes classifier is suitable for determining whether an individual has a particular disease or not, utilizing available data. This scenario typically represents a binary classification problem, where the task involves categorizing instances into two classes: presence or absence of the disease. Bernoulli Naïve Bayes is particularly well-suited for such cases, as it assumes that features are binary variables, making it effective for modeling binary outcomes in disease detection tasks. Different types of pancreatic tumor biomarkers, such as creatinine, LYVE1, REG1B, TFF1, are used for feature extraction and classification.

AdaBoost Decision Tree

In AdaBoost decision trees [19], individual base classifiers work on a dataset with adjusted weights for each instance. These weights are influenced by the performance of prior base classifiers on the same instances. When an instance is misclassified, its weight is boosted for subsequent models, while correct classifications maintain the instance's weight unchanged. At the outset, we evaluate the classification performance of samples by analyzing how well the variables categorize them. This entails constructing decision stumps, which are single-feature classifiers, for each feature, and subsequently assessing the Gini Index for each tree. The Gini Index serves as a metric for measuring the disorder or purity of a dataset within each tree. The decision stump with the minimum Gini Index is chosen as the initial stump. First, we import all the modules for Naïve Bayes and AdaBoost decision tree to work, secondly importing dataset for these algorithm. Printing the categorical features as a column. Presenting the count plots categorical feature. These four algorithms work to develop a robust and accurate system for diagnosing pancreatic tumors using CT images. The processing of CT images ensure the consistency and quality of input data, while the feature extraction and selection stage capture relevant information from the images. Finally, the classification and diagnostic module utilizes ML algorithms to interpret the extracted features, providing precise tumor classification and diagnoses suitable for clinical applications.

8 Result and Discussion

Significant features are extracted from the images using the different algorithms and different activation function. Dropout function is used to counter the problem of overfitting. Classification is conducted using CNN, Gaussian Naive Bayes, Bernoulli Naïve Bayes, and AdaBoost decision tree. This study utilizes four parameters, namely accuracy, precision, recall and F1-score, to assess and compare the performance of various algorithms.

According to Fig. 3, a model was developed using convolutional layer and pooling layer, with an increment in the number of filters for each subsequent layer. A kernel size of (3, 3) was chosen, and max pooling with a size of (2, 2) was applied with the use of ReLU activation function.

According to Fig. 4, there is a correlation between elevated levels of pancreatic tumor and an increase in creatinine biomarkers in the body. Creatinine, a byproduct of creatine phosphate breakdown in muscles, is typically generated at a steady rate based on muscle mass. It is commonly employed as an indicator of kidney function. Normal creatinine clearance rates range from 110 to 150 ml/min in males and 100–130 ml/min in females.

According to Fig. 5, there is a notable elevation in LYVE1 biomarkers when pancreatic tumor levels rise in the body. LYVE1 has garnered attention as a potential

```
Model: "sequential"
_________________________________________________________________
 Layer (type)                Output Shape              Param #
=================================================================
 conv2d (Conv2D)             (None, 222, 222, 16)      448

 conv2d_1 (Conv2D)           (None, 220, 220, 36)      5220

 max_pooling2d (MaxPooling2  (None, 110, 110, 36)      0
 D)

 conv2d_2 (Conv2D)           (None, 108, 108, 64)      20800

 max_pooling2d_1 (MaxPoolin  (None, 54, 54, 64)        0
 g2D)

 conv2d_3 (Conv2D)           (None, 52, 52, 128)       73856

 max_pooling2d_2 (MaxPoolin  (None, 26, 26, 128)       0
 g2D)

 dropout (Dropout)           (None, 26, 26, 128)       0

 flatten (Flatten)           (None, 86528)             0

 dense (Dense)               (None, 64)                5537856

 dropout_1 (Dropout)         (None, 64)                0

 dense_1 (Dense)             (None, 1)                 65

=================================================================
Total params: 5638245 (21.51 MB)
Trainable params: 5638245 (21.51 MB)
Non-trainable params: 0 (0.00 Byte)
_________________________________________________________________
```

Fig. 3 CNN connected neural layer output

focus for cancer treatment due to its increased expression observed in different tumor types such as breast, lung, and pancreatic cancer.

According to Fig. 6, it shows the increase in the level of REG1B biomarkers when pancreatic tumor is elevated in the body. REG1B concentrations were notably elevated in the blood of individuals diagnosed with pancreatic cancer when compared to both healthy individuals and patients with pancreatitis. Similarly, in the urine of pancreatic cancer patients, REG1B levels displayed an increase compared to both healthy individuals and those with chronic pancreatitis. Moreover, REG1A has been identified as a prognostic marker for pancreatic cancer, with higher REG1A levels correlating with worse outcomes. According to Fig. 7, it shows the increase in the level of TFF1 biomarkers when pancreatic tumor is elevated in the body. TFF1 levels were notably elevated in the blood of individuals with pancreatic cancer when compared to both healthy individuals and those with pancreatitis. Similarly, in the urine of pancreatic cancer patients, there was a marked rise in TFF1 levels compared to both healthy individuals and individuals with chronic pancreatitis (Fig. 8).

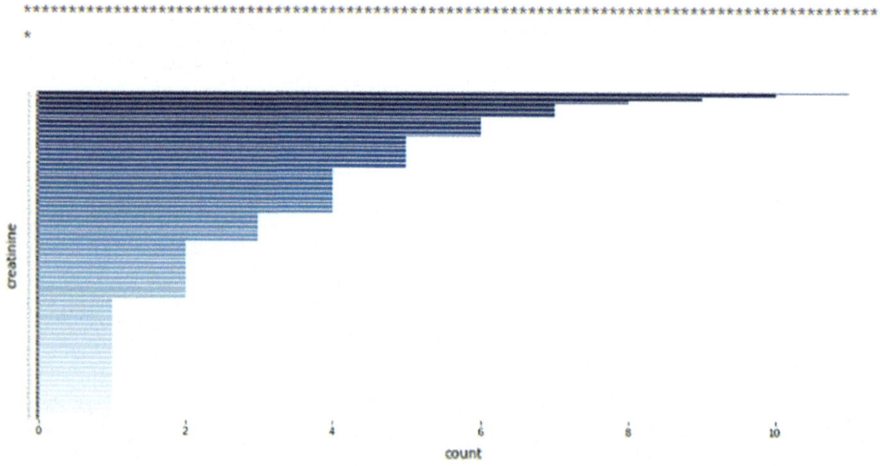

Fig. 4 Creatinine biomarkers

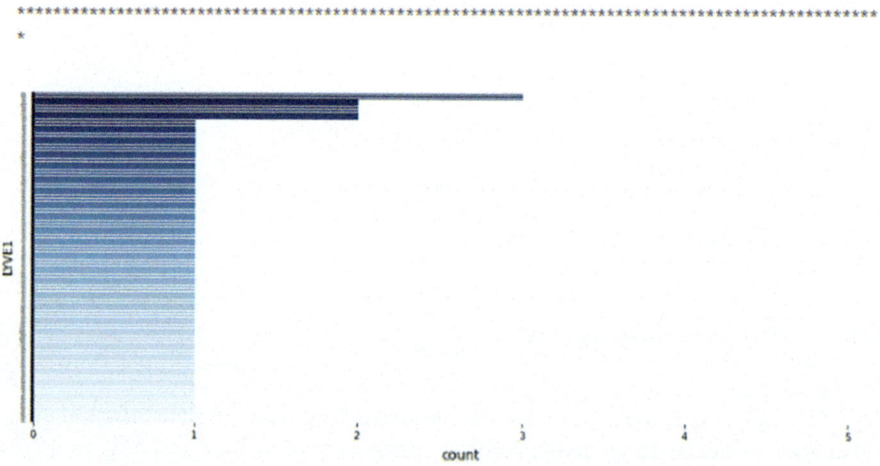

Fig. 5 LYVE1 biomarkers

Tables 1 and 2 show the accuracy, precision, recall, F1-score, and support of the Gaussian Naive Bayes and Bernoulli Naive Bayes algorithm, respectively. Accuracy measures how close results are to the true or known value. Recall signifies the capacity to identify all pertinent features of a class within a dataset. Precision denotes the ratio for input data points, which are recognized by model as belonging to the relevant class that were genuinely relevant. The F1-score is the harmonic mean of precision and recall, considering both metrics. It involves the use of TP = true positive, TN = true negative, FP = false positive and FN = false negative. In Fig. 9, the confusion matrix illustrates the predictions of non-cancerous and cancerous results for the image

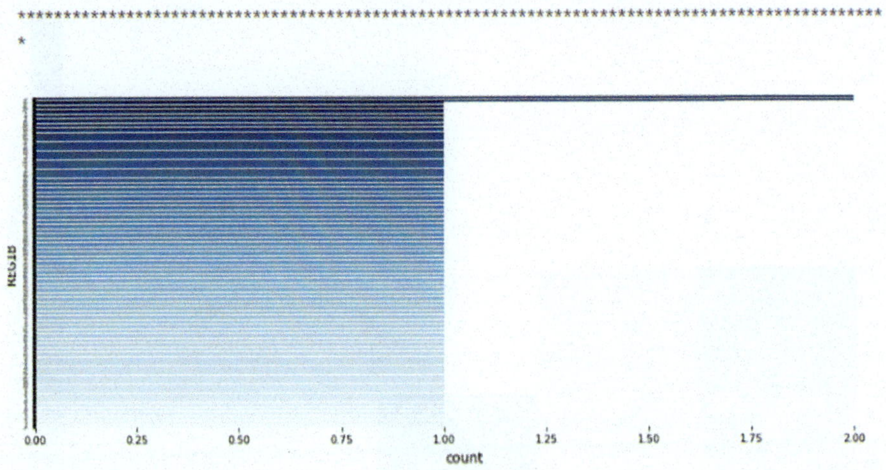

Fig. 6 REG1B biomarkers

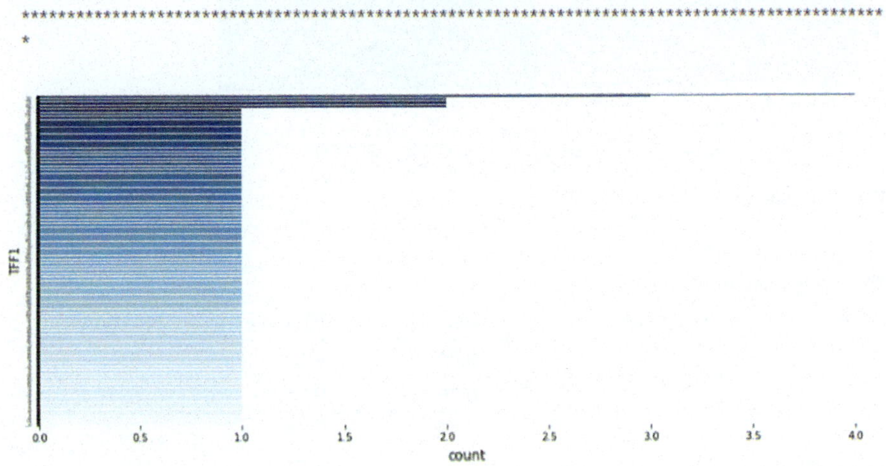

Fig. 7 TFF1 biomarkers

dataset. This matrix confirms that the CNN model achieves its highest accuracy of 96%, specifically on the CT image dataset. According to Table 3, it shows the result of the performance of four different algorithms. The accuracy, precision, recall, and F1-score are calculated for different classification algorithms, and from Table 3, it is clear that CNN produces better results than other algorithms. The highest accuracy is achieved by CNN, followed by AdaBoost, then Bernoulli Naive Bayes, and lastly Gaussian Naive Bayes. Figure 10 shows graph comparison of various algorithms. It is clear that the highest precision is achieved by CNN, followed by Bernoulli Naive Bayes, then Gaussian Naive Bayes, and lastly AdaBoost. The highest recall

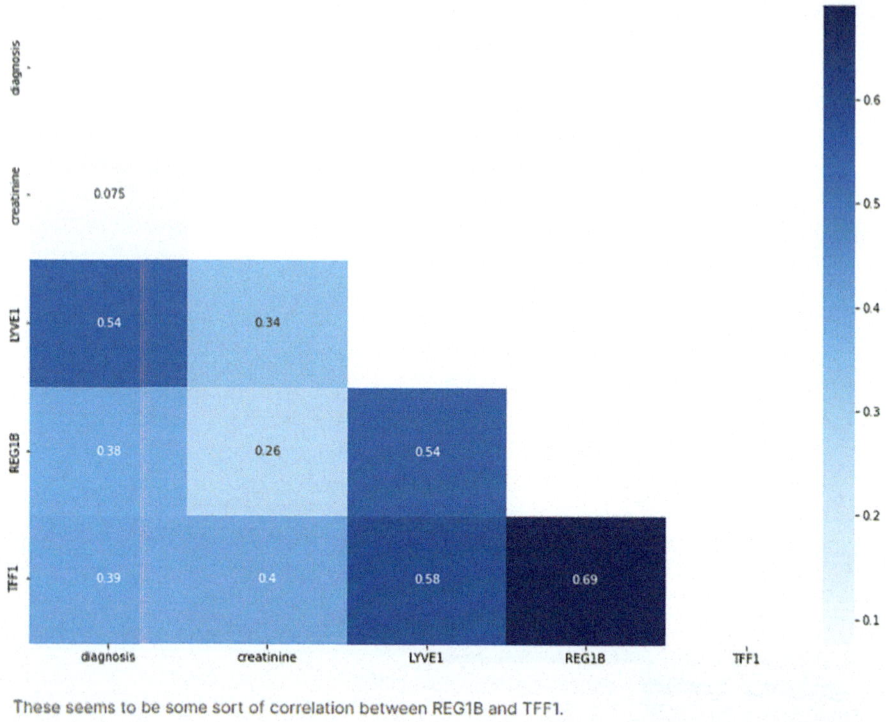

Fig. 8 Heatmap

Table 1 Gaussian NB result

Metrics	Precision	Recall	F1-score	Support
1	0.52	0.93	0.67	55
2	0.38	0.26	0.31	62
3	0.76	0.47	0.58	60
Accuracy			0.54	177
Macro-avg.	0.55	0.55	0.52	177
Weighted avg.	0.55	0.54	0.51	177

is achieved by AdaBoost, followed by CNN, then Gaussian Naive Bayes, and lastly Bernoulli Naive Bayes. The highest F1-score is achieved by CNN, followed by Gaussian Naive Bayes, then Bernoulli Naive Bayes, and lastly AdaBoost.

Table 2 Bernoulli NB result

Metrics	Precision	Recall	F1-score	Support
1	0.52	0.87	0.65	55
2	0.44	0.19	0.27	62
3	0.74	0.72	0.73	60
Accuracy			0.58	177
Macro-avg.	0.57	0.59	0.55	177
Weighted avg.	0.57	0.58	0.54	177

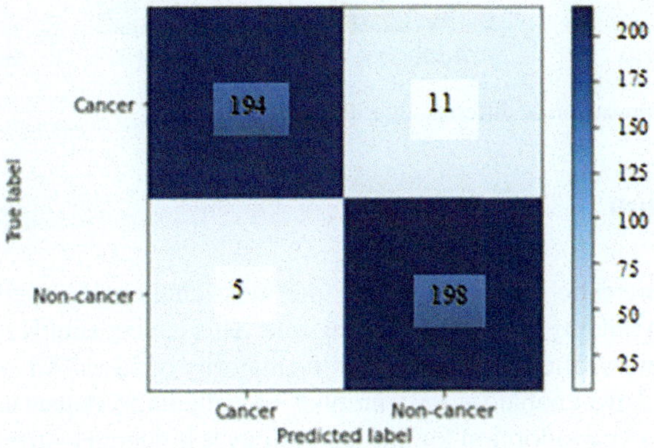

Fig. 9 Confusion matrix

Table 3 Performance metrics of various algorithms

	Accuracy	Precision	Recall	F1-score
CNN	96	94	97	95
AdaBoost	61	35	1	52
Bernoulli NB	58	52	87	65
Gaussian NB	54	52	93	67

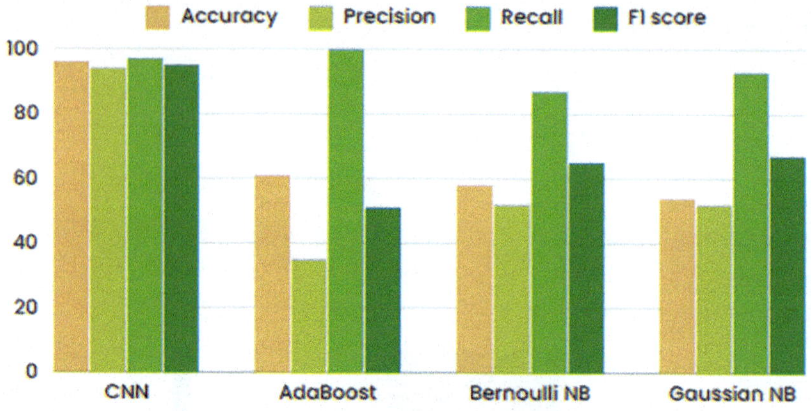

Fig. 10 Graph comparison of different algorithms

9 Conclusion

Diagnosing pancreatic cancer is among the most challenging medical conditions globally, with a notably low 5-years living rate. This cancer stands out as one of the most severe tumors, often responsible for the majority of cases. Advances in medical imaging scans have enabled a vast number of patients to detect at earlier stages. However, widespread adoption for this technology is hindered by the high costs associated with the necessary equipment and infrastructure, limiting accessibility. Here, we explores application of CNN, Naive Bayes, AdaBoost, and image processing to identify cancer in CT scan images of pancreas. The study employs different activation function, such as ReLU and sigmoid, and dropout function is used to counter overfitting in CNN algorithm. Different pancreatic biomarkers are analyzed in Gaussian and Bernoulli Naïve Bayes algorithms and AdaBoost for effective classification.

References

1. Chen PT, Wu T, Wang P, Chang D, Liu KL, Wu MS et al (2023) Pancreatic cancer detection on CT scans with deep learning: a nationwide population-based study. Radiology 306(1):172–182
2. Si K, Xue Y, Yu X, Zhu X, Li Q, Gong W et al (2021) Fully end-to-end deep-learning-based diagnosis of pancreatic tumors. Theranostics 11(4):1982
3. Qureshi TA, Gaddam S, Wachsman AM, Wang L, Azab L, Asadpour V et al (2022) Predicting PDAC using AI analysis of pre-diagnostic computed tomography images. Cancer Biomark 33(2):211–217
4. Karmazanovsky G, Gruzdev I, Tikhonova V, Kondratyev E, Revishvili A (2021) Computed tomography-based radiomics approach in pancreatic tumors characterization. La Radiol Med 126(11):1388–1395
5. Rhee H, Park MS (2021) The role of imaging in current treatment strategies for pancreatic adenocarcinoma. Korean J Radiol 22(1):23

6. Granata V, Grassi R, Fusco R, Galdiero R, Setola SV, Palaia R et al (2021) Pancreatic cancer detection and characterization: state of the art and radiomics. Eur Rev Med Pharmacol Sci 25(10)
7. Chu LC, Park S, Kawamoto S, Yuille AL, Hruban RH, Fishman EK (2021) Pancreatic cancer imaging: a new look at an old problem. Curr Probl Diagn Radiol 50(4):540–550
8. Miura S, Takikawa T, Kikuta K, Hamada S, Kume K, Yoshida N et al (2021) Focal parenchymal atrophy of the pancreas is frequently observed on pre-diagnostic computed tomography in patients with pancreatic cancer: a case-control study. Diagnostics 11(9):1693
9. Alabousi M, McInnes MD, Salameh JP, Satkunasingham J, Kagoma YK, Ruo L et al (2021) MRI vs. CT for the detection of liver metastases in patients with pancreatic carcinoma: a comparative diagnostic test accuracy systematic review and meta-analysis. J Magn Reson Imaging 53(1):38–48
10. Dhruv B, Mittal N, Modi M (2021) Early and precise detection of pancreatic tumor by hybrid approach with edge detection and AI techniques. EAI Endors Trans Perv Health Technol 7(28):el
11. Jasti V, Zamani A, Arumugam K, Naved M (2022) Computational technique based on ML and image processing for medical image analysis of breast cancer diagnosis. Secur Commun Netw 2022:7. https://doi.org/10.1155/2022/1918379.1918379
12. Chaudhury S, Krishna AN, Gupta S, Sankaran KS, Khan S, Sau K, Raghuvanshi A, Sammy F (2022) Effective image processing and segmentation-based ML techniques for diagnosis of breast cancer. Comput Math Methods Med 2022:6. https://doi.org/10.1155/2022/6841334.684 13346841334
13. Zamani AS, Anand L, Rane K, Prabhu P (2022) Performance of ML and image processing in plant leaf disease detection. J Food Qual 2022:7. https://doi.org/10.1155/2022/1598796.159 87961598796
14. Li X, Guo R, Lu J, Chen T, Qian X (2023) Causality-driven graph neural network for early diagnosis of pancreatic cancer in non-contrast computerized tomography. IEEE Trans Med Imaging 2023:1. https://doi.org/10.1109/TMI.2023.3236162
15. Ramesh JVN, Abirami T, Gopalakrishnan T, Narayanasamy K, Ishak MK, Karim FK, Mostafa SM, Allakany A (2023) Sparrow search algorithm with stacked deep learning based medical image analysis for pancreatic cancer detection and classification. IEEE Access 2023:10. https://doi.org/10.1109/ACCESS.2023.3322376
16. Miura F, Takada T, Amano H, Yoshida M, Furui S, Takeshita K (2006) Diagnosis of pancreatic cancer in HPB. Elsevier BV 8(5):337–342
17. Zavalsız MT, Alhajj S, Sailunaz K, Özyer T, Alhajj R (2023) Pancreatic tumor detection by CNN. IEEE 2023:1. https://doi.org/10.1109/ACIT57182.2022.9994181
18. Jagota V, Luthra M, Bhola J, Sharma A, Shabaz M (2022) A secure energy-aware game theory (SEGaT) mechanism for coordination in WSANs. Int J Swarm Intell Res (IJSIR) 13(2):1–16. https://doi.org/10.4018/ijsir.287549
19. Dodda K, Muneeswari G (2023) Pancreatic cancer detection through hyperparameter tuning and ensemble methods. Int J Adv Comput Sci Appl (IJACSA)

Combining Particle Swarm Optimization with Bacterial Foraging Optimization for Color Image Watermarking

Kavitha Soppari, N. Subhash Chandra, Swetha Sailaja Kuppa, and Mahendar Jinukala

Abstract Hybrid Meta-heuristic Color Image Watermarking (HMHCIW) algorithm is used to enhance the protection and authentication of digital color photographs. Traditional methods have primarily focused on grayscale images, but this study addresses the need for advanced techniques applicable to color images. The HMHCIW algorithm employs a combination of nature-inspired meta-heuristic techniques to optimize the embedding of watermarks, aiming to safeguard ownership rights effectively. In this study, the algorithm embeds an authenticated watermark into a color image and subsequently extracts the watermark to verify the authenticity of the image. The use of the BFPSO (Bacterial Foraging optimization and Particle Swarm Optimization) algorithm plays a crucial role in optimizing the coefficients within the transform domain, which is essential for the watermark embedding process. This optimization is intended to enhance the Peak Signal-to-Noise Ratio (PSNR), ensuring the watermarked image maintains high visual quality while being robust against attacks. By integrating these advanced optimization techniques, the proposed HMHCIW algorithm aims to improve the reliability and efficiency of digital rights management for color photographs, offering a significant advancement over previous methods focused on grayscale images. The study highlights the effectiveness of nature-inspired algorithms in addressing complex digital watermarking challenges, making a substantial contribution to the field of digital content protection.

Keywords Digital watermarking · Particle Swarm Optimization · Bacterial Foraging Optimization

K. Soppari (✉) · S. S. Kuppa
ACE Engineering College, Ankushapur, India
e-mail: kavithasoppari@aceec.ac.in

N. Subhash Chandra
CVR College of Engineering, Ibrahimpatnam, India
e-mail: subhashchandra@cvr.ac.in

M. Jinukala
MRCET, Secunderabad, India

© The Author(s), under exclusive license to Springer Nature Singapore Pte Ltd. 2025

H. Zen et al. (eds.), *Soft Computing and Signal Processing*, Lecture Notes in Networks and Systems 1200, https://doi.org/10.1007/978-981-97-9926-8_43

1 Introduction

Image watermarking involves embedding hidden information within an image to withstand various attacks and alterations, such as editing, compression, and noise [1]. This concealed data, known as a watermark, is used for purposes like safeguarding copyright, verifying authenticity, and retrieving images. Specifically, color image watermarking adds a watermark to color images while maintaining their quality [2]. As the use of digital photographs increases, there is a growing demand for more secure and reliable image watermarking solutions.

Optimization techniques like Bacterial Foraging Optimization (BFO) [3] and Particle Swarm Optimization (PSO) [4] are widely applied in fields such as engineering design, pattern recognition, and image processing. PSO is inspired by the social behavior of birds, while BFO is derived from the behavior of bacteria in nature. These methods are used in image watermarking to find the optimal position and size for the watermark within the image. However, individual optimization approaches can face challenges like local minima and slow convergence. To address these issues, hybrid meta-heuristic algorithms combine multiple optimization techniques, enhancing the efficiency and effectiveness of image watermarking.

The primary contribution of this paper is the development of the HMHCIW (Hybrid Meta-heuristic Color Image Watermarking) algorithm, designed to enhance the protection and authentication of digital color photographs. Unlike traditional methods that concentrate on grayscale images, the HMHCIW algorithm utilizes a combination of Bacterial Foraging Optimization (BFO) and Particle Swarm Optimization (PSO) to optimize watermark embedding in the transform domain. This dual optimization approach improves the Peak Signal-to-Noise Ratio (PSNR), ensuring that the watermarked images retain high visual quality while being resilient to various attacks. By implementing these advanced nature-inspired meta-heuristic techniques, the study significantly advances digital rights management for color photographs, offering a robust solution for complex digital watermarking challenges and making a notable contribution to digital content protection.

2 Literature Review

In 2018, Ali and associates [5] presented a reliable and effective photo watermarking method based on QIM and PSO. Discrete wavelet transforms the original image in the first step of the suggested method. A reliable and effective picture watermarking method based on QIM and PSO was presented by Bhajantri et al. [6] in 2018. The original image is first subjected to a DWT, and then the watermark is injected using QIM into the high-frequency DWT coefficients. PSO is used to determine the ideal places for watermark insertion and to maximize embedding strength [7]. This method's ability to resist conventional image processing techniques including filtering, noise addition, and compression is one of its advantages. The ideal places for watermark embedding are determined using PSO, which enhances watermarking performance. Additionally, the approach yields good imperceptibility, indicating that the watermarked image resembles the original image. Because of these drawbacks, a high embedding strength is necessary to ensure endurance, which could degrade how the image with the watermark looks. The program's color image quality hasn't been assessed.

In 2018, Liu and colleagues [7] proposed a reliable method of blind digital photo watermarking using SVD and DWT. The host image is divided into non-overlapping blocks in order to produce a watermark. Each block is then subjected to SVD and DWT. In this technique, the watermark is entrenched. After that, each block's diagonal SVD components receive the watermark. The extraction approach locates the watermark in the watermarked image by using a correlation detector. The experimental results show that the proposed method maintains a high degree of perceptual quality and is robust against various attacks such as JPEG compression, noise addition, and filtering. However, because embedding and removing watermarks demands a significant amount of computational effort, the suggested approach is less appropriate for real-time applications [8].

In 2017, Zhang and associates presented two effective picture watermarking methods: the DE method and SVD [9]. After SVD conversion, the watermark is appended to the high-frequency coefficients of the host image. The best places to include the watermark are determined using the DE approach. Comparing the suggested strategy to several existing watermarking techniques, the results demonstrate enhanced resilience and increased invisibility. This method's strength is its resistance to geometric attacks, standard image processing techniques, compression, and filtering. The drawback of this strategy is that it takes longer to embed than some other methods.

In 2017, Benmohammed and associates published a powerful picture watermarking technique that integrates the lifting wavelet transform (LWT), MOPSO method, and SVD [10]. The low-frequency coefficients of the LWT-SVD converted host picture are enhanced with the watermark. To improve the watermark embedding settings, the MOPSO algorithm is applied. The suggested approach is assessed using a sizable collection of reference images and contrasted with various watermarking methods currently in use. The outcomes demonstrate that it offers good resilience and outstanding invisibility. One benefit of this approach is that it yields reliable results when applied to several image processing operations, including noise addition, compression, and filtering. The drawback of this approach is that it can be vulnerable to particular geometric attacks and processing methods. The drawback of this strategy is that it takes longer to embed than some other methods.

A reliable photo watermarking technique utilizing the firefly algorithm in conjunction presented using SVD and DCT by Yang et al. in 2018 [11]. After transforming the host picture with DCT-SVD, the watermark is appended to its low-frequency coefficients. The firefly method is utilized to improve the parameters of watermark embedding. The outcomes demonstrate that, in comparison with several other watermarking methods already in use, the recommended approach offers greater invisibility and enhanced endurance. The benefit of this approach is that it generates strong defenses against a variety of image processing methods, including filtering, compression, and geometric attacks. The drawback of this approach is that noise addition assaults might be possible. Strong resistance to a variety of image processing methods, including filtering, compression, and noise addition. The drawback of this approach is that it can be vulnerable to particular geometric attacks.

In 2019, Abd El-Latif et al. [12] introduced a novel method of digital watermarking digital images. It employs a disorganized map. The suggested technique uses chaotic maps to produce random integers, which are then utilized to locate the embedding spots and jumble the watermark message. Experiments conducted on several standard photographs demonstrate how robust the system is against several types of attacks, such as noise addition, compression, and cropping. One benefit is having a strong defense against all types of attacks. Using chaotic maps to generate random numbers increases security. Because of its significant computational complexity, the suggested method might not be appropriate for real-time applications.

A unique adaptive digital image watermarking technique utilizing DWT, DCT, SVD, and chaotic map was proposed by Santhi et al. [13] in 2018. Depending on the picture attributes, the method dynamically selects the best combination of DWT, DCT, and SVD to achieve the best watermarking performance. The method was tested with several typical images, and the results show that it is very resilient to various attacks, such as compression, filtering, and cropping. Strong defense against various attacks and the option to select the best combination of DWT, DCT, and SVD depending on the image's characteristics are among the advantages. The recommended algorithm's shortcomings include the possibility that its high computational complexity makes it unsuitable for real-time applications.

A color image watermarking method using IWT and SVD was described in 2017 by Sudhakar and colleagues [14]. To obtain the LL sub-band, the procedure first employs IWT on the host image. This sub-band is then subjected to SVD in order to identify the locations where the watermark should be embedded. The method was tested on a number of common color photos, and the results show that it is very resilient to various attacks, such as compression, filtering, and cropping. Strong resistance to a variety of attacks is one of the advantages, and using SVD and IWT enhances security. The recommended method's shortcomings include the possibility that its high computational complexity makes it unsuitable for real-time applications.

A model that combines PSO and BFO methods to optimize the watermark embedding process was proposed by Abdalla et al. [15] in 2017. The watermark is added to the cover image by dividing it into various portions and using the best strength for each section. The advantages of the hybrid optimization approach make it possible for the watermark embedding process to be more successful and efficient, which increases resistance to various types of attacks. However, because of the computational complexity of the optimization process, the limitations of the proposed method might not be appropriate for real-time applications.

A hybrid photo watermarking technology using particle swarm optimization and bacterial swarm optimization was proposed by Qi et al. [16] in 2019. The watermark was implanted using bacterial swarm optimization and extracted using particle swarm optimization within the suggested framework. The testing's findings demonstrated that the suggested model outperformed a number of the currently used picture watermarking models. This method has the benefit of being highly resilient against a variety of attacks, including but not limited to cropping, noise addition, JPEG compression, and geometric distortions. The watermarked image appears to be nearly identical to the original due to the great degree of invisibility of the process. The system's overall efficiency is raised by combining particle swarm optimization and bacterial swarm optimization techniques to create a more effective watermark embedding process. Since the method is objective, it is possible to extract the watermark without having to comprehend the original image. One drawback of the suggested method is that it necessitates specifying a lot of parameters, which might make system optimization and fine-tuning challenging specific use cases. Due to higher computing costs, using two optimization algorithms may cause the process to move more slowly than with other watermarking methods. Because the approach does not support encryption or authentication, it might not be suitable for some uses that require a high level of security.

3 Proposed Color Image Watermarking Using Optimization by Hybrid Meta-heuristic Algorithm Based on BFO and PSO

3.1 Proposed Methodology

The reason for utilizing optimization in color image watermarking through a hybrid meta-heuristic algorithm based on BFO and PSO is due to the complexity of color images compared to grayscale images. As a result, more sophisticated methods are required to effectively embed a watermark without sacrificing the image's quality. Hybrid meta-heuristic algorithms [17] can also be used to improve the effectiveness of the watermarking process and help overcome the limitations of traditional optimization methods. The hybrid technique [18] combines the benefits of BFO and PSO [19] to discover the best position and scale of the watermark in the image, resulting in enhanced performance and resilience compared to utilizing only one optimization algorithm.

Enhancing Bacterial Foraging: The optimization method known as Bacterial Foraging Optimization (BFO) is based on the way bacteria forage for food. Using chemo taxis, reproduction, elimination, and dispersal processes to improve the solution, BFO simulates the movement of bacteria in the solution area. Picture watermarking is one of the numerous optimization problems for which BFO has been applied, and it has shown to be effective in locating the watermark's ideal location inside the image.

Particle Swarm Optimization: An optimization system known as PSO was inspired by fish schools and bird flocks. In order to improve the solution, PSO models the motion of particles inside the solution space by varying their position and velocity.

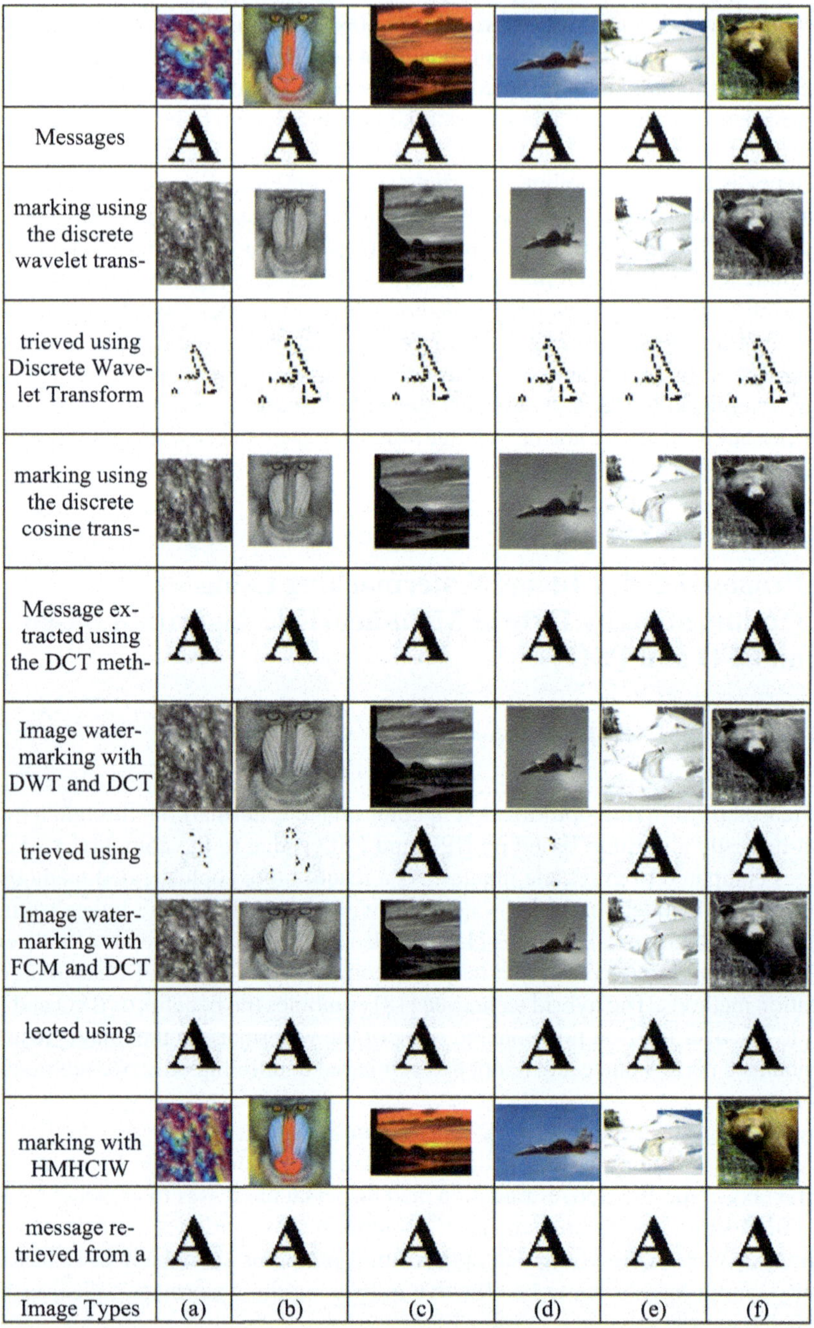

Fig. 1 The outcomes of the invented watermarking experiment using retrieved messages and conventional models like DWT, DCT, DWT + DCT, and FCM + DCT in addition to the proposed HMHCIW

Picture watermarking has made heavy use of PSO, which has been useful in figuring out the best watermark scale for the image.

Algorithm that Combines Meta-heuristic Techniques [20–24]: To determine the ideal location and scale of the watermark within the image, the hybrid meta-heuristic algorithm for color image watermarking combines the advantages of BFO and PSO. The ideal location and size for the watermark in the given image are determined by combining the benefits of two popular optimization methods, PSO and BFO, in this hybrid approach to color picture watermarking. This method's goal is to insert the watermark in a way that maximizes its invisibility while maintaining the watermarked image's quality in terms of PSNR. In the original color image, we select only blue channel to insert the watermark. DCT is applied on the blue channel pixel values, and the following embedding process is applied to identify the location of embedding. After embedding, inverse DCT is applied.

3.2 Watermark Embedding

Bacterial Foraging Optimization (BFO) Stage: This phase uses BFO to identify the best location for the watermark in the image. BFO employs a collection of microorganisms representing possible solutions, which move through the solution space via chemo taxis, reproduction, elimination, and dispersal. The fitness function, utilizing PSNR, evaluates the watermarked image's quality and imperceptibility to determine the optimal watermark location.

Particle Swarm Optimization (PSO) Stage: After BFO finds the optimal position, PSO determines the optimal watermark size. PSO uses a swarm of particles, each representing a potential solution, which improves their solutions by adjusting their positions and velocities based on individual and collective best solutions. The fitness function again evaluates the watermarked image using PSNR to find the ideal watermark size. Combining BFO's exploration capabilities with PSO's exploitation strengths, the algorithm refines particle positions and velocities to determine the optimal watermark size and location through iterative optimization.

Result and Insertion: The final output is the optimal watermark location and size. The watermark is embedded in the original image using these parameters, and the quality of the watermarking process is assessed using PSNR, SSIM, and NC on the watermarked image. This hybrid meta-heuristic method effectively determines the optimal placement and size of watermarks, ensuring high quality and imperceptibility, and can be adapted to various watermarking scenarios based on specific application requirements. Algorithm 3.1 provides a step-by-step explanation of the algorithm.

Algorithm 3.1: HMHCIW Watermark Embedding Input: Number of particles

(N), maximum number of iterations (MaxIter), watermark image (W), initial color picture (I), and optimization parameters ($c1$, $c2$).

a. To start with. Create N particles, each of which represents a possible resolution for the image's watermark's position and scale.
b. Randomly start each particle's position and velocity.
c. To determine the quality of the watermarked image, compare the original and watermarked image in terms of PSNR, SSIM, and NC. Maximizing PSNR is the objective function for each particle.
d. Fourth, adjust each particle's personal best solution in accordance with the objective function value.
e. Revise the overall best answer by taking into account each particle's own best solution.

Optimization Loop: Carry out the subsequent steps again until the required amount of iterations is reached or the convergence condition is satisfied.

a. Using the following formula, update each particle's velocity based on its current position, velocity, global best solution, and personal best solution: $w * vi(t) + c1 * \text{rand1} * (pi - xi) + c2 * \text{rand2} * (g - xi)$ equals $vi(t + 1)$, where t is the global best solution, g is the personal best solution, w is the inertia weight, vi is the particle's velocity, xi is its current position, $c1$ and $c2$ are the optimization parameters, and rand1 and rand2 are random values between 0 and 1.
b. Using the following formula, update each particle's position based on the revised velocity: $vi(t + 1) + xi(t)$ equals $xi(t + 1)$.
c. Using the updated position, evaluate each particle's objective function.
d. Using the updated objective function value as a basis, update each particle's personal best solution.
e. Make revisions to the overall optimal solution based on each particle's unique ideal solution.

Result: The optimal location and size of the watermark within the image are represented by the global top solution.

3.3 *Watermark Extraction*

Watermark extraction is a crucial phase in the watermarking process, involving the removal of the embedded watermark from the color image. This procedure begins by applying DCT on the blue channel of watermarked image and utilizing the watermarked image, along with the optimal position and scale information determined during the watermark embedding stage. The extraction process uses the optimal position (x_optimal) to locate the initial top-left corner of the watermark area within the watermarked image, and the optimal scale (s_optimal) to determine the final coordinates (bottom-right corner) of the watermark area, corresponding to where the watermark was initially embedded.

Once the watermark area is located, it is extracted from the watermarked image using the specified coordinates and resized to match the original watermark dimensions using suitable interpolation techniques, such as nearest-neighbor, bilinear, or bicubic interpolation. The pixel values of the extracted watermark are then normalized to fit within a valid range (e.g., 0–255 for an 8-bit grayscale image) to ensure accuracy and high quality. The resulting extracted watermark (W_extracted) is then compared to the original watermark (W) to evaluate the efficacy of the watermarking procedure, with the similarity between the two indicating the robustness and invisibility of the watermarking method. The outline for watermark extraction is detailed in Algorithm 3.2.

Algorithm 3.2: HMHCIW Watermark Extraction

1. Watermarked colored image (I_watermarked), Best position (x_optimal), Best scale (s_optimal), Watermark image (W), and Watermark size (W_size).
2. Change the watermarked image I_watermarked to grayscale if necessary (to match the format of the watermark picture W).
3. Create a blank picture W_extracted with the identical dimensions as W.
4. Specify the initial coordinates (x_start, y_start) of the watermark region, using x_optimal as a reference.
5. Using W_size and s_optimal, ascertain the watermark area's final coordinates (x_end, y_end).
6. Utilize the (x_start, y_start) and (x_end, y_end) coordinates to obtain the designated area from the marked image I_marked.
7. Adjust the size of the retrieved watermark area to match the dimensions of the original watermark picture W using an appropriate interpolation technique.
8. Adjust the pixel values of the recovered watermark to make sure they fall within the acceptable range (for example, between 0 and 255).
9. Save the standardized extracted watermark in the W_extracted picture.
10. Give back W_extracted.

4 Tests and Findings

Figure 1 shows the experimental results of watermarked photos and extracted images of various image types, including multi-design, classic, dark, fine detail, line, and smooth images.

We conducted trials on a set of color images, altering the watermarks' strengths and sizes, to evaluate the efficacy of a hybrid meta-heuristic technique for color image watermarking. The results, displayed in Fig. 1 and Table 1, show that in terms of PSNR, SSIM, NC, and robustness against various attacks, the hybrid meta-heuristic method outperformed the standalone BFO and PSO algorithms. The hybrid method produced higher PSNR, SSIM, and NC values and showed increased resilience to noise, compression, and image alteration, demonstrating its superior ability to

Table 1 Comparative examination of different images shown in Fig. 1 using HMHCIW clustering

Image type															
Error measures	DCT			DWT			DCT + DWT			FCM			HMHCIW		
	(a)	(b)	(c)	(a)	(b)	(c)	(a)	(b)	(c)	(a)	(b)	(c)	(a)	(b)	(c)
PSNR	41.41	41.41	41.41	23.1	23.1	24.5	42.5	42.4	43.0	42	42	42	44	43	44
SSIM	0.94	0.93	0.93	0.76	0.74	0.73	0.98	0.97	0.92	0.98	0.97	0.96	1.0	0.99	1.0
NC	0.98	0.956	0.964	0.984	0.956	0.964	0.984	0.956	0.964	0.984	0.956	0.964	0.984	0.956	0.964
	(d)	(e)	(f)	(d)	(e)	(f)	(d)	(e)	(f)	(d)	(e)	(f)	(d)	(e)	(f)
PSNR	41.41	41.41	41.41	23.27	23.1	23.6	42.41	42.41	42.41	42.41	42.41	42.41	44	42	42
SSIM	0.90	0.99	0.98	0.73	0.73	0.75	0.98	0.97	1.0	0.98	0.97	096	1.0	1.0	1.0
NC	0.98	0.956	0.964	0.984	0.956	0.964	0.984	0.956	0.964	0.984	0.956	0.964	0.984	0.956	0.964

determine the optimal position and size of the watermark compared to individual techniques.

The findings indicated that the hybrid algorithm reached higher quality solutions more quickly. The application of the hybrid PSO-BFO approach to watermarking was evaluated using metrics such as PSNR, SSIM, and NC. PSNR measures the quality of the watermarked image, with higher values indicating better quality. SSIM assesses the structural similarity between the original and watermarked images, with values closer to 1 indicating high similarity. NC measures the correlation between the original and watermarked images, with higher values indicating a higher degree of similarity.

5 Conclusion

In this study, we used Particle Swarm Optimization (PSO) and Bacterial Foraging Optimization (BFO) to develop a combined meta-heuristic strategy for color picture watermarking. The hybrid approach outperforms using a single optimization method by combining the advantages of BFO and PSO to determine the optimal location and size of the watermark within the image.

Based on the tests and findings, the hybrid meta-heuristic algorithm outperformed the standalone BFO and PSO algorithms in terms of resilience against various attacks, PSNR, SSIM, and NC. Compared to the individual methodologies, the hybrid method produced better PSNR, SSIM, and NC values and demonstrated greater robustness against noise, compression, and picture manipulation.

References

1. Cox IJ, Miller ML, Bloom JA (2002) Embedding digital marks. Morgan Kaufmann
2. Singh AK, Sharma AK (2018) An overview of digital watermarking: methods and uses. Multimed Tools Appl 77(14):17975–18006
3. Das S, Abraham A, Konar A (2013) Theoretical basis, analysis, and applications of the bacterial foraging optimization method
4. Kennedy J, Eberhart R (1995) Swarm intelligence optimization. In: Records of the IEEE international conference on neural networks, pp 1942–1948
5. Ali S, El-Melegy MA (2018) A strong and effective image watermarking method based on QIM and particle swarm optimization. Signal Image Video Process 12(6):1047–1054
6. Bhajantri NN, Uplaonkar MV (2018) An effective digital image watermarking method using DWT-SVD and firefly algorithm. Multimed Tools Appl 77(14):17991–18010
7. Liu J, Zhang H, Hu Z (2018) A strong blind digital image watermarking method based on singular value decomposition and DWT. Math Issues Eng 2018. https://doi.org/10.1155/2018/5459623
8. Shen J, Li X, Li W (2019) A strong combination watermarking system using the DWT-DCT-SVD and PSO. Multimed Tools Appl 78(5):5545–5566

9. Ye J, Zhang Y, Liu Y (2017) An effective image watermarking technique employing singular value decomposition and differential evolution algorithm. Multimed Tools Appl 76(20):21087–21101. https://doi.org/10.1007/s11042-017-4615-7
10. Benmohammed S, Boukrouche A, Bouguila N (2017) Strong picture watermarking utilizing LWT-SVD and multi-objective particle swarm optimization. Multimed Tools Appl 76(5):6455–6476. https://doi.org/10.1007/s11042-016-3844-4
11. Yang L, Xu Q (2018) Strong image watermarking using DCT-SVD with firefly algorithm. Multimed Tools Appl 77(5):5845–5861
12. Abd El-Latif SS, Ali NH (2019) Algorithm for embedding digital watermarks in digital images using a chaotic map. Multimed Tools Appl 78(22):31827–31852
13. Santhi SVNUN, Hemalatha M (2018) A novel flexible digital picture watermarking method using DWT-DCT-SVD and chaotic map. J Ambient Intell Humaniz Comput 9(2):357–367. https://doi.org/10.1007/s12652-017-0532-4
14. Sudhakar, Punithavalli (2017) An image watermarking approach for color images using the integer wavelet transform and singular value decomposition. Signal Image Video Process 11(7):1201–1208. https://doi.org/10.1007/s11760-017-1086-3
15. Abdalla MM, Fouad MM, Badr WM (2017) Model for watermarking hybrid images using bacterial swarm optimization and particle swarm optimization. J Ambient Intell Humaniz Comput 8(2):297–313
16. Qi F, Wang J, Wang X, Xu J (2019) A hybrid picture watermarking model using bacterial swarm optimization and particle swarm optimization. Multimed Tools Appl 78(15):22021–22039
17. Lin Q, Xiao L (2020) A technique for adding watermarks to images using a combination of a chaotic map and a hybrid optimization algorithm. Signal Process Image Commun 85:115881
18. Kumar V, Kumar R (2021) A new combination watermarking method for medical images using bacterial colony optimization and quantum particle swarm optimization. Int J Comput Intell Healthc Inform 11(1):1–22
19. Raman B, Ramalingam V (2020) Hybrid technique for embedding watermarks in color photos using discrete wavelet transform (DWT), discrete cosine transform (DCT), and singular value decomposition (SVD). Multimed Tools Appl 79(9):5557–5581
20. Chen Z, Zhang Y, Li Q (2020) A new image watermarking algorithm using a combination of DCT, DWT, and SVD for color images. Multimed Tools Appl 79(17):11855–11873
21. Chen Y, He X, Hu X (2021) Hybrid technique for adding watermarks to color images
22. Zhang C, Zhang Y, Cao Y, Jiang F (2019) A strong hybrid watermarking technique in curvelet and DCT domains utilizing artificial bee colony and particle swarm optimization. Signal Process Image Commun 76:157–172
23. Yang M, Zhang J, Lu L (2019) A hybrid watermarking approach for depth photos based on 3D DWT and particle swarm optimization. Multimed Tools Appl 78(17):24563–24578
24. Singh, Khanna (2020) Hybrid technique for embedding watermarks in color photos utilizing discrete wavelet transform (DWT), discrete cosine transform (DCT), and singular value decomposition (SVD) with particle swarm optimization (PSO). J Ambient Intell Humaniz Comput 11(3):1193–1204

Fertilizer and Crop Recommendation System using Machine Learning

M. Sucharitha, P. H. V. Sesha Talpa Sai, Munmun Gorai, Mithila Mahato, C. Adarsh, Vinodkumar Hiremath, Benu Pradhan, and Amiya Bhaumik

Abstract In agriculture, efficient and precise fertilizer management is crucial for optimizing crop yield, minimizing environmental impact, and ensuring sustainable farming practices. This paper presents a comprehensive study on fertilizer recommendation systems leveraging a machine learning model named random forest algorithm. The proposed model utilizes various agronomic parameters such as soil type, nutrient content, weather conditions, and crop type to predict the optimal fertilizer requirements. The machine learning algorithm employed demonstrates a high degree of accuracy in predicting fertilizer needs, providing farmers with personalized recommendations tailored to their specific agricultural conditions. By integrating the model into a web app, this approach aims to enhance the overall efficiency of fertilizer application, reduce resource wastage, and contribute to environmentally responsible farming practices. The research findings offer valuable insights into the potential of machine learning in revolutionizing fertilizer management, fostering precision agriculture, and ultimately contributing to global food security and sustainability.

Keywords Random forest · Machine learning · Fertilizers · Crop · Prediction

M. Sucharitha
School of Electronics Engineering (SENSE), VIT-AP University, Amaravati, Andhra Pradesh, India
e-mail: sucharitha.jackson@vitap.ac.in

P. H. V. Sesha Talpa Sai (✉)
Department of Research and Development, Malla Reddy University, Hyderabad, India
e-mail: polamrajusai@gmail.com

M. Gorai · M. Mahato · C. Adarsh · V. Hiremath · B. Pradhan
School of Electrical and Electronics Engineering, Lincoln University College, Petaling Jaya, Malaysia
e-mail: munmung@nttf.co.in

M. Mahato
e-mail: mithilam@nttf.co.in

A. Bhaumik
Lincoln University College, Petaling Jaya, Malaysia
e-mail: amiya@lincoln.edu.my

© The Author(s), under exclusive license to Springer Nature Singapore Pte Ltd. 2025
H. Zen et al. (eds.), *Soft Computing and Signal Processing*, Lecture Notes in Networks and Systems 1200, https://doi.org/10.1007/978-981-97-9926-8_44

1 Introduction

Predicting crop yields is crucial to the world's food production. Predicting crop production is essential to tackling new issues in food security, especially in an era of global climate change. To maintain national food security, policymakers depend on precise estimates to help them make timely decisions about import and export. Accurate crop projections help farmers make prudent financial and operational decisions and also contribute to the prevention of famine. Dealing with crop storage, transportation choices, and crop-related risk management are also crucial. The acronym NPK, which stands for nitrogen (N), phosphorus (P), and potassium (K), refers to the three main and crucial nutrients found in soil. Since soil is essential to agriculture, the quality of crops that grow is directly impacted by the nutrients in the soil.

It becomes necessary to give good recommendations to improve the fertility of soil. The designed system is an interface presented in which the following applications will be implemented, namely fertilizer recommendation and crop yield prediction. The objective is to develop a fertilizer recommendation system using machine learning and techniques for crop growth by taking factors like the nitrogen–phosphorus–potassium (NPK) values of the soil, etc., to build a simpler mechanism to recommend fertilizer that are suitable for that soil. To develop a prediction system for crop yield by taking factors like the year, location, crop type, etc., so that it will help farmers make informed economic and management decisions about the yield of the crop.

2 Related Work

The Indian economy now receives a significant boost from the agriculture sector. Developments in the agriculture sector are necessary to meet the needs of a nation like India, where the population growth has led to a constant increase in the need for food. For this reason, crop yield prediction is still a difficult challenge in this field. Crop output is influenced by a number of factors, including temperature, precipitation, pesticides, fertilizers, pH level, and other meteorological factors [1].

All farming record styles can be accommodated by the selection tree proposed by Wu et al. [2]. For agricultural data, a decision tree classifier was recommended. It can handle every record and uses up-to-date data. The 10-fold cross-validation approach is used to verify the horse-colic and soy datasets. Vardhan and Chowdary [3], an ID3 algorithm has been created to produce tomato crops more effectively. It makes use of CSV datasets and runs on the PHP framework. This study makes use of a number of factors, including humidity, temperature, area, and tomato crop output.

Gandhi et al. [4] presented the SVM for rice crop yield prediction. Numerous variables are included in the dataset used in this method, such as location, temperature, precipitation, and manufacturing. On this dataset, a sequential minimal optimization classifier has been applied. To construct a set of rules for the given dataset, they first prepared it using the Weka tool. The SVM method was used to generate the findings in Python.

Numerous scholars have illustrated how climatic anomalies affect crops by demonstrating the relationship between weather factors and crop productivity. A few of these studies have gone so far as to identify temperature, precipitation, and water thresholds that plants need to develop [5–8].

Tanmay et al. [9] processed the pest's image both continuously and discretely in order to administer the suggested insecticide. They did this by using the dual operators, transition probability function (TPF) and convolution neural network (CNN). The goal function was included in the mathematical model that was developed. Using CNN and machine vision, the system integrated two key components of farming: pest recognition and insecticide recommendation. Second, a soil NPK sensor is used for the soil nutrient analysis, and fertilizers are recommended based on the measured nutrient values.

3 Methodology

The suggested approach will use soil composition, temperature, humidity, soil pH, and rainfall data to forecast the best crop and fertilizer for a given plot of land. As illustrated in Fig. 1, the dataset is refined and then fed into the model so as to avoid any issues related to outliers.

3.1 Dataset Pre-processing

The dataset was meticulously curated by merging and augmenting existing datasets related to rainfall, climate, and fertilizer data, specifically tailored for the agricultural landscape of India. This Kaggle downloaded dataset serves as a foundation for developing machine learning models that can contribute to precision agriculture and

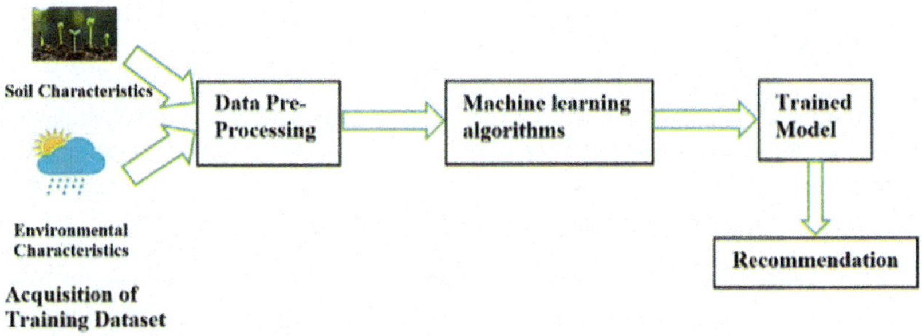

Fig. 1 Working of proposed system

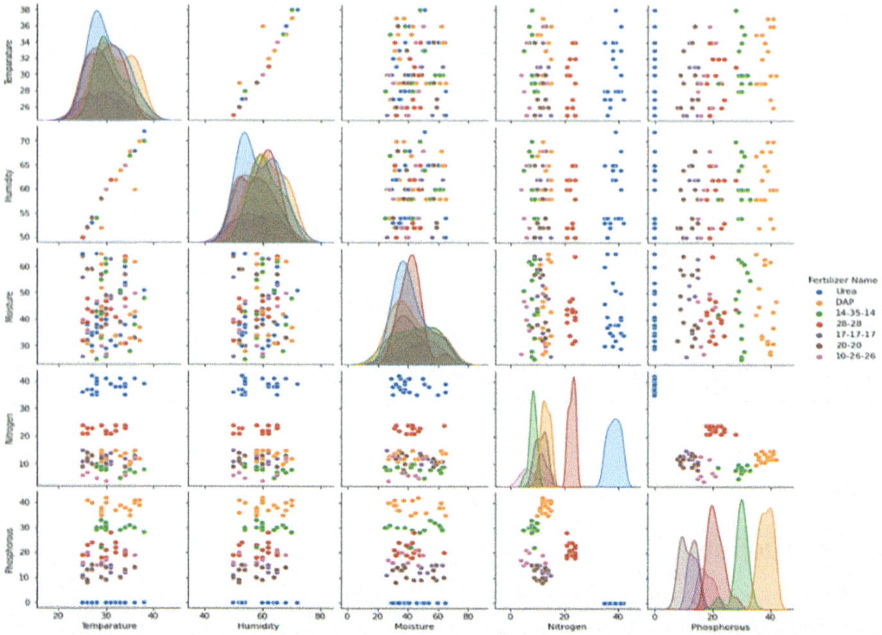

Fig. 2 Pair plot representing features

assist farmers in making optimal decisions regarding crop selection and resource management.

The dataset encompasses several key features critical to agricultural decision-making. These features include the ratios of nitrogen (N), phosphorous (P), and potassium (K) content in the soil—essential nutrients for plant growth. Additionally, the dataset includes temperature readings in degrees Celsius, relative humidity in percentage, soil pH values, and rainfall measurements in millimetres. These parameters as in Fig. 2 collectively represent a comprehensive set of variables that influence crop growth and development [10, 11].

3.1.1 Fertilizer Dataset

The fertilizer recommendation dataset is a type of agricultural dataset that contains information on the recommended fertilizer types for different crops based on various environmental factors.

The dataset typically contains the following columns:

1. Temperature: This column records the temperature of the environment in which the crop is grown. Temperature is an important factor that can affect the growth and yield of crops.

2. Humidity: This column records the humidity of the environment in which the crop is grown. Humidity is another important factor that can affect the growth and yield of crops.
3. Moisture: This column records the moisture level of the soil in which the crop is grown. Soil moisture is a critical factor that affects the growth and development of crops.
4. Soil type: This column records the type of soil in which the crop is grown. Different soil types have different nutrient and water holding capacities, which can affect the growth and yield of crops.
5. Crop type: This column records the type of crop that is being grown. Different crops have different nutrient requirements and thus require different types of fertilizers.
6. Nitrogen: The amount of nitrogen in the soil is noted in this column. Nitrogen is an essential nutrient for plant growth and is frequently the crop output bottleneck.
7. Potassium: This column records the level of potassium in the soil. Potassium is another important nutrient for plant growth and is often required in large amounts by crops.
8. Phosphorus: This column records the level of phosphorus in the soil. Phosphorus is a vital nutrient for plant growth since it is involved in both energy transmission and photosynthesis.
9. Fertilizer name: This column records the name of the recommended fertilizer for the given combination of environmental and crop factors.

The factors that most affect the crop and fertilizer are nitrogen, phosphorus, and potassium, these insights we were able to gather by using the plot that we designed Fig. 3.

Fig. 3 Factors influencing crop

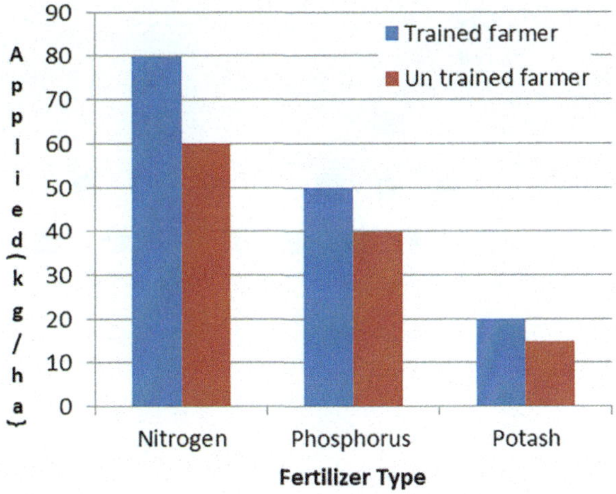

3.1.2 Crop Dataset

The data pertains to crop covered area (hectares) and production (tonnes) broken down by district, crop, season, and year.

Crop productivity, its contribution to the district, state, and nation, performance in each agro climatic zone, high yield production order for crops, crop growth patterns, and crop diversification are all being studied and analysed using the data.

3.2 Model Creation and Evaluation

Random forest algorithms are based on decision trees, which are used to make predictions by recursively splitting the data into subsets based on feature values, refer to Fig. 4. In a single decision tree, the tree is grown until it reaches its maximum size or until the predictions at each node are pure. However, single decision trees can be prone to overfitting, where they perform well on the training data but not on the test data. Random forest algorithm solves this problem by creating multiple decision trees, each trained on a different subset of the data and with a different subset of features. The final prediction is made by combining the predictions of all the trees in the forest. As a result, the model becomes less prone to overfitting and is more resilient and broadly applicable [12–15].

Terms Associated with Random Forest Algorithm

1. **Bootstrapping**: The data is sampled with replacement to create multiple training sets. Each training set will be used to train a separate decision tree.

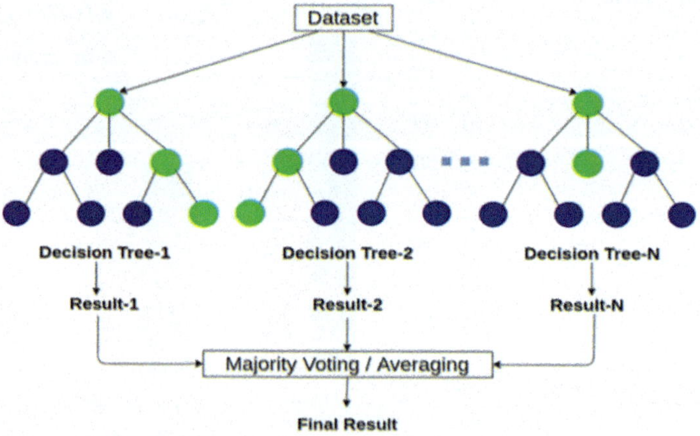

Fig. 4 Random forest algorithm

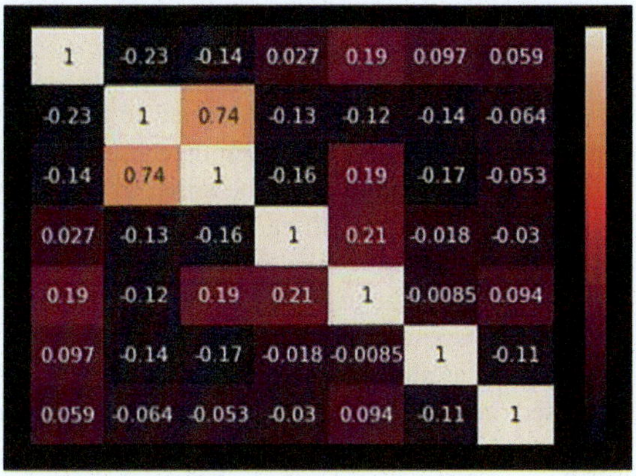

Fig. 5 Correlation matrix

2. **Feature Selection**: Only a random subset of the features in each tree is taken into account for splitting. This improves the diversity of the forest's tree population and lessens over fitting.
3. **Tree Growth**: Decision trees are grown from the bootstrapped training sets, considering only the random subset of features at each node. The trees continue growing until they reach a maximum size or until the predictions at each node are pure.
4. **Prediction**: The predictions of each tree are combined, typically by averaging for regression problems and voting for classification problems.

To understand the relation between features and make some valuable use of that dependency, we used the correlation matrix, which gave the results shown in Fig. 5.

3.3 Deployment

The user interface for the fertilizer recommendation and crop yield prediction system was built using HTML, CSS, and the Flask web framework. The standard markup language used to create web pages is called Hypertext Markup Language (HTML), and the style sheet language used to describe the appearance and formatting of an HTML document is called Cascading Style Sheets (CSS).

The Python web framework Flask offers resources and tools for creating online apps. The Flask web framework provides a way to route incoming requests to specific functions in the application using decorators. In this application, the Flask route decorator was used to create routes that correspond to the different functionalities of the application. For example, the route '/yield-predict' was used to route incoming requests for crop yield prediction, while the route '/fertilizer-predict' was used to route requests for fertilizer recommendation. After training the dataset and applying random forest classifier, we get classification accuracy of 95.8%.

4 Result Analysis

The proposed application is designed to address the issue of predicting crop yield through the use of machine learning techniques. The primary objective of this project is to develop a model that can be utilized to accurately predict crop yield and recommend the appropriate fertilizer to farmers. By doing so, this application assists farmers in making informed fertilizer-related decisions and provides them with an advance understanding of the expected crop yield.

The Web site consists of a form that needs to be filled by the user and provide basic details as shown in Fig. 6.

Based on the input provided by the user, the model calculates the precise fertilizer that most suits the features declared by the user in the form. According to the input provided, model produces results shown in Fig. 7.

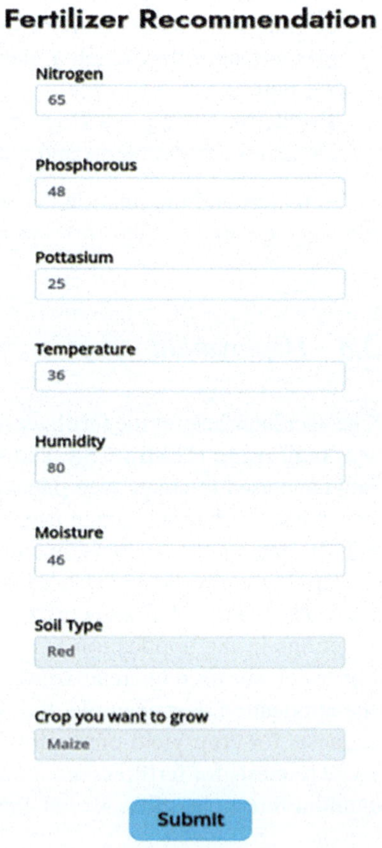

Fig. 6 Fertilizer recommendation input data form

Fig. 7 Annotations of the thumbs-up gesture

5 Conclusion

This project's purpose is to facilitate the acquisition of knowledge about various crops that can be employed to increase efficiency and profitability in the harvesting process. Accurately predicting different crop yields across various districts will enable farmers to make better decisions about their planting and harvesting strategies. This, in turn, will increase crop yield rates, ultimately contributing to the growth and development of the Indian economy. The suggested method calculates crop yields with accuracy by utilizing predictive modelling and machine learning techniques. The model is designed to analyse a wide range of variables, including weather conditions, soil type, irrigation, and crop type. By considering these variables, the model can provide recommendations on the most suitable fertilizers to be used to maximize crop yield. In summary, this application is an innovative solution to the age-old problem of predicting crop yield. By leveraging cutting-edge machine learning techniques, this project aims to empower farmers with the knowledge they need to make informed decisions that increase their crop yield rates and contribute to the overall growth of the Indian economy.

References

1. Gulati P, Jha SK (2020) Efficient crop yield prediction in India using machine learning techniques. Int J Eng Res Technol (IJERT) 8(10)
2. Wu J, Olesnikova A, Song C-H, Lee WD (2009) The development and application of decision tree for agriculture data. In: 2009 second international symposium on intelligent information technology and security informatics, Moscow, Russia, pp 16–20. https://doi.org/10.1109/IITSI.2009.10
3. IBM Research India (2020) AgroAI: AI-powered agriculture. https://www.research.ibm.com/artificialintelligence/projects/agro-ai/
4. Gandhi N, Armstrong LJ, Petkar O, Tripathy AK (2020) Rice crop yield prediction in India using support vector machines 2016. In: 13th international joint conference on computer science and software engineering (JCSSE), Khon Kaen, pp 1–5
5. Shakoor U, Saboor A, Ali I, Mohsin AQ (2011) Impact of climate change on agriculture: empirical evidence from arid region. Pak J Agric Sci 48:327–333
6. Javadinejad S, Eslamian S, Askari KOA (2021) The analysis of the most important climatic parameters affecting performance of crop variability in a changing climate. Int J Hydrol Sci Technol 11:1–25
7. Beillouin D, Schauberger B, Bastos A, Ciais P, Makowski D (2020) Impact of extreme weather conditions on European crop production in 2018: random forest—yield anomalies. Philos Trans R Soc B Biol Sci 375:20190510
8. Uleberg E, Hanssen-Bauer I, van Oort B, Dalmannsdottir S (2014) Impact of climate change on agriculture in Northern Norway and potential strategies for adaptation. Clim Change 122:27–39

9. Tanmay T, Patle BK, Kashyap SK (2023) Intelligent insecticide and fertilizer recommendation system based on TPF-CNN for smart farming. Smart Agric Technol 3:100114. ISSN 2772-3755
10. Rehman M, Agrawal SS, Jawandhiya PM (2020) A review on decision support system in smart farming. Int J Res Advent Technol 8(8):1–4. www.ijrat.org
11. Pivoto D, Waquil PD, Talamini E, Finocchio CPS, Dalla Corte VF, de Vargas Mores G (2018) Scientific development of smart farming technologies and their application in Brazil. Inf Process Agric 5(1):21–32
12. Rehman A, Liu J, Keqiu L, Mateen A, Yasin MQ (2020) Machine learning prediction analysis using IoT for smart farming. Int J Emerg Trends Eng Res 8(9):6482–6487
13. Astani M, Hasheminejad M, Vaghefi M (2022) A diverse ensemble classifier for tomato disease recognition. Comput Electron Agric 198:107054
14. Gokulnath BV (2021) Identifying and classifying plant disease using resilient LF-CNN. Ecol Inform 63:101283
15. Enkvetchakul P, Surinta O (2022) Effective data augmentation and training techniques for improving deep learning in plant leaf disease recognition. Appl Sci Eng Prog 15:3810

Plant Leaf Disease Detection System

P. H. V. Sesha Talpa Sai, M. Sucharitha, Nithul Prakash, Ezekiel Poulose, K. K. Ajesh, Aruldas Divakaran, and Amiya Bhaumik

Abstract An important part of the agricultural industry is crop production. In particular, tainted crops are largely blamed for the food loss, which naturally slows down development. In the world of agriculture, it is quite challenging to detect plant diseases. It is now feasible to produce enough food to meet society's demands thanks to modern technologies. But the security of the food and the harvests remained elusive. Numerous issues, such as plant disease, declining pollinators, and climate change, pose challenges for farmers. It is imperative to provide a solid foundation for these issues. The product's assembly and the market's economic value will both be severely harmed by inaccurate identification. This work aims at a novel approach to modelling and identification of plant disease progression through the categorization of leaf pictures and huge convolutional networks. To promote the program's simple execution in observance, a novel technique and method were used. The suggested approach can distinguish thirteen different forms of plant illnesses from healthy leaves by differentiating between diseased and healthy leaves. Agricultural consultants took all the required actions, starting with gathering photos to add information, to incorporate this disease recognition model area unit that was depicted during the project. To do the deep CNN technique, we choose to use PyCham and Python.

P. H. V. Sesha Talpa Sai (✉)
Department of Mechanical Engineering, Malla Reddy College of Engineering and Technology, Hyderabad, India
e-mail: polamrajusai@gmail.com

M. Sucharitha
School of Electronics Engineering (SENSE), VIT-AP University, Amaravathi, Andhra Pradesh, India
e-mail: sucharitha.jackson@vitap.ac.in

N. Prakash · E. Poulose · K. K. Ajesh · A. Divakaran
School of Electrical and Electronics Engineering, Lincoln University College, Petaling Jaya, Malaysia

A. Bhaumik
Lincoln University College, Petaling Jaya, Malaysia
e-mail: amiya@lincoln.edu.my

© The Author(s), under exclusive license to Springer Nature Singapore Pte Ltd. 2025
H. Zen et al. (eds.), *Soft Computing and Signal Processing*, Lecture Notes in Networks and Systems 1200, https://doi.org/10.1007/978-981-97-9926-8_45

Keywords Crops · Leaf diseases · Detection · Convolutional neural network · Python

1 Introduction

All people have access to food thanks to agriculture, even in situations where the population is growing quickly. To ensure there is food for everyone, it is advisable in the agricultural industry to anticipate plant diseases early on. Still, it is unfortunate to predict diseases during crop development's formative years. The study aims to inform farmers about new technologies that help reduce plant leaf diseases. A cutting-edge app called Dr. Plant analyses plant diseases quantitatively on plant parts like leaves. Users can compute the proportion of unhealthy tissue and gather or upload photos of sick plant organs. The program uses values entered by the user for the first eight hues of the photograph's healthy tissues. Each pixel's colour is then classified as either healthy or diseased based on how much it deviates from the healthy hues. Before calculating the percentage, users can shift the threshold bar until they are pleased with the accuracy with which sick tissues are represented. With the help of Dr. Plant, users may recognize and learn more about every type of plant found in the wild, including cactus, blooming plants, trees, grasses, conifers, ferns, vines, and wild salads. Though this is not the main objective of the application, it can also recognize a great deal of grown plants in parks and gardens. Users can observe plants in nature or in the middle of their vegetable gardens or on sidewalks in the cities.

An accurate identification of a plant will depend on how much visual information it has. Since many plants have similar appearances up close, sometimes it is the little things that set two species of the same genus apart. Photographing a species' most distinguishing organs, such as its flowers, fruits, and leaves, should come first. However, any additional information—such as hair on the stem, blooms, or thorns can be helpful. If there is a photo of the entire plant or tree, that is also very helpful information, though frequently insufficient for a trustworthy identification. The application employs advanced algorithms to evaluate the colour of each pixel in a photograph and assign a status of healthy or diseased. This approach is effective because it takes into account the range of healthy colours, and it allows users to adjust the threshold until they are satisfied that the results are accurate.

2 Review of Literature

Pre-processing techniques used by Krithika et al. [1] included colour-space conversion, contrast enhancement, and image scaling. K-means clustering is used to extract features and segment data using GLCM.

Abed et al. [2] applied scaling and stretching (min–max linear) to the raw data in order to improve the quality. After completing its development, the HIS model will be divided into many segments. K-means clustering and Euclidean distance techniques are combined to segment samples. For feature extraction and classification, respectively, the GLCM and SVM are employed.

A technique for classifying and identifying the various diseases that afflict plants is presented by Mrunalini et al. [3]. Because it saves time, money, and effort, a machine learning-based recognition system will be very beneficial to the Indian economy. This feature set extraction procedure is called the colour co-occurrence method. Using neural network technology, leaf disease detection is automated. The proposed approach can significantly improve the accuracy of leaf detection with minimal computational effort, and it seems to be an important approach for root and steam issues.

A process for accurate diagnosis of plant condition is presented by Kulkarni et al. [4] using a variety of image processing techniques and artificial neural networks (ANN). With a recognition rate of up to 91%, the suggested method yields better results because it is based on an "ANN classifier for classification" and a "Gabor filter feature extraction". To identify different plant diseases, an artificial neural network (ANN) classifier combines textures, colours, and other properties.

A CNN method was published by Jadhav et al. [5] to estimate plant condition. Using pre-trained CNN models, diseases in soybean plants were detected using this technique. Although the model performed worse in categorising diversity, better results were obtained when pre-trained transfer learning techniques such as AlexNet and GoogleNet were applied in the testing. Many existing models focus on identifying certain groups of plant diseases rather than developing a model to categorize various plant disorders. This is mostly due to the lack of databases with a wide range of plant species for training deep learning models.

A few others present a unique histogram transformation technique that, by creating synthetic image samples from poor test set images, improved the recognition accuracy of deep learning models. The goal of this work was to use a modified MobileNetV2 neural network model to apply motion blurring, over-exposure, resolution down-sampling, and Gaussian blurring to enhance photos quality in the dataset. They created synthetic images with altered colour value distributions in order to improve results and overcome the data scarcity that a data-hungry deep learning network encounters during training [6–10].

The objective is to develop techniques and algorithms that, when given images of leaves or other plant characteristics, can automatically recognize and classify plant diseases in agriculture. This can help farmers cope with the disease. After a thorough and critical review of several modern machine learning and deep learning-based algorithms created for plant disease detection and classification in the literature, the author has underlined a few major issues in crop disease detection and classification. Because of this, scientists will be able to investigate the factors that might have a significant influence on real-time systems for plant diagnosis and identification [11].

Wang et al. [12] have made great contributions to the use of neural networks in plant disease detection. They demonstrated the enhanced accuracy and efficiency of back propagation on neural networks (BPNNs) in the identification of plant diseases through the images. In 2012, the same scientists used principle component analysis

(PCA) in conjunction with neural networks to reduce the dimensionality of data while maintaining the features for enhanced accurate classification.

Similarly, the other scientists namely, Kai et al. [13] used back propagation on neural networks (BPNNs) to identify the crop diseases in maize. And their maize leaf image back propogation model effectively worked and also recognized the disease patterns, enhancing the potential of neural-based methods in agricultural disease detection and improving the detecting procedure.

A study in 2007 examined the use of fuzzy logic techniques for feature selection for cotton disease leaves in order to enhance the accuracy of disease detection and classification. This fuzzy logic was used to address the uncertainty present in the feature selection process, which also resulted in more accurate detection of disease signs in cotton leaves. Another study in 2002, compared different texture features obtained using Gabor filters. Their goal is to determine which features works best for texture analysis. Gabor filters are well known for their ability to capture both spatial and frequency information but have proven invaluable in differentiating various textures in plant leaves, which is very necessary in disease detection of plants. In 2013, some researchers aimed to detect the unhealthy areas and classification of plant leaf diseases using texture features. And their research effectively identified and categorized various leaf diseases by using advanced texture analysis and also demonstrated the importance of texture features in detection of diseased plants [14–16].

A comprehensive framework was introduced by the researchers in 2010, and this framework combined advanced image processing techniques with classification algorithms for the detection and classification of leaf and stem diseases. This framework mainly focused on the improvement of accuracy. It effectively analysed the visual symptoms in plant images, and for this it utilized both traditional image processing methods and modern image processing methods to enhance accurate detection. Similarly another technique was introduced to achieve fast and better accuracy in detection and classification of plant diseases. But it merged image processing methods with machine learning algorithms, resulting in high accuracy in detecting various plant diseases. This method was also suitable for real-time applications in agriculture due to its accuracy [17, 18].

Another study introduced hybrid intelligent system for colour imagery-based for grape leaf disease detection. This was basically introduced by integrating multiple machine learning methods and image processing methods to detect the symptoms of disease in grape leaves. Later it also combined various analytical techniques to increase its reliability and accuracy, showcasing the hybrid models in plant health management [19].

3 Methodology

3.1 Dataset Preparation

The dataset contains enormous amounts of unprocessed data that need to be trained in order to extract any valuable information. The dataset utilized by our system comprises an assortment of images depicting various plant leaves that either exhibit good health or fall under one of the illness categories: such as early spotted blight, spores of bacteria, and The TYLCV.

Often these types of datasets require pre-processing to improve quality and consistency. This step includes resizing, noise reduction, and colour normalization to ensure uniformity.

One needs a big database in order for our illness diagnostic algorithm to work as precisely and accurately as feasible. Each of the four conditions—bacterial spot, early blight, healthy, and TYLCV—has fifty leaf photographs. Each of these photos has been polished, scaled, and assigned a certain quality and dimension in order to produce a consistent dataset.

The images are converted from RGB to other colour spaces like HSV so that the effectiveness of certain types of image analysis and disease detection can be improved. We used histogram equalization to enhance the contrast of images so that it will be easy to prominently detect disease symptoms. Removing background elements to focus on leaf region enhanced the accuracy of our analysis.

This dataset governs the platform's earliest phases of digital photo processing and is used as our training set. This GitHub folder contains the original dataset. This collection, which is organized into 38 categories, has around 87 kb of photos of both healthy and damaged crop leaves. The entire dataset is split into training and validation collections with an 80/20 ratio while maintaining the directory structure. For prediction, a new directory with 33 test photos is subsequently created. Convolution ConvNets, also known as neural networks, are neural networks with shared parameters.

3.2 Work Flow

Layer of input has the raw input of the image to dimensions of $32 \times 32 \times 3$. The working of this neural network is demonstrated in Fig. 1. Convolution layer determines the output volume by computing the dot product between each filter and patching the image. Convolution layers work out the output volume by carrying out the dot product between every filter and patching over the image. The convolutional filters called as Kernels are applied to the input image by these layers. Each filter moves around the image, performing convolutional operations that result in feature maps being produced. These feature maps capture visual patterns such as edges, textures, or shapes which may indicate plant health or disease. Assume that 12 filters

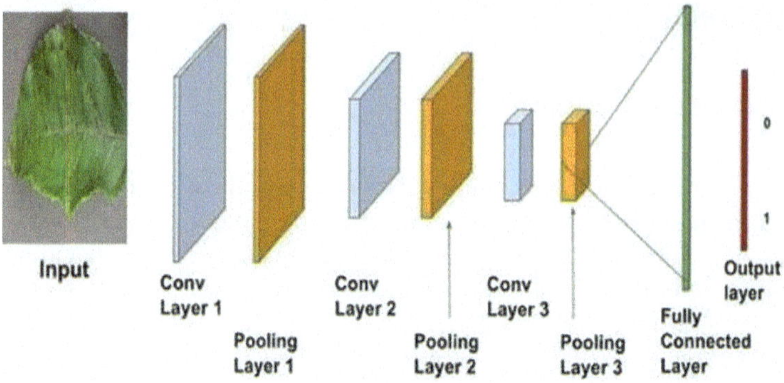

Fig. 1 Working of CNN

are used in total to obtain an output volume for this sheet that measure 32 × 32 × 12. Activation function layer enhances function of convolution layer by applying an element-wise activation function. Sigmoid: $1/(1 + e^{-x})$, Tanh, Leaky ReLU, and ReLU: max $(0, x)$ are a few examples of activation functions. Pooling layers such as max pooling reduce the spatial dimensions of the feature maps. These help to reduce computational complexity by down-sampling the data. For example, a 2 × 2 max pooling layer will reduce a 32 × 32 feature map to 16 × 16 by taking the maximum value in each 2 × 2 window.

3.3 Model Description

3.3.1 Detection with Image

In this Image-based detection, convolutional neural networks (CNNs) properly train and test the leaf if the image and its information is already available. Prediction.py is currently operating in the background as the picture is taken and assumptions are made about that leaf. The module now allows the detection, and the results are returned. As mentioned earlier, the user can only use this module if they have a picture of a leaf.

3.3.2 Detection with Camera

The camera module captures picture of leaf when we are missing it. Leaf information is applied through the script of CNN_predict.py. In accordance, CNN will check and train its data. Since this is a freshly obtained leaf image, the build module will create the leaf image. It works the same way as picture detection, but we used live detection to make it more convenient than picking leaves. In the backend, the script prediction.py is executed. The detection is now permitted by the module, and the outcomes are given back.

4 Results and Discussion

Leaf disease detection was established with the right preparation and direction. We have an admin page, as seen in Fig. 2. Here, we have three options: image detection, camera detection, and CNN model creation. The user must proceed in accordance with his or her needs. If the customer does not already have a picture of the leaf, it might also be able to identify the disease by taking a camera photograph of the leaf, as seen in Fig. 3. In the event that user is having image of the leaf, disease can be identified after choosing "image detection" at the admin page, as illustrated in Fig. 2, and carrying out the detection procedure, as illustrated in Fig. 4.

It can be seen that on the basis of input provided the model is able to generate results which can be seen in Fig. 5.

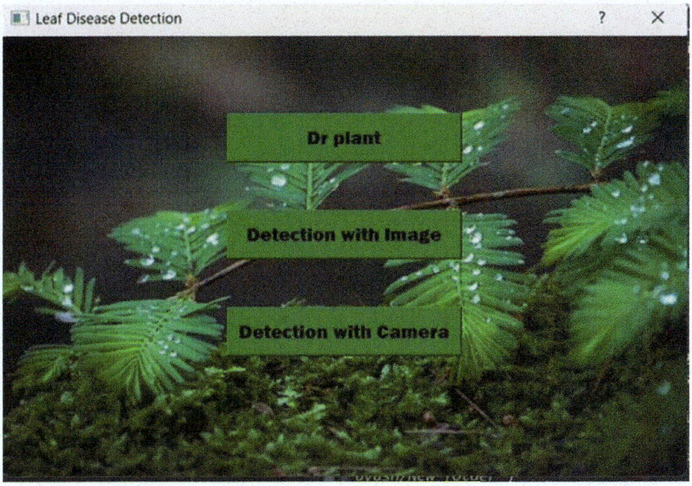

Fig. 2 Admin page

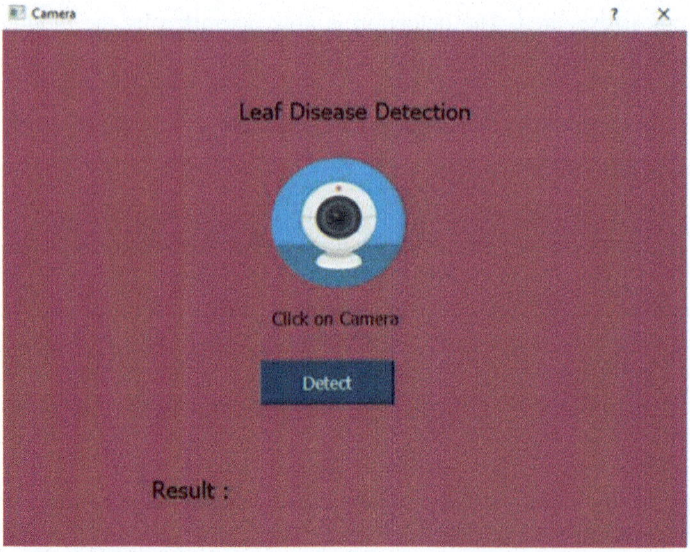

Fig. 3 Recording using camera

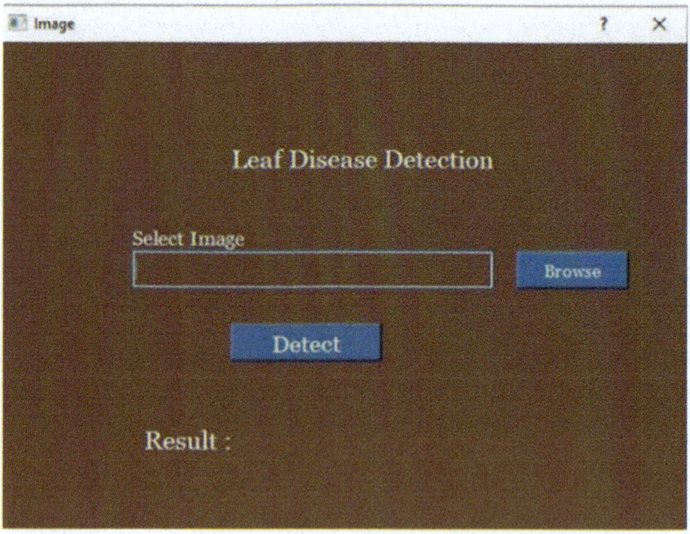

Fig. 4 Image uploading

```
Corn_(maize)___Common_rust_
1. Corn_(maize)___Common_rust_: 99.999511%
2. Apple___Apple_scab: 0.000317%
3. Corn_(maize)___Cercospora_leaf_spot Gray_leaf_spot: 0.000163%
4. Potato___Early_blight: 0.000005%
5. Apple___Cedar_apple_rust: 0.000003%
res= Corn_(maize)___Common_rust_
```

Fig. 5 Output data

5 Conclusion

In agriculture, crop monitoring—especially smart farming—is crucial to raising crop yield and quality. As a critical component that has given agriculture invaluable value and may support the maintenance of high-quality crops, smart farming has emerged. Thus, one crucial way that basic research will be applied to agriculture is in the creation of AI-based disease surveillance programmes. In the current work, a deep learning-based model for disease identification was built to resemble human disease recognition. Even in this field of study, there are different approaches to the detection and diagnosis of portable or machine-driven computer vision diseases. There are currently no commercial strategies available on the market other than the fact that plant species identification management was aided by leaf photos. This research proposed a novel deep learning victimization technique to automatically detect and identify plant diseases using leaf photos. The proposed model was able to identify 13 distinct diseases that can be visually identified, as well as distinguish between healthy leaves and the presence of the OFA virus. The entire approach involved drawing a clear distinction between the grouping of training and testing images, pre-processing and image augmentation, and, lastly, the deep CNN and fine-tuning work technique. Totally separate tests were carried out to determine the efficiency of the newly created model. The final accuracy of the qualified model was 96.3% overall. While augmentation has a bigger effect on yielding satisfactory results, further enhancement may be tried with more research.

References

1. Krithika P, Veni S (2017) Leaf disease detection on cucumber leaves using multiclass support vector machine. In: IEEE international conference on wireless communications, signal processing and networking, pp 1276–1281
2. Abed S, Esmaeel A (2018) A novel approach to classify and detect bean diseases based on image processing. In: IEEE symposium on computer applications & industrial electronics, pp 297–302
3. Badnakhe Mrunalini R, Deshmukh PR (2011) An application of K-means clustering and artificial intelligence in pattern recognition for crop diseases. In: International conference on advancements in information technology, vol 20, 2011 IPCSIT
4. Kulkarni AH, Ashwin Patil RK (2012) Applying image processing technique to detect plant diseases. Int J Mod Eng Res 2(5):3661–3664

5. Jadhav SB, Udupi VR, Patil SB (2021) Identification of plant diseases using convolutional neural networks. Int J Inf Technol 13:2461–2470
6. Abayomi-Alli OO, Damaševičius R, Misra S, Maskeliūnas R (2021) Cassava disease recognition from low-quality images using enhanced data augmentation model and deep learning. Expert Syst 38:e12746
7. Anh PT, Duc HTM (2021) A benchmark of deep learning models for multi-leaf diseases for edge devices. In: Proceedings of the 2021 international conference on advanced technologies for communications (ATC), Ho Chi Minh City, Vietnam, 14–16 Oct 2021, pp 318–323
8. Astani M, Hasheminejad M, Vaghefi M (2022) A diverse ensemble classifier for tomato disease recognition. Comput Electron Agric 198:107054
9. Gokulnath BV (2021) Identifying and classifying plant disease using resilient LF-CNN. Ecol Inform 63:101283
10. Enkvetchakul P, Surinta O (2022) Effective data augmentation and training techniques for improving deep learning in plant leaf disease recognition. Appl Sci Eng Prog 15:3810
11. Demilie WB (2024) Plant disease detection and classification techniques: a comparative study of the performances. J Big Data 11:5. https://doi.org/10.1186/s40537-023-00863-9
12. Wang H, Li G, Ma Z, Li X (2012) Image recognition of plant diseases based on back propagation networks. In: 5th international congress on image and signal processing, pp 894–900, Chongqing, China
13. Kai S, Zhikun L, Hang S, Chunhong G (2011) A Research of maize disease image recognition of corn based on BP networks. In: Third international conference on measuring technology and mechatronics automation, pp 246–249, Shenyang, China
14. Zhang Y-C, Mao H-P, Hu B, Li M-X (2007) Feature selection of cotton disease leaves image based on fuzzy feature selection techniques. In: Proceedings of the 2007 international conference on wavelet analysis and pattern recognition, pp 124–129, Beijing, China
15. Grigorescu SE, Petkov N, Kruizinga P (2002) Comparison of texture features based on Gabor filters. IEEE Trans Image Process 11(10):1160–1167
16. Arivazhagan S, Newlin Shebiah R, Ananthi S, Vishnu Varthini S (2013) Detection of unhealthy region of plant leaves and classification of plant leaf diseases using texture features. Commission Internationale du Genie Rural (CIGR) Journal 15(1):211–217
17. Al Bashish D, Braik M, Bani-Ahmad S (2010) A framework for detection and classification of plant leaf and stem diseases. In: 2010 International conference on signal and image processing, pp 113–118, Chennai, India
18. Al-Hiary H, Bani-Ahmad S, Reyalat M, Braik M, ALRahamneh Z (2011) Fast and accurate detection and classification of plant diseases. Int J Comput Appl 17(1):31–38
19. Menukaewjinda A, Kumsawat P, Attakitmongcol K, Srikaew A (2008) Grape leaf disease detection from color imagery using hybrid intelligent system. In: Proceedings of electrical engineering/electronics, computer, telecommunications and information technology (ECTI-CON), vol 1, pp 513–516, Krabi, Thailand

Creating an Error-Tolerant Adder for Image Processing Applications Using FPGA Technology

Bairi. Rohith Kumar, K. Niranjan Reddy, and E. John Alex

Abstract Image processing applications often require error-tolerant adders to ensure accurate computation despite potential hardware faults. Field-programmable gate array (FPGA) technology, coupled with very large-scale integration (VLSI) techniques, offers a versatile platform for implementing such adders. Contribution a new error-tolerant adder designed explicitly for image processing applications is proposed. Our methodology leverages the inherent reconfigurability of FPGAs to dynamically adapt to changing error patterns, thereby enhancing fault tolerance. By employing redundant computing resources and error detection/correction mechanisms at various stages of the adder architecture, we mitigate the impact of transient errors and aging-related faults. To validate the effectiveness of our approach, we conduct extensive simulations and synthesis experiments using XILINX software. The results demonstrate significant improvements in error tolerance compared to conventional adder designs, making our proposed architecture well-suited for demanding image processing applications in safety–critical environments. Overall, our work contributes to advancing the state-of-the-art error-tolerant computing for FPGA-based image processing systems.

Keywords Error tolerance · Emphasizing power consumption · Image processing · Circuit optimization

1 Introduction

In image processing, precision and accuracy are indispensable for obtaining reliable results. However, the computational hardware utilized in image processing tasks, including adders, is susceptible to errors that can significantly impact the outcomes. Consequently, there is a pressing need to develop error-tolerant adders (ETAs) that can mitigate such errors, ensuring the integrity of image processing applications.

Bairi. Rohith Kumar (✉) · K. Niranjan Reddy · E. John Alex
Department of Electronics and Communication Engineering, CMR Institute of Technology, Hyderabad, India
e-mail: rohithbairi.6754@gmail.com

© The Author(s), under exclusive license to Springer Nature Singapore Pte Ltd. 2025

H. Zen et al. (eds.), *Soft Computing and Signal Processing*, Lecture Notes in Networks and Systems 1200, https://doi.org/10.1007/978-981-97-9926-8_46

This study delves into the design and implementation of an error-tolerant adder tailored explicitly for image processing applications, employing. FPGAs provide a flexible and customizable platform for implementing digital circuits, while VLSI methodologies enable efficient integration and optimization of these circuits at the hardware level. Our approach involves a thorough investigation into the utilization of redundant computing resources and sophisticated error detection/correction mechanisms within the adder architecture. By redundantly computing critical operations and incorporating error detection techniques at various stages of computation, we aim to identify and rectify errors arising from transient faults or aging-related issues. Moreover, the integration of VLSI principles plays a pivotal role in optimizing resource utilization and minimizing power consumption. Through meticulous circuit partitioning, layout optimization, and the application of low-power design techniques, we strive to maximize the efficiency of FPGA resources while upholding high performance and reliability standards. To validate conduct extensive experiments utilizing both PC-based simulations and FPGA-based implementations [1]. By comparing the performance and error resilience of our design across these platforms, we aim to ascertain its efficacy in real-world scenarios.

In summary, this paper endeavors to advance the frontier of error-tolerant computing for FPGA-based image processing systems. By developing robust adder architectures capable of withstanding hardware faults, we aim to bolster the accuracy and reliability of image processing applications, thereby facilitating their widespread adoption in critical domains.

2 Error Tolerance

In digital systems, error tolerance is a critical concept aimed at ensuring the reliability and robustness of computational processes in the face of potential hardware faults or transient errors. Integrating error tolerance mechanisms is particularly essential in applications where accuracy and precision are paramount, such as image processing.

In the context of image processing, error tolerance serves as a safeguard against inaccuracies that may arise due to various factors, including electronic noise, manufacturing defects, or environmental disturbances. By implementing error tolerance strategies, such as redundant computation, error detection, and correction mechanisms, image processing systems can effectively identify and rectify errors, thereby preserving the integrity of the processed images. Moreover, error tolerance enables image processing applications to maintain operational continuity even in the presence of hardware faults or aging-related degradation [2]. This resilience is particularly crucial in safety–critical environments or mission-critical scenarios where the accuracy of image analysis directly impacts decision-making processes.

Furthermore, the incorporation of error tolerance techniques contributes to the longevity and reliability of image processing systems, as it helps mitigate the effects of hardware wear and tear over time. By proactively addressing potential sources of errors, error-tolerant designs can prolong the lifespan of hardware components and reduce the frequency of system downtime or maintenance interventions. Overall, the utilization of error tolerance in image processing applications plays a fundamental role in enhancing the accuracy, reliability, and longevity of digital systems [2]. By mitigating the impact of hardware faults and transient errors, error-tolerant designs ensure that image processing tasks can be performed with confidence, even in challenging operational environments.

An error-tolerant adder (ETA) arises from the necessity to balance between accuracy and efficiency in digital arithmetic operations, particularly in the face of potential errors. This involves dividing the operation into two distinct parts: the accurate part and the inaccurate part.

The accurate part of the ETA is responsible for ensuring precise arithmetic computations. It typically handles critical calculations where precision is paramount and errors must be minimized. This segment of the adder prioritizes accuracy over speed and may incorporate error detection and correction mechanisms to guarantee reliable operation [3]. By employing techniques such as redundant arithmetic or error-correcting codes, the accurate part mitigates errors and ensures that essential computations maintain high precision.

Conversely, the inaccurate part of the ETA is designed to prioritize efficiency and speed over absolute precision. This segment of the adder may utilize simplified or approximate arithmetic techniques to expedite calculations and reduce hardware complexity [4]. While the inaccurate part may introduce minor errors or inaccuracies in the computation, it does so within acceptable tolerances, balancing the trade-off between accuracy and efficiency. The inaccurate part is particularly beneficial for non-critical calculations or in scenarios where computational speed is paramount and small errors can be tolerated as shown in Fig. 1.

By partitioning the adder into accurate and inaccurate parts, the ETA achieves a synergistic balance between precision and efficiency, catering to the diverse requirements of different arithmetic operations and application scenarios. This approach ensures that critical computations maintain high accuracy while non-critical calculations are performed expediently, ultimately enhancing the overall efficiency and reliability of digital arithmetic operations.

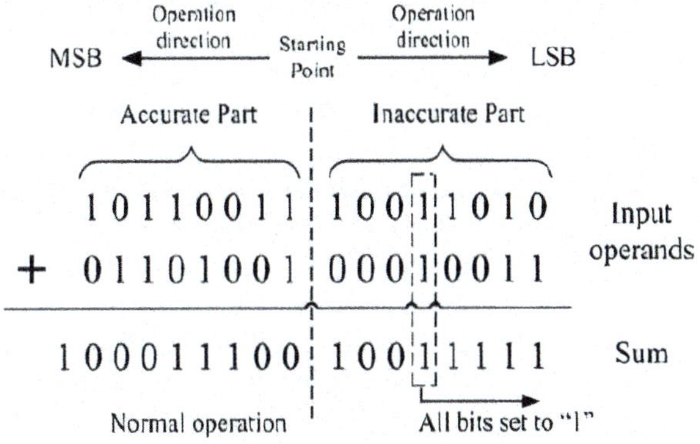

Fig. 1 Addition methodology of the error-tolerant adder

2.1 High Speed, Low Power, and Error-Tolerant Adder

- Vocabulary: Here are the definitions to a few of the words we use often, along with a few other details regarding the ETA.
- Over error (OE): OE = |Rc − Re|, where Rc is the right result, and Re is the result of the adder.
- In average, wolf agrees with others with an accuracy (ACC) of ~ 62% on the dataset; adder is only 52% agreement with the others.
- Minimum acceptable accuracy (MAA): The recognized accuracy of an output that a system demands to function as needed. That cut-off value is MAA.
- Acceptance probability (AP): A probability is denoted to an adder whose accuracy is greater than minimum allowable accuracy as acceptance probability [5].

2.2 Source of Power Consumption and Latency in DAC

The propagation chain of the carry from the least-significant bit (LSB) through to the most-significant bit (MSB) along the critical path is the cause for delay glitches, with carry propagation being attributed to large portion of the power dissipation of an adder in the case of conventional adder circuit. This results in reducing power as well as units critical path latency or, equivalently, size (can have less CLB/LUT) since speed performance and power consumption can be greatly increased if carry propagation is minimized or reduced to zero as possible [6].

2.3 Novel Proposed Addition Arithmetic Approach

Step-1: First, inaccurate part is the remaining lower order bits, and accurate part are many higher order bits by splitting the input operands. The length of each component does not have to be the same.

The Addition The act of adding works simultaneously in both opposite directions, starting from the (middle): the point of merging from two parts of the result structure. The best and worst piece of two 16-bit input operands are A = "1011001110011010" (45,978) and B = "0110100100010011" (26,899), divisible with the same portion slice of 8 bits [3].

Step-2: Accurate part addition process: For adding the higher order bits (accurate part) of the input operands from rightmost to leftmost (LSB to MSB), the normal addition method is applied. As higher order bits are more important than the lower order bits, this is done to preserve the correctness of multiplication.

Step-3: Generally in Parts addition part the lower order bits of input operant—the wrong portion will need a different adding process. To break the path of carry signal, no any carry signal is produced or received at any bit point. To decrease the total error

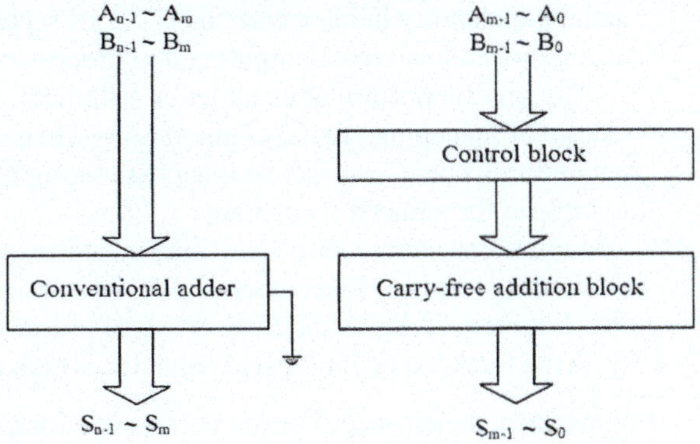

Fig. 2 Diagram depicting proposed the ETA

during the carry chain removal, we adopted a particular technique that generally can be described as follows;

(1) First check each bit position from left to right (MSB to LSB)
(2) The operation then leaves to the next bit position and if both the bit inputs are "0" or both the inputs are opposite, a usual one-bit addition is performed in this bit position as an output.
(3) If both the two input bits were 1, then the checking process was stopped, and all sum bits to the right were set to 1.

The example of a final result of "10001110010011111" (72,863) makes it easy to understand the addition technique explained. If standard arithmetic is used, the example provides a result in "10001110010101101" (72,877). One can calculate the total error produced using the formula OE = 14.

ETA utilizes the proposed addition arithmetic and is illustrated in Fig. 2. The sections may be two in this simplest design; a correct part and an accurate part. The correct part is just a combination of 4-bit conventional adders like RCA, CSK, CSL, CLA, etc. [2]. The carry-in of this adder is tied to the ground. The bad portion consists of two blocks—the control block and the carry-free addition block. The carry-free addition block follows the mode of operation (M) that is determined by the control signals that are generated by the control block.

2.4 Hardware Implementation

The development of a 32-bit fault-tolerant adder illustrates the hardware implementation of our proposed error-tolerant adder (ETA), which makes use of the addition arithmetic previously discussed. There are two primary parts to this simple structure:

(i) In the context of digital circuitry or signal processing, an "accurate part" is a segment or component inside a system that is intended to preserve precision and dependability in its operations [7]. This section is distinguished by its high accuracy and low error computing and processing capabilities.

The precise portion of an adder or arithmetic unit, for instance, would be in charge of managing crucial computations when accuracy is crucial. This could entail doing operations like adding or summing numbers while strictly adhering to criteria for numerical accuracy.

(ii) The term "inaccurate part" describes a section or part of a system that is prone to creating errors or inaccuracies in its operations; this term is frequently used in the context of digital circuitry or signal processing. This segment is deviating from the intended or anticipated behavior, which can result in inaccurate output.

For instance, the erroneous portion of an adder or arithmetic unit may use approximation methods or condensed circuits to execute computations faster or with less hardware complexity [7]. This method can increase productivity and efficiency, but it also has the potential to create tiny mistakes or divergences from the precise outcome.

The erroneous portion might be acceptable in some applications, such as an image or signal processing if the introduced faults fall within acceptable bounds and have little to no effect on the output's overall performance or quality. By accepting a specific amount of output error, the inaccurate portion acts as a trade-off between accuracy and performance, enabling the optimization of system resources.

We utilize a 32-bit case study in Sect. 3 to illustrate our design methodology and provide schematic illustrations of these two block types for the ETA circuit.

3 Development of a 32-Bit Fault-Tolerant Adder

The ability of a system to carry out its duties correctly even in the face of hardware or software failures is known as fault tolerance (FT). Any circuit implementation that can withstand more flaws than a non-defect-tolerant implementation is referred to as having defect tolerance (DT) [2]. The term "undefined" error tolerance has been observed in multiple texts recently. The phrase appears to allude to the reality that hardware might have flaws and that one must be able to reconfigure a device to utilize just the "good" circuitry in certain situations. This idea appears to be a combination of reconfigurable FT and DT concepts. We proposed a new definition of error tolerance as follows.

When it comes to an application or system, a circuit is considered error-tolerant (ET) if two things happen: (1) It has flaws that can lead to both internal and external failures, and (2) The system that uses it generates satisfactory results. An ET circuit differs from FT and DT in that it has the potential to yield false findings. The main reason for interest in ET is its ability to greatly increase a process's effective yield.

Chips will have billions of devices and thousands of flaws if VLSI scaling proceeds on its current course. It is possible that producing chips that are both functionally flawless and reasonably priced will require more than just FT and DT approaches. The 2001 International Technology Roadmap for Semiconductors (ITRS) makes a strong case for the idea. Error tolerance is one of the difficulties they talk about. Some applications like banking and nuclear power plant control need accurate computation. Many others don't either, which is a significant area where flawless functionality is crucial [8]. Functioning is not necessary for numerous DSP applications and certain multimedia applications. An MPEG motion detection encoder is an excellent illustration of an error-tolerant system, as it functions exceptionally well even when a significant portion of single stuck-at faults are present.

Separating the adder into two distinct pieces is the first stage in creating a suggested ETA. A guess-and-verify method underpins the splitting technique, which is based on many requirements like power, speed, and accuracy. First, we define time delay as the delay of the proposed adder. After defining this partitioning technique, we next determine whether the accuracy and performance of the adder satisfy the specifications given out by the designer or customer. This can be rapidly verified with certain software applications [2, 7]. If this condition is not satisfied, the checking procedure should be redone after shifting one bit from the erroneous to the accurate component.

Moreover, adding additional bits to this section results in greater power savings because of the streamlined circuit topology and the removal of switching activities in the erroneous part. After taking the aforementioned into account, we separated the 32-bit adder into 12 bits for the accurate component and 20 bits for the inaccurate part.

The faulty element in our suggested 32-bit ETA has 20 bits, whereas the accurate part uses 12 bits. Since the erroneous part determines the overall delay, a fast adder does not necessarily have to be the accurate part [9]. For the precise portion of the circuit, the most power-efficient conventional adder ripple carry adder has been selected. It affects the adder's accuracy, speed, and power consumption, and the erroneous portion of the suggested ETA is the most important one.

The control block and the carry-free addition block make up the erroneous portion. The twenty modified XOR gates that comprise the carry-free addition block are each responsible for producing a sum bit [5]. The improved XOR gate's schematic implementation and the carry-free addition block's block diagram are shown in Fig. 3.

The modified inverter is added to a standard XOR gate and three more transistors, M1–M3. A control signal (CTL) from the CB is used to set the operation mode of the crush. When CTL = 0, the input circuit behaves in the standard XOR mode, turning on M1 and M2 while turning off M3. Except that with CTL-1, M3 is on and collapses the output node to VDD providing the sum output = "1". M1 and M2 are both off. The role of control block is when both input bits are 1, then it will search for the first bit and enable the control signal at that position. This takes the people group on its right to high [3]. The structure is comprised of twenty control signal-generating cells (CSGCs), and each CSGC generates a control signal for the modified XOR. Carry-free addition block gate operates at the correct bit position to ensure accurate computation. Instead of the long chain of 20 cascading CSGCs, the control block is spliced into five groups' similar size, with additional connections between every contiguous pair of groups.

Two CSGC types I and II in Fig. 4a elaborate configurations are designed, and Fig. 4b shows the schematic implementations of these two CSGC kinds. The input

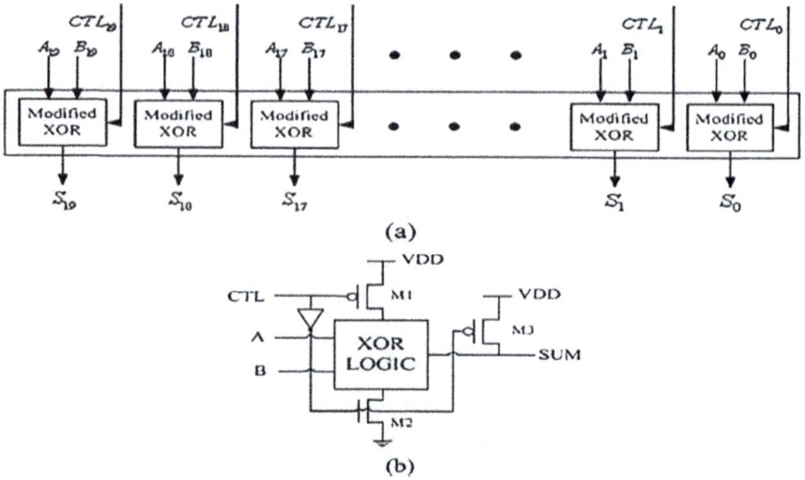

Fig. 3 Carry-less addition module. **a** Detailed configuration; **b** Schematic presentation of a modified XOR gate

of the leftmost cell in the following group is coupled to the control signal produced by left most cell in each group [2]. Transmitted high-control signal can "jump" from one group to another rather than going through all 20 cells because of additional connections. As a result, there are only ten cells on the worst-case propagation path (which is shaded in gray in Fig. 3a) (Fig. 5).

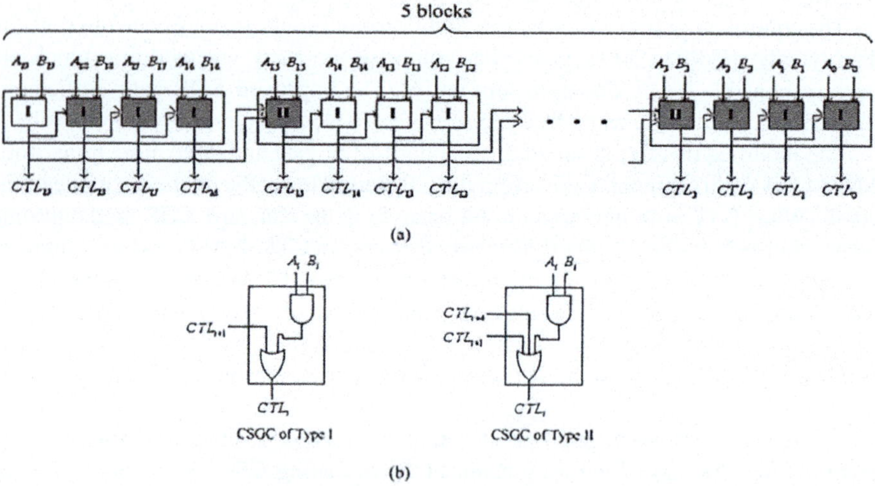

Fig. 4 Control block: **a** elaborate configuration; **b** CSGC implementation

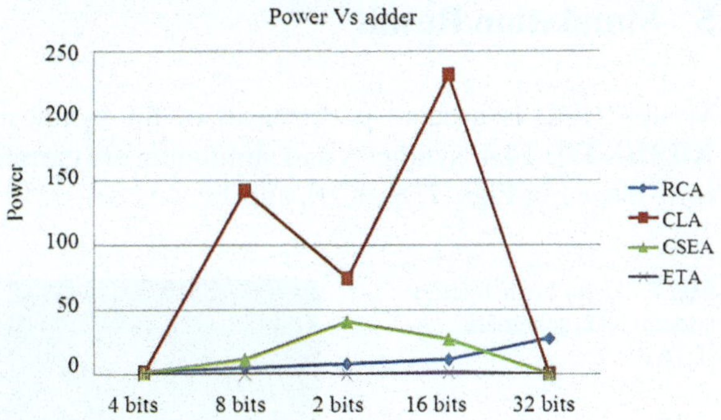

Fig. 5 32-bit ETA simulation graph

4 Block Diagram

To conduct simulations to contrast its performance with four conventional adders. Employing Xilinx software, we configured the input frequency to 30 MHz [10]. The outcomes of these simulations are succinctly presented as shown Fig. 6.

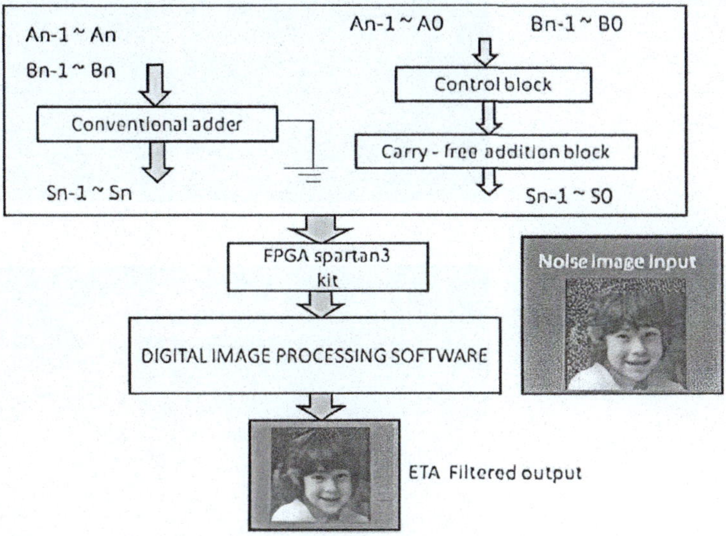

Fig. 6 Architecture of ETA

5 Simulation Result

Verilog HDL is used to perform all of the synthesis and simulation results. On XILINX ISE 14.4, synthesis and simulation are carried out. The simulation's output is displayed in Figs. 7, 8, 9, 10, and 11.

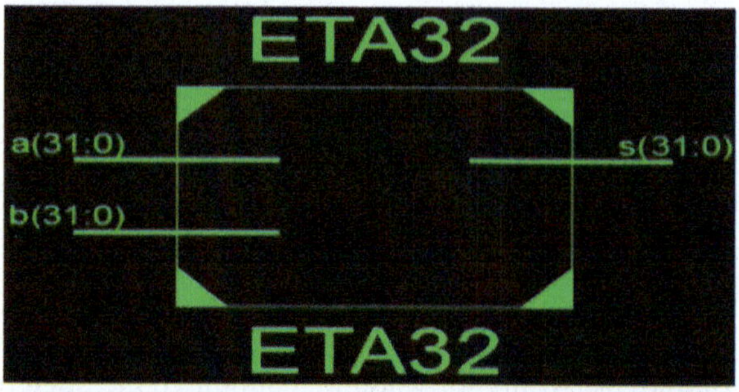

Fig. 7 32-bit error-tolerant adder's RTL schematic (ETA)

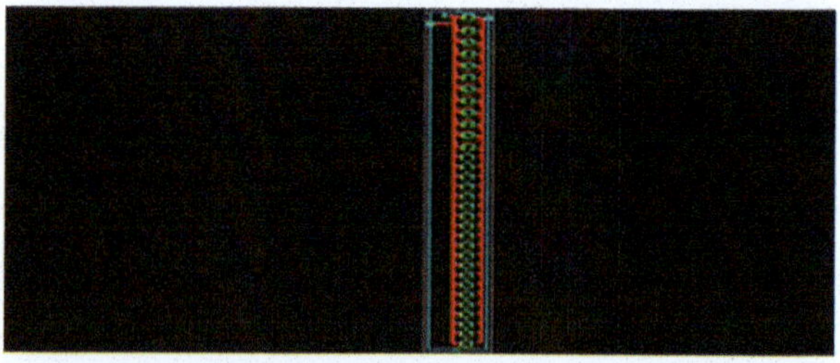

Fig. 8 RTL schematic of a 32-bit error-tolerant adder (ETA)

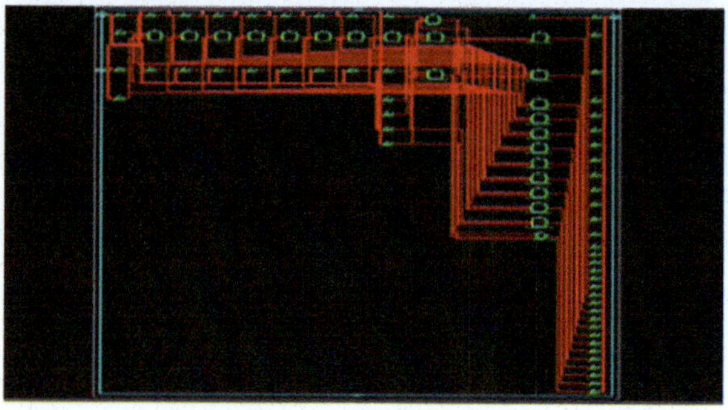

Fig. 9 Technology schematic of the 32-bit error-tolerant adder (ETA)

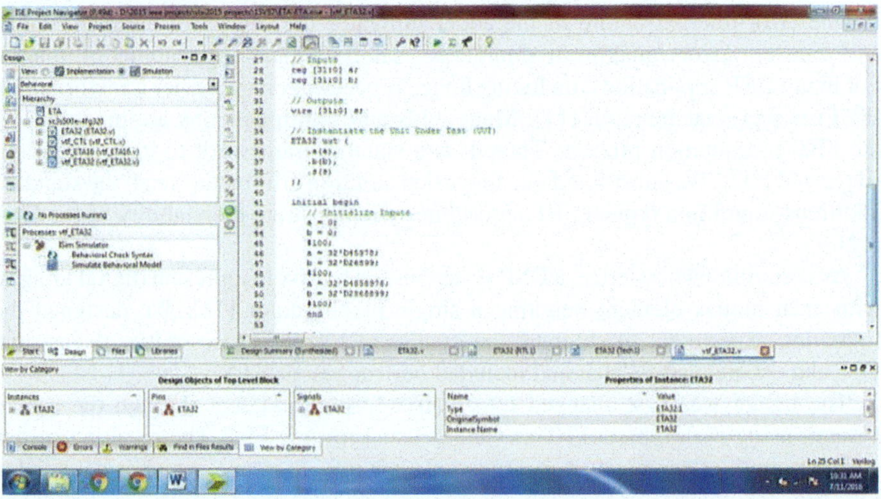

Fig. 10 Test bench of 32-bit error-tolerant adder

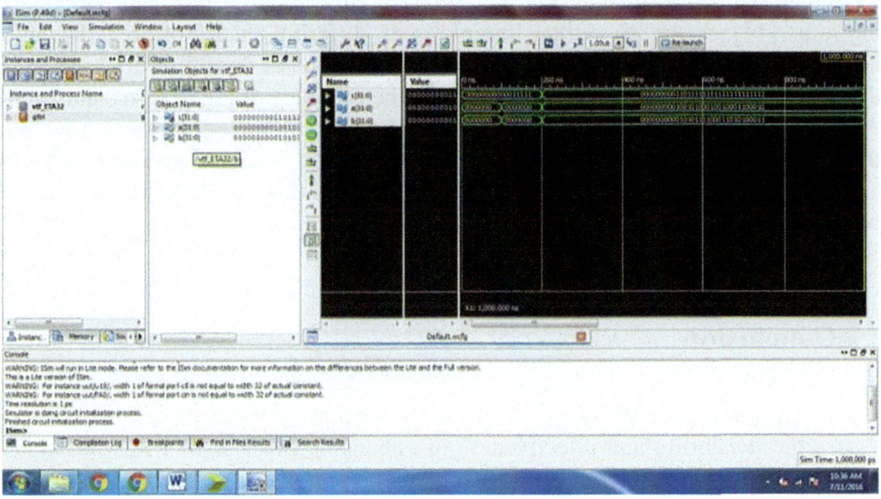

Fig. 11 32-bit error-tolerant adder simulation

6 Integration Error-Tolerant Adder Application in Digital Signal Processing

The error-tolerant adder is integrated into image processing workflows, offering key benefits such as enhancing the accuracy of computational operations and improving the robustness and reliability of image processing systems.

The error-tolerant nature of the adder mitigates the impact of hardware faults and transient errors, making it suitable for safety–critical applications in images. For many DSP applications, including image processing, fast Fourier transformation (FFT) is a crucial function [11]. Many adds and multiplications are made during the FFT computation process. Thus, it is a suitable framework to incorporate our suggested ETA. We substituted our suggested addition arithmetic for all the standard additions found in a typical FFT algorithm to demonstrate the viability of the ETA [12].

As is common knowledge, a DSP system uses a matrix to represent digital images, with each matrix element denoting a single pixel's color [13]. We designed the following experiment to compare the quality of images processed by the normal FFT and erroneous FFT that had included our suggested ETA.

The image was first changed into matrix format and sent through the carrier system, using both typical FFT as well as usual reverse FFT. The resulting matrix output from the system was then converted back into an image, which is depicted in Fig. 12a. The matrix for the same image was also processed within a system using the 32-bit ETA shown in Sect. 3 in perfect forward and imperfect backward FFTs. The processed master image is shown in this image Fig. 12b.

The two generated images (Fig. 12) had different results, despite being essentially the same [14]. A slightly darker line plot with differing shades of grey, displayed horizontally, indicates the variation in data. Application of this 95% MAA setting to the matrix of data/variables/results improves the accuracy of Fig. 12a to that Fig. AP 98.3% of b, and Fig. 12a. In short, the two images are compared to show how little the quality loss to the image should be using our ETA, and how good it will be for human acceptance [15]. This study has demonstrated that the ETA proposed in this work is feasible, based on its simulation results.

7 Conclusion

In summary, this research introduces a new concept in VLSI design that demonstrates the feasibility and effectiveness of enhancing error resilience while maintaining computational efficiency. The implementation of ETAs in image processing workflows offers several key benefits. Firstly, it enhances the accuracy of computational operations, ensuring that image transformations and manipulations maintain fidelity to the original data. This is critical in applications where precise pixel values are essential for maintaining image quality. Additionally, the error-tolerant nature of ETAs mitigates the impact of hardware faults and transient errors, improving the robustness and reliability of image processing systems. The research underscores the importance of error tolerance in image processing applications and highlights the potential of ETAs implemented on FPGA platforms to address this need effectively like smartphones and laptops.

Simulations were conducted using XILINX, utilizing the power estimator and analyzer to evaluate power consumption. While approximate adders may slightly

Fig. 12 Comparing images processed with conventional adder and recently introduced ETA for FFT and inverse FFT operations

(a)

(b)

compromise quality, they greatly enhance design metrics in these applications. Simulation results on XILINX validate that the ETA provides substantial improvements over precise adder architectures, with minimal errors observed in XILINX simulations. Consequently, the ETA can be effectively utilized in image and video processing applications.

References

1. Zhu N, Goh WL, Zhang W, Yeo KS, Kong ZH (2010) Design of low-power high-speed truncation-error-tolerant adder and its application in digital signal processing. IEEE Trans Very Large Scale Integr (VLSI) Syst 18(8)
2. Suganthi N (2015) Design and implementation of field programmable gate array based error tolerant adder for image processing application. In: 2015 2nd international conference on electronics and communication systems (ICECS). https://doi.org/10.1109/ecs.2015.7124821
3. Geetha S, Amritvalli P (2017) High speed error tolerant adder for multimedia applications. J Electron Testing
4. Breuer MA (2004) Intelligible test techniques to support error-tolerance. In: Proceedings of the Asian test symposium, Nov 2004, pp 386–393
5. Raheem MA, Gupta H, Fatima K, Adil O (2012) A high-speed reversible low-power error tolerant adder. In: 2012 Asia Pacific conference on postgraduate research in microelectronics and electronics
6. Palem KV (2005) Energy aware computing through probabilistic switching: a study of limits. IEEE Trans Comput 54(9):1123–1137
7. Jayanthi (2012) Design of an error tolerant adder. Am J Appl Sci 9(6):818–824. https://doi.org/10.3844/ajassp.2012.818.824
8. Jothin R, Peer Mohamed M, Vasanthanayaki C (2020) High performance compact energy efficient error tolerant adders and multipliers for 16-bit image processing applications. Microprocessors and Microsystem
9. Jothin T, Vasanthanayaki C (2017) High speed energy efficient static segment adder for approximate computing applications. J Electron Test
10. Breuer MA, Gupta SK, Mak TM (2004) Design and error-tolerance in the presence of massive numbers of defects. IEEE Des Test Comput 24(3):216–227
11. Patel H, Nandi A (2016) Modified error tolerant adder for digital signal processing application. In: 2016 international conference on micro-electronics and telecommunication engineering (ICMETE)
12. Melvin AB, Haiyang Z (2006) Error-tolerance and multi-media. In: Proceedings of 2006 international conference on intelligent information hiding and multimedia signal processing, pp 521–524
13. Paliwal S, Singh A, Kumar Y (2015) A programmable error tolerant adder for image and audio processing in modern day SoCs. In: 2015 IEEE UP section conference on electrical computer and electronics (UPCON)
14. Pareek M, Singhal M (2016) Low power high speed area efficient error tolerant adder using gate diffusion input method. In: 2016 3rd international conference on signal processing and integrated networks (SPIN)
15. Mendez T, Nayak SG, Vasanth Kumar P, Vijay SR, Vishnumurthy Kedlaya K (2022) Performance metric evaluation of error-tolerant adders for 2D image blending. Electronics

Predicting Customer Churn in Retail Environments Using Explainable Artificial Intelligence and BEE Colony Optimization Techniques

V. Kakulapati, P. Hasmita Reddy, Rutuja Shivputra Dhabale, and Rajagopal Jahnavi

Abstract Customer churn forecast is a vital undertaking in the retail sector, necessary for maintaining company growth and improving customer loyalty. Customer churn prediction is crucial for company performance in retail settings because of the higher expense associated with acquiring new customers compared to maintaining existing ones. This research examines the use of machine learning techniques, especially a combination of Random Forest and XGBoost classifiers, to forecast customer attrition in retail settings. This research utilizes a comprehensive dataset including various client characteristics collected from a shopping mall. It takes a methodical methodology that includes data preparation, feature engineering, model training, hyperparameter optimization, and performance assessment. The use of Explainable Artificial Intelligence methods, such as SHapley Additive exPlanations (SHAP) and LSTM, guarantees the interpretability of the model without compromising its prediction accuracy. The solid predictive performance attained by the ensemble model created using soft voting provided useful insights into consumer behavior and churn dynamics. Furthermore, the clarification of SHAP values enables a detailed analysis of individual forecasts, providing insight into the fundamental factors that contribute to customer turnover. Additionally, the use of bee colony optimization yields more efficient outcomes. These results have practical significance for retail practitioners as they provide them with the opportunity to create focused strategies to retain customers and ensure the long-term viability of their firm. This research adds to the expanding collection of literature on customer churn prediction in the retail industry. It provides valuable information on the retail customer journey and helps in making well-informed decisions in a highly competitive marketplace.

Keywords Customer · Churn · Prediction · Retail · Industry · SHAP · Feature engineering · Hyperparameters · Optimization · Performance · Voting · Retention · Sustainability

---

V. Kakulapati (✉) · P. Hasmita Reddy · R. Shivputra Dhabale · R. Jahnavi
Sreenidhi Institute of Science and Technology, Yamnampet, Ghatkesar, Hyderabad, Telanganaga 501301, India
e-mail: vldms@yahoo.com

© The Author(s), under exclusive license to Springer Nature Singapore Pte Ltd. 2025

H. Zen et al. (eds.), *Soft Computing and Signal Processing*, Lecture Notes in Networks and Systems 1200, https://doi.org/10.1007/978-981-97-9926-8_47

1 Introduction

Churners, or consumers who transfer platforms in the digital retailing period, largely from telecommunications services since 2000, now include online gaming, streaming, banking, ads, and insurance agencies [1].

Customer retention, also known as attrition, turnover, or defection, is the tendency of a company's customers to discontinue doing business with it. Businesses worldwide face a significant global challenge that leads to financial losses and negative effects on their operations [2]. This is a significant hurdle for many firms. Ensuring client loyalty is a significant priority for corporations since research suggests that maintaining existing customers is more financially advantageous than acquiring new ones [3]. Customer Relationship Management (CRM) primarily emphasizes the anticipation of customer loyalty, particularly in industries that rely on contractual agreements, such as telecommunications and banking. Machine learning algorithms have shown a high level of accuracy in predicting customer attrition [4, 5].

Machine learning and artificial intelligence are transforming a variety of industries, including business automation. ML improves software prediction skills by leveraging past data, whereas AI systems do complicated jobs like humans but at a higher cost and difficulty. Automated trading tactics have had a big influence on the US stock market, and AI applications employ input data, ML techniques, reinforcement, supervision, and unproven methodologies to increase prediction abilities [6].

The quantity of variables and computer power necessary for complex data analysis makes it difficult. With faster computers and parallel processing, efficient high-dimensional data analysis is critical. The analysis of data and methodologies is used to handle sets of features, using strategies such as feature reduction and subset selection methods to enhance classification accuracy while reducing computational expenses [7, 8].

The SHAP model conducts a thorough examination of characteristics and their impact on model output, employing embedded algorithms to calculate the model-independent SHAP and examine its effect on the intended parameter [10]. The study made use of K-means clustering for data segmentation due to its efficiency and simplicity. Each data point is assigned to the centroid that is closest to it after the algorithm clusters data points based on their similarity [11].

In the very competitive retail industry, predicting and preventing customer attrition is essential to long-term growth and customer loyalty. To do this, it is now important to use machine learning techniques, namely ensemble models like Random Forest and XGBoost. These models are well-suited to capturing the subtleties of churn patterns in retail contexts because they leverage the benefits of many algorithms to increase prediction accuracy.

The presented work aims to forecast customer attrition in retail contexts by utilizing an ensemble model, which is a combination of Random Forest (RF) and XG Boost (XGB) classifiers. These models are widely known for their exceptional adaptability and task-predictiveness. Our goal is to bring these components together to produce an accurate churn prediction model that helps retail businesses manage their customer relationships and allocate resources as efficiently as possible.

The method known as Explainable Artificial Intelligence (XAI) combines the advantages of straightforward and adaptable models to enhance the performance of quantitative analysis. It settles the performance vs. explainability trade-off [12].

Furthermore, we leverage Explainable Artificial Intelligence techniques, such as SHAP, to enhance the ability to comprehend and interpret models. By illuminating the factors that lead to churn estimates, our approach improves model transparency and enables stakeholders to make informed decisions. To provide practical insights for focused retention strategies, this study outlines a comprehensive approach for predicting customer attrition in retail settings. This framework uses a well-organized process that includes building models, preparing data, and evaluating their productiveness.

In [13] enhanced the artificial bee colony (ABC) metaheuristic optimization algorithm for energy forecasting by specifically addressing issues related to unpredictability and the presence of easily attainable local optimum solutions. They enhanced the performance of the method in many applications by improving the initialization and honey source update formulae, thereby avoiding unpredictability and the tendency to fall into local optimum solutions.

The organization of this paper is as follows: In part 2, the paper examines previous research on customer churn prediction, the techniques used in this field, and the process of choosing input data. It also demonstrates the use of statistical analysis approaches to predict customer attrition, including machine learning (ML), Explainable Artificial Intelligence (XAI) tools, and bee colony optimization strategies. The dataset is thoroughly described and the methods employed to build it are covered in the fourth section. Moreover, it explains Explainable Artificial Intelligence (XAI) and performance analytics. A thorough study of the recommended method is provided in the document's sixth section. The study's seventh section concludes with some closing observations and a discussion of possible directions for future investigation.

2 Related Works

Numerous studies have significantly improved our understanding of and application of machine learning techniques in predicting customer attrition in a wide range of industries. Based on earlier studies in this quickly expanding field, this study offers a unique and practical method for forecasting client attrition. By incorporating interpretability and explainability into the modeling process, the suggested approach improves customer churn prediction [14] in a variety of business sectors.

The study [15] uses cost-sensitive learning and sampling techniques to address data imbalances in the banking sector. Interval variables are used to impose penalties on minority misclassification and alter the class distribution, leading to random sample selection and higher noise tolerance. The method can be used in various service industries and technical applications, although it has been tried to predict attrition. The study shows how successful this tactic is at predicting consumer behavior.

The customer attrition data for a local telecom operator was categorized using Regression Stochastic Theory (RST) [16] in this study. When it came to accuracy, the evolutionary algorithm outperformed the other two reduction techniques. After multiple trials, the RST-based classifier performed the best. According to the study, RST may be used as a predictive system to identify client attrition and provides excellent methods for delivering better results.

The telecom industry employs AI and machine learning to predict customer attrition [17]. Data preprocessing, feature analysis, feature selection, data separation, and the use of well-known prediction models are the six steps of a suggested method. Hyperparameter optimization is accomplished by K-fold cross-validation. The examination of test results is done using an AUC curve and a confusion matrix. With Adaboost at 81.71% and XGboost at 80.8%, respectively, the Adaboost (AB) and XGB classifiers have the highest accuracy rates. Furthermore, their respective AUC scores of 84% are the highest.

The study offers a novel prediction model [18] that attempts to pinpoint potential customers who are likely to leave. Early warning signs for corrective action are provided by this paradigm. The Extreme Gradient Boosting "XGB" classifier is finally chosen as the best option after several machine learning techniques and algorithms are evaluated. To further improve the model's interpretability and visibility, SHapley values for possible feature combinations are generated.

The churn prediction feature selection model, which consists of a classification layer, two primary layers of GMDH-type neural networks, and an ensemble technique with a dynamic environment, is used [19]. This model exhibited superior performance compared to the other transfer learning techniques. A filter wrapper and an ensemble method [20] are employed to deal with unbalanced datasets in churn prediction. A neural network bagging was approach used [21] to test churn prediction and achieved 81% accuracy across many datasets. Forecast churn based on classification confidence was employed [22], and they discovered a direct link between distance and classification certainty. Churn detection algorithms discussed in [23] for attaining a high-profit model, which utilized ProfTree for classification and decision tree building. A new churn prediction approach [24] is used to maximize profit by utilizing LASSO and Tikhonov algorithms for regularization.

A churn prediction system for telecom operators has been developed [25] that employs machine learning in big data sets. They leveraged client social networks to extract SNA features. Jain et al. predicted customer churn Logit Boost and LR using WEKA ML with a real-world orange database.

Employ decision trees, C5.0, LR, and ANN models to forecast customer attrition, thereby improving customer management efficiency [26]. K-mean clustering and a rule-based algorithm, JRip, were used to forecast customer turnover in Nigeria. The data was processed and analyzed using the data mining tool, Weka, and applied to a dataset from a bank to identify clients who are likely to retain and design approaches to retain them [27].

The customer churn prediction method utilizes an enhanced balanced random forest model [28] that integrates balanced Random Forests and weighted Random Forests. This approach effectively tackles data distribution issues and exhibits superior accuracy when compared to DT (decision trees), support vector machines (SVM), and artificial neural networks (ANN).

Explainable Artificial Intelligence (XAI) has been used in many domains such as healthcare, medicine, and finance to improve the significance of features. This is achieved via the utilization of SHapley Additive exPlanations (SHAP) value and LIME methodologies [29]. Within the banking sector, the SHAP and LIME techniques have been effective in disrupting fraud detection programs, offering both broad and specific explanations for credit scoring model forecasts, and automating loan underwriting processes [30]. Nevertheless, there is an insufficient emphasis on the comprehensibility of machine learning models that forecast client attrition.

3 Methodology

Anticipating and averting client attrition is crucial for long-term expansion and consumer allegiance in the fiercely competitive retail sector [31]. Utilizing machine learning methods, namely ensemble models such as Random Forest and XGBoost, has become essential for this endeavor. These models are well-suited to capturing the subtleties of churn patterns in retail contexts because they leverage the benefits of many algorithms to increase prediction accuracy.

Using machine learning techniques, a forecasting framework for customer retention is developed and evaluated in this study as shown in Fig. 1.

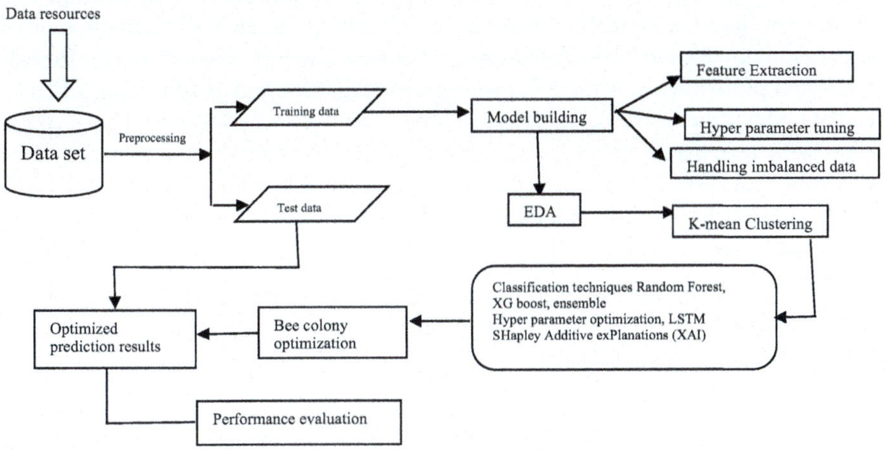

Fig. 1 Proposed framework for retail customer churn prediction

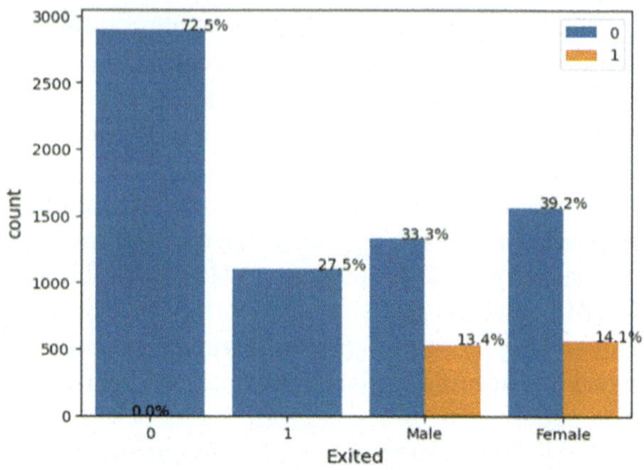

Fig. 2 Existing customer's percentage of gender wise

The dataset used in this study includes information on mall customers' socioeconomic status and frequency of visits. The dataset is the primary source of information used in the development and evaluation of models. This study analyzes past customer data from a retail mall to estimate client turnover using the supervised machine learning approach. The study plan comprises using ensemble learning techniques, specifically RF classifier and XGB, to build and evaluate forecasting models. The efficiency of this strategy is increased by utilizing the benefits of many algorithms.

3.1 Data Visualization

Understanding a dataset's features and distribution first requires visualization. The counterplot function in Seaborn was used to visually illustrate the distribution of the target variable "Exited," which indicates whether a client is churned or not. In addition, count plots were generated to investigate the distribution of additional variables related to the target variable, such as "Gender" and "VisitsPerMonth." These visualizations provide useful insights into the distribution of clients who churned vs those who remained, based on several category parameters (shown in Figs. 2 and 3). This aided in acquiring a basic understanding of the dataset's structure and completing subsequent analytic steps.

3.2 Ensemble Model

The ensemble model in this work combines the predictions of two separate classifiers, Random Forest and XGBoost, using soft voting. Random Forest is a common ensemble learning technique that creates many decision trees during training and then combines their predictions. XGBoost is a software implementation of the gradient

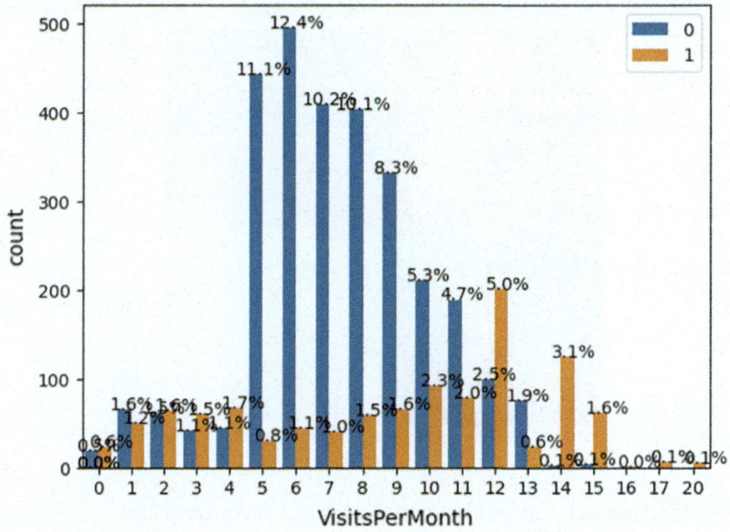

Fig. 3 Customer churn distribution

boosting approach that generates a series of decision trees step by step. Soft voting combines the projected probabilities from both models to select the class with the highest average probability as the final prediction.

By merging Random Forest with XGBoost [32], the ensemble model aims to increase forecast resilience and accuracy (shown in Table 1 and Fig. 4). This method provides a comprehensive approach to predicting customer attrition in retail settings, empowering businesses to make well-informed decisions to maximize customer retention strategies. The resampled dataset obtained through the application of the SMOTETomek approach was used to train the ensemble model.

Table 1 Performance measures of ensemble

Classification report for ensemble				
	Precision	Recall	Fl-score	Support
0	0.96	0.92	0.94	879
1	0.81	0.90	0.85	321
Accuracy			0.92	1200
Macro avg	0.88	0.91	0.90	1200
Weighted avg	0.92	0.92	0.92	1200
Accuracy score: 0.915				

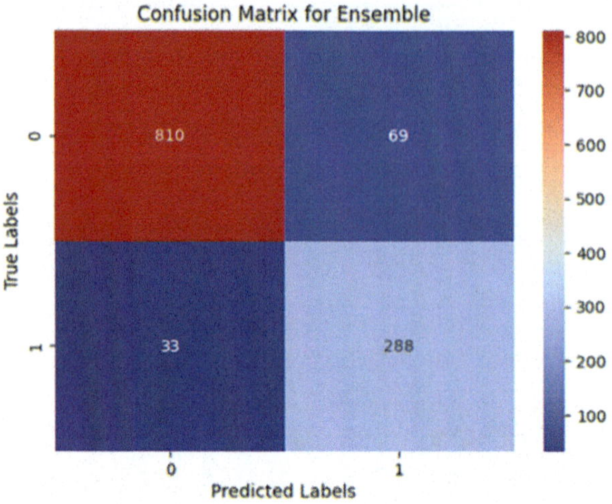

Fig. 4 Confusion matrix for ensemble model of customer churn prediction

3.3 Model Interpretability with SHAP Analysis

SHAP research provides important information about how the ensemble model, Random Forest, and XGBoost, two of its components, estimate customer turnover. By assigning a priority level to each element, SHAP sheds light on the factors that affect churn dynamics in retail settings. The optimal number of clusters (k) for classifying the resampled training data is determined using the K-means clustering approach. In SHAP analysis (shown in Fig. 5), this stage is crucial for producing background data. An approach known as the "elbow method" charts the sum of squared distances (SSDs) against the total number of clusters. It is employed to ascertain the threshold beyond which the SSD does not significantly improve with the addition of more clusters. Background data for SHAP analysis is then generated using the optimal value of k that has been found. As a result, by focusing on the most influential aspects of customer behavior, stakeholders can develop targeted retention strategies.

Developed by Lučić and Teodorović, the bee colony optimization (BCO) metaheuristic is a revolutionary approach in swarm intelligence that has proven effective in solving engineering and management problems. This paper provides an organized classification and analysis of the results that are produced when biogeography-based optimization (BCO) is used to simulate complex systems. The aim is to familiarize readers with the core ideas and identify potential applications in these fields [33].

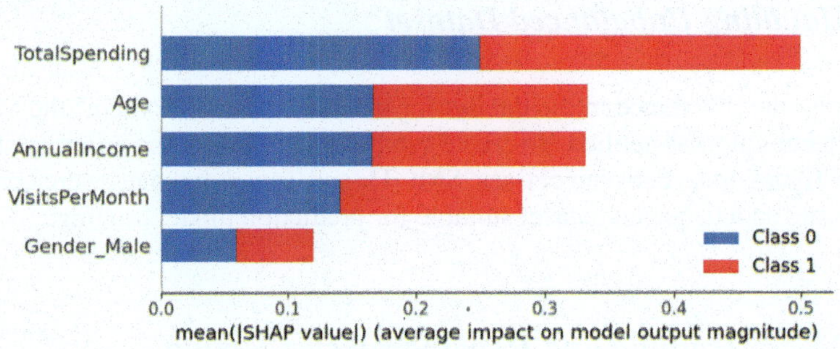

Fig. 5 Customer churn prediction SHAP Analysis

4 Implementation Analysis

Data is gathered from publicly accessible Kaggle, a large dataset taken from retail settings and specially designed for churn prediction of customers. The following features are included in the dataset: CustomerID, Gender, Age, Total Spending, VisitsPerMonth, and Exited.

4.1 Data Loading and Preprocessing

After the dataset was imported from Google Drive, extraneous data was removed by removing the "CustomerID" column. The aim variable "Exited" was visually represented using a Seaborn count plot, which was also utilized to investigate the relationship between "Gender," "VisitsPerMonth," and "Exited."

4.2 Data Splitting

The dataset is split into two distinct subsets, a test dataset comprising the remaining 30% of the data, and a training dataset comprising 70% of the data, to optimize the process of training and testing the model. The data used for training and testing are guaranteed to be distinct and non-overlapping by this segmentation. Next, the training dataset is used to train the proposed model, and the test dataset's unseen data is used to evaluate the model's performance.

4.3 Handling Unbalanced Dataset

To address the imbalance in the dataset for the aim variable "Exited," the SMOTE-Tomek technique was applied. To create synthetic samples for the minority class and remove Tomek links between classes, SMOTE and Tomek linkages were combined. This produced a balanced dataset suitable for prediction model training.

4.4 Model Training and Hyperparameter Tuning

GridSearchCV was used to optimize the RF and XGB hyperparameters during the training phase. The number of estimators, maximum samples, and maximum depth were the parameters that were changed for Random Forest. The learning rate, maximum depth, and number of estimators for XGBoost were the parameters that were changed.

4.5 Feature Importance

From the ensemble model, the individual base models are extracted to ascertain the features' relative importance/Next, a loop sequentially iterates over each base model to calculate their feature relevance. This significance is aggregated and tallied across all foundational models. Subsequently, the average feature significance is determined by dividing the total cumulative importance by the number of occurrences for each characteristic. Ultimately, the average feature importance is arranged in a decreasing order to provide a prioritized list of features according to their significance in the ensemble model. In addition to helping with feature selection and analysis, this process makes it easier to fully understand the combined characteristic contributions inside the ensemble model as shown in Fig. 6.

5 Performance Evaluation

The evaluation of model performance included the use of classification reports, accuracy scores, and confusion matrices for both the individual model and the ensemble model as shown in Figs. 7 and 8. In addition, several visualizations such as summary plots, heatmaps, pair plots, and individual prediction visualizations were used to effectively convey model insights. This ensured that these visualizations were readily accessible for generating real-time predictions in production scenarios.

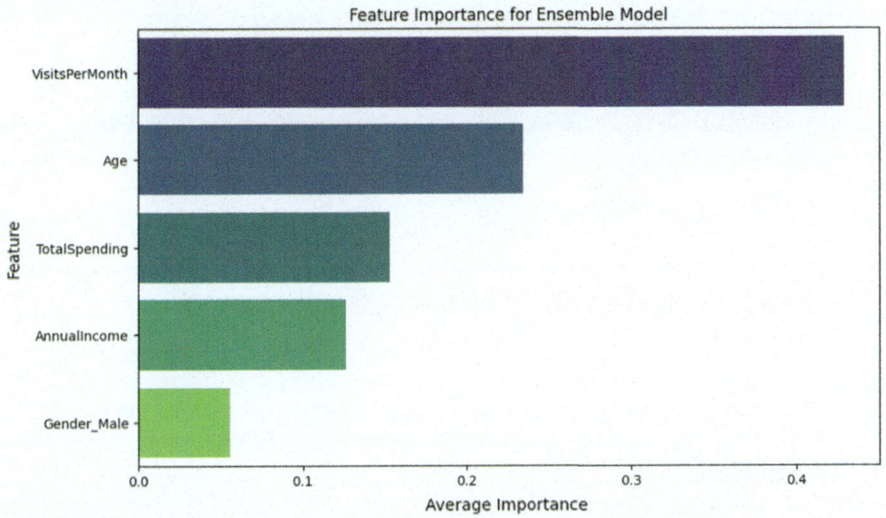

Fig. 6 Analysis of feature importance for the ensemble model

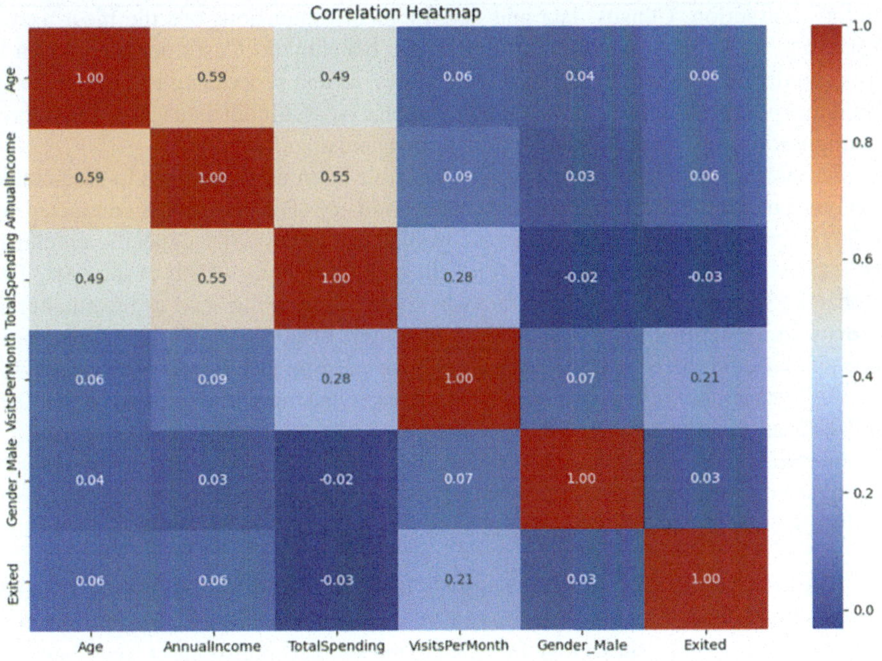

Fig. 7 Correlation heatmap

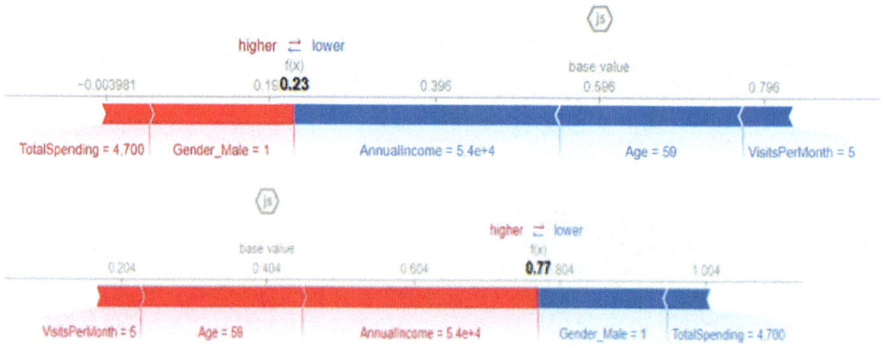

Fig. 8 Force plot

6 Discussion

CRM forecasts customer turnover based on transaction, location, and geodemographic data, with propensity influenced by product categories acquired after the initial transaction. Census data and geographical dimensions are the most useful variables, with spatial analyses serving as the benchmark. Customer impression is less significant than numerical and text reviews, as text processing is expensive and produces poor results. Successful integration necessitates input data validation, server configuration, and a continual assessment mechanism.

An analytical approach uses data and machine learning techniques to predict and analyze customer churn in the e-commerce industry. The analysis is conducted on a large-scale dataset obtained from a Swedish e-retailer, with a specific emphasis on aspects such as the recentness, frequency, and monetary worth of the data. The research [34] found that the churn rate was significantly influenced by the number of previous transactions, with over 75% of customers discontinuing after making only one purchase. Prediction models using machine learning techniques were created for the full dataset, as well as two specific subgroups: customers who completed either one or four transactions.

Interpretability and explainability of models improve customer trust and loyalty by using standardized automation methods that use machine learning models and Test-Driven Design. This will allow for precise predictions to be made at the beginning of customer interactions.

This work investigates explainability for four ML models by combining a model-explainable Ai method with the SHAP library and the bee colony optimization strategy. The results show that increasing the background dataset size causes a linear increase in runtime, which is correlated with the fact that SHapley values are computed via randomization.

7 Conclusion

In conclusion, an analytical strategy that takes advantage of data and machine learning methods. This study demonstrates how well the potent machine learning techniques of Random Forest and XGBoost may be combined to accurately forecast client attrition in the dynamic retail industry. To build a thorough framework, we gathered a variety of data from a crowded mall. Important steps in the development of this framework included data preparation, model training, parameter optimization, and a detailed assessment of the model's performance. Our strategy ensured the accuracy of our predictive models and provided insightful explanations for their forecasts by utilizing Explainable Artificial Intelligence techniques like SHapley Additive exPlanations (SHAP). This gave stakeholders a deeper understanding of how customers behave. Furthermore, adding RF and XGB to an ensemble model through soft voting yielded even better outcomes, improving our capacity for precise forecasting. Our findings have significant practical implications, giving merchants useful information to create tailored plans for customer retention and long-term financial sustainability.

8 Future Enhancement

In the future research, an analytical approach uses data and ML techniques. Possible improvements might be implemented by a comprehensive strategy targeting the advancement of several facets of customer churn prediction in retail settings. In addition, the use of a broader array of machine learning algorithms and ensemble approaches, beyond the limitations of RF and XGB, has the potential to cultivate a more varied and perhaps more precise ensemble model. Creating an interface or dashboard that is easy for users to navigate and understand and allowing them to interact with the models and their explanations will improve usability and make it easier to put the models into practical use. In addition, by incorporating continuous integration and deployment (CI/CD) pipelines, along with thorough unit testing and version control, it is possible to guarantee the stability and replicability of the entire machine learning process. This approach enables the development of scalable, dependable, and interpretable predictive models that can be applied in real-world scenarios.

References

1. Gunther CC, Tvete IF, Aas K, Sandnes GI, Borgan Ø (2014) Modelling and predicting customer churn from an insurance company. Scand Actuar J 1238(1):58–71
2. Chen K, Hu Y-H, Hsieh Y-C (2015) Predicting customer churn from valuable b2b customers in the logistics industry: a case study. Inf Syst e-Bus Manag 13:475–494

3. Dick AS, Basu K (1994) Customer loyalty: toward an integrated conceptual framework. J Acad Mark Sci 22:99–113
4. Bach MP, Pivar J, Jakovi´c B (2021) Churn management in telecommunications: hybrid approach using cluster analysis and decision trees. J Risk Financ Manag 14:544
5. Nie G, Rowe W, Zhang L, Tian Y, Shi Y (2011) Credit card churn forecasting by logistic regression and decision tree. Expert Syst Appl 38:15273–15285
6. Paschen U, Pitt C, Kietzmann J (2020) Artificial intelligence: building blocks and an innovation typology. Bus Horiz 63(2):147–155
7. Amin A, Shah B, Abbas A, Anwar S, Alfandi O, Moreira F (2019) Features weight estimation using a genetic algorithm for customer churn prediction in the telecom sector. In: Proceedings of the world conference on information systems and technologies, Galicia, Spain, Apr 2019
8. Han D, Zhao N, Shi P (2019) Gear fault feature extraction and diagnosis method under different load excitation based on EMD, PSO-SVM, and fractal box dimension. J Mech Sci Technol 33(2):487–494
10. Sina Mirabdolbaghi SM, Amiri B (2022) Model optimization analysis of customer churn prediction using machine learning algorithms with focus on feature reductions, discrete dynamics in nature and society. Hindawi. ISSN:1026-0226. https://doi.org/10.1155/2022/5134356
11. https://javilopezcastillo.medium.com/telecom-churn-analysis-using-clustering-techniques-for-customer-segmentation-4cdb7318f672
12. Biecek P, Tomasz B (2021) Explanatory model analysis: explore, explain, and examine predictive models. CRC Press, Boca Raton, FL, USA
13. Özdemir D, Dörterler S, Aydın D (2022) A new modified artificial bee colony algorithm for energy demand forecasting problem. Neural Comput Appl 34(20):17455–17471
14. Kakulapati V et al (2023) A Churn prediction of clinical decision support recommender system. ICAIHC 2022, Smart innovation, systems and technologies (SIST), pp 371–379. Springer Nature Singapore, @2023, ISBN-13: 9789811960673. https://doi.org/10.1007/978-981-19-6068-0_47
15. Xie Y, Li X, Ngai EWT, Ying W (2009) Customer churn prediction using improved balanced random forests. Expert Syst Appl 36(3), Part 1: 5445–5449. https://doi.org/10.1016/j.eswa.2008.06.121
16. Makhtar M, Nafis S, Mohamed MA, Awang MK, Rahman MN, Deris MM (2018) Churn classification model for local telecommunication company based on rough set theory. J Fundam Appl Sci 9:854. https://doi.org/10.4314/jfas.v9i6s.64
17. Lalwani P, Mishra M, Chadha J, Sethi P (2022) Customer churn prediction system: a machine learning approach. Computing 104:1–24. https://doi.org/10.1007/s00607-021-00908-y
18. Maan J, Maan H. Customer churn prediction model using explainable machine learning. Int J Comput Sci Trends Technol (IJCST) V11(1):33–38. ISSN: 2347–8578, https://doi.org/10.48550/arXiv.2303.00960
19. Xiao J, Wang Y, Wang S (2013) A dynamic transfer ensemble model for customer churn prediction. In: Proceedings of the 2013 6th international conference on business intelligence and financial engineering BIFE, vol 43, no 1, pp 115–119, Hangzhou, China, Nov 2013
20. Idris A, Khan A (2016) Churn prediction system for telecom using filter-wrapper and ensemble classification. Comput J 60(3), Article ID bxv123
21. Saghir M, Bibi Z, Bashir S, Khan FH (2019) Churn prediction using neural network based individual and ensemble models. In: Proceedings of the 2019 16th international Bhurban conference on applied sciences and technology, pp 634–639. IBCAST, Islamabad, Pakistan, Jan 2019
22. Amin A, Shah B, Abbas A, Anwar S, Alfandi O, Moreira F (2019) Features weight estimation using a genetic algorithm for customer churn prediction in the telecom sector. In: Proceedings of the world conference on information systems and technologies, Galicia, Spain, Apr 2019
23. Höppner S, Stripling E, Baesens B, vanden Broucke S, Verdonck T (2020) Profit-driven decision trees for churn prediction. Eur J Oper Res 284(3):920–933
24. Maldonado S, López J, Vairetti C (2020) Profit-based churn prediction based on Minimax probability machines. Eur J Oper Res 284(1):273–284

25. Ahmad AK, Jafar A, Aljoumaa K (2019) Customer churn prediction in telecom using machine learning in big data platform. J Big Data 6(1):24–24
26. Hudaib A, Dannoun R, Harfoushi O, Obiedat R, Faris H (2015) Hybrid data mining models for predicting customer churn. Int J Commun Netw Syst Sci 8:91–96
27. Oyeniyi AO, Adeyemo AB (2015) Customer churn analysis in banking sector using data mining techniques. Afr J Comput ICT 8(3):165–174
28. Xie Y, Li X, Ngai EWT, Ying W (2009) Customer churn prediction using improved balanced random forests. Expert Syst Appl 36:5445–5449
29. Psychoula I, Gutmann A, Mainali P, Lee SH, Dunphy P, Petitcolas FAP (2021) Explainable machine learning for fraud detection. arXiv
30. Sachan S, Yang JB, Xu DL (2020) Benavides DE, Li Y.: An explainable AI decision-support-system to automate loan underwriting. Expert systems with applications 144
31. Ahmad AK, Jafar A, Aljoumaa K (2019) Customer churn prediction in telecom using machine learning in the big data platform. J Big Data 6:28. https://doi.org/10.1186/s40537-019-0191-6
32. Kakulapati V et al XGBoost analyses of Classification of yoga is published in SSRN Elsevier digital library, id 394812. ISSN: 1556–5068. https://doi.org/10.2139/ssrn.3948127
33. Teodorović D (2009). Bee colony optimization (BCO). In: Lim CP, Jain LC, Dehuri S (eds) Innovations in swarm intelligence. Studies in computational intelligence, vol 248. Springer, Berlin, Heidelberg. https://doi.org/10.1007/978-3-642-04225-6_3
34. Lalwani RP, Mishra MK, Chadha JS, Sethi P (2022) Customer churn prediction system: a machine learning approach. Computing 104(2):271–294

Efficient Two-Bit Magnitude Comparator with Hybrid Logic Approach

R. Sindhuja, E. John Alex, and K. Niranjan Reddy

Abstract This research presents a design for a two-bit magnitude comparator utilizing various technologies including CCMOS (Conventional CMOS), TGL (Transmission Gate Logic), PTL (pass transistor logic), GDI (gate diffusion logic), and FinFET. By comparing FinFET with existing models, significant improvements in Power Delay Product (PDP) are observed. The FinFET-based comparator demonstrates superior performance in terms of power consumption, time delay, and transistor area. Utilizing Cadence Virtuoso, simulated waveforms depicting greater than, equal to, and less than scenarios are generated, highlighting the efficacy of FinFET technology in achieving greater control device efficiency and compactness. Comparing with CMOS, FinFET have more than 70% benefits for time delay and power consumption.

Keywords CCMOS · TGL · PTL · GDI · FinFET

1 Introduction

Several intelligent systems have been developed as a result of the assessment of contemporary, extremely complex electrical circuits and devices. High-performance VLSI circuits with rapid calculation and acceptable power loss are needed for these smart systems. The low power dissipation and acceptable speed requirements for portable devices make VLSI circuit performance crucial. Furthermore, in order to maintain their high functionality and high integration density, new microprocessors need extremely efficient circuits with little area usage. Consequently, there is a great deal of interest in VLSI circuit design approaches among researchers.

R. Sindhuja (✉)
M Tech Scholar, CMRIT, Hyderabad, India
e-mail: sindhuja6301@gmail.com

E. John Alex · K. Niranjan Reddy
Assosiate Professor, Department of ECE, CMRIT, Hyderabad, India

© The Author(s), under exclusive license to Springer Nature Singapore Pte Ltd. 2025
H. Zen et al. (eds.), *Soft Computing and Signal Processing*, Lecture Notes in Networks and Systems 1200, https://doi.org/10.1007/978-981-97-9926-8_48

An arithmetic component in VLSI circuits called the MC design determines whether a given binary number is equal to, greater than, or less than another binary number. To put it briefly, MC is employed in binary number comparison. Typically, MC receives two binary values as inputs and delivers the results of the binary comparison. Microprocessors employ MC extensively for sorting instructions. To ensure high performance, digital signal, and parallel processing processes require effective MC circuits. Because MC is used in high performance systems, it is imperative to use optimal MC design.

This work uses a hybrid logic language to demonstrate a two-bit MC architecture. In particular, CCMOS and TGL, two different CMOS VLSI circuit design methodologies, are used in this design. The performance comparison of the suggested MC with five alternative designs has been used to assess the design's efficiency. The MC used in this study maintained a low transistor count while demonstrating acceptable performance levels. These factors make the suggested MC very successful for contemporary VLSI systems.

A comparison is made between current technologies such as CMOS and FinFET. It also discusses the shortcomings of current technology and how FinFET, a new development, might address them. The benefits of FinFET significantly contribute to the improvement of VLSI. Transient response is obtained by simulating the two-bit magnitude comparator circuits. Values for power and delay are found for circuits that are less than, greater than, and equal to [1].

FinFETs are divided into two fundamental categories: There are two types of FinFET transistors: independent gate (IG) and short gate (SG). IG FinFETs have four terminals, whereas SG are so referred to as three terminal FinFETs [2]. Through shorted-gate (SG) mode, independent-gate (IG) mode with low-power (LP) mode is achieved using merging parallel transistors with separate inputs [3].

2 Two-Bit Magnitude Comparator

The magnitude comparator is a crucial tool that is generally employed in situations where it becomes necessary to compare two n-bit integers at some point. The two n-bit values, A and B, are accepted by the magnitude comparator's input nodes. The comparison of the two numbers is represented by three output nodes, which are labeled as greater than (»), equal to (=), and less than («) [4]. This principle is demonstrated in the block diagram as shown in Fig. 1.

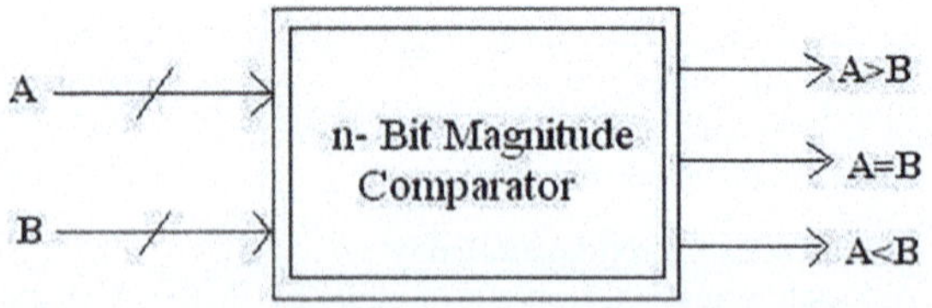

Fig. 1 Block diagram of two-bit MC

Using Karnaugh maps and the necessary comparator's Truth table, we can quickly and simply derive the equations for the three associated output nodes when implementing a two-bit magnitude comparator.

3 Prior Designs

Many design approaches for VLSI circuit implementation have been developed as a result of the quick changes in circuit design, including CCMOS, PTL, Transmission Gate Logic (TGL), Gate Diffusion Input Technique (GDI), etc. Different MC designs have been developed as a result. In PTL two-bit MC is implemented with 40 transistors. Voltage deterioration is a significant drawback of PTL-based designs. The issue of voltage degradation intensifies significantly in lower CMOS technology nodes. Because of this, there are currently very few designs that are only made using PTL and no swing restoring transistors. Weak driving power is the resultant significant drawback of this voltage decrease. New designs have arisen as a result of numerous PTL-related concerns [5].

Voltage deterioration and low drive power in PTL are resolved via CCMOS logic. Sixty-six transistors are used to build CCMOS two-bit MC. Nevertheless, the large number of transistors causes a variety of additional issues. These issues include large silicon area and high signal input impedance.

Because GDI-based circuit design may perform Boolean operations with fewer transistors, it has gained popularity in the present era. In contrast to PTL, however, the incapacity to provide a complete swing output leads to excessive DC power consumption, poor drive power, and voltage deterioration. Just 28 transistors are needed to implement the GDI two-bit MC design [6].

Another method that researchers employ to give VLSI circuits full swing output is TGL. TGL's drive power is inferior to CCMOS logic, although being more than PTL's. Sixty-six transistors are used to build the TGL two-bit MC architecture. Hybrid design methodology leverages many logic techniques to create Boolean functions. By utilizing the benefits of multiple design methods, hybrid design method implementation of Boolean functions has gained popularity. Two-bit MC is built with 42 transistors.

3.1 Two-Bit Comparator: CMOS Transistor Design

The two networks that make up the CMOS logic technology are the PMOS pull-up network and the NMOS pull-down network. The design of the CMOS logic circuit makes it obvious that the system is guiding simultaneously [7].

In actuality, CMOS logic is a multi-input stretching of CMOS inverters. The transistor performance of a traditional two-bit magnitude comparator is shown in Fig. 2.

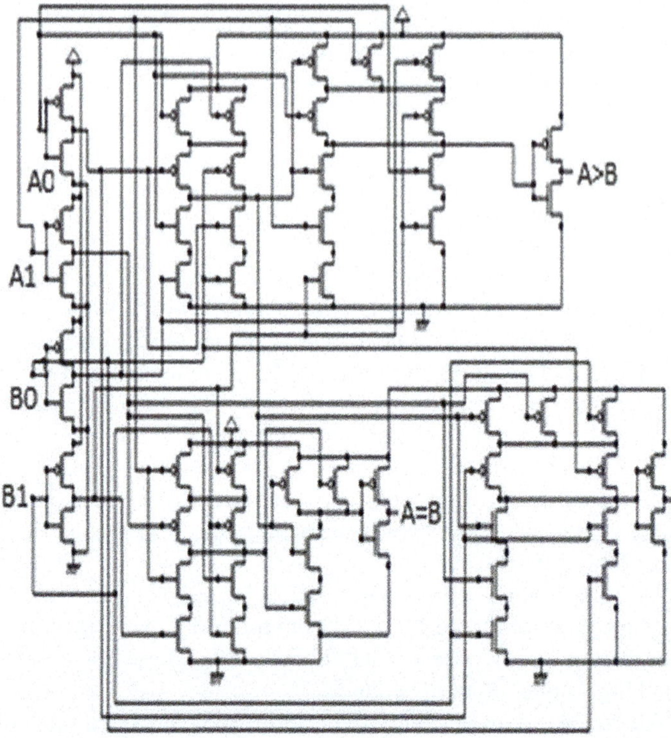

Fig. 2 Conventional CMOS logic two-bit comparator

3.2 Transmission Gate Logic Implementation of a Dual-Bit Comparator

Because the transistor stack height is tiny, fewer transistors are needed and commendatory input signals are not needed, transmission gate logic operates at a higher speed and consumes low power compared to standard CMOS. One NMOS and one PMOS transistor, connected in parallel, make up the transmission gate. Figure 3 displays the transmission gate's visual symbol.

It will function as a switch that, depending on the requirement, selectively blocks or passes a signal using complimentary biasing of the control gates, from the input to the output. The transmission gate logic truth table is shown in Table 1.

When both transistors are activated, they establish a low impedance pathway between the input and the output, facilitating the seamless transmission of signals when node g maintains a high logic level ($g = 1$). Conversely, if the voltage at node g drops below a certain threshold ($g = 0$), signal transfer from the input to the output is obstructed. Under such conditions, both transistors become deactivated, creating a high impedance channel between the input and the output, thus preventing signal propagation [8].

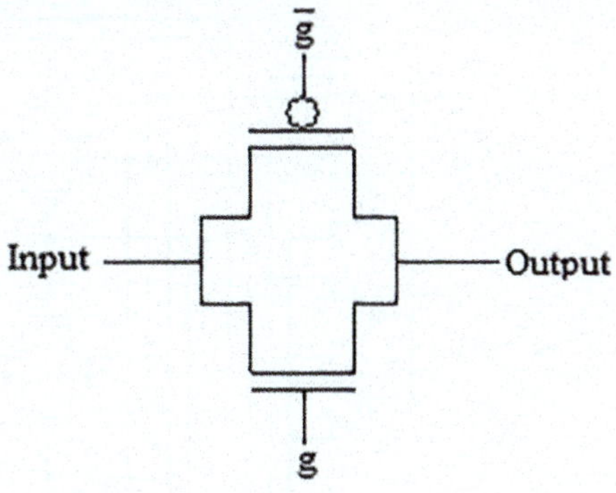

Fig. 3 Graphical representation of transmission gate

Table 1 Truth table of the transmission gate logic

Control input (g)	Input	Output
High logic level	High logic level	High logic level
High logic level	Low logic level	Low logic level
Low logic level	Don't care (x)	High impedance

The two-bit magnitude comparator built upon transmission gate logic incorporates a total of 74 transistors. It will have a lower packing density and require more power due to the huge number of transistors. However, in today's world, the three most important variables in a VLSI circuit's magnitude are power, speed, and area. This design is thus not thought to have better performance. Therefore, the focus of this study was on a two-bit magnitude comparator developed using modified transmission gates and only 30 transistors. The design that is being suggested reduces power usage while increasing speed and packing density. Every transmission gate in the suggested design functions as an AND gate, much like in the traditional gate design of the two-bit magnitude comparator.

It follows that the circuit needed the fewest transistors [9]. Figure 4 shows the suggested two-bit magnitude comparator based on a transmission gate.

4 PTL and GDI-Based Magnitude Comparator

The following subsections provide a description of the proposed two-bit MC's circuit design methodology.

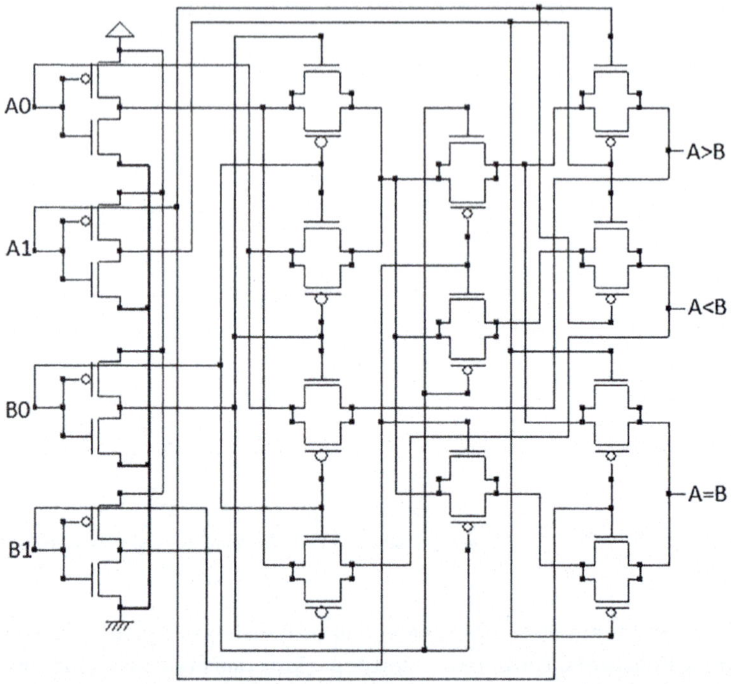

Fig. 4 Two-bit MC design using the TGL technique

4.1 Boolean Expression for Two-Bit MC

It is possible to represent the state and functionality of two-bit MC which is given in Table 2. The words A and B indicates inputs.

Table 2 Operation table of two-bit MC

Inputs		Outputs		
A1, B1	A0, B0	AGB	ALB	AEB
A1 = B1	A0 = B0	0	0	1
A1 = B1	A0 > B0	1	0	0
A1 = B1	A0 < B0	0	1	0
A1 > B1	X	1	0	0
A1 < B1	X	0	1	0
AGB: A > B (A is greater than B)				
ALB: A < B (A is smaller than B)				
AEB: A = B (A and B are equal)				
X = Don't care condition				

The outputs AGB, AEB, and ALB are indicated. The definitions of AGB, AEB, and ALB are given in Table 2. The Boolean functions presented below serve to define the operational principles of the two-bit magnitude comparator (MC).

$$\text{AEB} = \text{XN}_0\text{XN}_1 \tag{1}$$

$$\text{ALB} = \text{B}_1 + \text{B}_0\text{XN}_1 \tag{2}$$

$$\text{AGB} = \text{A}_1 + \text{A}_0\text{XN}_1, \tag{3}$$

where

XN_0 represents XNOR operation between A_0 and B_0.
XN_1 represents XNOR operation between A_1 and B_1.

Thus, it can be shown from the Boolean Eqs. (1)–(3) and the assertions XN_0 and XN_1 that designing AND and XNOR gates is necessary to build a two-bit MC. In the end, a complex VLSI circuit would be needed to provide outputs.

4.2 Preliminary Block Diagram for Dual-Bit Magnitude Comparator

Figure 5 displays the comprehensive block diagram of the suggested two-bit MC. Based on the findings shown in Fig. 5, it is evident that the suggested two-bit MC design utilizes a hybrid design approach, utilizing more techniques than VLSI logic circuit design. The XNOR circuit is designed using the PTL approach. Incorporating swing restorative transistors alongside PTL in the design aims to ensure complete swing output, boost circuit drive power, and reduce DC power loss stemming from the threshold voltage decline within PTL [10].

In order to deliver substantial drive power at the output terminals, CCMOS-based circuits are employed, even in scenarios where the input circuit employs PTL. The integration of PTL on the input side and CCMOS on the output side has led to a reduction in both transistor count and DC power loss, all while preserving the circuit's driving capabilities. This enhancement has notably boosted the circuit's adaptability [11].

The GDI technique necessitates only 2 transistors, while the CMOS technique demands 6 transistors. GDI technique has swing degradation problem, so that FinFET technology is introduced.

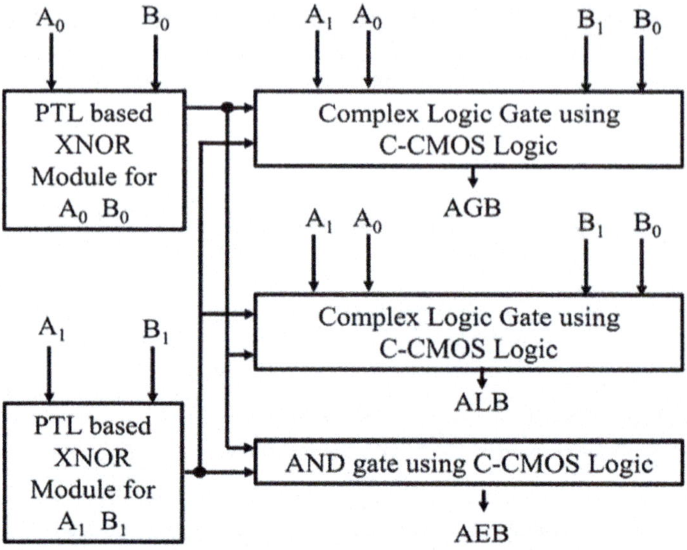

Fig. 5 Block diagram of two-bit MC of CMOS PTL

5 Proposed Model

5.1 Comparator Using FinFET Model

FinFET devices are extensively researched due to their potential utility in the production of CMOS integrated circuits. Deducing MOSFET size has a significant impact on electrostatic characteristic. The random differences in attributes result in a divergence effect, which is critical for design and manufacturing. At higher electric fields, FinFETs exhibit "volume inversion." The FinFET is distinguished by the fact that the conducting channel is encircled by a thin silicon "Fin" that serves as the device's body. The fin thickness (measured in the direction from source to drain) influences the device's effective channel length [12].

FinFET stands for Fin Field Effect Transistor. It's also known as multi-gate MOS transistors. With FinFET, the conducting channel is covered with a silicon wafer, which serves as the transistor's substrate. The device's thickness determines the channel length. It is not a planar transistor. It normally consists of two gates. This double gate transistor is made of either silicon wafers or silicon on insulator (SoI). The base in which it is constructed determines the kind of FinFET.

FinFET is the most recent technology in the VLSI area. It has the following advantages due to its multi-gate structure: Drive current is higher, channel length is shorter, implying a more compact design; power consumption is lower; performance is better; and speed is high. It is utilized in processors because it is small, allowing for the use of several FinFET transistors inside a given space. The CPUs of the Samsung

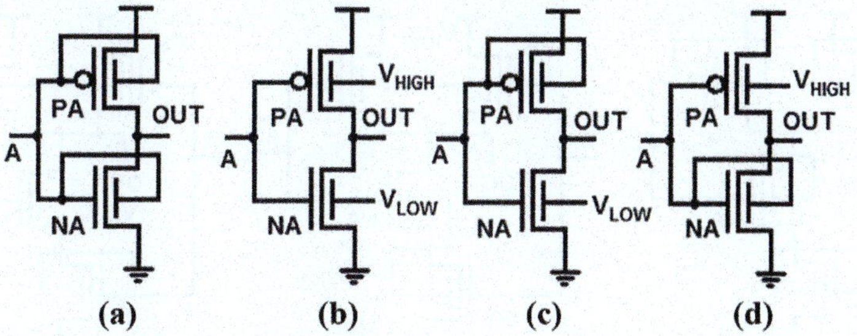

Fig. 6 INV gates. **a** SG, **b** LP, **c** IGn, **d** IGp

Galaxy S6 employ FinFET technology. In the future, the channel length will be reduced even more to FinFET 10 nm. The table shows the symbols, parameters, and values associated with double gate FinFETs.

Figure 6 illustrates four potential INV configurations utilizing SG/IG-mode FinFETs: SG-, low-power (LP-), IGn-, and IGp-INV. The SG-INV setup exclusively employs SG-mode n/p-FinFETs, resulting in a highly condensed structure [9]. In the LP-INV setup (Fig. 6b), the back gate of the pull-up (PA) and pull-down (NA) networks is biased to VHIGH and VLOW respectively, necessitating a more intricate design with a 36% larger footprint compared to the SG-INV of size X2. On the other hand, IGn-INV and IGp-INV occupy a comparable area to LP-INV, thanks to the multifin IG-mode FinFET back-gate connections.

Among the NAND2 gates, while SG-NAND2 exhibits the most condensed structure, LP-NAND2 consumes 27% more area than size X2 SG-NAND2 due to its staggered pull-up network of parallel FinFETs and shared back-gate contacts for the series pull-down FinFETs. Mixed-terminal (MT-) NAND2 matches LP-NAND2 in terms of area, with NB in SG mode (Fig. 7c). IG and IG2-NAND2 consolidate the parallel FinFETs of the pull-up network into a single p-FinFET, resulting in the same area as SG-NAND2. XT-NAND2, a variant of MT-NAND2, features both FinFETs of the pull-down network in SG mode, maintaining an identical layout area (not depicted). XT2-NAND2, another variant of MT-NAND2, features both parallel FinFETs of the pull-up network in SG mode, enabling a compact structure with the same area as SG-NAND2.

Employing numerous gates in a FinFET facilitates the control and reduction of off-state leakage current, while ensuring a significant current flow during the drive or ON state. Electrical channel control is also aided by these multigates. FinFET technology can help resolve the performance problems with planar transistors. In ULSI (Ultra Large-Scale Integrated Circuits) goods, FinFET is crucial. Inputs A0, A1, B0, B1 are used in this comparator, where A0, B0 represents the LSB bit and A1, B1 the MSB bit for the two values under consideration, A and B.

In the less than functional implementation of the two-bit magnitude comparator illustrated in Fig. 8, the N-FINFET logic signifies A < B. Conversely, in an equal-to-functional implementation for the two-bit magnitude comparator depicted in Fig. 9,

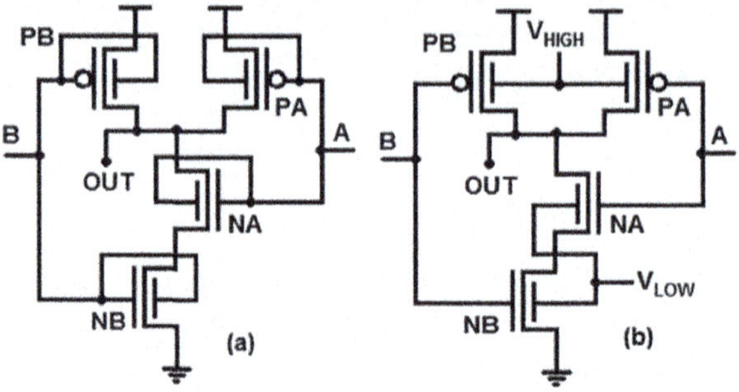

Fig. 7 NAND2 gates. **a** SG, **b** LP, **c** MT

N-FINFET logic denotes A = B. Moreover, in the greater than functional implementation of the two-bit magnitude comparator portrayed in Fig. 10, N-FINFET logic is utilized to represent A = B.

Assume that A and B are two-bit binary inputs, where the MSB bits are A0, B0, and the LSB bits are A1, B1. The phrases that follow are:

$$L = A1'B1 + (A0'B0B1) + (A1'A0'B0) \quad (4)$$

$$E = (A0B0 + A0'B0') \cdot (A1B1 + A1'1) \quad (5)$$

$$G = A1B1' + (A0B1'B0') + (A1A0B0') \quad (6)$$

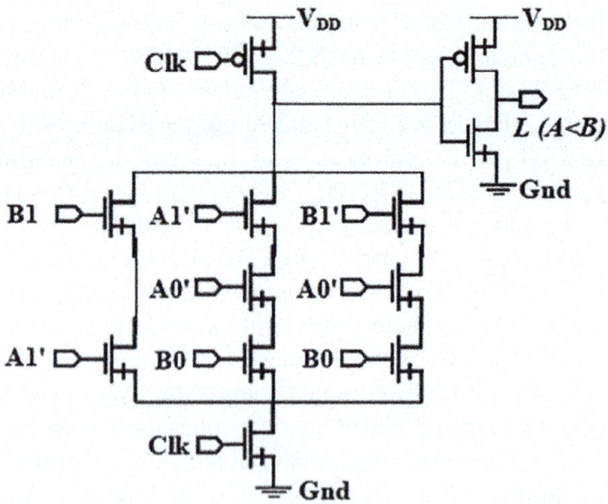

Fig. 8 Lesser than functional circuit of two-bit magnitude comparator

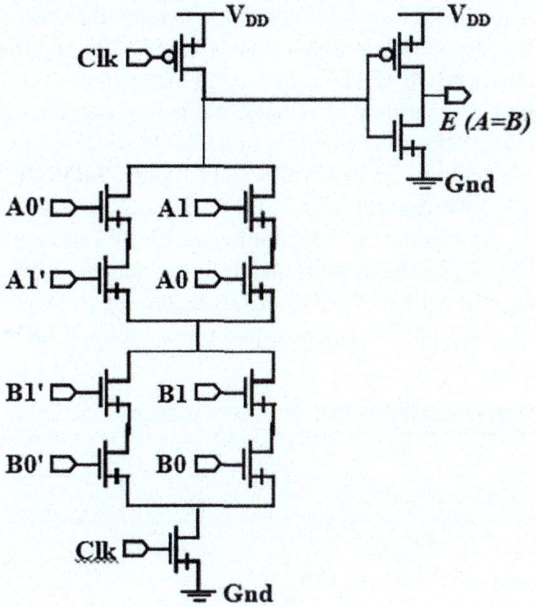

Fig. 9 Equal to functional circuit of two-bit magnitude comparator

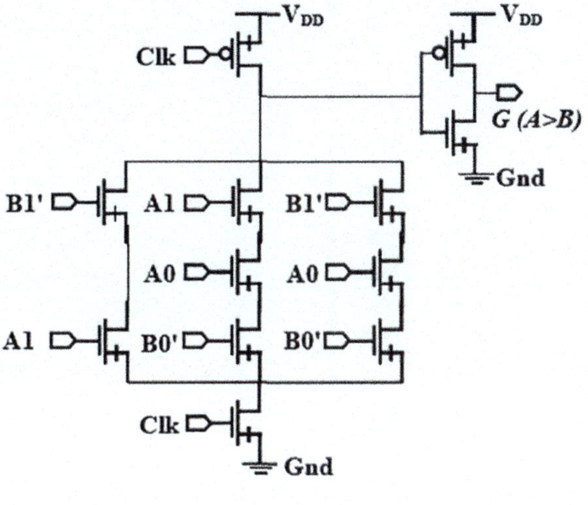

Fig. 10 Greater than functional circuit of two-bit magnitude comparator

6 Simulation Results

The simulation results for FinFET technology derived using Cadence Virtuoso tool with spectra models are shown in Figs. 11, 12, and 13. The comparator's entire functionality is represented by the transitory response mentioned above. The comparison table shown in Table 3 clearly shows that in FinFET technology the power consumption and propagation time delay is decreased as compared with CMOS comparators.

The FinFET technology demonstrates the lowest power consumption among the listed options, with a value of 44.64 μW, significantly lower than all CMOS variants. CMOS (PTL) exhibits the lowest power consumption among the CMOS variants, consuming 145.2 μW, indicating potential efficiency gains compared to other CMOS designs. CCMOS and CMOS (GDI) show relatively higher power consumption compared to CMOS (PTL) and CMOS (TGL), with values of 250 μW and 223 μW, respectively.

Among the listed technologies, FinFET demonstrates the lowest propagation delay of 7.15 ps showcasing its ability to achieve high-speed operation despite the lower supply voltage. CMOS (GDI) technology follows closely with a propagation delay of 7 ns, indicating fast signal propagation. CMOS (PTL) and CMOS (TGL) exhibit

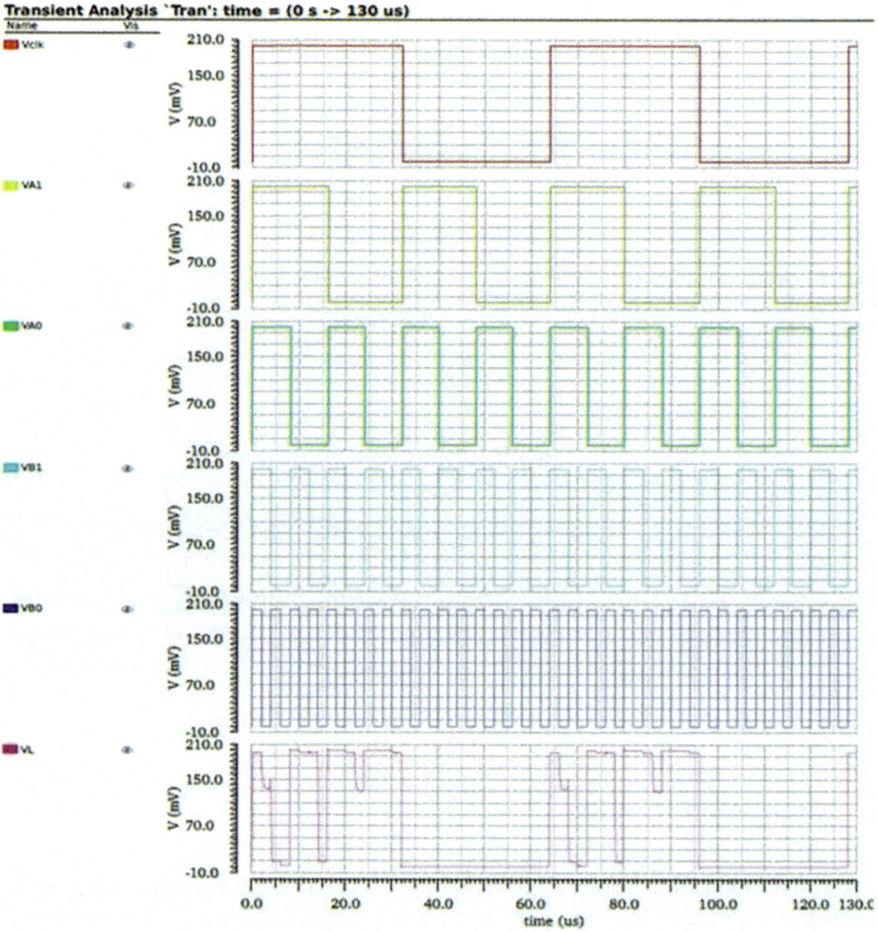

Fig. 11 Transient response of two-bit lesser than circuit

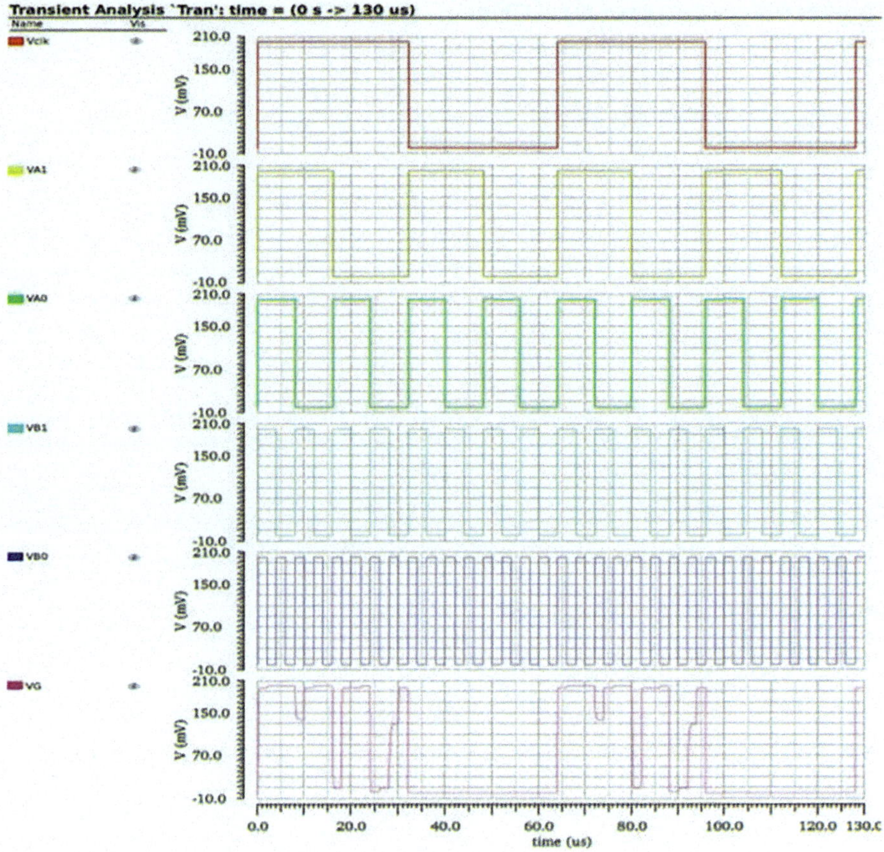

Fig. 12 Transient response of two-bit greater than circuit

propagation delays of 6 ns and 4.1 ns, respectively, indicating slower signal propagation compared to CMOS (GDI) and FinFET. CCMOS shows the highest propagation delay among the listed options, with a value of 1.679 ns, suggesting relatively slower signal propagation compared to other CMOS variants.

FinFET technology boasts the highest sampling frequency of 250 MHz, making it suitable for high-speed applications requiring rapid signal processing. CMOS (GDI) follows with a sampling frequency of 200 MHz, offering competitive performance in terms of signal processing speed. CMOS (PTL) and CCMOS share the same sampling frequency of 100 MHz, providing moderate performance suitable for various applications. CMOS (TGL) lags behind with a sampling frequency of 50 MHz, indicating lower performance in processing high-frequency signals compared to other options.

All technologies listed, including CMOS (PTL), CCMOS, CMOS (TGL), CMOS (GDI), and FinFET, have the same channel length of 32 nm. The FinFET technology operates at the lowest supply voltage of 0.8 V, indicating potential advantages in

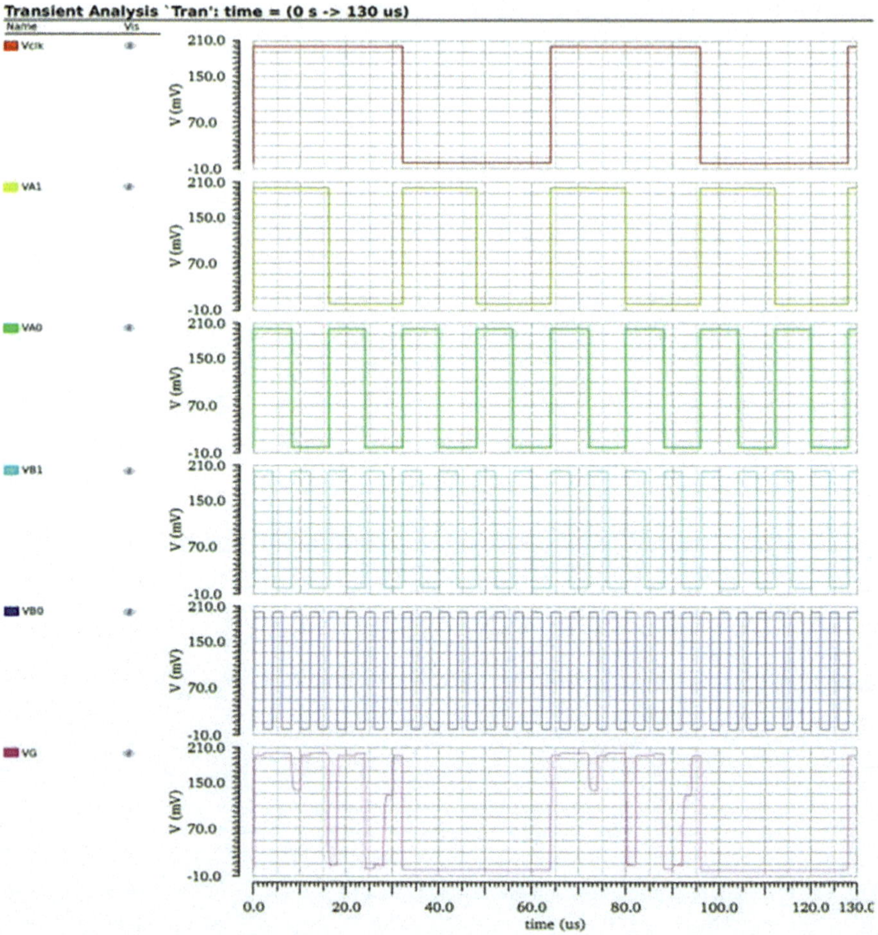

Fig. 13 Transient response of two-bit equal to circuit

power efficiency and reduced energy consumption compared to the CMOS variants. All CMOS variants, including CMOS (PTL), CCMOS, CMOS (TGL), and CMOS (GDI), operate at a supply voltage of 1.2 V, suggesting similar power supply requirements among these technologies. The difference in supply voltage between FinFET and CMOS technologies may impact their power consumption, with FinFET likely having lower power consumption due to its lower operating voltage.

Figure 14 gives the comparative analysis for FinFET and various CMOS as well Table 3 gives the detailed analysis of various technologies like channel length, supply voltage, power consumption, sampling frequency, and propagation delay implies in CMOS (PTL), CCMOS, CMOS (TGL), CMOS (GDI), and FinFET.

Table 3 Performance comparison among the proposed and prior comparator designs

Technology	CMOS (PTL)	CCMOS	CMOS (TGL)	CMOS (GDI)	FinFET (proposed)
Channel length (nm)	32	32	32	32	32
Supply voltage (V)	1.2	1.2	1.2	1.2	0.8
Power consumption (μW)	145.2	250	156.5	223	44.64
Sampling frequency (MHz)	100	100	50	200	250
Propagation delay	6 ns	1.679 ns	4.1 ns	7 ns	7.15 ps

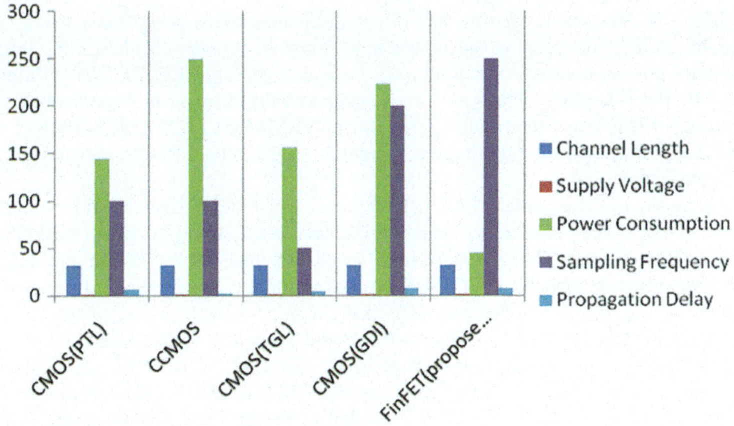

Fig. 14 Comparison analysis for FinFET and CMOS

7 Conclusion

CCMOS shows the highest propagation delay among the listed options, CMOS (TGL) lags behind with a sampling frequency among all. CMOS PTL is performing in average comparatively remaining CMOS, FinFET technology is more scalable, higher performance and better area efficiency than CMOS GDI. The FinFET technology operates at the lowest supply voltage of 0.8 V, indicating potential advantages in power efficiency in terms of lower static power consumption and reduced power dissipation during switching, and reduced energy consumption compared to the CMOS variants. Comparatively FinFET technology offers a significant advancement over traditional CMOS transistors, providing improved performance, lower power consumption, and enhanced reliability. Comparing with CMOS, FinFET have more than 70% benefits for time delay and power usage.

References

1. Rajalakshmi R, Aruna Priya P (2014) Design and analysis of a 4-bit low power universal Barrel-shifter in 16 nm FinFET technology. In: IEEE 2014 international conference on advanced communication, control and computing technologies (ICACCCT)
2. Rostami M, Mohanram K (2011) Dual-V_{th} independent-gate FinFETs for low power logic circuits. IEEE Trans Comput-Aided Des 30(3):337–349
3. Morad MJA, Talebiyan SR, Pakniyat E (2015) Design of new low-power high-performance full adder with new XOR-XNOR circuit. In: 2015 international congress on technology, communication and knowledge (ICTCK), Mashhad, pp 153–158
4. Sharma G, Arora H, Chawla J, Ramzai J (2015) Comparative analysis of a 2 bit magnitude comparator using various high performance techniques. In: 2015 international conference on communications and signal processing (ICCSP), Melmaruvathur, pp 0079–0083
5. Lubaba S, Faisal KM, Islam MS, Hasan M (2020) Design of a two-bit magnitude comparator based on pass transistor, transmission gate and conventional static CMOS logic. In: 2020 11th international conference on computing, communication and networking technologies (ICCCNT), Kharagpur, pp 1–5
6. Shekhawat V, Sharma T, Sharma KG (2014) 2-bit magnitude comparator using GDI technique. In: IEEE 2014 recent advances and innovations in engineering (ICRAIE). International conference on recent advances and innovations in engineering (ICRAIE-2014), Jaipur, India
7. Bhoj AN, Jha NK (2013) Design of logic gates and flip-flops in high-performance FinFET technology. IEEE Trans Very Large Scale Integr (VLSI) Syst 21(11):1975–1988
8. Anjuli SA (2013) Two-bit magnitude comparator design using different logic styles. Int J Eng Sci Invent 2(1):13–24
9. Pal RS, Sharma S, Dasgupta S (2017) Recent trend of FinFET devices and its challenges: a review. In: IEEE 2017 conference on emerging devices and smart systems (ICEDSS)
10. Hossain MM, Biswas SN (2019) A high speed and low power 32 nm FinFET dynamic latch comparator design. In: IEEE 2019 1st international conference on advances in science, engineering and robotics technology (ICASERT)
11. Hasan M, Saha UK, Hossain MS, Biswas P, Hossein MJ, Dipto MAZ (2019) Low power design of a two bit magnitude comparator for high speed operation. In: 2019 international conference on computer communication and informatics (ICCCI), Coimbatore, Tamil Nadu, India, pp 1–4
12. Mukherjee DN, Panda S, Maji B (2017) Design of low power 12-bit magnitude comparator. In: 2017 devices for integrated circuit (DevIC), Kalyani, pp 103–109

Alzheimer's Disease Multi-task Classification Using ResNet, MobileNet

Addagulla Mahalaxmi, Mynampati Meghana Reddy, Kandibanda Pramitha, Ch. Vijayabhaskar, and V. Kakulapati

Abstract Alzheimer's disease (AD) is a global disease that is in a serious and incurable stage. New approaches to AD diagnosis include a user-friendly platform for patient appointments and help doctors upload photos and medications accurately. Diagnosis and treatment of AD is a complex neurodegenerative disease that requires early diagnosis and appropriate treatment. The proposed system addresses this challenge by providing a seamless process for patients to request an appointment, eliminating the need for coordination, and reducing wait times. In this work, use the ResNet algorithm and MobileNet algorithm. The results of this model improve doctors' ability to diagnose patients more effectively. Doctors can also send medical notes for personalized care. The system aims to increase the efficiency and effectiveness of Alzheimer's diagnosis using technology. This approach holds great promise for using deep learning to advance AD classification, provide patients with timely access to treatment, and support clinicians in delivering optimal care.

Keywords Alzheimer · Disease · ML · Diagnosis · Pathological · Data · Imbalance · Technology · Doctor · Personalized · Care · Neurodegenerative

1 Introduction

AD is an incurable ailment that affects individuals worldwide, causing a loss in cognitive abilities, social functioning, and behavioral patterns. It is the primary etiology of dementia and significantly affects an individual's capacity for autonomy. Memory lapses and difficulties in remembering events or conversations are common in the

A. Mahalaxmi (✉) · M. M. Reddy · K. Pramitha · Ch. Vijayabhaskar · V. Kakulapati
Sreenidhi Institute of Science and Technology, Yamnampet, Ghatkesar, Hyderabad, Telangana 501301, India
e-mail: maha6092002@gmail.com

V. Kakulapati
e-mail: vldms@yahoo.com

© The Author(s), under exclusive license to Springer Nature Singapore Pte Ltd. 2025

H. Zen et al. (eds.), *Soft Computing and Signal Processing*, Lecture Notes in Networks and Systems 1200, https://doi.org/10.1007/978-981-97-9926-8_49

first stages of the disease. AD is a kind of dementia that has been defined by gradual cognitive and behavioral decline, often beginning in middle or late age. The pathological features include the existence of neurotic plaques in the brain and the degradation of certain brain cells. The symptoms often manifest gradually and reach a level of severity that hampers everyday functioning [1].

While old age is the primary risk factor, it is vital to note that AD is not exclusively a disease of old age. During the first phases, the memory loss is rather minor, however in the latter stages, the patient's verbal communication and responsiveness deteriorate significantly. While existing therapies are unable to halt the progression of AD, early detection may assist in mitigating the severity of the condition and enhancing the patient's quality of life [2].

AD is a multifaceted disorder that requires extensive knowledge acquisition for precise forecasting and representation [3, 4]. Medical professionals are progressively using deep learning methods to forecast and visualize diseases, facilitating customized treatment plans and aiding physicians in making treatment decisions, while also allowing health economists to examine costs. Ultimately, this enhances the lives of patients.

According to reports, the prevalence of AD is anticipated to intensify twofold in the next 20 years. By 2050, it is predicted that 1 out of every 85 adults will be afflicted. Therefore, the precise identification, particularly during the first phases of AD, is of utmost significance. People suffering from bacterial pneumonia often have symptoms including a cough that produces phlegm, a high temperature associated with freezing chills, problems with breathing, severe or sharp sensations in the chest while taking deep inhalations, and an increased rate of respiration.

Machine learning (ML) is a powerful tool for interpreting, analyzing, classifying, and modeling data in the elderly, enabling decisions not possible with traditional systems [5]. It is widely used in computer-aided diagnosis, medical image formation mining, retrieval, and detection of brain disease using CRT images and X-rays. However, the application of ML in AD prediction has been relatively late, with limited literature. Traditional measurements and human instincts are often insufficient, leading to the need for computational and non-traditional approaches like ML [6]. This shift toward personalized prescriptions is crucial for patients, physicians, and health economists, improving disease prediction, treatment decisions, and health outcomes.

Multitask learning (MTL) is an ML methodology that entails the simultaneous learning of numerous tasks to enhance the performance of the model. Instead of training separate models for each job, it utilizes shared knowledge across tasks. A wide range of learning tasks are included in MTL, such as online, active, supervised, unsupervised, semi-supervised, reinforcement learning, and multi-view learning. The concept encompasses the correlation between tasks and their description and may be used across different learning approaches [7].

The implementation of MTL is a technique that improves simplification by allowing parameters to be shared across many tasks [8]. This methodology diverges from conventional ML approaches, which concentrate on a solitary job and disregard interconnected information. Hard and soft sharing are two categories into which the sharing techniques may be placed. Hard sharing refers to the use of the network layer in a specific task together with different loss functions. On the other hand, soft sharing entails the use of different approaches and parameterization combinations [9].

The rest of the study is as follows. In Sect. 3, existing works are discussed. Section 4 presents the delves of the multi-task learning of the proposed work, highlighting the usage of MobileNet and ResNet for image-analysis. Section 5 provides the consequences and analysis of implementations are discussed. Section 6 provides the implications of our findings. Section 7 gives concluding remarks followed by prospects.

2 Background of the Study

AD [10] is a condition that affects the brain and is characterized by the buildup of certain proteins, which eventually affects the brain's size and finally results in dementia. In the United States, it affects around 6.5 million individuals, with 70% of those affected being 75 years old or older. Memory issues and trouble doing everyday chores are the first symptoms of the condition, which evolves from forgetting recent events to memory impairments. Individuals and their caregivers may get assistance from programs and services, while drugs may be able to bring symptoms under control. Unfortunately, there is currently no known cure for AD and significant loss of brain function may result in dehydration, starvation, or infection with the disease.

Cognitive impairment, dementia, and destruction of brain cells are all symptoms of AD, which is a neurodegenerative disorder that becomes more severe with time. For instances with a late beginning, detection takes 2.8 years [11], whereas for patients with an early onset, it takes 4.4 years. It is characterized by poor cognitive, social, and behavioral skills. To discover treatments and reduce medical costs, advances in diagnosis and diagnosis such as functional magnetic resonance imaging (fMRI), magnetic resonance imaging (MRI), computed tomography(CT), brain electroencephalogram EEG [12], and positron tomography (PET) scan is required. AD poses a heavy burden on society, and its incidence is expected to increase in the next two years [13].

MRI is a powerful tool that provides high resolution and non-invasiveness and is ideal for identifying brain changes in Alzheimer's patients [14]. It can also measure brain activity and activity by monitoring changes in the brain during exercise or stimulation using fMRI. The Mental Health Examination (MHE) and the Clinical Dementia Review (CDR) are two examples of cognitive assessment tools commonly used to evaluate cognitive impairment in AD and other types of dementia. Deep learning is a subfield of artificial intelligence that can perform complex tasks and extract information from images [15, 16].

3 Related Works

Numerous similar themes and gaps have been found after reviewing and assessing the previous research. One of the most noticeable developments is the significant increase in the use of ML approaches for detecting and predicting AD. One of the main issues was an uneven distribution of events with attributes, with few instances and an excessive number of attributes. Another problem was the use of unverified data sets, which introduced uncertainty in the results. There was also a class imbalance, with too few instances in one class and too many instances in another. Additionally, overtraining and the absence of external testing or validation were identified as shortcomings. However, the most well-designed and verified research has shown that ML approaches, when compared to normal statistical methods, may enhance the accuracy of AD forecasting [17].

The article examines many research endeavors that center on the use of ML techniques to identify and assess AD, a widespread neurological disorder. It emphasizes the increasing attention given to early diagnosis and treatment possibilities. Moreover, it delves into the challenges and potential benefits of using ML in this field.

Non-invasive techniques such as CT and MRI have replaced diagnostic tests such as pneumoencephalography and cerebral angiography. The analysis process consists of three steps: grayscale preprocessing, high-pass filtering, median filter, wavelet transform extraction, PCA, kernel support vector machine (KSVM) evaluation, and K-fold cross-validation [18]. The technology increases the effectiveness of KSVM in identifying and eliminating AD. The design has five convolutional layers for feature extraction and three full layers for AD/NC separation. This study shows very good performance with an accuracy rate of 98.74%, a detection rate of 100%, and a false alarm rate of 2.4%. Additionally, a comparison with seven existing methods confirmed its robustness [19].

This technology transforms brain data into details and advanced features and then uses two-dimensional convolutional neural networks (CNNs) for image classification [20]. The technology can provide global features from MRI and PET data that are important for future patient care and treatment. This project aims to improve ML using multiple measurements to evaluate neuroimaging quantitatively and computerized diagnosis in AD.

The study [21] titled "Integrating global and local features of the hippocampus using three-dimensional densely connected convolutional networks and image analysis." By a combination of local visual and global image cues, this approach improves classification accuracy, making it a more powerful disease biomarker. Improved neural network design to identify AD by analyzing two-dimensional MRI slices to diagnose the disease.

This model uses the difference between difference, RELU activation function, and the SE module to improve feature extraction performance. Experiments indicate that the precision of categorization of AD increases by 4.18% compared with CN, while the classification accuracy of MCI increases by 6% compared with CN [22].

4 Methodology

AD has a worldwide impact, affecting millions of people, and unfortunately, there is now no known therapy for it. Timely diagnosis is essential for mitigating symptoms and enhancing quality of life. Conventional diagnostic techniques, such as evaluations of medical history, neurological examinations, cognitive assessments, and brain imaging scans, may not consistently provide correct detection of the condition and proposed work is shown in Fig. 1.

In addition, ML is crucial in predicting and determining the future course of AD. To address these constraints, a suggested model aims to accurately diagnose the early stages of AD. When analyzing the data set that has been confirmed to include pathological evidence, the suggested model includes a preprocessing phase to address the problem of class imbalance. The use of ML in attribute selection is crucial in mitigating the issue of insufficient examples and excessive characteristics, sometimes referred to as the curse of dimensionality [23]. The dataset is partitioned into training and testing data by the model. When training data is based on a small amount of testing data, it might result in a phenomenon known as over-training [24]. Therefore, it is important to use training data that covers a representative portion of the real data. The model uses association rule mining to do classification, using minimal support and minimum confidence thresholds.

Fig. 1 Flow of proposed work

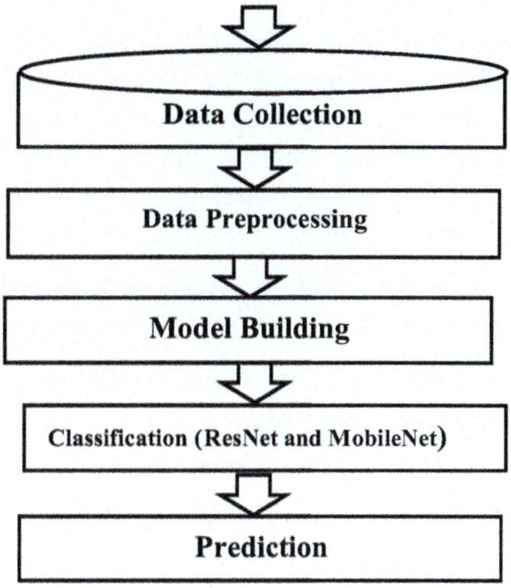

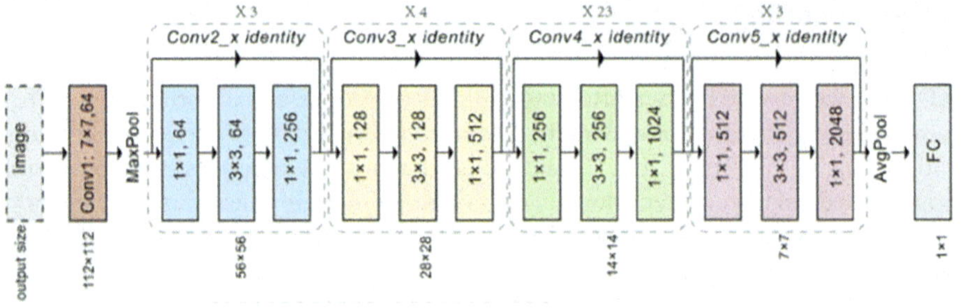

Fig. 2 Architecture of the ResNet [R]

4.1 ResNet

To enable gradient flow without vanishing gradients, ResNet (shown in Fig. 2), a CNN design, combines residual blocks, convolutional layers, batch normalization, activation function, pooling layers, and fully connected layers [25].

4.2 MobileNet

With mobile and edge devices in mind, Google's MobileNet (shown in Fig. 3) is a deep learning architecture for object identification and picture categorization. The model attains a trade-off between precision and effectiveness by using depth-wise separable convolutions, which minimize computational requirements without compromising model accuracy. It may be customized and is not heavy [26].

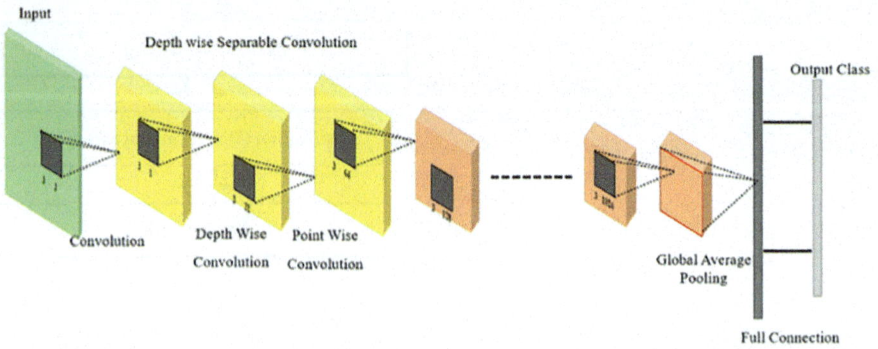

Fig. 3 Architecture of the MobileNet [27]

5 Implementation Analysis

The data set is obtained from the open-access Kaggle data source. The illness prediction dataset is partitioned into two subsets: the training dataset, which is used for instructing the model, and the testing dataset, which is employed to assess its efficacy.

5.1 Data Preprocessing

Images in a uniform format are part of the training procedure for the illness diagnosis model. The dataset is then inputted into the model, whereby it assimilates knowledge from the photos and their corresponding labels to cultivate prediction capabilities. Whether the symptoms of the disease or is within normal limits. User modules allow users to access images for classification and models to analyze images to provide predictions. Users can access results that provide information about diseases or common conditions. The user can access the findings, which provide information on the illness or normal state. This information is crucial to the diagnostic process and informed decision-making.

The metrics are evaluation of ResNet (shown in Figs. 4, 5, 6, and 7):

Specificity: 98%
Sensitivity: 98%
Accuracy: 83%.

The metrics are evaluation of MobileNet shown in Figs. 8, 9, and 10:

Specificity: 98%
Sensitivity: 98%
Accuracy: 89%.

```
154/154 [==============================] - 1079s 7s/step - loss: 0.6311 - accuracy: 0.7650 - val_loss: 96.0539 - val_accuracy: 0.2304
Epoch 2/15
154/154 [==============================] - 996s 6s/step - loss: 0.3693 - accuracy: 0.8646 - val_loss: 3.9509 - val_accuracy: 0.2365
Epoch 3/15
154/154 [==============================] - 3797s 25s/step - loss: 0.2398 - accuracy: 0.9121 - val_loss: 3.9209 - val_accuracy: 0.2365
Epoch 4/15
154/154 [==============================] - 1290s 8s/step - loss: 0.2029 - accuracy: 0.9300 - val_loss: 3.9857 - val_accuracy: 0.2365
Epoch 5/15
154/154 [==============================] - 1062s 7s/step - loss: 0.1501 - accuracy: 0.9521 - val_loss: 9.9343 - val_accuracy: 0.2304
Epoch 6/15
154/154 [==============================] - 1153s 7s/step - loss: 0.1242 - accuracy: 0.9596 - val_loss: 2.0377 - val_accuracy: 0.3901
Epoch 7/15
154/154 [==============================] - 940s 6s/step - loss: 0.0898 - accuracy: 0.9704 - val_loss: 1.5163 - val_accuracy: 0.5133
Epoch 8/15
154/154 [==============================] - 827s 5s/step - loss: 0.0898 - accuracy: 0.9697 - val_loss: 1.6806 - val_accuracy: 0.5643
Epoch 9/15
154/154 [==============================] - 812s 5s/step - loss: 0.1073 - accuracy: 0.9629 - val_loss: 2.0730 - val_accuracy: 0.4684
Epoch 10/15
154/154 [==============================] - 940s 6s/step - loss: 0.0715 - accuracy: 0.9759 - val_loss: 2.8154 - val_accuracy: 0.3635
Epoch 11/15
154/154 [==============================] - 885s 6s/step - loss: 0.0389 - accuracy: 0.9883 - val_loss: 4.5962 - val_accuracy: 0.4000
Epoch 12/15
154/154 [==============================] - 823s 5s/step - loss: 0.0521 - accuracy: 0.9837 - val_loss: 2.7834 - val_accuracy: 0.4700
Epoch 13/15
154/154 [==============================] - 859s 6s/step - loss: 0.0808 - accuracy: 0.9727 - val_loss: 7.5733 - val_accuracy: 0.2730
Epoch 14/15
154/154 [==============================] - 839s 5s/step - loss: 0.0595 - accuracy: 0.9801 - val_loss: 6.4778 - val_accuracy: 0.3072
Epoch 15/15
154/154 [==============================] - 841s 5s/step - loss: 0.0688 - accuracy: 0.9772 - val_loss: 4.2568 - val_accuracy: 0.4707
```

Fig. 4 ResNet training epochs, train accuracy, loss, and validation accuracy

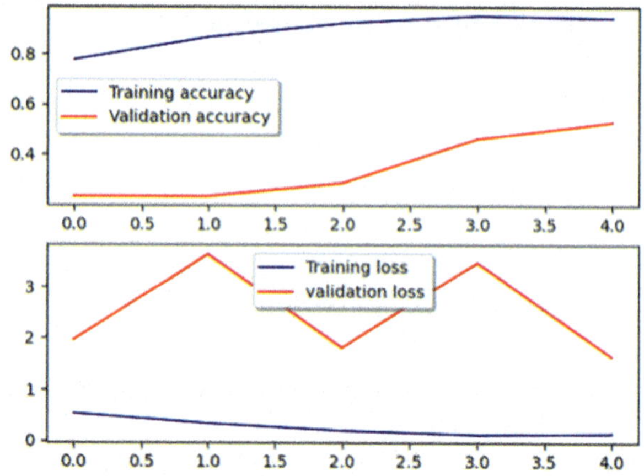

Fig. 5 ResNet train and validation accuracy, loss

```
Layer (type)                Output Shape              Param #
=================================================================
resnet50 (Functional)       (None, 8, 8, 2048)        23587712

global_average_pooling2d (  (None, 2048)              0
GlobalAveragePooling2D)

dense (Dense)               (None, 64)                131136

batch_normalization (Batch  (None, 64)                256
Normalization)

dropout (Dropout)           (None, 64)                0

dense_1 (Dense)             (None, 4)                 260

=================================================================
Total params: 23719364 (90.48 MB)
Trainable params: 23666116 (90.28 MB)
Non-trainable params: 53248 (208.00 KB)
```

Fig. 6 Summary of model ResNet

6 Discussion

It has been shown that deep learning methods, such as multimodality, TL-based approaches, and CNN algorithms, can accurately diagnose and classify AD utilizing a range of neuroimaging modalities. The research used a publically available dataset of magnetic resonance (MR) images and employed a DenseNet model for data augmentation and feature extraction. Evidence [28, 29] shows that the accuracy of the study in classifying AD reaches 96.05%. DenseNet models surpassed prior methods, including VGG-19, SqueezeNet, and MobileNet, with precision rates fluctuating from 77.60 to 83.97%. The transfer learning strategy used the DenseNet architectures [30, 31] to extract generic features, while the top layers of deep CNNs

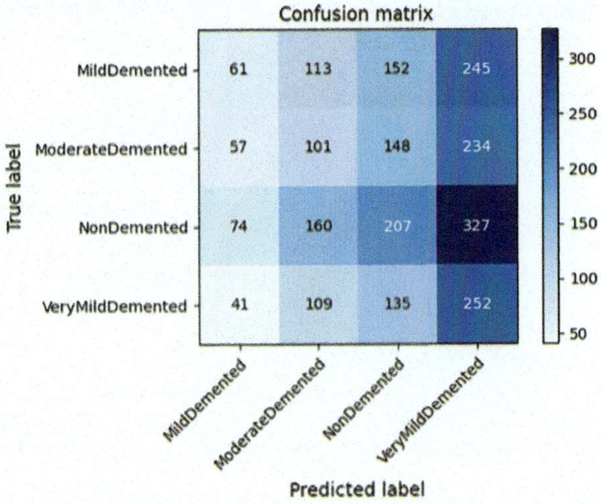

Fig. 7 ResNet confusion matrix for different classes of AD

```
Layer (type)                 Output Shape              Param #
================================================================
mobilenet_1.00_224 (Functi   (None, 8, 8, 1024)        3228864
onal)

global_average_pooling2d_1   (None, 1024)              0
 (GlobalAveragePooling2D)

dense_2 (Dense)              (None, 64)                65600

batch_normalization_1 (Bat   (None, 64)                256
chNormalization)

dropout_1 (Dropout)          (None, 64)                0

dense_3 (Dense)              (None, 4)                 260

================================================================
Total params: 3294980 (12.57 MB)
Trainable params: 3272964 (12.49 MB)
Non-trainable params: 22016 (86.00 KB)
```

Fig. 8 Summary of model MobileNet

were utilized to collect domain-specific data. This approach demonstrated the efficacy of DenseNet models in extracting appropriate features from MRI images.

Imaging study of AD categorization was done using the MobileNet model, which achieved ResNet 83% accuracy and 89% accuracy, respectively, confirming the effectiveness of the proposed technique. Other models proceeded less well than the MobileNet model. The ResNet18 and MobileNet models were exposed to multi-task learning, and their performance was evaluated based on a system of measurement such as accuracy, precision, recall, and F1 score. MobileNet determined

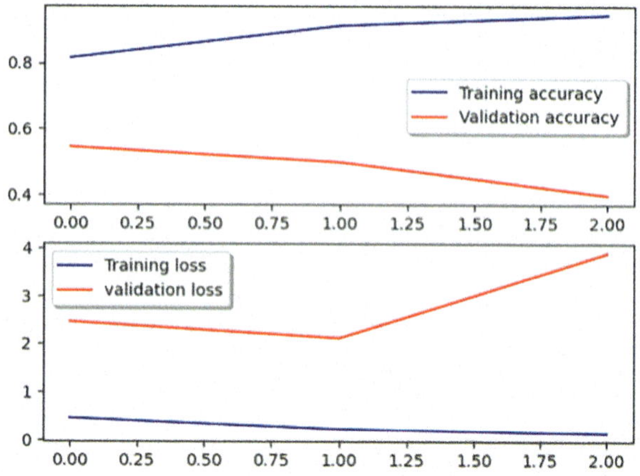

Fig. 9 MobileNet train and validation accuracy, loss

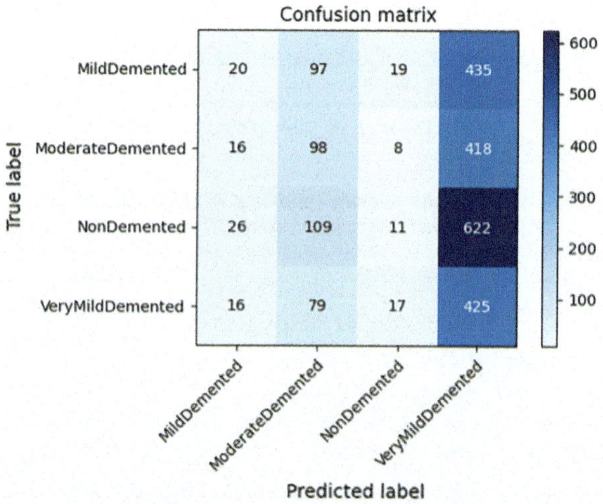

Fig. 10 ResNet confusion matrix for different classes of AD

outstanding effectiveness, with a precision of 89% in detecting dementia. In contrast, ResNet exhibited the shortest training time and a comparable level of performance. Depth-wise convolutions in MobileNet demonstrated superior performance in sound classification and the identification of lung illnesses.

7 Conclusion

AD is a worldwide health problem. AD treatment, on the other hand, continues to be symptomatic and has not improved the prognosis. Diet and exercise consistently are regarded to be the first line of defense against the condition. The deep learning algorithms used in this study effectively categorized MRI pictures into four distinct types: Minor Demented, Reasonable Demented, Non Demented, and Very Mild Demented. The dataset included four dissimilar categories of MRI images, which were trained using a Modified CNN using the ResNet and MobileNet architectures. The categorization was evaluated after the upload of the images.

8 Future Enhancement

In future, ResNet, a sophisticated neural network structure, will streamline the categorization of infections, facilitating the identification of cases at an early stage and allowing for immediate management. This technology improves patient outcomes by allowing healthcare personnel to promptly detect infections and perform timely treatments.

References

1. DeTure MA, Dickson DW (2019) The neuropath logical diagnosis of Alzheimer's disease. Mol Neurodegener 14:32. https://doi.org/10.1186/s13024-019-0333-5
2. Jahn H (2013) Memory loss in Alzheimer's disease. Dialogues Clin Neurosci 15(4):445–454. https://doi.org/10.31887/DCNS.2013.15.4/hjahn
3. Saleh AW, Gupta G, Khan SB, Alkhaldi NA, Verma A (2023) An Alzheimer's disease classification model using transfer learning DenseNet with embedded healthcare decision support system. Decis Anal J 9:100348. ISSN 2772-6622. https://doi.org/10.1016/j.dajour.2023.100348
4. Brookmeyer R, Johnson E, Ziegler-Graham K, Arrighi HM (2007) Forecasting the global burden of Alzheimer's disease. Alzheimer's Dement 3(3):186–191. https://doi.org/10.1016/j.jalz.2007.04.381
5. Speiser JL, Callahan KE, Houston DK, Fanning J, Gill TM, Guralnik JM, Newman AB, Pahor M, Rejeski WJ, Miller ME (2021) Machine learning in aging: an example of developing prediction models for serious fall injury in older adults. J Gerontol A Biol Sci Med Sci 76(4):647–654. https://doi.org/10.1093/gerona/glaa138
6. Rekha S, Rekha G, Sudha C (2020) Alzheimer's detection model using machine learning. Int Res J Eng Technol 7. ISSN 2395-0056
7. Zhang Y, Yang Q (2018) An overview of multi-task learning. Natl Sci Rev 5:30–43. https://doi.org/10.1093/nsr/nwx105
8. Sun TX, Shao YF, Li XN, Liu PF, Yan H et al (2020) Learning sparse sharing architectures for multiple tasks. In: Proceedings of the AAAI, New York, NY, pp 8936–8943
9. Chen Z, Badrinarayanan V, Lee CY, Rabinovich A (2018) Gradnorm: gradient normalization for adaptive loss balancing in deep multitask networks. In: Proceedings of the ICML, Stockholm, Sweden, pp 794–803
10. https://www.mayoclinic.org/diseases-conditions/alzheimers-disease/symptoms-causes/syc-20350447
11. van Vliet D, de Vugt ME, Bakker C, Pijnenburg YAL, Vernooij-Dassen MJFJ, Koopmans RTCM, Verhey FRJ (2013) Time to diagnosis in young-onset dementia as compared with late-onset dementia. Psychol Med 43(2):423–432

12. Sperling RA, Aisen PS, Beckett LA, Bennett DA, Craft S, Fagan AM, Iwatsubo T, Jack CR Jr, Kaye J (2011) Toward defining the preclinical stages of Alzheimer's disease: recommendations from the National Institute on Aging-Alzheimer's Association workgroups on diagnostic guidelines for Alzheimer's disease. Alzheimer's Dement 7(3):280–292
13. Wimo A, Jönsson L, Bond J, Prince M, Winblad B (2013) The worldwide economic impact of dementia 2010. Alzheimer's Dement 9(1):1–11
14. Ghosh S, Hazarika AP, Chandra A, Mudil RK (2021) Adaptive neighbor constrained deviation sparse variant fuzzy c-means clustering for brain MRI of AD subject. Vis Inform 5(4):67–80
15. Shah SA, Tahir A, Ahmad J, Zahid A, Pervaiz H, Shah SY et al (2020) Sensor fusion for identification of freezing of gait episodes using Wi-Fi and radar imaging. IEEE Sens J 20(23):14410–14422
16. Abbasi SF, Abbas A, Ahmad I, Alshehri MS, Almakdi S, Ghadi YY et al (2023) Automatic neonatal sleep stage classification: a comparative study. Heliyon
17. Wang Y, Liu L, Wang C (2023) Trends in using deep learning algorithms in biomedical prediction systems. Front Neurosci 9(17):1256351. https://doi.org/10.3389/fnins.2023
18. Kakulapati V et al (2023) A novel method of early detection of AD (Alzheimer's disease). Acta Sci Comput Sci 5(6):36–41
19. Bäckström K, Nazari M, Gu I, Jakola A (2018) An efficient 3D deep convolutional network for Alzheimer's disease diagnosis using MR images, pp 149–153. https://doi.org/10.1109/ISBI.2018.8363543
20. Cheng D, Liu M (2017) CNNs-based multi-modality classification for AD diagnosis, pp 1–5. https://doi.org/10.1109/CISP-BMEI.2017.8302281
21. Cui R, Liu M (2019) Hippocampus analysis by combination of 3-D DenseNet and shapes for Alzheimer's disease diagnosis. IEEE J Biomed Health Inform 23(5):2099–2107. https://doi.org/10.1109/JBHI.2018.2882392. Epub 20 Nov 2018
22. Zhang X, Gao L, Wang Z, Yu Y, Zhang Y, Hong J (2024) Improved neural network with multitask learning for Alzheimer's disease classification. Heliyon 10(4):e26405. ISSN 2405-8440. https://doi.org/10.1016/j.heliyon.2024.e26405
23. Khan A, Usman M (2019) Alzheimer's disease prediction model using demographics and categorical data. Int J Online Biomed Eng (iJOE) 15:96. https://doi.org/10.3991/ijoe.v15i15.11472
24. Nguyen QH, Ly H-B, Ho LS, Al-Ansari N, Le HV, Tran VQ, Prakash I, Pham BT (2021) Influence of data splitting on performance of machine learning models in prediction of shear strength of soil. Math Probl Eng 2021:1–15. Article ID 4832864. https://doi.org/10.1155/2021/4832864
25. Borawar L, Kaur R (2023) ResNet: solving vanishing gradient in deep networks. https://doi.org/10.1007/978-981-19-8825-7_21
26. Dong K, Zhou C, Yihan R, Li Y (2020) MobileNetV2 model for image classification, pp 476–480. https://doi.org/10.1109/ITCA52113.2020.00106
27. Islam MM, Alam KMR, Uddin J, Ashraf I, Samad MA (2023) Benign and malignant oral lesion image classification using fine-tuned transfer learning techniques. Diagnostics 13:3360. https://doi.org/10.3390/diagnostics13213360
28. Ajagbe SA, Amuda KA, Oladipupo MA, Afe OF, Okesola KI (2021) Multi-classification of Alzheimer's disease on magnetic resonance images (MRI) using deep convolutional neural network (DCNN) approaches. Int J Adv Comput Res 11(53):51–60. https://doi.org/10.19101/IJACR.2021.1152001
29. Ghosh T, Palash MIA, Yousuf MA, Hamid MA, Monowar MM, Alassafi MO (2023) A robust distributed deep learning approach to detect Alzheimer's disease from MRI images. Mathematics 11(12):2633. https://doi.org/10.3390/math11122633
30. Odusami M, Maskeliūnas R, Damaševičius R (2023) Pixel-level fusion approach with vision transformer for early detection of Alzheimer's disease. Electronics 12(5):1218. https://doi.org/10.3390/electronics12051218
31. https://www.ikomia.ai/blog/mastering-resnet-deep-learning-image-recognition

Classification And Prediction of Parkinson's Disease Using Machine Learning Algorithms

Nagesh Deevi, Sai Geethika, Deekshitha, A. Vyshnavi, and B. Sruthi

Abstract Parkinson's Disease (PD) is a nervous system disorder that affects movement and have motor symptoms such as tremors, rigidity, and non-motor symptoms like cognitive impairment and depression. Early diagnosis and prediction of PD are crucial for effective management and intervention strategies. In this work, we work on the utilization of machine learning algorithms that best classifies and predicts the Parkinson's Disease, aiming to provide valuable insights for doctors to take the best decisions regarding patients' health. The main aim of our project is to predict the Parkinson's disease of a patient using machine learning algorithms in early stage, so that the affected person can be treated without any delay. Comparative study of performances of various machine learning algorithms will be done through calculating various terms like accuracy, precision, F1score, Recall etc. Hence, this work delivers valuable findings by evaluating and contrasting various computational methods and machine learning algorithms employed in the detection and classification of Parkinson's Disease.

Keywords Extra tree classifier · XGBoost · SVM classifier · Parkinson's disease · Machine learning

1 Introduction

Parkinson's disease, is a chronic nervous system disorder that mainly affects the central nervous system. The symptoms usually start slowly, and as the seriousness of disease increases, they become more common. The most general and early symptoms are rigidity, slowness of movement, shivering in voice and difficulty with walking. Behavioral problems may also occur with depression and anxiety in many people who suffer from this disease. These patients can also have problems like sleep-lessness and improper functioning of sensory systems.

N. Deevi (✉) · S. Geethika · Deekshitha · A. Vyshnavi · B. Sruthi
Department of ECE, BVRIT HYDERABAD College of Engineering for Women, Hyderabad, Telangana, India
e-mail: 21wh5a0409@bvrithyderabad.edu.in

© The Author(s), under exclusive license to Springer Nature Singapore Pte Ltd. 2025
H. Zen et al. (eds.), *Soft Computing and Signal Processing*, Lecture Notes in Networks and Systems 1200, https://doi.org/10.1007/978-981-97-9926-8_50

Even though the Parkinson's disease is a chronic and a long-term disorder, however-er the actual causes for this disorder are unknown. But it is believed that, a combination of genetic and some other external factors may lead to the development of this disease. If a family member gets affected by parkinson's disease, then there are high chances of diagnosing this disease in their offspring's. Being exposed to pesticides for a very long period of time, having traumatic head injuries, being a tobacco smoker etc. are also few other factors that causes parkinson's disease. The fact that most of the people doesn't know is, there is no cure for this, but treatment helps to reduce the effects of the symptoms.

1.1 Literature Survey

In [1], proposed SVM Approach that identifies Parkinson's Disease Using Gait Analysis. In this approach 12 Gait features were analyzed by using Gaussian Radial Basis Function (RBF) & Kernel Based SVM Classifier. In [2], they put efforts to focus on the specific Gait characteristics which would help in differentiating Parkinson's Disease from other diseases that affects nervous system like Amyotrophic Lateral Sclerosis (ALS) etc. The dataset utilized in this study was sourced from 'The National Institutes of Health-sponsored Research Resource'. The SVM classifier exhibited a notably high accuracy of 83.33%, coupled with a commendable detection rate of 75% for identifying Parkinson's disease patients in contrast to other neurodegenerative conditions.

In [3], a study proposed the use of machine learning for predicting Freezing of Gait in Parkinson's Disease patients. The analysis focused on a dataset having 237 individual Freezing of Gait events across 8 patients, with 18 features. Various algorithms including Support Vector Machines with polynomial kernel, XGBooster, K-Nearest Neighbors, Naive Bayes, and Random Forest were applied to the Daphnet Freezing of Gait Dataset obtained from the UCI Machine Learning Repository. In [4], A comparative study was proposed to evaluate existing machine learning approaches for the detection of Parkinson's Disease. In this study, a dataset comprising PD patients with General Voice Disorders was employed to differentiate Parkinson's disease patients. Various machine learning algorithms, including Multi-Layer Perceptron (MLP), Multi-Layer Feed Forward Neural Network, K-Nearest Neighbor, and Support Vector Machine, were utilized to analyze the voice dataset. The aim of this research paper [5] is to perform an extensive survey and comparison of computational intelligence methods for detecting Parkinson's Disease. The goal is to pinpoint the most effective classifier for this purpose.

In [6], a technique for early detection of Parkinson's disease through machine learning algorithms was introduced. This diagnostic method encompasses both feature selection and classification procedures. Feature Importance and Recursive Feature Elimination techniques were employed for selecting relevant features. In the experimental phase, Artificial Neural Networks and Support Vector Machines were utilized to classify Parkinson's patients. In [7], experimentation involved the utilization of the Speech Signal Dataset, which includes 22 voice features and 195 instances sourced from the UCI Machine Learning Repository. In [8], proposed Parkinson's Disease Detection Using Isosurfaces-Based Features and Convolutional Neural Networks. This work proposes the use of isosurfaces as a way to reduce huge amount of information that leads to complex CNN architectures keeping the most relevant in-formation. These isosurfaces are then used to implement a classification system which uses two of the most well-known CNN architectures, LeNet and Alex Net. Dataset used in this paper was obtained from PPMI (Parkinson's Progression Markers Initiative). Convolutional Neural Networks algorithms were used to get an accuracy of 89%.

The [9], emphasizes the application of machine learning methods in telemedicine for the early detection of Parkinson's Disease (PD). The study conducted research on MDVP audio data from 30 Parkinson's patients and healthy individuals, utilizing four machine learning models during training. In [10], comparison of classification results from Support Vector Machine (SVM), Random Forest, K-Nearest Neighbors (KNN), and Logistic Regression models identified the Random Forest classifier as the most effective technique for PD detection. The Random Forest classifier model achieved a detection accuracy of 91.83% and a sensitivity of 0.95.

In [11] performed spatial–temporal analysis of brain MRI scans. They implemented decision trees, random forest and KNN to detect Mild Cognitive Impairment (MCI) in PWP. However, dataset was small and artificial data augmentation [12] was needed. In [13], they trained a ResNet model on images of audio data, instead of training the model on the nuances of the frequency of audio. In [14], they aimed to remove subjectivity of doctors in prediction of PD using an unbiased ML model, however their results achieved peak accuracy of 85% only [15].

2 Design

2.1 Dataset

The features listed below extracted from voice recordings, which we used as input dataset to assess and diagnose Parkinson's disease. Let's break down what each fea-ture represents:

- MDVP: Fo (Hz) (Mean Frequency): This is the average vocal fundamental frequency, representing the average pitch of the voice.
- MDVP: Fhi (Hz) (Maximum Frequency): The highest frequency of the voice signal.
- MDVP: Flo (Hz) (Minimum Frequency): The lowest frequency of the voice signal.
- MDVP: Jitter (%): Jitter is a measure of the frequency variation of consecutive periods of a voice signal. High jitter may indicate voice instability.

- MDVP: Jitter (Abs): Absolute jitter measures the absolute difference in the length of consecutive periods in the voice signal.
- MDVP: RAP (Relative Amplitude Perturbation): Similar to jitter, RAP measures the perturbation in the amplitude of consecutive periods.
- MDVP: PPQ (Pitch Period Perturbation Quotient): It quantifies the variation in pitch over short periods.
- Jitter: DDP: This is the difference between consecutive cycle lengths, which also reflects pitch variation.
- MDVP: Shimmer: Shimmer measures the variation in amplitude between consecutive cycles of the voice signal.
- MDVP: Shimmer(dB): Shimmer in decibels, which provides a more perceptually relevant measure.
- Shimmer: APQ3 (Amplitude Perturbation Quotient): It measures the amplitude variation over the first three cycles.
- ShimmerAPQ5 (Amplitude Perturbation Quotient): Similar to APQ3, but over the first five cycles.
- MDVP: APQ (Amplitude Perturbation Quotient): It quantifies the overall amplitude variation in the voice signal.
- Shimmer: DDA (Amplitude Perturbation Quotient): It is the difference between consecutive amplitudes within the first three cycles.
- NHR (Noise-to-Harmonics Ratio): It measures the ratio of noise to tonal components in the voice signal.
- HNR (Harmonics-to-Noise Ratio): The ratio of tonal to noise components in the voice signal.
- Status: This binary variable typically indicates the presence (1) or absence (0) of Parkinson's disease.
- RPDE (Recurrence Period Density Entropy): A measure of the complexity of the voice signal.
- DFA (Detrended Fluctuation Analysis): It quantifies the fractal-like temporal correlations in the voice signal.
- spread1, spread2, D2, PPE: These are nonlinear measures related to the distribution of data points in multidimensional space.

2.2 *Proposed Methodology*

Designed a machine learning model to predict Parkinson's disease. It involves different algorithms to analyze relevant data and make predictions based on patterns and voice features. Here's how we approached this task:

- Data Collection: We gathered a dataset containing features of size 815 × 23 related to Parkinson's disease. These features are associated with voice and speech analysis.

- Data Pre-processing: In this stage, we handled with missing data by imputation, encoded categorical variables, normalized numerical features as different features fall into different ranges and explored the data to understand its characteristics.
- Model Selection: Then we choose appropriate machine learning models after referring various research papers. They are:
- Support Vector Machines (SVM)
- XGBoost Classifier
- Extra Tree Classifier
- Model Training: We splitted 70% of dataset into training and the other 30% into testing sets and trained our models on the training set.
- Model Evaluation: We evaluated the models on the testing set using metrics like accuracy, precision, recall, F1 score, and area under the ROC curve.
- Validation and Interpretability: We validated our model on new, unseen data to check whether model's predictions are interpretable and explainable, especially in a medical context.
- Deployment: Once the model's performance is satisfied, we deployed it by using gradio as user interface.

2.3 *Block Diagram*

Figure 2 shows the block Diagram of Proposed Method. Data transformation techniques are applied to the raw dataset and processed initially. Then, the model is trained with 70% of the processed dataset. Later machine learning algorithms like XGBoost, SVM classifier and Extra Tree classifier are used to design the prediction model. After training and evaluating the model with 30% of labelled data, gradio user interface is designed. This user interface acts as a medium, which takes voice features as input and predicts the result (Fig. 1).

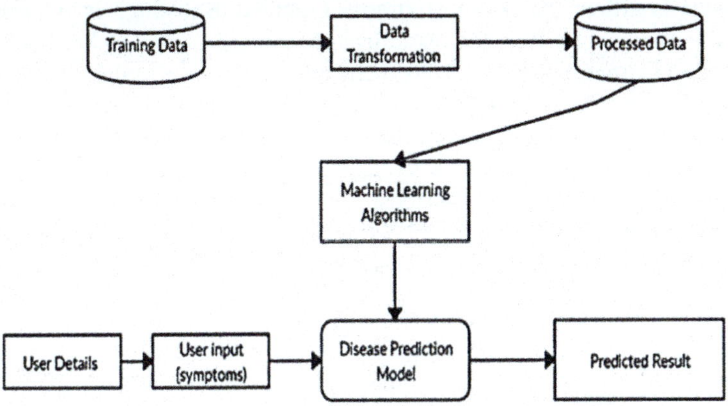

Fig. 1 A Block diagram of proposed method

Fig. 2 XGBoost output

```
Training Accuracy: 0.9708588957055214
Testing Accuracy: 0.9877300613496932
Mean Squared Error: 0.012269938650306749
Classification Report:
              precision    recall  f1-score   support

           0       0.98      0.98      0.98        49
           1       0.99      0.99      0.99       114

    accuracy                           0.99       163
   macro avg       0.99      0.99      0.99       163
weighted avg       0.99      0.99      0.99       163
```

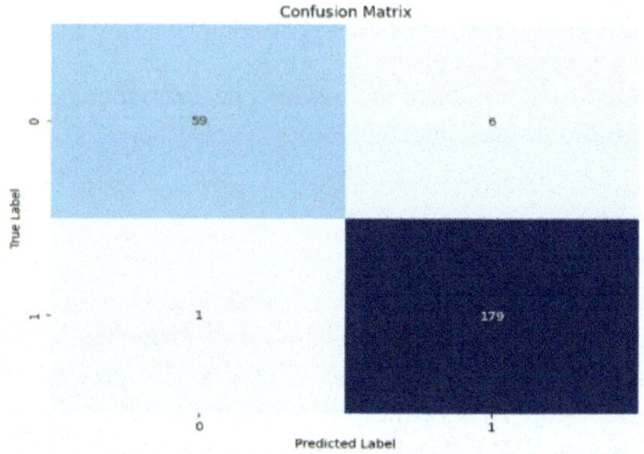

2.4 *Algorithms Used*

XGBoost Classifier: XGBoost (Extreme Gradient Boosting) is a mostly-used machine learning algorithm for classification and regression tasks. It belongs to the ensemble learning family. XGBoost has great efficiency, scalability, and performance and being widely adopt-ed in various industry settings. The important features of XGBoost includes handling large datasets efficiently, flexible in handling different types of data, and its robustness. It uses a gradient boosting framework, optimizing a differentiable loss function. XGBoost incorporates regularization techniques to control model complexity and prevent overfitting. It is an effective algorithm suitable for a wide range of predictive modeling tasks.

Support Vector Machine Classifier: Support Vector Machine (SVM) is a supervised learning algorithm. In classification, SVM tries to find the optimal hyperplane which separates different classes in the feature space in the best way, and tries to maximize the margin between the classes. It reduces the risk of overfitting. Additionally, it can also handle both linearly separable and non-linearly separable data through the use of kernel functions, which map the input data into higher-dimensional spaces. SVM has high effectiveness in high-dimensional spaces, robustness to overfitting, and ability to handle complex datasets. Its performance may be sensitive to the choice of kernel function and regularization parameters.

Extra Tree Classifier: Extra Trees Classifier is an ensemble learning method based on decision trees. It builds multiple decision trees using random subsets of features and training data. Unlike traditional decision trees, Extra Trees introduces randomness at two levels: it randomly selects feature subsets at each split and uses random thresholds to partition the data. This randomness helps to reduce overfitting and improves generalization performance. Additionally, Extra Trees Classifier is computationally efficient, as it requires less computational resources compared to other ensemble methods like Random Forests. It is particularly useful for high-dimensional datasets with many features, making it a popular choice in applications such as classification, regression, and anomaly detection. Extra Trees Classifier is implemented in various machine learning libraries such as scikit-learn in Python, providing an easy-to-use and versatile tool for building robust predictive models.

3 Results and Discussion

The three different classifiers or algorithms that we used for designing this model, predicted the output with different accuracies. XGBoost Classifier predicted with an accuracy of 97% (see Fig. 2.), Support Vector Machine with an accuracy of 89% (see Fig. 3.) and Extra Tree Classifier was 99% (see Fig. 4.) accurate. We also performed a comparative analysis with various machine algorithms that were used in existing approaches for this problem. Accuracy and efficiency of our model are comparatively high. And taking voice features as dataset is another advantage of our model, as analyzing and extracting voice features would be far easier than taking any other symptoms into consideration as early prediction and diagnosis is needed for this disease.

Table 1 shows a brief comparison of accuracy of algorithms we implemented in our work with existing algorithms.

Fig. 3 Support vector machine output

```
Training Accuracy: 0.6964912280701754
Testing Accuracy: 0.8938775510204081
Mean Squared Error: 0.10612244897959183
Classification Report:
              precision    recall  f1-score   support

           0       0.80      0.85      0.82        71
           1       0.94      0.91      0.92       174

    accuracy                           0.89       245
   macro avg       0.87      0.88      0.87       245
weighted avg       0.90      0.89      0.89       245
```

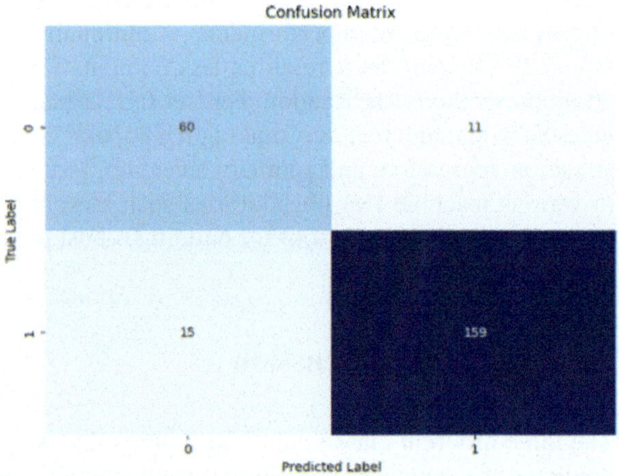

4 Results and Discussion

Following an extensive evaluation, XGBoost and Extra Tree classifiers were identified as the most effective models for predicting Parkinson's Disease using Machine Learning. These classifiers consistently delivered high accuracy in their predictions. We selected three classifiers based on their performance and applied them to a dataset obtained from the UCI Machine Learning repository. By assessing the classification performance of these algorithms on the same dataset and comparing results in terms of misclassification and correct classification rates, it was evident that both XGBoost Classifier and Extra Tree Classifier Surpassed Support Vector Machine and other existing algorithms in performance.

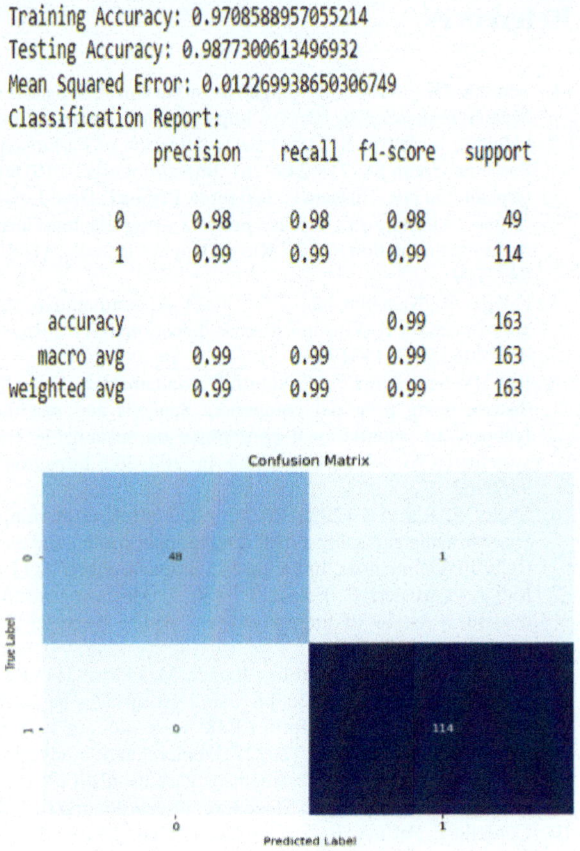

Fig. 4 ExtraTree output

Table 1 Comparison table with existing algorithms

Algorithms	Accuracy
Naive Bayes [16]	64%
Perceptron [17]	86%
Logistic regression [18]	89%
XGBoost (Proposed)	97%
SVM (proposed)	89%
Extra tree (proposed)	99%

Additionally, more reliable and accurate algorithms, including advanced learning techniques such as Deep Learning, can be considered for further enhancement. By integrating these methods, predictions of Parkinson's Disease chances can be achieved with greater efficiency and accuracy in less time, rendering the project more robust and dependable.

References

1. Senturk ZK (2020) Early diagnosis of Parkinson's disease using machine learning algorithms. Med Hypotheses 138:109603. https://doi.org/10.1016/j.mehy.2020.109603
2. Govindu A, Palwe S (2023) Early detection of Parkinson's disease using machine learning. Procedia Comp Sci 218:249–261. https://doi.org/10.1016/j.procs.2023.01.007
3. Orphanidou NK, Hussain A, Keight R, Lishoa P, Hind J, Al-Askar H (2018) Predicting freezing of gait in Parkinson's disease patients using machine learning. In: IEEE congress on evolutionary computation (CEC). Rio de Janeiro, Brazil, pp 1–8. https://doi.org/10.1109/CEC.2018.8477909
4. Pahuja G, Nagabhushan TN (2018) A comparative study of existing machine learning approaches for Parkinson's disease detection. IETE J Res 67(1):4–14. https://doi.org/10.1080/03772063.2018.1531730
5. Rao DV, Sucharitha Y, Venkatesh D, Mahamthy K, Yasin SM (2022) Diagnosis of Parkinson's disease using principal component analysis and machine learning algorithms with vocal features. In: International conference on sustainable computing and data communication systems (ICSCDS). Erode, India, pp 200–206. https://doi.org/10.1109/ICSCDS53736.2022.9760962
6. Shetty S, Rao YS (2022) SVM based machine learning approach to identify Parkinson's Disease using gait analysis. In: International conference on inventive computation technologies (ICICT). Coimbatore, India, pp 1–5. https://doi.org/10.1109/INVENTIVE.2016.7824836
7. Mei J, Desrosiers C, Frasnelli J (2021) Machine learning for the diagnosis of Parkinson's disease: a review of literature. Front Aging Neurosci 13:633–752. https://doi.org/10.3389/fnagi.2021.633752
8. Ortiz A, Munilla J, Martínez-Ibañez M, Górriz JM, Ramírez J, Salas-Gonzalez D (2019) Parkinson's disease detection using isosurfaces-based features and convolutional neural networks. Front Neuroinform 13:48. https://doi.org/10.3389/fninf.2019.00048
9. Prabhavathi K, Patil S (2022) Tremors and bradykinesia. In: Arjunan SP, Kumar DK (eds) Techniques for assessment of Parkinsonism for diagnosis and rehabilitation. Series in bioengineering. Springer, Singapore. https://doi.org/10.1007/978-981-16-3056-9_9
10. Cordella F, Paffi A, Pallotti A (2021) Classification-based screening of Parkinson's disease patients through voice signal. In: IEEE international symposium on medical measurements and applications (MeMeA). Lausanne, Switzerland, pp 1–6. https://doi.org/10.1109/MeMeA52024.2021.9478683
11. Trivedi D, Jaeger H, Stadtschnitzer M (2019) Mobile device voice recordings at king's college london (MDVR -KCL) from both early and advanced Parkinson's disease patients and healthy controls
12. Wang W, Lee J, Harrou F, Sun Y (2020) Early detection of Parkinson's disease using deep learning and machine learning. In: IEEE access, vol 8, pp 147635–147646. https://doi.org/10.1109/ACCESS.2020.3016062
13. Huang F, Xu H, Shen T, Jin L (2021) Recognition of Parkinson's disease based on residual neural network and voice diagnosis. IEEE 5th information technology, networking, electronic and automation control conference (ITNEC). Xi'an, China, pp 381–386. https://doi.org/10.1109/ITNEC52019.2021.9586915
14. Moradi S, Tapak L, Afshar S (2022) Identification of novel noninvasive diagnostics biomarkers in the Parkinson's diseases and improving the disease classification using support vector machine. BioMed Res Int 5009892. https://doi.org/10.1155/2022/5009892
15. Kumar, Kiran C, Sethi G, Rawal K (2022) An effective approach for object detection using deep convolutional networks. In: 2022 International conference on breakthrough in heuristics and reciprocation of advanced technologies (BHARAT), pp 37–42. https://doi.org/10.1109/BHARAT53139.2022.00019
16. Yadav D, Jain I (2022) Comparative analysis of machine learning algorithms for Parkinson's disease prediction. In: 2022 6th International conference on intelligent computing and control

systems (ICICCS). Madurai, India, pp 1334–1339. https://doi.org/10.1109/ICICCS53718.2022.9788354
17. Alkhatib R, Diab MO, Corbier C, Badaoui ME (2020) Machine learning algorithm for gait analysis and classification on early detection of Parkinson. IEEE Sens Lett 4(6):1–4. https://doi.org/10.1109/LSENS.2020.2994938
18. Gupta I, Sharma V, Kaur S, Singh AK (2022) PCA-RF: an efficient parkinson's disease prediction model based on random forest classification. arXiv:2203.11287

Hardware Implementation for Advanced Arrythmia Detection of Multi-channel ECG Signals

V. Priyanka Brahmaiah, R. Bharath Vardhan Reddy, Veda Mouktika Boddanapalli, Srinivas Talasila, and Y. Padma Sai

Abstract Irregular heartbeats known as arrhythmia can be detected by studying the ECG signals in which RR intervals are calculated to find out the beats per minute (BPM) (Yeh and Wang in Comput Methods Progr Biomed 91:245–254, 2008 [1]). These BPM numbers help in identifying eight different types of arrhythmias. Each phase represents different levels of irregularity in cardiac rhythm that assist doctors to diagnose and plan for treatment. Real-time monitoring with precise classification is made possible through machine learning methods together with advanced algorithms used by ECG analysis systems while dealing with arrhythmias. This aids in early intervention as well as personalized management approaches necessary to avert risks like congestive heart failure or stroke. There is promise for better outcome among patients with arrhythmia managed continuously due to improvements brought about by technological advancement in cardiac monitoring devices. Timely identification and intervention are required when it comes to dealing with arrhythmia. In this case, ECG signals measure RR intervals which determine the number of times the heart beats within one minute (BPM), and therefore, they can be grouped into eight stages of arrhythmias depending on their rate. However, a correct diagnosis needs care. Advancements in technology and healthcare systems are important for better diagnosis of arrhythmia and improving patient results since they help in early detection which leads to immediate medical attention, hence cutting down on complications such as heart failure or stroke.

Keywords Arrythmia · Verilog · ECG · Data compression

V. Priyanka Brahmaiah (✉) · R. Bharath Vardhan Reddy · V. Mouktika Boddanapalli · S. Talasila · Y. Padma Sai
Department of ECE, VNR Vignana Jyothi Institute of Engineering and Technology, Bachupally, Hyderabad, Telangana, India
e-mail: priyanka.veeramosu@ieee.org

© The Author(s), under exclusive license to Springer Nature Singapore Pte Ltd. 2025

H. Zen et al. (eds.), *Soft Computing and Signal Processing*, Lecture Notes in Networks and Systems 1200, https://doi.org/10.1007/978-981-97-9926-8_51

1 Introduction

There are several causes why people should detect arrhythmia. Some arrhythmias may result in severe problems that include heart failure, stroke, or even sudden cardiac arrest if they are not treated early enough. In addition to this, it allows for timely intervention which could save life. It can be difficult to diagnose many arrhythmias because they do not manifest any symptoms at all or show mild signs unless monitored properly over time, hence requiring frequent detection. Thus, routine recognition of abnormal heart rhythms becomes essential especially among those who are more vulnerable due to old age bracket, family background, or other related heart diseases which might have occurred before their time. In general terms, recognizing these irregular beats helps protect the heart through quick response towards treatment and control when necessary. Arrhythmia is a medical condition that messes with the rhythm of your heart. Normally, your heart's arrhythmia, this electrical system goes haywire, causing your heart to beat too fast (tachycardia), too slow (bradycardia), or just all over the place. Various forms of a rapid heartbeat, faintness, chest pains, short breathes, and a possibility of losing consciousness are the different ways in which you may experience arrhythmias. They may be transient or take long periods of time; it might be non-life threatening to life-threatening ones depending on their type and severity.

Doctors use a bunch of tests to diagnose arrhythmias, but the main one is called an electrocardiogram (or ECG or EKG for short). Treatment options vary depending on what is causing your arrhythmia and your specific circumstances. It could be anything from taking meds to doing procedures like cardioversion, catheter ablation, or getting a pacemaker or defibrillator put in.

On the whole, arrhythmias would definitely affect your heart's well-being significantly as well as the quality of life you lead. Therefore, it is very crucial to detect and diagnose them early enough and also have healthcare professionals help you manage the condition. Normally, arrhythmia is a heart rhythm disorder that interrupts this organ's normal working. Your heart's chambers contract and expand in harmony due to electrical impulses that travel along them, but this is not the case with tachycardia, bradycardia, or chaotic atrial rhythms where heartbeats become too fast, slow, or irregular, respectively, causing poor blood pumping ability.

It may appear in different ways such as when you feel your heart beating fast, dizziness, difficulty in breathing, chest pain, or even fainting. They could be transient or persistent, posing a threat so slight to severe risk based on type and severity of arrhythmias, respectively. There are many tests that doctors use to diagnose arrhythmias, but the most common one is called electrocardiogram (or EKG or ECG). Depending on what is causing your arrhythmia and your personal circumstances, treatment options differ. This might involve anything from medications to procedures like cardioversion, catheter ablation, or having a pacemaker/implantable cardioverter-defibrillator installed. On the whole therefore, arrhythmias will have great significance towards the quality of life as well as general health of your heart. This is why it is extremely important for such conditions to be caught early before they do much harm; hence early detection and accurate diagnosis by health professionals who can guide you through the management process.

2 Types of Arrythmia

Arrhythmia occurs when incredible things are going wrong with your heartbeat rhythm or beats that might not sound right at all. Basically, there are electrical impulses which normally make chambers in your heart contract and then relax in unison. That way, blood can flow through your body smoothly. However, this electrical system will go amok leading to a high heart rate (tachycardia), slow heart rate (bradycardia), or irregular heartbeat. Arrhythmias manifest in various ways such as palpitations, loss of consciousness, shortness of breath, and chest pain among others. Sometimes they occur intermittently while at other times become persistent with gravity ranging from harmless to potentially fatal depending on the type and severity shown in Table 1.

To ascertain if someone has an arrhythmia, doctors can perform many tests, but among them the most common is electrocardiogram abbreviated as EKG or ECG. They come in different forms like medication or procedures including cardioversion, catheter ablation, pacemaker implantation, or implantable defibrillators [2]. The determination will be based on your kind of arrhythmia and how it affects you.

Table 1 Description of arrhythmia

Type of arrhythmia	Description	BPM range
Tachyarrhythmias	Abnormally fast heart rates	> 100
Supraventricular tachycardia	Rapid heart rate originating above the ventricles	> 100
Ventricular tachycardia	Rapid heart rate originating in the ventricles	> 100
Bradyarrhythmia's	Abnormally slow heart rates	< 60
Sinus bradycardia	Slow heart rate originating in the sinus node	< 60
Heart block	Delay or blockage in electrical signals	Variable
Atrial fibrillation	Irregular and rapid atrial activity	> 100
Atrial flutter	Rapid but organized atrial contractions	> 100

3 Data Compression Algorithm

This is one great signal processing technique known as multi-channel linear prediction that we use to predict future samples of a given signal using previous ones. In essence it is just about using past samples of any signal to make highly accurate predictions for multiple input channels simultaneously. As opposed to single-channel linear prediction [3], which relies on a single input signal, multi-channel linear prediction uses many such signals to predict more accurately future samples. We frequently employ this method in audio processing, speech recognition as well as telecommunications where it is often useful for tasks like noise reduction [4], echo cancellation, and speech enhancement. By considering information from multiple channels, multi-channel linear prediction boosts prediction accuracy especially in situations where signals are complex or the environment varies. This eventually results into better performance and high quality of signal processing. Design flow is shown in Fig. 1.

Adaptive linear prediction is a technique used in signal processing in order to forecast future samples of a given signal. Here, however, the predictive model is adaptive not static as observed with linear models whereby its parameters continuously adjust according to characteristics of an input data stream being monitored. Such adaptability leads to accurate predictions even under the highly dynamic and non-stationary nature of such signals Adaptive linear prediction has various applications that ranges from speech and audio processing through telecommunication systems up to system identification. The model parameters are constantly tuned so as to improve the accuracy and robustness of the predictions, hence making it a very promising adaptive linear prediction method.

Golomb–Rice coding is a lossless data compression method used in very efficient encoding of lengthy sequences of positive integers. It mingles components of Golomb coding with simplified binary representation and is hence particularly appropriate

Fig. 1 Procedural Flow

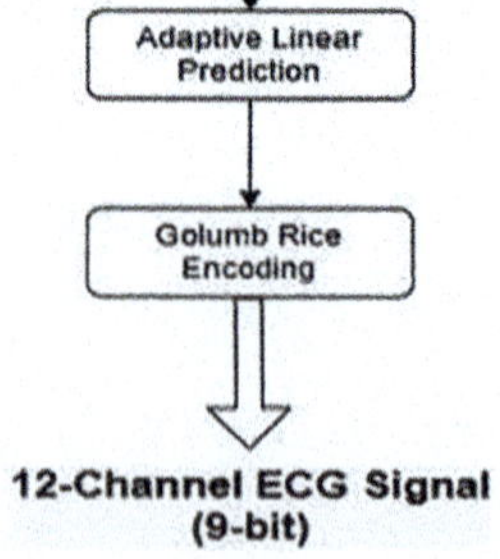

for geometrically distributed data. Two parameters are involved in this process: one being the divisor, m, chosen based on statistical distribution of input values, and the other being the quotient remainder arising from the division of each value in the set by m. To encode each number, it is necessary to represent its quotient using a unary code which has a set of 1 s followed by zero for each value and to encode its reminder with fixed length binary code that needs as much bits as the divisor does. Consequently, Golomb–Rice coding can take into account features inherent to source information, and therefore, it works well when compressing sparse data types such as run lengths in bitmap images or residual values in audio and video signals' predictive coding. Because it is simple and highly effective for compression certain kinds of sparse data like picture image run lengths or residual values in predictive coding audio/video signals. It also makes Golomb–Rice codes popular among designers working within restricted computational resource environment or where simple but effective compression technique is required due to its simplicity and high efficiency. RTL schematic and output pulses are shown in Figs. 2 and 3.

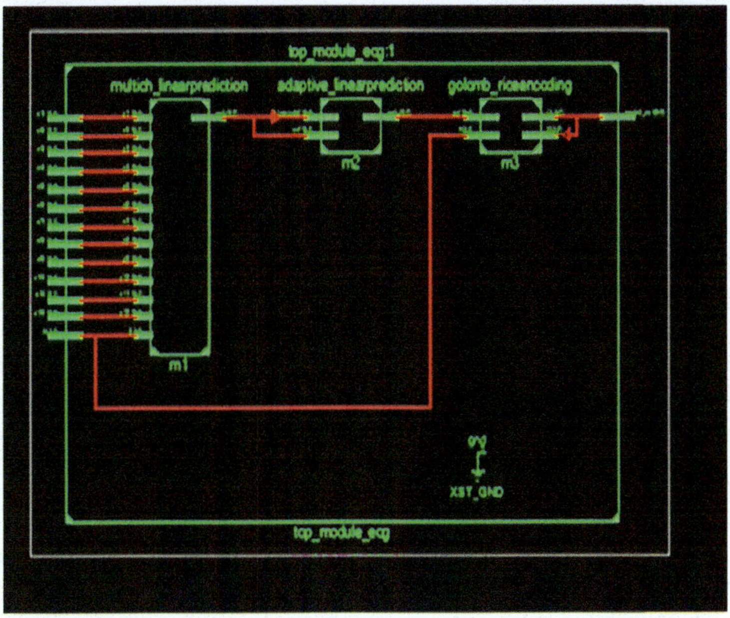

Fig. 2 RTL schematic

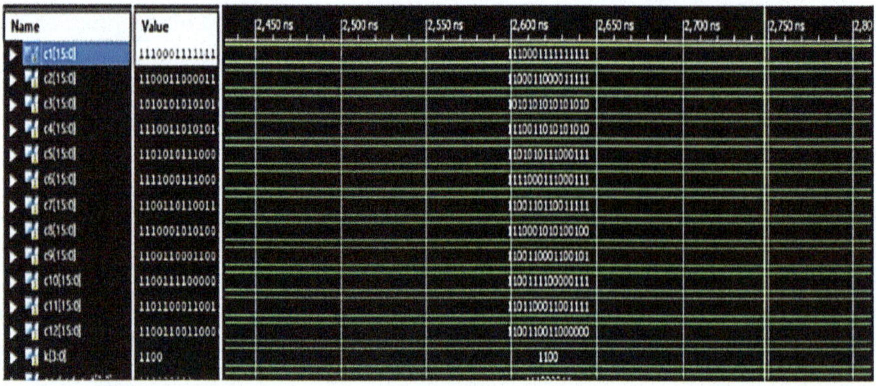

Fig. 3 Output pulses

4 Proposed Method

The process begins by using wavelet decomposition that disintegrates an ECG signal into separate frequency components. This helps since ECG signals can be contaminated with noise from different sources [5]. The next two steps are aimed at eliminating various types of noise. High-frequency noise is removed by discarding certain wavelet coefficients (D1 and D2) from the decomposed signal [4]. Then, low-frequency noise is eliminated by discarding another set of wavelet coefficients (A10). After that, the flow chart chooses some wavelet coefficients (D3 and D3′) as if for the following step in its stage of processing. However, the flow chart then proceeds to a classification stage, whereas such information is not available in the picture. The next step involves detecting R peaks—these are prominent upward deflections seen on a normal ECG signal which correspond to ventricular contractions of the heart [6]. Feature extraction follows next whereby feature extraction refers to identifying and picking out major characteristics of a signal. Relevant examples include amplitude or duration time taken during R peaks in ECG processing. Finally, there is a threshold applied for phasor detection within the signal. A phasor in ECG processing is a mathematical representation of the magnitude and phase of a periodic signal. It could be a specific value on which the threshold is likely based to distinguish between signal and noise. In general, it can be observed from the flow chart that wavelet decomposition is used as well as other processing steps in order to remove noise and detect key components of an ECG signal. Design flow and expected output are shown in Figs. 4 and 5.

Atrial fibrillation (AFib) is a common arrhythmia marked by rapid and irregular heartbeats in the atria, raising the risk of stroke and other heart issues. Management includes regular heart rhythm monitoring, limiting alcohol and caffeine, managing blood pressure, following a heart-healthy diet, and proper hydration.

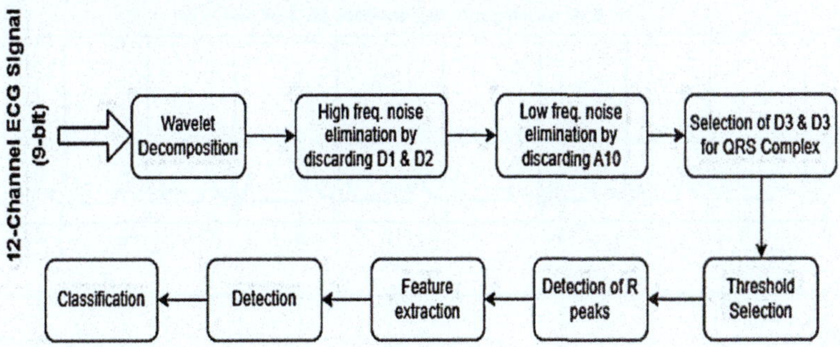

Fig. 4 Design flow

Fig. 5 Expected output

```
database
live ecg recording
enter mode 1
choose a data base
Heart Rate=66

ans =

    '66'

NORMAL SINUS RYTHM
>>
```

Atrial Flutter has some similarities with AFib, but involves more organized atrial beats. The condition requires regular heart monitoring, possibly medication for heart rate or rhythm control, and sometimes catheter ablation.

Ventricular tachycardia (VT) is a dangerous condition when instant cardiac death may occur because HR rises sharply in ventricular chambers. To manage VT: avoid stimulants; follow treatment plan; update your knowledge about this disease; check if implanted devices work properly. Ventricular fibrillation (VFib) is an emergency where fast, erratic heartbeats prevent the heart from pumping blood effectively making immediate medical action like CPR and defibrillation necessary. Wearing a medical alert device and avoiding known triggers are recommended for those at risk of this condition.

Bradycardia is a condition marked by abnormally low heart rate, which could result in such symptoms as dizziness and weariness. It may be necessary to monitor for symptoms, modify medications, or place a pacemaker to control the heart rate.

5 Results

Figures 6, 7, and 8 show the results.

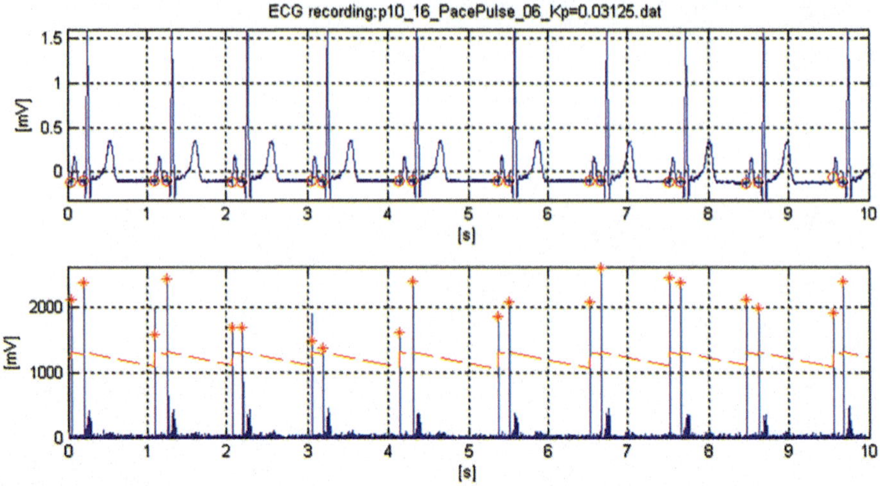

Fig. 6 R peak interval

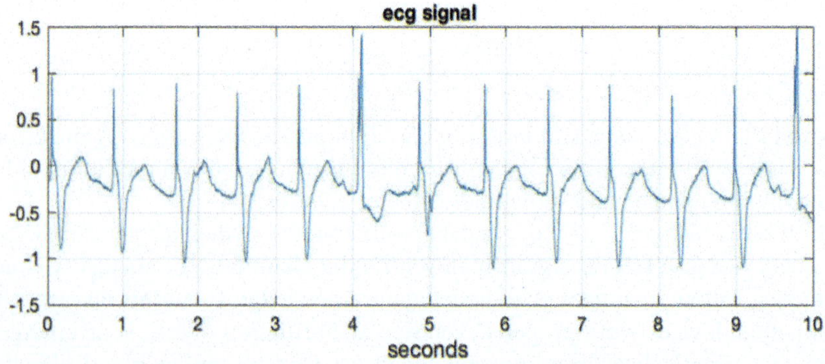

Fig. 7 ECG plot

```
Command Window
>> val

val =

  Columns 1 through 14

   -40   -40   -40   -40   -40   -40   -40   -40   -38   -35   -38   -41   -42   -4

  Columns 15 through 28

   -39   -42   -46   -47   -49   -47   -46   -46   -49   -54   -52   -48   -45   -4

  Columns 29 through 42

   -45   -45   -45   -45   -44   -47   -52   -52   -52   -52   -51   -51   -52   -5

  Columns 43 through 56

   -53   -50   -48   -51   -53   -56   -52   -51   -50   -49   -50   -51   -49   -4

  Columns 57 through 70

   -50   -49   -53   -54   -50   -50   -49   -50   -51   -51   -48   -49   -53   -5

  Columns 71 through 84

   -54   -56   -56   -53   -53   -52   -53   -51   -49   -46   -46   -47   -46   -4
```

Fig. 8 ECG sensor values

6 Conclusion

This design is implemented using Verilog language and Arduino Ide, data compression of ECG signal from 16 to 8 bit. An efficient method of arrhythmia detection has been developed based on heart rate. After enhancement, R peak detection, the most relevant feature of an ECG waveform, is done using threshold methodology. Once the R peak is detected, RR interval is calculated to estimate the heart rate for arrhythmia detection. The design is implemented in Arduino mega using a heartbeat sensor. It can compute the heart rate of the signal and detect the arrhythmia. The performance of the proposed method is evaluated using different arrhythmia data records of MIT-BIH databases. Till now, the analysis is done with R peak but in the future, work will go on with the other peaks too as some heart arrhythmias depend not only on the R-R interval but also P, T waves. Our system cannot detect those kinds of arrhythmias. It can be used for the processing of real-time ECG signals. The algorithm can be modified further to calculate all the ECG parameters automatically at the output.

References

1. Yeh Y-C, Wang W-J (2008) QRS complex detection for ECG signal: the difference operation method. Comput Methods Progr Biomed 91:245–254; Rangayyan RM, Biomedical Signals

2. Deep-learning-based arrhythmia detection using ECG signals: a comparative study and performance evaluation
3. Bannajak K, Theera-Umpon N, Signal acquisition-independent lossless electrocardiogram compression using adaptive linear prediction
4. Al Mahamdy M, Bryan Riley H (2014) Performance study of different denoising methods for ECG signals. Procedia Comput Sci 37:325–332
5. Kumar P, Sharma VK, Detection and classification of ECG noises using decomposition on mixed codebook for quality analysis
6. Pal S, Mitra M (2012) Empirical mode decomposition-based ECG enhancement and QRS detection. Comput Biol Med 42

Principle Study of Nanosheet Field-Effect Transistors with Transition Metal Dichalcogenide Channel Materials

Aruru Sai Kumar, V. Bharath Sreenivasulu, Srinivas Talasila, Venkat Jukuru, Thanvitha Valluru, and D. P. S. S. S. K. Vamsi

Abstract This research work comprehensively investigates the performance characteristics of nanosheet field-effect transistors (FETs) employing transition metal dichalcogenides (TMDs) as channel materials. Molybdenum disulfide (MoS_2), molybdenum diselenide ($MoSe_2$), and molybdenum ditelluride ($MoTe_2$) are all members of the transition metal dichalcogenide (TMD) family. These materials are captivating researchers due to their unique electronic properties. Each TMD offers a distinct combination of band gap, carrier mobility, and intrinsic qualities, making them ideal candidates for the development of 2D transistor channels. Leveraging simulations on Nanohub.org, we systematically explore the effects of varying gate lengths and dielectric materials, including HfO_2, Si_3N_4, and SiO_2, on device performance metrics. Specifically, we calculate the on-current (I_{ON}), off-current (I_{OFF}), and I_{ON}/I_{OFF} ratio to evaluate the transistor's operational efficiency and switching behavior. Our findings reveal intricate relationships between gate length scaling, dielectric selection, and device performance, offering valuable insights for optimizing nanosheet FETs with TMD channels. This research contributes to advancing the understanding of nanoelectronic device design and paves the way for developing high-performance transistors for future technological applications.

Keywords Nanosheet FET · Transition metal dichalcogenides (tmd) · On-current (I_{ON}) · Off-current (I_{OFF}) · Electron desity per sheet

A. S. Kumar (✉) · S. Talasila · V. Jukuru · T. Valluru · D. P. S. S. S. K. Vamsi
Department of ECE, VNR Vignana Jyothi Institute of Engineering and Technology, Telangana, India
e-mail: asaikumar.nitw@gmail.com

V. B. Sreenivasulu
Department of Electronics and Communication Engineering, Manipal Institute of Technology Bengaluru, Manipal Academy of Higher Education, Manipal 576104, Karnataka, India
e-mail: vbharathsree818@gmail.com

© The Author(s), under exclusive license to Springer Nature Singapore Pte Ltd. 2025

H. Zen et al. (eds.), *Soft Computing and Signal Processing*, Lecture Notes in Networks and Systems 1200, https://doi.org/10.1007/978-981-97-9926-8_52

1 Introduction

In recent years, the exploration of novel materials and device architectures has been pivotal in advancing the field of nanoelectronics. Among these materials, transition metal dichalcogenides (TMDs) have emerged as promising candidates for channel materials in field-effect transistors (FETs) due to their unique electronic properties and two-dimensional nature [1]. In parallel, the development of nanosheet FETs has garnered significant attention for their potential to overcome limitations associated with conventional FET structures. Nanosheet FETs offer several advantages over traditional planar and FinFET architectures. Their thin-layered structure enables superior electrostatic control and reduced short-channel effects, making them highly suitable for scaled-down devices. Furthermore, the two-dimensional geometry of nanosheets facilitates efficient gate control over the channel region, leading to enhanced device performance and reduced power consumption [2–4]. Nanosheets represent the next evolutionary stage in transistor technology, poised to supersede FinFET devices for technology nodes below 7 nm and beyond. As a member of the multi-gate transistor family, the nanosheet gate-all-around (GAA) FET embodies a specialized transistor configuration, distinguished by its nanoscale channel thickness. This design refinement enables nanosheet field-effect transistors (NSFETs) to exert exceptional electrostatic control over the channel area, thereby mitigating short-channel effects (SCEs) characteristic of conventional transistors [5–7].

In this study, we focus on evaluating the performance of nanosheet FETs with TMD materials as channel materials. Specifically, we investigate the influence of varying gate lengths and dielectric materials on device characteristics. The choice of gate length and dielectric plays a crucial role in determining key metrics such as the on-current (I_{on}), off-current (I_{off}), and I_{on}/I_{off} ratio, which are pivotal indicators of device performance. The on-current (I_{on}) represents the maximum current flowing through the device when the transistor is in the on-state, reflecting its capability to conduct charge carriers effectively. Conversely, the off-current (I_{off}) signifies the leakage current when the transistor is in the off-state, which directly impacts the power consumption and overall efficiency of the device. The I_{on}/I_{off} ratio serves as a measure of the transistor's switching efficiency, indicating how effectively it can switch between ON and OFF states.

This research paper is further arranged as follows: Sect. 2 explains the literature survey, Sect. 3 provides the proposed methodology which explains the monolayer nanosheet field-effect transistor (NSFET). Section 4 provides the result analysis, and the conclusion of this research is explained in Sect. 5.

2 Related Work

Sreenivasulu et al. [8] implemented a Junctionless Vertically Stacked Nanowire Gate-All-Around Field-Effect Transistor (JVS NW GAA-FET) incorporating spacer technology and variable gate lengths and spacer materials. It investigates how spacer length variations impact the performance of the GAA nanowire FET (NSFET). The study reveals that while changing spacer length helps to some extent, it isn't the most effective method for mitigating short-channel effects (SCEs). However, an optimized asymmetric spacer design demonstrates the best results in terms of on-current (I_{on}), subthreshold swing (SS), and the I_{on}/I_{off} ratio. This optimized design with asymmetric nanowires achieves low SS, low off-current (I_{off}), and a high I_{on}/I_{off} ratio. This translates to minimal standby power consumption, making it ideal for applications demanding ultra-low power. Additionally, the proposed device exhibits strong currents and superior gate control even at very small gate lengths (LG). Liang et al. [9] introduced an improved SPICE model for the negative capacitance field-effect transistor (NCFET). The model captures the interaction between the MOSFET's gate charge and the ferroelectric layer's stored charge. It also incorporates parasitic capacitance and resistance, ensuring adherence to the principle of charge conservation. The model is then used to analyze two key NCFET characteristics: negative resistance (NR) and drain-induced barrier lowering (DIBL). Finally, the impact of these properties on analog circuit performance is explored using classic examples like current mirrors and latch comparators. The study demonstrates that the NR property can simultaneously mitigate DIBL and channel-length modulation (CLM) effects, leading to improved current mirroring accuracy and faster latch comparator speeds.

Xie et al. [10] focused on junctionless double-gate (JL DG) MOSFETs. The study develops analytical formulas to calculate two-dimensional potential and subthreshold current, considering the dependence on channel length. These formulas are then validated using TCAD simulations. The analysis reveals that JL DG MOSFETs suffer more significantly from short-channel effects (SCEs) compared to undoped MOSFETs. This is primarily caused by the impact of the pre-exponential factor in the calculations. Sung et al. [11] investigated the electrical properties of Gate-All-Around stacked Silicon Nanosheet MOSFETs (GAA NSFETs) using computational analysis. The focus is on maximizing channel count and its impact on digital circuits. Without metal sidewall source/drain (S/D) regions, the analysis is limited to three channels due to decreasing electrostatic potential from top to bottom contacts. However, including metal sidewalls (often tungsten) significantly improves on-state current (I_{on}) by enabling more than three channels. This enhancement is attributed to the low resistivity of tungsten, which strengthens the operational efficiency of stacked GAA NSFETs with higher channel counts. These findings offer valuable insights for designing source and drain structures in stacked GAA NSFETs for future device technologies.

3 Proposed Methodology

The proposed work aims to comprehensively evaluate the performance of nanosheet field-effect transistors (FETs) utilizing transition metal dichalcogenide (TMD) materials (MoS_2, $MoSe_2$, $MoTe_2$) as channel materials, with a focus on varying gate lengths and dielectric materials (HfO_2, Si_3N_4, SiO_2) as depicted in Fig. 1. Leverag-

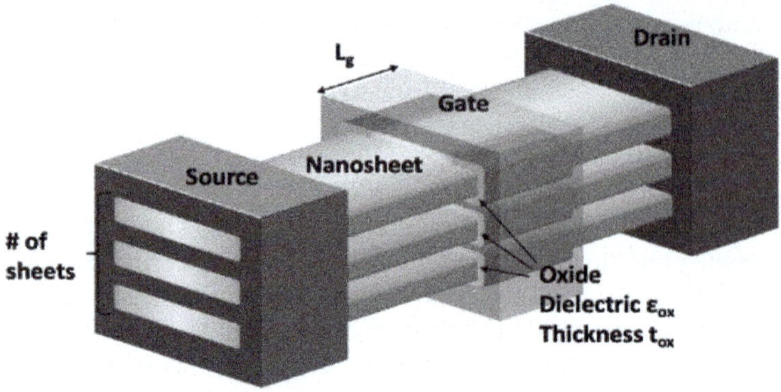

Fig. 1 Monolayer nanosheet field-effect transistor (NSFET)

ing simulation capabilities on Nanohub.org, the study will calculate the on-current (I_{on}), off-current (I_{off}), and I_{on}/I_{off} ratio for each combination of TMD material, gate length, and dielectric material, facilitating a comprehensive analysis of device performance [12–17]. Optimization strategies will be explored to identify optimal combinations of parameters, considering potential trade-offs between gate length scaling, dielectric selection, and TMD material choice. Through comparative analysis and discussions on future directions, the research aims to advance understanding in the field and provide insights for the design of high-performance nanosheet FETs tailored for advanced nanoelectronic applications [18–22].

4 Result Analysis

We have analyzed the performance of nanosheet FET using TMD's as channel materials with different dielectrics. The transfer characteristics (Id versus Vgs) and the switching ratio(I_{on}/I_{off}) were calculated and concluded that 2DFET with TMD's as channel materials using HfO_2 as dielectric is performing better when compared to Air, SiO_2 as dielectric materials. Nanohub.org, the open online simulation tool has been used to run the simulations of nanosheet FET by varying the channel materials for different dielectrics.

4.1 Energy Band Profile

The energy band profile of a nanosheet field-effect transistor (NSFET) at different gate voltages (Vg) as depicted in Fig. 2. The different lines in the graph correspond to different gate voltages. At low gate voltages (Vg = 0.0V and 0.1V), the Fermi level (Ef) is close to the valence band (Ev), which means that the channel is in the depletion

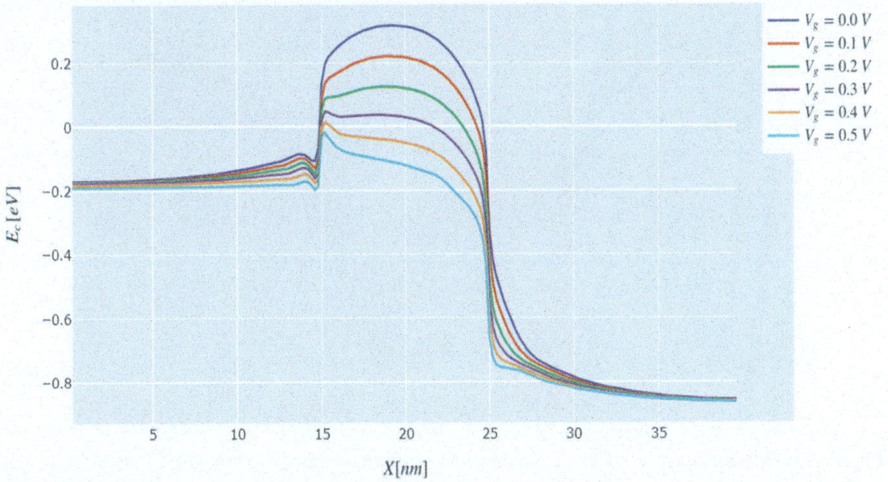

Fig. 2 Energy band profile of a nanosheet field-effect transistor (NSFET) at different gate voltages

mode and has low conductivity. As the gate voltage increases (Vg = 0.2V, 0.3V, 0.4V, and 0.5V), the Fermi level moves closer to the conduction band (Ec), and the channel starts to conduct more current. This is called the accumulation mode. The band profile of the nanosheet FET is an important factor in its performance [23–27]. It affects the threshold voltage, subthreshold swing, and on-current of the transistor. These parameters are important for designing low-power and high-performance electronic devices.

4.2 Electron Density per Sheet

The energy density per sheet of a nanosheet field-effect transistor (NSFET) at different gate voltages (Vg) as depicted in Fig. 3. The different lines in the graph correspond to different gate voltages. As the gate voltage increases, the energy bands shift upwards. This indicates that the gate voltage is controlling the conductivity of the channel by affecting the energy bands of the nanosheet [28–30]. In a nanosheet FET, the concept of electron density is more complex than a single value due to the sheet-like nature of the channel. Electrons aren't uniformly spread, but rather concentrate near the center.

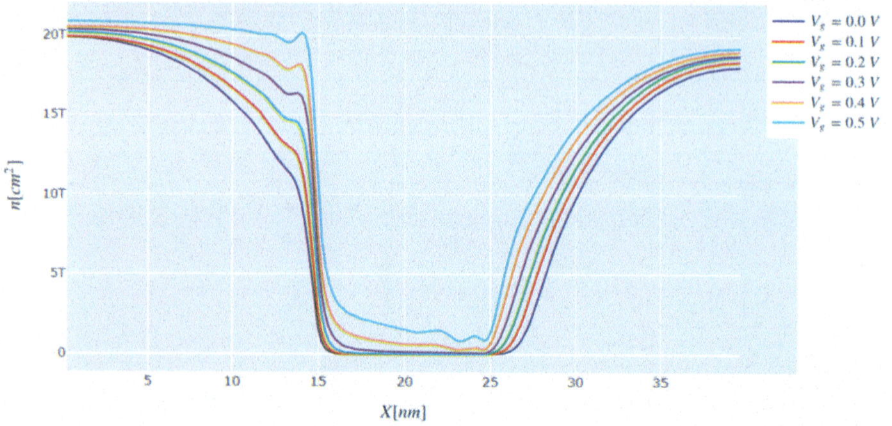

Fig. 3 Energy density per sheet of a nanosheet field-effect transistor (NSFET) at different gate voltages

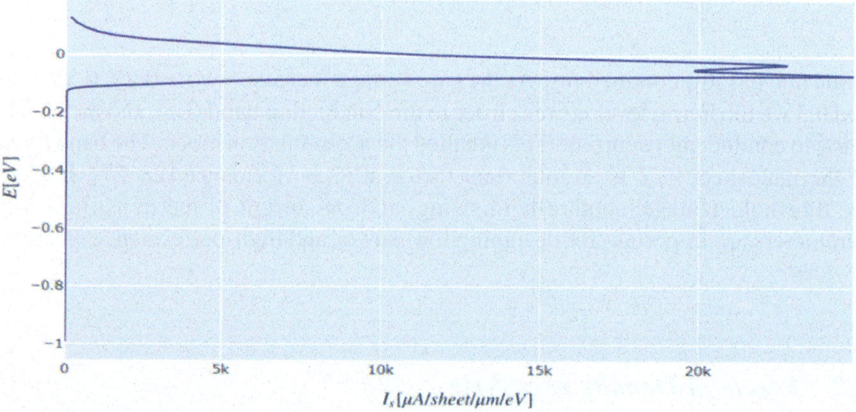

Fig. 4 Current spectrum graph of a nanosheet field-effect transistor (FET)

4.3 Current Spectrum

This type of transistor uses a nanosheet, which is a thin layer of material with a thickness of just a few atoms, as its channel. The current through the nanosheet increases as the gate voltage becomes more positive as depicted in Fig. 4. This is because a positive gate voltage attracts electrons to the channel, which increases the conductivity of the channel. The current starts to increase significantly at around −0.4 eV and reaches a maximum value at around 0.2 eV. This indicates that the threshold voltage for this particular nanosheet FET is around −0.4 eV. The current spectrum graph of a nanosheet FET is a useful tool for characterizing the electrical properties of the device.

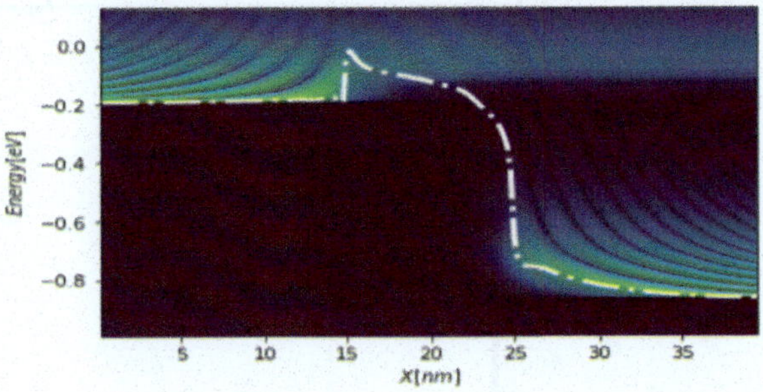

Fig. 5 Density spectrum graph of a nanosheet field-effect transistor (FET)

4.4 Density Spectrum

A density spectrum in a nanosheet FET reveals the energy levels of electrons within the material, depicted on the X-axis, ranging from the valence band to the conduction band. Peaks and valleys on the Y-axis denote the density of states, highlighting available energy states per unit volume as depicted in Fig. 5. Features such as valence and conduction bands, Dirac peaks (for graphene-based nanosheets), and doping effects are discernible, offering insights into electronic band structure, carrier concentration, and device performance optimization. Understanding these characteristics aids researchers in tailoring nanosheet FETs for enhanced functionality and efficiency.

4.5 I_D Versus V_{GS}

The drain current (Id) is high for a gate length of 10 nm when compared with others. Figure 6 shows the relation between the drain current (Id) and gate-source voltage (Vgs) for the field-effect transistor (FET) and serves as a fundamental characterization of the device's behavior. This curve illustrates the device response to varying gate-source voltages while maintaining a constant drain-source voltage. This graph typically exhibits distinct regions such as the cutoff region, saturation region, and linear region, each representing different operational states of the transistor.

4.6 I_D Versus V_{DS}

The Id versus Vds (drain current versus drain-source voltage) is referred to as the "output characteristic" of the field-impact transistor (FET) because this graph essentially depicts the output behavior of the FET in response to versions inside the implemented drain-source voltage as depicted in Fig. 7. It showcases how the transistor responds to outside load and voltage changes, presenting a complete knowledge of its overall performance characteristics beneath distinct operating conditions.

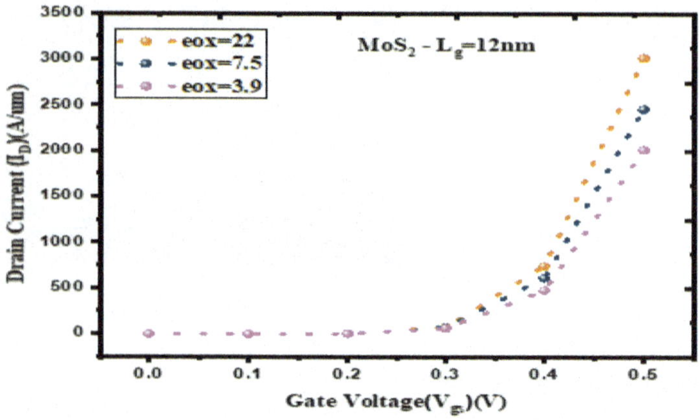

Fig. 6 Id versus Vgs for various dielectric materials while the channel materials are MoS_2

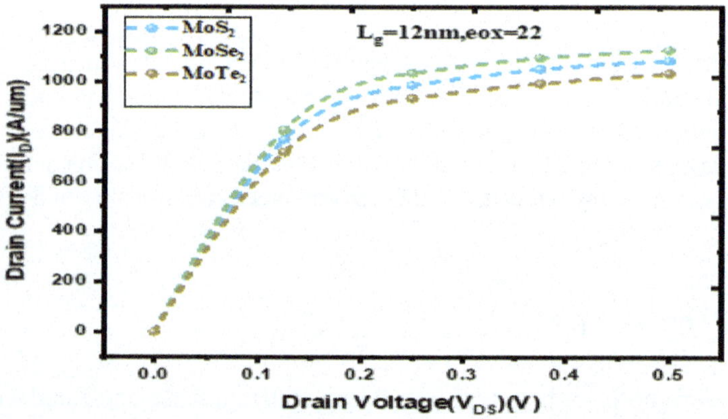

Fig. 7 Id versus Vds for various channel materials while the dielectric is HfO_2 (eox = 22) and gate length is 12 nm

4.7 I_{ON} and I_{OFF}

The I_{ON} is a critical parameter in field-effect transistors (FETs) that represents the contemporary flowing between the drain and source terminals while the transistor is in the ON state typically happening when the device is within the saturation region of operation. It is calculated with the aid of applying a sufficient gate-supply voltage (Vgs) to turn the transistor ON and measuring the resulting drain contemporary (Id) below unique operating situations. The I_{OFF} is a vital parameter in field-effect transistors (FETs) that signify the leakage present day among the drain and supply terminals while the transistor is inside the OFF state. It is calculated by using applying

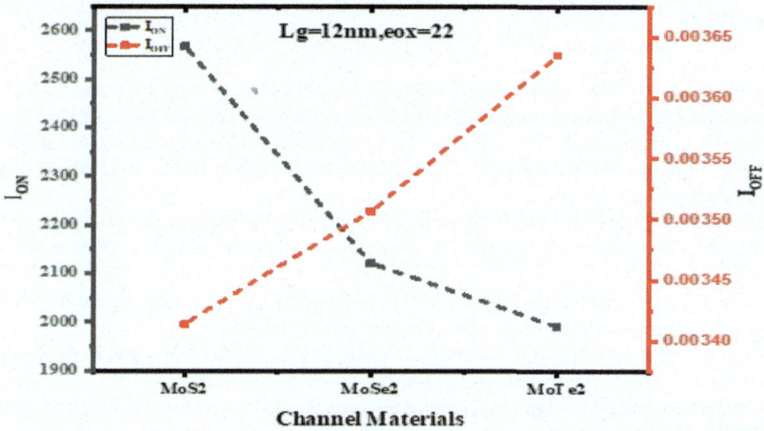

Fig. 8 I_{ON} and I_{OFF} for various channel materials while the dielectric is HfO_2 (eox = 22) and gate length is 12 nm

a gate-source voltage (Vgs) that guarantees the transistor is fully grown to become off, and then measuring the resulting drain modern-day (Id) under precise running conditions as depicted in Fig. 8.

5 Conclusion

This research study on nanosheet field-effect transistors (FETs) with transition metal dichalcogenide (TMD) channel materials has provided valuable insights into their performance characteristics. Through variations in gate length and dielectric materials, we have comprehensively evaluated key metrics including the on-current (I_{on}), off-current (I_{off}), and I_{on}/I_{off} ratio. Furthermore, our investigation into dielectric materials highlights their crucial role in determining the overall performance of nanosheet FETs. Dielectrics such as HfO_2, Si_3N_4, and SiO_2 influence the device's gate capacitance, leakage current, and subthreshold behavior, thereby directly impacting I_{on}, I_{off}, and I_{on}/I_{off} ratios. This study underscores the importance of optimizing gate length, dielectric selection, and channel materials for maximizing the performance of nanosheet FETs. By understanding and manipulating these factors, we can pave the way for the development of next-generation nanoelectronic devices tailored for advanced applications in computing, communication, and beyond.

Acknowledgements We sincerely acknowledge Nanohub.org for the experimental tool support. We also acknowledge Jing Guo, Ning Yang, and Qimao Yang, University of Florida, for the simulation tool support.

References

1. Sreenivasulu VB, Narendar V (2022) Design insights of nanosheet FET and CMOS circuit applications at 5-nm technology node. IEEE Trans Electron Devices 69(8):4115–4122
2. Kumari NA, Sreenivasulu VB, Vijayvargiya V, Upadhyay AK, Ajayan J, Uma M (2024) Performance comparison of nanosheet FET, CombFET, and TreeFET: device and circuit perspective. IEEE Access 12:9563–9571
3. Kumar AS, Rao TVKH (2019) Efficient core mapping on customization of NoC platforms, 2019 IEEE international symposium on smart electronic systems (iSES) (Formerly iNiS), pp 57–62
4. Kumar AS, Hanumantha Rao TVK (2021) An adaptive core mapping algorithm on NoC for future heterogeneous system-on-chip. Comput Electr Eng 95
5. Kumar AS, Rao KN, Sujith A, Dhanuja T, Vinay MVS (2023) Design and implementation of 1KB SRAM array in 45 nm technology for low-power applications, 2023 3rd international conference on advances in computing, communication, embedded and secure systems (ACCESS), India, pp 245–250
6. Kumar AS et al (2023) Nanosheet field effect transistor device and circuit aspects for future technology nodes. ECS J Solid State Sci Technol 12(8)
7. Kumar AS, Siddhesh U, Sai kiran N, Bhavitha K. Design of high speed 8-bit Vedic multiplier using Brent Kung adders, 2022 13th international conference on computing communication and networking technologies (ICCCNT), Kharagpur, India, pp 1–5
8. Sreenivasulu VB, Narendar V (2021) Characterization and optimization of junctionless gate-all-around vertically stacked nanowire FETs for sub-5 nm technology nodes. Microelectron J 116:105214
9. Liang Y, Li X, Gupta SK, Datta S, Narayanan V (2018) Analysis of DIBL effect and negative resistance performance for NCFET based on a compact SPICE model. IEEE Trans Electron Devices 65(12):5525–5529
10. Xie Q, Wang Z, Taur Y (2017) Analysis of short-channel effects in junctionless DG MOSFETs. IEEE Trans Electron Devices 64(8):3511–3514
11. Sung WL, Li Y (2021) Characteristics of stacked gate-all-around Si nanosheet MOSFETs with metal sidewall source/drain and their impacts on CMOS circuit properties. IEEE Trans Electron Devices 68(6):3124–3128
12. Sreenivasulu VB, Neelam AK, Panigrahy AK, Vakkalakula L, Singh J, Singh SG (2024) Benchmarking of multi-bridge-channel FETs toward analog and mixed-mode circuit applications. IEEE Access 12:7531–7539
13. Sreenivasulu VB, Neelam AK, Kola SR, Singh J, Li Y (2023) (2023) Exploring the performance of 3-D nanosheet FET in inversion and junctionless modes: device and circuit-level analysis and comparison. IEEE Access 11:90421–90429. https://doi.org/10.1109/ACCESS.2023.3306050
14. Kumar AS et al (2023) Device analysis of vertically stacked GAA nanosheet FET at advanced technology node, 2023 3rd international conference on advances in computing, communication, embedded and secure systems (ACCESS), India, pp 274–279
15. Kumar AS et al (2023) Nanosheet field effect transistor device and circuit aspects for future technology nodes. ECS J Solid State Sci Technol 12(8)
16. Guo J, Yang N, Yang Q (2023) Nanosheet FET. https://nanohub.org/resources/nanosheetfet; https://doi.org/10.21981/1R27-EN52
17. Liu L, Lu Y, Guo J (2013) Monolayer MoS_2 field-effect transistors at the scaling limit. IEEE Trans Electron Devices 60:4133
18. Bharath Sreenivasulu V et al (2024) Spacer engineering on multi-channel FinFET for advanced wireless applications. AEU—Int J Electron Commun 178
19. Gowthami U et al (2024) Performance improvement of spacer-engineered N-type tree shaped NSFET toward advanced technology nodes. IEEE Access 12:59716–59725
20. Panigrahy AK et al (2024) Spacer dielectric analysis of multi-channel nanosheet FET for nanoscale applications. IEEE Access 12:73160–73168

21. Naresh Kumar Reddy B, Kumar AS (2024) Evaluating the effectiveness of bat optimization in an adaptive and energy-efficient network-on-chip routing framework. J Parallel Distrib Comput 188
22. Kumar AS, Naresh Kumar Reddy B (2023) An efficient real-time embedded application mapping for NoC based multiprocessor system on chip. Wirel Pers Commun 128:2937–2952
23. Kumar AS, Hanumantha Rao TVK (2020) Scalable benchmark synthesis for performance evaluation of NoC core mapping. Microprocess Microsyst 79
24. Kumar AS, Hanumantha Rao TVK, Kumar Reddy BN (2020) Exact formulas for fault aware core mapping on NoC reliability. 2020 IEEE 17th India council international conference (INDICON), pp 1–5
25. Sarangam K, Kumar AS, Reddy BNK (2024) Design and investigation of the 22 nm FinFET based dynamic latched comparator for low power applications. Trans Electr Electron Mater 1–14
26. Jaisawal RK (2023) Self-heating and interface traps assisted early aging revelation and reliability analysis of negative capacitance FinFET. 7th IEEE electron devices technology and manufacturing conference (EDTM). Korea, Republic of, Seoul, pp 1–3
27. Sreenivasulu VB, Kumari NA, Lokesh V et al (2023) Design of resistive load inverter and common source amplifier circuits using symmetric and asymmetric nanowire FETs. J Electron Mater 52:7268–7279
28. Sreenivasulu VB, Narendar V (2022) Circuit analysis and optimization of GAA nanowire FET towards low power and high switching. Silicon 14:10401–10411
29. Sreenivasulu VB, Narendar V (2022) Junctionless gate-all-around nanowire FET with asymmetric spacer for continued scaling. Silicon 14:7461–7471
30. Sreenivasulu VB, Narendar V (2022) Design and temperature assessment of junctionless nanosheet FET for nanoscale applications. Silicon 14:3823–3834

Quantum Transport Analysis of Nanosheet FETs Utilizing 2D Semiconductor Materials

Aruru Sai Kumar, V. Bharath Sreenivasulu, Chinthakindi Kiran Kumar, Kondeti Keerthi, Shivani Kethapelly, and Ragidimilli Sai Abhinav

Abstract This paper presents a comprehensive investigation into the ballistic I-V characteristics of nanosheet field-effect transistors (FETs) employing 2D semiconductor materials. By concurrently solving the quantum transport equation in conjunction with the Poisson equation, we offer a rigorous analysis of device performance. Specifically, we examine the case of FINFETs, where the number of sheets equals one, serving as a benchmark for our study. The channel materials under scrutiny consist of monolayer sheets of van der Waals (vdW) 2D semiconductors, demonstrating their potential in advanced device architectures. This research work comprehensively investigates the performance characteristics of nanosheet field-effect transistors (FETs) employing transition metal dichalcogenides (TMDs) as channel materials. Molybdenum disulfide (MoS_2), molybdenum diselenide ($MoSe_2$), and molybdenum ditelluride ($MoTe_2$) are all members of the transition metal dichalcogenide (TMD) family. Our computational framework yields multifaceted insights, including I-V characteristics, potential and charge profiles, as well as quantum transmission and local density of states. Through self-consistent numerical simulations, we elucidate the intricate interplay between quantum mechanical effects and device operation, thereby elucidating key design considerations for future nanoscale electronics. This research contributes to advancing the understanding of nano-electronic device design and paves the way for developing high-performance transistors for future technological applications.

Keywords Nanosheet FET · Transition metal dichalcogenides (TMD) · On-current (I_{ON}) · Off-current (I_{OFF}) · I_{ON}/I_{OFF} ratio

A. S. Kumar (✉) · K. Keerthi · S. Kethapelly · R. S. Abhinav
Department of ECE, VNR Vignana Jyothi Institute of Engineering and Technology, Hyderabad, Telangana, India
e-mail: asaikumar.nitw@gmail.com

V. Bharath Sreenivasulu
Department of Electronics and Communication Engineering, Manipal Institute of Technology Bengaluru, Manipal Academy of Higher Education, Manipal 576104, Karnataka, India

C. K. Kumar
Malla Reddy College of Engineering and Technology, Hyderabad, India

© The Author(s), under exclusive license to Springer Nature Singapore Pte Ltd. 2025

H. Zen et al. (eds.), *Soft Computing and Signal Processing*, Lecture Notes in Networks and Systems 1200, https://doi.org/10.1007/978-981-97-9926-8_53

1 Introduction

In recent years, the field of electronics and photonics has undergone a significant shift with the rise of two-dimensional (2D) materials as leading contenders for technological advancement. Their incredibly thin atomic structure, lack of surface defects, and unique physical properties make them highly attractive for future electronic and photonic applications [1]. As the demand for smaller, more efficient, and high-performing devices continues to grow, researchers and semiconductor industries have focused on scaling down metal-oxide-semiconductor field-effect transistors (MOSFETs) in line with Moore's Law. However, the relentless pursuit of miniaturization has brought about substantial challenges such as leakage currents, drain-induced barrier lowering (DIBL) effects, and short-channel effects [2].

Miniaturization of transistors has led to problems like leakage currents and short-channel effects. To overcome these limitations and maintain high performance with low power consumption, researchers are exploring 2D materials as replacements for conventional silicon. Graphene, the first discovered 2D material, initially attracted attention due to its conductivity, but its lack of a bandgap hinders its use in transistors [3]. This has shifted focus to transition metal dichalcogenides (TMDs) like MoS_2 and WS_2, which possess desired optical, electrical properties, and a bandgap. The ultra-thin structure of 2D materials like MoS_2 effectively reduces short-channel effects in scaled devices due to superior electrostatics. However, these materials exhibit limitations in on-current ((I_{on}) due to a low saturation velocity, even at minimal channel lengths. Additionally, their low saturation velocity makes increasing mobility less impactful on on-current. Two main carrier transport mechanisms exist: scattering and ballistic transport. These determine how carriers move within a material, affecting device performance. The specific transport mode in 2D FETs depends on the device's configuration and characteristics. Ballistic transport, ideal for high-speed FETs, minimizes carrier collisions within the channel [4–6].

In the realm of nanosheet FETs utilizing TMD materials as channels, understanding the intricacies of leakage transport mechanisms is crucial for shaping device performance. TMDs like MoS_2, $MoSe_2$, WS_2, and WSe_2, known for their atomically thin films, present a unique set of challenges and opportunities in charge carrier mobility and scattering phenomena. Various scattering mechanisms, including phonon scattering induced by lattice vibrations, impurity scattering due to defects, and surface roughness scattering, significantly influence charge carrier mobility. Quantum confinement effects in nanoscale TMD layers further complicate matters by affecting energy levels and carrier transport properties. The choice of dielectric material, typically HfO_2, also plays a critical role, with the interface between the TMD channel and the dielectric influencing scattering phenomena [7].

This research paper is further arranged as follows: Sect. 2 explains the literature survey, Sect. 3 provides the proposed methodology which explains the monolayer nanosheet field-effect transistor (NSFET). Section 4 provides the result analysis and the conclusion of this research is explained in Sect. 5.

2 Related Work

As modern electronics continue to advance, understanding the advantages and limitations of various dielectrics for achieving low-power operation becomes increasingly crucial. Examining existing literature provides valuable insights into dielectric choices. Liu et al. [8] conducted a comprehensive review on the maximum scaling limit of monolayer thin-body double-gate molybdenum disulfide (MoS_2) field-effect transistors (FETs). They found that even at a short gate length of 10 nm, scattering significantly reduces the on-current, which is essential for accurately estimating the intrinsic delay of MoS_2 FETs. Self-consistent quantum transport simulation in the presence of phonon scattering examines and compares the ultimate scaling limit of double-gate molybdenum disulfide (MoS_2) field-effect transistors (FETs) with a monolayer thin body to ultrathin-body Si FETs. It can correctly evaluate monolayer MoS_2 FET performance by modeling phonon scattering, quantum mechanical effects, and self-consistent electrostatics. Monolayer MoS_2 FETs had 52% lower DIBL and 13% lower SS than 3-nm-thick-body Si FETs at 10 nm channel length with the same gating.

Sreenivasulu et al. [9] implemented a Junctionless Vertically Stacked Nanowire Gate-All-Around Field-Effect Transistor (JVS NW GAA-FET) incorporating spacer technology and variable gate lengths and spacer materials. It investigates how spacer length variations impact the performance of the GAA nanowire FET (NSFET). The study reveals that while changing spacer length helps to some extent, it isn't the most effective method for mitigating short-channel effects (SCEs). However, an optimized asymmetric spacer design demonstrates the best results in terms of on-current (I_{ON}), subthreshold swing (SS), and the I_{ON}/I_{OFF} ratio. This optimized design with asymmetric nanowires achieves low SS, low off-current (I_{OFF}), and a high I_{ON}/I_{OFF} ratio. Additionally, the proposed device exhibits strong currents and superior gate control even at very small gate lengths (LG).

Due to its suitable direct band gap, mature synthesis techniques, and ambient stability, 2D MoS2 is the most extensively studied and utilized 2D semiconductor. Ni et al. [10] explored the performance upper limit of sub-10 nm monolayer MoS2 transistors, showing promising potential for next-generation nanoelectronic devices. They highlighted that conventional bulk silicon FETs struggle to perform reliably at sub-10 nm channel lengths due to increasingly apparent short-channel effects, which degrade device performance and challenge Moore's Law. Sung et al. [11] investigated the electrical properties of Gate-All-Around stacked Silicon Nanosheet MOSFETs (GAA NS-FETs) using computational analysis. The focus is on maximizing channel count and its impact on digital circuits. Without metal sidewall source/drain (S/D) regions, the analysis is limited to three channels due to decreasing electrostatic potential from top to bottom contacts. However, including metal sidewalls (often tungsten) significantly improves on-state current (I_{ON}) by enabling more than three channels. This enhancement is attributed to the low resistivity of tungsten, which strengthens the operational efficiency of stacked GAA NS-FETs with higher channel counts.

3 Proposed Methodology

The proposed work examines a monolayer double-gated nanosheet FET configuration using MoS2, MoSe2, MoTe2, WSe2, and WS2 as the five distinct types of transition metal dichalcogenide (TMD) channel materials. The main goal is to investigate the impacts of different spacer thicknesses-which can range from one to twenty-two layers-and comprehend how they affect important device properties. The inquiry is further complicated by the fact that the gate insulator is made of three distinct compounds [12–17]. (I_{on} current (I_{ON}), off-state current (I_{OFF}), switching ratio (I_{ON}/I_{OFF}), electron density, band profile, current spectrum and finally density spectrum are among the variables that are being examined. While off-state current measures leakage while the transistor is in the off state, Ion current indicates the transistor's on-state behavior. Changes in the spacer thickness are anticipated to impact carrier transport qualities by altering the gate's electrostatic control over the channel [18–24].

The monolayer nanosheet field-effect transistor(NSFET) is as depicted in Fig. 1. This circuit allows for the exploration of different 2D channel materials and dielectrics to better understand their influence on the FET's performance, which is crucial for designing transistors with specific threshold voltages [25–28]. This property is important to minimize the power consumption of electronic devices in idle or standby mode. Using their special qualities, TMDs are frequently employed as channel materials in field-effect transistors (FETs). TMDs' atomically thin structure allows for superior electrostatic channel control, which enhances gate control and lowers leakage currents. Additionally, TMDs are compatible with high-k dielectrics in transistor designs. This compatibility improves gate control, further reduces power consumption, and makes TMD-based transistors promising candidates for energy-efficient applications [29, 30].

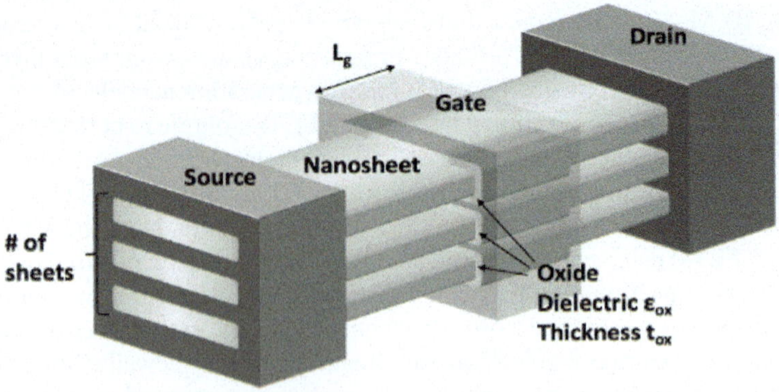

Fig. 1 Monolayer nanosheet field-effect transistor (NSFET)

4 Result Analysis

We have analyzed the performance of nanosheet FET using TMD's as channel materials with different dielectrics. The transfer characteristics (Id versus Vgs) and the switching ratio (I_{ON}/I_{OFF}) were calculated and concluded that 2DFET with TMD's as channel materials using HfO_2 as dielectric is performing better when compared with air, SiO_2 as dielectric materials. Nanohub.org, the open online simulation tool has been used to run the simulations of nanosheet FET by varying the channel materials for different dielectrics [16, 17].

4.1 I_D Versus V_{GS} for Mos2

The transfer characteristics (Id versus Vgs) for different gate lengths and different spacer materials are depicted in Fig. 2. The drain current (Id) is high for a gate length of 10 nm when compared with others. The graph that shows the relation between the drain current (Id) and gate-source voltage (Vgs) for the field-effect transistor (FET), serves as a fundamental characterization of the device's behavior. This curve illustrates the device response to varying gate-source voltages while maintaining a constant drain-source voltage. As the channel length decreases the drain current, i.e., conductivity of the NSFET increases.

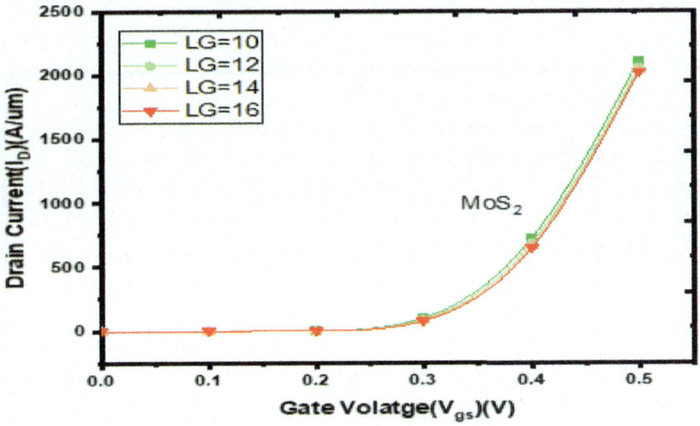

Fig. 2 Id versus Vgs for various channel lengths while the channel materials is MoS2

4.2 I_D Versus V_{GS} For Various Dielectric Materials and Channel Material Is Fixed as Mos2

The ID versus VGS (drain current vs gate-source voltage) plot is commonly referred to as the "transfer characteristics" of a transistor as depicted in Fig. 3. This graph illustrates how the transistor's output current (ID) responds to varying input gate-source voltage (VGS) as depicted in the above figure, by keeping the channel length fixed at 10nm and channel material as MoS2. We obtained a higher drain current while using HfO_2 as the spacer material whose dielectric constant is around 22. HfO_2 is helping to reduce gate leakage compared to traditional dielectric materials.

4.3 I_D Versus V_{GS} for Various Channel Material and LG = 10 nm

The five materials shown in the graph are: MoS2, MoSe2, MoTe2, WS2, and WSe2. The legend indicates that all of the materials have a channel length (L) of 10 nm as depicted in Fig. 4. In all the TMDCs, the drain current increases as the gate voltage increases. This suggests that these materials are n-type semiconductors, which means that the majority carriers are electrons. The rate of increase of the drain current with gate voltage is different for the different TMDCs. MoS2 has the highest rate of increase, while MoTe2 has the lowest rate of increase. This suggests that MoS2 may be a better conductor than MoTe2.

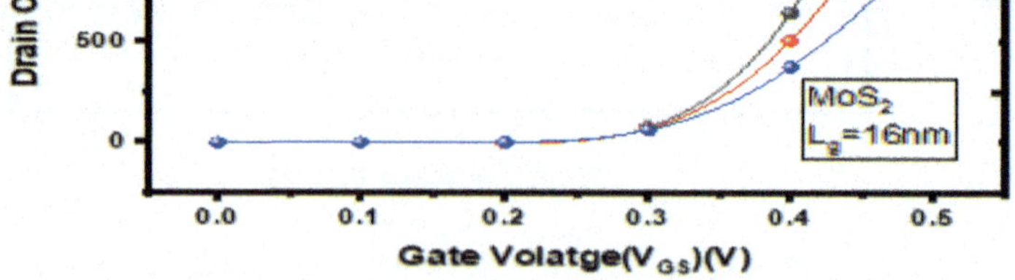

Fig. 3 ID versus VGS for various dielectric material and Lg = 16 nm and channel material is MoS2

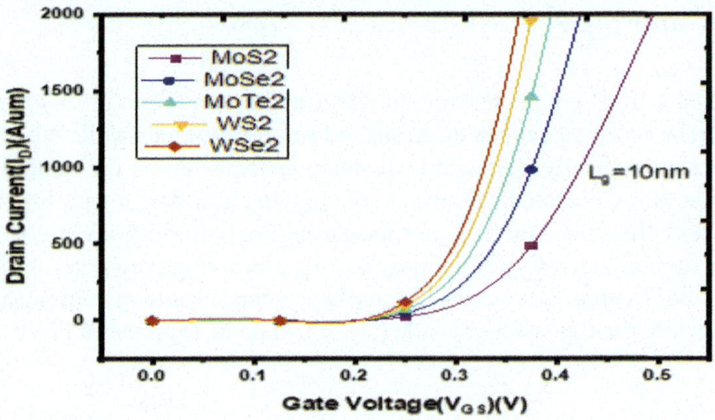

Fig. 4 Id versus Vgs for various channel material and Lg = 10 nm

4.4 I_D Versus V_{DS}

The output traits(Id vs Vds) for exceptional dielectric materials as depicted in Fig. 5. The Id versus Vds (drain current versus drain-source voltage) is referred to as the "output characteristic" of the field-impact transistor (FET) because this graph essentially depicts the output behavior of the FET in response to versions inside the implemented drain-source voltage. It indicates how the drain present day(Id) acts because of the output of the transistor, modifications concerning the applied voltage throughout its terminals.

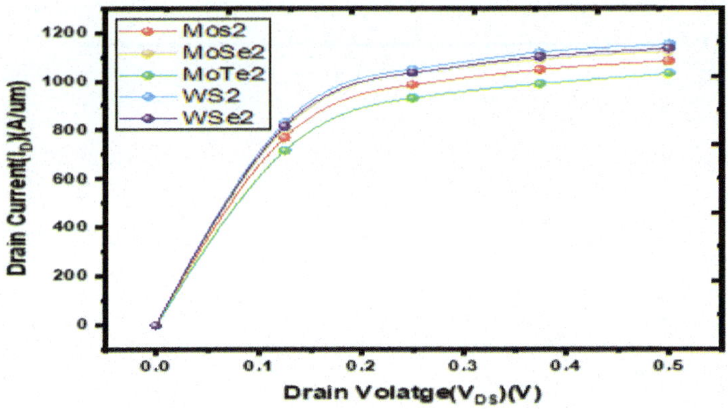

Fig. 5 Id versus Vds for various channel materials while the dielectric is HfO_2(eox = 22) and gate length is 12 nm

4.5 I_{ON} and I_{OFF}

The I_{ON} is a critical parameter in field effect transistors (FETs) that represents the contemporary flowing between the drain and source terminals while the transistor is in the ON state typically happening when the device is within the saturation region of operation. It is calculated with the aid of applying a sufficient gate-supply voltage (Vgs) to turn the transistor ON and measuring the resulting drain contemporary (Id) below unique operating situations. The I_{OFF} is a vital parameter in field-effect transistors (FETs) that signifies the leakage in present day among the drain and supply terminals while the transistor is inside the OFF state as depicted in Fig. 6.

4.6 I_{ON}/I_{OFF} (Switching Ratio)

The I_{ON}/I_{OFF} ratio, often termed the "switching ratio," is a critical parameter in field-effect transistors (FETs) that quantifies the difference between the ON current (I_{ON}) and the OFF current (I_{OFF}) as depicted in Fig. 7. It is calculated by dividing the ON current by the OFF current, providing a measure of the transistor's ability to switch between ON and OFF states effectively. This ratio is essential for evaluating the performance of FETs as it directly reflects the device's switching efficiency and power consumption characteristics. A higher switching ratio signifies a more significant distinction between the ON and OFF states, indicating better control over the flow of current and reduced leakage when the transistor is in the OFF state. The switching ratio serves as a crucial metric for assessing the overall performance and reliability of FET-based devices.

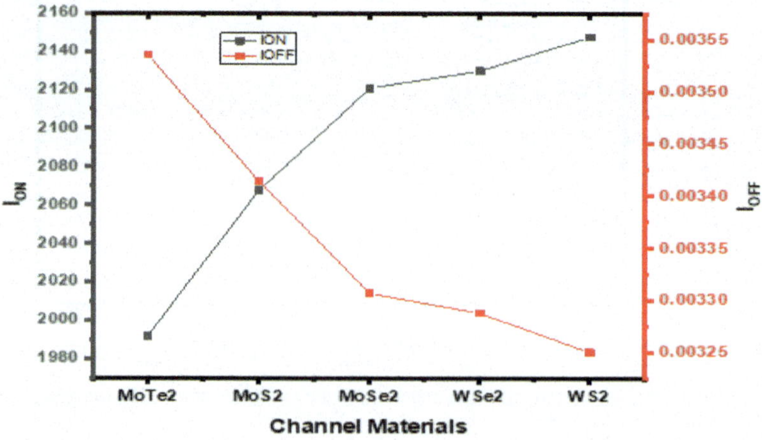

Fig. 6 I_{ON} and I_{OFF} for various channel materials while the dielectric is HfO_2(eox = 22) and gate length is 12 nm

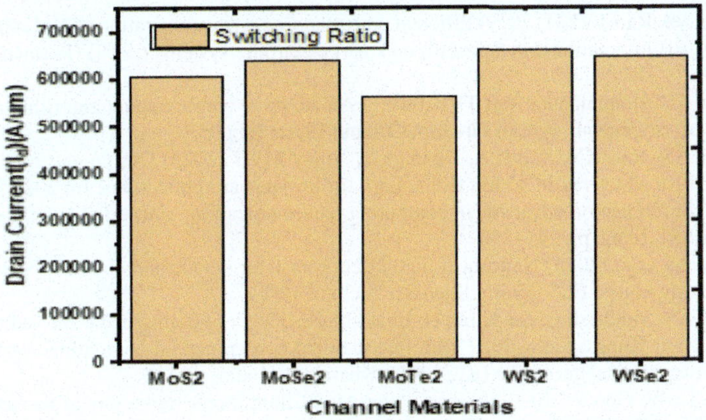

Fig. 7 I_{ON}/I_{OFF} (switching ratio) for various channel materials while the dielectric is HfO_2(eox = 22) and gate length is 12 nm

5 Conclusion

This research study on nanosheet field-effect transistors (FETs) with transition metal dichalcogenide (TMD) channel materials has provided valuable insights into their performance characteristics. In summary, this research utilized self-consistent quantum transport simulations to examine the ballistic I-V characteristics of nanosheet FETs employing a variety of 2D semiconductor channel materials, including MoS2, MoSe2, MoTe2, WS2, and WSe2. The findings underscored the potential of these materials for future nanoelectronics applications. Notably, HfO_2 emerged as a promising dielectric option due to its superior ability to minimize gate leakage current when compared with SiO_2 and air. This study underscores the importance of optimizing gate length, dielectric selection, and channel materials for maximizing the performance of nanosheet FETs. By understanding and manipulating these factors, we can pave the way for the development of next-generation nano-electronic devices tailored for advanced applications in computing, communication, and beyond.

Acknowledgements We sincere acknowledge nanohub.org for the experimental tool support. We also acknowledge Jing Guo, Ning Yang, Qimao Yang, University of Florida for the simulation tool support.

References

1. Sreenivasulu VB, Narendar V (2022) Design insights of nanosheet FET and CMOS circuit applications at 5-nm technology node. IEEE Trans Electron Devices 69(8):4115–4122
2. Kumari NA, Sreenivasulu VB, Vijayvargiya V, Upadhyay AK, Ajayan J, Uma M (2024) Performance comparison of nanosheet FET, CombFET, and TreeFET: device and circuit perspective. IEEE Access 12:9563–9571

3. Kumar AS, Rao TVKH (2019) Efficient core mapping on customization of NoC platforms. In: 2019 IEEE international symposium on smart electronic systems (iSES) (Formerly iNiS), pp 57–62
4. Kumar AS, Hanumantha Rao TVK (2021) An adaptive core mapping algorithm on NoC for future heterogeneous system-on-chip. Comput Electr Eng 95
5. Kumar AS, Rao KN, Sujith A, Dhanuja T, Vinay MVS (2023) Design and implementation of 1 KB SRAM array in 45 nm technology for low-power applications. In: 2023 3rd international conference on advances in computing, communication, embedded and secure systems (ACCESS), India, pp 245–250
6. Kumar AS et al (2023) Nanosheet field effect transistor device and circuit aspects for future technology nodes. ECS J Solid State Sci Technol 12(8)
7. Kumar AS, Siddhesh U, Sai Kiran N, Bhavitha K. Design of high speed 8-bit Vedic multiplier using Brent Kung adders. 2022 13th international conference on computing communication and networking technologies (ICCCNT), Kharagpur, India, pp 1–5
8. Liu L, Lu Y, Guo J (2013) On monolayer MoS2 field-effect transistors at the scaling limit. IEEE Trans Electron Devices 60(12):4133–4139
9. Sreenivasulu VB, Narendar V (2021) Characterization and optimization of junctionless gate-all-around vertically stacked nanowire FETs for sub-5 nm technology nodes. Microelectron J 116:105214
10. Ni Z et al (2016) Performance upper limit of sub-10 nm monolayer MoS2 transistors. Adv Electron Mater 2(9):1600191
11. Sung WL, Li Y (2021) Characteristics of stacked gate-all-around Si nanosheet MOSFETs with metal sidewall source/drain and their impacts on CMOS circuit properties. IEEE Trans Electron Devices 68(6):3124–3128
12. Sreenivasulu VB, Neelam AK, Panigrahy AK, Vakkalakula L, Singh J, Singh SG (2024) Benchmarking of multi-bridge-channel FETs toward analog and mixed-mode circuit applications. IEEE Access 12:7531–7539
13. Sreenivasulu VB, Neelam AK, Kola SR, Singh J, Li Y (2023) Exploring the performance of 3-D nanosheet FET in inversion and junctionless modes: device and circuit-level analysis and comparison. IEEE Access 11:90421–90429. https://doi.org/10.1109/ACCESS.2023.3306050
14. Kumar AS et al (2023) Device analysis of vertically stacked GAA nanosheet FET at advanced technology node. In: 2023 3rd international conference on advances in computing, communication, embedded and secure systems (ACCESS), India, pp 274–279
15. Kumar AS et al (2023) Nanosheet field effect transistor device and circuit aspects for future technology nodes. ECS J Solid State Sci Technol 12(8)
16. Guo J, Yang N, Yang Q (2023) Nanosheet FET. https://nanohub.org/resources/nanosheetfet, https://doi.org/10.21981/1R27-EN52
17. Liu L, Lu Y, Guo J (2013) Monolayer MoS2 field-effect transistors at the scaling limit. IEEE Trans Electron Devices 60:4133
18. Bharath Sreenivasulu V et al (2024) Spacer engineering on multi-channel FinFET for advanced wireless applications. AEU—Int J Electron Commun 178
19. Gowthami U et al (2024) Performance improvement of spacer-engineered N-type tree shaped NSFET toward advanced technology nodes. IEEE Access 12:59716–59725
20. Panigrahy AK et al (2024) Spacer dielectric analysis of multi-channel nanosheet FET for nanoscale applications. IEEE Access 12:73160–73168
21. Naresh Kumar Reddy B, Kumar AS (2024) Evaluating the effectiveness of bat optimization in an adaptive and energy-efficient network-on-chip routing framework. J Parallel Distrib Comput 188
22. Kumar AS, Naresh Kumar Reddy B (2023) An efficient real-time embedded application mapping for NoC based multiprocessor system on chip. Wireless Personal Commun 128:2937–2952
23. Kumar AS, Hanumantha Rao TVK (2020) Scalable benchmark synthesis for performance evaluation of NoC core mapping. Microprocess Microsyst 79

24. Kumar AS, Hanumantha Rao TVK, Kumar Reddy BN (2020) Exact formulas for fault aware core mapping on NoC reliability. In: 2020 IEEE 17th India council international conference (INDICON), pp 1–5
25. Sarangam K, Kumar AS, Reddy BNK (2024) Design and investigation of the 22 nm FinFET based dynamic latched comparator for low power applications. Trans Electr Electron Mater pp 1–14
26. Jaisawal RK (2023) Self-heating and interface traps assisted early aging revelation and reliability analysis of negative capacitance FinFET. 7th IEEE electron devices technology and manufacturing conference (EDTM). Korea, Republic of, Seoul, pp 1–3
27. Sreenivasulu VB, Kumari NA, Lokesh V et al (2023) Design of resistive load inverter and common source amplifier circuits using symmetric and asymmetric nanowire FETs. J Electron Mater 52:7268–7279
28. Sreenivasulu VB, Narendar V (2022) Circuit analysis and optimization of GAA nanowire FET towards low power and high switching. Silicon 14:10401–10411
29. Sreenivasulu VB, Narendar V (2022) Junctionless gate-all-around nanowire FET with asymmetric spacer for continued scaling. Silicon 14:7461–7471
30. Sreenivasulu VB, Narendar V (2022) Design and temperature assessment of junctionless nanosheet FET for nanoscale applications. Silicon 14:3823–3834

Enhancing Disaster Response: A Dual-Drone Approach for Efficient Area Scanning and Life Detection

A. Mummoorthy, P. Balamurugan, and Anitha Patibandla

Abstract The paper proposes a new methodology for utilizing drones and UAVs during natural disasters such as floods and earthquakes. Natural disasters represent some of the most devastating events imaginable, underscoring the critical importance of preparedness to swiftly aid affected individuals. While natural disasters occur with varying frequency across the globe, their occurrence can severely impede human communication and mobility within affected regions. Providing assistance and procuring necessary equipment become significantly challenging, compounded by the constrained time available for response efforts. The initial 72 h following a disaster is pivotal for rescue operations. Developing an optimal rescue plan from ground-level assessments proves impractical. This is where drones and Unmanned Aerial Vehicles (UAVs) come into play. With advancements in drone technology, it becomes imperative to devise intelligent strategies for their deployment during natural disasters. This paper endeavors to propose an intuitive methodology aimed at analyzing affected areas, providing optimized routes, and predicting safe zones through modern machine learning algorithms. The proposed methodology seeks to expedite the analysis of current conditions and enhance the efficiency and safety of rescue operations. Central to this project is the establishment of reliable communication between the two drones and the server, along with the development of efficient task allocation strategies for each drone. Additionally, the methodology includes an interactive user interface housing all the information communicated and computed by the system. In conclusion, the proposed methodology aims to streamline rescue operations during natural disasters by harnessing the capabilities of drones and UAVs. Through improved efficiency and informed decision-making,

A. Mummoorthy (✉)
Department of Information Technology, Malla Reddy College of Engineering and Technology, Hyderabad, India
e-mail: a.mummoorthy@mrcet.ac.in

P. Balamurugan
School of Computing, SRM Institute of Science and Technology, Kattankulathur, Tamil Nadu 603203, India

A. Patibandla
Department of ECE, Malla Reddy College of Engineering and Technology, Hyderabad, India

© The Author(s), under exclusive license to Springer Nature Singapore Pte Ltd. 2025

H. Zen et al. (eds.), *Soft Computing and Signal Processing*, Lecture Notes in Networks and Systems 1200, https://doi.org/10.1007/978-981-97-9926-8_54

this approach endeavors to enhance the effectiveness of disaster response efforts, ultimately contributing to saving lives and mitigating the impact of natural disasters.

Keywords Drones—natural disasters · Unmanned Aerial Vehicles (UAVs)

1 Introduction

Natural disasters pose significant challenges to emergency responders, often hindering their ability to assess the situation and deliver timely assistance to affected populations. In the face of these challenges, leveraging advanced technologies such as drones offers a promising solution to enhance disaster response efforts. This paper presents a novel methodology that harnesses the capabilities of two drones to improve disaster response operations during natural disasters, focusing specifically on floods and earthquakes. By integrating satellite imagery, real-time data processing, and smart routing algorithms, our proposed [1, 2] approach aims to provide responders with actionable insights and efficient routes for search and rescue missions in disaster-affected areas. The devastating impact of natural disasters underscores the critical need for effective disaster management and response strategies. From earthquakes that topple buildings and infrastructure to floods that submerge entire communities, these events can cause widespread destruction, displacement, and loss of life. In the aftermath of such disasters [3], the window of opportunity for rescue and relief efforts is often narrow, requiring responders to act swiftly and decisively to save lives and mitigate further harm. Traditional methods of disaster response often rely on ground-based assessments and human intervention, which can be impeded by the scale, terrain, and hazards present in disaster-affected areas. Moreover, the chaotic and dynamic nature of disasters can make it challenging to accurately assess the situation and prioritize resources effectively. In recent years, advances in technology have opened up new possibilities for enhancing disaster response capabilities, with drones emerging as a versatile tool for aerial reconnaissance, data collection, and situational awareness. Our proposed methodology builds upon these technological advancements, leveraging the unique capabilities of drones to streamline and optimize disaster response operations. Central to our approach is the use of two drones, each equipped with distinct functionalities tailored to specific tasks within the response workflow. The first drone serves as a stationary platform for data processing and analysis, while the second drone is deployed for dynamic reconnaissance and search missions along optimized routes. The first drone, equipped with enhanced computing power, is tasked with conducting initial assessments of the disaster-affected area. By leveraging satellite imagery and real-time data feeds, this drone generates smart routes and identifies safe areas for rescue operations. The process begins with the drone ascending to an optimal altitude to provide broad coverage of the affected area. From this vantage point, the drone analyzes satellite maps of the area during normal conditions, comparing them to real-time imagery to identify changes and assess the extent of damage. Using advanced image processing and machine learning algorithms, the first drone identifies viable routes for search and rescue missions, taking into account factors such as road conditions, terrain features, and potential hazards. These smart routes are calculated to maximize coverage of the affected area while minimizing travel time and exposure to risks. Additionally, the drone identifies safe areas within the disaster zone where rescued individuals can be brought for medical treatment and temporary shelter. Once the smart routes and safe areas have been determined, the first drone transmits this information to a centralized server, which then relays the data to the second drone. The second drone, equipped with lightweight and agile capabilities, is tasked with executing search and rescue missions along the designated routes. As it navigates through the disaster zone, the second drone utilizes onboard sensors and cameras to detect signs of life and identify individuals in need of assistance. If the second drone detects any individuals within its field of view, it promptly sends their location coordinates back to the central server, allowing responders to prioritize their rescue efforts and dispatch ground teams to the identified locations. Meanwhile, the second drone continues along the smart route, systematically scanning the area for additional survivors [4] and relaying relevant information to the response team in real-time. By leveraging the complementary capabilities of the two drones and integrating them into a cohesive response framework, our proposed methodology aims to improve the efficiency and effectiveness of disaster response operations. The combination of aerial reconnaissance, data analysis, and dynamic routing enables responders to quickly assess the situation, identify critical needs, and deploy resources with precision and agility.

In addition to its immediate impact on disaster response efforts, our proposed methodology also holds long-term potential for enhancing overall disaster preparedness and resilience. By providing responders with actionable insights and real-time situational awareness, drones can help inform decision-making processes and improve resource allocation strategies [5]. Furthermore, the data collected during disaster\re response missions can be used to inform future planning and mitigation efforts, allowing communities to better anticipate and mitigate the impact of future disasters. In the following sections of this paper, we will provide a detailed overview of our proposed methodology, including the technical specifications of the drones, the algorithms used for data processing and analysis, and the operational procedures for executing search and rescue missions. We will also present a case study or simulation demonstrating the effectiveness of our approach in a real-world disaster scenario, as well as a discussion of the potential benefits, limitations, and future directions for research and application [6].

2 Related Works

Several studies have investigated the utilization of drones and UAVs in disaster response and management, contributing valuable insights to the field. Here are some noteworthy examples:

Drone Talk: An Internet-of-Things-Based Drone System for Last-Mile Drone Delivery (2022): Proposes a methodology enabling drones to fly autonomously in mixed indoor–outdoor environments [7].

An Image Processing Based Classifier to Support Safe Dropping for Delivery-by-Drone (2022): The paper proposes a simple and fast classifier to assess the safety of a designated dropping zone before and during the dropping operation, using a single onboard camera.

Exploration and Object Detection via Low-Cost Autonomous Drone (2023): Proposes autonomous drone path-planning solution to locate and inspect wind turbines in a wind farm.

The application of autonomous drones in the environment of overhead lines (2018): The paper advocates for autonomous drones over remote-controlled ones to inspect overhead lines efficiently by following conductor curves, ensuring electric network safety [8]. Design and Development of an Autonomous Pesticides Spraying Agricultural Drone (2020): This paper introduces an autonomous agricultural drone for pesticide spraying, featuring variable altitude [9], cruise speed, and payload capacity, aimed at enhancing efficiency and reducing health risks for farmers.

Autonomous drone control within a Wi-Fi network (2020): The paper introduces a multiple drone network using commercially available drones to extend coverage area and overcome quadcopter drones' short flight time limitation through autonomous communication and coordination [10].

A Path Planning Drone Solution to Safe Pesticide Usage in Arable Lands (2022): The paper proposes a path-planning drone solution to minimize pesticide overuse and human exposure by autonomously spraying approved amounts only in arable areas, promoting satellite farming with autonomous drones.

The drone-following models in smart cities (2018): The paper examines drones' pivotal role in smart cities, highlighting the need for secure operational environments and government regulations while presenting two drone-following models for intelligent transportation systems.

FPV drone with GPS used for surveillance in remote areas (2017): The paper outlines an energy-efficient drone prototype for surveillance, emphasizing increased flight time, reduced noise, and affordability for various agencies, with future potential for additional features and customization based on specific criteria.

Optimization and Representation of a Network of Food Delivery Drones in Simulation (2023): The paper explores the central coordination of a multi-node drone network for efficient food delivery in urban areas, utilizing a genetic algorithm for route planning and heuristic optimization simulation to address logistical challenges and ensure stable drone performance under various conditions.

Drone Forensics: A Case Study on a DJI Mavic Air (2019): The research delves into drone forensics, specifically investigating the DJI Mavic Air using an iOS-based smartphone, revealing significant forensic data within the control application and extracting valuable information from the drone, external MicroSD card, and associated smartphone device [11].

These studies underscore the diverse applications of drone technology in disaster response scenarios and provide valuable insights for the development of effective methodologies and strategies.

3 Methodology

Our proposed methodology aims to leverage the capabilities of two drones to enhance disaster response operations during natural disasters, focusing specifically on floods and earthquakes. The methodology involves a coordinated approach that combines aerial reconnaissance, data processing, and dynamic routing to optimize search and rescue missions in disaster-affected areas [12]. To implement this methodology, a combination of hardware and software solutions is required as shown in Figs. 1 and 2. Drones equipped with high-resolution cameras, Lidar sensors, GPS receivers, and communication systems serve as the primary hardware components. These drones are capable of capturing detailed imagery, generating 3D maps, navigating accurately, and communicating in real-time with the central server. On the software side, various algorithms and software packages play crucial roles in different aspects of the methodology:

1. **Image Processing Algorithms**: These algorithms analyze the collected imagery to assess the extent of damage, identify critical areas, and detect signs of life. Techniques such as object detection, image segmentation, and feature extraction can be employed to extract valuable information from the imagery.
2. **Machine Learning Algorithms**: Modern machine learning algorithms, including convolutional neural networks (CNNs), recurrent neural networks (RNNs), and support vector machines (SVMs), are utilized for tasks such as

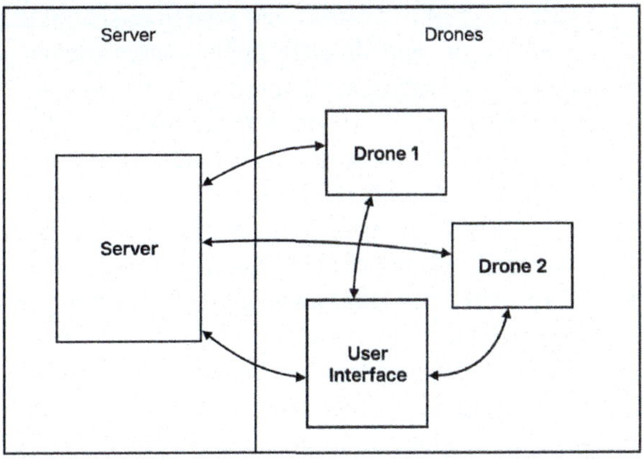

Fig. 1 Architecture diagram

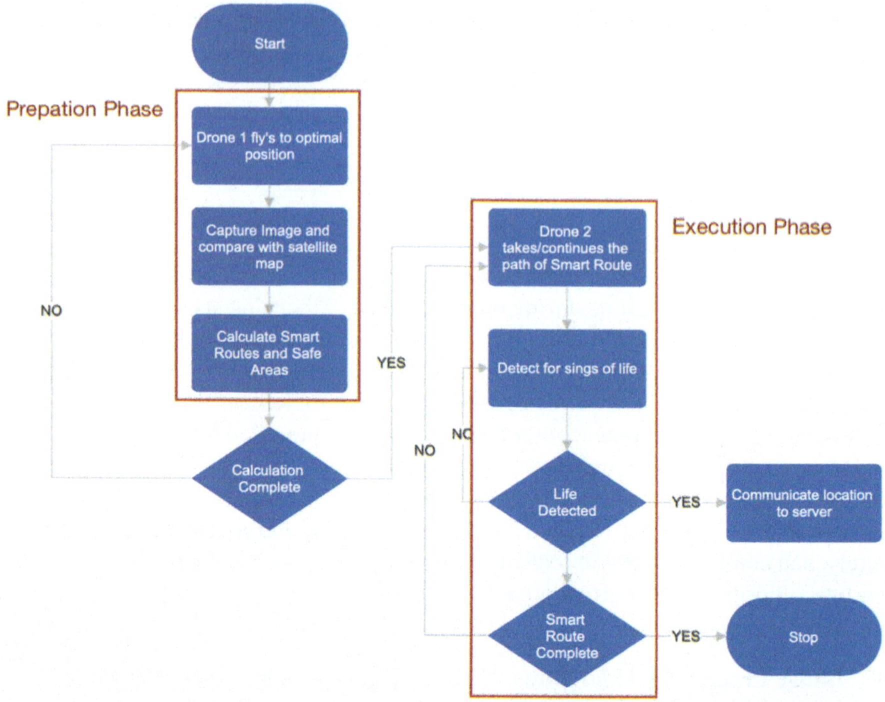

Fig. 2 Flowchart

route optimization, safe zone prediction, and survivor detection. These algorithms learn patterns from historical data and make predictions based on the current situation, improving the efficiency and accuracy of response efforts.

3. **Geospatial Analysis Software**: Geospatial analysis software, such as ArcGIS or QGIS, is used to process and visualize spatial data, including satellite imagery, terrain maps, and disaster impact assessments. These tools enable responders to analyze geographic patterns, plan response strategies, and make informed decisions based on spatial information.
4. **Communication Protocols**: Efficient communication protocols, such as TCP/IP or MQTT, facilitate real-time data exchange between drones and the central server. These protocols ensure reliable communication, even in challenging environments, and enable seamless coordination of response efforts.
5. **Graphical User Interface (GUI) Software**: Interactive GUI software provides a user-friendly interface for responders to monitor drone operations, visualize collected data, and make command decisions [13]. These interfaces display real-time information, such as drone positions, detected hazards, and safe zones, allowing responders to react quickly to changing conditions.

By integrating these hardware components, algorithms, and software solutions, the proposed methodology aims to optimize disaster response operations, improve situational awareness, and ultimately save lives in disaster-affected areas. The proposed methodology consists of two main phases: Preparation Phase: In this phase, the first drone conducts initial assessments of the disaster-affected area, generates smart routes, and identifies safe areas for rescue operations. Execution Phase: Once the necessary data has been processed and analyzed, the second drone is deployed to execute search and rescue missions along the designated routes, identifying survivors and relaying relevant information to the response team in real-time [14].

4 Preparation Phase

During the preparation phase, the first drone, equipped with enhanced computing power and aerial reconnaissance capabilities, performs the following tasks: Initial Assessment: The first drone ascends to an optimal altitude to provide broad coverage of the affected area. Using onboard sensors and cameras, the drone captures high-resolution imagery of the disaster zone, which is then processed and analyzed to assess the extent of damage and identify critical areas of interest. Data Processing and Analysis: Using advanced image processing and machine learning algorithms, the first drone analyzes the collected data to generate smart routes and identify safe areas within the disaster zone [15]. This involves overlaying satellite maps of the area during normal conditions and comparing them to real-time imagery to identify changes and assess the impact of the disaster. Route Optimization: Based on the results of the data analysis, the first drone calculates optimized routes for search and rescue missions, taking into account factors such as road conditions, terrain features, and potential hazards. These smart routes are designed to maximize coverage of the affected area while minimizing travel time and exposure to risks. Safe Area Identification: In addition to generating smart routes, the first drone also identifies safe areas within the disaster zone where rescued individuals can be brought for medical treatment and temporary shelter. These safe areas are selected based on criteria such as accessibility, proximity to resources, and suitability for temporary habitation [16–19].

5 Execution Phase

Once the necessary data has been processed and analyzed during the preparation phase, the second drone is deployed to execute search and rescue missions along the designated routes. The execution phase involves the following steps: Route Execution: The second drone [15], equipped with lightweight and agile capabilities, follows the smart routes generated by the first drone, systematically scanning the area for signs of life and identifying individuals in need of assistance. Survivor Detection:

Using onboard sensors and cameras, the second drone detects signs of life within its field of view and promptly sends the location coordinates of any survivors back to the central server. This allows responders to prioritize their rescue efforts and dispatch ground teams to the identified locations. Real-Time Communication: Throughout the execution phase, the second drone maintains real-time communication with the central server, relaying relevant information about the disaster zone and any survivors detected during the mission. This enables responders to make informed decisions and adapt their strategies in real-time based on the latest available data.

6 Equipment and Sensors

Drones

Drone 1: The first drone serves as the primary aerial reconnaissance platform and is equipped with enhanced computing power and payload capacity. It is typically a larger drone capable of carrying heavier equipment and staying aloft for extended periods. The first drone is responsible for conducting initial assessments of the disaster-affected area, generating smart routes, and identifying safe areas for rescue operations. Due to its computing power, it can process complex algorithms for image analysis and route optimization.

Drone 2: The second drone is a smaller, more agile aerial vehicle designed for dynamic reconnaissance and search missions. It is equipped with lightweight sensors and cameras for detecting signs of life and relaying real-time data to the central server. The second drone follows the smart routes generated by the first drone and systematically scans the area for survivors. Its agility allows it to navigate through tight spaces and challenging terrain, maximizing coverage of the disaster zone and sensor.

Cameras

Both drones are equipped with high-resolution cameras capable of capturing detailed imagery of the disaster-affected area. These cameras provide visual data used for assessing damage, identifying survivors, and monitoring response operations. Infrared cameras may also be utilized to detect heat signatures and identify survivors in low visibility conditions or at night.

Light Detection and Ranging (Lidar)

Lidar sensors may be mounted on the drones to generate detailed 3D maps of the terrain and infrastructure. Lidar data can be used for terrain mapping, obstacle detection, and structural integrity assessment, providing valuable insights for response operations.

Global Positioning System (GPS)

Both drones are equipped with GPS receivers for navigation and precise positioning. GPS data allows the drones to follow predefined routes accurately and transmit their location coordinates to the central server in real-time.

Obstacle Avoidance Sensors

Both drones may be equipped with obstacle avoidance sensors, such as ultrasonic or laser sensors, to detect and avoid obstacles during flight. These sensors enhance the safety and autonomy of the drones, allowing them to navigate through complex environments without colliding with obstacles or structures. By leveraging advanced equipment and sensors, the drones in the proposed methodology are capable of conducting comprehensive aerial reconnaissance, detecting signs of life, and relaying real-time data to support decision-making in disaster response operations. The integration of these technologies enhances the efficiency, effectiveness, and safety of response efforts, ultimately saving lives and mitigating the impact of natural disasters [15].

7 Challenges and Limitations

While the proposed methodology utilizing drones for disaster response offers numerous benefits, it also faces several challenges and limitations that must be addressed to ensure its successful implementation. Here are some of the key challenges and limitations:

Limited Payload Capacity: Drones, especially smaller models, have limited payload capacities, which restrict the types and quantities of equipment and sensors they can carry. This limitation may hinder their ability to carry essential supplies or advanced sensor systems necessary for comprehensive disaster assessment and response.

Battery Life and Flight Endurance: Battery life and flight endurance are significant limitations for drones, particularly in large-scale disaster scenarios that require prolonged aerial surveillance and search missions. Short battery life may necessitate frequent recharging or battery swaps, leading to interruptions in response operations and reduced coverage area.

Weather Dependence: Drones are highly susceptible to adverse weather conditions, such as high winds, heavy rain, or fog, which can affect their flight stability and performance. Inclement weather may limit the operational window for drone deployment and compromise the safety of response personnel.

Regulatory Restrictions: Regulatory restrictions, including airspace regulations, licensing requirements, and privacy laws, may pose challenges to the deployment of drones in disaster response operations. Compliance with regulations and obtaining necessary permits may delay or restrict drone operations, particularly in urban or densely populated areas.

Communication Reliability: Communication reliability is essential for maintaining real-time connectivity between drones and the central server, particularly in remote or disaster-affected areas with limited infrastructure. Poor communication infrastructure or interference may disrupt data transmission and hamper response coordination.

Obstacle Avoidance and Navigation: Drones must navigate through complex and dynamic environments, including urban landscapes, debris-filled areas, and confined spaces, without colliding with obstacles or structures. Limited obstacle avoidance capabilities or navigation errors may result in accidents or mission failures.

Data Processing and Analysis: Processing and analyzing the vast amounts of data collected by drones require advanced computational resources and algorithms. Inadequate data processing capabilities or lack of expertise in data analysis may limit the effectiveness of drone-based disaster assessment and response efforts.

Humanitarian Considerations: Ethical and humanitarian considerations, such as privacy concerns, consent, and cultural sensitivities, must be carefully addressed when deploying drones in disaster-affected areas. Respect for human rights and dignity, as well as community engagement and consultation, are essential for building trust and acceptance of drone technology.

Integration with Existing Systems: Integrating drone technology into existing disaster management systems and workflows requires coordination among multiple stakeholders, including emergency responders, government agencies, and humanitarian organizations. Ensuring seamless integration and interoperability with existing systems may pose logistical and organizational challenges.

Cost and Sustainability: The cost of acquiring, operating, and maintaining drones, as well as training personnel in their use, can be significant barriers to adoption, particularly for resource-constrained organizations or communities. Sustainable funding models and cost-effective solutions are essential for the long-term viability of drone-based disaster response initiatives.

8 Conclusion

In conclusion, our methodology presents a significant advancement in disaster response operations through the integration of drones and UAVs. By leveraging cutting-edge technology, we aim to overcome challenges posed by natural disasters, enhancing communication, mobility, and timely assistance to affected populations. Through rapid aerial reconnaissance and smart route generation, our methodology enables efficient resource allocation and decision-making, maximizing coverage of affected areas. Despite challenges such as regulatory compliance and ethical considerations, we strive to continuously improve the effectiveness and safety of disaster response efforts, ultimately contributing to saving lives and building resilience in the face of natural disasters.

References

1. Drone talk: an internet-of-things-based drone system for last-mile drone delivery. In: IEEE transactions on intelligent transportation systems (2022)
2. An image processing based classifier to support safe dropping for delivery-by-drone. In: 2022 IEEE 5th international conference on image processing applications and systems (IPAS) (2022)
3. Exploration and object detection via low-cost autonomous drone (2023)
4. The application of autonomous drones in the environment of overhead lines. In: 2023 intermountain engineering, technology and computing (IETC) (2018)
5. In: 2018 IEEE 18th international symposium on computational intelligence and informatics (CINTI)
6. Design and development of an autonomous pesticides spraying agricultural drone. In: 2020 IEEE region 10 symposium (TENSYMP) (2020)
7. Autonomous drone control within a Wi-Fi network (2020)
8. In: 2020 12th international symposium on communication systems, networks and digital signal processing (CSNDSP)
9. A path planning drone solution to safe pesticide usage in arable lands (2022)
10. In: 2022 2nd international conference on advanced research in computing (ICARC)
11. The drone-following models in smart cities (2018)
12. In: 2018 IEEE 59th international scientific conference on power and electrical engineering of Riga Technical University (RTUCON)
13. FPV drone with GPS used for surveillance in remote areas (2017)
14. In: 2017 third international conference on research in computational intelligence and communication networks (ICRCICN)
15. Optimization and representation of a network of food delivery drones in simulation (2023)
16. Drone forensics: a case study on a DJI Mavic Air. In: 2019 IEEE/ACS 16th international conference on computer systems and applications (AICCSA) (2019)
17. Priyadharshini SP, Balamurugan P (2023) An in-sight analysis of cyber-security protocols and the vulnerabilities in drone communication. In: Karthikeyan P, Kumar S, Anbarasu V (eds) Drone data analytics in aerial computing. Transactions on computer systems and networks. Springer Nature, Singapore Pte Ltd. ISSN 2730-7492. (electronic) ISBN 978-981-99-5056-0
18. Priyadharshini SP, Balamurugan P (2023) Empirical analysis of packet-loss and content modification for secure flying ad-hoc networks (FANETs). In: 2023 international conference on networking and communications (ICNWC), 5–6 April 2023. IEEE Xplore. https://doi.org/10.1109/IC-NWC57852.2023.10127499. ISBN: 979-8-3503-3600-9. (Indexed date: 25 May 2023)
19. Priyadharshini SP, Balamurugan P (2022) Unmanned aerial vehicle in smart farming systems: types, applications, and cyber-security threats. In: International conference on innovative computing, intelligent communication and smart electrical systems, ICSES2022, pp 1–9, 15–16 July 2022. IEEE Xplore. ISSN 978-166547413-9. https://doi.org/10.1109/IC-SES55317.2022.9914070. (14 Oct 2022)

Author Index

© The Editor(s) (if applicable) and The Author(s), under exclusive license to Springer Nature Singapore Pte Ltd. 2025

H. Zen et al. (eds.), *Soft Computing and Signal Processing*, Lecture Notes in Networks and Systems 1200, https://doi.org/10.1007/978-981-97-9926-8

The manufacturer's authorised representative in the EU is Springer Nature Customer Service Centre GmbH, Europaplatz 3, 69115 Heidelberg, Germany. If you have any concerns regarding our products, please contact ProductSafety@springernature.com

Printed and bound by CPI Group (UK) Ltd, Croydon, CR0 4YY
01/10/2026
02225901-0014